Technology, Crafting and Artisanal Networks in the Greek and Roman World

Technology, Crafting and Artisanal Networks in the Greek and Roman World

Interdisciplinary Approaches to the Study of Ceramics

Edited by
Diego Elia, Eleni Hasaki and Marco Serino

DE GRUYTER

This volume has received funding from the European Union's Horizon 2020 research and innovation programme under the Marie Skłodowska-Curie grant agreement no. 893629 – A.G.A.T.H.O.C.L.E.S. – The 'Archaeology of Gesture': Apprenticeship, Tools, Hands, Organization, Collaborations, Learning Experience and Social Network Analysis.

The volume was supported by the following institutions:

ISBN 978-3-11-115461-9
e-ISBN (PDF) 978-3-11-118963-5
e-ISBN (EPUB) 978-3-11-119058-7

Library of Congress Control Number: 2024942041

Bibliographic information published by the Deutsche Nationalbibliothek
The Deutsche Nationalbibliothek lists this publication in the Deutsche Nationalbibliografie; detailed bibliographic data are available on the internet at http://dnb.dnb.de.

Cover image: Fragment of a red-figured, Attic kylix showing a man decorating a kylix, attributed to the Antiphon Painter, ca. 480 B.C., with the permission of © Boston, Museum of Fine Arts, Henry Lillie Pierce Fund, Accession Number 01.8073
Typesetting: Integra Software Services Pvt. Ltd.
Printing and binding: CPI books GmbH, Leck

www.degruyter.com

Acknowledgements

A number of people and institutions helped us complete this volume which includes the proceedings of the conference held at the University of Turin in October 20–22, 2022 and organized by the A.G.A.T.H.O.C.L.E.S project, the University of Turin, and the University of Arizona. Our deepest thanks to all institutions which supported us in numerous way with organizing the conference and publishing this volume: the European Commission Marie Sklodowska-Curie Global Fellowship for the A.G.A.T.H.O.C.L.E.S project (Grant Agreement no. 893629), the University of Turin (Department of Historical Studies) and the University of Arizona (School of Anthropology, Department of Religious Studies and Classics, and Provost's Author Support Fund).

We would like to thank immensely all the members of the Scientific Advisory Committee that supported us in this initiative. Following an alphabetic order, they are: Peter Acton (University of Melbourne), Martin Bentz (Bonn University), Monica Gulmini (University of Turin), Pirjo Hamari (University of Helsinki), Mario Iozzo (Museo Archeologico Nazionale di Firenze), Maria Chiara Lebole (University of Turin), Kathleen Lynch (University of Cincinnati), Ian McPhee (La Trobe University), Valeria Meirano (University of Turin), Dimitris Paleothodoros (University of Thessaly), Claude Pouzadoux (Paris-Nanterre University), Ted Robinson (University of Sydney), Susan Rotroff (Washington University in St. Louis), Francesca Silvestrelli (Università del Salento), Mark Stansbury O'Donnell (University of St. Thomas, Minnesota), and Athena Tsingarida (Université libre de Bruxelles). They peer-reviewed both the conference abstracts and this volume's essays and they generously offered advice to the volume editors on several aspects.

Our warmest thanks also to the session chairs who enriched the discussion during the two-day conference in Turin: Mario Iozzo, Kathleen Lynch, Valeria Meirano, and Ted Robinson. Many thanks also to Elisa Ercolin and Simone Guion, both at the University of Turin, who ably handled several administrative duties as conference secretaries.

We sincerely express our appreciation to the contributing authors who helped us meet a tight timetable by responding quickly to our requests. The roster of authors in this volume draws both on conference participants and presenters in the A.G.A.T.H.O.C.L.E.S International Webinar Series (two editions) held on Spring 2022 and Spring 2023. We greatly acknowledge the oral presentations by Sanchita Balachandran (Smithsonian Museum), Jared Benton (Old Dominion University), Sheramy Bundrick (University of South Florida), Shawn Graham (Carleton University), Caroline Jeffra (Immersive Heritage Studio), Martin Langner (Universität Göttingen), Christopher F. Motz (University of Richmond), David Saunders (J. Paul Getty Museum) and Angela Ziskowsky (Coe College), who could not participate in this volume.

Our warmest thanks to Paul Psoinos and Gil Renberg for assisting the editors with copy-editing of the text. Finally, we are greatly indebted to Jessica Bartz, Mirko Vonderstein and Jens Lindenhain (De Gruyter) for their steadfast support and their attentive production of this volume.

https://doi.org/10.1515/9783111189635-202

https://doi.org/10.1515/9783111189635-203

Contents

Introduction

Marco Serino, Diego Elia and Eleni Hasaki

1 The A.G.A.T.H.O.C.L.E.S. Project and Beyond: Archaeology of Gesture and Material Consciousness in Ceramic Studies

The hand is the window on to the mind (Immanuel Kant)[1]

People make objects, but objects make people (Daniel Miller)[2]

In Kant's quote above on how the gestures of hand is a window into the creative process of humans and on Miller's statement on how creation is a consciously dialectic process between objects and people, the concepts of gesture and material consciousness take center stage, as are in the current volume. More than words, a creative adjustment in a famous Renaissance statue help us visualize such mental processes. In 1564, more than twenty years after completing the statue of Moses which still stands in San Pietro in Vincoli in Rome, the famous master Michelangelo Buonarroti (better known simply as Michelangelo) revisited his work, making a major change: he turned the head (!) of Moses from its original frontal position so that it would face left (Fig. 1). Thus, what we can see today does not correspond to the great sculptor's original version of the statue. Both recent conservation activities on the statue and personal correspondence from Michelangelo's circle of friends have confirmed independently this artistic modification,[3] allowing us to conclude that, as this prominent example demonstrates, artistic creation and recreation are dynamic processes that sometimes continue even when an artifact is finished.

The selection of a Renaissance marble statue to introduce this book on ceramics from the ancient Greek and Roman worlds may at first seem an odd choice, but the shared idea of crafting and re-crafting an artifact is central to the conceptualization of this volume as well. It helps us understand the continuous intellectual spectrum of archaeology of gesture and how artifacts acquire a certain material consciousness as they move through production, distribution, and consumption. These two concepts—archaeology of gesture and material consciousness—are useful for anchoring the genesis of the conference and the proceedings that are presented here.

The concept of "material consciousness", proposed by R. Sennett, refers to the "craftsmen's proper conscious domain" and argues that "people invest thought in things they can change." According to him, such thinking revolves around three key issues that are related, to a certain extent, to the concept of movement: metamorphosis, presence, and anthropomorphosis.[4] In his analysis, he argues that "the long history of crafting clay shows three ways of becoming aroused consciously by materials, in altering, marking, or identifying them with ourselves".[5]

Artifacts from all periods and places are intrinsically related to the concept of movement in different ways: crafting procedures, human manipulation and gestures on them, displacements in certain locations and spaces (e.g., productive, domestic, ritual), import/export, and restorations or modifications over time. Thereby, movements can generate networks and crucial relationships: links among different geographical areas, cultures, and ideas; links between artisanal knowledge, dynamics of apprenticeship, and artistic collaborations; and links between specific or general contexts of learning experience. All these movements and networks in antiquity need to be investigated through an interdisciplinary approach that aims to focus on a (reconsidered) "material consciousness" involving the exploration of ancient materiality from multiple perspectives.

Even if we deal with a distant world—chronologically, geographically, materially—the technical adjustments of Michelangelo's modified Moses could be considered as a proof to invisible to the naked eye 'movements' of raw material and of artisanal gestures involved in rendering the new position of the head, as well as evidence of 'movements' of

1 Tallis (2003) 4.
2 Miller (2005) 18.
3 Frommel (2014) 54–58. The statue was begun in 1506, later modified in 1532, and completed in 1544. The reference to the Moses' turned head was discovered in a letter dated in 1564 and written by Tommaso Cavalieri. It was addressed to Giorgio Vasari since he was collecting memories of Michelangelo's friends for the second edition of his book on the lives of the artists: see Vasari (1550). Concerning the crafting and technological aspects related to the modification of the Moses' head, see Forcellino (2014) 295.
4 Sennett (2008) 120.
5 Sennett (2008) 144.

https://doi.org/10.1515/9783111189635-001

Fig. 1: Statue of Moses by Michelangelo Buonarroti in San Pietro in Vincoli, Rome (Wikimedia Commons / Jörg Bittner Unna, "'Moses' by Michelangelo JBU140", CC BY 3.0; https://it.m.wikipedia.org/wiki/File:%27Moses%27_by_Michelangelo_JBU140.jpg [last access 15.07.2024]).

Michelangelo's own cultural, philosophical, and religious beliefs.[6] Moreover, Michelangelo's Moses is extremely fascinating as inspiration for understanding the "embodied humanity"[7] of ancient artisans, their techniques, their movements, their invisible—at least, for us today—gestures, and their relationship with the finished object itself.

Archaeology of artisanal gesture is central to the project A.G.A.T.H.O.C.L.E.S., the acronym for "**A**rchaeology of **G**esture: **A**pprenticeship, **T**ools, **H**ands, **O**rganization, **C**ollaborations, **L**earning **E**xperience and **S**ocial Network Analysis". This three-year project has been funded by the Marie-Sklodowska Global Individual Program at the European Commission and performed at the University of Turin (beneficiary) and the University of Arizona (host for outgoing phase).[8] A.G.A.T.H.O.C.L.E.S. was the driving force for the international conference entitled *Technology, Crafting and Artisanal Networks in the Greek and Roman World: Interdisciplinary Approaches to the Study of Ceramics*, which was held in hybrid format in Turin in 2022, and which led to the creation of this volume.

Our main goal in this volume is to present a number of important case studies relating to ceramics and the people behind them,[9] with a special focus on new methodological approaches, their efficacy, and how they can be combined to better understand ancient craft manufacturing and the societies that lay behind these crafting dynamics. We have included most of the papers presented during the conference and in the international webinar series held in 2022 and 2023.[10] These two types of settings for scholarly exchange, conference and webinar series, had a shared objective: to

6 On the Michelangelo's religious beliefs and his connection with the Neoplatonism, see Frommel (2014) 22.

7 Dobres (2014) 203.

8 Grant Agreement no. 893629: Principal Investigator: Marco Serino, University of Turin (Supervisor: Diego Elia, Department of Historical Studies); University of Arizona (Supervisor: Eleni Hasaki, School of Anthropology and Department of Religious Studies and Classics).

9 Inspired by the title of Stissi's (2002) dissertation: *Pottery to the People. The Production, Distribution and Consumption of Decorated Pottery in the Greek World in the Archaic Period (650–480 BC).*

10 The A.G.A.T.H.O.C.L.E.S. International Webinar Series—organized by Eleni Hasaki and Marco Serino—– were entitled, respectively: *Technology, Craft Connections and Digital Humanities in Classical Archaeology* (March-April 2022), and *Present and Future of Red-figure Pottery Studies: Thematic and Methodological Approaches* (February-March 2023).

address all of the topics mentioned above, and to create fruitful occasions for discussing past and current methodological approaches and how it might be possible to combine them all in order to better understand craftspeople of the ancient world. This volume centers on *tēchne*, the powerful ancient Greek word which emphasizes the inseparability of art, skill, craft, process, knowledge, understanding, and awareness, as well as material consciousness.

The intensely revised scholarship of red-figured production of Southern Italy and Sicily in recent decades has provided a beneficial gestation period for the A.G.A.T.H.O.C.L.E.S. project. The analysis of archaeological context became – finally! – increasingly important for better understanding functions, symbolical meanings, workshop dynamics, and the possible migration of artisans.[11] In addition to a renewed attention to the stylistic aspects of artifacts, their iconographic themes, and their distribution patterns, the technological features of the raw materials for these artifacts have now been recognized as essential elements within current scholarship.

A.G.A.T.H.O.C.L.E.S. focuses on craft traditions related to red-figured vase production in Southern Italy during the 5th and 4th centuries B.C. It examines the manual skills and gestures which are the most powerful learning-medium provided by those members of society who would transmit their knowledge through the expert use of their hands. The acronym of the project, A.G.A.T.H.O.C.L.E.S., encompasses the various angles for studying South Italian ceramics, but it is also a reference to Agathocles, the tyrant of Syracuse and self-proclaimed *basileus* of Sicily (between 316 and 289/288 B.C.).[12] But he was not only the tyrant of the Greek polis Syracuse: according to the ancient literary sources, he was the son of a potter named Carcinus of Rhegion, another Greek polis in Southern Italy. The project therefore refers to Agathocles as a son of a craftsman, and the acronym aims to evoke the curiosity and the typical behavior of children, who are used to learning by observing the intangible actions of their elders.

In this project material science is embedded from a network perspective exploring links between ancient people and landscapes in various ways, such as raw materials and artisans, apprenticeship between masters and pupils, links between individuals and communities. From the individual level the project moves towards a macroscale economic dynamic.[13] It serves more broadly as a model case study to better visualize and embrace the effectiveness of such methodological intersections. But data from technology studies, anthropological approaches, archaeometric analyses, and computational imaging can become even more useful when integrated into a wider picture -- and this picture is more fully captured as a network. As M.-A. Dobres has argued, only when the "communities of craftspeople [. . .] are taken as central to understanding how technical activities unfolded" can gestures and artisanal relationships become visible in the archaeological record.[14]

It is from precisely these premises that the idea for developing an archaeology of (artisanal) gesture was born, supported by the realization that we could overcome some serious challenges (as we cannot witness an ancient artisan performing such gestures) thanks to new diagnostic modern technologies. These can really help and support us in our efforts to find the material clues – often invisible to the naked eye – to the various steps behind the completed item, as a result of specific sequences of artisanal gestures.

Scholars have increasingly focused on technology in Classical Archaeology in the last few decades, adopting the concept of *chaîne opératoire* promoted by anthropological studies,[15] and combining it with the advances in archaeometry, especially chemical and petrographic analyses of clay.[16] Archaeometric studies focused their attention not only on the clay provenance, but also on the technological dynamics related to the quality of black gloss, to the kiln temperature reached during the three-stage firing process, and to some crucial issues related to the conservation of these archaeological materials.[17]

---

11 For some general and recent overviews of early South Italian red-figured workshops, see Denoyelle/Iozzo (2009) 97–136, 165–170; Todisco (2012) vol. II. To mention some post-Trendall studies related to Sicilian workshops: de Cesare (2009); Madella (2010); Barresi (2013), (2018); Elia (2012), (2014), (2019); Serino (2014), (2019); Santostefano (2020). For a brief overview of the last two decades of studies, see Soleti (2012) 66–71.

12 Polyb. 15.35.2; Diod. 19 (*passim*), Diod. 20.63.4.; Plut. *Regum et imperatorum apophthegmata*, 176, and *De laude ipsius*, 544; Iustin. 22.1.1; Amm. 14. 11.30; Aus. Epigr. 2, 5–6.

13 Dobres (2014) 203: "As knowledgeable but imperfect agents, people must never be an afterthought in technology studies, no matter how hard it is to find them in the archaeological record".

14 Dobres (2014) 201.

15 Delage (2017).

16 For a complete and comprehensive recent overview, see Roux (2019) 1–14.

17 Concerning archaeometric studies related to black gloss, kiln temperatures and other technological, see some recent papers (with previous bibliography): Aloupi-Siotis (2020); Gliozzo (2020); Jones (2021).

Recently, the relationship between archaeology and technology has also been investigated both from an ontological point of view and from that of the history of technology. M.-A. Dobres has emphasized the "mutual becoming" of people and products, focusing on the distinction between "cultural reason" and "practical reason" and highlighting how "people" need to be the ontological starting point of concrete research on ancient technologies.[18] A.M. Pollard and C. Gosden have more recently explored from a diachronic and theoretical perspective how technology is a social product as it "represents a socially embedded set of activities sustained by the transmission of relevant bodies of knowledge and practices".[19]

The strong, interdisciplinary frameworks of A.G.A.T.H.O.C.L.E.S characterize both the academic institutions where the project was performed and also the broader scholarly trends in the study of ceramic production in the Greek and Roman world. The outgoing phase of the project was performed at the University of Arizona, the School of Anthropology and its Laboratory for Traditional Technology (LTT) together with the Department of Religious Studies and Classics.[20] At the LTT ongoing projects by E. Hasaki and her collaborators pertaining to craft apprenticeship, technology and organization of ancient workshops, experimental archaeology, and the social networks of ancient communities of practice provided a stimulating environment for A.G.A.T.H.O.C.L.E.S. to further refine its research angle with regard to the South Italian vase industry.[21] The LTT offers semester-long instruction on potting techniques through its Ceramics Resident Instructor program, while also hosting short-term residencies by local and international ceramic artists. These artists instruct students and also participate in the laboratory's research projects concerning forming, decorating, and firing techniques within the A.G.A.T.H.O.C.L.E.S. framework, such as that of Roberto Paolini in 2023.[22]

Similarly, during the incoming phase at the University of Turin, in its Department of Historical Studies, the A.G.A.T.H.O.C.L.E.S. project capitalized on a long tradition of archaeological and material culture studies—started by M. Barra Bagnasco and C. Preacco Ancona, and continued by D. Elia, especially in the field of figured pottery—[23] which has significantly expanded interdisciplinary collaborations in a wide range of research areas in recent years, both within and beyond the Classical world. At the University of Turin, an important institution for collaboration and research planning has been the Structure in Sciences for Conservation, Restoration, Valorization of Cultural Heritage (SUSCOR): within its framework, training and research activities have been developed in relation to the Master's Degree in Conservation and Restoration of Cultural Heritage, created in agreement with the Conservation and Restoration Center "La Venaria Reale". This institution involves a close collaboration between colleagues from the Department of Historical Studies – archaeologists and art historians – and those from the Departments of Chemistry, Physics, Earth Sciences, Life Sciences and Systems Biology. Although the focus is on restoration and conservation, the interest in materials and techniques has led to other synergies and collaborations on various subjects related to cultural heritage, from archaeology (e.g., the House of the Ancient Hunt in Pompeii),[24] to contemporary art.

Advancing interdisciplinary perspectives in the field of cultural heritage studies, the University of Turin has also launched Tech4Culture (T4C), an innovative research and training doctoral program in Heritage Sciences.[25]

18 See Dobres (2010), with previous bibliography.

19 Pollard/Gosden (2023) 1.

20 See ltt.arizona.edu (last access 24.04.2024).

21 Hasaki (2016), (2019), (2021), (2024); Hasaki/Harris Cline (2020); Harris Cline/Hasaki (2023).

22 Hasaki this volume; Serino this volume; Hasaki/Serino/Elia this volume.

23 For the first archaeometric analyses of Western fine wares carried out at the University of Turin, see Mirti *et al.* (1995), (1998), (2004a), (2004b). An early application of RTI is shown in Bovero *et al.* (2012). For the use of microscopy, see Elia *et al.* (forthcoming).

24 The project *Da Pompei a Venaria. Per un progetto di conoscenza, valorizzazione, divulgazione: la Casa della Caccia Antica*, directed by D. Elia, adopts an interdisciplinary perspective with a multidisciplinary team to promote the knowledge, conservation and valorization of a Pompeian domus through archaeological, historical, scientific and technical studies. The results from the first phase of this project were presented to the public in the exhibition *Pompeiana Fragmenta. Conoscere e conservare (a) Pompei*: see Elia/Meirano (2018); a second phase of the project is now underway.

25 See https://www.tech4culture2020.unito.it/do/home.pl (last access 01.05.2024). This program (Technology Driven Sciences: Technologies for Cultural Heritage), which involves 35 national and international partners, both academic and non-academic, aims to train the next generation of Cultural Heritage professionals, to equip them with technological solutions and services for the restoration, protection, diagnostics and valorization of cultural heritage, while also offering a complementary scientific training in socio-economic disciplines and humanities.
A non-Mediterranen project, *BeArchaeo* (2019–2023), focusing on the archaeological exploration of the Tobiotsuka Kofun and other Kofun burial mounds in Japan (see *Beyond Archaeology*: https://www.facebook.com/BeArchaeoProject/?locale=it_IT [last access 01.05.2024]), has also contributed significantly to the scientific growth of researchers at the University of Turin, as well as the development of methodological and instrumental approaches: many collaborations started in the framework of this project continue to this day, supporting the study of archaeological sites and materials (ceramics, glass, metal and stone artifacts).

At the University of Turin, the experimental archaeological sessions in the framework of the A.G.A.T.H.O.C.L.E.S. project continued as along with extensive analyses using RTI (Reflectance Transformation Imaging) and photogrammetry. Moreover, a large number of microscopic, archaeometric, SEM (Scanning Electron Microscope) and Laser Ablation analyses have been undertaken in collaboration with the Department of Chemistry and the University "Aldo Moro" in Bari.[26]

Both the Universities of Turin and the University of Arizona have very strong traditions in the study of ceramics from different but highly complementary perspectives. All of the research environments noted above have created a perfect combination for allowing the project to integrate different methodological approaches. The resulting interdisciplinary study is characterized by the integration of traditional studies with innovative diagnostic techniques in search of ancient artisanal gestures and the craftspeople behind them.

Some of the key methodologies employed by the A.G.A.T.H.O.C.L.E.S. project and by several contributing authors in this volume are visual archaeology, experimental archaeology, and network studies. Visual Archaeology, especially the computational imaging techniques, when applied to the study of technological and artisanal procedures developed in ancient times, is deeply changing the way archaeological studies can approach the technological aspects of ancient materials.[27] Computational imaging has emerged as an incredible tool in archaeology and cultural heritage. It can capture, process, and analyze images, unlocking hidden details and restoring artifacts virtually. It enables us to better follow the gestures, skills, and movements of raw materials, as well as the 'movements' of peculiar artisanal *savoir-faire*.

Experimental studies have progressed both for Bronze Age archaeology in the Mediterranean and for Classical studies, and there is great potential for further integration of experimental projects in the investigation of craft technologies.[28] Prehistoric figurine marble sculpting, stone cutting technology, and textile technology have been studied alongside making attempts to replicate Roman hairstyles and Roman concrete.

Network Studies, both in the broader sense of interconnectedness and in its Social Network Analysis variation, have progressed significantly, but the combination between technology and networks is far from being systematically explored, especially for the ancient Greek and Roman worlds.[29] In particular, Social Network Analysis—mostly applied in "pure" historical studies, and less in archaeology—in the last few years has been revealed to be a very useful instrument for investigating the materiality of the ancient world.[30] Social Network Analysis, as applied on red-figured vases, can also shed new light on the mobility[31] of ancient craftspeople and the possibility of following their movements through their specific technological procedures.[32]

But the case study of red-figured pottery is just one of the instances where such interdisciplinary approaches can be applied in the ceramic world. Thus, we decided to go beyond the A.G.A.T.H.O.C.L.E.S. project itself, and include a large spectrum of ceramic artifacts and to focus on some key-concepts: technology, crafting procedures, and artisanal networks as anthropological, historical and archaeological evidence for both material and cultural metamorphosis in artisanal productions related to the Greek and Roman worlds. Taking inspiration from the concept of "material metamorphosis", we intended to investigate how "material consciousness" developed for the ceramics of the Greek and Roman world.[33]

---

26 Thanks to the collaboration of Annarosa Mangone, (Department of Chemistry, University "Aldo Moro" of Bari).

27 See the annual *Proceedings of the International Conference on Cultural Heritage and New Technologies*, held in Vienna, Austria (https://ojs.chnt.at/index.php/proceedings [last access 24.04.2024]) and the *Annals of the Photogrammetry, Remote Sensing, and Spatial Information Sciences* by the International Society for Photogrammetry and Remote Sensing (https://www.isprs.org/publications/annals.aspx [last access 24.04.2024]). For some case studies applied on clay materials, see: Bechtold *et al.* (2010); Karl *et al.* (2018); Mara/Portl (2013); Kipke *et al.* (2022). More specifically, concerning RTI applied to red-figured pottery see: Artal Isbrand/Klausmeyer (2013); Saunders/Collmann/Borda (2017); Balachandran (2018).

28 For Early Cycladic figurines, see Papadatos and Venieris (2017); for the Mycenaean pendulum saw, see Blackwell (2018), and for the Bronze Age textile technology, see Jones (2015); for Classical archaeology especially of the Roman world one can mention the replication of Roman hairdos—Stephens (2008)—or the technology of the Roman concrete—Brandon *et al.* (2014). See also some recent attempts of experimental archaeology related to figured pottery in the Classical world: Balachandran (2019) on firing process; Aloupi-Siotis (2020) on coatings and painted parts; Neth/Hasaki (2021) on the potter's wheel; all with previous bibliography.

29 Attempts of combining technology and network perspectives have been adopted in various studies on the Bronze and Iron Ages: see, for example, Brysbaert (2011), and Rebay-Salisbury/Brysbaert/Foxhall (2014). In particular, they focused on material crossover and enhanced how cross-craft can help to find more nodes among people, their artisanal processes and performances, and their interactions with materials and spaces.

30 See Mills/Fowles (2017); Mills/Peeples (2019); Brughmans *et al.* (2023).

31 Horden/Purcell (2000) 342–400; Rouillard (2007), (2009), (2010); Jockey (2009); Archibald (2011) 53–55; Isayev (2017).

32 For artisanal mobility in the prehistoric Mediterranean, see Kiriatzi/Knappett (2017); for the Classical world, see Armstrong/Cohen (2022).

33 See *supra*, note 4.

In this volume, essays by more than fifty authors representing different fields of study – such as archaeology, chemistry, physics, geology, computational imaging techniques and digital humanities, conservation science, network studies, and statistics – highlight how interdisciplinary approaches on ceramics can be crucial to better defining artisanal and social networks in the Greek and Roman world. Each section has two different parts: longer essays come first and correspond to oral presentations delivered at the Turin conference and the webinar series. The second part consists of shorter essays, which correspond to poster presentations at the Conference.

Case studies presented in **Section I** focus on the different shapes of clay and their related crafting local traditions. Starting from the raw material, archaeological research can investigate in many ways how ancient technology and craft procedures have shaped the clay, creating artifacts with specific functions in everyday life. Thus, those papers focus on workshop organization, artisanal skills, the relationship between different types of pottery, and the movements of these ceramics based on some specific technological features.

Section II is about the communities of the ancient pottery workshops and their relationship with workspace, workflow, timing, qualitative and quantitative standards, and tools.

Section III collects archaeometric and petrographic studies on ceramics and shows how they can be combined with other approaches to better define specific workshops' recipes, workflow protocols, and technological solutions.

Section IV is related to ceramics and the invisible features that can become visible: diagnostic analyses of artisanal gestures through computational imaging techniques, development of ceramic databases and the ontology behind them, and new ways of visualizing ceramics networks thanks to network studies.

Section V focuses on craftspeople's mobility and knowledge transfer, especially in relation to the new advances in distribution analyses and statistics, which are often generated from well-developed databases as the ones treated in Section IV.

The book certainly does not claim to be exhaustive and complete, but it is meant to offer stimulating insights for future research on related topics, especially by encouraging the use of new technologies and renewed hermeneutic approaches in relation to ancient material culture, in an integrated analytical framework. All contributions taken together emphasize how crafting dynamics allow us to deal with "[. . .] the issue of technique – but technique considered as a cultural issue rather than as a mindless procedure [. . .]".[34] Investigating artisanal gestures, transfer of knowledge, and the adoption of specific tools[35] can really pave the way to re-humanizing the craftspeople's world and the entire cultural system behind them.

Abbreviation

CVA Corpus Vasorum Antiquorum.

Bibliography

Aloupi-Siotis (2020): Eleni Aloupi-Siotis, "Ceramic Technology: How to Characterise Black Fe-Based Glass-Ceramic Coatings", in: *Archaeological and Anthropological Sciences* 12, 191–206.

Armstrong/Cohen (2022): Jeremy Armstrong and Sheira Cohen (eds.), *Production, Trade, and Connectivity in Pre-Roman Italy*, London/New York.

Artal Isbrand/Klausmeyer (2013): Paula Artal Isbrand and Philip Klausmeyer, "Evaluation of the Relief Line and the Contour Line on Greek Red-figure Vases Using Reflectance Transformation Imaging and Three-dimensional Laser Scanning Confocal Microscopy", in: *Studies in Conservation* 58, 338–359.

Balachandran (2018): Sanchita Balachandran, "Uncovering Ancient Preparatory Drawings on Greek Ceramics". *The Iris*. Online: http://blogs.getty.edu/iris/uncovering-ancient-preparatory-drawings-on-greek-ceramics (last access 15.07.2024).

Barresi (2013): Sebastiano Barresi, "Sicilian Red-figure Vase Painting", in: Claire Lyons, Michael Bennett and Clemente Marconi (eds.), with the collaboration of Alexandra Sofroniew, *Sicily. Art and Invention between Greece and Rome* (Exhibition Catalogue, J. Paul Getty Museum, Los Angeles 2013), Los Angeles, 210–218.

34 Sennett (2008) 8.

35 Knappett (2006) 241. He even points out that tools are used to such an extent that they are almost part of the individual, as an organism, an agent and a person.

Barresi (2018): Sebastiano Barresi, "Il Gruppo di Locri in Sicilia: analisi ed elementi di riflessione", in: Martine Denoyelle, Claude Pouzadoux and Francesca Silvestrelli (eds.), *Mobilità dei pittori e identità delle produzioni*, Cahier du Centre Jean Bérard 25, Naples, 41–63.

Bechtold *et al.* (2010): Sebastian Bechtold, Susanne Krömker, Hubert Mara and Bettina Kratzmüller, "Rollouts of Fine Ware Pottery using High Resolution 3D Meshes", in: Alessandro Artusi, Morwena Joly-Parvex, Genevieve Lucet, Alejandro Ribes and Denis Pitzalis (eds.), *11th International Symposium on Virtual Reality, Archaeology and Cultural Heritage VAST 2010*, Eurographics Association, 79–86. https://doi.org/10.2312/VAST/VAST10/079-086.

Blackwell (2018): Nicholas G. Blackwell, "Experimental Stone-cutting with the Mycenaean Pendulum Saw," in: *Antiquity* 92, 217–232.

Bovero *et al.* (2012): Alessandro Bovero, Marco Demmelbauer, Diego Elia, Maria Gargano, Tommaso Poli and Carla Scilabra, "Studio e intervento di restauro su un cratere apulo a figure rosse", in: *Lo Stato dell'arte 10* (Atti del X Congresso Nazionale del Gruppo Italiano dell'IGIIC International Institute for Conservation, Roma, Accademia Nazionale di San Luca, 22–24 novembre 2012), Rome, 181–188.

Brandon *et al.* (2014): Christoph J. Brandon, Robert L. Hohlfelder, Marie D. Jackson and John P. Oleson, *Building for Eternity: The History and Technology of Roman Concrete Engineering in the Sea*, Oxford.

Brysbaert (2011): Ann Brysbaert (ed.), *Tracing Prehistoric Social Networks through Technology. A Diachronic Perspective on the Aegean*, London/New York.

Brughmans *et al.* (2023): Tom Brughmans, Barbara J. Mills, Jessica Munson and Matthew A. Peeples, *The Oxford Handbook of Archaeological Network Research*, Orford.

de Cesare (2009): Monica de Cesare, "Il pittore della Scacchiera e la nascita della ceramica figurata siceliota", in: Carmine Ampolo (ed.), *Immagine e immagini della Sicilia e di altre isole del Mediterraneo antico* I, Pisa, 277–294.

Delage (2017): Christophe Delage, "Once Upon a Time. . .the (Hi) Story of the Concept of the *Chaîne Opératoire* in French Prehistory", in: *World Archaeology* 49, 158–173.

Denoyelle/Iozzo (2009): Martine Denoyelle and Mario Iozzo, *La céramique grecque d'Italie Meridionale et de Sicile. Productions coloniales et apparentées du viiie au iiie siècle av. J.-C.*, Paris.

Dobres (2010): Marcia-Anne Dobres, "Archaeologies of Technology", in: *Cambridge Journal of Economics* 34, 103–114.

Dobres (2014): Marcia-Anne Dobres, "Concluding and Future Thoughts on Material Crossovers", in: Rebay-Salisbury/Brysbaert/Foxhall (2014) 199–204.

Elia (2012): Diego Elia, "Birth and Development of Red-figured Pottery between Sicily and South Calabria", in: Dimitris Paleothodoros (ed.), *The Contexts of Painted Pottery in the Ancient Mediterranean World (Seventh-Fourth Centuries BCE)*, BAR International Series 2364, Oxford, 101–116.

Elia (2014): Diego Elia, "Local Production of Red-figure Pottery at Locri Epizephyrii: A Synthesis on the Last Decade of Studies", in: Stine Schierup and Victoria Sabetai (eds.), *The Regional Production of Red-figure Pottery: Greece, Magna Graecia and Etruria*, Gösta Enbom Monographs 4, Aarhus, 279–290.

Elia (2019): Diego Elia, "Produzione, distribuzione e consumo da Locri a Caulonia nella prima metà del IV sec. a.C.", in: *Produzioni e committenze in Magna Grecia* (Atti del LV Convegno di Studi sulla Magna Grecia, Taranto, 24–27 settembre 2015), Taranto, 553–567.

Elia/Meirano (2018): Diego Elia and Valeria Meirano (eds.), *Pompeiana Fragmenta. Conoscere e conservare (a) Pompei*, Catalogo della Mostra (Venaria Reale, 12 novembre-21 dicembre 2018), Turin.

Elia *et al.* (forthcoming): Diego Elia, Patrizia Davit, Marco Demmelbauer, Silvia Gobbato and Monica Gulmini, "Le produzioni di ceramica figurata dell'Italia meridionale: problemi e prospettive di ricerca", in: Alfonsina Benincasa, Eliana Mugione and Claude Pouzadoux (eds.), *Le produzioni di ceramica figurata dell'Italia meridionale: problemi e prospettive di ricerca*, forthcoming.

Forcellino (2014): Antonio Forcellino, "Le statue di Michelangelo per la tomba di Giulio II in S. Pietro in Vincoli", in: Christoph L. Frommel (ed.), *Michelangelo. Il marmo e la mente. La tomba di Giulio II e le sue statu*e, Milan, 289–296.

Frommel (2014): Christoph L. Frommel, "La tomba di papa Giulio II: genesi, ricostruzione e analisi", in: Christoph L. Frommel (ed.), *Michelangelo. Il marmo e la mente. La tomba di Giulio II e le sue statu*e, Milan, 19–69.

Gliozzo (2020): Elisabetta Gliozzo, "Ceramic Technology. How to Reconstruct the Firing Process", in: *Archaeological and Anthropological Sciences* 12, 260.

Harris Cline/Hasaki (2023): Diane Harris Cline and Eleni Hasaki, "Assortative Mixing in the Social Networks of Athenian Potters and the Search for Communities of Practice", in: *Journal of Historical Network Research* 8, 21–55.

Hasaki (2016): Eleni Hasaki, "Craft Apprenticeship in Ancient Greece. Reaching Beyond the Masters", in: Willeke Wendrich (ed.), *Archaeology and Apprenticeship: Body Knowledge, Identity and Communities of Practice,* revised edition, Tucson, 171–202.

Hasaki (2019): Eleni Hasaki, "Potters and their Wheels in Ancient Greece: Skills and Secrets in Communities of Practice", in: Mario Denti and Mathilde Villette (eds.), *Archéologie des espaces artisanaux. Fouiller et comprendre les gestes des potiers,* Rennes, 297–314.

Hasaki (2021): Eleni Hasaki, *Potters at Work in Ancient Corinth: Industry, Religion, and the Penteskouphia Pinakes*, Hesperia Supplement 51, Princeton N.J.

Hasaki (2024): Eleni Hasaki, "Handling Greek Vases", in: Nikos Dimakis and Eirene Dimitriadou (eds.), *Καλλίνικος: Honorary Volume for Professor Panos Valavanis*, Athens, 25–36.

Hasaki/Harris Cline (2020): Eleni Hasaki and Diane Harris Cline, "Social Network Analysis and Connoisseurship in the Study of Athenian Potters' Communities", in: Eleni Hasaki and Martin Bentz (eds.), *Reconstructing Scales of Production in the Ancient Greek World: Producers, Processes, Products, People* (Panel 3.4, Proceedings of the 19th International Congress of Classical Archaeology, vol. 8, Cologne/Bonn, May 22–26, 2018), Heidelberg, 59–80. https://doi.org/10.11588/propylaeum.639.c9977.

Isayev (2017): Elena Isayev, *Migration, Mobility and Place in Ancient Italy*, Cambridge.

Jones (2015): Bernice R. Jones, *Ariadne's Threads: The Construction and Significance of Clothes in the Aegean Bronze Age*, Aegaeum 38, Leuven-Liège.

Jones (2021): Richard Jones, "The Decoration and Firing of Ancient Greek Pottery: A Review of Recent Investigations", in: *Advances in Archaeomaterials* 2, 67–127.

Karl *et al.* (2018): Stephan Karl, Paul Bayer, Hubert Mara and András Márton, "Advanced Documentation Methods in Studying Corinthian Black-figure Vase Painting", in: *Proceedings of the 23rd International Conference on Cultural Heritage and New Technologies, CHNT 23* (Vienna, November 2018). https://doi.org/10.11588/heidok.00025189.

Kipke *et al.* (2022): Marta Kipke, Lukas Brinkmeyer, Souaybou Bagayoko, Lars Schmidt-Thieme and Martin Langner, "Deep Level Annotation for Painter Attribution on Greek Vases Utilizing Object Detection", in: *Proceedings of the 4th ACM International workshop on Structuring and Understanding of Multimedia Heritage Contents* (SUMAC '22, Lisboa, October 10th), 23–31. https://dl.acm.org/doi/proceedings/10.1145/3503161#heading16 (last access 15.07.2024).

Kiriatzi/Knappett (2017): Evangelia Kiriatzi and Carl Knappett (eds.), *Human Mobility and Technological Transfer in the Prehistoric Mediterranean*, Cambridge 2017.

Knappett (2006): Carl Knappett, "Beyond Skin: Layering and Networking in Art and Archaeology", in: *Cambridge Archaeological Journal* 16, 239–251.

Madella (2010): Paolo Madella, "Nuove osservazioni su un cratere del pittore della Scacchiera del Museo Paolo Orsi", in: Filippo Giudice and Rosalba Panvini (eds.), *Il greco, il barbaro e la ceramica attica* I, Rome, 77–92.

Mara/Portl (2013): Hubert Mara and Julia Portl, "Acquisition and Documentation of Vessels Using High-Resolution 3D-Scanners", in: Elisabeth Trinkl (ed.), *CVA* Österreiches Beiheft 1. *Interdisziplinäre Dokumentations- und Visualisierungsmethoden*, Vienna, 25–40.

Miller (2005): Daniel Miller, *Materiality*, Durham/North Carolina.

Mills/Fowles (2017): Barbara J. Mills and Severin Fowles (eds.), *The Oxford Handbook of Southwest Archaeology*, New York.

Mills/Peeples (2019): Barbara J. Mills and Matthew A. Peeples, "Reframing Diffusion through Social Network Theory", in: Karen G. Harry and Barbara Roth (eds.), *Interaction and Connectivity in the Greater Southwest*, Boulder, 40–62.

Mirti *et al.* (1995): Piero Mirti, Antonella Casoli, Marcella Barra Bagnasco and Maria Cristina Preacco Ancona, "Fine Ware from Locri Epizephiri: A Provenance Study by Inductively Coupled Plasma Emission Spectroscopy", in: *Archaeometry* 37, 41–51.

Mirti *et al.* (1998): Piero Mirti, Maurizio Aceto and Maria Cristina Preacco Ancona, "Campanian Pottery from Ancient Bruttium (Southern Italy): Scientific Analysis of Local and Imported Products", in: *Archaeometry* 40, 311–329.

Mirti *et al.* (2004a): Piero Mirti, Monica Gulmini, Marco Pace and Diego Elia, "The Provenance of Red Figure Vases from Locri Epizephirii (Southern Italy): New Evidence by Chemical Analysis", in: *Archaeometry* 46, 183–200.

Mirti *et al.* (2004b): Piero Mirti, Monica Gulmini, Alessandra Perardi, Patrizia Davit and Diego Elia, "Technology of Production of Red Figure Pottery from Attic and Southern Italian Workshops", in: *Analytical and Bioanalytical Chemistry* 380, 712–718.

Neth/Hasaki (2021): Brandon Neth and Eleni Hasaki, "The Ancient Greek Potter's Wheel: Experimental Archaeology and Web Applications for Velocity Analysis", in: *Interdisciplinaria Archaeologica. Natural Sciences in Archaeology* 12, 115–125.

Papadatos/Venieris (2017): Yannis Papadatos and Epaminondas Venieris, "An Experimental Approach to the Manufacture of Cycladic-type Figurines with Folded Arms: Preliminary Observations", in: Marisa Marthari, Colin Renfrew and Michael J. Boyd (eds.), *Early Cycladic Sculpture in Context*, Oxford, 483–490.

Pollard/Gosden (2023); Alan Mark Pollard and Chris Gosden, *An Archaeological Perspective on the History of Technology*, Elements, in Archaeological Perspectives on Materials and Technologies, Cambridge.

Rebay-Salisbury/Brysbaert/Foxhall (2014): Katharina Rebay-Salisbury, Ann Brysbaert and Lin Foxhall (eds.), *Knowledge Networks and Craft Traditions in the Ancient World. Material Crossovers*, London/New York.

Rieck/Mara/Krömker (2013): Bastian Rieck, Hubert Mara and Susanne Krömker, "Unwrapping Highly-Detailed 3D Meshes of Rotationally Symmetric Man-Made Objects", in: *ISPRS Annals of the Photogrammetry, Remote Sensing and Spatial Information Sciences*, Volume II-5/W1, (XXIV International CIPA Symposium, Strasbourg, 2–6 September), 259–264. https://isprs-annals.copernicus.org/articles/II-5-W1/259/2013 (last access 15.07.2024).

Roux, Valentine (ed.), in collaboration with Marie-Agnès Courty (2019): *Ceramics and Society: A Technological Approach to Archaeological Assemblages*, Switzerland.

Roux, Valentine (2020): "*Chaîne Opératoire*, Technological Networks and Sociological Interpretations", in: *Cuadernos de Prehistoria y Arqueología de la Universidad de Granada* 30, 15–34.

Santostefano (2020): Antonella Santostefano, "Riferimenti cronologici per la ceramica a figure rosse siceliota tra V e IV sec. a.C.", in: *Quaderni di Archeologia. A cura dell'Università degli Studi di Messina* X, nuova serie, 75–108.

Sennett (2008): Richard Sennett, *The Craftsman*, New Haven/London.

Serino (2014): Marco Serino, "The Beginnings of Sicilian Red-figured Pottery and its Relationship with Early South Italian Productions: A Reappraisal through the Case-Study of the Himera Painter Workshop", in: Stine Schierup and Victoria Sabetai (eds.), *The Regional Production of Red-figure Pottery: Greece, Magna Graecia and Etruria*, Gösta Enbom Monographs 4, Aarhus, 247–267.

Serino (2019): Marco Serino, *La Bottega del Pittore di Himera e le altre Officine Protosiceliote. Stile, Iconografia, Contesti, Cronologia*, Fecit te, Rome.

Soleti (2012): Vita Maria Soleti, "Storia degli studi", in: Todisco (2012) 37–71.

Spigo (2023): Umberto Spigo, "Per una rilettura del Gruppo di Maron e delle «tendenze apulizzanti» della ceramica siceliota", in: *Mare Internum. Archeologia e culture del Mediterraneo* XV, 31–62.

Stephens (2008): Janet Stephens, "Ancient Roman Hairdressing: On (Hair)pins and Needles", in: *Journal of Roman Archaeology* 21, 110–130.

Stissi (2002): Vladimir Stissi, *Pottery to the People. The Production, Distribution and Consumption of Decorated Pottery in the Greek World in the Archaic Period (650–480 BC)*, PhD Dissertation, University of Amsterdam.

Tallis (2003): Raymond Tallis, *The Hand: A Philosophical Inquiry in Human Being*, Edinburgh.

Todisco (2012): Luigi Todisco (ed.), *La ceramica a figure rosse della Magna Grecia e della Sicilia*, Vols. I-III. Rome.

Vasari (1550): Giorgio Vasari, *Le vite de' più eccellenti pittori, scultori e architettori*, (original edition), Florence.

Section I: **The Shapes of Clay: Manufacturing and Local Traditions**

Gudrun Klebinder-Gauß

2 What Makes Aeginetan Cooking Pots So Special? Technological Choices in Ancient Aegina's Pottery Production

Abstract: Aegina's reputation as a producer of ceramic vessels for use with fire is attested by a large number of finds from outside the island, but also by ancient written sources. Intensive interdisciplinary research during the last few years has provided a comprehensive picture of Aeginetan cookingware production from prehistory to the Hellenistic period, of its stylistic and technical characteristics, and of the dynamics of its production and supply. This paper examines the question of what made Aeginetan cookingware so attractive to the many consumers abroad. It discusses technical choices and potential physical advantages of Aeginetan cooking ware items, but also considers the impact of non-technical aspects on the decision to buy an Aeginetan product, such as style, cultural connotations and, last but not least, the given socio-economic and political conditions.

Introduction

Through the Late Archaic and Early Classical periods, cooking pottery from the Greek island of Aegina was a highly regarded and widely traded megaseller.

The reputation of Aeginetan cooking pottery far beyond the island was noted in ancient written sources, where Aegina was called χυτρόπωλιν, "pot seller."[1] The term "chytra" refers generally to an earthenware pot for everyday use, and in particular to a cooking pot.[2] Aegina's designation as a 'pot seller is striking, because information about potters and their products is generally rare in ancient literature, especially in relation to everyday, undecorated pottery.[3] Therefore, Aegina's association with cooking pots was considered noteworthy, and Aegina gained a reputation for producing ceramic vessels for use with fire.

That Aeginetan cooking pottery was well known is also supported by its wide distribution beyond the island. In the Late Archaic and Early Classical periods Aeginetan cooking pottery was traded not only with nearby Athens and Corinth but also farther afield—to Crete, the Black Sea region, North African Cyrenaica, Italy, and the southern coast of France.[4] Its popularity is particularly evident at the Athenian Agora, where, according to S. Rotroff, Aeginetan products account for over 70% of the cooking pots discarded there between c. 525 and 480 B.C.[5]

Such wide distribution invites the question what made Aeginetan cooking pottery so attractive for so many consumers abroad. In other words, did Aeginetan cooking ceramics have specific qualities that made them superior to alternative products? Scholarship that assesses what makes a good cooking pot is extensive, as is the bibliography on this topic. Our study of cooking ware production on Aegina can contribute significantly to the question,[6] as intensive research has provided a comprehensive picture of production from prehistoric times to the Hellenistic period. By combining systematic stylistic examination with scientific analysis and geological sampling on the island, we were able to link Aeginetan cooking ware to specific technologies and types of raw materials, and we could reconstruct the dynamics of its production and supply.[7]

---

1 Poll. 7.197; see also Weilhartner (2012) 209–210 for the reading of χυτρόπωλιν as "pot-seller", in contrast to the reading χυτρόπλοιν "town that produces cooking pots" as is sometimes found in current studies.

2 Weilhartner (2012) 208 with further references.

3 Scheibler (1995) 72–73.

4 For further references, see Klebinder-Gauß (2012) 198 and Klebinder-Gauß/Gauß (2015) 74–75 and fig. 8.

5 Rotroff (2015) 183.

6 For the term 'cooking ware', see e.g., Sparkes/Talcott (1970) 34; see also Klebinder-Gauß/Strack (2015) 79 for the use of the term 'ware' for defining a group of vessels by typological, technological, and compositional data.

7 For Aeginetan pottery production in the Bronze Age and for geological sampling on the island see especially Gauß/Kiriatzi (2011) 71–92; for the historic period, see Klebinder-Gauß (2012); Pentedeka *et al.* (2012).

https://doi.org/10.1515/9783111189635-002

Cooking Ware Production on Archaic and Classical Aegina

Aeginetan cooking ware production in the historical period can be traced back at least to the early 7th century B.C. We currently lack securely dated finds from earlier phases of the Iron Age, most probably because the evidence for this time is incomplete; we can only assume that cooking ware was produced on Aegina also in the Protogeometric and Geometric periods.[8] Aegina certainly boasted flourishing cooking ware production throughout the Bronze Age, and its products have been found in contexts also beyond the island dating from the Middle Bronze Age until the Mycenean postpalatial period of the 12th century B.C.[9] It is interesting that scientific analysis has revealed that Archaic/Classical cooking ware fabric closely resembled the fabric used throughout the Middle and Late Bronze Age, reflecting the use of similar types of raw materials and even similar sources, though differing in the better sorting of the Archaic/Classical fabrics.[10] It is, however, currently still rather difficult to close the gap between the Late Helladic IIIC Early and the Archaic period.[11]

By the 6th century B.C., Aeginetan potters had attained a high level of expertise. Between the Late Archaic and Early Classical period they considerably expanded their repertoire, and they enhanced their manufacturing techniques. Also productivity increased significantly, satisfying needs not only for local consumption but also for overseas trade. During this period, Aeginetan cooking ware reached its widest distribution across the Aegean and beyond. The heyday of Aeginetan cooking ware production coincides with a phase of great prosperity on the island, which was at that time politically independent and extraordinarily wealthy because of its success in long-distance sea trade.[12] At the end of the 3rd quarter of the 5th century B.C., however, pottery production decreased dramatically, most probably aligning with the historical developments of the time. The capture of the island by the Athenians in 456 B.C. and, above all, the later expulsion of the local population in 432 B.C. cemented the loss of Aegina's independence and outstanding position as a trading power.

The Repertoire of Aeginetan Cooking Ware

The range of shapes of Aeginetan cooking ware expanded during the Archaic and Classical periods.[13] In the 7th and 6th centuries B.C., all-purpose globular chytrai with one or, less commonly, two vertical strap handles were the only cooking pots used (Fig. 1).

Then, in the late 6th and early 5th century new types joined the repertoire, such as the lidded globular chytra, the round-bottomed shallow lopas, various kinds of pans, and a bell-shaped cover, as well as the barrel-shaped brazier and the eschara. The repertoire also included specific implements for braising and stewing, grilling, frying, parching, spit-roasting, baking, and heating. Containers for dipping and for carrying water, such as jugs and amphoroid jars, were made from the same fabric.

The introduction of more specialized equipment for cooking can also be observed elsewhere in the late 6th and earlier 5th century, for example in the increasing elaboration of Corinthian mortaria at that time.[14] This phenomenon probably coincided with changes in diet and food processing, as well as more complex recipes and a finer sophistication in cooking techniques, which may have arisen from the spread of symposium culture.[15] The fact that the Aeginetans took the lead in including these specific shapes into their repertoire and introduced them to places like Athens may be because Aeginetan cooking pottery production was already well established by this time and known far beyond the island. Savvy Aeginetan potters possibly recognized the demand for more sophisticated cooking methods and specific implements and realized their capacity to cover this demand. The exact source of inspiration for the Aeginetan potters is, however, difficult to assess. There is no indication of an influx of new people with distinct cooking traditions on Archaic Aegina or of imports of 'foreign' lopades, lidded chytrai, barrel-shaped braziers, or escharai that may have served as models for Aeginetan

8 Klebinder-Gauß/Strack (2015) 81.
9 Klebinder-Gauß/Gauß (2015) 79–80.
10 Gauß/Kiriatzi (2011) 239–240 and Pentedeka *et al.* (2012) 111.
11 Gauß *et al.* (2017) 51.
12 Klebinder-Gauß (2012) 196–197; Klebinder-Gauß/Gauß (2015) 71–77.
13 For a detailed overview on the range of shapes of Aeginetan cooking ware, see Klebinder-Gauß (2012) 182–193.
14 Villing/Pemberton (2010) 624–625.
15 Trapichler (1995) 74–75; Dibble (2010) 107–108, 131–132, 138; Villing/Pemberton (2010) 625 with further references.

Fig. 1: Aeginetan cooking ware chytra from a Late Archaic/Early Classical context. © Gudrun Klebinder-Gauß.

products.[16] Rather, early experimental versions of these types, made in well-known Aeginetan cooking ware fabric and using traditional manufacturing techniques, attest that their production is deeply rooted in the Aeginetan pottery tradition.[17] Aeginetans may have come into contact with and found inspiration in exotic spices, food preparation practices, and implements during trade activities. In any case, the new Aeginetan products quickly made their way into kitchens beyond the island, as for example in Athens, where the very first lopades, lidded chytrai, barrel-shaped braziers, escharai, and other new Aeginetan types were part of local culinary practices.

Fabric and Manufacturing Techniques

The particular quality and appearance of Aeginetan cooking ware products—thin walls, light weight, and a dense, lustrous surface—come not only from the raw materials used but also from a set of specific shaping and finishing techniques.

Aeginetan cooking ware of the Archaic/Classical period is made of a characteristic red-brown fabric that macroscopic examination has characterized as sandy with flakes of golden biotite mica and lathlike glossy black inclusions. Petrographic and elemental analysis has confirmed the origin of this fabric from Aegina and has located the raw material sources in the volcanic central and southern part of the island, most possibly in an area about four kilometers south of Kolonna in a restricted zone along the west coast of the island. Aeginetan cooking ware fabric is thus associated with fragments of volcanic rocks of intermediate composition which most probably are the products of natural weathering.[18]

Until the 7th century B.C. Aeginetan cooking pots were handmade, but from the 6th century onward potters also used a rotational device.[19] The technique for building large closed vessels such as globular chytrai and flat-based jugs is most characteristic. The lower body and the shoulder-neck section were usually made in two steps. In this way, the interior of the lower body remained accessible for finishing with a flexible scraping tool such as a thin piece of wood or bark (Fig. 2). Only small pots whose interior was less easily accessible for that kind of tool appear to have been made in one piece with the help of a rotational device, as indicated by spiral markings running all the way from the interior base upwards (Fig. 3).

16 See e.g., the case of Medieval Corinth where a change in cooking-pot typology can be convincingly associated with the arrival of new people: Joyner (2007) 188–189.

17 Klebinder-Gauß (2012) 184–185, 188, 191 on Aeginetan prototypes of lidded chytrai, lopades, barrel-shaped braziers, and escharai; see also Sparkes/Talcott (1970) 227.

18 Klebinder-Gauß (2012) 89–90; Pentedeka *et al.* (2012) 106–115; Kiriatzi *et al.* (2011) 93–99, 131–134 and geological map on fig. 32.

19 Klebinder-Gauß/Strack (2015) 81–82; for earlier handmade Aeginetan cooking pots, see Klebinder-Gauß (2012) 283–284, nos. 484–485, figs. 48. 111; for an overview of the shaping and finishing techniques used with Aeginetan cooking ware see Klebinder-Gauß (2012) 178–182 and Klebinder-Gauß/Strack (2015) 81–83.

Fig. 2: Tool marks from scraping at the interior of an Aeginetan cooking ware jug. © Gudrun Klebinder-Gauß.

Fig. 3: Wheel marks on the interior base of a small Aeginetan cooking ware chytra. © Gudrun Klebinder-Gauß.

This suggests that the lower part of larger vessels made in two separate stages were formed on the wheel also, even though the thorough finish of the surfaces usually obliterates evidence of the primary forming process. For the forming of the rounded base of globular vessels the clay may have been pressed into a mold while rotating on a wheel.[20] Most probably the upper shoulder-neck part of vessels that were formed in two separate steps was added to the lower one on the wheel by applying a coil of clay to the upper edge of the lower part and then drawing it up.[21]

The smoothing of the lower part's interior surface with a scraping tool was done when the clay was still wet or slightly dried. The edge of this tool left angular impressions perpendicular to the direction of its movement, occasionally extending from the bottom to the wall of the vessel (Fig. 2). Moved in various directions up to the upper edge, the scraper generated irregularly overlapping stripes of parallel striations from dragging the larger particles along the surface. This technique

20 For the use of molds on the wheel, see e.g., Rice (2005) 129, 132–133; Rye/Evans (1976) 32–33, fig. 6, pl. 21.
21 For forming pots in several steps, see, among others, Hampe/Winter (1962) 93; Rye/Evans (1976) 33.

probably did not generate much displacement of material but primarily created an even wall thickness throughout by obliterating irregularities and smoothing wheel marks.[22]

Wide round-bottomed shapes such as lopades were probably built by forming the lower body part in a mold and then adding the upper part on the wheel, as is indicated by the presence of regular horizontal grooving on the interior shoulder and on the rim. Molding was probably also used for wide shallow pans and the wide thick-walled bowls of escharai. Large and heavy flat-bottomed pans with a low vertical rim and a diameter up to 48 cm most likely were completely hand-built.

Exterior surfaces of closed shapes and both the interior and exterior surfaces of open shapes were thoroughly wet-smoothed in order to create a smooth, dense coating from a finer layer of clay that was largely free of inclusions. Then, the leather-hard or dry surface was burnished by drawing with a hard, well-smoothed tool, such as a pebble or the rounded end of a bone, more or less closely spaced parallel or zigzag lines whose compacted and at least partly vitrified surface created a specific luster (Fig. 4).[23]

Fig. 4: Burnishing marks on an Aeginetan cooking ware vessel. © Gudrun Klebinder-Gauß.

Fig. 5: Glaze wash on an Aeginetan cooking ware lopas. © Gudrun Klebinder-Gauß.

22 For the use of scraping tools for finishing a pot, see e.g., Rice (2005) 137; for ethnoarchaeological studies on the scraping operations, see e.g., Gosselain (2008) 36. For the use of this technique in Aeginetan Early Helladic II pottery production, see Berger (2010) 32: '(Grob) verstrichen/ gewischt'.

23 Rice (2005) 138; Rutter (1995) 56–57.

The burnishing marks usually cover the entire lower exterior of closed shapes, often including the bottom, and the interior of open shapes such as lopades or pans, and also the rims of escharai or the exterior side of lids. Throughout the Archaic and Early Classical periods burnishing is among the most characteristic features of Aeginetan cooking ware.

After the mid–5th century, such burnishing was gradually replaced by a thin clay-based wash applied with a sponge or brush, creating a dense surface with streaky bright red to dark red hues (Fig. 5).

Organization and Location of Aeginetan Cooking Ware Production

Unfortunately our understanding of the organization and scale of Aeginetan cooking ware production is incomplete. Our detailed knowledge of its technical and stylistic characteristics and of the distribution pattern of Aeginetan cooking ware suggests that its production was fairly specialized and prolific, aiming at both local consumption and export trade. The potters' high level of expertise and their specialized methods of ceramic production are indicated not only by the consistent homogeneity in stylistic and morphological properties, manufacturing techniques, and the composition of the fabric but also by the repertoire's inclusion of specific implements for specific cooking methods, as well as by the widespread consumption of Aeginetan cooking ware products around the Aegean and beyond.[24]

Production in more than one workshop is suggested by variations in the execution of particular morphological features within the same conceptual scheme, such as for example the curve of the rim of lopades.[25] The increase in specialization and standardization of Aeginetan cooking ware production through the Late Archaic and Early Classical periods corresponds to an increase in production and a surge in exports during a phase of intensifying external contacts within the Aeginetan long-distance trade network. Increased trade may even have been the driving force behind the intensification of Aeginetan cooking ware production and the wide distribution of Aeginetan products. A possible mode of Aeginetan cooking ware production may thus be D. Peacock's model of the nucleated workshop that also includes the involvement of middleman traders and their wide distribution networks.[26]

With the exception of a waster whose exact find spot is unknown,[27] no direct evidence about a production site for Aeginetan cooking ware dated to the Archaic or Classical periods has so far been discovered (Fig. 6).[28] Ethnographic studies reveal that, although proximity to the raw material sources was desirable, water and fuel availability, as well as easy access to markets, was at least as important.[29] In this respect the volcanic central and southern part of the island, with its unfavorable environmental conditions, was probably considered less attractive for the location of pottery workshops. Therefore, workshops should rather be anticipated in the area of the island's capital, in the flat and fertile northwest of the island, where the majority of Aegina's inhabitants lived and where the large trading port was located. Moreover, a supply of fresh water was probably better provided here, as the Aeginetans depended primarily on the strict management of rainwater, with an underground aqueduct, wells, cisterns, and tanks.[30] The use of sea water in the Aeginetan cooking ware production has not yet been attested.[31] A possible hypothesis is that the pottery workshops were located in the broad area between the main settlement and the western foothills of the volcanic mountainous landscape, at a reasonable distance from the raw material sources.[32]

24 Pentedeka *et al.* (2012) 106–111, 170 on the homogeneous composition of the fabric; for the discussion of a link between standardization and specialized modes of ceramic production see, among others, Rice (2005) 202–204 and Kotsonas (2014), with further references.

25 For variation in the execution of the rim of lopades of the Classical period, see Klebinder-Gauß (2012) 186.

26 Peacock (1982) 9; see also Rice (2005) 184, 187.

27 For this waster from the 2nd half of the 4th century B.C., which consists of numerous cooking vessels of various types, and for its origin on Aegina, see Klebinder-Gauß (2012) 204–206, nos. 488–505.

28 For this information I would like to thank Sophia Michalopoulou, Ephorate of Antiquities of Piraeus and Islands.

29 Hampe/Winter (1962) 87–88, 103; Scheibler (1995) 107. See also Rice (2005) 177–180 on the various factors governing the location of pottery-making communities, particularly proximity to resources and markets.

30 For the Classical city plan of Aegina, see Welter (1938) 39, fig. 36 and map.

31 My thanks to Evangelia Kiriatzi for this comment. For the use of sea water in pottery production, in particular to reduce lime-popping in calcite-tempered ceramics, see e.g., Rice (2005) 119; Tite/Kilikoglou (2002) 3.

32 Gauß/Kiriatzi (2011) 249.

Fig. 6: Waster of Aeginetan cooking ware pottery, 2nd half of 4th century B.C. © Gudrun Klebinder-Gauß.

Trade in Cooking Pottery

How can we explain the widespread and persistent popularity of Aeginetan cooking ceramics among the many consumers beyond the island, and why did consumers prefer Aeginetan products over their own local products or alternative foreign wares? In general, the trade of cooking pottery is not exceptional. Although past scholars thought that only fine-ware pottery was widely traded, it is clear that plain, utilitarian household pottery was also traded.[33] The wide distribution of Aeginetan cooking ware across the Mediterranean and beyond bears witness to the trade of highly specialized coarse household pottery. Aeginetan cooking ware mainly appears along coastlines and in areas where ancient written sources place Aeginetans, often in connection with trade activities. Thus, trade of Aeginetan cooking pottery is closely linked to the routes of the Aeginetan mercantile fleet,[34] which supports the hypothesis that trade in plain, utilitarian household pottery took place primarily in colonial or diaspora contexts.[35] At Elea, for example, a Greek city on the coast of the Tyrrhenian Sea, Greek-style cooking pots far outnumber Italic-style ones; in non-Greek settlements on the same coast the opposite is true.[36]

It is interesting that foreign consumers of Aeginetan cooking pottery wanted only the vessels intended for use with fire, and not the water jars made of the same fabric and with the same manufacturing techniques. Moreover, Aeginetan products were also used in places with a firmly established local cooking pottery tradition, such as in Athens, where the production of local cooking implements eventually stopped because consumers preferred Aeginetan imports.[37] This discontinuance suggests

33 For this discussion see, among others, Gassner (2003) 220–225; Villing/Pemberton (2010) 555–556, 575, 625.
34 Klebinder-Gauß/Gauß (2015) 74–75, fig. 8.
35 Villing (2006) 40.
36 Trapichler (1995) 75–76.
37 Klebinder-Gauß/Strack (2015) 77.

that Aeginetan cooking ware was deemed superior to alternative products in heating and cooking performance, and that it was traded and consumed as a highly specialized commodity because of its high functionality and its specialized range of shapes.[38]

Performance Characteristics of Aeginetan Cooking Ceramics

The choice of which clay vessel to use for cooking and heating was certainly influenced by various factors. The performance of these specialized implements, which was affected by a complex range of variables, was probably of chief significance. Cooking pots ideally should meet certain requirements to guarantee efficient heating and durability. Most important is good thermal behavior: the ability to transfer heat efficiently throughout the vessel wall to its contents and the ability to withstand the stress of a rapid change in temperature as well as repeated heating and cooling. A number of studies have evaluated the physical properties of ceramics exposed to heat, such as strength, toughness, thermal-shock resistance, and thermal conductivity. These studies are often contradictory, as they variously assess the influence of forming techniques, surface treatments, clay-paste preparation, firing, and morphological features on the performance of cooking implements, as well as how various technological choices can each contribute to different physical properties.[39]

Certainly, Aeginetan cooking ware boasts some of the properties that are generally thought to be best for efficient cooking. For example, rounded shapes are considered advantageous for regular and efficient heat transfer.[40] Furthermore, rounded vessels better withstand the stress of rapid changes in temperature, and they resist cracks and vessel failure better than vessels with sharp angles at the junction of bottom and wall.[41] Globular or round-bottomed shapes, such as plain and lidded chytrai, lopades, and roasting pans, therefore dominate the Aeginetan repertoire of cooking pots. It is not surprising that other production centers of the period throughout the Mediterranean, as well as ethnographic studies of contemporary pottery production, indicate a wide use of rounded shapes for cooking pots.[42]

Other key properties of Aeginetan cooking pots include thin walls and smooth surfaces. Thin walls transfer heat more quickly and thus save fuel, and they are thought to better resist the stress caused by rapid temperature changes.[43] Smooth, even surfaces—in Aeginetan cooking ware achieved by a smooth, dense coat of a finer layer of clay on the exterior and by scraping the interior of closed shapes—are likewise deemed important for thermal-stress resistance.[44]

Various surface treatments affect the performance of cooking ceramics. Dense surfaces, for example, impact porosity and permeability, and reduced porosity and permeability are particularly desirable for containers used to store and transport liquids.[45] Dense surfaces are also thought to have a positive effect on heat conductivity.[46] Burnished surfaces or surfaces with a thin glaze wash are not only less permeable but also harder and thus more resistant to external mechanical impacts such as stirring the contents or moving the pot.[47] In Archaic/Classical Aeginetan cooking ware production dense surfaces are achieved by wet-smoothing and subsequent burnishing or the application of a thin glaze wash, both for vessels used with fire and for vessels for water storage. Because burnishing marks and wash appear on the underside of vessels, it is likely that these surface treatments not only served as decoration but also had a practical function. Although burnishing appears in local Aeginetan pottery since the Bronze Age, it is interesting that at that time it was rarely used for vessels intended for use with fire.[48] Burnishing marks may, however, also have served a function beyond possible physical

38 Similarly Gassner (2003) 222–223 on Late Archaic Elea where the cooking ceramics used with fire were mostly imported, e.g., from as far away as Sicily or North Africa.

39 For a comprehensive overview of the discussion with further references, see e.g., Whitbread (2015).

40 Hein *et al.* (2015) 52, fig. 4.3.

41 Rice (2005) 229; Tite/Kilikoglou (2002) 1.

42 E.g., Aydemir (2005) 87–89 for an Archaic East Greek type of cooking pot; Ben-Shlomo *et al.* (2008) 225–226 for Canaanitan-style cooking pots from the Late Bronze and Iron Age; Joyner (2007) 189, notes 33–34.

43 Rice (2005) 227, 369; Tite/Kilikoglou (2002) 1.

44 Rice (2005) 232, 368–369.

45 Henrickson/McDonald (1983) 632–633 on an ethnographic study on pottery production in the Zagros highlands of central western Iran where burnishing and glazing are used more often for certain liquid storage vessels than for any other functional groups.

46 For the complex effects of reduced permeability on cooking pots see, among others, Schiffer (1990) 377–379; Rice (2005) 231, 367–368.

47 Henrickson/McDonald (1983) 633; Rutter (1995) 56; Rice (2005) 231–232, 353–355.

48 Berger (2010) on the emergence of burnishing in Aeginetan pottery production at the very end of Early Helladic II; Gauß/Kiriatzi (2011) 247–248, 250.

advantages: distinct patterns—vertical or horizontal stripes, zigzags, or vertical stripes with a zone of horizontal stripes at the greatest circumference—may have differentiated the works of individual potters or production units.[49]

The methods of clay-paste preparation as well as the different types, shapes, and amounts of temper, have an effect on a pot's performance, especially its thermal-shock resistance. Large amounts of temper material, in particular, appear to increase thermal-shock resistance.[50] Any advantage of volcanic inclusions, a characteristic component of Aeginetan cooking ware fabric, is difficult to assess.[51] Low firing temperatures below or close to the onset of vitrification, as is the case with Aeginetan cooking ware, are thought to support thermal-shock resistance in cooking ceramics.[52]

Why Choose an Aeginetan Cooking Pot?

Returning to the question posed at the start of this chapter—Why would a foreign consumer choose an Aeginetan cooking pot?—our instinct suggests that functional qualities must have been paramount. We have credited the Aeginetan potters with selecting functionally specific raw material and technological choices for cooking pots to produce durability and good thermal behavior. Indeed, a second principal fabric type on Aegina, made from a light-fired calcareous clay from deposits in the northwest of the island, was not used for the production of cooking pottery, neither in the Archaic/Classical nor in the prehistoric periods.[53]

It is, however, difficult to assess how well these various potential physical advantages of Aeginetan cooking ceramics were understood by consumers. Foreign consumers may (also) have been attracted by other factors, such as their light weight, the range of their repertoire, their style, and their widespread availability. The light weight of Aeginetan cooking pots, due to their thin walls, and thus their easy handling, may have been reasons for preferring them to alternative products.[54] Their varied repertoire, with its wide range of specialized shapes, possibly opened up new, previously unknown ways of food preparation and thus contributed to the great popularity of Aeginetan cooking ware.[55] Other, nonfunctional factors, such as the style, the fashion, and the pleasant visual appearance or touch of the vessels, should also be considered. Aeginetan cooking ceramics may have yielded a specific flavor experience.[56] Cultural connotations may also have played a role in the decision to choose an Aeginetan cooking pot. One final factor most likely played an important role in the wide distribution of Aeginetan cooking ware, namely its ready availability (also of replacements), through the far-reaching Aeginetan trade networks.[57]

In closing, no single explanation for why Aeginetan cooking pottery became a popular megaseller is sufficient. Rather, many factors—functional but also sensory, cultural, and the particular historical (socioeconomic-political) conditions—contributed to its wide acceptance and to Aegina's being called a '*chytropolis*'.

Bibliography

Aydemir (2005): Ahmet Aydemir, "Funde aus Milet. XX. Kochgeschirr und Küchengeräte aus dem archaischen Milet", in: *Archäologischer Anzeiger* 2005/2, 85–101.

Ben-Shlomo *et al.* (2008): David Ben-Shlomo, Itzhaq Shai, Alexander Zukerman and Aren M. Maier, "Cooking Identities: Aegean-Style Cooking Jugs and Cultural Interaction in Iron Age Philistia and Neighboring Regions", in: *American Journal of Archaeology* 112, 565–580.

49 See the example of an ethnoarchaeological project in a community of Philipino potters: London (1991) 193.

50 E.g., Müller *et al.* (2014) 267.

51 Klebinder-Gauß (2012) 176; Whitbread (2015) 32.

52 Müller *et al.* (2014) 267; Pentedeka *et al.* (2012) 202 on the original firing temperatures of Aeginetan cooking ware around/below 800°C.

53 For this fabric type, see Gauß/Kiriatzi (2011) 49–50, 99–104; Klebinder-Gauß (2012) 90; Pentedeka *et al.* (2012) 115–124.

54 See Whitbread (2015) 31 for an ethnographic study in the Kalinga village of Guina-ang, Philippines, where consumers justify their preference for particular cooking pots because of their light weight.

55 See above page 12.

56 Sillar (2001) 175–176.

57 Klebinder-Gauß/Gauß (2015) 74–75 fig. 8; see also Whitbread (2015) 32–33.

Berger (2010): Lydia Berger, “Zur Terminologie und Definition der Oberflächenbehandlung anhand gebrannter Gefäße in der prähistorischen Keramikforschung”, in: Barbara Horejs, Reinhard Jung and Peter Pavuk (eds.), *Analysing Pottery- Processing – Dlassification – Publication*, Bratislava, 29–37.
Dibble (2010): William F. Dibble, *The Archaeology of Food in Athens: The Development of an Athenian Urban Lifestyle*, Master thesis, University of Cincinnati. http://rave.ohiolink.edu/etdc/view?acc_num=ucin1274209972 (last access 16.04.2024).
Gassner (2003): Verena Gassner, *Materielle Kultur und kulturelle Identität in Elea in spätarchaisch-früklassischer Zeit*, Velia-Studien 2, Vienna.
Gauß *et al.* (2017): Walter Gauß, Evangelia Kiriatzi, Michael Lindblom, Bartlomiej Lis and Jerolyn E. Morrison, “Aeginetan Late Bronze Age and Early Iron Age Cooking Pottery”, in: Julie Hruby and Debra Trusty (eds.), *From Cooking Vessels to Cultural Practices in the Late Bronze Age Aegean*, Oxford, 46–56.
Gauß/Kiriatzi (2011): Walter Gauß and Evangelia Kiriatzi, *Pottery Production and Supply at Bronze Age Kolonna, Aegina. An Integrated Archaeological and Scientific Study of a Ceramic Landscape*, Ägina-Kolonna. Forschungen und Ergebnisse 5, Vienna.
Gosselain (2008): Olivier P. Gosselain, “Ceramics in Africa”, in: Helaine Selin (ed.), *Encyclopaedia of the History of Science, Technology, and Medicine in Non-Western Cultures*, Berlin, 32–44.
Hampe/Winter (1962): Rudolf Hampe and Adam Winter, *Bei Töpfern und Töpferinnen in Kreta, Messenien und Zypern*, Mainz.
Hein *et al.* (2015): Arno Hein, Noémi Müller and Vassilis Kilikoglou, “Heating Efficiency of Archaeological Cooking Vessels: Computer Models and Simulations of Heat Transfer”, in: Spataro/Villing (2015) 49–54.
Henrickson/McDonald (1983): Elizabeth F. Henrickson and Mary M. A. McDonald, “Ceramic Form and Function: An Ethnographic Search and an Archeological Application”, in: *American Anthropologist* 85, 630–643.
Joyner (2007): Louise Joyner, “Cooking pots as Indicators of Cultural Change. A Petrographic Study of Byzantine and Frankish Cooking Wares from Corinth”, in: *Hesperia* 76, 183–227.
Kiriatzi *et al.* (2011): Evangelia Kiriatzi, Myrto Georgakopolou and Areti Pentedeka, “Pottery Production and Importation at Bronze Age Kolonna: The Ceramic Fabrics and the Island’s Landscape”, in: Walter Gauß and Evangelia Kiriatzi (eds.), *Pottery Production and Supply at Bronze Age Kolonna, Aegina. An Integrated Archaeological and Scientific Study of a Ceramic Landscape*, Vienna, 69–156.
Klebinder-Gauß (2012): Gudrun Klebinder-Gauß, *Keramik aus klassischen Kontexten im Apollon-Heiligtum von Ägina-Kolonna*, Ägina-Kolonna, Forschungen und Ergebnisse 6, Vienna.
Klebinder-Gauß/Gauß (2015): Gudrun Klebinder-Gauß and Walter Gauß, “Opportunity in Scarcity”, in: Achim Lichtenberger and Constance von Rüden (eds.), *Multiple Mediterranean Realities. Current Approaches to Spaces, Resources and Connectivities*, Paderborn, 67–91.
Klebinder-Gauß/Strack (2015): Gudrun Klebinder-Gauß and Sara Strack, “An Aeginetan Potters’ Workshop in Athens?”, in: Walter Gauß, Gudrun Klebinder-Gauß and Constance von Rüden (eds.), *The Transmission of Technical Knowledge in the Production of Ancient Mediterranean Pottery*, Vienna, 77–95.
Kotsonas (2014): Antonis Kotsonas, “Standardization, Variation, and the Study of Ceramics in the Mediterranean and Beyond”, in: Antonis Kotsonas (ed.), *Understanding Standardization and Variation in Mediterranean Ceramics Mid 2nd to Late 1st Millennium BC*, Leuven, 7–23.
London (1991): Gloria Anne London, “Standardization and Variation in the Work of Craft Specialists”, in: William A. Longacre (ed.), *Ceramic Ethnoarchaeology*, Tucson, 182–204.
Müller *et al.* (2014): Noémi S. Müller, Vassilis Kilikoglou, Peter M. Day and George Vekinis, “Thermal Shock Resistance of Tempered Archaeological Ceramics”, in: M. Martinón-Torres (ed.), *Craft and Science: International Perspectives on Archaeological Ceramics*, DOH, Qatar http://dx.doi.org/10.5339/uclq.2014.cas.ch28.
Peacock (1982): David P. S. Peacock, *Pottery in the Roman world: An Ethnoarchaeological Approach*, London.
Pentedeka *et al.* (2012): Areti Pentedeka, Evangelia Kiriatzi and Myrto Georgakopoulou, “Understanding Local Products and Exploring Sources of Imports: Petrographic and Chemical Analysis of Classical Pottery from Kolonna, Aegina”, in: Klebinder-Gauß (2012) 102–170.
Rice (2005): Prudence M. Rice, *Pottery Analysis. A Sourcebook*, Chicago.
Rotroff (2015): Susan I. Rotroff, “The Athenian Kitchen from the Early Iron Age to the Hellenistic Period”, in: Spataro/Villing (2015) 180–189.
Rutter (1995): Jeremy B. Rutter, *The Pottery of Lerna IV*, Lerna III, Princeton N.J.
Rye/Evans (1976): Owen S. Rye and Clifford Evans, *Traditional Pottery Techniques of Pakistan: Field and Laboratory Studies*, Washington.
Scheibler (1995): Ingrid Scheibler, *Griechische Töpferkunst*, 2nd edition, Munich.
Schiffer (1990): Michael B. Schiffer, “The Influence of Surface Treatment on Heating Effectiveness of Ceramic Vessels”, in: *Journal of Archaeological Science* 17, 373–381.
Sillar (2001): Bill Sillar, “Strength,Toughness and Thermal Shock Resistance of Ceramics. Comment III: Technological Choices and Experimental Archaeology”, in: *Archaeometry* 45, 173–181.
Sparkes/Talcott (1970): Brian A. Sparkes and Lucy Talcott, *Black and Plain Pottery of the 6th, 5th and 4th Centuries BC*, Agora XII, Princeton N.J.
Spataro/Villing (2015): Michaela Spataro and Alexandra Villing (eds.), *Ceramics, Cuisine and Culture. The Archaeology and Science of Kitchen Pottery in the Ancient Mediterranean*, Oxford.
Tite/Kilikoglou (2002): Michael Tite and Vassilis Kilikoglou, “Do We Understand Cooking Pots and Is There an Ideal Cooking Pot?”, in: Vassilis Kilikoglou, Arno Hein and Yannis Maniatis (eds.), *Modern Trends in Scientific Studies on Ancient Ceramics*, BAR International Series 1011, Oxford, 1–8.
Trapichler (1995): Maria E. Trapichler, “*Chytra, Kakkabe, Lopas, Tagenon*. Zur Entwicklungsgeschichte velinischer Kochtöpfe von spätarchaischer bis in hellenistische Zeit”, in: Barbara Brandt, Verena Gassner and Sabine Ladstätter (eds.), *Synergia. Festschrift für Friedrich Krinzinger* 2, Vienna, 69–76.
Villing (2006): Alexandra Villing, “‘Drab Bowls’ for Apollo. The Mortaria from Naukratis and Exchange in the Archaic Eastern Mediterranean”, in: Alexandra Villing and Udo Schlotzhauer (eds.), *Naukratis. Greek Diversity in Egypt*, London, 31–46
Villing/Pemberton (2010): Alexandra Villing and Elizabeth G. Pemberton, “Mortaria from Ancient Corinth. Form and Function”, in: *Hesperia* 79, 555–638.

Weilhartner (2012): Jörg Weilhartner, "Die Keramikproduktion auf Ägina in historischer Zeit nach Aussage der Literarischen Quellen", in: Klebinder-Gauß (2012) 207–211.
Welter (1938): Gabriel Welter, *Aigina*, Berlin.
Whitbread (2015): Ian Whitbread, "Material Choices in Utilitarian Pottery: Kitchen Wares in the Berbati Valley, Greece", in: Spataro/Villing (2015) 28–36.

Kathleen M. Lynch

3 Production Relationships between Attic Black-glazed and Figured Pottery

Abstract: Excellent and innovative research on the logistics of Attic pottery production has contributed greatly to our understanding of the scale of output, workshop relationships, and energetics of production, but these studies largely focus on figure-decorated pottery as their data. Using a methodology for close study of vessel profiles pioneered by H. Bloesch, this paper presents evidence that Athenian potters produced vessels for figural decoration and plain black-glazed decoration simultaneously. Much of this evidence has been known, but it is useful as a reminder that production estimates based only on figured pottery are not the full picture of activity in the Athenian Kerameikos.

Introduction

> um neben dem Vasenmaler auch dem Töpfer den ihm gebührenden Platz im Strome des unendlich reich durchpulsten antiken Lebens zu sichern.
>
> *[The aim of the book is to give] not only the vase painter but also the potter the place he deserves in the stream of infinitely richly pulsing ancient life.]*

The quote is from H. Bloesch's study of Athenian potters:[1] he describes the aim of his study of cup profiles as "securing not only the vase painter but also the potter the place he deserves in the stream of infinitely richly pulsing ancient life." His poetic statement meaningfully connects the products of craftspeople to the "pulse of life" because whatever the craft, whatever the product, it existed only to fulfill a human need.

The prevalence of figure decorated pottery has structured the study of ancient Athenian pottery production. The work of Sir J. D. Beazley importantly organized the vast body of Attic black-figured and red-figured pottery, but his emphasis on vase painters continues to influence our vision of the ancient industry.[2] J. D. Beazley structured Athenian figure decorated pottery by identifying "hands" or, rather, individuals—often anonymous artisans—behind handmade products through careful study of formal details. The assumption is that individual painters will draw the same thing in slightly different but identifiable ways while aiming for a generally homogenous style. As discussed below, scholars like H. Bloesch have extended Beazley's methodology to identify anonymous potters through analysis of idiosyncratic details of their products. Just as vase-painters aimed for a general "Athenian" style of form and decoration, so too the potters aimed to produce pottery forms with common features; yet in both cases the painters and potters achieved these ideals in their own ways, creating idiosyncratic details that we can now recognize. Beyond identifying "hands," this paper considers what attention to potting details may tell us about the organization of the pottery industry in Late Archaic Athens.

Potter and Painter in Ancient Athens

It is well known that Athenian painters and potters of figured wares occasionally but inconsistently signed their work, although we continue to debate what exactly ἔγραπσεν and ἐποίεσεν mean.[3] These two verbs used in signatures, however, lead us to envision two separate individuals in the *chaîne opératoire* of Attic vase production. Dual signatures, the same

1 Bloesch (1940) xi.
2 *ARV*², *ABV* in particular, but also Beazley (1986). P. Sapirstein in this volume demonstrates how Beazley's study can be used to address new questions about the potting industry, although limited to figure decorated pottery.
3 Bolmarcich/Muskett (2017) 155–156 for a recent discussion with references.

Note: All P numbers in this chapter are from Athens, Agora Museum.

https://doi.org/10.1515/9783111189635-003

person signing as potter and painter, indicate that the same person could do both.[4] R. Osborne sketched out the bewildering relationships between the individuals who signed as potters and those who signed as painters.[5] And E. Hasaki and D. Harris Cline transformed this data and more through the visualization of social network analysis.[6] As the visualizations illustrate, the relationship between signed potter and painter is quite "fluid" or, in Osborne's words, "promiscuous."[7] P. Sapirstein has pointed out that the equipment requirements of the potter make his side of production more likely to be fixed, so it is more likely the painters who are moving around than the potters.[8] But note that these schemes consider only figured pottery, because for the Late Archaic and Classical periods, totally black pottery rarely receives signatures.[9] Exceptional are the few examples of olpai from the Kriton Group with an *epoiesen* signature in a reserved field on the body.[10] The Nikosthenes Potter also signs the reserved edge of a Type C cup variant with odd foot and undecorated phialai with tongues around the omphalos,[11] and an otherwise unknown potter named Salax signs a banded plate from the Athenian Agora.[12] Signatures, therefore, provide limited help in understanding the relationship of figure-decorated and plain black-glazed pottery.

Because we lack direct evidence for workshops—no contemporary textual records, very little archaeological evidence, and no first-hand accounts—our primary methodology is to infer relationships from the output, the pottery vessels themselves, to reconstruct the production relationships.[13] Further complicating things is that J. D. Beazley used terms like "Penthesilea Workshop," or "followers of Douris," to characterize stylistic affinities of individual painters; but at the same time, Beazley's terms evoke a physical setting, one that has no direct contemporary archaeological or textual evidence.

The Relationship of Black-glazed and Figured Wares

How can plain black Athenian fineware pottery provide new insights into the production structure of the Athenian potting industry? B. Sparkes and L. Talcott in their publication of undecorated black-glazed fineware from the excavations of the Athenian Agora recognized potting relationships between individual plain black pots and between plain black pots and figure decorated pots, and they even recognized potter relationships between different shapes. To them, these shared features pointed to products of a single "workshop" or potter. Here I use the term "workshop" as B. Sparkes and L. Talcott did, to mean specifically a physical location of production (vs. Beazley's more stylistic use of the word).[14] The methodology to identify potter relations also uses a Beazley-inspired methodology. Similarly idiosyncratic details of potting, such as the way a particular potter finishes the interior face of a foot or the way he attaches handles, can identify the hand of an individual potter. The pioneer of this type of potter "attribution" was the Swiss scholar H. Bloesch, whose study *Formen*

---

4 For example, Exekias signs as painter and potter on Berlin, Antikensammlung, inv. no. 1720, *ABV* 143.1; and Douris on Brussels Museum, inv. no. A 718, *ARV*² 445, 256. Sophilos signs as potter (Athens, Acropolis Museum, inv. no. 2035.1, 2) and painter (Athens, Acropolis Museum, inv. no. Akr.1.587) separately. Dual painter/potters are also discussed in Sapirstein's contribution to this volume.

5 Osborne (2004) 90, fig. 6.8.

6 Harris Cline/Hasaki (2019); Hasaki /Harris Cline (2020) 64 and fig. 3.

7 Osborne (2004) 92; Williams (2016) 64.

8 Sapirstein (2013) 500, (2014), and Sapirstein in this volume.

9 *Agora* XII, 13, note 18 lists potters who signed vases without decoration, but many of these are Little Master cup painters whose signature appears between palmettes, so technically with decoration. To my knowledge there are no true Little Master cups without any subsidiary black-figured decoration.

10 *ABV* 446, four olpai assembled into "The Kriton Group," which J. D. Beazley associates with olpai by the Amasis Painter. Three potters' names appear: Lysias, Kriton, and Priapos. The inscription on the example Paris, Louvre, inv. no. F 339, *ABV* 446, 1, also states "hemichonei[mi]" perhaps indicating that the *epoiesen* inscription confirms capacity.

11 Cup: Rome, Vatican Museum, inv. no. G74; *ABV* 233, 19. The bowl of the cup is a Type C with concave lip, but the foot is more like a type A with very pronounced fillet at the junction with the bowl. There is a tongue pattern around the handle attachments. Phialai: London, British Museum inv. nos. 1852, 0522.6; Würzburg, Martin von Wagner Museum, inv. no. 429; Paris, Cabinet des Medailles, inv. no. 334. Sotades also signs a phiale as potter: London, British Museum, inv. no. 1894, 0719.2.

12 P 1386, *Agora* XXIII, 270, no. 1407, ca. 500 B.C.

13 On archaeological evidence for workshops: Hasaki (2011); Monaco (2000); Baziotopoulou-Valavani (1994); and the online resource: "Mapping Ancient Athens," www.dipylon.org (last access 28.09.2023), which allows you to search for pottery production remains discovered in excavations in Athens. Stissi (2016) discusses the process of inference from products to production.

14 *Agora* XII, 13. Also Stissi (2016).

attischer Schalen von Exekias bis zum Ende des strengen Stils classified potter relationships among cups using potter signatures as a starting point. He focused on cups with figure decoration but also included black-glazed versions when they exist, which are rare for cups of Types A and B. Bloesch's study appeared in 1940, and he lamented that, at that time, few museum collections included plain black-glazed cups, thus limiting his study.[15] He did include the then-recently published Type C cup feet found in a well on the North Slope of the Athenian Acropolis used as ostraka against Themistokles. He grouped the Themistokles cup feet with red-figured cups with similar foot profiles.[16] In this way, H. Bloesch introduced the idea that the same potters made vessels for decoration with figures or in plain black-glaze.

Many other scholars have followed Bloesch's method for identifying the hands of potters of figure-decorated vessels. For example, in his study of the Achilles Painter, Oakley identified five potters all of whom also potted for the Achilles Painter and his teacher, the Berlin Painter.[17] Oakley identified hands of different potters in the feet of pelikai. The P potter preferred a concave interior face of the foot and recessed underside, whereas the A potter's feet have more a squared profile. B. Kreuzer also argued for potter identifications as a key to unraveling the Leagros Group.[18] In these studies, the potter identifications add further support to the painter relationships that J. D. Beazley identified, but they do not consider plain black-glazed production by the same potters.

B. Sparkes and L. Talcott, writing more than 30 years after Bloesch, document potter relationships among vessels from the excavations of the Athenian Agora and other collections. They note the common occurrence of "twins," two vessels of the same shape clearly made at the same time by the same potter and found together.[19] In our study of the pottery from the Crossroads Enclosure from the Excavations of the Athenian Agora, S. Rotroff and I have found not just twins but many multiples of shapes. A distinctive example is a set of oversized black-glazed Corinthian skyphoi from the same potter or at least the same workshop.[20] This skyphos shape, as found in the Crossroads Enclosure, combines an Attic-type rim and handles and a Corinthian-type foot and lower wall.

Perhaps the most interesting of Sparkes and Talcott's observations relates to the remarkable red-figured Delicate Class stemless cup from the Celtic burials at Klein-Aspergle attributed to the Amphitrite Painter by J. D. Beazley (KAS 113).[21] Sparkes and Talcott saw a parallel to the profile of a plain black-glazed stemless cup from the Agora excavations (P 5332, Fig. 1).[22] The two share careful moldings on the underside and an unusual projecting cone at the center of the underside. The exterior profile of the foot in two degrees and the delicate offset of the rim on the interior also unite the pair.[23] B. Sparkes and L. Talcott and E. Böhr connect a plain black-glazed stemless cup from the same Celtic grave (KAS 114) to another stemless cup from the Agora excavations (P 23661), but the foot profiles are different enough to cast doubt on the workshop relationship (Fig. 2).[24] On the other hand, it is entirely possible that one Athenian workshop produced all large Delicate Class stemless cups as well as other drinking vessels with finely articulated details and incised decoration, such as light-walled cup-skyphoi.[25] In the absence of exact parallels, it is best to be conservative.

15 Bloesch (1940) 111.
16 Bloesch (1940) 124–126; Broneer (1938), Well M, 228–242.
17 Oakley (1997) figs. 28, 29, 30.
18 Kreuzer (2016) 103.
19 *Agora* XII, 13.
20 Deposit J 5:2, Shear (1973) 366, fig. 67e; for a preliminary report, see Rotroff /Lynch (2022).
21 *ARV*^{2} 831, 25; Böhr (1988) 176–177, no. 5.1, fig. 96, pl. 29, with previous publications.
22 *Agora* XII, 103; no. 484, fig. 5; discussed further by Böhr (1988) 178–184.
23 Böhr (1988) 180, note 16 and 181–183 connects the Agora and Kleinaspergle stemless cups with others from Stryme. The stemless cup, (Komotini, Archaeological Museum, inv. no. 20), features the same reserved leaf decoration on the interior of the lip; however, the foot and moldings on the underside do not match those of the Amphitrite Painters' stemless cup (see figs. 103, 105).
24 Böhr (1988) 177–178, figs. 97, 102; *Agora* XII, 103, no. 483, fig. 5.
25 *Agora* XII, 110–111, esp. no. 581.

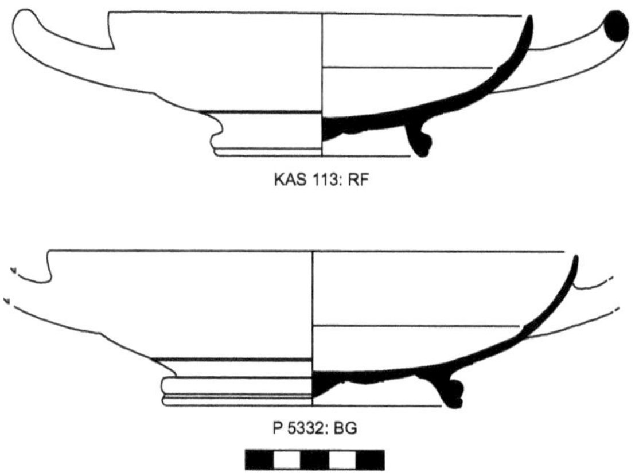

Fig. 1: Stemless cups compared: Stuttgart, Württembergisches Landesmuseum inv. no. KAS 113, (red-figured; attributed to the Amphitrite Painter; from a burial at Kleinaspergle); P 5332 (black-glazed; from the Athenian Agora). After Böhr (1988) fig. 96 and *Agora* XII, no. 484, fig. 5.

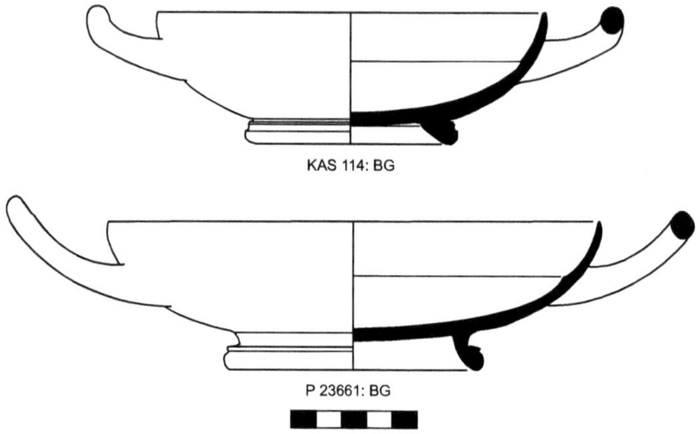

Fig. 2: Black-glazed stemless cups compared: Stuttgart, Württembergisches Landesmuseum inv. no. KAS 114 (from a burial at Kleinaspergle); P 23661 (from the Athenian Agora). After Böhr (1988) fig. 97 and *Agora* XII, no. 483, fig. 5.

Late Archaic pottery from deposits formed after the Persian sack of Athens in 479 B.C. provide additional evidence for shapes produced for both plain black-glazed and figured decoration. The excavations of the Athenian Agora excavated 23 deposits of debris, most of which consisted of decommissioned wells filled with trash.[26] The Stoa Gutter Well, Q 12:3, published by S. Roberts, represents the stock of a pottery boutique, a sales shop.[27] Of the 489 inventoried vessels from the Stoa Gutter Well, only 17 are red-figured, 3% of the whole deposit, and household wares are always underrepresented in the inventory; therefore red-figured pottery comprised an even smaller portion of the deposit.[28] However, a large number of Late Archaic black-figured lekythoi from the Athens 581 workshop and multiples and different-sized lekythoi by the same painter's hand probably stood on the shelves at the time of the sack.[29]

S. Roberts tried to identify red-figured and black-glazed kylix cups potentially produced by the same pottery workshop.[30] She connects the profiles of cups from the Stoa Gutter Well with the potters whom H. Bloesch identified as associated with the workshop of Kachrylion, which includes the Hermione Group, along with the painter Phintias.[31] For example, she compares the profile of a red-figured Type C cup with concave rim attributed to Epiktetos (P 24131) convincingly with two plain black-glazed cups (P 24595, P 24120; Fig. 3).[32] The profiles share similar fillets on the stem, thickness and angle of the foot plate, rounded exterior face of the foot, and thick floor. The concave rims do vary a little bit, especially with the stronger offset on the interior of the red-figured example.

Roberts' assessment is less convincing in another case. Of the three Type C cups with plain rim (Fig. 4), the top profile is a red-figured cup signed by the painter Phintias and in the collection of The Johns Hopkins University, in Baltimore, MD; and the bottom two are plain black-glazed cups from the Stoa Gutter Well (P 24601 and P 24602) Roberts states that the black-glazed cups from the Agora "...parallel the profile of the [red-figured] Baltimore cup."[33] I would argue that the profile of the red-figured Baltimore cup is appreciably different from the two black-glazed cups. The red-figured cup has

26 Shear (1993) to which should be added: Lynch (2011a); Camp/Martens (2020) 625–626.
27 Roberts (1986) 4.
28 Lynch (2011b) 71 has suggested that the small number of red-figured kylikes in this deposit and in the Rectangular Rock-Cut Shaft (G 6:3) suggest that they were not for sale but may have been the property of the shop owner. A similar number of both red-figured and black-glazed kylikes in a domestic deposit from the Agora, well J 2:4 (discussed below) adds merit to this hypothesis.
29 Lynch (2011b) 73.
30 Roberts (1986) 3–9.
31 Roberts (1986) 3.
32 Roberts (1986) 5, fig. 2.
33 Roberts (1986) 3.

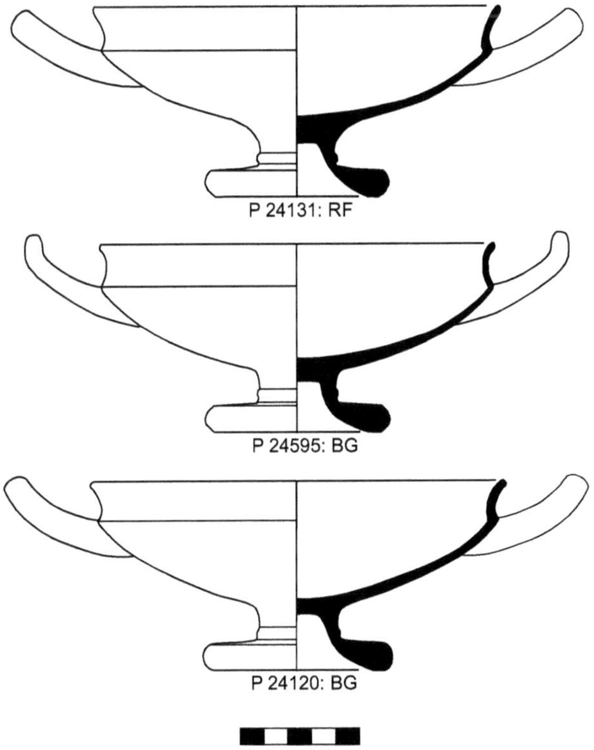

Fig. 3: Type C cups compared, all from Athenian Agora deposit Q 21:3, the Stoa Gutter Well: P 24131 (red-figured; attributed to Epiktetos); P 24595 and P 24120 (black-glazed). After Roberts (1986) fig. 2.

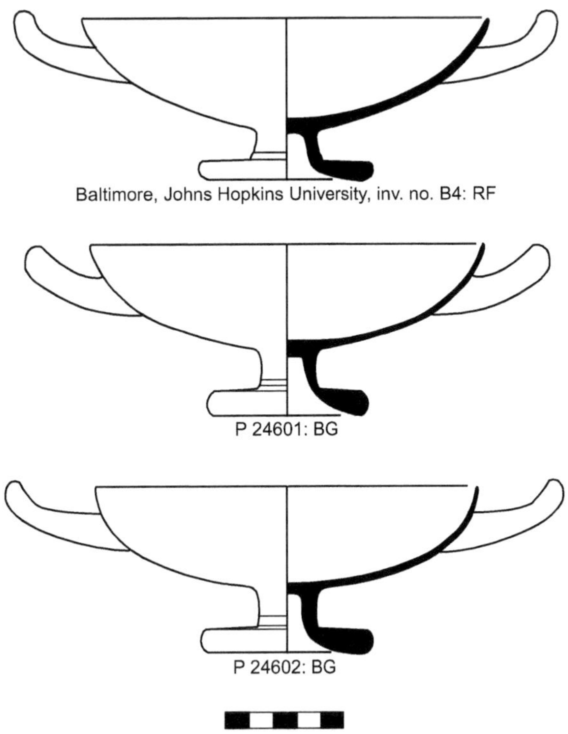

Fig. 4: Type C cups compared: Baltimore, MD, Johns Hopkins University, inv. no. B4 (red-figured; signed by Phintias); P 24601 and P 24602 (black-glazed; from Athenian Agora deposit Q 21:3, the Stoa Gutter Well). After Roberts (1986) fig. 7.

a clear fillet, and the two black-glazed cups have only incised lines marking out the concept of a fillet. The footplates of the black-glazed Agora cups are thicker. This example of potters producing cups for decoration in black-glaze or red-figure is less persuasive than I would like. The two Agora black-glazed cups, however, are convincingly similar and point to the same potter.

Following Roberts' lead, we can identify both black-glazed and red-figured cups by the same potter from other Athenian Agora deposits (Fig. 5). A Late Archaic period domestic deposit, well J 2:4, contained a set of Type C cups with both black-glazed and red-figured members.[34] The profile of a red-figured Type C cup with decoration in the tondo and attributed to the Euergides Painter (P 32419) matches two black-glazed cups (P 32475; P 32470). A stemmed dish from the same deposit (P 32465) shares the same foot profile, so it could also come from the same potter. A red-figured Type C cup with concave rim with the unusual decoration of a wheel in its tondo (P 32431) again matches the profile of two black-glazed cups (P 32396, P 32755; Fig. 6). Cup-skyphoi from the same deposit also reveal pottery connections, this time with black-figured and black-glazed decoration. In contrast to the hasty silhouette-style black-figured decoration, the potting of the cup-skyphoi is quite good and allows us to identify similarities between the black-figured (P 32474) and black-glazed vessels (P 32482, P 32481; Fig. 7).

Other scholars have recorded products of the same potter decorated with red-figured and in plain black. K. Kathariou's fine studies of the Jena Painter workshop and products illustrated very convincingly that the same potter produced plain black-glazed and red-figured stemless cups (among other shapes) within the Jena Painter workshop.[35]

34 Lynch (2011a) nos. 87–95 (red-figured) and 128–136 (black-glazed).

35 Kathariou (2016) 150–153.

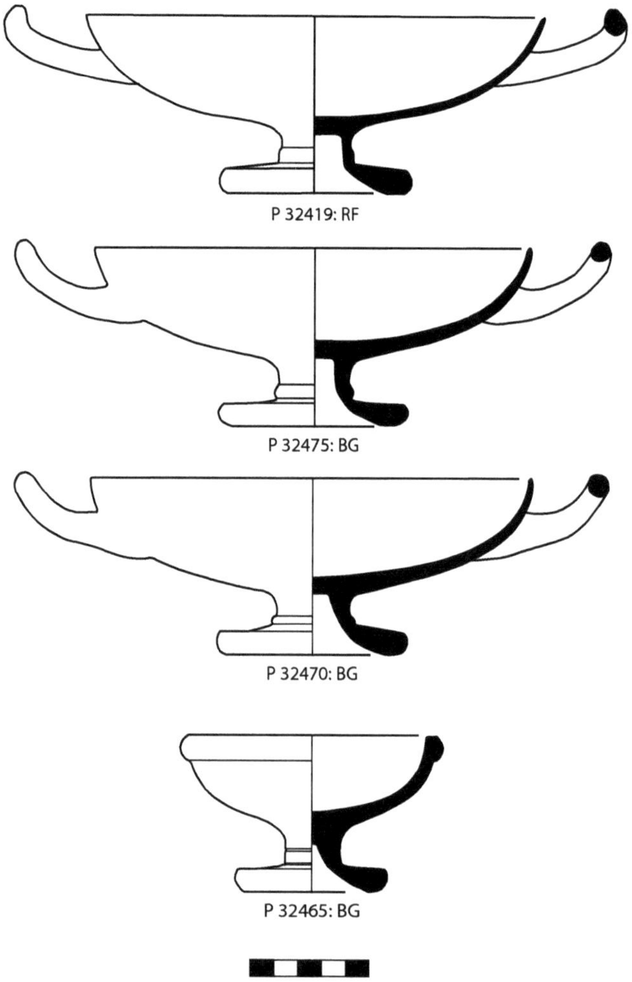

Fig. 5: Type C cups and dish compared, all from Athenian Agora deposit J 2:4: P 32419 (red-figured, attributed to the Manner of the Euergides Painter); P 32475, P 32470 (black-glazed cups), P 32465 (black-glazed stemmed dish). After Lynch (2011) figs. 88, 112.

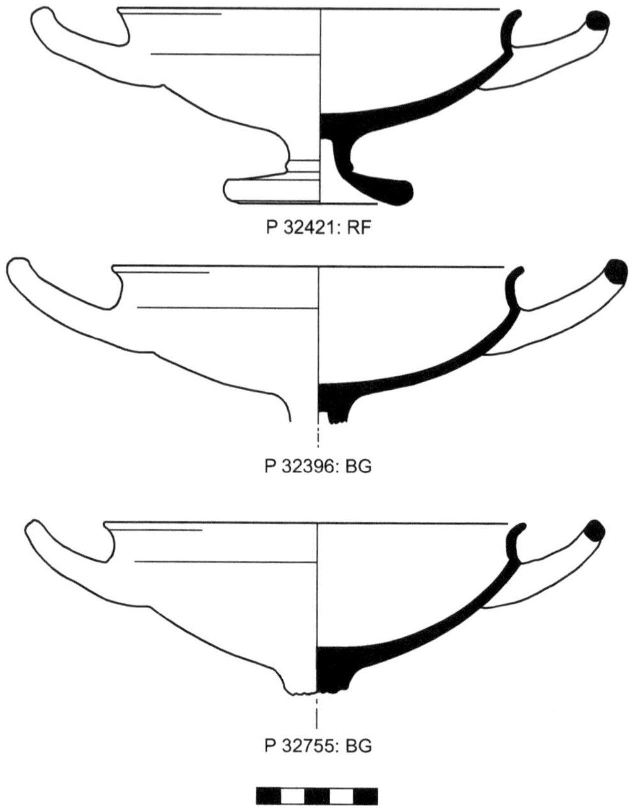

Fig. 6: Type C cups compared, all from Athenian Agora deposit J 2:4 : P 32421, red-figured unattributed; P 32396 and P 32755. After Lynch (2011) figs. 90, 110.

Conclusions

What do the potter relationships between figured and black-glazed vessels tell us? First, Attic potters expanded their repertoire to include undecorated versions of their products in the late 6th century. I have suggested elsewhere that this sudden increase in pottery production may reflect a revolution in the symposium, as more men participated in the drinking event than ever.[36] It also tells us that consumers in Athens looking for new sympotic equipment shopped at pottery boutiques that featured both figured and plain black vessels produced by the same potter. The red-figured and black-glazed cups from well J 2:4, connected by potting relationship and all dating closely together, indicate that the homeowner purchased the red-figured and the black-glazed cups at the same time, or over a brief period. The stock of Athenian pottery boutiques represented by the Stoa Gutter Well and the Rock-Cut Shaft, which I have not discussed here, already confirmed that they carried a mix of decorative styles: black-figure in aisle 1, black-glaze in aisle 2, and maybe red-figured in an exclusive

36 Lynch (2011a) 169–173, (2020) 129–131.

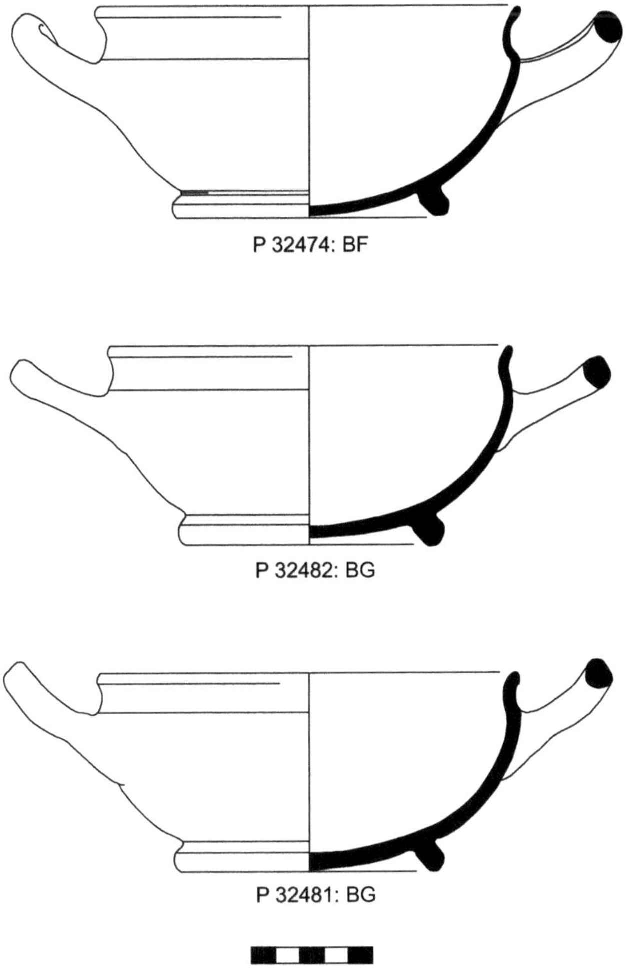

Fig. 7: Cup-skyphoi compared, all from Athenian Agora deposit J 2:4: P 32474 (black-figured); P 32482 and P 32481 (black-glazed). After Lynch (2011) figs. 55, 107.

display in the front.[37] Notably, these deposits do not represent the full production line of pottery available in Athens, providing further evidence that we should think of them as "boutiques," representing the wares of a particular workshop or several individual workshops, not a large department store. There are no loutrophoroi, for example, although we know these were being produced and sold in Athens at the same time.

From about 525 B.C. the production of undecorated black-glazed fineware, largely for use in Athenian symposia, needs to be considered alongside the more visible production of figured wares. Prior to 525, some shapes existed in both black-glazed and black-figured, but the figured versions vastly outnumbered the plain. versions. For example, the neck amphora exists in both black-figured and black-glazed Athenian production, but the plain ones are rare.[38] In contrast, the shapes invented after the introduction of red-figured, such as the pelike, appear equally with plain black and figured decoration ca. 500 B.C. Stemmed cups had existed in black-figure since the early 6th century Komast cup. A few undecorated black-glazed kylix forms exist, such as the Class of Athens 1104, an analog to black-figured Komast cups, but all black versions of Little Master cups, are rare at best.[39] To be clear, this means that black-figured cups accounted for almost all

37 Stoa Gutter Well: Roberts (1986); Rectangular Rock-Cut Shaft: Vanderpool (1946).
38 Black-glazed, P 24986, *Agora* XII, 236, no. 1, pl. 1; black-figure: P 23200, *Agora* XXIII, 128–129, no. 209, pl. 24.
39 Class of Athens 1104: P 24975, *Agora* XII, 262, no. 380, pl. 18, relationship to Komast cups discussed 88–89.

symposia cups of the first three quarters of the 6th century B.C. Only in the final quarter of the 6th century, closer to 500, do black-glazed kylikes become much more abundant, as the cups from deposit J 2:4, discussed here, reveal. After 525 B.C., plain black-glazed kylikes are more numerous than their figured—red-figured or black-figured—brothers.

The introduction of undecorated black-glazed analogs of black-figured and red-figured vessels comes at a time of great innovation and experimentation in the arts: M. Meyer has called it "two crucial generations," acknowledging the Late Archaic seeds of the Classical style.[40] Our Late Archaic Athenian potters were businessmen, and market demand prompted them to expand their production significantly in the last quarter of the 6th century B.C. An efficient way to do so was to produce undecorated versions of normally figured shapes—a smart move to increase output. The Attic pottery workshops producing small Late Archaic black-figured shapes like lekythoi took a different approach: for maximum efficiency, they opted for hastier and faster decoration. Another move to meet market demand, surely.

What can the change in production strategies tell us about the social life of the Late Archaic city of Athens? As I noted, the potters chose to make black-glazed versions of sympotic shapes. In the last quarter of the 6th century B.C., Athenian potters also introduced new forms to accompany the symposium: saltcellars and small bowls, but only in black-glazed and rarely with figures. What Athenian potters did not produce provides an even more interesting insight. The Athenian potters did not produce individual plates and bowls: that is, dining equipment. They directed their energies instead into expanding and articulating equipment for symposia. For me, this is the smoking gun pointing to the motivation for late 6th century developments in the Kerameikos: an expansion of the symposium to include men from new status groups meant the potters needed to increase production. The entrepreneurial potters met the demand in creative, efficient ways.[41]

This paper has tried to foreground a significant missing piece in the discussion of pottery production at Athens, the production of black-glazed vessels. That the same potter might produce a vessel for decoration in figured or plain black is an important observation because it means that potter may not have known how the piece would be decorated. He was told to make cup-skyphoi, so he did, and passed them onto a painter to finish. I hope that scholars concerned with estimating potter/painter output and the energetics of the pottery industry at Athens, some present in this volume, can include black-glazed pottery production as part of their equations. Our potters were savvy businessmen, and the full range of their products reflects their market strategies.

Abbreviations

Agora XII	Brian Sparkes and Lucy Talcott, *Black and Plain Pottery of the 6th, 5th and 4th Centuries B.C.*, Princeton N.J., 1970.
Agora XXIII	Mary Moore and Mary Zelia P. Philippides, *Attic Black-figured Pottery*, Princeton N. J., 1986.
ABV	John Davidson Beazley, *Attic Black-figure Vase-painters*, Oxford, 1978.
*ARV*²	John Davidson Beazley, *Attic Red-figure Vase-painters*, 2nd edition, Oxford, 1963.

Bibliography

Baziotopoulou-Valavani (1994): Effie Baziotopoulou-Valavani, "Ανασκαφές σε Αθηναϊκά Κεραμικά Εργαστήρια Αρχαϊκών και Κλασικών Χρόνων", in: William D. E. Coulson, Olga Palagia, Theodore L. Shear Jr., H. Alan Shapiro and Frank. J. Frost (eds.), *The Archaeology of Athens and Attica Under the Democracy*, Oxbow Monograph 37, Oxford, 45–54.

Beazley (1986): John Davidson Beazley, *The Development of Attic Black-Figure*, revised edition, Berkeley.

Bloesch (1940): Hansjorg Bloesch, *Formen attischer Schalen von Exekias bis zum Ende des strengen Stils*, Bern.

Böhr (1988): Elke Böhr, "Die griechischen Schalen", in: *Das Kleinaspergle: Studien zu einem Fürstengrabhügel der frühen Latenezeit bei Stuttgart*, Stuttgart, 176–195.

Bolmarcich/Muskett (2017): Sarah Bolmarcich and Gina Muskett, "Artists' Signatures on Archaic Greek Vases from Athens", in: Peter Schultz and Kristen Seaman (eds.), *Artists and Artistic Production in Ancient Greece*, Cambridge, 154–176.

Broneer (1938): Oscar Broneer, "Excavations on the North Slope of the Acropolis, 1937", in: *Hesperia* 7, 161–263.

Camp/Martens (2020): John McK Camp II and Brian Martens, "Recent Excavations in the Athenian Agora, 2013–2019", in: *Hesperia* 89, 593–657.

40 Meyer (2020) 20–22.

41 Discussed further in Lynch (2020).

Eschbach/Schmidt (2016): Norbert Eschbach and Stefan Schmidt (eds.), *Töpfer-Maler-Werkstatt: Zuschreibungen in der griechischen Vasenmalerei und die Organisation antiker Keramikproduktion*, *CVA* Deutschland Beihefte 7, Munich.

Harris Cline/Hasaki (2019): Diane Harris Cline and Eleni Hasaki, "The Connected World of Potters in Ancient Athens: Collaborations, Connoisseurship, and Social Network Analysis", in: *CHS Research Bulletin* 7, http://nrs.harvard.edu/urn-3:hlnc.essay:ClineD_and_HasakiE.The_Connected_World_of_Potters.2019 (last access 02.07.2024).

Hasaki (2011): Eleni Hasaki, "Crafting Spaces: Archaeological, Ethnographic, and Ethnoarchaeological Studies of Spatial Organization in Pottery Workshops in Greece and Tunisia", in: Mark Lawall and John Lund (eds.), *Pottery in the Archaeological Record: Greece and Beyond* (Acts of the International Colloquium held at the Danish and Canadian Institutes in Athens, June 20–22, 2008), Aarhus, 12–28.

Hasaki/Cline (2020): Eleni Hasaki and Diane H. Cline, "Social Network Analysis and Connoisseurship in the Study of Athenian Potters' Communities", in: Eleni Hasaki and Martin Bentz (eds.), *Reconstructing Scales of Production in the Ancient Greek World: Producers, Processes, Products, People* (Panel 3.4, Proceedings of the 19th International Congress of Classical Archaeology, vol. 8, Cologne/Bonn, May 22–26, 2018), Heidelberg, 59–80. https://doi.org/10.11588/propylaeum.639.c9977.

Kathariou (2016): Kleopatra Kathariou, "On the Quest for the Missing Link in Late Classical Athenian Kerameikos: A Study of the Jena Painter's Workshop", in: Eschbach/Schmidt (2016) 149–161.

Kreuzer (2016): Bettina Kreuzer, "Töpfer und Maler, Klasse und Gruppe. Beazley und die Leagrosgruppe", in: Eschbach/Schmidt (2016) 96–106.

Lynch (2011a): Kathleen M. Lynch, *The Symposium in Context: Pottery from a Late Archaic House near the Athenian Agora*, *Hesperia* Supplement 46, Princeton N.J.

Lynch (2011b): Kathleen M. Lynch, "Depositional Patterns and Behavior in the Athenian Agora: When Disaster Strikes", in: Mark L. Lawall and John Lund (eds.), *Pottery in the Archaeological Record: Greece and Beyond* (Acts of the International Colloquium held at the Danish and Canadian Institutes in Athens, June 20–22, 2008), Aarhus, 67–74.

Lynch (2020): Kathleen M. Lynch, "Attic Fineware at the Turn of the 5th Century: Shapes and Images", in: Marion Meyer and Gianfranco Adornato (eds.), *Innovations and Inventions in Athens, c. 530 to 470 BCE—Two Crucial Generations*, Vienna, 125–133.

Meyer (2020): Marion Meyer, "Joint Efforts", in: Marion Meyer and Gianfranco Adornato (eds.), *Innovations and Inventions in Athens, c. 530 to 470 BCE—Two Crucial Generations*, Vienna, 11–22.

Monaco (2000): Maria C. Monaco, *Ergasteria: Impianti Artigianali Ceramici ad Atene ed in Attica dal Protogeometrico alle soglie Dell'ellenismo*, Rome.

Oakley (1997): John H. Oakley, *The Achilles Painter*, Mainz.

Osborne (2004): Robin Osborne, "Workshops and the Iconography and Distribution of Athenian Red-Figure Pottery: A Case Study", in: Simon Keay and Stephanie Moser (eds.), *Greek Art in View*, Oxford, 78–94.

Roberts (1986): Sally R. Roberts, "The Stoa Gutter Well, a Late Archaic Deposit in the Athenian Agora", in: *Hesperia* 55, 1–74.

Rotroff/Lynch (2022): Susan Rotroff and Kathleen M. Lynch, "The Crossroads Enclosure in the Athenian Agora: A Preliminary Report", in: Jenifer Neils and Olga Palagia (eds.), *From Kallias to Kritias: Art in Athens in the Second Half of the Fifth Century B.C.*, Berlin/Boston, 155–172.

Sapirstein (2013): Philip Sapirstein, "Painters, Potters, and the Scale of the Athenian Vase-painting Industry", in: *American Journal of Archaeology* 117, 493–510.

Sapirstein (2014): Philip Sapirstein, "Demographics and Productivity in the Ancient Athenian Pottery Industry", in: John H. Oakley (ed.), *Athenian Potters and Painters* III, Oxford, 175–186.

Shear (1973): Theodore L. Shear Jr., "The Athenian Agora: Excavations of 1972", in: *Hesperia* 42, 359–407.

Shear (1993): Theodore L. Shear Jr., "The Persian Destruction of Athens: Evidence from Agora Deposits", in: *Hesperia* 62, 383–482.

Stissi (2016): Vladimir Stissi, "Minor Artisans, Major Impact?" in: Eschbach/Schmidt (2016) 47–53.

Vanderpool (1946): Eugene Vanderpool, "The Rectangular Rock-Cut Shaft: The Upper Fill", in: *Hesperia* 15, 265–336.

Williams (2016): Dyfri Williams, "Peopling the Athenian *kerameia*: Beyond the Master Craftsmen", in: Eschbach/Schmidt (2016) 54–68.

Francesca Silvestrelli

4 Potters in South Italian Red-Figured Workshops: An Overview

Abstract: Shapes and potters' work have received only occasional attention in the study of South Italian red-figured ware. While the personalities of the painters working in South Italian workshops are to some degree established, there is little if any information available on the potters working within the same contexts. The absence of signatures and the reduced number of published profiles (an essential tool for recognizing work by an anonymous potter) have hindered the progress of research, and potters are as a consequence still shadowy figures. Through an analysis of the transformations of the morphological repertoires produced in the workshops operating in the main production centers, this chapter aims to provide new elements to highlight the role played by potters in the development of red-figured pottery in southern Italy. Case studies will be chosen among Metapontine workshops, where the large output and the variety of shapes produced both in red-figured and in black gloss suggest the potters' originality and capacity for innovation.

Introduction

Potters and their exact role in pottery workshops have received scant attention in the study of South Italian ware. Signatures (a basic element in the construction of corpora) occur only in a few cases from the Archaic to the Early Hellenistic period, and as far as figured pottery is concerned, they mainly relate to painters.[1] Furthermore, the identification of workshop practices, if not those of individual potters, via the morphological study of shapes and the analysis of potters' work details and idiosyncrasies, is a line of research that has virtually no tradition. In the case of South Italian red-figured pottery, the approach to classification adopted by A. D. Trendall was consciously focused on style,[2] partly because of the nature of the available documentation. While J. D. Beazley dealt with material relating to a single production site, the workshops established in settlements along the Ionian Coast and in Campania and Sicily from the third quarter of the 5th century B.C. onward gave rise to traditions that are highly distinctive not only in terms of style, technique, and iconography but also in their morphological repertoire. This complexity was the result of the variety of consumers, who included both Greeks of the poleis and the inhabitants of the hinterland. This interaction between the workshops and the communities they supplied, which in the second half of the 4th century B.C. would bring about the apparent progressive decentralization of production (by then also active in Daunia, Peucetia and inland Lucania),[3] underscores a complexity unfolding on both the diachronic and the synchronic levels.

The similarity between Attic and South Italian vases in terms of shape, especially with regard to early Apulian[4] and Metapontine production, is cited by A. D. Trendall as a means of establishing a relative chronology; it is highlighted when remarkable consistencies can support a workshop connection.[5] These similarities are merely pointed out, however, and are never discussed in detail.

---

1 D'Agostino (2003); Rucco/Tagliamonte (2007); Lippolis (2018). On signatures in black gloss vessels produced in Teano cf. Manzini (2011), with bibliography.

2 As stated in *LCS*, vii: "Nor are shapes discussed at any length, though their fuller study would undoubtedly complement much of what is said here and shed further light upon the relations between potters and painters in the various workshops. I hope that the present classification may provide a basis for work in both these fields".

3 Lippolis (2008); Gadaleta (2012).

4 *RVAp* I, 4–5.

5 Cf., for example, *RVAp* I, 170 (on the Painter of Ruvo 1364): "The three situlae [. . .] form a very compact group, similar in shape and decoration, all with bead and reel ornament around the rim"; *RVAp* I, 223, no. 260 (*A volute-krater depicting Achille and Penthesilea):* "The mascaroons on the obverse take the form of the usual female masks, with flesh in added white, applied to the roundels; on the reverse, however, they are painted on, a very rare practice, for which a parallel will be found on Naples Stg. 31 (no. 100), by a close associate of the Iliupersis Painter. On this vase, too, there are no swan-heads on the shoulder at the handle-joins, and the two are so alike in shape that they might well be the work of the same potter, which would provide further evidence for the connection of the Basel krater with the Iliupersis workshop."

https://doi.org/10.1515/9783111189635-004

An example is provided by two hydriai from the Tomb of the Policoro Painter in Herakleia (Policoro, Matera), the first by the Amykos Painter and the second by the Creusa Painter in his initial phase. The morphological similarities led A. D. Trendall to suggest that the vases were by the same potter, confirming the collaboration between the two painters and his training within the workshop of the Amykos Painter, which had already been established on a stylistic basis.[6]

Nevertheless, a prevailing focus on vase painting resulted in the relative neglect of the support, sometimes creating inconsistencies in the classification itself. An example is the amphora of the Panathenaic type from Tomb 1-1974 in Gravina–Botromagno,[7] which was attributed by A. D. Trendall to a painter of the York Group (380–360 B.C.).[8] However, the amphora's decorative pattern, style, and shape find parallels with the workshop of the Sisyphus Painter.[9]

Subsequent scholars have shown a similar lack of appetite for this topic, and papers on individual shapes tend to focus on their distribution, their functions in specific contexts, and the relationship between individual shapes and their decoration.[10] This lack of interest results in a failure to reconstruct the unity of the two fundamental processes involved in the manufacture of a vase, namely potter's work and decorating. It thus prevents us from assessing the relationship between potters and painters in the development of red-figured pottery in southern Italy.

Notwithstanding the increasing attention paid to shapes in recent studies,[11] the documentation currently available is still insufficient to identify individual potters in accordance with the methods developed for Attic pottery.[12] It does, however, enable us to examine workshop and production-site traditions, which can be identified by means of a synchronic and diachronic analysis of the shapes.

Shapes at Metapontum

Because of a lack of literary and epigraphic sources, the issue needs to be addressed by looking at the archaeological documentation. An interesting case study is Metapontum, where the first South Italian red-figured workshops were set up just after the middle of the 5th century B.C. The personalities of the painters working in the city until the first quarter of the 4th century B.C. are well established, and information on their workshops is available.

Metapontum's red-figured workshops are characterized by an outstanding output and a similarly outstanding fineware morphological repertoire, wider than the one in use in the first half of the 5th century.[13] In the period when the Pisticci Painter's workshop was active, the range of vessels produced in the city included vases of large dimensions for containing and mixing liquids (red-figured kraters, especially bell-kraters, hydriai, amphorae of Panathenaic type, and red-figured, patterned, and black gloss pelikai), drinking vases,[14] and vessels for perfumed oils (both red-figured and black gloss). With a few exceptions (such as the Type A and C skyphoi), this repertoire is composed of new shapes mostly derived from the Attic repertoire (but seldom if ever based on imports) or shapes already in use in the city that were reinvented or updated.

Just as the Pisticci Painter's style and iconography are built from a plurality of references,[15] the shapes introduced by his workshop also originate from a variety of models. For example, the rare lug-handled krater type is related to a shape elaborated in the Niobid Painter's workshop,[16] as is the bell-krater with narrow fascia between the rim and the body at the

6 *LCS*, 84.

7 The funerary assemblage includes South Italian and Attic vases (a kantharos by the Eretria Painter and a lekane lid by the Achilles Painter) and has been dated to the last quarter of the 5th century B.C.: Ciancio (2005).

8 *RVAp* I, 95, no. 221 pl. 33, 1–2.

9 Ciancio (1997) 181–186, (2005) 54.

10 Cassimatis (1993) (lebetes gamikoi); Gaunt (2002) (volute-kraters); Lucchese (2012); Schierup (2014); Silvestrelli (2019a) (loutrophoroi); Gadaleta (2022) (fish-plates).

11 Hoffmann (2002); Barresi/Giudice (2011) 54; Serino (2019) 147–160; *CVA* Italia 82 – Angelo Bottini and Lucia Lecce, *Matera, Museo Archeologico Nazionale "Domenico Ridola" 2. Vasi italioti a figure rosse*, 11–13.

12 Discussed in the contribution by Kathleen Lynch in this volume.

13 Cf. Silvestrelli (2016b) 177–180, fig. 1–3, (2018) 136–141, figs. 1–2 and (2019b) 537–538, fig. 1.

14 Denoyelle (2002).

15 Denoyelle (2007), (2013).

16 Jircik (1992) 68.

Metropolitan Museum of New York, originally attributed to the Niobid Painter himself.[17] The only volute krater decorated by the Pisticci Painter, produced in cooperation with the Cyclops Painter,[18] is based on Polygnotan prototypes,[19] as are his calyx and bell-kraters.[20] His squat lekythoi resemble vessels of the third quarter of the 5th century B.C. or later.[21]

Nonetheless, shapes were to a certain degree reinterpreted and adapted, establishing a local pottery tradition that would be adopted and developed by later workshops, as a few examples illustrate.

Bell-kraters

Most of the bell-kraters by the Pisticci Painter (and the Cyclops Painter) are very consistent in shape: the lip, with a minor overhang, extends just behind the slightly inturned handles. The body, slightly square in appearance, with the height roughly coincident with the diameter of the rim, curves inward slightly at the top and tapers gently to a low stem supported by a disk foot, whose outer edge is reserved (Fig. 1a–b).

Vessels of the late phase have a narrower body and a slightly taller stem,[22] also seen in the bell-kraters painted by the Cyclops and the Amykos Painters (Fig. 1b–c) and later, with some development in the proportions of the shape and in the type of foot (now in two degrees) by the Dolon Painter.[23]

Hydriai

Rarely attested among imports[24] and only occasionally locally produced around the mid-5th century B.C. in banded ware and black gloss,[25] the hydriai were intended for both internal consumption and export.[26] They represent another form introduced by the Pisticci Painter (Fig. 2), who painted 37 of them, all decorated on the body. Produced in a range of sizes, these hydriai present a certain morphological homogeneity: mouth with overhanging lip in two degrees, the upper step glazed and the lower step circled by an egg band; a short, concave neck ending in a plump body with a rounded shoulder tapering to an ogee foot. The body reaches its maximum diameter, normally roughly equivalent to its height, immediately below the handles.

The later workshops operate within the tradition codified by the Pisticci Painter, albeit with constant updating of decoration and development in morphology.

Black Gloss and Red-Figured Shapes: Pelikai and Lebetes Gamikoi

A third shape, the pelike, deserves some attention. Pelikai are popular at Metapontum, where they appear in funerary contexts of the mid-5th century B.C. in the form of black gloss vessels with an outturned rim angled downward on the outer side, flat handles, base rings, and the lower part of the body reserved. This type is still attested, with some morphological

17 New York, Metropolitan Museum of Art, inv. no. 2011.604.2.2398: Denoyelle (1997) 399–400, figs. 6–7.

18 Taranto, Museo Archeologico Nazionale, inv. no. 141384, from Rutigliano, Contrada Purgatorio T 11/1976: Denoyelle (2013) 180–181, figs. 8–9.

19 J. Gaunt, pers. comm. 2023. Cf. also Gaunt (2002).

20 Denoyelle (1997) 396.

21 Regarding the squat-lekythos from Pantanello, Tomb 136 in Burn (1998) 617, cf., for instance, a lekythos in Berlin: *CVA* Germany 62 – Irma Wehgartner, *Corpus Vasorum Antiquorum. Berlin, Antikensammlung 8*, 60, pl. 42, 6–8 (profile in fig. 21,2).

22 Cf., for example, *LCS* III, 7, no. 43a.

23 Silvestrelli (2016b) 184, fig. 6.

24 Hydria by the Pan Painter, *BAPD* 206341, from Crucinia, Tomb 11/1952: Lo Porto (1966) 203–204, fig. 57. Sabouroff Painter, *BAPD* 43135: Lo Porto (1988–89) 331, fig. 9, from a tomb excavated in 1911 in Contrada Ricotta.

25 Elliott (1998) 655–656.

26 Metapontum, Sant'Angelo Vecchio and Saldone: *LCS* III, 5–6, nos. 7a, 24a, 25a; Metapontum, Pantanello necropolis: Burn (1998) 616, T 88–1; Metapontum, Deposit 475 Prop. Vitelli and Tomb 697 Prop. Tarulli: Silvestrelli (2014) 100, WF044. Pisticci (Matera): *LCS*, 18, 25 and *LCS* I, 5 and *LCS* III, 7, nos. 56a–b; Rutigliano (Bari): *LCS* III, 6, no. 24a.

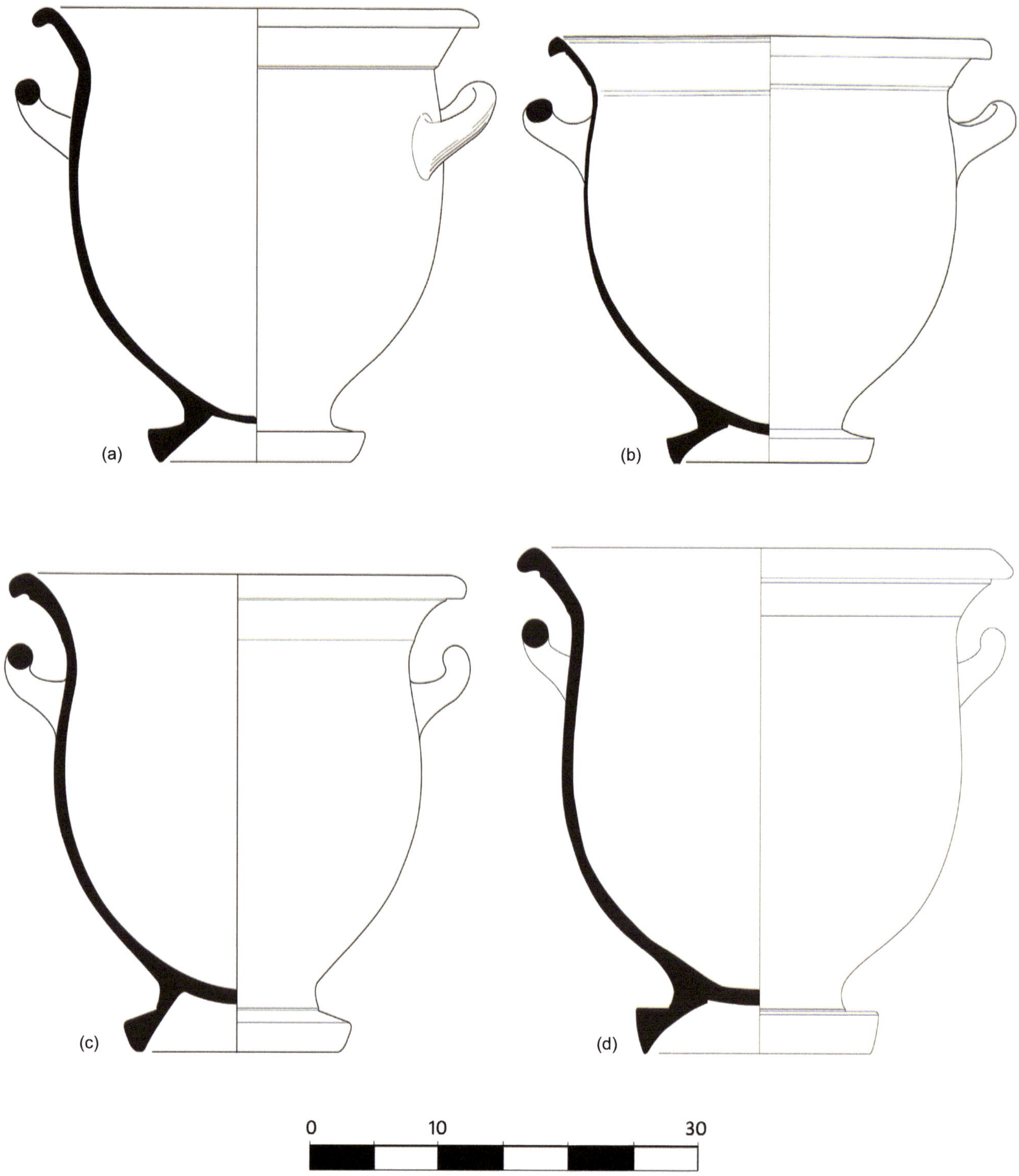

Fig. 1: (a) Bell-krater by the Pisticci Painter, © London, British Museum, inv. no. 1849,0518.12 (E500). After *LCS*, 17 no. 16, Drawing: Kate Morton; **(b)** Bell-krater by the Pisticci Painter, Metapontum, MAN, from Metapontum, Tomb 415 Prop. La Torre. After De Siena (2007) 440, pl. XIII, 1 © Ministero della Cultura – Direzione Regionale Musei Basilicata. Drawing: Giulia Gioia. Ink: Fabiola Malinconico; **(c)** Cyclops Painter, Sydney, Nicholson Museum, inv. no. NM 48.2. After *CVA* Australia 2–Michael Turner and Alexander Cambitoglou, *The Nicholson Museum, The University of Sydney. Red Figure and Over-Painted Pottery of South Italy*, 7–8, pl. 7–8; 13–15, pl. 11–12; **(d)** Amykos Painter, Sydney, Nicholson Museum, inv. no. 46.53. After *CVA* Australia 2–Michael Turner and Alexander Cambitoglou, *The Nicholson Museum, The University of Sydney. Red Figure and Over-Painted Pottery of South Italy*, 7–8, pl. 7–8; 13–15, pl. 11–12.

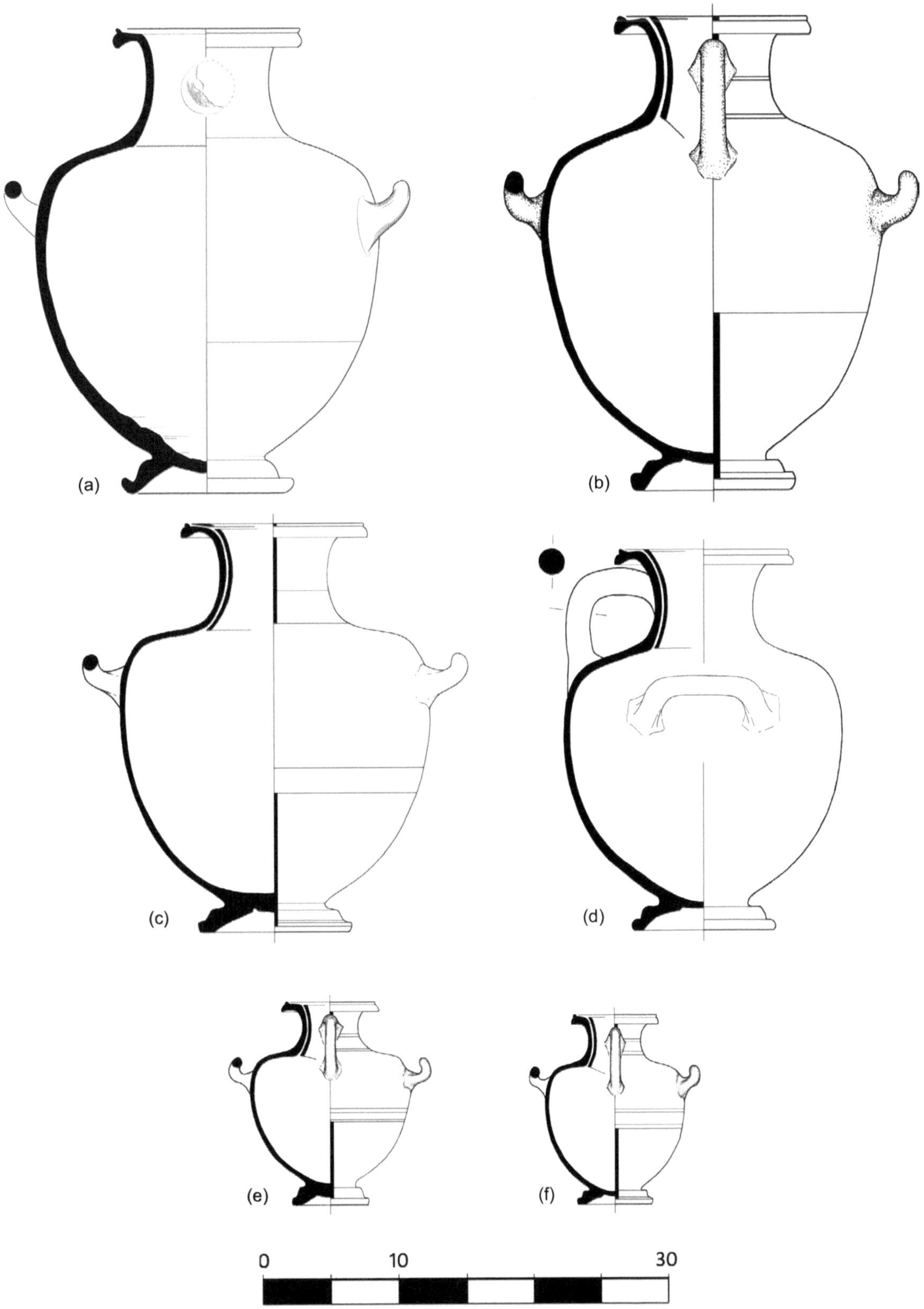

Fig. 2: Hydriai by the Pisticci Painter **(a)** London, British Museum, inv. no. 1867,0508.1128 (E223). After *LCS*, 15 no. 6. © London, British Museum, Drawing: Kate Morton; **(b)** Metapontum, MAN, inv. no. 326742, from Metapontum, Deposit 475 Prop. Vitelli. After Silvestrelli (2014) 100; **(c)** Metapontum, MAN, from Metapontum, Tomb 697 Prop. Mutinati. After Silvestrelli (2014) 100, WF044; **(d)** Athens, Benaki Museum inv. no. 35302. After Tzannes (2012); **(e–f)** Metapontum, MAN, inv. nos. 324220 and 324221, from Metapontum, Tomb 55 Prop. Riccardi. After Nava (2003) 667, pl. XXXIII,1. © Ministero della Cultura – Direzione Regionale Musei Basilicata. Drawing: Giulia Gioia and Fabiola Malinconico.

evolution, in the 4th century B.C.[27] At roughly the same time, a second type appears,[28] characterized by an everted rim, molded handles, and disc foot, again introduced by the Pisticci Painter.[29] The 24 vases of this shape that were decorated by him include some important variants, for example, the Matera vase and the patterned pelike from Tomb 259 of Pantanello (Fig. 3a–b). Both vases have a thick torus rim, short neck, pirifom body, disk foot reserved on the outside, and molded handles sometimes decorated with inverted palmettes.[30] They can be compared with Attic vases of the central decades of the 5th century.[31] A second variant, exemplified here by a pelike in the British Museum, differs from the previous one in having a thinner, overhanging lip, longer neck, and rounder body (Fig. 3c). This is the model of the Metapontine pelikai, as seen in examples by the Cyclops and Amykos Painters and further developed in the Dolon Painter's workshop (Fig. 4a–c).

The shape and subsidiary decoration (egg band on the neck and below the scene, which features an owl) of the small pelike from Tomb 55 Riccardi (Fig. 3d) find parallels with vases by the Washing Painter,[32] further confirming the eclecticism of the Pisticci Painter's workshop.

The archaeological evidence demonstrates that the *ergasteria* producing Metapontine red-figured vessels throughout the 4th century B.C. also manufactured black gloss, banded, plain, and kitchen ware.[33] Although the Pisticci and Amykos workshops, active in the second half of the 5th century, have not yet been discovered, they are believed to have operated in a similar way at least as far as black gloss is concerned. Potters operating in the Pisticci and Amykos workshops were probably responsible for both red-figured and black gloss vases, at least in some shapes: the elaborate pelikai of the second variant were certainly produced in black gloss, as were the glaukes (Fig. 5).[34]

Also produced using both techniques were the lebetes gamikoi, one of the most popular shapes in the Metapontine repertoire, part of the funerary assemblages of the Achaean colony since the Archaic period. This Late Archaic and Early Classical form was an adaptation of Corinthian models (Fig. 6b); the shape created by the Pisticci Painter for Metapontine consumers was partly Attic in appearance (Fig. 6a), and his sources of inspiration may have included the rare Type 2 from the Washing Painter's workshop.[35] Clearly original, however, was the subsidiary decoration (tongues on the shoulder, palmette fans below the handles, and meanders), which would become canonical in Metapontine lebetes.[36] This was also seen in the Apulian version of the shape, which would become popular only with the Long Overfall Painter at end of the first quarter of the 4th century B.C.[37] With many variants, the new version of the shape was also produced in black gloss ware (Fig. 6c–g).[38]

In the workshops of the early 4th century B.C., the close interaction between red-figured and black gloss pottery becomes even more evident, thanks to the greater quantity of known material and to the discovery of the workshops where both were manufactured.[39]

Stamped decoration (among the main innovations in Metapontine black gloss production, introduced just before the mid-5th century B.C.)[40] is also present in red-figured vases, albeit rarely. It is seen in a lebes gamikos from Pantanello that can be attributed to the Dolon and Creusa Painters' workshop.[41] Mixed decoration occurs in a fragmentary *pinax* from the Sanctuary of Apollo with a lion between two bulls beneath an ivy branch with black-figured corymbs (Fig. 7a). The scene can be compared to the black-figured decoration on the rims of Metapontine column kraters of the second half of the 5th

27 Elliott (1998) 671 (Plain pelike).
28 Elliott (1998) 670–671 (Elaborate pelike).
29 In contrast, pelikai would become popular in Apulian red-figured pottery with the Tarporley Painter in the late 5th–early 4th centuries.
30 Cf. *LCS*, 22, no. 58; *LCS* III, 6, no. 30a, pl. I,4.
31 Cf. Bochum, Ruhr-Universität, Kunstsammlungen, inv. no S1060 (*BAPD* 6974): *CVA* Germany 81– Norbert Kunish, *Bochum, Kunstsammlungen der Ruhr-Universität 2*, 13–15, fig. 2.1, pls. 4.1–2, 5.1–4, 6.1–5, 9.5–6 (Niobid Painter) and Matheson (1995) 183–184 (Early Polygnotan pelikai).
32 Lezzi-Hafter (1988) 277–280, fig. 91.
33 Silvestrelli (2016a), with bibliography.
34 On glaukes, cf. *CVA* Italy 86 – Giada Giudice and Elvia Giudice, *Collezione Jatta 6. Ruvo di Puglia. Ceramica attica a figure rosse e a figure nere*, 135–137, with previous bibliography.
35 *CVA* Denmark 8 – Knud Friis Johansen, *Copenhague, Musée National 8,. Collection des antiquités classiques*, 266–267, pl. 345,1a-b (*BAPD* 11167); *CVA* Greece 9 – Victoria Sabetai, *Athens Benaki Museum 1*, 38–39, pls. 31–32, fig. 17.
36 Cf., for example, *LCS*, 29, no. 103, pl. 7,6 (Cyclops Painter); *LCS* III, 46, no. C21 (Creusa Painter).
37 *RVAp* I, 86, nos. 154–161.
38 Elliott (1998) 659–663.
39 Silvestrelli (2016a), with previous bibliography.
40 Silvestrelli (2016b) 179.
41 Burn (1998) 617, T110-1.

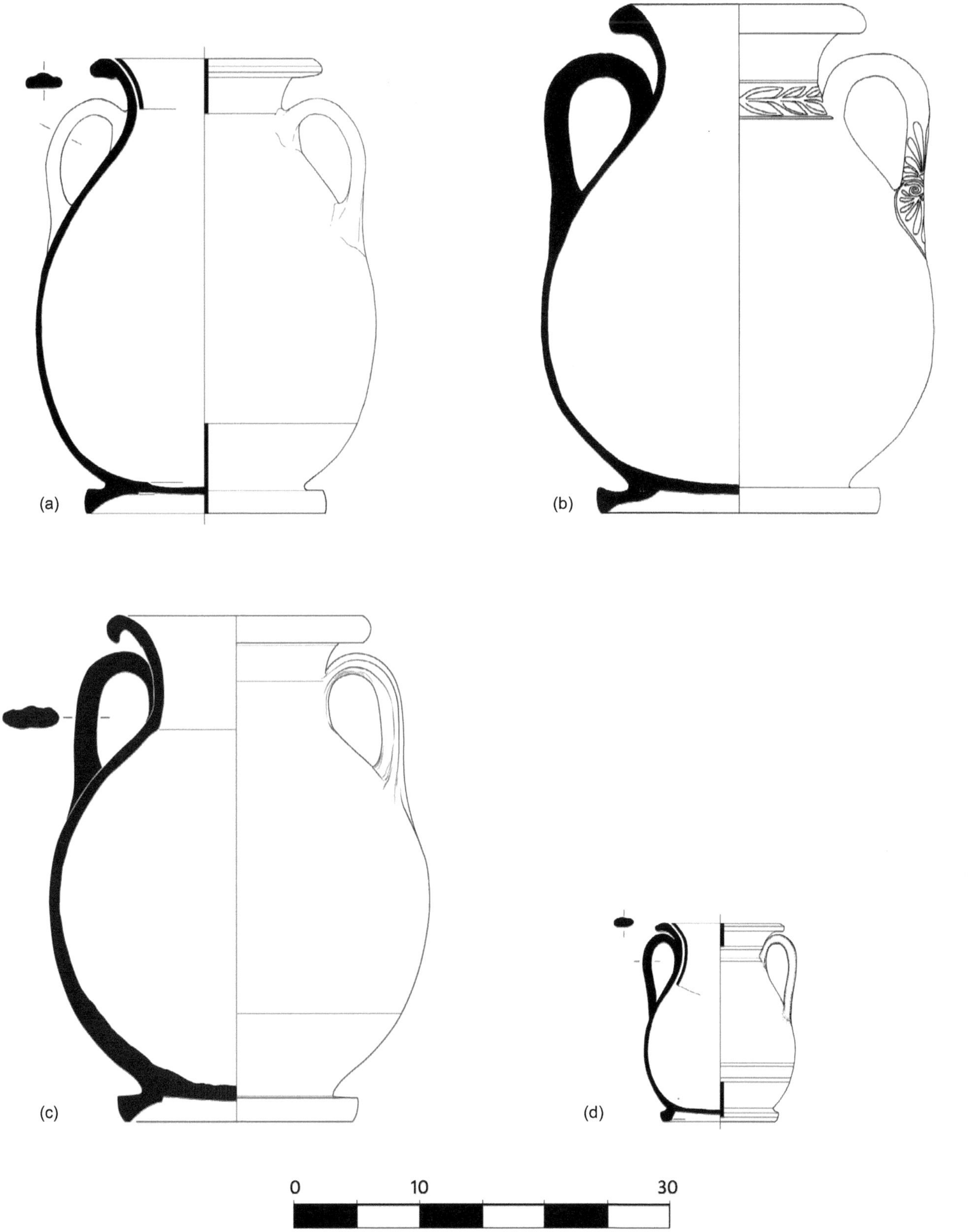

Fig. 3: Pelikai by the Pisticci Painter **(a)** Matera, MAN, inv. no. 172312. After *CVA* Italia 82 – Angelo Bottini, and Lucia Lecce, *Museo Archeologico Nazionale "Domenico Ridola"* 2. *Vasi italioti a figure rosse*, 17–18, pl. 1, fig. 1; **(b)** Metapontum, MAN, from Pantanello, Tomb 259. After Elliott (1998) 670, P27 (T 259–2); **(c)** London, British Museum, inv. no. 1859,0216.111 (E 423), *LCS*, 22 no. 59. © London, British Museum. Drawing: Kate Morton; **(d)** Metapontum, MAN, inv. no. 324299, from Metapontum, Tomb 55 Prop. Riccardi, site M. After Nava (2003) 667, pl. XXXIII,1. © Ministero della Cultura – Direzione Regionale Musei Basilicata. Drawing: Giulia Gioia. Ink: Fabiola Malinconico **(a–b, d)**.

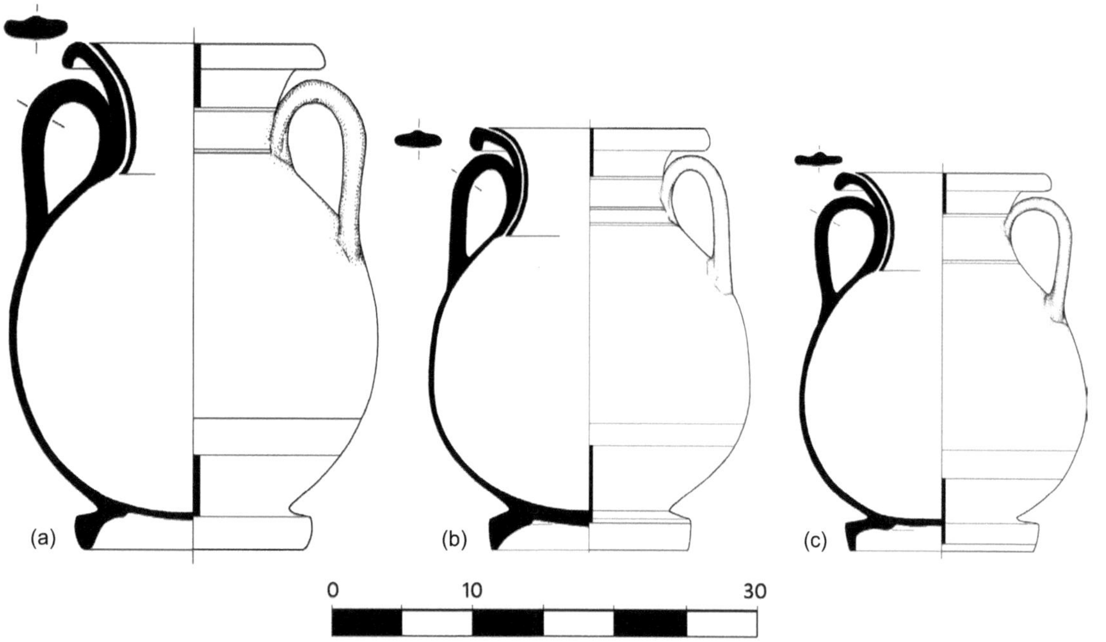

Fig. 4: Pelikai: **(a)** Metapontum, MAN, inv. no. 326119, by Cyclops Painter, from Metapontum, Tomb 296 Prop. Corrado. After Silvestrelli (2014) 100; **(b)** Metapontum, MAN, by Amykos Painter, from Metapontum, Tomb 759 Prop. Gallotta. After Silvestrelli (2014) 113; **(c)** Metapontum, MAN, by Dolon Painter, from Metapontum, Tomb 548 Prop. Mutinati. After Silvestrelli (2014) 107. © Ministero della Cultura – Direzione Regionale Musei Basilicata – Drawing: Fabiola Malinconico.

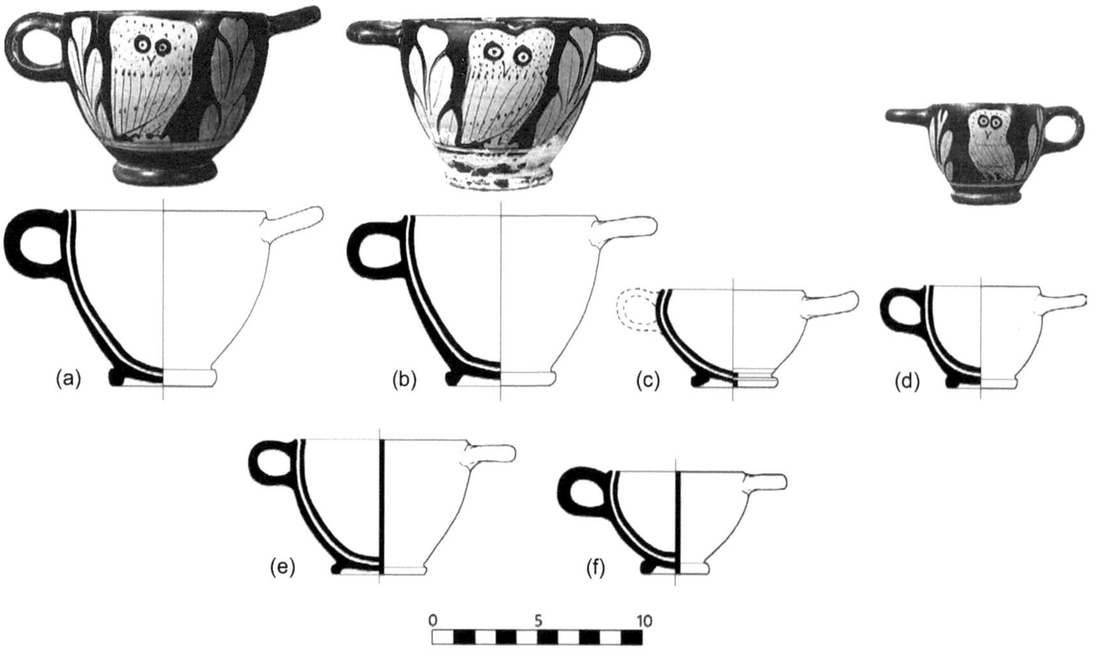

Fig. 5: Metapontum, MAN. Glaukes **(a–d)** and black gloss B-Type skyphoi **(e–f)**: **(a)** Glaux by Pisticci Painter, Metapontum, MAN, inv. no. 128555, from Saldone Tomb 12. After *LCS* III, 9 no. OS1, Burn (1998) 638, SS 12-2; **(b)** Glaux by Pisticci Painter, Metapontum, MAN, inv. 128552. After *LCS* III, 9 no. OS2; Burn (1998) 638, SS 12–3; **(c)** Glaux by Pisticci–Amykos workshop, Metapontum, MAN, from Metapontum, Pantanello, Tomb 87 (end 5th-beginning 4th century B.C.). After Burn (1998) 616, T87–2; **(d)** Glaux by Pisticci–Amykos workshop, Metapontum, MAN, inv. no. 128562, from Metapontum, Saldone, Tomb 19 (end 5th beginning 4th century B.C.). After Burn (1998) 638, SS 19–2; **(e)** black gloss B-Type skyphos, from Metapontum, Pantanello, Tomb 259 (second half 5th century B.C.). After Elliott (1998) 679; 681, S45 (T 259 – 3); **(f)** black gloss B-Type skyphos, from Metapontum, Pantanello, Tomb 350 (second half 5th century B.C). After Elliott (1998) 679; 681, S64 (T 350–25). © Ministero della Cultura – Direzione Regionale Musei Basilicata and Institute of Classical Archaeology at Austin – Drawing: Susan Langston. Ink: Fabiola Malinconico.

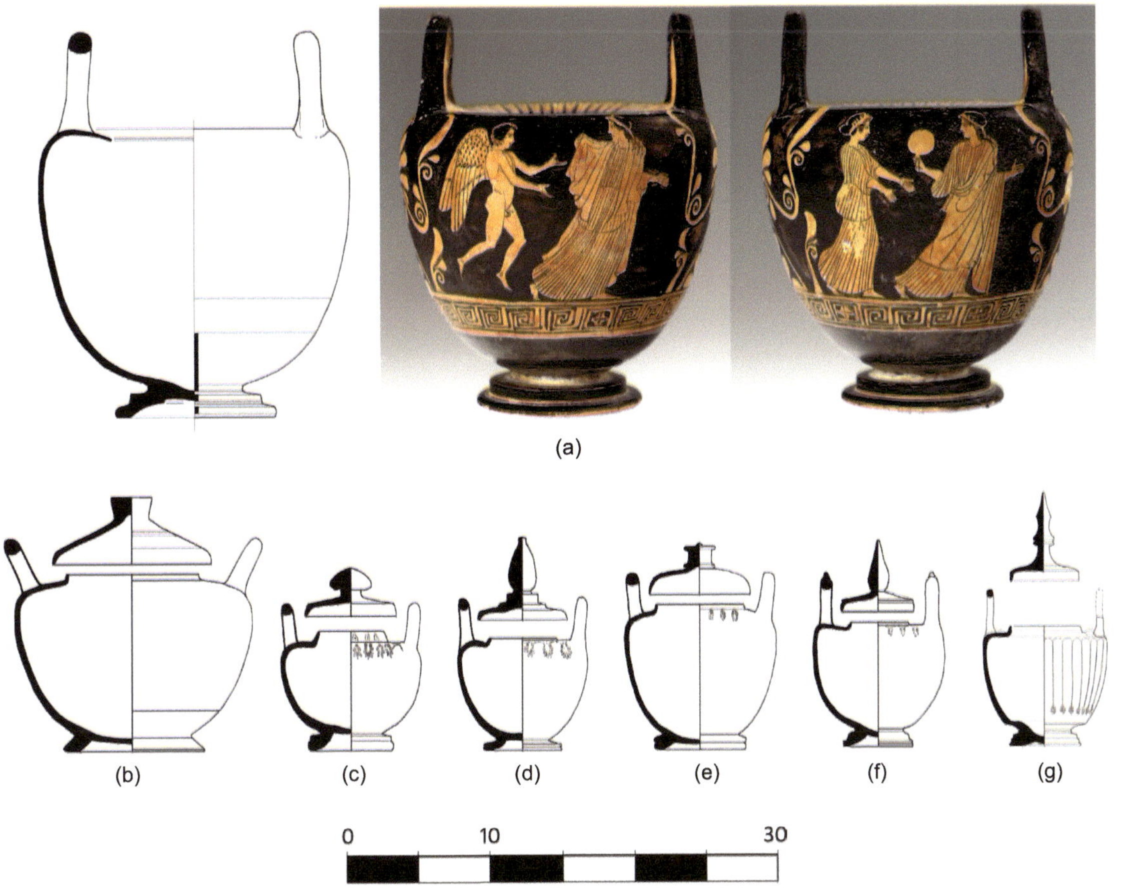

Fig. 6: Metapontum, MAN. Red-figured and black gloss lebetes gamikoi: **(a)** inv. no. 327034, from Metapontum, Tomb 597 Prop. Mutinati (Pisticci Painter). After Silvestrelli (2014) 100 WF045a-b © Ministero della Cultura – Direzione Regionale Musei Basilicata – Drawing: Giulia Gioia and Fabiola Malinconico **(b–g)** Black gloss lebetes gamikoi from the Pantanello necropolis. After **(b)** Elliott (1998) 659–660, LG 13 (T 86–4); **(c)** Elliott (1998) 661, LG 55 (T350–32); **(d)** Elliott (1998) 661, LG 54 (T 350–18); **(e)** Elliott (1998) 661, LG 31 (T 209–8); **(f)** Elliott (1998) LG18 (T 136–2) **(g)** Becker et *al.* (2016) 102, T04–01. © Ministero della Cultura – Direzione Regionale Musei Basilicata and Institute of Classical Archaeology at Austin – Drawing: Susan Langston.

and the beginning of the 4th century B.C.,[42] confirming its production in red-figured pottery workshops. Below the scene is a chain of stamped palmettes and lotus flowers that recalls similar decorations on black gloss vases found in association with vases by the Dolon and Anabates Painters (Fig. 7b).[43]

Furthermore, the rope pattern, a distinctive feature with oblique lines used on the ridges of Panathenaic amphorae and nestorides by the Dolon Painter (Fig. 8a), is also seen in a Xenon Group kantharoid vase and in black gloss vessels from this workshop (Fig. 8b).

Further study may also help to identify individual potters; four squat lekythoi (Fig. 9) by the Dolon, Creusa, and Anabates Painters (who cooperated in the same workshop for at least part of their careers, as established on the basis of finds recovered from the Metapontine potter's quarter) show strong similarities in the details of the rims, molded feet, and handles, and they may have been made by the same potter.

In conclusion, the broad and eclectic Metapontine morphological repertoire, established by the Pisticci Painter's workshop and continued by later *ergasteria*, highlights the potters' originality and capacity for innovation. Its variety was a response to the needs of the Metapontine workshops' multifarious clientele, and its complexity becomes even more evident when black gloss ware is considered in addition to red-figured pottery. Indeed, the two classes of materials were found together not only in the city, but also in the hinterland: of Metapontine production are the black gloss lebes gamikos and

42 Cf. *LCS*, 18, no. 23 pl. 4,1 (Pisticci Painter); 102, no. 535, pls. 51,8 and 53,1 (Dolon Painter).
43 Lebes gamikos from Pantanello, Tomb 306: Elliott (1998) 661 (in association with a pelike of the Dolon Painter's workshop).

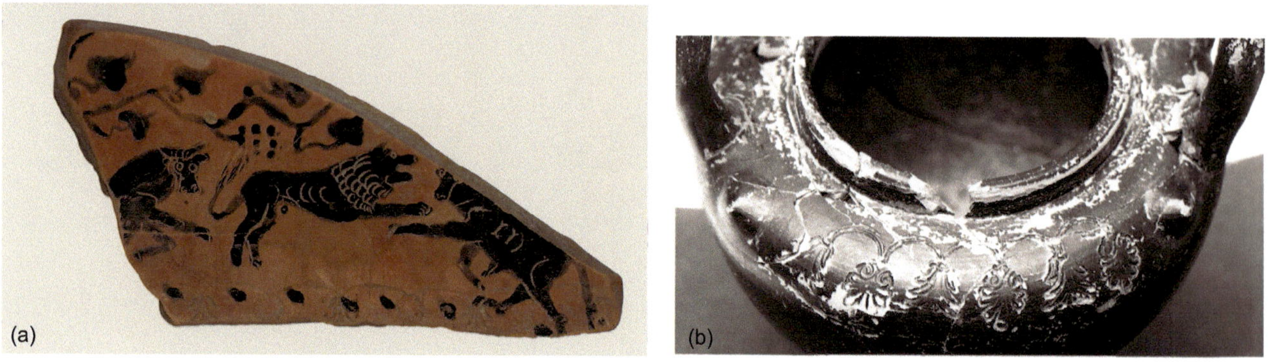

Fig. 7: (a) Metapontum, MAN, Terracotta placque inv. no. 147148, from Metapontum, Sanctuary of Apollo. After Adamesteanu (1975), fig. 50b. Photo: Francesca Silvestrelli; **(b)** Lebes gamikos with stamped decoration from Pantanello, Tomb 306, Metapontum, MAN. After Elliott (1998) 662, LG 41 (T306–2). © Ministero della Cultura – Direzione Regionale Musei Basilicata and Institute of Classical Archaeology at Austin – Drawing: Susan Langston.

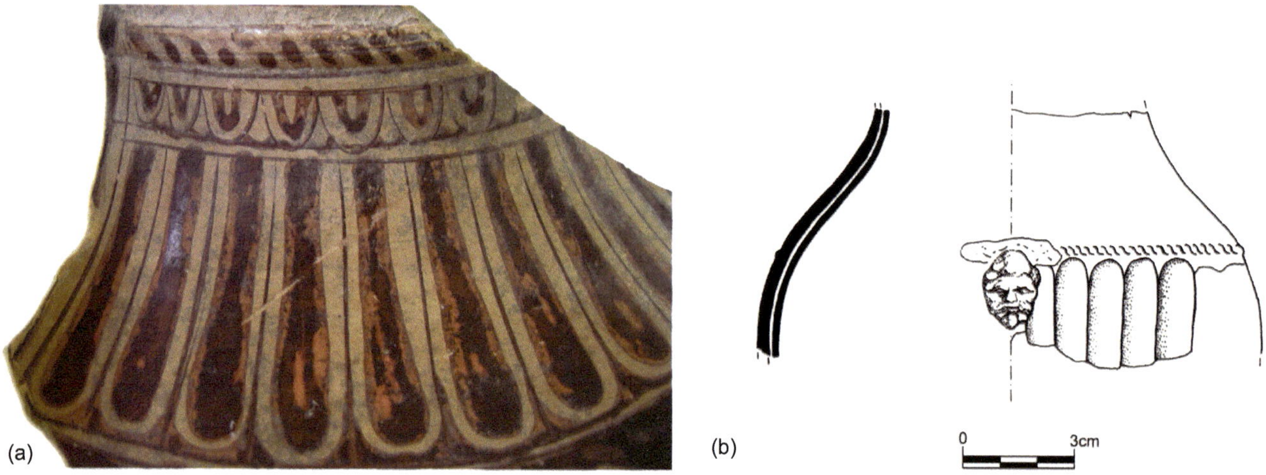

Fig. 8: (a) Amphora of Panathenaic type. Metapontum, MAN, inv. no. 29059 from Metapontum, Kerameikos, Deposit no. 1 (Dolon Painter), *LCS* III, 61 no. D 34. Photo: Francesca Silvestrelli; **(b)** Xenon Group kantharoid vase. Metapontum, MAN inv. no. 132911, from Metapontum, Kerameikos, Deposit no. 1, after Robinson (2013) 39, no. 77, pl. 10,77. © Ministero della Cultura – Direzione Regionale Musei Basilicata– Drawing: Fabiola Malinconico.

olpe found in Tomb 68 of Serra di Vaglio in association with a bell-krater with a pursuit scene by the Pisticci Painter;[44] the black gloss mugs and squat lekythoi from Tomb 57 of Baragiano, Toppo Sant'Antonio (Potenza) accompanying a hydria with youths exchanging gifts with women attributed to the Amykos Painter;[45] and the black gloss vessels (for example, the cylindrical mug with horizontal ridges)[46] that, together with the owl skyphos and a stemless kylix, both by the Anabates Painter, are part of the funerary assemblage of Tomb 8 at Ascoli Satriano, Valle Castagna, in Daunia.[47]

---

44 Greco (2014) 291–328.
45 Bruscella/Pagliuca (2013) 298–303.
46 Elliott (1998) 667, M15 (T330–10) and Silvestrelli (2018) 146 fig. 6,4. On the Metapontine origin of the shape, also confirmed by finds from the local Kerameikos, cf. Williams (2000).
47 Liseno (2012) 171–188.

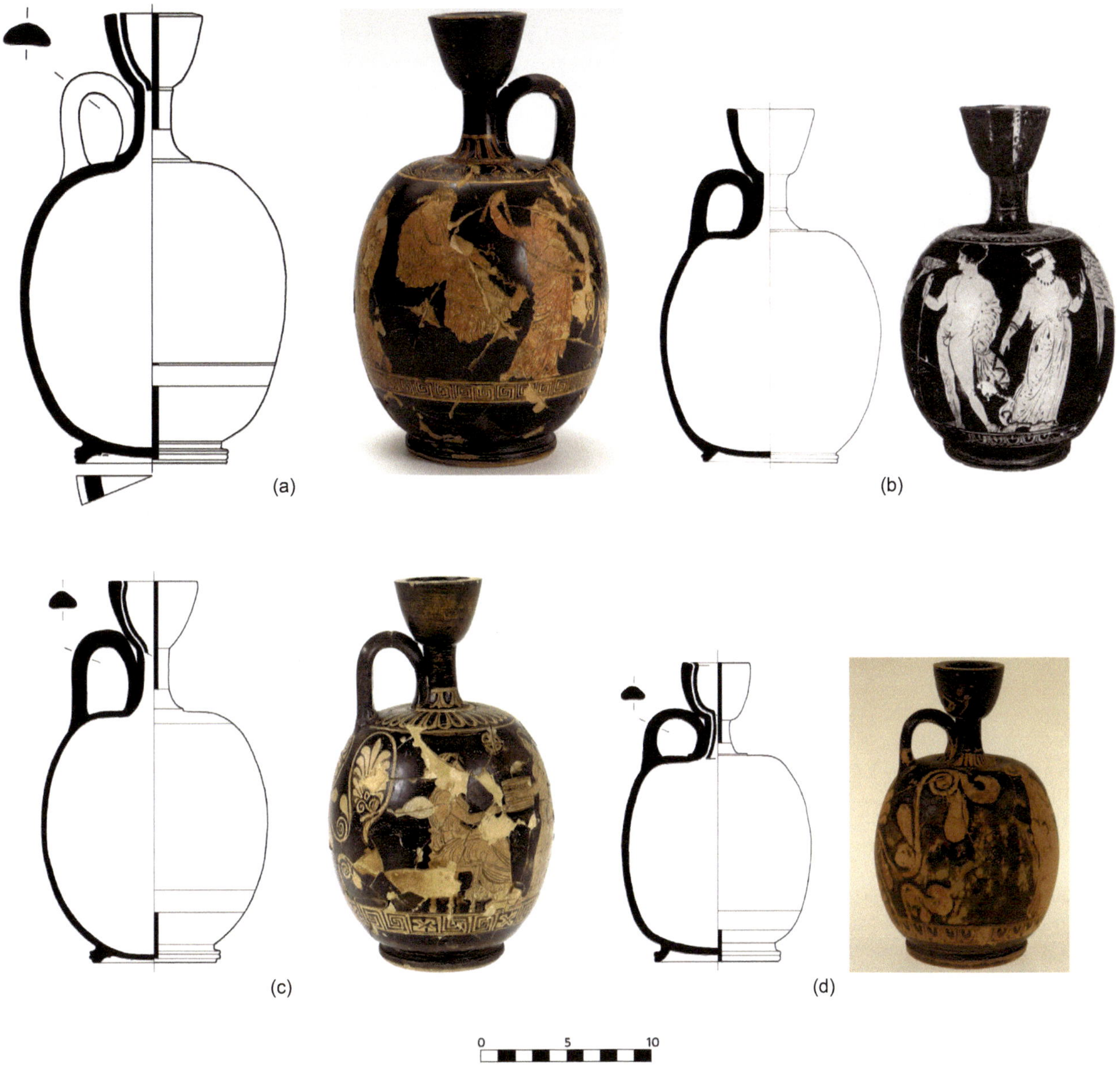

Fig. 9: Metapontine Red-figured squat-lekythoi: **(a)** Metapontum, MAN, inv. no. 325772. From Metapontum, Torre di Mare, Tomb 100 (Dolon Painter). After De Siena (2018) 103–107, figs. 5–9: 116, fig. 19b; **(b)** Taranto, MArTA inv. no. 8282. From Ceglie del Campo (Dolon Painter), *LCS* III, 63 no. D57. After *CVA* Italia 35 – Ciro Drago, *Taranto, Museo Nazionale 3*, IV D, 7, pl. 16; **(c)** Metapontum, MAN. From Metapontum, Tomb 369 Prop. La Torre (Creusa Painter). After Silvestrelli (2014) 106, WF 052, a–c; **(d)** Metapontum, MAN. From Metapontum, Tomb 546 Prop. Mutinati (Anabates Painter). After Silvestrelli (2014) 106, fig. 4.6 © Ministero della Cultura – Direzione Regionale Musei Basilicata – Drawing: Giulia Gioia and Fabiola Malinconico.

Abbreviations

BAPD Beazley Archive Pottery Database, https://www.carc.ox.ac.uk/carc/pottery (last access 02.07.2024).
CVA *Corpus Vasorum Antiquorum*.
LCS Arthur Dale Trendall, *The Red-figured Vases of Lucania, Campania and Sicily*, Oxford.
LCS I Arthur Dale Trendall, *The Red-figured Vases of Lucania, Campania and Sicily. First Supplement*, London.
LCS III Arthur Dale Trendall, *The Red-figured Vases of Lucania, Campania and Sicily. Third Supplement (consolidated)*, London.
MAN Museo Archeologico Nazionale.
MArTA Museo Archeologico Nazionale di Taranto.
RVAp I Arthur Dale Trendall and Alexander Cambitoglou, *The Red-Figured Vases of Apulia. Early and Middle Apulian*, Vol. I, Oxford.

Bibliography

Adamesteanu (1975): Dinu Adamesteanu, "Il Santuario di Apollo e urbanistica generale, in: Dinu Adamesteanu, Dieter Mertens and Francesco D'Andria (eds.), *Metaponto I*, Rome, 5–312.

Barresi/Giudice (2011): Sebastiano Barresi and Giada Giudice, "Attico, non attico: Camarina, il pittore di Mesagne ed il problema della definizione delle prime produzioni ceramografiche d'Occidente" in: Giada Giudice and Elvia Giudice (eds.), *Ἀττικὸν . . . κέραμον. Veder greco a Camarina dal principe di Biscari ai nostri giorni*, 2, Catania, 43–64.

Becker *et al.* (2016): Marshall Joseph Becker, Ingrid E.M. Edlund-Berry, Massimo Limoncelli, Edward G.D. Robinson, Francesca Silvestrelli and Serena Viva, "The Tombs at Sant'Angelo Vecchio", in: Francesca Silvestrelli and Ingrid Edlund-Berry (eds.), *The Chora of Metaponto 6. A Greek Settlement at Sant'Angelo Vecchio*, Austin, 85–128.

Bruscella/Pagliuca (2013): Antonio Bruscella and Salvatore Pagliuca, *Baragiano. Le tombe 53, 37, 57*, in: Massimo Osanna and Mariasilvia Vullo (eds.), *Segni del potere. Oggetti di lusso dal Mediterraneo nell'Appennino Lucano di età arcaica*, Venosa, 272–303.

Burn (1998): Lucilla Burn, "Figured Vases", in: Joseph Coleman Carter (ed.), *The Chora of Metaponto. The Necropoleis*, Austin, 592–640.

Cassimatis (1993): Hélène Cassimatis, *Le lébès à anses dressées italiote: à travers la collection du Louvre*, Naples.

Ciancio (1997): Angela Ciancio, *Silbíon, una città tra greci e indigeni: la documentazione archeologica dal territorio di Gravina in Puglia dall'ottavo al quinto secolo a.C.*, Bari.

Ciancio (2005): Angela Ciancio, "Recenti acquisizioni di ceramica italiota da Gravina in Puglia", in: Martine Denoyelle, Enzo Lippolis, Marina Mazzei and Claude Pouzadoux (eds.), *La céramique apulienne: bilan et perspectives* (Actes de la Table Ronde, Naples, 30 novembre–2 décembre 2000), Naples, 47–57.

D'Agostino (2003): Bruno d'Agostino, "Scrittura e artigiani sulla rotta per l'Occidente", in: Silvana Marchesini and Paolo Poccetti (eds.), *Linguistica è storia – Sprachwissenschaft ist Geschichte. Scritti in onore di Carlo De Simone – Festschrift für Carlo De Simone*, Pisa, 75–84.

Denoyelle (1997): Martine Denoyelle, "Attic or non- Attic? The Case of the Pisticci Painter", in: John H. Oakley, William D.E. Coulson and Olga Palagia (eds.), *Athenian Potters and Painters* I, Oxford, 395–405.

Denoyelle (2002): Martine Denoyelle, "Style individuel, style local et centres de production: retour sur le cratère des "Karneia", in: *Mélanges de l'École française de Rome: Antiquité* 114, 587–609.

Denoyelle (2007): Martine Denoyelle, "La ceramica: appunti sulla nascita delle produzioni italiote", in: *Atene e la Magna Grecia dall'età arcaica all'ellenismo* (Atti del XLVII Convegno di studi sulla Magna Grecia, Taranto, 27–30 settembre 2007), Taranto, 339–350.

Denoyelle (2013): Martine Denoyelle, "Spina: un avant-poste de la céramique Italiote en Etrurie padane?" in: Athena Tsingarida and Didier Viviers (eds.), *Pottery Markets in the Ancient Greek World: (8th – 1st Centuries B.C.)* (Proceedings of the International Symposium, Bruxelles, 19–21 June 2011), Brussels, 203–211.

De Siena (2007): Antonio De Siena, "L'attività archeologica in Basilicata nel 2006", in: *Passato e futuro dei convegni di Taranto* (Atti del XLVI Convegno di Studi sulla Magna Grecia, Taranto, 29 settembre – 1 ottobre 2006), Taranto, 407–463.

De Siena (2018): Antonio De Siena, "La tomba 100 della necropoli di Metaponto in località Torre di Mare ed il suo contesto", in: Martine Denoyelle, Claude Pouzadoux and Francesca Silvestrelli (eds.), *Mobilità dei pittori e identità delle produzioni*, Cahier du Centre Jéan Berard 25, Naples, 99–133.

Elliott (1998): Maria Elliott, "Black-Glazed Pottery", in: Joseph Coleman Carter (ed.), *The Chora of Metaponto. The Necropoleis*, Austin, 643–693.

Gadaleta (2012): Giuseppina Gadaleta, "Provenienze e contesti", in: Luigi Todisco (ed.), *La ceramica a figure rosse della Magna Grecia e della Sicilia*, II, Rome, 77–109.

Gadaleta (2022): Giuseppina Gadaleta, "ΙΧΘΟΠΙΝΑΚΙΑ. Riflessioni sui piatti da pesce a figure rosse", in: *Sicilia Antiqua. An International Journal of Archaeology* 19, 65–70.

Gaunt (2002): Jasper Gaunt, *The Attic Volute-krater*, PhD Dissertation, Institute of Fine Arts, New York University.

Greco (2014): Giovanna Greco, "Intorno al complesso funerario-cultuale della tomba 68 di Serra di Vaglio: la ricostruzione del contesto", in: Bianca Ferrara and Giovanna Greco (eds.), *Segni di appartenenza e identità di comunità nel mondo indigeno* (Atti del seminario di Studi, Napoli, 6–7 luglio 2012), Naples, 291–328.

Hoffmann (2002): Andreas Hoffmann, *Grabritual und Gesellschaft. Gefässformen, Bildthemen und Funktionen unteritalisch-rotfiguriger Keramik aus der Nekropole von Tarent*, Leidorf.

Jircik (1992): Nancy Rogers Jircik, *The Pisticci and Amykos Painters: The Beginning of Red-Figured Vase Painting in Ancient Lucania*, PhD Dissertation, University of Texas at Austin.

Lezzi-Hafter (1988): Adrienne Lezzi-Hafter, *Der Eretria-Maler. Werke und Weggefährten*, Mainz.

Lippolis (2008): Enzo Lippolis, "Modelli attici e artigianato artistico in Magna Grecia", in: *Atene e la Magna Grecia* (Atti del XLVII Convegno di Studi sulla Magna Grecia, Taranto, 27–30 settembre 2007), Taranto, 351–403.

Lippolis (2018): Enzo Lippolis, "La mobilità del ceramografo dalla formazione alla produzione. Problemi generali e un caso di studio: il Pittore di Dario e il suo ambiente artigianale", in: *Archeologia Classica* 69, 73–111.

Liseno (2012): Maria Grazia Liseno, "Tomba 8 di Valle Castagna", in: Marisa Corrente (ed.), *Lo spreco necessario. Il lusso nelle tombe di Ascoli Satriano*, Foggia, 171–188.

Lo Porto (1966): Felice Gino Lo Porto, "Metaponto. Scavi e ricerche archeologiche", in: *Notizie degli scavi di antichità*, 136–231.

Lo Porto (1988–89): Felice Gino Lo Porto, "Metaponto (Matera): rinvenimenti nella città antica e nel suo retroterra ellenizzato", in: *Notizie degli scavi di antichità*, 299–441.

Lucchese (2012): Claudia Lucchese, "Forme vascolari", in: Luigi Todisco (ed.), *La ceramica a figure rosse della Magna Grecia e della Sicilia*, II, Rome, 129–152.

Manzini (2011): Ilaria Manzini, "La ceramica a vernice nera di Teano: nuovi dati sulle caratteristiche della produzione locale" in: Gloria Olcese (ed.), *Immensa Aequora. Workshop. Ricerche archeologiche, archeometriche e informatiche per la ricostruzione dell'economia e dei commerci nel bacino occidentale del Mediterraneo (metà IV sec. A.C. – I sec. D.C.)* (Atti del Convegno, Roma, 24–26 gennaio 2011), Rome, 201–208.

Matheson (1995): Susan Burke Matheson, *Polygnotos and Vase Painting in Classical Athens*, Madison, Wisconsin.

Nava (2003): Maria Luisa Nava, "L'attività archeologica in Basilicata nel 2002", in: *Ambiente e paesaggio in Magna Grecia* (Atti del LII Convegno di studi sulla Magna Grecia, Taranto, 5–8 ottobre 2002), Taranto, 653–717.

Robinson (2013): Edward G.D. Robinson, "New Pixe-Pigme Analyses for South Italian Pottery", in: *Mediterranean Archaeology* 26, 15–42.

Rucco/Tagliamonte (2007): Gabriella Rucco and Gianluca Tagliamonte, "Stenio Pupidio, ceramografo campano", in: *Oebalus* 2, 135–151.

Schierup (2014): Stine Schierup, "Patterns of Use in Early Metapontine Red-Figure Pottery: Distribution, Shapes and Iconography", in: Stine Schierup and Victoria Sabetai (eds.), *The Regional Production of Red-figure Pottery: Greece, Magna Graecia and Etruria*, Gösta Enbom Monograph 4, Aarhus, 191–216.

Serino (2019): Marco Serino, *La bottega del Pittore di Himera e le altre officine protosiceliote. Stile, iconografia, contesti, cronologia*, Fecit te, Rome.

Silvestrelli (2014): Francesca Silvestrelli, "Red-Figure Vases from Metaponto: The Evidence from the Necropoleis along the Coast Road", in: Thomas H. Carpenter, Kathleen M. Lynch and Edward G.D. Robinson (eds.), *The Italic People of Ancient Apulia. New Evidence from Pottery for Workshops, Markets, and Customs*, Cambridge, 96–115.

Silvestrelli (2016a): Francesca Silvestrelli, "Ceramic Production at Sant'Angelo Vecchio and in the Metapontine Chora, in: Francesca Silvestrelli and Ingrid Edlund-Berry (eds.), *The Chora of Metaponto 6. A Greek Settlement at Sant'Angelo Vecchio*, Austin, 129–142.

Silvestrelli (2016b): Francesca Silvestrelli, "Potters and Painters in Metapontine Red-Figure Workshops: Some Preliminary Observations", in: Norbert Eschbach and Stefan Schmidt (eds.), *Töpfer-Maler-Werkstatt: Zuschreibungen in der griechischen Vasenmalerei und die Organisation antiker Keramik production*, *CVA* Beiheft Deutschland 7, Munich, 176–189.

Silvestrelli (2018): Francesca Silvestrelli, "Trasformazioni diacroniche del repertorio morfologico di Metaponto", in: Martine Denoyelle, Claude Pouzadoux and Francesca Silvestrelli (eds.), *Mobilità dei pittori e identità delle produzioni*, Cahier du Centre Jéan Berard 25, Naples, 135–160.

Silvestrelli (2019a): Francesca Silvestrelli, "Loutrophoroi. Forma e funzione", in: Claude Pouzadoux and Federica Giacobello (eds.), *Savoir-faire antichi e moderni tra Ruvo di Puglia e Napoli. Il caso del cratere dell'Amazzonomachia e di una loutrophoros con il mito di Niobe* (Giornata di Studio internazionale, Napoli, 13 Giugno 2013), Naples, 183–204.

Silvestrelli (2019b): Francesca Silvestrelli, "La *polis* di Metaponto e il suo territorio dalle prime esperienze alla metà del IV secolo a.C.", in: *Produzioni e committenze in Magna Grecia* (Atti LV Convegno di Studi sulla Magna Grecia, Taranto, 24–27 settembre 2015), Taranto, 536–555.

Tzannes (2012): Maria-Christina Tzannes, "Eros Pursuing a Maiden on an Early Lucanian Hydria in the Benaki Museum", in: Eurydike Kefalidou and Despoina Tsiafaki (eds.), *Kerameos Paides: Studies Offered to Professor Michalis Tiverios by his Students*, Thessaloniki, 218–227.

Williams (2000): Dyfry Williams, "Cista a cordoni-bicchiere a cordoni", in: David Ridgway, Francesca R. Serra Ridgway, Mark Pearce, Edward Herring, Ruth D. Whitehouse and John B. Wilkins (eds.), *Ancient Italy in its Mediterranean Setting. Studies in Honour of Ellen Macnamara*, London, 257–269.

Valeria Meirano

5 Clay Crafting, Hand Modeling: A Coroplastic Perspective. Case-Studies from Sicily and Magna Graecia

Abstract: Ancient clay handicraft has been extensively investigated in recent decades, and has also benefited significantly from modern multidisciplinary approaches that have considerably improved our knowledge of ancient technology, craftspeople apprenticeship, *savoir-faire,* and workshop organization. With regard to terracotta figurines and related artifacts, molding techniques and their processes (from prototypes to generations) have been the subject of important studies, whereas hand modeling has received much less attention in the research on ancient technology. This is probably due to its apparent simplicity and presumed association with early stages of terracotta crafting and/or less technologically advanced production contexts. On the contrary, hand modeling is a long-lasting technique: thanks to its versatility, it enabled craftspeople to combine a variety of forming methods and to produce composite artifacts, made up of hand-modeled, moldmade, and wheel-made parts. Focusing on some case-studies from Western Greece (handmade figurines in eastern Sicily and terracotta cave-nymphaeum models at Locri Epizephyrii), this paper aims to show the wide range of techniques used by coroplasts in clay manipulation and assembly strategies, and also addresses how the adoption of hand modeling, along with iconographic and typological choices, may reflect regional trends and specific, local market demands related to ritual necessities.

Introduction

The production of terracotta figurines has been extensively investigated in recent years; in particular, molding techniques and related manufacturing procedures have been the subject of important studies concerning both small and large artifacts. Scholars have shed light on various skills of ancient craftspeople in specific contexts and chronological periods, and have been able to tentatively reconstruct the organization of the workshops and the roles played by the artisans (and perhaps the artists) in the ateliers, and their mutual interactions. The dynamics that led from the creation of the orginal pieces to the production of replicas – let us say from prototypes to generations and series – are the crucial objectives of this investigation, and were the nodal point of an output that was mainly aimed at seriality and standardization.[1]

On the contrary, hand modeling has received much less attention in coroplastic studies, at least until the last few decades.[2] This lack of visibility was probably due to the apparent simplicity of hand-modeled figurines compared to their moldmade counterparts, their alleged association with 'primitive' terracotta crafting or less technologically advanced production contexts, and – last, but not least – to their limited stylistic implications and value in determining chronology. Finally, they were considered 'immature' products, often attributed to children, on the microscale of human life,[3] and relegated to early steps of craftsmanship, in the macroscale of ancient material culture.

It is important to notice, however, that manual modeling is definitely a long-lasting technique. Despite its traditional association with a predominant use in 'idol' making,[4] as well as its perceived reference to a primitive, non-Greek/'indigenous' concept of the human body in earlier studies,[5] recent research has pointed out that this technique continued to be used even after the introduction of molding and in fact coexisted with it. The production of animal statuettes, probably

1 Muller (1997); Caubet (2009); Muller (2014); Muller/Lafli (2016b); Muller (2017–2018); Favier (2022) 94–96.
2 Cf. Muller (2017–2018) 155; Aurigny/Rohaut (2022b). Two recent conferences have drawn attention to hand modeling in a Mediterranean perspective (Aix-en-Provence 2019: Aurigny/Rohaut 2022a), even in relation to bronze casting (Paris 2016: Descamps-Lequime/Jeammet 2023).
3 See *infra*.
4 Fourrier (2022) 194.
5 See P. Orsi's statements, *infra*, note 45; P. Orlandini's remarks on a piece from Metapontum, commented by Parisi in Parisi/Albertocchi (2019) 484–485; E. Pottier's reflections on terracotta "horribles fétiches", recalled in Aurigny/Rohaut (2022b) 6.

https://doi.org/10.1515/9783111189635-005

the largest corpus of hand-modeled terracottas since the Iron Age,[6] demonstrates the *longue durée* of this technique into the Hellenistic period and beyond. Other common categories of hand-modeled terracottas are the 'daily life' compositions showing female figures engaged in cereal processing, food preparation and other activities,[7] as well as the large and heterogeneous group of miniature models (food, offering trays and plates, boats, etc.).[8]

"A macchietta"[9] Figurines: An Overview of Hand-Modeled Terracottas from Eastern Sicily

Some areas of Western Greece have proven to be rich sources of hand-modeled terracottas, but these have never been examined systematically or from a broad perspective.[10] Without claiming comprehensivity, I will focus on some contexts of eastern Sicily – from Messina to Camarina, between the 6th century B.C. and the Hellenistic period – where numerous hand-modeled statuettes are known from necropoleis and sanctuaries that also yielded moldmade products.

Although terracottas are generally rare among grave goods of the Archaic and Classical periods,[11] several handmade figurines have been found in the Syracuse necropoleis of Fusco[12] and Giardino Spagna (Fig. 1a–d),[13] as well as in Lentini (Fig. 1e),[14] Francavilla (Fig. 1f)[15] and Ragusa (Fig. 1g).[16] The funerary areas of Megara Hyblaea (Fig. 2a–f)[17] and Camarina (Fig. 2g–n)[18]

6 Poplin (2003): for technical aspects, see especially Muller (2003); for Archaic and Classical Greece, Huysecom (2003); for Athens, Georgaka (2014); for Crete, Pautasso (2022).

7 Pisani (2003) provides a detailed overview of this class, which is mainly attested in funerary contexts of Greek motherland (Boeotia, Athens, Argos and other Peloponnesian sites, Cyprus) and Western Greece, especially Sicily (Lipari; for other sites see *infra*), between the second half of the 6th and the first quarter of the 5th century B.C.; see also Tsoukala (2009).

8 For offering plates and food: Meirano (2017), (2021) with previous bibliography; see also *infra*; for boat miniatures: Basile (1991), (1993); Tigano (1991), (2002) 118–120, 133–137; Tigano (2021); Gianfrotta (2014) and *infra*.

9 The definition as "a macchietta" figurines was used by P. Orsi for the handmade figurines he found in Sicilian excavations: cf. Orsi (1904) 798–799, (1918) 566, 568.

10 For Magna Graecia cf. the examples from funerary and sacred contexts (Saturo, Metapontum, Francavilla Marittima, Crotone, Paestum) belonging to the earliest phases, discussed by Parisi in Parisi/Albertocchi (2019) 475–486, figs. 7–10; see also Bilbao Zubiri (2022) for Metapontum.

11 Meirano (2012a) 113–116.

12 In tomb CCX a crouched "mostriciattolo" was found, accompanied by several animal figurines (a dog, two turtles, a dove, a hen on a plinth), at least some of which are handmade; in tomb CDLIX, two 'a nastro' female figurines with molded face: Orsi (1895) 133–134, fig. 11; 177–178, fig. 76; see also Meola (1971) 77, pl. IV,b (wrong provenance); Albertocchi (2012) 94–95 (said to be probably Corinthian; first half of the 6th century B.C.).

13 Tomb 121 contained ten figurines, both molded and hand-modeled: the latter group consisted of a dog (?), a "scimmiotto" riding a horse and three "quadrupedi indefinibili": Orsi (1925) 304, fig. 59. A statuette of a duck with two ducklings on a plinth is reported from tomb II: Cultrera (1943) 45, h, fig. 12; F. Fouilland in *La Sicilia greca* (1989) 144, 220, no. 227 (said to be Rhodian, dated 575–525 B.C.); see also: Gabba/Vallet (1980) pl. C,124.

A sporadic, probably hand-modeled dog (?) comes from the Archaic funerary area investigated near Viale Paolo Orsi: Gentili (1951) 323, no. 12; a monkey (?) from tomb H: Gentili (1954a) 315, no. 5, fig. 13,2 (Archaic); a probably handmade fawn (?) from tomb 35 in the area near Piazza della Vittoria: Gentili (1956a) 127, no. 4, fig. 7h (Archaic).

14 One male figurine is known from tomb 93 of the Piscitello necropolis, dating to the second quarter of the 5th century B.C.: Frasca (1982) 57, no. 7, pl. XXI,11.

15 A handmade satyr with other terracottas (a dove and a statuette) comes from tomb 1, dated around 450 B.C.: Spigo/Rizzo (1993–1994) 1053, no. 10, pl. CVIII, fig. 2; E. D'Amico in Spigo *et al.* (2008) 133, 139, no. 11; E. D'Amico in Tigano/Vanaria (2022) 203, no. 26.

16 From the Rito necropolis, woman on a plinth, sitting in front of the oven: F. Fouilland in *La Sicilia greca* (1989) 149, 224, no. 252; Pisani (2003) 7, 16, no. 88, fig. 17a-b (end of the 6th-beginning of the 5th century B.C.).

17 Western necropolis: two satyrs were found in tomb 401: Gras *et al.* (2004b) 116, note 116; a horseman in tomb LXXI: Orsi/Cavallari (1889) 834; a hydrophora, with molded face and a probably wheel-turned hydria, and a female figurine at the oven were found in tomb 413; another, similar figurine also comes from the site: Sparkes (1965) 163, no. 79A, pl. XXX,4, (1981) 176, pl. 4b; Gabba/Vallet (1980) pl. XCI,96; F. Fouilland in *La Sicilia greca* (1989) 149, 224, no. 253; Pisani (2003) 7, 17, nos. 98–99, figs. 6–7, 18; Chillemi (2015) 641, nos. 10–11, 13: they are all dated to the end of the 6th-beginning of the 5th century B.C. A bird, several female figurines and a "rozzissimo" ithyphallic satyr were also recovered from the same funerary area: Orsi/Cavallari (1889) 910, pl. VI,2.

Southern necropolis: an ithyphallic three-legged satyr with molded face comes from tomb 72: Gentili (1954b) 400, no. 4, fig. 9 (Archaic); a woman at the oven from tomb B81: Cébeillac Gervasoni (1976–1977) 592, pl. CXXX (early 6th century B.C.?). For general remarks on the *femmes au travail* and satyrs in the necropoleis of Megara Hyblaea, see also Bérard (2017) 228–229.

18 From the Orsi excavations in the Passo Marinaro necropolis, a handmade singing "mostriciattolo" with moldmade head accompanied by a miniature vase and pieces of furniture: Orsi (1904) 798–799, fig. 21 (tomb 142); a satyr with molded face: 835–836, fig. 47 (found outside tomb

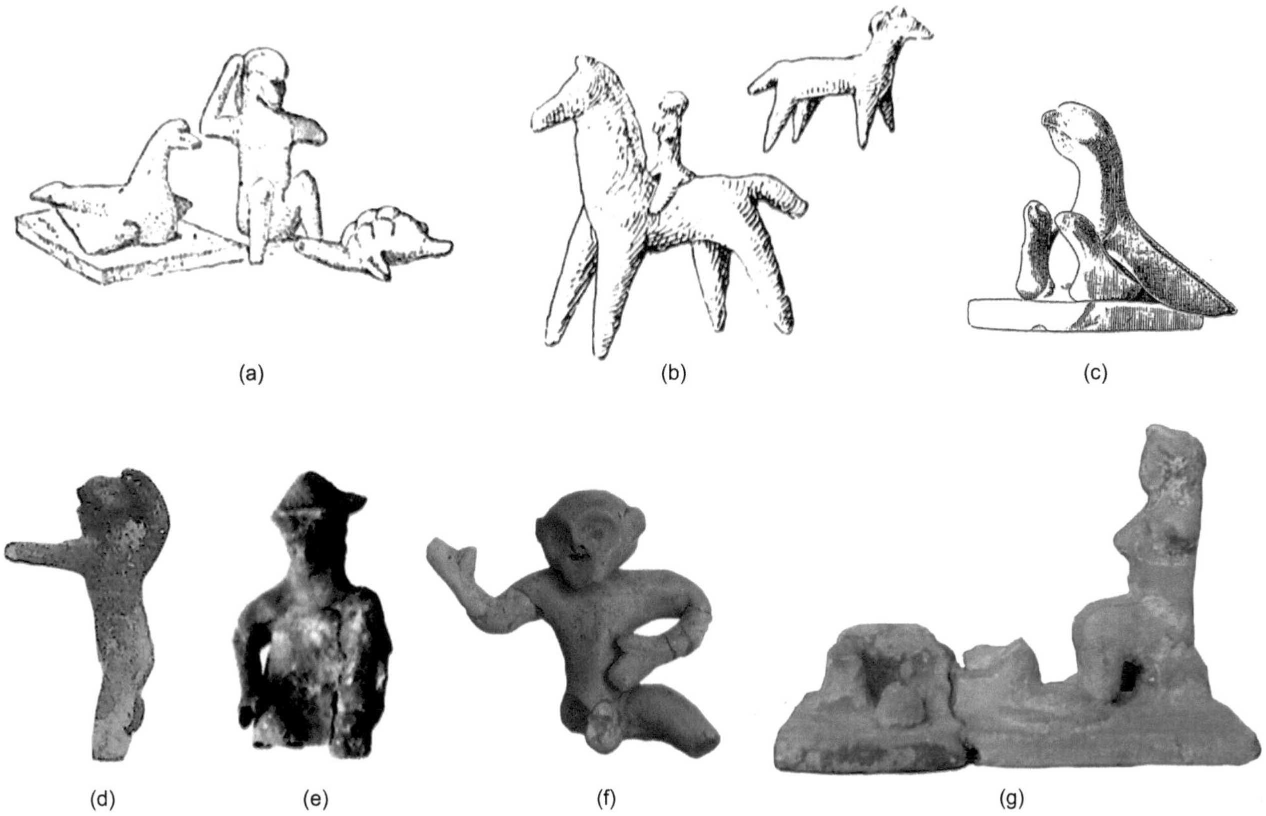

Fig. 1: Handmade terracotta figurines from necropoleis in eastern Sicily: **(a)** Syracuse-Fusco. After Orsi (1895), fig. 11; **(b–d)** Syracuse-Giardino Spagna. After Orsi (1925), fig. 59; Cultrera (1943), fig. 12; Gentili (1954a), fig. 13,2; **(e)** Lentini-Piscitello. After Frasca (1982), pl. XXI,11; **(f)** Francavilla. After Tigano/Vanaria (2022), no. 26; **(g)** Ragusa-Rito. After Pisani (2003), fig. 17b.

have provided numerous examples: it is noteworthy that in these two sites, in particular, excavations brought to light workshops which produced hand-modeled terracottas of the same types as those known from the necropoleis (Fig. 3).[19]

The types attested in funerary areas include animals – both real and mythical –, satyrs, 'monsters' with rudimentary bodies, articulated 'dolls' and other anthropomorphic figures, many of which correspond to *sujets de genre* showing cereal processing and food preparation scenes. Similarly to other categories of terracottas, recent research has shown that these occur mainly in burials belonging to children and young women.[20] In the past, scholarship has emphasized their possible identification as toys, despite their typological heterogeneity and various associations (e.g. satyrs, 'monsters' and 'every-

318); two jointed 'dolls' with hand-modeled limbs: 868, fig. 73 (tomb 444); several animals, gathered both inside (a turtle from tomb 509, birds from tombs 308 and 314: *ibidem* 835, 884, 948) and outside (a sphinx, a dog, a horse, a pig, a dove: *ibidem* 921) the tombs; an unusual three-legged figurine with male attributes but with moldmade female face, holding a dove: Orsi/Lanza (1990) 41–42, no. 2, pl. XVII (tomb 711; *ca* 450 B.C.); two three-legged satyrs: 123, pl. LXXXV (tomb 1248; 450 B.C.) and 149, pl. XCVIII (tomb 1451; 460–450 B.C.); other sporadic finds (a dove, a satyr, and a 'monster' sitting on his queue: 83, nos. 13–15, pl. XLVII).

From later excavations in the same funerary area, a dog, found on the *a cappuccina* cover of tomb 3: Salibra (2016) 51, 249, no. 3, pl. XXXIV (second half of the 5th century B.C.); the head of a bird and a jointed 'doll' arm (outside tomb 297: 175, 249, pl. LXXX,x–y); a sporadic (hand-modeled?) cock (249, 309, pl. CVIII).

From the Rifriscolaro necropolis, one 'a nastro' female figure: Pelagatti (1973) 144–145, no. 429, pl. XLIII (tomb 70); Albertocchi (2012) 95.

19 At Megara Hyblaea the Archaic kilns in the area of the southern fortifications yielded fragments of satyrs, partly hand-modeled female figures and animal statuettes (birds, dogs, bulls, an ox): Tréziny (2004) 175–177, 181, figs. 198, 201; satyrs are also attested in the inhabited area: Gras *et al.* (2004b) 116.

The Provide kiln near the Ippari river in Camarina, in use between the third quarter of the 5th and the middle of the 4th century B.C., has yielded handmade animals (dogs, fawns, a sow, a panther, a dove) and three-legged figures with molded heads: Pisani (2008) 81–84, nos. 172–196, fig. 15, pl. XIV; 72–76, nos. 162–167, figs. 13–14, pl. XII. Other hand-modeled terracottas from the same area belong to the phases prior to the kiln installation (an ithyphallic figurine, pieces of '*genre* scene' statuettes and other fragments: 28–29, 32, nos. 1–6, pl. V).

20 Hermary/Dubois (2012).

Fig. 2: Handmade terracotta figurines from necropoleis in eastern Sicily: **(a–e)** Megara Hyblaea-western necropolis. After Orsi/Cavallari (1889) 834 and pl. VI,2; Pisani (2003), figs. 6–7 and 18; **(f)** Megara Hyblaea-southern necropolis. After Gentili (1954b), fig. 9; **(g–n)** Camarina-Passo Marinaro. After Orsi (1904) figs. 21 and 47; Orsi/Lanza (1990), no. 2, pl. XVII; no. 123, pl. LXXXV; no. 149, pl. XCVIII; nos. 13–14, pl. XLVII; Salibra (2016), no. 3, pl. XXXIV.

day-life' statuettes). More recently, a wider range of interpretations has emerged, concerning the symbolic and ritual values that terracotta figurines in general could have assumed in funerary contexts.[21] Another important issue is the way in which these small artifacts were involved in funerary ceremonies, both during the burial and later. Their precise find places can provide valuable clues as to the roles they played in specific funerary customs: inside tombs together with other grave goods, outside tombs, sometimes on their covers (like a dog at Camarina),[22] or scattered across the ground of the necropolis.[23]

21 Cf. Graepler (1997); Dasen (2012); Huysecom-Haxhi *et al.* (2012); Meirano (2012a); Pisani (2012); Scilabra (2012).

22 Cf. *supra*, note 18.

23 Meirano (2018).

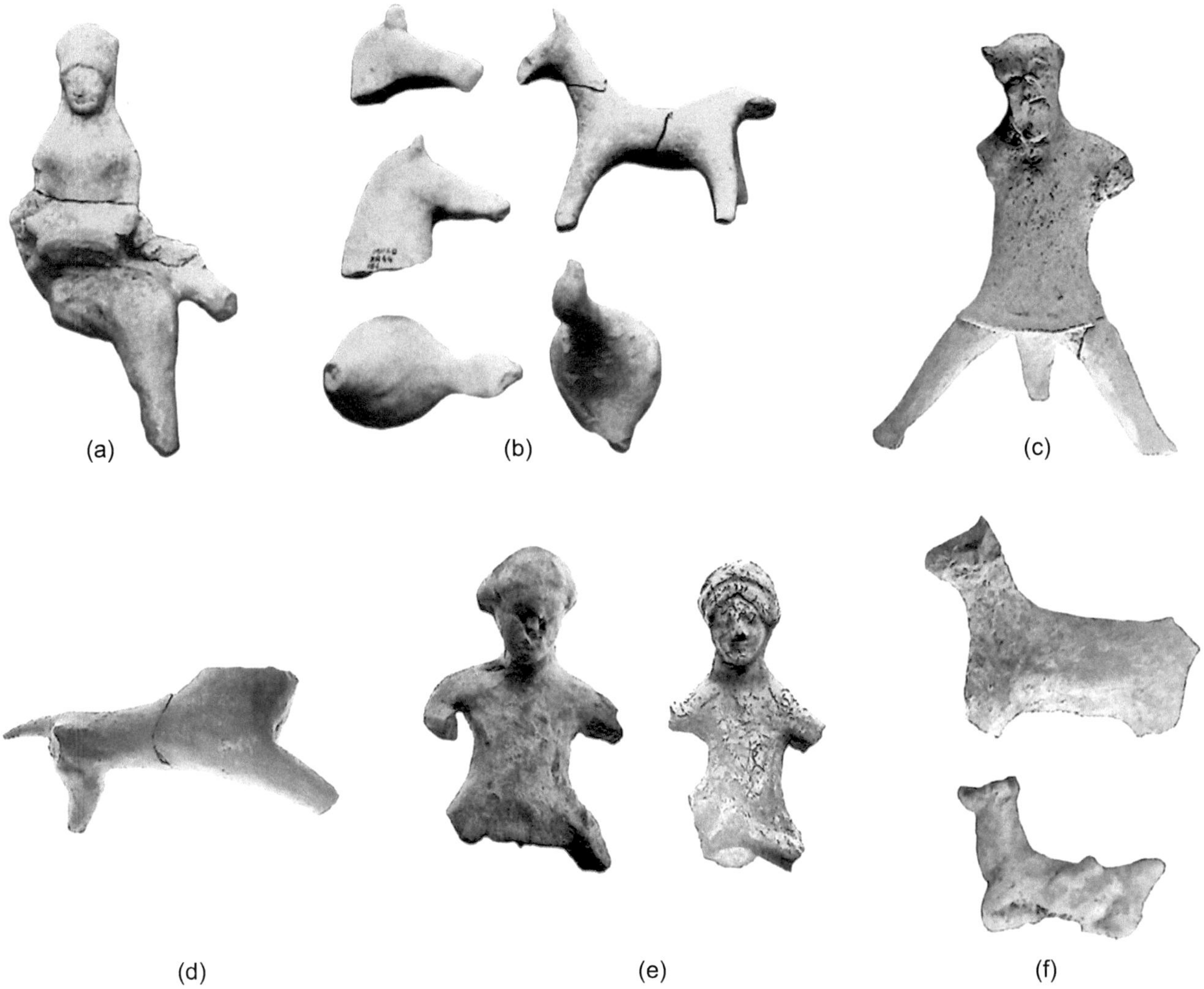

Fig. 3: Handmade terracotta figurines from production contexts in eastern Sicily: **(a–b)** Megara Hyblaea. After Tréziny (2004), figs. 198 and 201; **(c–f)** Camarina. After Pisani (2008), figs. 13 and 15, pls. XII and XIV.

Hand-modeled figurines have also been found in sacred areas at various sites, mainly in contexts related to goddesses and female worship, for example in Naxos (Fig. 4a),[24] Lentini (Fig. 4b–c),[25] Francavilla,[26] Syracuse (Fig. 4d–e)[27] and Megara Hyblaea (Fig. 4f).[28] However, the largest quantity of these artifacts, approximately 800 items, was discovered in the area of Piazza San Francesco in Catania (Fig. 4g–j), which represents one of the most impressive votive deposits in Sicily. According to the archaeological evidence, the chronology of this sacred area extends from the early 6th to the late 4th century B.C., and a connection with the religious sphere of Demeter has been documented, that dates back at least to the 5th century B.C.[29]

24 A small seated figure comes from the urban sanctuary at the mouth of the Santa Venera stream, dedicated to a female goddess: Gentili (1956b) 331, fig. 12.
25 Four birds and a small 'monkey' come from the Alaimo sacred area, probably dedicated to Artemis in the period in which ritual activities are archaeologically documented (660-first decades of the 6th century B.C.): Grasso (2008) 130, 152–156, nos. 622–626, p. 130, pl. LXII.
26 Handmade kourotrophos from the sanctuary of Via don Nino Russotti: Pedrucci (2022) 355, Fran 1 (second half of the 6th century B.C.).
27 From the Athenaion area, an anthropomorphic seated figure and an animal with spotted coat: Orsi (1918) 566, 568, figs. 155, 158.
28 Fragments of ithyphallic satyrs, female figures on horses (?), three ducks: Orsi/Cavallari (1889) 932, nos. 115–116, pl. VIII,7 and 9; 933, no. 118, pl. VIII,2; 935, nos. 132–133; 937, nos. 183–185. A hand-modeled kourotrophos from an indetermined context is reported in Pedrucci (2022) 381, MH 4 (early 5th century B.C.).
29 On the context and cult implications: Rizza (1960); Pautasso (2010), (2023).

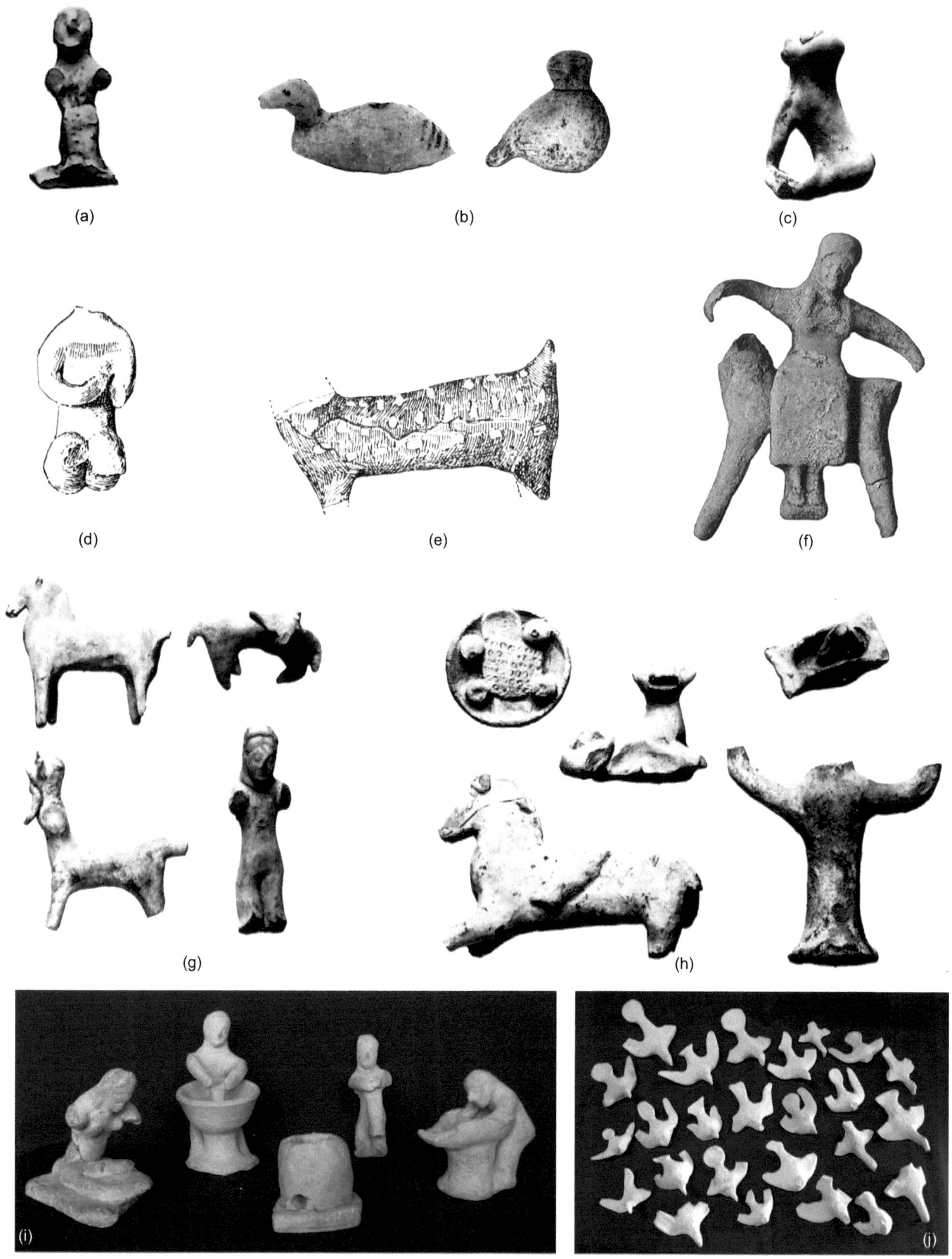

Fig. 4: Handmade terracotta figurines from sacred areas in eastern Sicily: **(a)** Naxos-Santa Venera. After Gentili (1956b), fig. 12; **(b–c)** Lentini-Alaimo. After Grasso (2008), nos. 622–623 and 626, pl. LXII; **(d–e)** Syracuse-Athenaion. After Orsi (1918), figs. 155 and 158; **(f)** Megara-Hyblaea. After Orsi/Cavallari (1889), pl. VIII,2; **(g–j)** Catania-San Francesco. After Rizza (1960), figs. 17–18; Chillemi (2010), figs. 1–2.

The handmade terracottas from Piazza San Francesco depict various iconographies, some of which correspond to the types also attested in eastern Sicilian funerary areas. Apart from a series of artifacts reproducing offering plates and trays (Fig. 4h, top left),[30] the majority of the statuettes features anthropomorphic and zoomorphic subjects. The former group includes standing and seated figures, articulated 'dolls', subjects with animals, horsemen, dancing groups and a figure in a bathtub. The latter consists of mythological beings such as centaurs, satyrs and three-legged satyrs, other so-called 'monsters', and animals (fawns, horses, dogs and felines, wild boars, pigs, sheep, goats, turtles, frogs, and especially swans, doves, cocks, hens, and other winged creatures).[31] The body is usually made from a clay cylinder to which the limbs and head are affixed, according to the 'snow-man assembly technique'. Despite the summary representation of the details ("più o meno rozzamente plasmate a mano"),[32] these terracottas are highly expressive.

Specific studies have recently been devoted to the '*genre* scene' figurines, which represent individuals or groups performing daily tasks such as grinding, pounding, kneading and baking, accompanied by their tools (portable ovens, basins and mortars, saddle-querns, pestels, etc.).[33] Information on the hybrid production technique used for these statuettes is now available: the body is generally done freehand, while the heads are moldmade. The hair is created by modeling a strip of clay and attaching it to the head; occasionally, details are added through incised lines, and the nose is sometimes shaped by pinching the clay between the fingers. Mortars and other containers are wheel-made.[34]

Since their discovery in the 1950s, the handmade materials from San Francesco have been shown to have parallels among terracottas found in Greece (at Corinth, Megara, Argo, Tyrinth, Sparta, and especially in Boeotia for the 'daily-life' figurines), and have been linked to sub-Mycenaean and Geometric legacies, as well as to the persistence of tradition before the diffusion of the molding technique.[35] Recent studies have expanded on identifying stylistic influences from Greece,[36] and shed new light on the interpretation of the statuettes, emphasizing the symbolic meanings with which they may have been invested in the ritual contexts, rather than the more secular readings that have often been suggested in the past.[37]

Hand-modeled figurines seem to be less attested in eastern Sicily for the Hellenistic period; anthropomorphic and animal statuettes, however, are documented, and, above all, boat models with handmade parts (e.g. figurines of rowers and helmsmen) from necropoleis, sanctuaries and profane contexts (Messina,[38] Francavilla,[39] Lentini,[40] Syracuse,[41] Avola[42]), corresponding to a typical creation of coroplasts from Syracuse and other sites in this part of the island.[43]

30 Rizza (1960) 258, fig. 17,1; Pautasso (2020) 63–66, fig. 2.

31 Rizza (1960) 252–254, 258, figs. 17, 18,1 and 4–6, 19, 20,1–4 and 7; see also Albertocchi (2012) 91, 94–95 about the possible Corinthian production (or imitation) of the statuettes at figs. 17,4, 18,6 and 20 in Rizza (1960); Chillemi (2010).

32 Rizza (1960) 252.

33 Cf. Sparkes (1965) 163, 85A-B; Pisani (2003) 7, 9, 12, 14, nos. 41 and 52; and especially Chillemi (2010), (2011); V. Chillemi in Ursino (2012) 228–229; Chillemi (2015).

34 Chillemi (2011), (2015).

35 Rizza (1960) 253 ("... plastica fittile la cui produzione ancora sfuggiva all'uso livellatore della matrice.").

36 Especially for '*genre* scenes' in San Francesco and for three-legged satyrs in other Sicilian sites: Chillemi (2011), (2015); Lanza in Orsi/Lanza (1990) *passim*; Pisani (2003), (2008) 28–29, 72–75; Salibra (2016) *passim*.

37 Chillemi (2015).

38 Boat models: Tigano (1991) 62, 64, notes 47–48, 73. The inhabited area of ancient Zankle has also yielded some late Archaic hand-modeled terracottas: G. Zavettieri in Bacci/Tigano (1999) 145, 147, S116, S121 (horse and male torso).

39 A hand-modeled female figure has been found in the residential area: E. D'Amico in Spigo *et al.* (2008) 211, no. 4; E. D'Amico in Tigano/Vanaria (2022) 135–136, no. 17.

40 From the house in Contrada Crocefisso: Frasca (1995) 6, 16–17, nos. 69–73, fig. 25, pl. IV,f (fragments of boat models); 6, 15, nos. 54–55, fig. 20 (a horse, a ram); Basile (1991) 25, no. 14, fig. 14.

41 From the Fusco necropolis, tomb 346: Basile (1991) 11–17, figs. 1, 3–6; Basile (1993) 70–75, figs. 5–8; Frasca (1995), pl. IV,g; Bell (2012) 197, pl. X,3; Gianfrotta (2014) *passim*, figs. 2, 4–5. Other examples come from the necropoleis of Fusco and Scala Greca (tomb XLI), the 'Archimedes Tomb', the amphitheater area, the Hellenistic sacellum in the Santa Lucia catacombs area, a bothros in the sanctuary of Piazza della Vittoria, well 16 at Ortigia, the villa Maria and Vigna Cassia areas; one is in a private collection: Orsi (1891) 386–387, (1897) 499, (1915) 190; Fallico (1971) 587, 628, fig. 47; Basile (1991) 19–27, nos. 2–9, 11–13, 16–17, figs. 8–13, 16–17, (1993) with complete catalogue and discussion.

42 From a votive deposit: Basile (1991) 23, no. 10; Basile (1993) 78, no. 10, fig. 17.

43 *Supra*, note 8.

Even if some of the statuettes discussed above were imported,[44] the majority have been[45] and are considered to be local output. Moreover, production in the eastern part of the island is supported by the discovery of workshops that created various types of handmade figurines: despite the influences from Greece from an iconographic and stylistic point of view,[46] their elaboration is local. Their small size, limited concern for anatomical and other details, and technical 'simplicity' compared to their moldmade counterparts probably contributed to their affordability, even for less wealthy clients ("[. . .] mondo delle "piccole cose" che ruotava intorno alla vita del grande santuario"[47]).

Although the aforementioned observations are generally true, some artifacts seem to be less 'simple'. While most handmade figurines are very small (starting from 2 cm in height/length), some are considerably larger (up to 12 or even 16 cm), suggesting a greater level of skill on the part of the artisan.[48] The hybrid products which combine different techniques, such as the potter's wheel and molds (e.g. the hydrophora from Megara Hyblaea – Fig. 2d[49] – and some '*genre* scenes'), demonstrate a higher level of accuracy and precision in shaping and assembling. This often results in unique and original pieces (e.g. the three-legged male figurine with a female face from Camarina: Fig. 2i)[50] compared to moldmade statuettes.

When large corpora are available, systematic studies can provide particularly valuable information allowing, for example, further investigation of the relationship with contemporary moldmade production through different approaches. These include evaluating the possible use of different clayey mixtures,[51] and studying the ways in which the manufacturing

44 For example, a few statuettes from Syracuse and Catania mentioned *supra*, notes 12, 13, 31.
As for the 7th century B.C., handmade figurines imported from Corinth are attested at Megara Hyblaea, together with local examples, both coming from votive deposits and from dwellings: cf. Vallet/Villard (1964) 27, 30, 33–38, nos. 1–5, 8; Albertocchi (2012) 89, fig. 5. Imported products are also known from Gela: Meola (1971) 58–59, no. 8384, pl. VII,b; Panvini/Sole (2009) 73, VII/11; 78, VII/24: said to be Rhodian; see also Albertocchi (2016) for a general overview of Greek-oriental terracottas in southern Sicily and in particular 100, fig. 10 and 98–99, fig. 5, with a different interpretation compared to the previous bibliography.
At Himera, a male figurine is paralleled to artifacts from Olympia and considered to be imported from the Peloponnese: Vassallo (2005) 131, fig. 243; see also Parisi/Albertocchi (2019) 478, note 8.
For the 6th century B.C., oriental-Greek animals are known from the sacred area of Bitalemi at Gela: Bertesago (2022) 153, 335, no. 984, pl. CXCIV,i (560–520 B.C.); 110, 335, dep. 2584, CLXXII,f (560–540 B.C.). The same context also yielded Corinthian products such as 'a nastro' female statuettes, a dancer and a female figurine on a horse: Meola (1971) 53–54, 60, 65–66, 75, pls. IIe-III; Albertocchi (2012) 95, pl. VI,3, (2022b) 344–345, 51, dep. 743, pl. LXXXVIII,b (early 6th century B.C.); a male statuette: Meola (1971) 67, pl. XIX,b *contra* Albertocchi (2012) 91, (2022b) 35, 344, dep. 250, pl. LX,g (first half of 6th c. B.C.); and a horse: Albertocchi (2022b) 48, 346, dep. 636, pl. LXXXV,a (first half of the 6th century B.C.).
45 P. Orsi notes on the handmade terracottas from Megara Hyblaea: "[. . .] aggiungasi ancora che i prodotti veramente locali sembrano essere quelli assai rozzi, plasmati a mano libera, alcuni saggi dei quali derivano pure dai sepolcri. [. . .] In taluni esemplari solo la testa è rilevata a stampo, mentre il corpo è modellato rozzamente a mano libera; ed è qui che apparisce la deficenza artistica dell'operaio coroplasta, incapace di produrre alcunché di naturale ed artistico, se non lavora a mano obbligata; quindi forme pesanti, anormali, movimenti inesatti; ben si vede che siamo molto lontani ancora dagli abilissimi coroplasti del IV secolo, i quali anche a mano libera seppero cavare dalla creta forme così vive, parlanti e belle. [. . .] Di stile libero a Megara non abbiamo indizio nelle terrecotte, e quando il foggiatore di esse ardisce allontanarsi dalle forme consuete, allora abbiamo prodotti infantili che non meritano nemmeno il nome di artistici."
Translation by author: ("[. . .] still, it should be added that the truly local products seem to be the very rudimentary ones, made freehand, some examples of which also come from the tombs. [. . .] In some examples only the head is molded, while the body is handmade; and here the artistic deficiency of the coroplast appears, incapable of producing anything natural and artistic, if he does not work by forced hand; hence heavy, abnormal forms, inaccurate movements; it is clear that we are still very far from the very skillful coroplasts of the 4th century, who even by freehand knew how to extract from clay such lively, expressive and beautiful forms. [. . .] Of free style at Megara we have no evidence in the terracottas, and when the craftsman dares to depart from the usual forms, then we have childish products that do not even deserve the name of artistic."). Orsi/Cavallari (1889) 917–918.
46 See Albertocchi (2012).
47 Chillemi 2010 on the San Francesco materials.
48 Cf. the remarks about the size of terracotta handmade animals in Muller (2003) 13.
49 *Supra*, note 17.
50 *Supra*, note 18.
51 Macroscopic and archaeometric analyses have been carried out on the terracotta figurines from San Francesco in Catania, which have also allowed the differentiation between local and oriental-Greek specimes, on the basis of the presence of volcanic temper in the fabrics: Pautasso *et al.* (2022). A microscopy study on samples taken from Classical female piglet carriers – Ferlito (2020) – has shown a general homogeneity, despite limited differences appearing from a technological point of view. On the contrary, the fabrics of the Classical figurines are clearly different from those of the Archaic. It would be interesting to extend this approach to handmade specimens in order to evaluate similarities or discrepancies compared to the molded materials.

procedures were carried out in workshops, selecting and/or combining different hand modeling techniques (e.g. hand assembly, coiling and slab construction).[52] In the case of intensively iterated artifacts, a systematic classification can help in order to reconstruct the manual work of ancient craftspeople, the sequence of gestures and the methods of assembling the parts together.[53] The study of the handmade doves of San Francesco has led to the suggestion that they were shaped by children:[54] if this could be confirmed, it would contribute to the reconstruction of the apprenticeship and work dynamics in the ateliers and to the knowledge about the involvement of sub-adults (and young women?) in this type of production. The alleged childish allure of these items, which has been considered as negative in previous literature,[55] could be viewed in a new perspective.[56]

Returning to more general observations, hand modeling enabled craftspeople to combine a variety of shaping methods to produce composite artifacts made up of freehand, moldmade, and wheel-made parts, and even medium and life-size sculptures.[57] This versatility, often associated with artisans's skill levels, their interest in experimentation and their familiarity with gestures, led to a wide range of possibilities in clay manipulation, assembly strategies and compositional solutions. In some contexts, coroplasts developed a particular *savoir-faire* that sometimes also gave birth to serial production, even in the Classical and Hellenistic periods. Moreover, clay and wax hand modeling are the first steps in the creation of moldmade terracottas and cast bronzes, also in relation to large-scale statues, as recent studies have helped to further illuminate through different approaches.[58]

Cave-nymphaeum Terracotta Models: An Insight into Locrian Handicraft

Another issue I would like to explore is the relationship between hand-modeled production and local consumption. In other words, the extent to which the adoption of hand modeling techniques, together with typological and iconographic choices, could meet specific, local trends and market demands, mainly linked to ritual necessities.

The recent study of a votive architectural roof model from the sanctuary of Poseidon Helikonios at Nikoleika, ancient Helike in Achaea, has illuminated, in light of experimental archaeology, how hand-modeling, combined with painted and impressed decoration, contributed to the creation of an outstading dedication, a special commission from a small group of élite members of the local community.[59]

This perspective takes us to the Greek colony of Locri Epizephyrii on the Ionian coast of southern Calabria, famous in Magna Graecia for its flourishing and refined craftsmanship: outstanding bronzes, of course, but also pottery – even

52 Cf. for example Serwint (2022).

53 Cf. Georgaka (2013), (2022).

54 Pautasso *et al.* (2022) 33–34, figs. 6–7. The possible participation of children in the manufacture of terracotta figurines from San Francesco has also been suggested for some molded specimens. However, due to the difficulties in identifying children and women fingerprints, a dactyloscopic study, althoug rarely conclusive, should be attempted. The bibliography on dactyloscopic studies applied to archaeology is vast: for terracottas, see Muller (2023), especially 12–16 for the uncertainty of sex and age determination.

55 Cf. *supra* note 45 Orsi's remarks on the handmade figurines from Megara Hyblaea; Aurigny/Rohaut (2022b) 6 on Pottier's considerations on Greek material; P. Orlandini's reflections reported in Parisi/Albertocchi (2019) 484–485.

56 Muller and Klinger have pointed this out for terracottas: Muller (1994) 177, (2000) 96–97, (2014), (2022) 253–254; Klinger (2021) 9; on pottery and other objects: Hasaki (2012) 175–176; Langdon (2015); on the possible role of miniature objects in apprenticeship processes: Meirano (2012a) 123, note 70, with further bibliography.

57 Detailed analyses of technical features and mixed techniques based on the evidence from different sites are provided in Aurigny/Rohaut (2022a): see in particular Georgaka (2022) on Athens; Philippa-Touchais (2022) on Argos; Muller (2022) and Huysecom-Haxhi (2022) on Thasos; Fourrier (2022) and Raptou (2022) on Cyprus; Bilbao Zubiri (2022) on Metapontum.

58 See Devogelaere (2022); Jeammet (2022); Descamps-Lequime (2022), (2023); Queyrel (2023); Moustacha (2023).

59 The coil-built roof, dated to 720–700 B.C., was intended to fit on a clay building model, now missing. Five additional examples have been found in Achaean sanctuaries, suggesting the existence of a specialized manufacture in the area: cf. Gadolou (2011), (2015) 271–274, figs. 4–8; Gadolou (2022).

figured –,[60] terracotta figurines, architectural terracottas, and so on. In every field of craftsmanship, Locrian workshops gave birth to original creations with distinctive stylistic features, and to refined products.[61]

As far as coroplastic art is concerned, an extraordinary example is the renowned *Cavaliere* on a sphinx (Fig. 5), dated to 420-400 B.C.[62] The group was part of the decoration (probably an *akroterion*) of the Doric temple of Casa Marafioti. The subject of a long and complicated restoration after its discovery, this work was the focus of a new conservation intervention and a diagnostic investigation in 2015, allowing detailed observations of technical aspects and polychromy, which confirmed local craftspeople's skills in shaping and firing.[63]

Fig. 5: Terracotta *Cavaliere* on a sphinx, from Locri, Casa Marafioti. After Pugliese Carratelli (1983), fig. 497.

On a smaller scale, Locrian coroplasts also produced special, large hand-modeled votive offerings, such as the misfired Archaic architectural model of a probably Ionic prostyle temple, recovered in fragments from a kiln in the Centocamere area (Fig. 6).[64]

60 See Elia *et al.* in this volume.

61 Cf. Elia/Meirano (2022) for a synthesis on Locrian craftsmanship in the Archaic and Classical periods; Meirano (2012b) on the archaeological evidence of kilns and production units.

62 Orsi (1911) 39–50, figs. 33–37; Orsi (1925–1926). The group is 132 cm high; the identification of the subject depicted is still controversial (Dioscorus?). See Agostino/Paoletti (2016), with complete bibliography.

63 Guido/Mantella (2016). For the technical issues related to the production of large size terracottas, see Moustacha (2023) 153–155.

64 The fragments recovered belong to the roof, the *oikos* wall and the pavement of the building: Barra Bagnasco (1975) 333; Gullini (1980) 38–39, pl. VII,4–5; Costamagna/Sabbione (1990) 117, fig. 152. Two clay models of temples also come from the Locrian colony of Medma: Orsi (1913) 68–69, figs. 75–76; Demangel (1946–1948) 17–22, figs. 1–7; Ferri (1948).

Fig. 6: Misfired roof of an Archaic terracotta temple model from Locri, Centocamere. After Costamagna/Sabbione (1990), fig. 152.

A case study of extraordinary interest with regard to handmade artifacts can be found in the so-called cave-nymphaeum models (Fig. 7). Almost all of the examples were found in the rupestrian sanctuary of Grotta Caruso[65] dedicated to the Nymphs and associated with female rites of passage, that was in use from the late Classical to the Hellenistic period. The archaeological context was investigated in the 1940s and then suddenly destroyed by a landslide.[66]

Cave, nymphaeum and fountain models are quite rare in the Greek world.[67] In Western Greece, only seven examples are known besides the Locrian ones: two models come from tombs in Tarentum, with completely different morphological features,[68] two terracotta fountains were found at the Locrian sub-colony of Medma, modern Rosarno on the Tyrrhenian coast,[69] two model fragments come from Kaulonia[70] and Reggio[71] respectively, and a cave model from Agrigentum.[72]

The Locrian specimens constitute an unparalleled corpus in the ancient world so far. I will not expand on the iconographic, cultic and ritual implications of these artifacts, but rather will focus on the morphology and the variety of solutions adopted in the realization of these terracottas. They were the subject of a systematic study at the beginning of the 1990s by Francesca Martorano,[73] who developed a typology distinguishing the models into four groups, from those that closely imitate natural caves to the more artificial ones, which were provided with architectural elements and façades, and inspired by nymphaea: in the words of Henri Lavagne: "[. . .] certaines de ces terres cuites s'en tiennent résolument à l'imitation de l'antre naturel, alors que d'autres évoluent plus nettement vers un type d'édifice construit, où la grotte n'est

65 In addition to the ten examples and large fragments found in the sanctuary of Grotta Caruso, four models and nine fragments come from other contexts within the site (the theater, the Centocamere and Cusemi areas) or are of unknown provenance: Meirano in Elia/Meirano (2016) 425, note 25 and *infra*; a further example was discovered during recent excavations: see *infra*.

66 Arias (1946); Costabile (1991).

67 Martorano (1991a).

68 Mustilli (1932–1933) 231, note 10; Martorano (1991a) 50–51, figs. 70–71; Graepler (1997) 221–222, 261–262, no. 17, figs. 245–246 (tomb 46); Danner (2000) 54–57, nos. 13–14, figs. 23–28.

69 One of the two specimens from Medma is exhibited in the Naxos Antiquarum: Foti (1978) 149–150, fig. 2; Sabbione (1979) 389–390, pl. XVIII; Pelagatti (1984–1985) 288–292, figs. 29–30; Martorano (1991a) 51–52, note 197, figs. 90, 315–317; Danner (2000) 47–52, nos. 10–11, figs. 11–19.

70 Danner (2000) 35–36, no. 1, fig. 1.

71 Martorano (1991a) 51–52, figs. 72–73; Danner (2000) 52–54, no. 12, figs. 20–22.

72 From the filling of a sacred pool in the area of the sanctuary of Zeus Olympios, dismissed at the end of the 4th century B.C.: de Cesare/Portale (2017) 83, fig. 8, (2019) 16–21, (2020) 109–110, fig. 7. See also the "edicola gemina" from the *Antro* of the Chiesa Madre at Centuripe: Portale (2012) 231–232, fig. 6.

73 Martorano (1991a) 45–50, (1991b); see also Danner (2020) 36–47, nos. 2–9, figs. 2–10.

Fig. 7: Cave-nymphaeum terracotta models from Locri, Grotta Caruso: (**a-d**): A1.4, B 1.2, B 1.3, D 1.1 in Martorano's classification. Photo: Valeria Meirano.

plus qu'un souvenir ou une lointaine transposition".[74] Scholarly debate is open about the origin of these creations: 'fantastic places' imagined by the coroplasts or, rather, inspired by caves, grottoes and nymphaea existing at Locri or elsewhere in the Greek world.[75]

It should be noted, however, that Martorano's classification does not reflect an evolutionary development in the creation of these objects, nor a chronological sequence.

Although the models can be traced back to these four types, each artifact is unique in terms of the solutions adopted, from a planimetric, volumetric and iconographic point of view. Unlike the examples from Tarentum, which are basically wheel-turned, the Locrian models are made up of different hand-modeled pieces that were assembled and joined together; in one case, a curved tile positioned vertically serves as the back wall of the cave.[76]

The walls are decorated with many hand-modeled and moldmade elements; among the *appliqués* pieces are: parapets, partitions and architectural elements such as small, Ionic semi-columns and caryatids; antefixes and lion-head spouts, sometimes connected to small reservoirs; stalactites and *pastillage* additions, which imitate the roughness of the cave walls (the so-called "scodellette"); shells evoking the presence of water; nymphs and satyrs (both heads and whole figures), *thyrsi* and other subjects decorating the surfaces, and so on.

74 Lavagne (1988) 152.
75 Lavagne (1988) 153–154; Martorano (1991a).
76 Martorano (1991b) 64.

A metal tool or wooden stick was used to engrave large pebbles or polygonal blocks on the surfaces, and stamps were used to define other details, such as protomes and ovolo friezes.

The publications dedicated to these objects do not focus on technical aspects and make only limited reference to the hand modeling of the structural or decorative parts; past restorations have not provided any observations on the techniques of execution.

Two models, however, have been the subject of recent reappraisals, although they have not been studied from a technical point of view.

One, a previously unpublished example, found during an imprecise rescue excavation, shows long finger traces suggesting the roughness of the surfaces, abundant red color, and several moldmade *appliqué* motifs, including lion-head spouts and figures representing satyrs.[77]

The second, a well-known model from the Caruso cave (B 1.2 in Martorano's classification: Fig. 7b), was the subject of a new restoration in 2015. The intervention showed that its surface is a kind of palimpsest, the result of a complex stratigraphic sequence of successively added elements, including an unexpected and previously unknown inscription.[78]

The careful naked-eye analysis revealed that the model was first covered with a red coating applied directly on the clay; then, on the front surface, hand-modeled protruding clayey strings were added in order to evoke the rough façade of a cave, and subsequently the inscription was written with a stick on the clay strings while they were still soft, before firing. During its long lifespam, the model was also the object of an ancient restoration, as evidenced by the traces of lead clamps; then, probably after a long period of use and exposure, another white coating was added, which finally covered the inscription. The discovery of these 'hidden' incised letters provided a new and important confirmation concerning religious and ritual aspects, but no further studies were carried out to better understand the way the model was made as a whole.

During the recent excavations of our archaeological mission in the central sector of the urban plan of ancient Locri, we discovered a new ritual area, the only one known so far in this part of the city.[79] The context, dating from the mid- 4th to the end of the 3rd century B.C., is related to female worship and rites of passage. The building and the adjacent spaces feature ritual infrastructures and devices that allow us to reconstruct the ceremonies and rituals that took place in the area. Water played an important role in these performances, as shown by the presence of terracotta basins, bathtubs and hydraulic installations. Numerous votive deposits have been found, as well as other rich archaeological evidence for the activities and rituals performed in the complex (Fig. 8).

At the end of its life, the building was dismantled and, in the layer belonging to this phase, numerous pieces of a cave model were intentionally broken and scattered during an abandonment rite. We unearthed the fragments in subsequent, extended campaigns (Fig. 9).[80] The newly found model is the only one known so far to have come from a clear ritual context other than that of Grotta Caruso.

With the authorization of the Archaeological Superintendency, the fragmentary model was transferred to the laboratories of the Centro Conservazione e Restauro 'La Venaria Reale', where the activities of the Master Course in Conservation and Restoration of Cultural Heritage of the University of Turin take place. Here, the artifact has been – and still is – included in a project aimed at achieving its restoration and an in-depth study of its technological features and manufacturing process, also thanks to an articulated diagnostic plan.[81] Several non-invasive techniques have been used during the restoration of the object, some on the individual fragments and others on the temporarily reconstructed model, including optical microscopy, optical profilometry, CAT (Computerized Axial Tomography), SEM (Scanning Electron Microscopy), XRF (X-ray Fluorescence Analysis), UVF (UV-induced Visible Fluorescence) and a micro-invasive technique, Fourier-transform Infrared Spectroscopy (FT-IR), in order to study the manufacturing techniques, and to evaluate the causes of degradation of the surface and thus derive the best choices of materials and strategies in the restoration intervention.

77 Pizzi (2017) 96–99, figs. 11–13, 15.

78 Costabile (2022) 166–174, figs. 7–9.

79 For the ongoing excavation programs of the University of Turin at Locri (*Concessione* by the Italian Ministry of Culture) and the ritual context see: Elia/Meirano (2016); Elia (2019), (2020); Elia/Meirano (2020) with further bibliography.

80 For a preliminary presentation of the first unearthed parts of the model: Meirano in Elia/Meirano (2016) 424–425.

81 The project is directed by the author together with Marco Demmelbauer (Centro Conservazione e Restauro 'La Venaria Reale', University of Turin); Cecilia Cittadini participated in the restoration project in the frame of her Master Degree Thesis in Conservation and Restoration of Cultural Heritage.

Fig. 8: Finds from one of the votive deposits recently excavated at Locri, Casino Macrì area. Photo: Diego Elia.

Fig. 9: Fragments of the terracotta cave-nymphaeum model discovered at Locri, Casino Macrì, emerging during the excavation; one of the fragments. Photo: Diego Elia.

Thanks to this varied approach (Optical Microscopy and CAT), it is possible to identify the shaping sequence, mainly consisting of the so-called 'a colombino' or coil technique. This allows us to trace back the various steps in the construction of the model and the gestures of the artisan(s), from the assembly of the morphological pieces and decorations to the treatment of the surfaces (Fig. 10). Naked-eye analysis, combined with microscopy and CAT, makes it possible to reconstruct the detailed sequence in which each piece, whether freehand done or molded, was added.

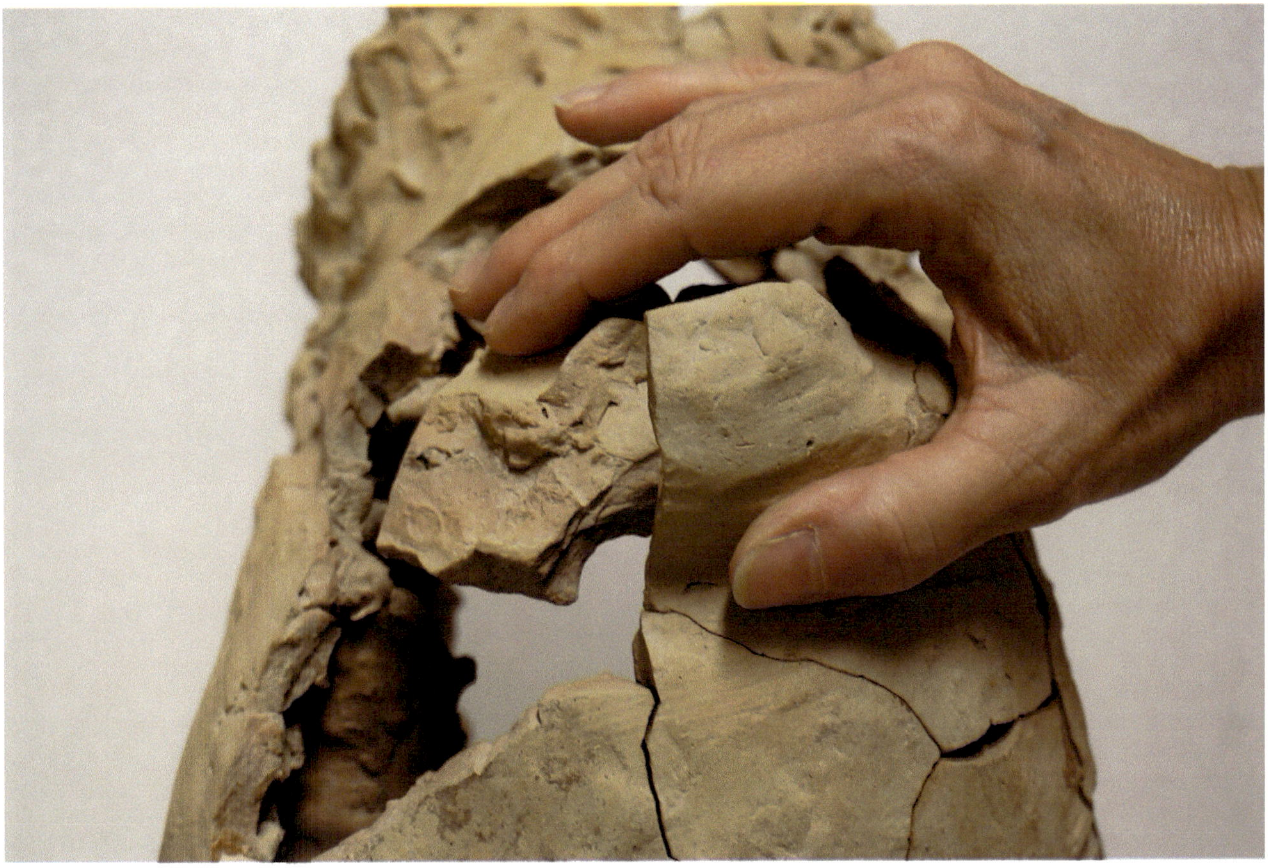

Fig. 10: Terracotta cave-nymphaeum model discovered at Locri, Casino Macrì: retracing the gestures of the artisan. Photo: Diego Elia.

Other important aspects of the project that are still in progress concern the study of the fingerprints on the surface, using a dactyloscopic approach, and further analysis of the hand-modeled reservoir.[82] Like the majority of Locrian examples, our model is also equipped with such a device. Water was poured into these reservoirs in order to create humidity or even spurts on the front surface by means of various solutions. Some items show complicated hydraulic systems, consisting of a reservoir, a channel and an outflow with basins and holes in order to keep the water level constant.

In conclusion, the case-study of the Locrian cave models – a unique corpus in which each example is itself unique – demonstrates once again, also for the Hellenistic period, that clay hand modeling, due to its versatility, enabled artisans to create and develop a new coroplastic type in order to meet local demand, strictly driven by ritual needs.

Abbreviation

La Sicilia greca (1989) *La Sicilia greca* (Catalogo della Mostra, Malmö-Rooseum, 7 ottobre–15 dicembre 1989), Palermo.

82 The publication of the results of the project is in preparation; it will also provide the opportunity for a detailed technical re-examination of some specimens from the Grotta Caruso cave.

Bibliography

Agostino/Paoletti (2016): Rossella Agostino and Maurizio Paoletti, "Giovane cavaliere (Dioscuro?) sorretto da una sfinge, detto 'Cavaliere Marafioti' ", in: Bertelli/Bonsanti (2016) 49–55.

Albertocchi (2012): Marina Albertocchi, "Dalle origini all'età arcaica: importazioni e rielaborazioni locali", in: Albertocchi/Pautasso (2012) 85–111.

Albertocchi (2016): Marina Albertocchi, "La coroplastica arcaica greco-orientale nella Sicilia meridionale. Vecchi problemi e nuove acquisizioni", in: Muller/Lafli (2016a) 95–111.

Albertocchi (2022a): Marina Albertocchi (ed.), *Gela. Il Thesmophorion di Bitalemi. La fase arcaica. Scavi P. Orlandini 1963–1967,* Rome.

Albertocchi (2022b): Marina Albertocchi, "Coroplastica di altre fabbriche", in: Albertocchi (2022a) 343–347.

Albertocchi/Pautasso (2012): Marina Albertocchi and Antonella Pautasso (eds.), *Philotechnia. Studi sulla coroplastica della Sicilia greca*, Catania.

Arias (1946): Paolo Enrico Arias, "Locri. Scavi archeologici in contrada Caruso-Polisà (aprile-maggio 1940), in: *Notizie degli Scavi di Antichità*, 138–161.

Aurigny/Rohaut (2022a): Hélène Aurigny and Laura Rohaut (eds.), *Quand on a la terre sous l'ongle. Le modelage dans le monde grec antique*, Aix-en-Provence.

Aurigny/Rohaut (2022b): Hélène Aurigny and Laura Rohaut, "Introduction", in: Aurigny/Rohaut (2022a) 5–9.

Bacci/Tigano (1999): Giovanna Maria Bacci and Gabriella Tigano (eds.), *Da Zancle a Messina. Un percorso archeologico attraverso gli scavi*, I, Messina.

Barra Bagnasco (1975): Marcella Barra Bagnasco, "Lo scavo di Locri Centocamere 1974", in: *Orfismo in Magna Grecia* (Atti del XIV Convegno di Studi sulla Magna Grecia, Taranto, 6–10 ottobre 1974), Naples, 325–335.

Basile (1991): Beatrice Basile, "Modellini fittili di imbarcazioni dalla Sicilia orientale", in: *Atti della IV Rassegna di archeologia subacquea, IV Premio Franco Papò* (Giardini Naxos, 13–15 ottobre 1989), Messina, 1–50.

Basile (1993): Beatrice Basile, "Modellini fittili di imbarcazioni dalla Sicilia orientale", in: *Bollettino di Archeologia Subacquea*, 1–1993, 69–101.

Bell (2012): Malcolm Bell III, "Terracottas in Hellenistic Sicily", in: Albertocchi/Pautasso (2012) 187–209.

Bérard (2017): Reine Marie Bérard, *Mégara Hyblaea 6. La nécropole méridionale de la cité archaïque, 2. Archéologie et histoire sociale des rituels funéraires*, Rome.

Bertelli/Bonsanti (2016): Carlo Bertelli and Giorgio Bonsanti, *Restituzioni 2016. Tesori d'arte restaurati*, Venice.

Bertesago (2022): Silvia Martina Bertesago, "Coroplastica greco-orientale", in: Albertocchi (2022a) 332–343.

Bilbao Zubiri (2022): Eukene Bilbao Zubiri, "Le modelage dans les pratiques artisanales de Métaponte aux VIIe et VIe s. av. J.-C.", in: Aurigny/Rohaut (2022a) 269–280.

Caubet (2009): Annie Caubet, "Les figurines antiques de terre cuite", in: *Perspective. Actualité en histoire de l'art*, 1–2009, 43–56. https://doi.org/10.4000/perspective.1251.

Cébeillac Gervasoni (1976–1977): Mireille Cébeillac Gervasoni, "Une étude systématique sur les nécropoles de Megara Hyblaea: l'exemple d'une partie de la nécropole méridionale", in: *Kokalos* XXII-XIII, 587–597.

Chillemi (2010): Vanessa Chillemi, "Coroplastica modellata a mano in Sicilia. Problemi di produzione e diffusione", in: *CSIG News. Newletter of the Coroplastic Studies Interest Group*, 4–2010, 6–7. http://www.coroplasticstudies.org (last access 31.01.2013).

Chillemi (2011): Vanessa Chillemi, "Soggetti di vita quotidiana?", in: Guido Guarducci and Stefano Valentini (eds.), *Il Futuro nell'Archeologia. Il Contributo dei Giovani Ricercatori* (Atti del IV Convegno Nazionale dei Giovani Archeologi, Tuscania, 12–15 Maggio 2011), Rome, 248–256.

Chillemi (2015): Vanessa Chillemi, "Handmade Terracotta Figurines: Subjects of Daily Life", in: Pietro Maria Militello and Hakan Öniz (eds.), *SOMA 2011, Proceedings of the 15th Symposium on Mediterranean Archaeology* (Catania, 3–5 March 2011), BAR International Series 2695, Oxford, 639–649.

Costabile (1991): Felice Costabile (ed.), *I ninfei di Locri Epzefiri. Architettura, culti erotici, sacralità delle acque*, Soveria Mannelli.

Costabile (2022): Felice Costabile, "Euthymos e le ninfe: l'iscrizione nascosta sul modello di grotta con simulacro di Eros dagli scavi Arias a Locri Epizefiri", in: Ilenia Achilli, Giuseppe Mariotta, Salvo Micciché and Anna Maria Seminara (eds.), *Heorté. Studi in onore di Michele R. Cataudella in occasione del suo 80° compleanno*, Rome, 155–178.

Costamagna/Sabbione (1990): Liliana Costamagna and Claudio Sabbione, *Una città in Magna Grecia. Locri Epizefiri. Guida archeologica*, Reggio Calabria.

Cultrera (1943): Giuseppe Cultrera, "Siracusa. Scoperte nel Giardino Spagna", in: *Notizie degli Scavi di Antichità*, 33–126.

Danner (2000): Peter Danner, "Tonmodelle von Brunnenanlagen aus der Magna Graecia", in: *Jahreshefte des Österreichischen Archäologischen Institutes in Wien* 69, 35–75.

Dasen (2012): Véronique Dasen, "Cherchez l'enfant! La question de l'identité à partir du matériel funéraire", in: Hermary/Dubois (2012) 9–22.

de Cesare/Portale (2017): Monica de Cesare and Elisa Chiara Portale, "Le ricerche dell'Università di Palermo nel santuario di Zeus Olympios ad Agrigento", in: Natascha Sojc (ed.), *Akragas: Current Issues in the Archaeology of a Sicilian Polis*, Leiden, 81–94.

de Cesare/Portale (2019): Monica de Cesare and Elisa Chiara Portale, "Il santuario di Zeus Olympios nel quadro urbano dell'antica Akragas" in: *Archeologia Classica* LXX, 1–28.

de Cesare/Portale (2020): Monica de Cesare and Elisa Chiara Portale, "Il santuario di Zeus Olympios ad Agrigento: al di là del tempio monumentale", in: de Cesare *et al.* (2020) 99–124.

de Cesare *et al.* (2020): Monica de Cesare, Elisa Chiara Portale and Natascha Sojc (eds.), *The Akragas Dialogue. New Investigations on Sanctuaries in Sicily*, Berlin.

Demangel (1946–1948): Robert Demangel, "Anecdota dorica I", in: *Annuario della Scuola archeologica di Atene e delle missioni italiane in Oriente*, nuova serie VIII-X, XXXIV-XXXVI, 17–23.

Descamps-Lequime (2022): Sophie Descamps-Lequime, "Polyclète était doublement modeleur. . .", in: Aurigny/Rohaut (2022a) 99–112.

Descamps-Lequime (2023): Sophie Descamps-Lequime, "De l'argile au bronze: au commencement était le *plastès*", in: Descamps-Lequime/Jeammet (2023) 13–48.

Descamps-Lequime/Jeammet (2023): Sophie Descamps-Lequime and Violaine Jeammet (eds.), *Pratiques d'ateliers dans la Grèce et l'Égypte anciennes. Du coroplathe au bronzier*, Paris-Villenueve d'Ascq.
Devogelaere (2022): Jonathan Devogelaere, "Empreintes et modelage de la cire: les traces de la céroplastie sur les bronzes antiques", in: Aurigny/Rohaut (2022a) 13–25.
Elia (2019): Diego Elia, "La gestione delle acque a Locri Epizefiri. Criticità e soluzioni, dalle origini alla conquista romana", in: Sophie Bouffier, Oscar Belvedere and Stefano Vassallo (eds.), *Gérer l'eau en Méditerranée au premier millénaire avant J.-C.*, Aix en Provence, 87–108.
Elia (2020): Diego Elia, "Locri Epizefiri: la città e la gestione delle acque in età arcaica", in: Elisabetta Bianchi and Matteo D'Acunto (eds.), *Opere di regimentazione delle acque in età arcaica. Roma, Grecia e Magna Grecia, Etruria e mondo italico*, Rome, 185–201.
Elia/Meirano (2016): Diego Elia and Valeria Meirano, "Il sacro e l'acqua a Locri Epizefiri: osservazioni alla luce delle scoperte recenti", in: Alfonsina Russo and Francesca Guarneri (eds.), *Santuari mediterranei tra Oriente e Occidente. Interazioni e contatti culturali* (Atti del Convegno Internazionale, Civitavecchia-Roma, 2014), Rome, 419–434.
Elia/Meirano (2020): Diego Elia and Valeria Meirano, "Locri Epizefiri: in città e oltre. A cinquant'anni dall'avvio delle ricerche UNITO", in: Diego Elia (ed.), *Chiedi alla terra. Scavi e ricerche archeologiche del Dipartimento di Studi Storici dell'Università di Torino*, Turin, 142–157.
Elia/Meirano (2022): Diego Elia and Valeria Meirano, "Produzioni artigianali, contesti e forme di consumo a Locri Epizefiri. Appunti e riflessioni, tra periodo arcaico ed età classica", in: *Orizzonti. Rassegna di archeologia* XXIII, 39–66.
Fallico (1971): Anna Maria Fallico, "Siracusa. Saggi di scavo nell'area della Villa Maria", in: *Notizie degli Scavi di Antichità*, 581–639.
Favier (2022): Éléonore Favier, "La coroplathie dans l'historie des collaborations artisanales. Historiographie entre la fin du XIXe siècle et aujord'hui", in: Aurigny/Rohaut (2022a) 89–98.
Ferlito 2020: Flavio Ferlito, "La produzione coroplastica a Catania. Osservazioni preliminari sugli impasti delle statuette votive dal deposito di Piazza San Francesco (V-IV sec. a.C.)", in: *Facem* (version December/06/2020). http://www.facem.at/project-papers.php (last access 31.03.3024).
Ferri (1948): Silvio Ferri, "I "tempietti" di Medma e l'origine del triglifo", in: *Rendiconti dell'Accademia Nazionale dei Lincei, Classe di Scienze morali, storiche e filologiche*, serie VIII, III, 402–413.
Foti (1978): Giuseppe Foti, "Attività della Soprintendenza archeologica della Calabria": in: *Klearchos* 77–80, 147–157.
Fourrier (2022): Sabine Fourrier, "Tours de main et styles: le moulage dans la coroplastie chypriote archaïque", in: Aurigny/Rohaut (2022a) 193–202.
Frasca (1982): Massimo Frasca, "Leontini. necropoli di Piscitello: campagna di scavi 1977–78", in: Massimo Frasca and Dario Palermo, *Scavi nelle necropoli di Leontini (1977–1982)*, in: *Cronache di Archeologia* XXI, 37–60.
Frasca (1995): Massimo Frasca, "Leontini. Piccola coroplastica ellenistica da una abitazione rupestre di contrada Crocefisso", in: *Bollettino d'Arte* XCI, serie VI, 1–23.
Gadolou (2011): Anastasia Gadolou, "A Late Geometric Architectural Model with Figure Decoration from Ancient Helike, Achaea", in: *Annual of the British School at Athens* 106, 247–273.
Gadolou (2015): Anastasia Gadolou, "Narrative Art and Ritual in the Sanctuary of Poseidon Heliconius in Ancient Helike, Achaea", in: Vlachou (2015) 267–276.
Gadolou (2022): Anastasia Gadolou, "Technique and Craftmanship in the Manufacture of Early Iron Age Achaean Architectural Models. A Votive Roof from the Nikoleika Sanctuary", in: Aurigny/Rohaut (2022a) 27–36.
Gabba/Vallet (1980): Emilio Gabba and Georges Vallet (eds.), *La Sicilia antica I,2. Le città greche di Sicilia*, Naples.
Gentili (1951): Gino Vinicio Gentili, "Siracusa. Scoperte nelle due nuove arterie stradali, la Via di Circonvallazione, ora Viale P. Orsi, e la Via Archeologica, ora Viale F.S. Cavallari", in: *Notizie degli Scavi di Antichità*, 261–334.
Gentili (1954a): Gino Vinicio Gentili, "Siracusa. Saggio di scavo a sud del Viale Paolo Orsi, in predio Salerno Aletta", in: *Notizie degli Scavi di Antichità*, 302–333.
Gentili (1954b): Gino Vinicio Gentili, "Megara Hyblaea. Scoperta di nuove tombe arcaiche della necropoli meridionale", in: *Notizie degli Scavi di Antichità*, 390–402.
Gentili (1956a): Gino Vinicio Gentili, "Siracusa. Contributo alla topografia dell'antica città", in: *Notizie degli Scavi di Antichità*, 94–164.
Gentili (1956b): Gino Vinicio Gentili, "Naxos alla luce dei primi scavi", in: *Bollettino d'Arte* XLI, serie IV, 326–332.
Georgaka (2013): Vassiliki Georgaka, "Typological Classification of the Archaic, Handmade Figurines of the Athenian Acropolis", in: *CSIG News. Newletter of the Coroplastic Studies Interest Group*, 9–2013, 5–6. http://www.coroplasticstudies.org (last access 31.01.2014).
Georgaka (2014): Vassiliki D. Georgaka, "Handmade Terracotta Animal Figurines from the Athenian Acropolis", in: *Les Carnets de l'ACoSt* 11. https://doi.org/10.4000/acost.435.
Georgaka (2022): Vassiliki D. Georgaka, "Modeling Technique of a New Coroplastic Type of Archaic Handmade Terracotta Figurines from the Athenian Acropolis and from the Sanctuary of the Nymph at the South Slope of the Acropolis Hill", in: Aurigny/Rohaut (2022a) 137–147.
Gianfrotta (2014): Piero Alfredo Gianfrotta, "Barchette fittili siceliote e prove di immaginarie navigazioni", in: *Topografia Antica, 3. Tradizione, tecnologia e territorio*, 2, Acireale-Rome, 183–193.
Graepler (1997): Daniel Graepler, *Tonfiguren im Grab. Fundkontexte hellenistischer Terrakotten aus der Nekropole von Tarent*, Munich.
Gras/Tréziny/Broise (2004a): Michel Gras, Henri Tréziny and Michel Broise, *Mégara Hyblaea, 5. La ville archaïque. L'espace urbain d'une cité grecque de Sicile orientale*, Rome.
Gras/Tréziny/Broise (2004b): Michel Gras, Henri Tréziny and Michel Broise, "Recherches topographiques sur l'habitat du plateau Sud (1977–1983)", in: Gras/Tréziny/Broise (2004a) 53–155.
Grasso (2008): Lorenza Grasso, *La stipe del santuario di Alaimo a Lentini. Un'area sacra tra la chora e il mare*, Rome.
Guido/Mantella (2016): Sante Guido and Giuseppe Mantella, "Giovane cavaliere (Dioscuro?) sorretto da una sfinge, detto 'Cavaliere Marafioti' ", in: Bertelli/Bonsanti (2016), with no p. nos.
Gullini (1980): Giorgio Gullini, *La cultura architettonica di Locri Epizefirii: documenti e interpretazioni*, Taranto.

Hasaki (2012): Eleni Hasaki, "Craft Apprenticeship in Ancient Greece. Reaching Beyond the Masters", in: Willeke Wendrich (ed.), *Archaeology and Apprenticeship. Body Knowledge, Identity, and Communities of Practice*, Tucson, 171–202.
Hermary/Dubois (2012): Antoine Hermary and Céline Dubois (eds.), *L'enfant et la mort dans l'antiquité III. Le matériel associé aux tombes d'enfants* (Actes de la table ronde internationale, Aix-en-Provence 20–22 janvier 2011), BiAMA 12, Paris.
Huysecom (2003): Stéphanie Huysecom, "Terres cuites animales dans les nécropoles grecques archaïques et classiques du bassin méditerranéen", in: Poplin (2003) 91–103.
Huysecom (2022): Stéphanie Huysecom, "Techniques et usage du modelage dans la coroplathie thasienne archaïque", in: Aurigny/Rohaut (2022a) 229–239.
Huysecom-Haxhi/Pautasso (2022): Stéphanie Huysecom-Haxhi and Antonella Pautasso (eds.), *'Craft Economy' and Terracotta Figurines. Approaching Systems of Production through Coroplastic Studies*, in: Archaeology and Economy in the Ancient World 16, Panel 3.14 (Proceedings of the 19th International Congress of Classical Archaeology, Cologne-Bonn, 22–26 May 2018), Heidelberg.
Huysecom-Haxhi *et al.* (2012): Stéphanie Huysecom-Haxhi, Irini-Despina Papaikonomou and Stratis Papadopoulos, "Les figurines en terre cuite dans les sépultures d'enfants en Grèce ancienne: le cas des jeunes filles nues assises", in: Hermary/Dubois (2012) 343–366.
Jeammet (2022): Violaine Jeammet, "Du modeleur au coroplathe: tentatives de définitions", in: Aurigny/ Rohaut (2022a) 49–61.
Klinger (2021): Sonia Klinger, *Corinth XVIII.8. The Sanctuary of Demeter and Kore. Miscellaneous Finds of Terracotta*, Princeton N.J.
Langdon (2015): Susan Langdon, "Geometric Pottery for Beginners: Children and Production in Early Greece", in: Vlachou (2015) 21–36.
Lavagne (1988): Henri Lavagne, *Operosa Antra. Recherches sur la grotte à Rome de Sylla à Hadrien*, Rome.
Martorano (1991a): Francesca Martorano, "I modellini fittili di ninfei", in: Costabile (1991) 45–56.
Martorano (1991b): Francesca Martorano, "Catalogo dei modelli locresi", in: Costabile (1991) 63–92.
Meirano (2012a): Valeria Meirano, "Les terres cuites dans les sépultures d'individus immatures en Grèce d'Occident: types, contextes, significations", in: Hermary/Dubois (2012) 111–130.
Meirano (2012b): Valeria Meirano, "Productions et espaces artisanaux à Locres Epizéphyrienne", in: Arianna Esposito and Giorgos Sanidas (eds.), *Quartiers artisanaux en Grèce ancienne. Une perspective méditerranéenne* (Actes du Symposium international, Lille, décembre 2009), Villeneuve d'Ascq, 257–279.
Meirano (2017): Valeria Meirano, "Offerte incruente in Magna Grecia. Un approccio iconografico per lo studio di dolci e pani in contesto rituale", in: *Scienze dell'Antichità* 23.3, 351–371.
Meirano (2018): Valeria Meirano, "Fuori dalla tomba. Terrecotte nelle necropoli di Magna Grecia e Sicilia: spunti di riflessione", in: Simonetta Bonomi and Carmelo Malacrino (eds.), *Ollus leto datus est. Architettura, topografia e rituali funerari nelle necropoli dell'Italia meridionale e della Sicilia fra Antichità e Medioevo, I. Dalla preistoria all'ellenismo* (Atti del Convegno Internazionale, Reggio Calabria, 22–25 ottobre 2013), Reggio Calabria, 133–138.
Meirano (2021): Valeria Meirano, "Offrandes alimentaires en Italie du sud: le témoignage des terres cuites en milieu funéraire", in: Dominique Frère, Barbara Del Mastro, Priscilla Munzi and Claude Pouzadoux (eds.), *Manger, boire, se parfumer pour l'éternité. Rituels alimentaires et odorants en Italie et en Gaule du IXe siècle avant au Ier siècle après J.-C.*, Collection du Centre Jean Bérard 53, Naples, 51–61.
Meola (1971): Emma Meola, "Terrecotte orientalizzanti di Gela ('Daedalica Siciliae III')", in: *Monumenti Antichi dei Lincei* XLVIII, s.m. I, I, 7–88.
Moustacha (2023): Aliki Moustacha, "La statuaire en terre cuite: entre bronze et marbre?", in: Descamps-Lequime/Jeammet (2023) 151–163.
Muller (1994): Arthur Muller, "La coroplathie: un travail de petite fille? Les figurines de terre cuite, de l'atelier à la publication: question de méthode", in: *Revue Archéologique*, n.s. 1, 177–187.
Muller (1997): Arthur Muller (ed.), *Le moulage en terre cuite dans l'antiquité. Création et production dérivée, fabrication et diffusion*, Paris.
Muller (2000): Arthur Muller, "Artisans, techniques de production et diffusion: le cas de la coroplathie", in: Francine Blondé and Arthur Muller (eds.), *L'artisanat en Grèc ancienne. Les productions, les diffusions* (Actes du Colloque, Lyon, 10–11 décembre 1998), Lille, 91–106.
Muller (2003): Arthur Muller, "Fabriquer des figurines d'animaux dans l'antiquité: techniques et matériaux", in: Poplin (2003) 11–16.
Muller (2014): Arthur Muller, "L'atelier du coroplathe: un cas particulier dans la production céramique grecque", in: *Perspective. Actualité en histoire de l'art*, 1–2014, 63–82. https://doi.org/10.4000/perspective.4295.
Muller (2017–2018): Arthur Muller, "Coroplastic Studies: What's New?", in: *Archaeology in Greece 2017–2018*, in: *Archaeological Reports* 64, 153–169.
Muller (2022): Arthur Muller, "Du pire au meilleur. Le modelage à Thasos: techniques et artisans du IVe siècle", in: Aurigny/Rohaut (2022a) 241–256.
Muller (2023): Arthur Muller, "Les doigts dans la terre. L'exploitation des dactylotypes de coroplathes: questions, exemples et perspectives", in: *Pallas* 121, 23–41.
Muller/Lafli (2016a): Arthur Muller and Ergün Lafli (eds.), *Figurines en terre cuite en Méditerranée grecque et romaine, volume 1. Production, diffusion, étude, Bulletin de Correspondance Hellénique* Supplément 54.
Muller/Lafli (2016b): Arthur Muller and Ergün Lafli, "Introduction", in: Muller/Lafli (2016a) 3–13.
Mustilli (1932–1933): Domenico Mustilli, "La necropoli tirrenica di Efestia", in: *Annuario della Scuola archeologica di Atene e delle missioni italiane in Oriente* XV-XVI, 1–278.
Orsi (1891): Paolo Orsi, "Siracusa. Nuove scoperte di antichità siracusane", in: *Notizie degli Scavi di Antichità*, 377–416.
Orsi (1895): Paolo Orsi, "Siracusa. Gli scavi nella necropoli del Fusco a Siracusa nel giugno, novembre e dicembre del 1893", in: *Notizie degli Scavi di Antichità*, 109–192.
Orsi (1897): Paolo Orsi, "Siracusa. Di alcune necropoli secondarie di Siracusa", in: *Notizie degli Scavi di Antichità*, 471–504.
Orsi (1904): Paolo Orsi, "Camarina. Campagne archeologiche del 1899 e 1903", in: *Monumenti Antichi dei Lincei* XIV, 757–952.
Orsi (1911): Paolo Orsi, "Rapporto preliminare sulla quinta campagna di scavi nelle Calabrie durante l'anno 1910", in: *Notizie degli Scavi di Antichità*, Supplemento, 3–26.

Orsi (1913): Paolo Orsi, "Rosarno (Medma). Esplorazione di un grande deposito di terrecotte ieratiche", in: *Notizie degli Scavi di Antichità*, Supplemento, 55–144.

Orsi (1915): Paolo Orsi, "I. Siracusa", in: *Notizie degli Scavi di Antichità*, 175–208.

Orsi (1918): Paolo Orsi, "Gli scavi intorno all'Athenaion di Siracusa negli anni 1912–1917", in: *Monumenti Antichi dei Lincei* XXV, 353–762.

Orsi (1925): Paolo Orsi, "Siracusa. Necropoli greco-arcaica nel predio ex-Spagna", in: *Notizie degli Scavi di Antichità*, 296–321.

Orsi (1925–1926): Paolo Orsi, "Il gruppo equestre fittile di Locri Epizeiri", in: *Dedalo* VI, 345–365.

Orsi/Cavallari (1889): Paolo Orsi and Francesco Saverio Cavallari, "Megara Hyblaea. Storia, topografia, necropoli e *anathemata*", in: *Monumenti Antichi dei Lincei* I, 689–950.

Orsi/Lanza (1990): Paolo Orsi and Maria Teresa Lanza, *La necropoli di Passo Marinaro a Camarina. Campagne di scavo 1904–1909*, Rome.

Panvini/Sole (2009): Rosalba Panvini and Lavinia Sole (eds.), *La Sicilia in età arcaica. Dalle apoikiai al 480 a.C.* (Catalogo della Mostra, Caltanissetta, 12 Giugno–12 Agosto 2006, Catania, 26 Ottobre–7 Gennaio 2007), Palermo.

Parisi/Albertocchi (2019): Valeria Parisi and Marina Albertocchi, "Coroplastica: produzioni per santuari, abitati, necropoli", in: *Produzioni e committenze in Magna Grecia* (Atti del LV Convegno di Studi sulla Magna Grecia, Taranto, 24–27 settembre 2015), Taranto, 473–501.

Pautasso (2010): Antonella Pautasso, "Santuari lungo le rotte. Per una storicizzazione della stipe votiva di Piazza San Francesco", in: Maria Grazia Branciforti and Vincenzo La Rosa (eds.), *Tra lava e mare. Contributi all'*Archaiologhia *di Catania* (Atti del Convegno, Catania), Catania, 109–118.

Pautasso (2020): Antonella Pautasso, "Dedicants, Offerings, and Sacrifice. The Value of the Images", in: de Cesare *et al.* (2020) 59–78.

Pautasso (2022): Antonella Pautasso, "Les figurines zoomorphiques modelées dans la Crète de l'Âge du Fer. Remarques sur la technique et les interactions entre pratiques artisanales", in: Aurigny/Rohaut (2022a) 125–134.

Pautasso (2023): Antonella Pautasso, "La città "invisibile". Il deposito di piazza San Francesco", in: Daniele Malfitana and Antonino Mazzaglia (eds.), *Catania. La città antica e quella del futuro. Archeologia, topografia, urbanistica per la riqualificazione dello spazio urbano*, Rome, 225–240.

Pautasso *et al.* (2022): Antonella Pautasso, Vanessa Chillemi, Ambra Pace and Lighea Pappalardo, "Né fornaci né matrici. Evidenza indiretta e produzione coroplastica: il caso di Katane", in: Huysecom-Haxhi/Pautasso (2022) 25–40.

Pedrucci (2022): Giulia Pedrucci, *Votive Statuettes of Adult/s and Infant/s in Ancient Italy from the end of the 7th to the 1st c. BCE: A New Reading, Volume II. Campania, Magna Graecia and Sicily*, Rome.

Pelagatti (1973): Paola Pelagatti, "Camarina", in: Paola Pelagatti and Giuseppe Voza (eds.), *Archeologia nella Sicilia sud-orientale*, Naples, 133–158.

Pelagatti (1984–1985): Paola Pelagatti, "Naxos (Messina). Gli scavi extraurbani oltre il Santa Venera (1973–1975)", in: *Notizie degli Scavi di Antichità*, 253–304.

Philippa-Touchais (2022): Anna Philippa-Touchais, "Les figurines modelées d'un sanctuaire archaïque péri-urbain à Argos (Péloponnèse, Grèce)", in: Aurigny/Rohaut (2022a) 163–178.

Pisani (2003): Marcella Pisani, "Vita quotidiana nel mondo greco tra il VI e il V secolo a.C. Un contributo per la classificazione delle rappresentazioni fittili", in: *Bollettino d'Arte* CXXIII, serie VI, 3–24.

Pisani (2008): Marcella Pisani, *Camarina. Le terrecotte figurate e la ceramica da una fornace di V e IV secolo a.C.*, Rome.

Pisani (2012): Marcella Pisani, "Iconografia e funzione delle terrecotte figurate nelle sepolture ellenistiche d'infanti e di adolescenti della necropoli nord-orientale di Tebe", in: Hermary/Dubois (2012) 367–385.

Pizzi (2017): Francesca Pizzi, "I volti del sacro a Locri. Nuovi dati sulla Ninfe da scavi di magazzino", in: *Atti e memorie della Società Magna Grecia* serie V, II, 85–104.

Poplin (2003): François Poplin (ed.), *Figures animales des mondes anciens*, *Anthropozoologica* 38.

Portale (2012): Elisa Chiara Portale, "Busti fittili e ninfe: sulla valenza e la polisema delle rappresentazioni abbreviate in forma di busto nella coroplastica votiva siceliota", in: Albertocchi/Pautasso (2012) 227–252.

Pugliese Carratelli (1983): Giovanni Pugliese Carratell (ed.), *Megále Hellás. Storia e civiltà della Magna Grecia*, Milan.

Queyrel (2023): François Queyrel, "Modelage de l'argile (Pline l'Ancien, Histoire Naturelle, XXXV, 152) et fabrication des grands bronzes (Posidippe, 63 A-B)", in: Descamps-Lequime/Jeammet (2023) 87–99.

Raptou (2022): Eustathios Raptou, "Technique de fabrication des terres cuites: le cas de Marion (Chypre) à l'époque classique", in: Aurigny/Rohaut (2022a) 213–226.

Rizza (1960): Giovanni Rizza, "Stipe votiva di un santuario di Demetra a Catania", *Bollettino d'Arte* XLV, serie IV, III, 247–262.

Rizza (2008): Giovanni Rizza, "Demetra a Catania", in: Carmela Angela Di Stefano (ed.), *Demetra. La divinità, I santuari, il culto, la leggenda* (Atti del I Convegno Internazionale, Enna, 1–8 luglio 2004), Rome-Pisa, 187–191.

Sabbione (1979): Claudio Sabbione, "Ricerche archeologiche nei territori di Locri e delle sue subcolonie", in: *Gli Eubei in Occidente* (Atti del XVIII Convegno di Studi sulla Magna Grecia, Taranto 8–12 ottobre 1978), Taranto, 382–398.

Salibra (2016): Roberta Salibra, *La necropoli di Passo Marinaro a Camarina. Campagna di scavo 1972–1973*, Rome.

Scilabra (2012): Carla Scilabra, "*Veneri pupa negata*. Giocattoli in tomba: casi di studio dall'Occidente greco", in: Hermary/Dubois (2012) 387–402.

Serwint (2022): Nancy Serwint, "The Terracotta Sculpture from Ancient Marion: Evidence of the Coroplast's Craft", in: Huysecom/Pautasso (2022) 5–24.

Sparkes (1965): Brian A. Sparkes, "The Greek Kitchen: Addenda", in: *Journal of Hellenic Studies* 85, 162–163.

Sparkes (1981): Brian A. Sparkes, "Not Cooking, but Baking", in: *Greece & Rome* 28.2, 172–178.

Spigo/Rizzo (1993–1994): Umberto Spigo and Cettina Rizzo, "Ricerche a Francavilla di Sicilia 1989–1991", in: *Kokalos* XXXIX–XL, 1039–1057.

Spigo *et al.* (2008): Umberto Spigo, Cettina Rizzo, Elvira D'Amico and Maria Grazia Vanaria (eds.), *Francavilla di Sicilia. L'anonimo centro di età greca, l'area archeologica e l'Antiquarium*, Soveria Mannelli.

Tigano (1991): Gabriella Tigano, "Modellini fittili di imbarcazioni dalla necropoli ellenistica di Milazzo", in: *Atti della IV Rassegna di archeologia subacquea, IV Premio Franco Papò* (Giardini Naxos, 13–15 ottobre 1989), Messina, 51–70.

Tigano (2002): Gabriella Tigano (ed.), *Le necropoli di Mylai (VIII-I sec. a.C.)*, Milazzo.

Tigano (2021): Gabriella Tigano, "I "modellini" fittili di imbarcazioni. Nuovi spunti di riflessione", in: Valeria Patrizia Li Vigni (ed.), *Sebastiano Tusa. L'uomo, lo studioso, l'archeologo*, Castelvetrano-Selinunte, 55–58.

Tigano/Vanaria (2022): Gabriella Tigano and Maria Grazia Vanaria (eds.), *M.A.FRA. Museo archeologico Francavilla di Sicilia. Nuovi studi e guida all'esposizione*, Saponara.

Tréziny (2004): Henri Tréziny, "Sondages sur la fortification méridionale (1978–1982)", in: Gras/Tréziny/Broise (2004a) 157–235.

Tsoukala (2009): Victoria Tsoukala, "Cereal Processing and the Performance of Gender in Archaic and Classical Greece: Iconography and Function of a Group of Terracotta Statuettes and Vases", in: Çiğdem Özkan Aygün (ed.), *SOMA 2007. Proceedings of the XI Symposium on Mediterranean Archaeology* (Istanbul, 24–29 April 2007), BAR International Series 1900, Oxford, 386–395.

Ursino (2012): Michela Ursino (ed.), *Da Evarco a Messalla. Archeologia di Catania e del territorio dalla colonizzazione greca alla conquista romana* (Catalogo della mostra, Catania 21 dicembre 2012–10 marzo 2013), Palermo.

Vallet/Villard (1964): Georges Vallet and François Villard, "Mégara Hyblaea VIII. Remarques sur la plastique du VIIe siècle", in: *Mélanges de l'Ecole française de Rome. Antiquité* 76, 25–42.

Vassallo (2005): Stefano Vassallo, *Himera città greca. Guida alla storia e ai monumenti*, Palermo.

Vlachou (2015): Vicky Vlachou (ed.), *Pots, Workshops and Early Iron Age Society: Function and Role of Ceramics in Early Greece* (Proceedings of the International Symposium, Bruxelles, 14–16 November 2013), Brussels.

Dario Giuliano

6 The Manufacture of "Thickened-edge" Roof Tiles between Unintentional Traces and Intentional Marks: The Case of the Hellenistic-Roman Temple in Agrigento

Abstract: The study of materials from the excavation of the Hellenistic-Roman temple in Agrigento provided an opportunity to conduct an in-depth analysis of a class of materials that are notoriously underappreciated, namely roof tiles. Those presented here pertain to a specific type found in practically all Roman-era sites in Sicily, the so-called thickened-edge roof tiles. This type derives from the Laconic roofing system, characterized in Sicily by large, concave pan-tiles and curved cover tiles, replacing the previously attested "hybrid" system characterized by large, flat pan tiles and curved cover tiles. This analysis allowed to reconstruct all the phases of the production chain of these artifacts, recognizing both involuntary and intentional traces. Among the latter, there are a considerable number of marks, and it has thus been our aim to understand their function, by advancing a hypothesis regarding their purpose.

Introduction

In this paper we discuss a smaller part from a larger project that focuses on the roofing systems used in the ancient city of Agrigento, with particular attention to the different types of roof tiles from their inception up to the Roman conquest. We have chosen to present a case study of the Hellenistic-Roman temple of Agrigento, which, following recent archaeological excavation, has yielded a significant quantity of a specific type of roof tile, commonly referred to as "thickened-rim" roof tiles. These were extensively used in roof coverings during the Roman period in Sicily. The study conducted has allowed us to carry out a detailed analysis of these artifacts and reconstruct their production chain through the unintentional and intentional traces present.

The sanctuary complex being studied is located north of the city's Agora and was first investigated by E. de Miro from 1995 to 2005. Since 2013, research has resumed under the direction of the Polytechnic University of Bari, led by L. Caliò, M. Livadiotti, and R. Belli. Recent investigation has provided a better understanding of the monument's phases of construction, which began in the mid-2nd century B.C., but was interrupted for a long period and resumed during the 1st century A.D., when the sacred building was completed and a portico was added that surrounded the temple on three sides. The final appearance was that of a podium temple of the *rostratum* type, located in a porticoed square, which classifies the building as a *templa cum porticibus.*[1]

The study focused on all the fragments recovered from the area during the recent excavation campaigns. This lot comprises 903 fragments of various types, covering a chronological period from the Late Archaic period to the abandonment that occurred during the 5th century A.D. Despite the considerable quantity involved, selection among all the identified fragments was carried out directly in the field by the excavators. Most of the fragments come from secondary deposits, except for trench 3, where a portion of the collapsed roof of the northern portico was discovered;[2] this had been constructed using the roof tiles under study.

1 I would like to thank Prof. Luigi Caliò for entrusting me with the study of the roof tiles in this context. At the Archaeological and Landscape Park Valle dei Templi di Agrigento, I express my gratitude to Drs. Valentina Caminneci, Maria Concetta Parello and Maria Serena Rizzo, for allowing me to conduct this study with the utmost freedom. On the results of recent investigations, see Caliò *et al.* (2016); Livadiotti/Fino (2017).

2 SSUU 2337, 2352, 2459.

https://doi.org/10.1515/9783111189635-006

The "Thickened-edge" Roof Tiles and their Manufacture

The roof tiles that have been designated "thickened-edge" constitute the most commonly attested type at the Agrigento complex,[3] just as in the Roman-era buildings constructed on the island. They are wide, curved tiles, associated with narrower cover tiles, which also have a semicircular section. The term "thickened-edge" derives from the intentional thickening of the edge on the short front side, which was achieved by the craftsman during the production phase and exhibits a varied profile, commonly referred to as "almond-shaped." These roof tiles, classified by R. Wilson as type B,[4] mark a break from the previous roof covering system prevalent in Sicily, known as the "hybrid system."[5] The hybrid system consisted of a combination of large, rectangular pan tiles with flanges on their long sides and semicircular cover tiles to cover the contact lines between two adjacent pan tiles. The new roof covering system replaced the pan tiles with wide, curved roof tiles following the model of the "Laconic system." Wilson had already hypothesized the derivation from the Laconic model for this type of roof tile,[6] which was widespread in Sicily during the Roman era. In our opinion, it would be preferable to employ the definition adopted by P. Hamari[7] and introduced for the first time by Skoog,[8] i.e. a "Roman Laconic-style" system or, even better, a regional variant of this "style" with distinct local characteristics. We fully support Hamari's statement during the Ghent conference that "We should advance from generic 'Laconian' or 'Corinthian' to a more nuanced classification of roof-tiles."[9]

In Sicily, the use of thickened-edge tiles extended over a long period. While their use is reliably attested until the third quarter of the 5th century A.D., based on the stratigraphic excavation of the Campanaio settlement,[10] we have less data regarding their introduction, which we generally attribute to the 2nd century B.C.[11]

During this long period of use, the roof tiles did not undergo substantial morphological changes. The excavation carried out at the Hellenistic-Roman temple in Agrigento has allowed the recovery of some near-intact examples, which have provided insights into their morphological characteristics and dimensions (Fig. 1). The wider, curved tiles have a slightly trapezoidal shape with an average length of 72.5 cm and a width of 32.6 cm at the front edge and 28.5 cm at the rear edge. The height is also variable but decreases from 10.2 cm[12] to 8.3 cm at the rear edge. The average thickness of the body is 1.7 cm, and the estimated weight is 7.5 kg. The cover tiles have practically the same length,[13] but a lesser width, measuring 19.4 cm at the front edge and 17 cm at the rear edge. The height is also variable, but decreases from 8.7 cm just beyond the rim to 6.8 cm at the other end. The body has a thickness of 1.5 cm, and the estimated weight of a single element is 3.8 kg.

Study of the nearly complete roof tiles and fragments has led us to reflect on their manufacturing process, partly based on the unintentional traces present on their surfaces. As regards their production, we might imagine the use of a presumably trapezoidal-shaped mold from which a slab would be removed and placed on a semicylindrical wooden form. Evidence of the latter remains on the lower surface, where the negative impressions of wood grain are often visible (Fig. 2a).

---

3 This accounts for 41% of the total. To this value, a significant portion of the curved tile bodies (35%) should be added. The counting of these fragments was performed separately because it is not always easy to distinguish the fragments related to Roman era curved roof tiles from those from earlier periods.

4 Wilson (1979) 20.

5 Akerström (1966) 198; Wikander (1988) 214.

6 Wilson (1979) 20.

7 Hamari (2019) 30.

8 Skoog (1998) 129.

9 Hamari (2019) 115.

10 This refers to the rural settlement of Campanaio (Agrigento), which underwent a violent destruction around 460 A.D., probably because of Vandal raids on the island: see Wilson (2000) 365.

11 We take this opportunity to provide an update on the proposal presented in Giuliano (2022). In that study, the introduction of tiles with a thickened rim in the 3rd century B.C. was hypothesized based on the existing literature. However, upon direct examination of the materials mentioned in that contribution (currently under study), slight differences have been observed that set them apart from the roof tiles with a "thickened rim". Therefore, a date of 2nd century B.C. is proposed for the introduction of "thickened rim" roof tiles.

12 Measurement taken just beyond the edge.

13 There are two preserved examples measuring 71.6 cm and 67 cm, respectively. The shorter value is due to the roof tile deforming during firing, causing it to curve upwards.

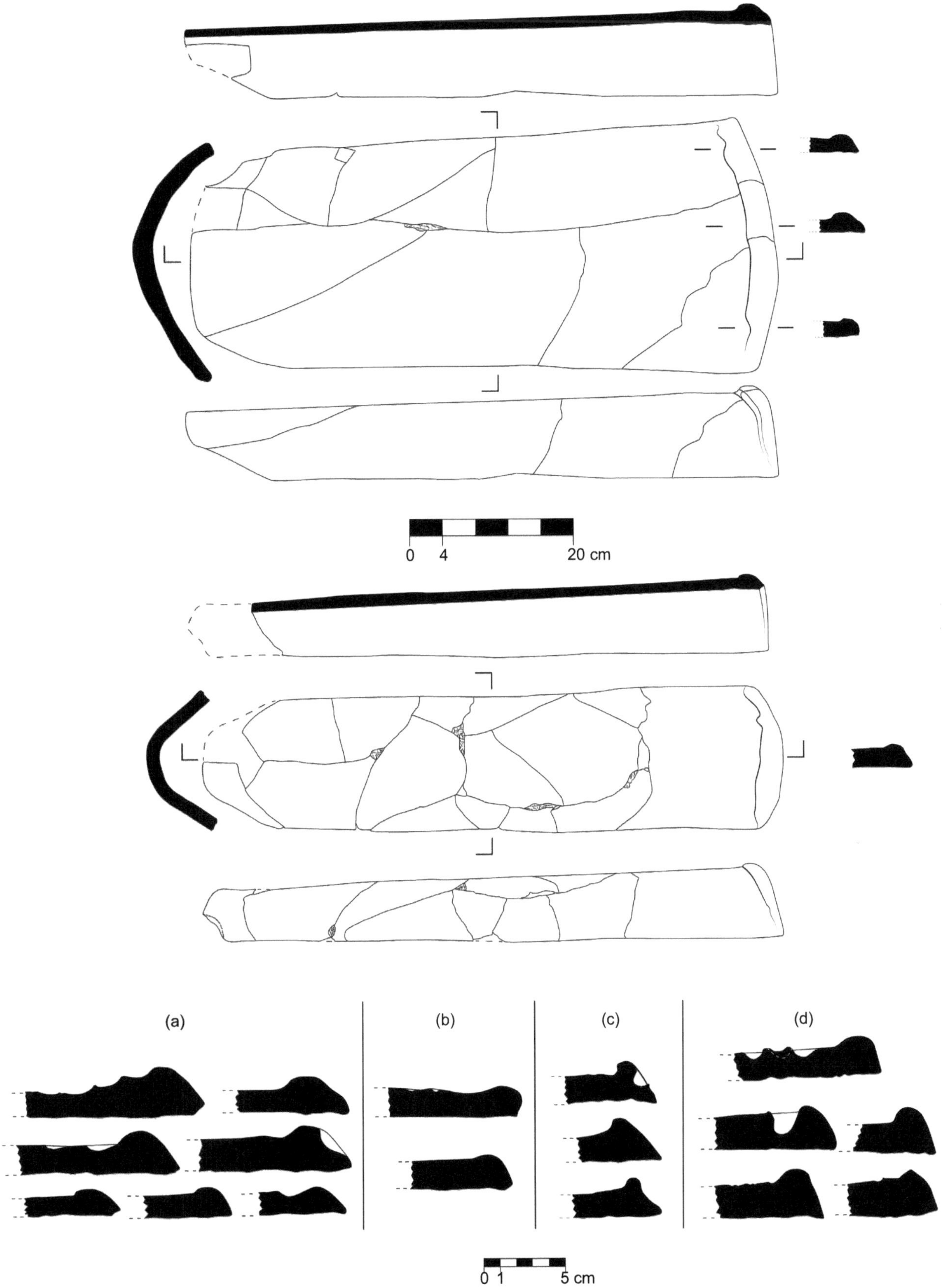

Fig. 1: Above: the thickened-edge roof tiles. Below: the different types of edge: **(a)** Almond-shaped edge; **(b)** The slightly indicated profile; **(c)** The crest profile; **(d)** The massive profile. Drawings: Dario Guiliano.

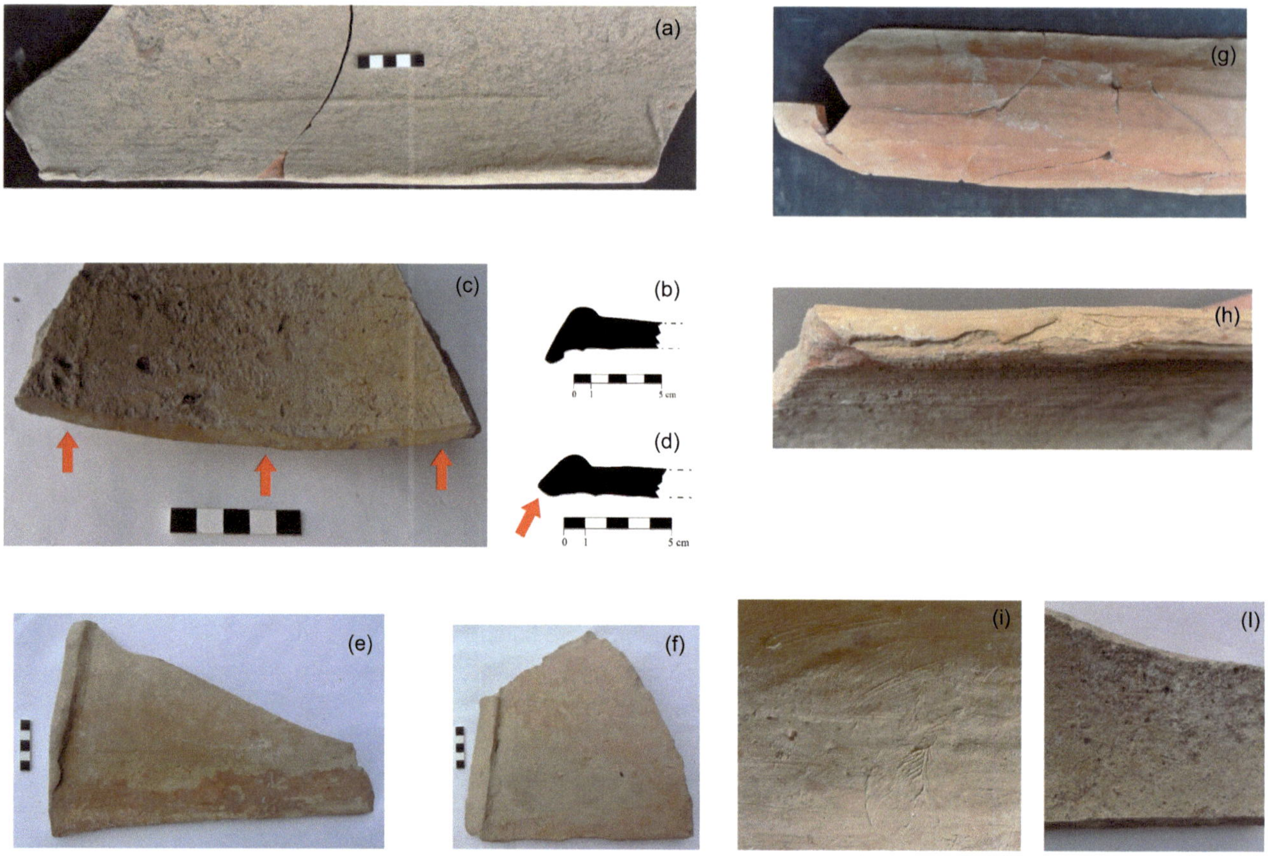

Fig. 2: (a) Negative impressions of wood grain visible in the lower surface; **(b)** Edge with clay smudge; **(c–d)** Anterior edge with cutting traces of a clay smudge; **(e–f)** The receded anterior margin due to the molding of the edge. Treatment of the upper surface: **(g)** Longitudinal streaks; **(h)** Clay smudges on the lateral margins; **(i)** Impressions left by fingers or palms of the hands; **(l)** "Orange peel" lower surface. Drawings: Dario Guiliano.

Once removed from the mold, the edge had to be formed by folding the clay of the front edge onto itself, shaping it to the desired form. This operation left several traces, including the formation of clay smudges on the lower surface, which were sometimes removed with a blade (Fig. 2b–d). The front edge also receded, sometimes by several centimeters, giving it a more or less pronounced concave shape (Fig. 2e–f). Lastly, various profiles were created for the edges. The most common is "almond-shaped" (Fig. 1a), characterized by a pointed end and an elongated and rounded profile. Beyond the point of maximum height, the edge gradually tapers off, with a short, sinuous section connecting to the body of the roof tile. The "barely hinted at" profile (Fig. 1b) is rarer and the simplest among those observed, with a slight rounded swelling that is a little more developed than the body. The "crest" (Fig. 1c) profile is more distinctive, featuring a short, either straight or slightly concave profile with a short and rounded upper end that is sharply separated from the body. Lastly, the "massive" profile (Fig. 1d) has the anterior face developing almost vertically, or with a slight inclination, and a wide, rounded profile, clearly separated from the body of the tile. This variability is the result of the artisan's skill and the speed with which the edge was fashioned, often in a very cursory manner.

The rear edge is characterized by the absence of the two corners, which were removed with a blade, resulting in two oblique sections (Fig. 1). This action created a short "tail" that was functional for their installation.

The final phase for these roof tiles was the treatment of the upper surface. The artisan would pass his wet hands over it along its length, or with circular movements across the entire surface. This created a slip (a sort of engobe) with a characteristic beige color, observed in almost all the fragments. That this is not engobe can be deduced from the fact that this slip is not distinct from the ceramic body, but is part of it. This operation left evident traces, such as clay smudges on the lateral margins, which were sometimes removed (Fig. 2h),[14] as well as unintentional fingerprints, in the form of longitudinal streaks and impressions left by fingers or palms (Fig. 2g, i). The function of this slip was to render the upper

14 Also in this case, with the use of a blade.

surface smoother and improve its impermeability. Its use is not limited to the production of these roof tiles, but has ancient origins, also being attested in the Archaic and Classical periods.[15] A noteworthy element of the slip used at Agrigento was the addition of common salt (sodium chloride, NaCl). This could be achieved using seawater[16] or brackish clay, which is common in Sicily due to the island's geology.[17] During the drying phase of the artifact, the salt would rise to the surface and during firing would give it a white or ivory color.[18] It has been suggested that this technique was employed in Gela in the production of Western Greek amphorae[19] and protomes.[20] During the Hellenistic period, this *slip* is attested in some productions in Lilibeo[21] and Selinous.[22] This technical feature was never abandoned, as Hampe and Winter noted the use of salt as an additive in roof tile production.[23] The two scholars emphasize that this is a peculiarity of Sicilian artisans and does not find parallels in other areas of the ancient Mediterranean.[24] Such a practice is still employed today, as recently reported by G. Montana, following direct observation of the work of various ceramists in different locations in central-western Sicily.[25]

The lower surface of the "thickened-edge" roof tile did not undergo any treatment and was left rough and irregular, characterized by a peculiar surface that could be described as "orange peel-like" (Fig. 21).

Intentional Marks

The study conducted enabled us to observe the presence of intentional marks on 88 fragments,[26] which can be classified into 11 different types (Figs. 3–5). Their intentional nature is evident because they are all located in the same area of the roof tile, approximately at the center of the front edge, directly on or just beyond the edge. All these marks are clearly visible, taking the form of finger impressions, grooves or letters, and were made either with tools or one to three fingers. The most common mark is the "X" which is traced using a finger or a tool, with the latter also being used to create straight lines with hooked ends. Numerous finger impressions can be distinguished, some made directly on the edge and others slightly beyond it. Among these marks, an interesting one is a set of three finger impressions made with the thumb, index, and middle finger of a boy's hand. This particular mark appears five times, four of which were found on roof tiles from the roof collapse of the northern portico.[27] Considering the uniformity in execution, position, and characteristics,[28] there seems to be no doubt that they were made by the same person.

These observations allow us to make some considerations about their function, which is still being debated,[29] especially considering that similar marks are already known on roof tiles produced in Etruria in the 6th century B.C.[30] While we do not intend to draw comparisons with such geographically and chronologically distant contexts, it is worth mentioning that one interpretation suggests they were placement markers or guidelines for laying the tiles on the roof.[31] While this

15 As was possible to ascertain following the study conducted by the author on the roof tiles from those epochs.
16 Spagnolo *et al.* (2018) 52.
17 It specifically refers to the gypsum-sulfur series, which is also characterized by the presence of extensive underground halite deposits that are crossed by aquifers, bringing salt to the surface, and flowing into rivers.
18 Cuomo di Caprio (2007) 311–314.
19 Spagnolo *et al.* (2018) 46.
20 Uhlenbrock (1988) 31.
21 It corresponds to fabric 2 in Bechtold's classification: Bechtold (1999) 136.
22 Fourmont (1992) 60–61.
23 Hampe/Winter (1965) 103.
24 Hampe/Winter (1965) 176.
25 Montana/Polito/Sulli (2011). The workshops where the scholar was able to observe this practice are located in Racalmuto, Mussomeli, Paceco, Marsala, and Mazzara del Vallo.
26 In this count, all intentional marks, including those partially preserved, have been considered. The latter group consists of thirteen fragments, eleven of which appear to represent a portion of the "X" mark. However, due to their fragmented nature, it was preferred to count them within the category of "unidentifiable letters."
27 The fifth one comes from the dump from the 5th century A.D.
28 Same measurements as the impressions in various instances, in some cases, the mark left by the thumbnail is also visible.
29 A comprehensive summary of all interpretive hypotheses about their function is compiled in Hamari *et al.* (2019) 2, pl. 1.
30 Wikander (2017) 201–226.
31 Wikander (2017) 205.

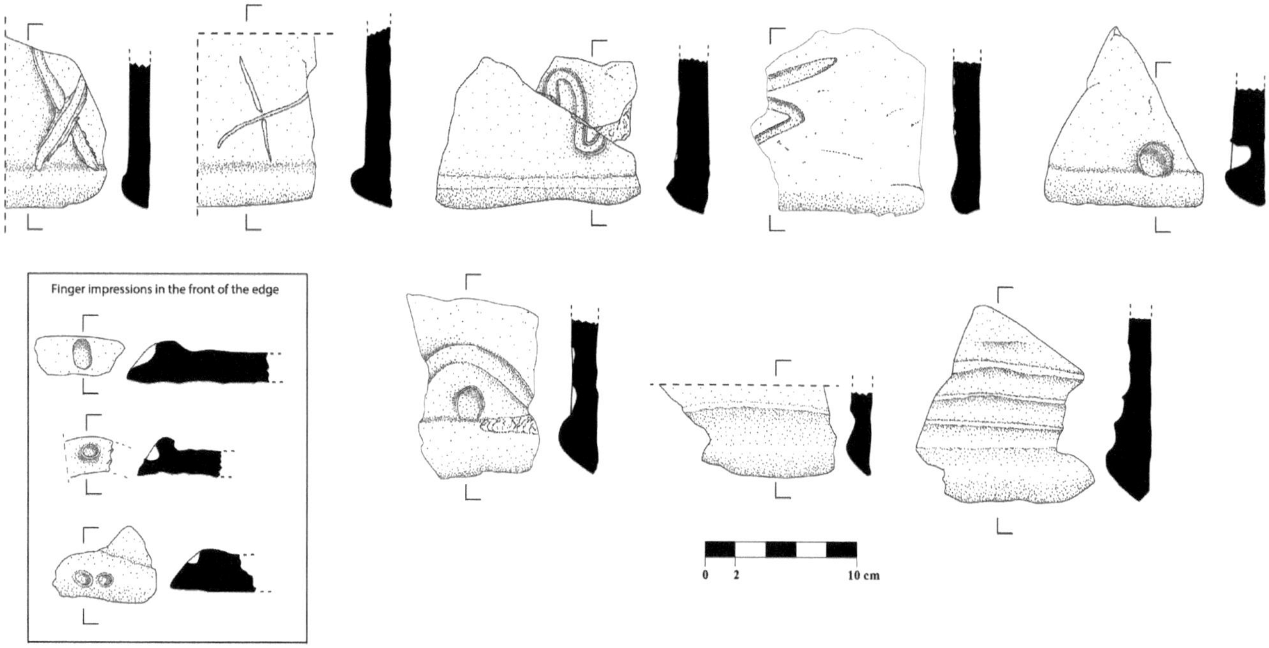

Fig. 3: The marks on the roof tiles. Drawings: Dario Guiliano.

hypothesis seems plausible for decorated, architectural terracotta,[32] for which precise placement was crucial for obtaining uninterrupted decorative friezes;[33] it may not hold true for the roof tiles we are analyzing, which can be considered "standardized" products, manufactured "in series." Another interpretation of these marks is that they served as dryness tests. This hypothesis, put forth again by E. J. Shepherd[34] specifically because of the semicircular grooves made with one or more fingers, suggests that they were "tests to assess the drying stage, possibly performed repeatedly."[35] However, as E. J. Shepherd herself points out, "the impressions should have been lighter due to a higher degree of drying [. . .] a phenomenon that does not seem to occur regularly".[36] Therefore, this interpretation also encounters some difficulties. Additionally, it is evident that some marks on our roof tiles were made when they were clearly still wet, as indicated by clay smudges or the depth of the grooves.

In light of the above, it seems plausible to interpret these marks as symbols that workers in the same *figlina* would place on the last roof tile at the end of each batch. These marks would serve both as distinctive symbols for batches produced by individual craftsmen and as counters for tallying the number of pieces, hence serving a function related to daily wage payment.[37] This hypothesis aligns with the theory put forward by P. Warry.[38] B. Clément likewise sees the grooves as distinctive marks of each craftsman, useful for quality control of the produced materials and for counting the manufactured items for daily wage payment.[39] The latter hypothesis is also mentioned by S. Alegiani, based on the provisions of

32 Simas and revetment plaques.
33 Averna (2019) for a recent study on marks in Himera.
34 Shepherd (2006) for the roof tiles from the Vingone workshop.
35 Shepherd (2006) 175. This interpretation is also mentioned in Alegiani (2016) 56.
36 Shepherd (2006) 175.
37 It follows that the position in which the marks are made can be an indication of production by a different craftsman. As is the case of a single fingerprint made in some examples directly on the front part of the edge, and in others just beyond it.
38 Warry (2006) 14–16.
39 In their study, the author examines the semi-circular and ogival grooves found in Roman flat roof tiles in central-eastern Gaul: see Clément (2013) 67.

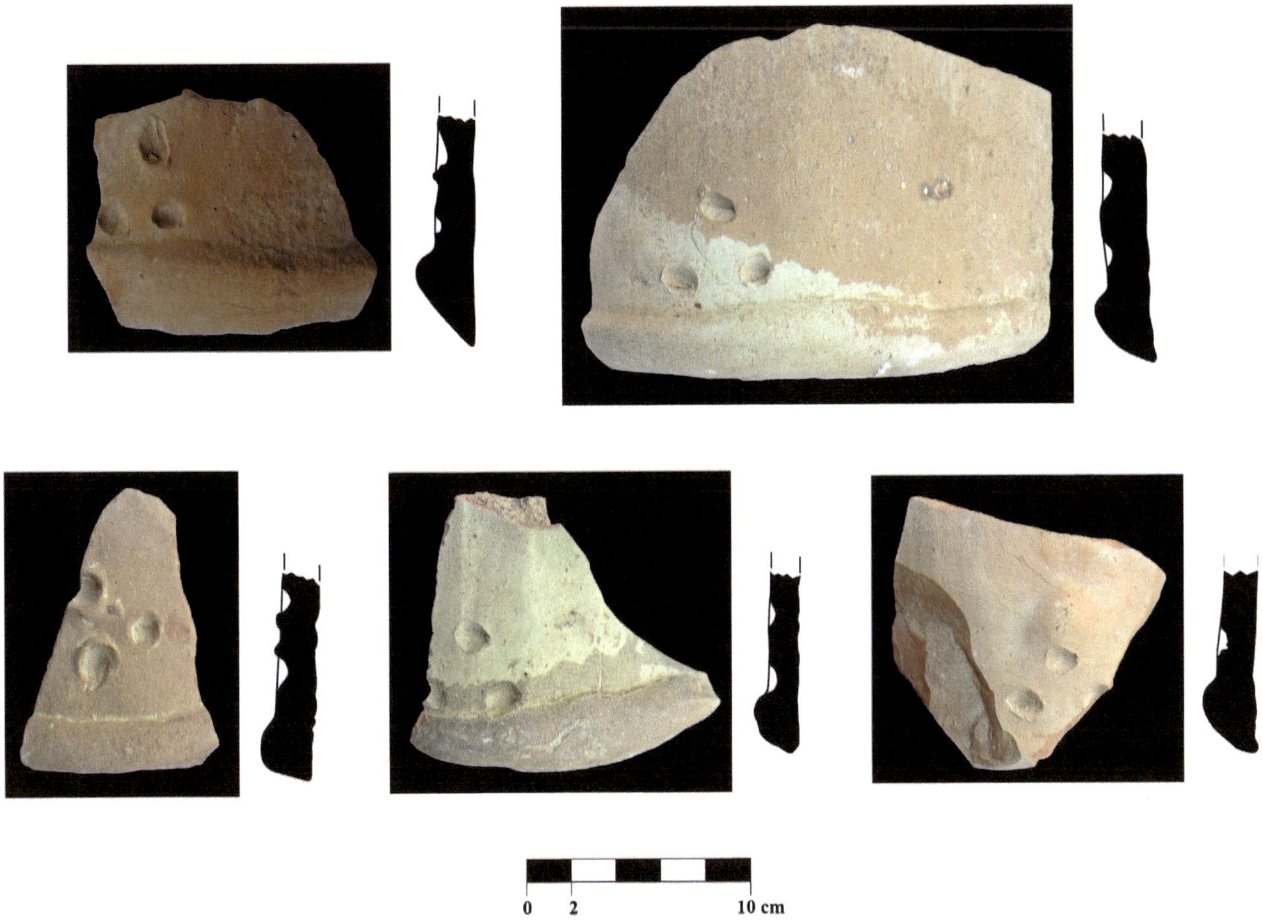

Fig. 4: Photos and drawings of the three finger impressions. Drawings: Dario Guiliano.

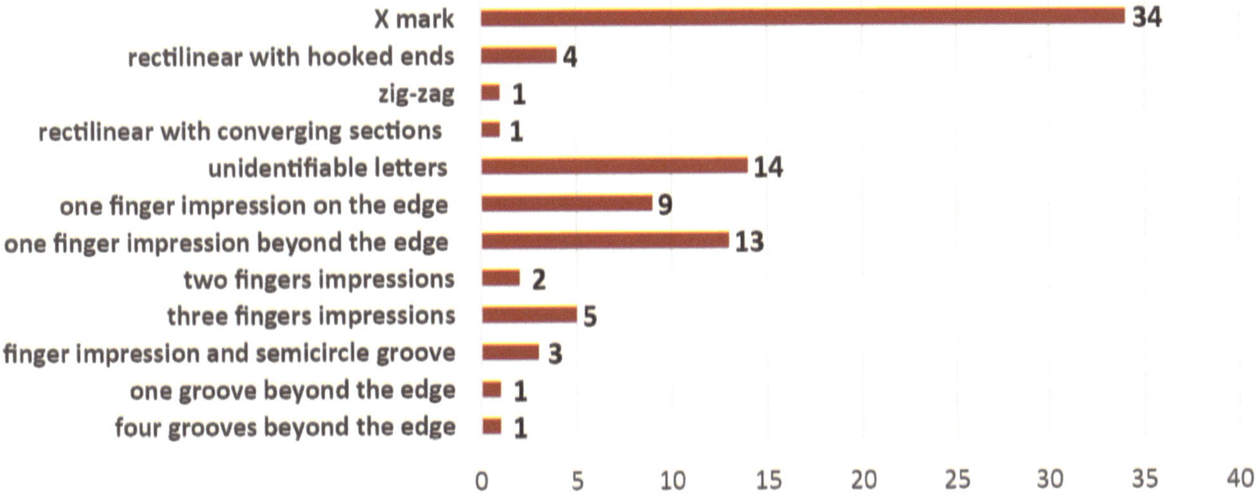

Fig. 5: Marks attested in the roof tiles. © Dario Guiliano.

Diocletian's Edict, which states that "workers' payment [involved in brick production ed.] usually occurred on a piecework basis."[40] Recently, this hypothesis has been put forward as one of the possible scenarios suggested by the presence of different marks on roof tiles originating from the same roof, plausibly produced in the same workshop.[41]

If it is assumed that a specific mark is associated with a worker's production, this information can be used to address another question: whether it is possible to create a chrono-typological seriation of these specific roof tiles based on the profile of the edge. By comparing the profiles of edges marked with the three-finger mark (Figs. 3–4), and therefore attributable to the same craftsman, a substantial difference can be observed; this implies that the same artisan did not produce identical edge profiles. The significant variability in the manufacturing of edges can also sometimes be observed within the same edge. To complicate any attempt at chronological-typological seriation based on the edge, it must be pointed out how the edge profile of the same roof tile is not consistent throughout its development, but varies due to its cursive execution; in fact, by making the design at three different points within the same edge, substantially different profiles will be obtained (Fig. 1). As a result, due to the absence of any standardization in its shape, any attempt at chrono-typological seriation of these roof tiles based on the edge would be less effective.

Conclusions

The examination of numerous fragments from the excavation of the Hellenistic-Roman temple in Agrigento has allowed us to reconstruct the production chain of the so-called "thickened-edge" roof tiles, which remained substantially unchanged for centuries during the Roman rule of the island. After molding the slab, the treatment of the short sides followed, with the corners being removed on the back side with a blade, while the front side was manipulated to produce the thickening of the edge, a tectonic peculiarity from which this roof tile's name derives. Finally, the smoothing of the upper surface was carried out, done directly by hand on the wet surface to create an outer slip that, during firing, acquired a characteristic, light color. It seems plausible that the formation of this slip was facilitated by means of clays naturally containing salt, which was used precisely for its bleaching properties.

The wide variability of the edge profile necessitated the roof tiles' classification into types – almond-shaped, barely defined, crested, and massive – which helps to show that their variation is not indicative of different chronologies. In fact, the most substantial lot of artifacts analyzed comes from the same context, namely the collapse of the roof of the northern portico which did not undergo modifications or renovations during its centuries-long period of use. Another clue supporting the hypothesis of approximate contemporaneity of the various profiles comes from the study of the marks present on the roof tiles; indeed, among the eleven types identified, the one bearing three fingerprint impressions would, in our opinion, have been made by the hand of the same craftsman, since it appears on roof tiles with different edge profiles.

The presence of different marks on the tiles could be interpreted as signs left by artisans operating within the same workshop to indicate the specific artifacts produced; in this way, counting and consequently paying workers on a piecework basis would have been facilitated.

Bibliography

Akerström (1966): Ake Akerström, *Die Architektonischen Terrakotten Kleinasiens*, Lund.
Alegiani (2016): Silvia Alegiani, "Un esempio di catalogazione di bolli laterizi anepigrafici a Roma", in: Evelyne Bukowiecki, Rita Volpe and Ulrike Wulf-Rheidt (eds.), *Il laterizio nei cantieri imperiali. Roma e il Mediterraneo* (Atti del I workshop "Laterizio", Roma, 27–28 novembre 2014), Florence, 54–59.
Averna (2019): Andrea Averna, "Markers di cantiere e segni di numerazione nei rivestimenti architettonici del tempio B di Himera", in: Patricia Lulof, Ilania Manzini and Carlo Rescigno (eds.), *Deliciae Fictiles V. Networks and Workshops: Architectural Terracottas and Decorative Roof Systems in Italy and Beyond* (Proceedings of the 5th International Conference, Napoli, 15–17 Marzo 2008), Havertown, 208–216.
Bechtold (1999): Babette Bechtold, *La necropoli di Lilybaeum*, Trapani.

40 Alegiani (2016) 56.
41 Hamari *et al.* (2019).

Caliò *et al.* (2016): Luigi Maria Caliò, Gian Michele Gerogiannis, Fernando Giannella, Monica Livadiotti, Antonello Fino, Marina Albertocchi and Francesca Leoni, "Il santuario Ellenistico Romano di Agrigento. Lo scavo, l'inquadramento urbano, l'architettura", in: Maria Concetta Parello and Maria Serena Rizzo (eds.), *Paesaggi urbani tardoantici. Casi a confronto* (Atti delle VIII Giornate Gregoriane, Agrigento, 29–30 novembre 2014), Bari, 295–318.
Clément (2013): Benjamin Clément, *Les couvertures de tuiles en terre cuite en Gaule du Centre-Est (IIe s. av.-IIIe s. ap. J.-C.)*, Drémil Lafage.
Cuomo di Caprio (2007): Ninina Cuomo di Caprio, *La ceramica in archeologia 2*, Rome.
Fourmont (1992): Martine H. Fourmont, "Les ateliers de Sélinonte (Sicile)", in: Francine Blondé and Jacques Y. Perreault (eds.), *Les ateliers de potiers dans le monde grec aux époque géométrique, archaïque et classique, Bulletin de Correspondance Hellénique* Supplément 23, 57–68.
Giuliano (2022): Dario Giuliano, "Tegole e coppi", in: Luigi Maria Caliò, Gian Michele Gerogiannis, Francesca Leoni and Giulia Raimondi (eds.), *Agrigento 2. Il santuario Ellenistico-Romano. Scavi 2013–2017. I materiali*, Rome, 565–590.
Hamari (2019): Pirjo Hamari, *Roman-period Roof Tiles in the Eastern Mediterranean. Towards Regional Typologies*, PhD Dissertation, University of Helsinki.
Hamari *et al.* (2019): Pirjo Hamari, Despoina Tsiafaki, Nikolaos Kazakis and Nestor Tsirliganis, "Tracing Ancient Fingerlines: Applying Micro-XRF to Signatures on Laconian Roof Tiles in Late Antique Arethousa, Paliambela (Greece)", in: *Journal of Archaeological Sciences: Reports* 26, 1–11.
Hampe/Winter (1965): Roland Hampe and Adam Winter, *Bei Töpfern und Zieglern in Süditalien Sizilien und Griechenland*, Mainz.
Livadiotti/Fino (2017): Monica Livadiotti and Antonello Fino, "Il complesso porticato a Nord dell'Agorà", in: Luigi Maria Caliò, Valentina Caminneci, Monica Livadiotti, Maria Concetta Parello and Maria Serena Rizzo (eds.), *Agrigento. Nuove ricerche sull'area pubblica centrale*, Rome, 97–110.
Montana/Polito/Sulli (2011): Giuseppe Montana, Anna M. Polito, Attilio Sulli, *Le "argille ceramiche" della Sicilia occidentale*, Enna.
Shepherd (2006): Elizabeth J. Shepherd, "Laterizi da copertura e da costruzione", in: Elizabeth J. Shepherd, Gabriella Capecchi, Giuliano De Marinis, Fabio Mosca and Anna Patera (eds.), *Le fornaci del Vingone a Scandicci. Un impianto produttivo di età romana nella valle dell'Arno*, Rassegna di archeologia 22/B, Florence, 165–200.
Skoog (1998): Victoria Nevius Skoog, *The Laconian-style Roof: Development, Distribution, and Technologies*, PhD Dissertation, University of Missouri-Columbia.
Spagnolo *et al.* (2018): Grazia Spagnolo, Germana Barone, Paolo Mazzoleni and Erica Aquilia, "Le anfore da trasporto e la ceramica medio-grossolana di Gela in età greca: caratterizzazione e tecniche produttive", in: Caterina Ingoglia (ed.), *Risorse ambientali. Impianti produttivi a Gela*, Bari, 43–56.
Uhlenbrock (1988): Jaimée P. Uhlenbrock, *The Terracotta Protomai from Gela: A Discussion of Local Style in Archaic Sicily*, Rome.
Warry (2006): Peter Warry, *Tegulae. Manufacture, Typology and Use in Roman Britain*, BAR British Series 417, Oxford.
Wikander (1988): Örjan Wikander, "Ancient Roof-tiles – Use and Function", in: *Opuscula Atheniensia* XVII, 203–216.
Wikander (2017): Örjan Wikander, *Roof-tiles and Tile-roofs at Poggio Civitate (Murlo)*, Stockholm.
Wilson (1979): Roger Wilson, "Brick and Tiles in Roman Sicily", in: Alan McWhirr (ed.), *Roman Brick and Tile*, BAR International Series 68, Oxford, 11–44.
Wilson (2000): Roger Wilson, "Rural Settlement in Hellenistic and Roman Sicily Excavations at Campanaio (AG), 1994–8", in: *Papers of the British School at Rome* 68, 337–369.

Alessandra De Natale, Laura Di Siena, Giovanni Murro and Sine Grove Saxkjær

7 Teanum Ware Pottery from Loc. Cappella Marcella (Castrocielo, Frosinone)

Abstract: This chapter presents Teanum Ware pottery from Loc. Cappella Marcella, in southern Latium. Although our understanding of this Sidicinian ceramic ware remains insufficient, we take a closer look at the production cycle through observations of processing signs and traces. Based on the assemblage from Loc. Cappella Marcella, we have established six production phases that include the shaping of the vessel, several stages of decoration, the application of black gloss coating, and the firing of the vessel. In terms of decoration, the impressed decoration is done before the application of black gloss coating, the incised and painted decoration is done before the firing, while details in red paint are applied after the firing. Although the study is not exhaustive, it clearly illustrates that the Teanum Ware pottery from Loc. Cappella Marcella differs in appearance and manufacturing techniques both from the traditional Roman black gloss pottery and from other assemblages of Teanum Ware pottery.

Introduction

In the 4th century B.C., the Sidicinians expanded their territory northward into southern Latium, leading to a wider distribution of their distinctive Teanum Ware pottery. The pottery is named after the Sidicinian capital Teanum Sidicinum in northern Campania, which also constitutes the main find spot for this type of pottery. So far, with no remains of kilns or production waste in Teanum Sidicinum, and with no archaeometric analyses, our technical knowledge on the production of Teanum Ware remains incomplete.

The traces of manufacture on the Teanum Ware vessels from Loc. Cappella Marcella can, however, provide an insight into the operational sequences of Sidicinian pottery. In what follows, we present a preliminary study of the Teanum Ware pottery from the Loc. Cappella Marcella necropolis,[1] dating to the 4th –3rd centuries B.C. The necropolis, located close to the northwestern suburban area of the ancient city of Aquinum,[2] was excavated in 2015–2016 in connection with public construction and included a total of 44 burials dating from the 5th century B.C. to the 1st century A.D. Of these, nine adult burials could be ascribed to the Sidicinian culture and contained Teanum Ware pottery (Fig. 1).

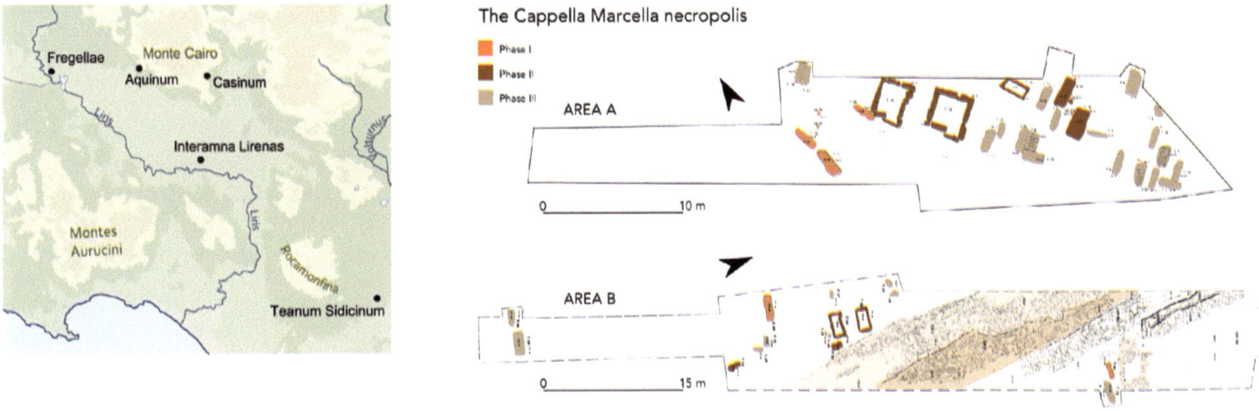

(a) (b)

Fig. 1: (a) Map of the Liri Valley; **(b)** Plan of the Loc. Cappella Marcella necropolis with its three cultural phases indicated. © Società Land s.r.l.

1 The excavations at Loc. Cappella Marcella were carried out between November 2015 and March 2016 by Società Land s.r.l., coordinated by archaeologists Gian Matteo Matullo, Giovanni Murro, and Mauro Treglia.
2 For a general overview on the site of ancient Aquinum, see Ceraudo/Murro (2018). On Cappella Marcella's necropolis, see also Bellini/Murro (2019).

https://doi.org/10.1515/9783111189635-007

Teanum Ware Pottery Production

Teanum Ware pottery is characterized by a technique which combines impressed, incised, and painted decorations.[3] The impressed decorations include a varied repertoire of stamps and stamp combinations; the incised decorations most often consist of a wavy, continuous vegetable branch in a series of concentric circles, less frequently with other details, such as ovals, arches, or crosses. The painted decoration is marginal, and often serves as accessory to the incised, impressed one, consisting mostly of leaves and berries adorning the vegetable branches or as a series of dots surrounding the stamps.

When comparing the material from Loc. Cappella Marcella with that from Loc. Gradavola,[4] which currently constitutes the most homogenous assemblage of Teanum Ware, one notices a significant technical and aesthetic difference. Among the Teanum Ware examples from Loc. Gradavola, the execution of the decoration on the vessels is very precise, often occupying the entire surface of the vessel. The overall aesthetic impression and the repertoire of decorations is roughly homogeneous within the Loc. Cappella Marcella assemblage, but its technological and decorative quality is clearly inferior to that from Loc. Gradavola. In fact, there are several imperfections in the execution of the vessels, both in the molding of the vessels on the wheel and in the decoration, painting, and firing of the vessels.

Based on observations of the processing signs, the production cycle of the Teanum Ware assemblage from Loc. Marcella Cappella can be categorized into the six phases described below (Fig. 2).

Fig. 2: A selection of the Teanum Ware pottery from T.13. © Alessandra De Natale.

3 Scholars first attempted to define this pottery class date in the early 20th century, following the discovery of an extensive Hellenistic necropolis at Gradavola (Gabrici 1910). On the characteristics of the pottery and its production sphere, see e.g., Morel (1965) 57–58, (1985) 172–173; Pedroni (2001) 123–129.

4 Gabrici (1910).

The Phases of Production Based on the Processing Signs

1st Phase: Shaping of the Vessel

In general, the molding of the vessels on the potter's wheel is done at a high technical level. The few vessels that are misformed are large open and closed shapes: plates (Fig. 3A.1) and oinochoai. In this phase, imperfections can be attributed to a general lack of attention to surface finishing. This is less the result of inexperienced craftspeople and more a strategy to save time and costs. Observations of the processing traces include:

- Marks from turning: the concentric grooves of the forming phase are rarely smoothed or refined (Fig. 3A.2).
- Traces of tooling: Often grooves are left by tools used to regulate and straighten the horizontal level of the wide rims of plates or cups; several traces of scraping are clearly visible on the undersides of ring bases.
- Assembling signs: Small imperfections, dents, and finger impressions are frequently observed in the attachment of handles, the tall vertical feet of the cups (which are formed separately), and other decorative elements such as heads and pyxides of kernoi (Fig. 3A.3).

2nd Phase: Division of the Overall Decorative Scheme and Impressed Decoration

After a partial drying phase, the vessel is placed back on the potter's wheel. The first stage of decoration can be accomplished in various ways, such as:

- Concentric sequences of rouletting and stamps (Fig. 3.B.1).
- Decoration by means of a roulette with multiple grooves (Fig. 3B.2).
- Impression of stamps on the center of the tondo (Fig. 3B.3).
- Concentric sequences of small stamps (Fig. 3B.4).

After this, the vase is left to dry completely (Tab. 1).

3rd Phase: Coating with Black Gloss

Generally, the vessel is held by the hand—two or three fingers are used for smaller vessels, four or five for heavier or more complex vessels. For the coating, which covers the impressed and incised decoration an open vessel is dipped upside down while a closed vessel is tilted sideways. Immediately after, the vessel is turned over with rapid movements in various patterns to ensure that the coating covers the surface completely and uniformly (Tab. 2; Fig. 3C.1–C.4).

4th Phase: Adding Incised and Painted Decoration

Once the painting is finished, the second decorative cycle of incisions and overpainting occurs. Previously, it had been assumed that both the incised vegetable branches and the accessory decorations (e.g., ivy leaves, berries, running waves, dots) were created after firing, with vitrified varnish graffiti and white pigment paint. However, analyses of the incised decoration and its characteristics (type of incisions and post firing effects) have made it possible to place this decoration phase after the painting but before the firing of the vessel (Tab. 3; Fig. 3D.1–D.4).

5th Phase: Firing of the Vessel

Leaving aside aspects of the firing method that relate to the loading of the kiln and the placement of the vessels inside the chamber (as seen in the marks left by the spacers and stacking rings), it is important to consider the impact of the firing on the color of the paint, the level of sintering, and the adhesion to the ceramic body. The color of the slip undergoes during

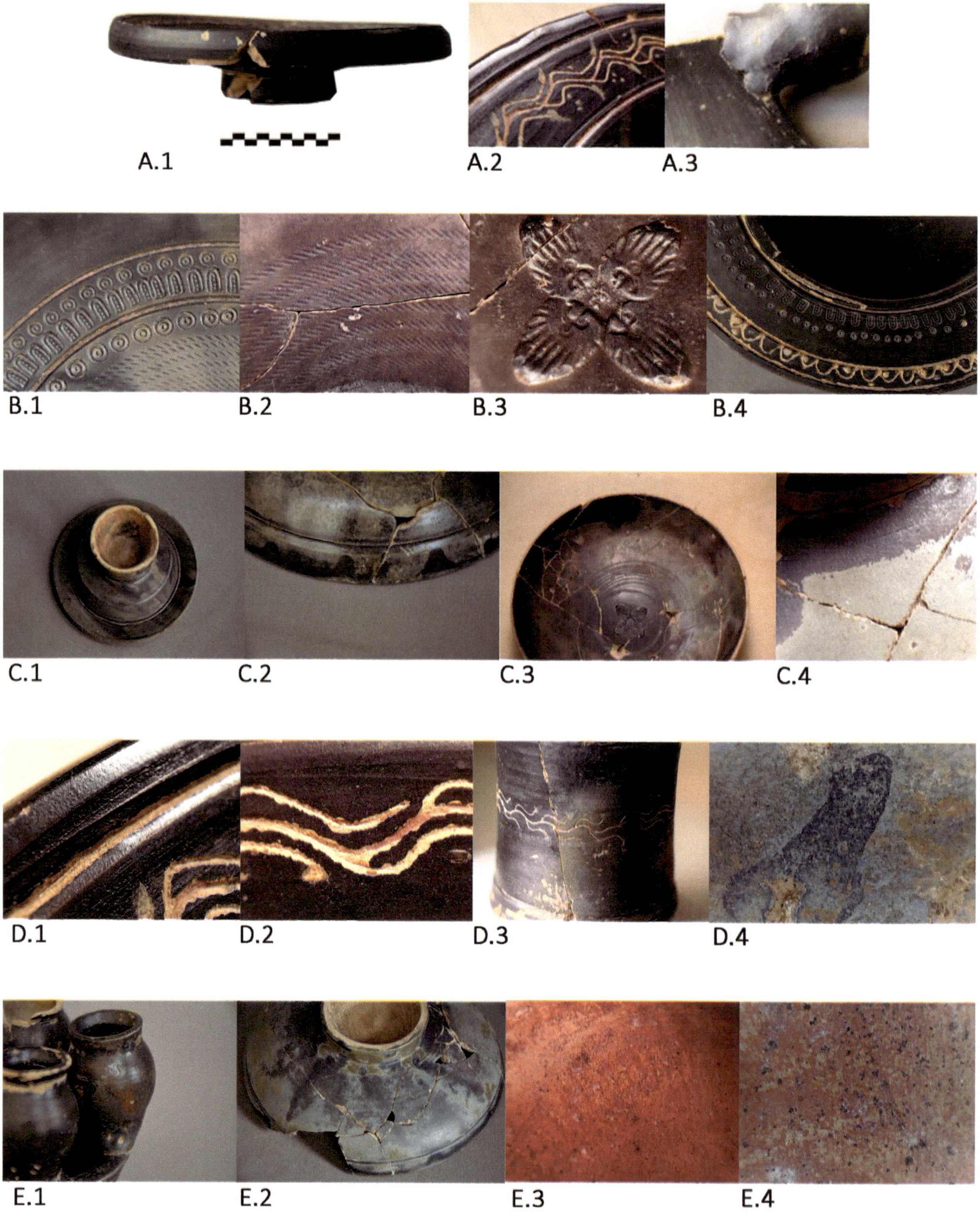

Fig. 3: Overview of observed production-related features and processing signs. © Alessandra De Natale.

Tab. 1: Observations related to the 2nd phase: Impressed decoration. © Alessandra De Natale/Laura Di Siena.

Action of the potter and type of decoration	Operating the potter's wheel	Technological characteristics	Processing signs or imperfections	Cause – observations
1- INCISED DECORATION Partition of decoration schemes with concentric circles.	Potter's wheel operated in slow motion, continuous.	Deep V-shaped groove, made with a thin pointed instrument.	– Eccentric position of the circles; – Uncertain connection and closure of the circle, decentralized and/or staggered.	– Inaccurate centring of the vessel on the potter's wheel – Absence of a stable fulcrum in the pointing of the instrument.
2- IMPRESSED DECORATION Decoration with concentric "rotellature".	Potter's wheel operated in slow motion, continuous.	Shallow grooves, organized in continuous concentric bands and in sequence; made with a rotating instrument equipped with a head with multiple sets of teeth.	– Irregular and/or uneven size of the instrument's teeth; – Regular parallelism between independent grooves; – Involuntary overlapping between sets of grooves.	– Insecure grip of the instrument and poor pressure; – Tool head with teeth organized in multiple and parallel sets; – Incorrect repositioning of the instrument, absence of fulcrum for the hand.
3- STAMPED DECORATION Decoration of the central tondo with a composition of stamps.	Potter's wheel stopped or operated with few rotations.	Use of different stamps to create the composite decoration of the tondo; Sometimes shallow but generally clear and defined stamps.	– Decentralization of the tondo; – Overlapping of stamps; – Unclear details.	– Inexperience in visual centring; – The speed of the decoration action in sequence, lack of expertise or the need to reduce time; – Worn stamps or low pressure.
4- IMPRESSED DECORATION Decoration with concentric series of stamps.	Potter's wheel operated with few rotations.	Concentric series of impressions organized in parallel registers: – Sequence of vertical arcs with central dots; – Simple dice eyes with band and centre point.	– Progressive inclination of the arc impressions; – Staggering of the correspondence between suprajacent impressions and overlapping of impressions.	– Fast execution, unfavourable direction for the hand due to the minimization of rotations of the potter's wheel; – Fast execution, poor skill or need for contraction of time, absence of fulcrum for the hand.

Tab. 2: Observations related to the 3rd phase: Application of black gloss coating. © Alessandra De Natale/Laura Di Siena.

Action of the potter	Processing signs or imperfections	Causes – observations
1- TAKING THE ARTIFACT The potter seizes the vessel to dip it into the paint bath.	– Fingerprints at the gripping point.	– The character of the imprints is not always clear or 'legible', but always made evident by the chromatic difference of the paint in these places or by the accumulations of the paint itself.
2- TOWARDS IMMERSION The potter quickly immerses the vessel in the clayey suspension.	– Placement of the gripping points.	– The grip by the foot reveals an immersion of the artifact 'upside down' in the paint bath. This is always the case for open shapes, while for closed ones the vessel is probably almost always seized by the foot but immersed in paint in an inclined position to prevent the coating from going inside the vessel; – Saving of paint? – A need to leave uncoated parts for firing?
3- SPREADING OF THE COATING The potter takes the vessel out of the paint bath and start to swirl it to distribute the paint on the inner and outer surface.	– Drips and accumulations of paint scattered on the surface with irregular movements; – Uneven surface distribution and frequent dripping.	– Excessive fluidity of the clayey suspension and immediate hydration of the dried product do not leave much time to spread the paint evenly over the entire ceramic body; – Minimize the possibility of localized accumulations on the bottoms; – Reduce the risk of excessive rehydration of the clay and consequent deformations or cracks; – Excessive fluidity of the clayey suspension.

Tab. 2 (continued)

Action of the potter	Processing signs or imperfections	Causes – observations
4- DRYING The potter puts the painted vessel away to allow the paint and the rehydrated vessel to dry.	– The paint drips from the edge towards the foot; – Accumulations of paint on the bottoms of open forms.	– The frequent drips of fluid paint always have a preferential direction that proceeds from the top of the shape downwards and which indicate that the shape has been repositioned on the surface in its 'natural' direction (i.e. on the foot); – The excess paint slowly descends from the vertical walls and accumulates on the bottom before drying.

Tab. 3: Observations related to the 4th phase: Incised and painted decoration. © Alessandra De Natale/Laura Di Siena.

Action of the potter	Operating the potter's wheel	Technological and morphological characteristics	Processing signs or imperfections	Causes – observations
1- RE-INCISING Removal of the paint from the grooves of the concentric circles.	The potter's wheel operated in slow motion, continuous.	The V-shaped profile of the circles has sloping walls covered with paint while the coating just applied on the bottom is removed to recreate the orange/black chromatic play.	– Attachment errors between the beginning and end of the circles remain covered with paint; – The walls of V-shaped groove are covered with paint, while the groove is uncovered at the bottom.	– In the presence of incision errors in the 2nd phase, only the exact part of the circle is rectified; – The second incision removes the paint only at the bottom of the circle, leaving the sloping walls of the V-shaped groove covered with coating.
2- INCISED DECORATION Execution of the incised decorations with vegetable branches, arcs or other types of motifs.	Potter's wheel operated in slow motion, continuous.	Generally curved incisions, in particular that of the branch which includes a continuous main branch on which the short secondary branches, always wavy, are grafted.	– The incision is less deep than the circles performed in the 2nd phase, the profile of the groove is no less clear and clean, the curvilinear course is smooth and fluid but, in some places, it has a stiffness or slight edges; – The connection between the main and secondary branches is soft and legible; – The paint on the edges of the groove remains well adhered and only small fraying can be seen.	– The incision tool is thin, it acts on an already dried and partially hardened material and the line is fluid if the hand is still and works with precision, it is harsher/more angular where the hand is hasty or if the part of the vessel is delicate or the hand unsteady; – First the continuous wavy line of the main branch is executed and then the secondary ones. At the point of attachment between the two, the connection of the incision is soft, a sign that the material still had such a consistency as to allow a good penetration of the instrument and a discreet cleaning of the graffiti; – If the incisions had been performed after firing, the hardness of the sintered paint would not have allowed a such a selective and precise removal of the coating, which would have flaked off on the edges and led to much more jagged edges.

Tab. 3 (continued)

Action of the potter	Operating the potter's wheel	Technological and morphological characteristics	Processing signs or imperfections	Causes – observations
3- PAINTED DECORATION IN WHITE Execution of the over-painted decorations in white on the black-glossed surface.	Potter's wheel operated with few rotations.	Small accessory decorations (ivy leaves, berries, dots, wavy meanders, etc.) made by brush with a very dense and opaque white pigment (probably kaolin clay), applied directly over the black-gloss coating.	– In the places where the painted decoration has vanished (post-depositional period), we often notice the negative impression of the decoration in contrast with the chromatically altered paint; – In some cases, the white pigment changes chromatically to blue in vessels where the paint is also blue-azure.	– If the decoration had been applied to the coating after firing, no negative trace of its shape would remain: this occurs in all the vessels that have a paint coating with colour changes that allow it to be seen (see below); – When those phenomena that alter the black colour of the coating occur during firing, even the white pigment seems to undergo a chromatic alteration towards the azure colour range.

firing is a remarkable transformation caused by the composition of the slip itself in combination with the management of temperature and the reduction phase inside the kiln.[5]

Frequently, the black gloss surfaces of Teanum Ware pottery display a wide variation in hues of the fired slip, with intense contrasts in color or shades that can go from black or reddish brown to blue and gray gradations with areas of intense blue and shiny greenish patches. This phenomenon, which can be further accentuated by nonhomogeneous distribution of the black gloss and its poor coverage, strongly affects the appearance of the final product, making it—at least aesthetically—very different from the traditional Roman black gloss pottery. The interplay of colors, furthermore, is often associated with differentiated gloss effects, contrasting opaque areas, generally in reddish brown or blue, and more compact and bright areas, blue-black or gray. There are also green-yellow areas with a glassy appearance, very oily to the touch and with a greater thickness, which sometimes seem to constitute a distinct layer in their own right. The vessels decorated with mixed techniques are less affected by these chromatic anomalies, which affect more vases with impressed decoration or with no decoration. Even more important, the phenomenon is not limited to the Cappella Marcella assemblage but has also been observed among the material from the western necropolis at the Casilina Est service area[6] and, above all, among the finds from Teanum Sidicinum, especially among those from the Loc. Loreto sanctuary.[7] It is perhaps safe to assume that the chromatic anomalies are not the results of depositional effects of the vessels in the ground, but rather a combination of the mineralogical composition of the various clay bodies used for the black gloss as well as the conditions and range of temperatures during the firing, in particular those developed during the two phases,of reduction and reoxidization.[8] As demonstrated by numerous scientific studies, the critical element is generally represented by the degree of oxidation of ferrous compositions present in the initial mixture: if reddish-brown colors result from high presence of hematite exposed to poor or too-short reductive processes, gloss with a bluish appearance can be attributed to high contents of hercynite and magnetite, which can only be formed at higher temperatures, usually higher than 950° C.[9] (Fig. 3C.4, D.3, D.4, E.1, E.2).

5 Maniatis/Aloupi/Stalios (1993); Gliozzo *et al.* (2004); De Vito *et al.* (2014).
6 Bellini/Trigona (2011).
7 Johannowsky (1963).
8 Gliozzo *et al.* (2004); De Vito *et al.* (2014).
9 De Vito *et al.* (2014) 211.

Coloration is a central feature of Teanum Ware production as it relates both to visual impact and to its presence in vessels from different contexts and areas. Further analyses may establish whether this effect was an intentional choice by its producers to imitate silver-plated metal artifacts.[10]

6th Phase: Post-Firing Application of Red Paint

The last phase of the production cycle is the application of the red pigment after the firing on the parts that had previously received no decoration: the external base of the foot and the incised grooves. We have only two types of evidence for this post firing decoration:

- Transparency, bright tone, and material delicacy of the red pigment: this layer of paint is very has a red/coral-pink color, is transparent, and spread in a very thin layer on the ceramic surface. Its appearance and the touch suggest that the application may have been done with an organic binder, perhaps protein, lending it this sheen. Furthermore, it is extremely delicate, which causes some organic solvents to weaken or to detach from the substrate under minimal mechanical stress; these are typical characteristics of a pigment that is not fixed by firing but dispersed in a binder.
- The so-called black spots only present in this pigment; those spots can be small alteration points of oxides and manganese hydroxides that are frequently formed after the deposition phase on substrates rich in iron oxides (red pigment) or rich in organic material (the binder of the paint; Fig. 3D.2, E.3, E.4).[11]

Conclusions

While our observations on the manufacturing traces have elucidated various stages of the production cycle of the Teanum Ware pottery from Loc. Cappella Marcella, many aspects of the pottery production remain unclear, some of which will require targeted archaeometric investigations in the future. However, with the pottery from Loc. Cappella Marcella it has become clear that we deal with an assemblage which differs in execution and appearance not only from traditional Roman black gloss production but also from Teanum Ware assemblages from other Sidicinian sites, such as Loc. Gradavola.

Bibliography

Bandini/Diana/Guidi (1989): Giovanna Bandini, Silvio Diana and Giolj Guidi, "Rimozione di macchie nere da reperti ceramici, vitrei e ossei", in: *Estratto dal Notiziario dell'Enea/Energia e Innovazione* 35, 1–10.

Bellini/Murro (2019): Giovanna Rita Bellini and Giovanni Murro, "Paesaggi agrari di Aquinum. Resti e contesti nella trasformazione diacronica del territorio", in: *Lazio e Sabrina* 12, Rome, 157–166.

Bellini/Trigona (2011): Giovanna Rita Bellini and Simon Luca Trigona, "La necropoli occidentale di Aquinum: IV sec. a.C. – IV sec. d.C.", in: *Lazio e Sabina* 7, 479–492.

Ceraudo/Murro (2018): Giuseppe Ceraudo and Giovanni Murro, *Aquinum. Guida ai monumenti e all'area archeologica*, Foggia.

Cottier-Angeli/Duboscq/Harari (1997): David Cottier-Angeli, Bertrand Duboscq and Maurizio Harari, "La couleur de l'argent. Une enquête archéométrique autour des poteries à placage", in: *Antike Kunst* 40, 124–133.

De Vito *et al.* (2014): Caterina De Vito, Laura Medeghini, Silvano Mignardi, Diletta Orlandi, Lorenzo Nigro, Federica Spagnoli, Pier Paolo Lottici and Danilo Bersani, "Technological Fingerprints of Black-Gloss Ware from Motya (Western Sicily, Italy)", in: *Applied Clay Science* 88–89, 202–213.

Gabrici (1910): Ettore Gabrici, "Necropoli di età ellenistica a Teano dei Sidicini", in: *Monumenti antichi, Accademia Nazionale dei Lincei* XX, 5–152.

Gliozzo *et al.* (2004): Elisabetta Gliozzo, Ian W. Kirman, Emanuel Pantos and Isabella Turbanti Memmi, "Black Gloss Pottery: Production Sites and Technology in Northern Etruria. Part II: Gloss Technology", in: *Archaeometry* 46, 227–246.

Johannowsky (1963): Werner Johannowsky, "Relazione preliminare sugli scavi di Teano", in: *Bollettino d'arte* XLVIII, 131–165.

Maniatis/Aloupi/Stalios (1993): Yannis Maniatis, Eleni Aloupi and A.D. Stalios, "New Evidence for the Nature of the Attic Black Gloss", in: *Archaeometry* 35, 23–34.

10 Massa Pairault (1985) 77; Cottier-Angeli/Duboscq/Harari (1997) 124; Michetti (2003) 17.

11 Bandini/Diana/Guidi (1989).

Manzini (2013): Ilaria Manzini, "La ceramica a vernice nera di Teano: nuovi dati sulle caratteristiche della produzione locale", in: Gloria Olcese (ed.), *Immensa Aequora Workshop - Ricerche archeologiche, archeometriche e informatiche per la ricostruzione dell'economia e dei commerci nel bacino occidentale del Mediterraneo (metà IV sec. a.C.-I sec. d.C.)* (Atti del Convegno, Roma, 24–26 gennaio 2011), Rome, 201–208.

Massa Pairault (1985): Françoise Héléne Massa Pairault, *Recherches sur l'art et l'artisanat étrusco-italiques à l'époque hellénistique*, Rome.

Michetti (2003): Laura Maria Michetti, "Le ceramiche argentate e a rilievo in Etruria nella prima età ellenistica", in: *Monumenti Antichi, Accademia Nazionale dei Lincei* VIII, Rome.

Morel (1965): Jean Paul Morel, *Céramique à vernis noir du Forum romain et du Palatin, Mélanges d'archéologie et d'histoire*, Supplément 3, Paris.

Morel (1985): Jean Paul Morel, "La ceramica e altre merci d'accompagno da e per Roma in età repubblicana," in: Rolando Bussi and Vittorio Vandelli (eds.), *Misurare la Terra. Centuriazione e Coloni nel Mondo Romano. Città, Agricoltura, Commercio. Materiali Da Roma e dal Suburbio*, Modena, 172–179.

Pedroni (2001): Luigi Pedroni, *Ceramica calena a vernice nera. Produzione e diffusione*, Città di Castello.

Section II: **Populating the Ancient Pottery Industry**

Eleni Hasaki

8 Timing Euxitheos and Euphronios: Energetics and the Scale of Production in the Athenian Potters' Quarters

Dedicated to Toby Schreiber (1928–2023)

Abstract: Scholars have utilized various proxies to estimate the organization and size of workforce employed in the workshops of figured pottery, a niche sector of the entire ceramic industry. In this study an energetics approach is used to estimate the time Athenian potters and painters invested in forming and decorating major types of shapes. The focus piece for this approach is the red-figured Sarpedon krater by the potter Euxitheos and the painter Euphronios, produced in Archaic Athens (ca. 515 B.C.). Drawing on evidence from ethnoarchaeological data to semi-structured interviews with active vase-painters in Greece and Italy, I estimated a total time for potting, painting, and firing this labor-intensive vessel. Energetics invite us to carefully examine all steps included in the potting and painting processes and to appreciate the embodied knowledge of potter and painter in terms of forming techniques, tools for decoration, slip preparation, positioning of the vessel on and off the wheel, and the careful monitoring of firing. Required drying times are of crucial importance, accounting for more than half of the production time. Workshop energetics can shed light on the organization of production, shared specialized knowledge of the workforce, the transfer of knowledge in apprenticeships, and supply-and-demand estimates, so that we can formulate a more balanced view of this sector of the ancient economy.

Introduction

Scholars of Athenian decorated ceramics have been successful in identifying specific hands and styles, but they still struggle to estimate the overall scale of this industry, which was undoubtedly a niche industry, alongside workshops specializing in gloss ceramics, architectural terracottas, plainwares, rooftiles, and pipelines. Scholar's individual specialties and vantage points have guided the discussion toward survival rates, attribution rates, the demographics of ancient cities and their ceramic needs, the spatial footprint of workshops, kiln capacity, and economics and competitive advantage.[1] In this paper I propose that a calibrated approach to all these angles can augment the strengths of each while also limiting their disproportionate impact on the whole.

To the existing mix of approaches I add the energetics and individual worker output, by providing an estimate of time investment for potting and painting Athenian figure-decorated ceramics. For this exercise I focus on one of the largest red-figured vessels from Athens, the red-figured Sarpedon krater by Euxitheos and Euphronios. I developed my data through observations and structured interviews with practicing potters and painters who produce exact replicas of ancient vases using techniques similar to those in antiquity. After developing a detailed step-by-step energetics approach

1 Survival rate: Stissi (2016), (2020); attribution rate and demographics of potters' quarters: Sapirstein (2013), (2014), (2020); demographics of ancient cities and their needs of ceramics: Bentz (2020); Chapman (2022); workshop spatial organization: Hasaki (2011), (2021); Stissi (2012); economics and competitive advantage: Acton (2014).

Note: My deep appreciation to the masters of potting and painting: Thanassis Katsaras (Eleusis) who generously shared his expertise, and his mother Barbara Katsara, Dima Romanciuc (Ancient Corinth), Roberto Paolini and Mario Patarino (both at Cerveteri). My gratitude to Roberto Paolini who kindly explained to me the subtleties of timing and project management and to Yannis Nakas who worked tirelessly with me on drawing the potting and painting sequences in Tables 1 and 2. I also express my sincere gratitude to Alan May, Bryan Burns, Kathleen Lynch, Ben Russell, Alan Shapiro, and Mark Stansbury O'Donnell for their advice and support. Financial support for this project was provided by the College of Humanities, Faculty Research Grant, the College of Social and Behavioral Sciences, Faculty Small Grant, and the University of Arizona Provost's Author Support Fund. I welcomed the initiatives that Marco Serino enthusiastically developed at the University of Arizona during the outgoing phase of the A.G.A.T.H.O.C.L.E.S. project, funded through a Marie Sklodowska-Curie Global Postdoctoral Fellowship. Finally, my sincere thanks to the School of Anthropology (Director: Diane Austin) and the Department of Religious Studies and Classics at the University of Arizona (Head: Karen Seat) which supported the A.G.A.T.H.O.C.L.E.S. project and my supervision of its outgoing phase in several important ways.

https://doi.org/10.1515/9783111189635-008

for an individual artist, I proceed to estimate the seasonal and annual output of a workshop in Archaic Athens. Finally, I turn to the consumer's perspective to explore how we can build an equilibrium of producer's supply with consumer's demand.

Chaînes opératoires, Energetics and Ancient Communities of Practice

The *chaînes opératoires* (manufacturing sequences) have been long established for a variety of crafts, and informative charts and drawings take the reader through the extraction of a block to the production of a sarcophagus (Fig. 1a). M. Korres, the current director of the Acropolis Restoration, outlines the fictional journey of a column capital for the Parthenon from the quarry to the Acropolis in a deeply personal style narrated by a mason working on the Acropolis;[2] we follow how it was quarried and shaped almost to almost its final dimensions at the Penteli quarries before being transported through the Athenian countryside and refined on the Acropolis, only to be found flawed (because of internal fissure) and then abandoned. M. Korres estimates two months from the quarrying of the original block to the discovery of its flaws. His illustrated journey is based on his expertise working with both the ancient monument and the modern sculptors who were making new pieces to complete the original fragmentary pieces. Moving to studies of energetics in Roman sculpture B. Russell and his collaborators have undertaken several interesting studies, including study of the energetics of shipwrecks that highlights how many person-hours are lost when a cargo of semi-worked sculptural pieces sinks.[3]

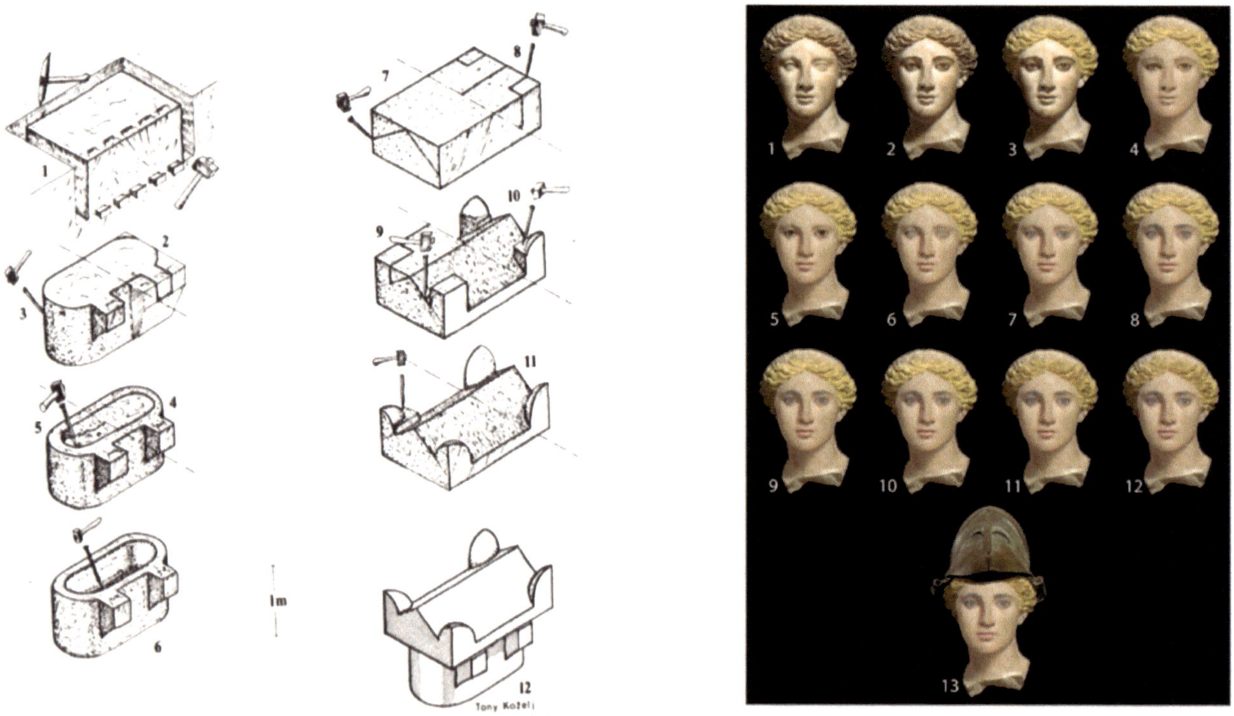

(a) (b)

Fig. 1: (a) The *chaîne opératoire* of a sarcophagus and its lid. After Kozelj (1988) pl. 10; **(b)** The 13-stage sequences of coloring the Treu Head (London, British Museum, inv. no. 1884,0617.1). After Verri/Opper/Lazzarini (2014) 174, fig. 17.

2 Korres (2001). The book was originally published in Greek in 1995.

3 Russell/Leidwanger (2020); for a critical discussion of labor figures for Roman stone working, see Barker/Russell (2012); and general discussion of labor economics and Roman stone working in Russell (2014).

The scope of sculptural energetics has broadened recently to include the important but often overlooked lengthy process of painting the carved sculpture. In recent years, a careful reconstruction of the steps involved in the painting of marble statues highlighted the complexity and skill of statue painters:[4] a research team who tried to reconstruct the coloring process for the Treu Head identified at least 13 stages, starting with the application of a transparent sealant to the marble head, and then continuing to the complex work of depicting the eyes and balancing skin tones (Fig. 1b). The transformation is slow, requiring painterly skills as well as the ability to achieve the right composition of pigments. The authors did not calculate time, but this detailed presentation of each step greatly complements earlier, noteworthy, efforts to juxtapose a colored reconstruction next to the original statue.[5]

Energetics in archaeology is the methodology for translating the process of completing an artifact/structure/project into its person/hours.[6] In other words, it inserts time into the *chaîne opératoire* and divides the total time according to average workdays of individuals and teams. It appeared first as architectural energetics, which "seeks to quantify past architectural remains in terms of the labor force involved in construction projects, or labor 'cost' ".[7] Its first applications tended to be large-scale undertakings, so that community involvement and the wide ranges of colossal structures could be more easily appreciated in crew schedules. For prehistoric societies, architectural energetics also served as an index of social stratification, power hierarchy, and complexity. For later periods and for monumental projects, its use expresses in concrete terms the sheer quantity of raw materials, complexity of construction, craft specialization, and size of the workforce, skilled and unskilled alike.

Architectural energetics had been successfully applied in Classical Archaeology since the late 1990s to estimate the labor time and the size of the crew for building the largest baths in Rome (the Baths of Caracalla), considering the time required, for example, to quarry stones, lay the foundation, erect the building, plaster the walls, and apply any revetment.[8] It entered Greek archaeology with estimating the labor required and community involvement for constructing the palaces and elite houses on Minoan Crete, the labor-intensive chamber and tholos tombs of Mycenaean Greece, and Cyclopean fortifications.[9] Another application treats Archaic domestic architecture in Sicily, where digital renderings helped estimate the total surfaces that need to be built.[10] More recently, scholars have employed this approach to estimate labor for rooftiles, from Bronze Age to Archaic-period structures.[11]

Specifically for pottery making, the step-by-step illustrations by Toby Schreiber of about 30 Athenian shapes, along with her detailed comments, have been instrumental for archaeologists to notice seams between different parts in composite shapes, or adjustments of the shape's profile. Some shapes were illustrated with as few as 3–4 stages (e.g., fish plate, alabastron) while the construction of others required as many as 14 (e.g., amphora, column krater). For the calyx krater, our focus shape in this study, Schreiber illustrates two sequences: one for a krater shorter than 40 cm and one for a krater taller than 40 cm, the difference being that the body of the shorter krater can be thrown from one piece, while the taller krater requires two separate thrown parts for its body, as the untempered clay cannot be pulled up too high without collapsing.[12] As an accomplished potter herself (with a decorative repertoire including mostly outlines of land and sea animals in relief), Schreiber emphasized the virtuosity of the potter, concluding each entry with the line "The Krater then awaited the painter's brush."[13]

Pallecchi also proposed a timetable of production from clay quarrying to firing in the Albinia kilns of the Roman Republican Roman period. The 20-day timetable to produce 150–200 amphorae may be a bit short, but it is a worthy attempt to reconstruct the sequence and spatial patterning of a workshop (Fig. 2a).[14] More recently, Sapirstein applied energetics to the

4 Verri/Opper/Lazzarini (2014).
5 Brinkman/Koch-Brinkman/Piening (2017).
6 The seminal article for architectural energetics is Abrams (1989).
7 Abrams/McCurdy (2019) 3.
8 DeLaine (1997); (2015); McCurdy/Abrams (2019).
9 Devolder (2012), (2013) for Minoan Crete; Fitzsimons (2020) for Mycenaean tombs; Brysbaert *et al.* (2018), for an edited volume of various cases of labor economics of buildings; Harper (2016) and Boswinkel (2021) for Mycenaean monumental architecture; Pakkanen *et al.* (2020) for using photogrammetry to assess labor costs.
10 Lancaster (2019), (2021).
11 For prehistoric rooftiles, see Jazwa (2020), (2024); for the tiles on the Archaic temples in the Corinthia, see Sapirstein (2013) with emphasis on the experimental replication process, (2020) with emphasis on energetics.
12 Schreiber (1999) 138–140.
13 Schreiber (1999) 143.
14 Pallecchi (2008).

Archaic rooftiles from the temple of Apollo at Corinth/Isthmia, building upon his earlier experimental replication project (Fig. 2b). He estimated that a crew of 6 workers could produce all 1500 tiles for the temple in about 6 months. He provided estimates for a worker producing tiles alone versus two workers. He also calculated time for collecting water, fuel, and firing the kiln. His combined methodology of energetics studies and experimental archaeology underscores how intrinsically linked and mutually beneficial these two approaches could be when undertaken in tandem. In a similar vein, an energetics-inspired schedule has been recorded during the building of a replica of the tallest clay statue from Bronze Age Kea, the largest surviving sculpture from the Greek Bronze Age.[15] A preliminary dataset of potting times for Athenian vessels has also been developed based on the throwing data from one potter with no lengthy experience of throwing these shapes, and without timing the waiting intervals and it would be interesting to compare it with the timings of more specialized potters.[16]

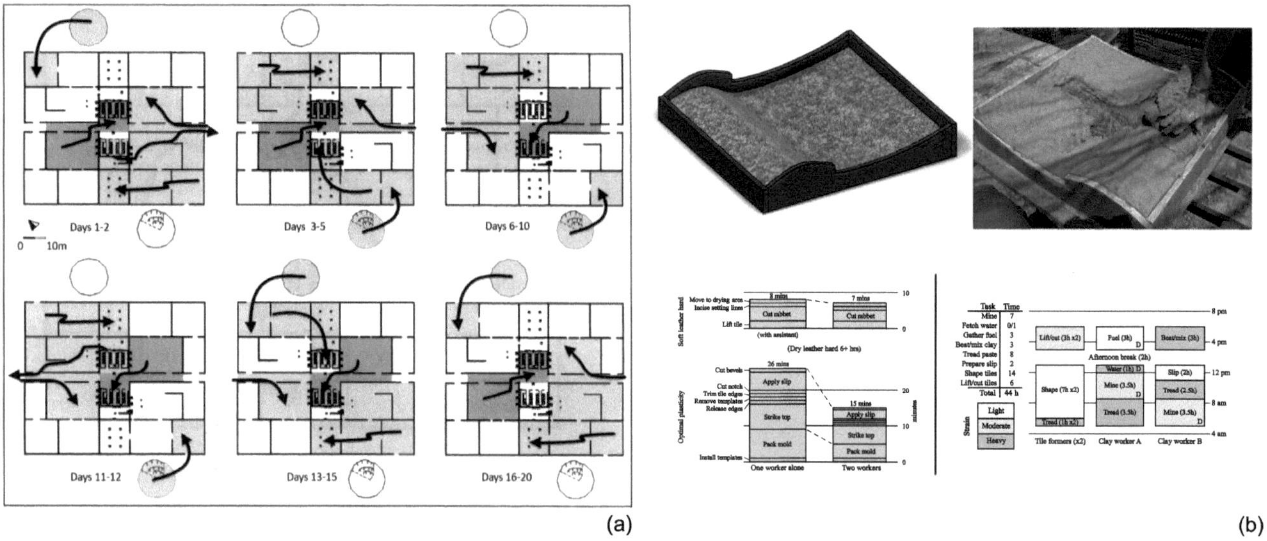

Fig. 2: Selection of sequences, timetables, and energetics in ceramics: **(a)** The production and firing of Roman amphorae in Albinia. After Pallecchi (2008) 335, fig. 6. Courtesy: Silvia Pallechhi; **(b)** The energetics for the composite ceramic tiles of the Archaic Temple of Apollo at Ancient Corinth. After Sapirstein (2020) 114–115, fig. 4.3, tables 4.2–4.3. Courtesy: Philip Sapirstein.

Labor in craft production at times has been addressed indirectly through economic studies of demand and demographics. P. Acton, in his study of different craft industries in Classical Athens, briefly touched upon energetics (without using this term) through output in economic terms; his discussion of supply analysis reflects his work's focus on economics and consumer use of certain items.[17] For ceramics, he used kiln capacity to estimate how many minutes it might have taken an ancient potter to throw different size of vessels to fill a kiln with 2–3 layers of stacked vessels.[18] Combining kiln capacity and consumer use, he states, “the Greek single kiln could have supplied a tall thin vase or ornament to every citizen's household in Athens every 40 days”.[19] Although the specifics of such a calculation are difficult to reconstruct, if we interpret a “tall thin vase” to be a lekythos, we could estimate that 50 workshops in Ancient Athens, each producing 1200 lekythoi (what one could consider a tall thin vase) during 10 firings in a year, could have provided all 60000 households with a lekythos in a year. P. Acton's analysis is very brief and at times does not take into account the required times for assembling, trimming, or drying different parts of a vessel.[20] It does represent, however, a noteworthy attempt to consider in tandem with pot-

15 Hasaki/DeLozier (2018).

16 Neth/Hasaki (2021).

17 Acton (2014) 94, Table 3.2; 316–317, Table A10.

18 Acton (2014) 95. Acton's time estimates rely on a single observation of a potter throwing in Rajasthan, and they can therefore be finetuned when a larger number of sessions is undertaken for replicating a broader variety of Greek shapes. The capacities for different size of vessels also need to be reconsidered with specific shapes and sizes described, as the tall thin shape (which could be a lekythos); an average kiln of 1 m in diameter, stacked on four levels, amount to ca. 120 vessels, not 440 per level as calculated.

19 Acton (2014) 96.

20 Other misconceptions also exist in the overall approach. For example, Greek pots were not regularly bisqued and potters rarely fire half-full

ter's output, kiln capacity, and consumer demand. Overall extrapolations from the production-line perspective need to be adjusted by calculating the consumption rate and use life/breakage rate of a vessel on consumer level.[21]

These detailed manufacturing sequences enhance our appreciation of the ancient artisans, especially in cases where two distinct phases of shaping an object (potting/sculpting) and decorating it (vase-painter/sculpture painter) were involved, even if it was likely that one individual could complete both phases.

The Athenian Red-figured Sarpedon Krater by Euxitheos and Euphronios

In 2022, I launched a new project, the *Energetics of Potters and Painters in Ancient Greece*, with the aim of calculating the time that specialists spent on creating a wide spectrum of vessels. As the first test case I chose to apply the approach to an iconic vase of Archaic Athens–indeed, one of the largest open vessels from Archaic and Classical times. The Sarpedon calyx krater, produced in Athens ca. 515 B.C, was signed by Euxitheos as potter and Euphronios as painter (Fig. 3).[22] It measures 45.7 cm in height and 55.1 cm in rim diameter. At this size, it is one of the tallest and widest shapes in the Greek pottery repertoire, surpassed only by some Panathenaic amphorae, which could reach up to 80 cm.[23] The krater weighs 13 kg[24] and can hold up to 45 L of wine.

Fig. 3: The Sarpedon Krater by Euxitheos and Euphronios: From left to right: profile, side A, handle A palmette, side B, handle B palmette. Courtesy: Cerveteri Archaeological Museum.

In terms of its decorative elements, the krater contains a total of 8 distinct areas, two main scenes (one on each side), two elaborate palmette systems, one above each handle, and a total of four additional motives framing the scene and delineating the foot. The main scenes are framed by a continuous framing motif on the top (sloping palmettes, with the ones on side having less detail in the leaves than the ones on side A), while the lower framing motifs, almost twice as high as the top motif, differs from side A (lotus chain) to side B (sloping palmettes). At the junction of foot and lower part to the body, a band of tongues surrounds the vessel. On side A the slain Trojan hero Sarpedon is carried away by Hypnos and Thanatos; side B shows warriors. The figural zone measures ca. 20 cm in height.

Both makers' signatures appear on side A: the *egraphsen* signature of Euphronios is placed behind the wings of Hypnos, and the Euxitheos' *epoiesen* signature is placed behind the wings of Thanatos. It is one of 19 kraters (17 calyx and 2 volute) prepared by Euphronios, the only krater bearing his signature. We know of Euxitheos only from full or partial signatures on a handful of other vases; he is paired by signatures once with Euphronios and twice with Oltos, on two cups.[25]

The vessel was produced in Athens, and later exported to Etruria. In recent times, it was illegally excavated and subsequently sold for $1 million to the Metropolitan Museum of Art in New York City, where it stayed until 2004.[26] It has since

---

kilns, so kiln capacity estimates need to be adjusted for some sizes: Acton (2014) 95, 316. P. Acton confirmed that his preliminary calculations need to be readjusted by scholars of ceramic studies (pers. comm. April 2024).

21 See below: "Consumption of Calyx Kraters and Uselife of Vessels in Domestic Assemblages".

22 Denoyelle (1990); Wehgartner (1992); Spivey (2020).

23 Bentz (1998).

24 Kind information supplied by Alan Shapiro.

25 For signatures, see Bolmarcich/Muskett 2017; Hurwit (2015). Calyx krater: Paris, Louvre Museum, inv. no. G33 (*Euxi*)*theos epoiesen*; amphora: London, British Museum, inv. no. 1836,0224.124 (Euchsitheos *epoiesen*; *BAPD* 200436); 2 cups by Euxitheos as potter and Oltos as painter: Berlin, Antikensammlung inv. no. F2264 (*BAPD* 200457); cup: New York, Metropolitan Museum of Art /Tarquinia, Museo Nazionale di Tarquinia, inv. no. L2008.1.1 (*BAPD* 200502).

26 For the tumultuous history of the vase, see Watson/Todeschini (2007).

been returned to Italy and is now permanently exhibited in the Cerveteri Archaeological Museum (Cerveteri, Archaeological Museum, inv. no. 145139).

Estimating Potting and Painting Energetics: An Energetics Approach with Modern Practitioners

For the calculations below I conducted a study merging ethnoarchaeology and experimental archaeology with practicing ceramicists. The inspiration of breaking down the stages came from Scheibler's detailed drawings of potting the major Athenian vases in her seminal 1999 study. I interviewed and videotaped four artists with significant experience in the making and decorating of replicas of ancient Greek and Italian vases: the painters Thanassis Katsaras from Eleusis, Dima Romanciuc from Ancient Corinth, as well as Roberto Paolini and the potter Mario Patarino both from Cerveteri.[27] The Greek potters use acrylic paint to decorate the already-fired vessels, while Paolini prepares a slip from local clays in the vicinity of Cerveteri that he uses to decorate unfired vessels. All painters have decorated both replicas of the Sarpedon krater or vessels of similar shape and complexity of composition.

The Sarpedon krater, like all pottery vessels, underwent several extreme transformations during production. Potters and painters had to anticipate such changes in size and weight from the wheel to the kiln and in slip appearance during the firing.[28] At all times both the potter and the painter would flip the vase upside down, hold it on their legs (perhaps with the help of a soft cushion), and handle it in ways that their contemporary customers and modern scholars never do. A large vessel—even one as heavy as 19 kg, as the Sarpedon vessel must have been prior to its firing—was no exception. The daring holding poses of Athenian potters at work represent the reality of the workshop and underscore the tactile nature of a clay vessel.

At the outset, it needs to be emphasized that all the artisans I consulted are full-time specialists, at the peak of their careers. The energetics presented, therefore, are those of masters, not apprentices or even mid-level practitioners.[29] If the same people were interviewed ca. 5–15 years earlier the times would be different, so any longitudinal study should probably use an average time, not their fastest time. Roberto Paolini, whose painting schedule we follow in this study, spent ca. 25 days when he was 30 years old, 30% more time his current time of 18 days, at age 36 (as of 2024).

Moreover, one has to consider energetics for free-designed decoration vs. energetics of replication. When asked to reproduce a replica of the Euphronios krater, the artists emphasized how difficult it is to make a replica of an existing piece. The desire for accuracy severely slows down the process, as they have to consult the original, often replicating oversights or hasty work by the ancient artist. Paolini noted that if he had to make a freehand copy he could finish a krater of similar size and difficulty in approximately two weeks, or 33% less time than he currently needs to finish an exact replica of an ancient vessel. The prolonged process that results when adhering to a prototype is also encountered in sculpting, when it is estimated that making a statue freehand takes about 40% less time than when using plaster casts and copying exactly all elements from the original.[30]

For this energetics exercise, it is calculated that it takes about 21 days to form the krater, give it some time to dry, trim the excess clay, add the handles, and allow it to reach a leather-hard stage for the painter to start applying the slip. Drying times of course would have fluctuated according to the seasons. The active time working on the pot is about 7 hours over 21 days, so a thorough knowledge of drying requirements is of paramount importance. For the painting sequence, the painter needs ca. 18 days with an active time of about 72 hours.

Both in the potting and painting sequences, when the potter and painter have finished the krater-related work they can start or resume work on other pieces—perhaps of lesser difficulty, as the strenuous activity on the krater prevents them from resuming work on equally demanding pieces.

---

27 See https://www.amasis-painter.gr; https://www.pithosriproduzioniceramiche.it (both last access 20.05.2024).

28 For standard references on *the chaîne opératoire* and physical transformations in Greek pottery, see Rice (1987); Noble (1988); Cuomo di Caprio (2017); Hasaki (2019), (2021), (2024).

29 For craft apprenticeship, see Hasaki (2016).

30 Landwehr (2010) for the energetics of creating an exact replica versus a similar copy for a marble statue of an Amazon.

Euxitheos' Potting Schedule

The energetics for potting the Sarpedon krater are provided below (Tab. 1; Fig. 4). It is important to note the dimensions of the krater. With a height of 45 cm it is ranked as high, along with contemporary hydriai and amphorae, but what is even more challenging is its diameter of 55.1 cm, perhaps the widest opening of any Greek vase. That dimension alone suggests that Euxitheos had years of experience in potting, including open vessels, and must have been one of the most accomplished potters in that category of vessels in the Athenian Kerameikos. For the body of a large krater (over 40 cm in height), the potter has to prepare three parts on the wheel, the upper and lower halves of the body and the rim; he needs to let each of them dry for a short period of time and to shave off excess clay from the base. Then the pot will dry for a week or so before it is assembled: coordinating the appropriate drying stage for all parts which are often of different sizes and thickness also requires experience. The potter will score the surfaces where the parts join and aim for an adhesion strong enough that could withstand size and weight reduction before and during firing. In the meantime the potter has to shape handles of the right size and dry them in the correct inclination before attaching them to the krater at the right time while the vase is still upside down; the handles are formed and left to dry to match the drying stage of the krater before they are attached.[31] The total estimated time for making one krater of the exceptional proportions of the Sarpedon krater is approximately three weeks (21 days) with an active time of 7 hours.

Tab. 1: Forming Stages and Duration. © Eleni Hasaki.

<table>
<tr><th>DAY</th><th>Time (hrs)</th><th>Stages</th><th>Activity</th></tr>
<tr><td rowspan="3">1</td><td rowspan="3">2 hrs</td><td>1</td><td>Shaping foot</td></tr>
<tr><td>2</td><td>Shaping lower half of body</td></tr>
<tr><td>3</td><td>Shaping upper half of body</td></tr>
<tr><td>2
3
4
5</td><td>2 hrs</td><td>4</td><td>Drying (short-term)
Connecting two halves of body and foot</td></tr>
<tr><td>6</td><td>1 hr</td><td>5</td><td>Preparing and drying handles</td></tr>
<tr><td>7</td><td>2 hrs</td><td>6</td><td>Attaching handles to body</td></tr>
<tr><td>8–21</td><td></td><td>7</td><td>Drying (long-term)</td></tr>
</table>

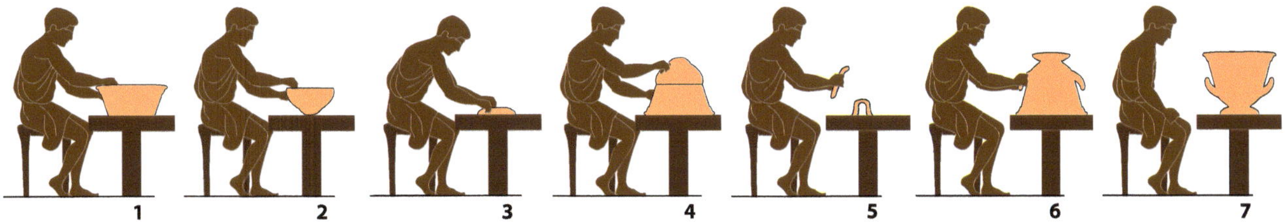

Fig. 4: Stages and times of forming the Sarpedon krater. © Eleni Hasaki. Drawing: Yannis Nakas.

Euphronios' Painting Schedule

Euphronios' painting schedule was equally lengthy and demanding. With the help of several experts, especially Roberto Paolini, a set of 41 stages was reconstructed over 18 days (Tab. 2; Fig. 5). I highlight below some major principles that guide the organization of painting.

31 Schreiber (1999).

Tab. 2: Forming Stages and Duration. © Eleni Hasaki.

DAY	Time (hrs)	Stages	Activity	Side A	Side B
1	3 hrs	1	boundary lines	✓	✓
		2	lowest tongue frieze: sketching and slip	✓	✓
		3	upper palmette: sketching	✓	✓
2	4 hrs	4	upper palmette: relief/contour/background slip	✓	✓
3	5 hrs	5	lower reclining palmettes B: sketching		✓
		6	lower reclining palmettes B: relief/contour/background slip		
4	5 hrs	7	main scene B: sketching		✓
		8	handle palmette B: sketching		✓
		9	main scene B: contour/relief lines		✓
		10	main scene B: contour/relief lines		✓
		11	handle palmette B: relief lines, contour lines		
5	7 hrs	12	lower palmettes A: sketching	✓	
6	6 hrs	13	lower palmettes A: slip	✓	
		14	main scene A and handle palmette A: sketching	✓	
		15	handle palmette A: sketching		
7	4 hrs	16	main scene A: relief lines	✓	
8	4 hrs	17	main scene A: contour lines	✓	
9	4 hrs	18	handle palmette A: relief lines, contour lines	✓	
10	5 hrs	19	blobs for hair/beards/shield *emblemata* B		✓
		20	frontal view layer of handles B: 1st layer of slip main scene B: 1st layer of slip		✓
		21	handle palmette B: 1st layer of slip		✓
11	4 hrs	22	blobs for hair/beards/shield *emblemata* A	✓	
		23	frontal view layer of handles A: 1st layer of slip main scene A: 1st layer of slip	✓	
		24	handle palmette A: 1st layer of slip	✓	
12	2 hrs	25	main scene and handle palmette B: 2nd layer of slip		✓
		26	handle palmette B: 2nd layer of slip		✓
13	2 hrs	27	main scene and handle palmette A: 2nd layer of slip	✓	
		28	handle palmette A: 2nd layer of slip	✓	
14	2 hrs	29	frontal view of handles: 2nd layer of slip	✓	✓
		30	foot: 1st layer of slip	✓	✓
15	2 hrs	31	foot: 2nd layer of slip	✓	✓
		32			
16	2 hrs	33	polishing slipped vessel	✓	✓
		34			
		35	accent slips/labels/signatures	✓	
		36			
17	1 hr	37	interior: 1st layer of slip	✓	✓
		38	rim: 1st layer of slip		✓
18	1 hr	39–40	interior: 2nd layer of slip rim: 2nd layer of slip	✓	✓

Fig. 5: Stages and times of painting the Sarpedon krater. © Eleni Hasaki. Drawing: Yannis Nakas.

Banding as the foundational skill: Placing the vessel upside down on the wheel and painting the boundary zones marks the fundamental skill of all painters, as they divide their painting surface. All painters emphasized this part and how a well-balanced pot shaped by the potter facilitates their work tremendously.

Entire vessel vs. side-specific approaches: For some parts the vessel is approached as a whole, for others according to each side. The painter completes both sides to the same stage: both sides will reach at the same time the sketching stage, then the relief/contour, and then the slipping of the background. The two layers of slip, which need to be applied a day apart for optimal drying, dictate a longer production time. Moreover, for areas that are viewed together and on a similar plane (e.g., the main scene, the frontal view on the handles and the lower foot), the painter must use the same batch of slip, for a homogeneous appearance. Painters know from experience that if they slip those surfaces in two different sessions, they run the risk of the same slip behaving differently. The wet slip always has to be respected and the painter had to carefully avoid any smudging as he or other workers handle the vase. Every person who contributes labor has to be trusted to do his best and not jeopardize the integrity of the vase.

Division of labor: The painter works 3 days on framing motifs, 4 days on side B, which is less complex, 7 days on side A, which is more complicated, and the remaining 4 days on adding accents and slipping the interior and the rim. The active time is about 72 hours, of which almost half the time, 34 hours (or 47%), is spent on side A and handle palmette, 23 hours (32%) for side B and handle palmette, and 15 hours (21%) on work on the entire vessel at the beginning and at the end of the process. The painter works 1.5 times longer on side A than side B. The painter finishes both sides to the same stage before proceeding to the next one, so a pot will look the same all around at any one completed stage. In the past scholars often suggested a labor division for certain vases that promoted a hierarchy of skill where side A is more advanced compared to the less advanced skills exhibited on side B. But for several areas the vase is decorated all at once. Instead of adopting the rigid division of side A and B, perhaps we can divide the different stages (side-specific or not) in two major groups, one requiring more advanced skills and a second one completed with less advanced skills. If we then value the main scene and its framing motives on side A as the most difficult one, and all other drawings as less demanding then the division of time is equitably divided between the more difficult tasks on side A (34 hours) and the tasks relating to side B and to the vessel as a whole inside and outside (15 + 23 = 38 hours). There are of course several scenarios, from one artist doing everything (as our consultants in this study), to a workshop that can successfully and smartly employ two painters who according to their skill and age could contribute in a minor or major way, even equaling half of the entire time required for the completion of a vessel.

Artisans' specialization: The artisans who provided information for this energetics project work alone, with no assistant, so it is not evident which parts of the pot could have been assigned to a less-advanced painter. On the Caputi hydria (Fig. 6e–h) we see a younger worker working on the lowest border motif, with the top finished (presumably by him) but the main scene area left blank; it is impossible to know whether this is because the main scene will be painted later, or because any its details would be illegible in such a small area. I have discussed elsewhere how framing motifs could be assigned to assistants to hone their skills in designing a band, evenly spreading out the motif, practicing elaborate strokes, and perfecting the right amount of slip, holding of tools, and hand position.[32] Still, we must consider the possibility that for some very intricate vessels, such as the Sarpedon krater, the main painter may have undertaken all decoration, main and secondary, himself, to ensure a high standard of execution.

Timing: Some tasks are more physically strenuous than others; sketching the main scene and the handle palmette can take up to six hours, while relief and contour lines, although shorter in duration (ca. 4–5 hours), are more taxing on the painter's positioning relative to the vase and his vision. He cannot, therefore, move on to another task that is similarly labor-intensive. It is also important to note that the painters need to "warm up" their hands before undertaking a difficult task. For example, a painter may do some practice slipping on practice pieces, which also serves to test the consistency of the slip, but these repetitive movements–whether on practice pieces, easier stages, or simpler vases–warm up his painterly hand, so he can switch to the more demanding detailed work on more complicated scenes. Of the 18 days required for the Sarpedon krater, half of them is labor intensive, while the rest require shorter interactions with the pot. The physical strain of some stages on the painter, coupled with long drying intervals for the slip, force the painters to organize (or delegate) their tasks carefully to maximize the daily workflow of the workshop.

32 Hasaki (2016).

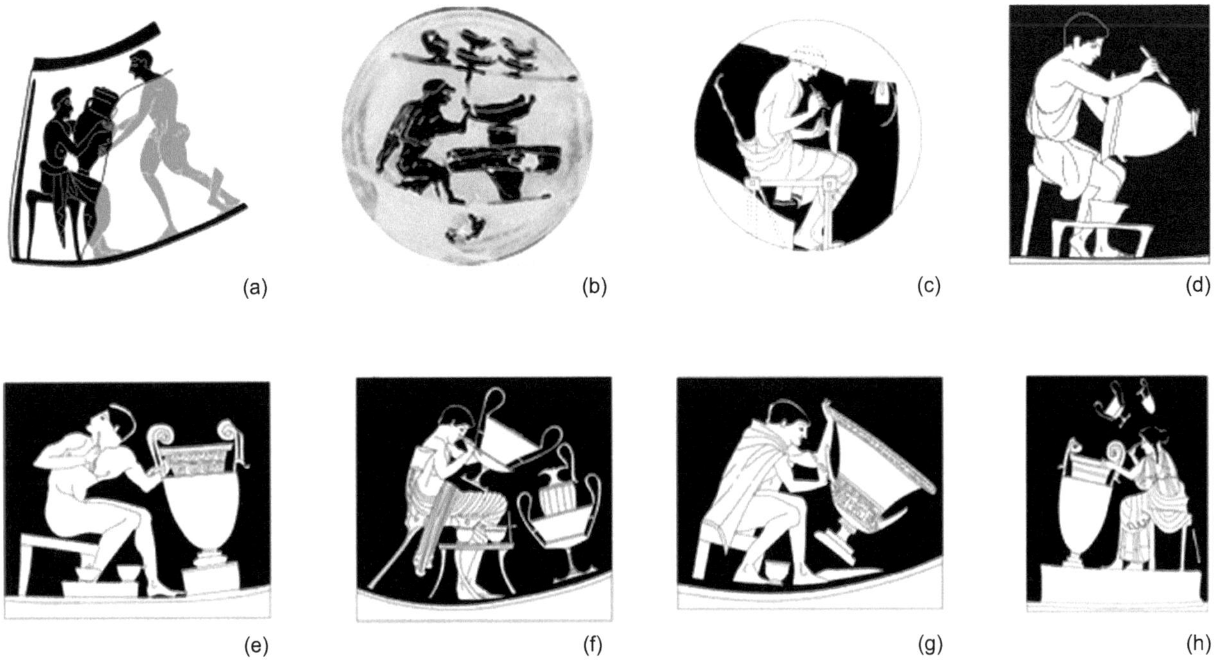

Fig. 6: Athenian vases with potters and painters at work. **(a)** Black-figured hydria (Munich, Museum antiker Glyptothek und Kleinkunst, inv. no. 1847,1125.18); **(b)** Red-figured kylix (London, British Museum, inv. no. B 432); **(c)** Red-figured kylix (Boston, Museum of Fine Arts, inv. no. 01.8073); **(d)** Red-figured krater (Oxford, Ashmolean Museum, inv. no. 526); **(e–h)** Red-figured hydria (Vicenza, Banca Intesa Collection, inv. no. C 278). Drawings: Yannis Nakas.

It is plausible but unlikely that the painting of a vase would be so subdivided as to have one painter for the border motifs, a second for side B, and a third master painter for side A. Of course, anyone painting on a vase, even subsidiary motifs, was skilled enough to be assigned an important vase. The primary testing pieces might have been small plaques of clay, either flat or slightly curved, where the budding painter would have perfected his technique, before being given responsibility for an important vase.[33]

Constant repositioning of the vessel: It is important to note that at the beginning and end the vessel is handled in its entirety, while in the middle phases, focus is directed to each side. The pot is flipped and changes positions several times to receive the full spectrum of its decoration. The first stages are upside down on the wheel for sketching and slip; there is also longer phase when the vessel is off the wheel: the painter holds the vessel closer to him at an oblique position perhaps with the help of a pillow or slanting it against support (see Caputi hydria; Fig. 6e–h); and at the end for the interior slip and for the slipping of the foot the pot goes back on the wheel. The potter uses different consistencies of the same base slip for different purposes: the slip is thicker for the contour and for the plastic 3D blobs on beards, hair, or on the details of the shield *emblemata*. The painter strategically slips last the areas, such as the rim, that are frequently used stabilize the pot as the last to be slipped.

Iconographical evidence of vessel positioning: As we consider the various positions in which modern practitioners hold the vessel while in production, it may be useful to revisit well-known Athenian scenes of potters at work. A number of Athenian depictions of potters at work convey a sense of intimacy with the objects, as the vase is always kept very close to the painter.[34] On a black-figured hydria (attributed to the Leagros Group, 520 B.C.; Fig. 6a), the potter/painter holds an entire amphora, which at that pre-firing point may weigh much as 23–25 kgs, on his thighs, perhaps for polishing. On a black-figured kylix tondo (500 B.C.; Fig. 6b) the potter leans toward the wheel to attach a handle. Again on the tondo of a red-figured kylix (attributed to the Antiphon Painter, 480 B.C.; Fig. 6c) the painter rests the kylix on his thigh and proceeds

33 Hasaki (2021), where I argue for a multi-purpose function of the Penteskouphia pinakes and for similarly-shaped objects in pottery workshops.
34 Stissi (2002); Chatzidimitriou (2003); Williams (2009); Hasaki (2019), (2021), (2024).

to paint it carefully with his brush and with his anatomically correct extending pinky. On the scene depicted on a red-figured bell krater, the painter holds a bell krater on his thigh, and with a thick brush covers with slip its lower part (Fig. 6d; attributed to Komaris Painter, 430 B.C.). Such scenes provide conclusive evidence that potters and the painters had to turn over the vessel on multiple occasions during forming and painting—as well as during drying, to ensure that all parts of the vessel were uniformly dry.

Four separate scenes on the Caputi hydria (460 B.C.; Fig. 6e–h) show four painters in a busy workshop, working on three different types of vessels, but only two of them have been entrusted with working on the main body. From left to right the first seated painter has completed the decorative motif on the neck of a volute krater which stands on a small base; next a second painter is about to start decorating the lower area of a delicate kantharos which is resting on his thighs; the third painter is tilting a calyx krater on its base to paint the simplest petal-and-dot motif around the lower attachment area of the handles. Both top and bottom decorative areas framing the main (undecorated area) have been completed; finally, to the extreme right of the scenes, one seated woman painting the filling ornament on the handles of a volute krater while no other surface on the krater is decorated. The painter of this hydria has depicted the brushes of both painters with great precision as well the correct anatomical details of holding the brush in each position. The painter has also paid great attention to indicate which areas of the krater has been completed, but it is not possible to determine whether the painter faced space limitations and could not depict the central scene which was already painted, or if indeed the central scene was not yet painted.

Overall color scheme: For the duration of the painting process, the painter was dealing only with different shades of orange—from the leather-dry surface to the wet orange slip used for boundaries, fillings of drawn surfaces, and painting solid bands. The relief-line slip and diluted lines also fell within the orange spectrum. The only color deviation came from details highlighted in white clay slip or deep red/purple earth. It is important to consider the appearance of those vases on which no hint of black/dark brown color was visible to the painter prior to placing the pots in the kiln.

Delineating the painting stages of a demanding shape invites us to project the acquisition of skills backwards, and to appreciate how many different skills an established painter like Euphronios would have to master: steadily dividing the vase into distinct zones, painting the framing motifs around the vessel, then focusing on the design of the scene with preliminary drawings, engaging in the actual drawing of the scene using different thicknesses of slips and tools, then carefully filling in the scene's backgrounds with slip, and finally filling all other external and internal surfaces with thick slip. Every time the painter applies the brush(es) to the vessel, he always runs the risk of uneven application of the slip.[35] The constant challenge of attaining the desired thickness everywhere during the painting phase, coupled with the unpredictability of the interaction of the flames with the vase during the firing process, easily explain why most Athenian masterpieces are unevenly fired. The painter's artistic skill, his positioning of the pot, his toolkit, and his project management, as well as the workshop's mastery of slip preparation and its viscosity, constitute a rather extensive spectrum of the painter's knowledge and the workshop's competitive edge.

Labor Investment and Workshop Schedule

Turning to the workshop schedule, I have estimated that a potter can start working on about four calyx kraters a week. They would all need three weeks to be ready for the painter, who could finish them by week 6. By week 7 they will be ready for firing. An average kiln of this period (ca. 1.00–1.20 m in diameter) could hold up to four kraters of the size of the Sarpedon krater. With time allotted for the gradual firing and cooling of the kiln we would easily reach the 7-week mark.[36] It is unlikely, however, that the workshop would risk losing all 4 kraters in a single firing, so it was more prudent to place one or two in different firings during the summer. This estimate assumes a crew of 4–6 people (one potter, one painter, and 2–4 assistants).

This production timeline could produce false impressions, implying that the potter seems to be active in the first part and the painter in the second part. The workshop always has additional tasks to be completed, such as collecting

35 Vases with uneven application of slip layers are ubiquitous. See for example, a black-figured amphora attributed to the Manner of Lydos (New York, Metropolitan Museum of Art, inv. no. 51.11.3; ca. 540 B.C.; H. 55.7 cm). For a recent discussion on how slip may have been produced from a different clay than the clay paste, see Chiavara/Aloupi-Siotis (2016). For the different processes occurring during firing, and the multi-sensory effects, see Balachandran (2019).

36 Hasaki (2021); See also WebAtlas of Greek Ceramic Kilns (atlasgreekkilns.arizona.edu; last access 15.04.2024).

and preparing clay, producing sufficient quantities of slip for painting, moving pots during drying time, or collecting fuel. Moreover, if the workshop maintained a diversified production line with symposium vessels, such as amphorae, psykters, stamnoi, or cups (all within Euphronios' general shape repertoire),[37] its crew would operate in full-time mode producing other symposium shapes, while also shielding the business from abrupt financial strain if the firing of the kraters was not successful.

In the firing schedule, we must question whether a second firing was routine. Based on observations conducted on single fragments, a more complex protocol with two firings has recently been suggested.[38] In the first firing, the pot, without the relief-lines, is going through the tripartite stage of oxidization-reduction-reoxidization (ORO). Then the painter simply adds the relief-lines on the already-fired pot which then undergoes a second, three-stage firing a second, lengthy, firing with the same three-stage (ORO) protocol. A double firing would have incurred additional fuel costs to the workshop and would have involved additional risk for the products being fired. It seems to me that subjecting pots to two firings without a guaranteed success of a second firing would not have been financially or technologically savvy.

Moving from seasonal to annual production, we could hypothesize that a workshop with the appropriate specialists in potting and painting could produce ca. 20–30 kraters and distribute them evenly over 6–12 firings consisting of 1–2 kraters per load.[39] For an annual production of 20 kraters the workshop would need to have 540 kg (or 1,188 lbs) of clay, levigated and ready for throwing on the wheel. Given the considerable investment of material and of effort by both potter and painter, it seems unlikely that the workshop would have stockpiled kraters; instead, it would have carefully calculated the profitability of this expensive line of production. With a required lead time of two months, export traders and other interested buyers must have placed their requests in advance, rather than casually visiting workshops in hopes of immediately acquiring one of these highly prized pieces or having a big selection to choose from. In this way, energetics and operations management invite us to consider whether Athenian workshops could afford to stockpile labor-intensive shapes.[40]

Consumption of Calyx Kraters and Uselife of Vessels in Domestic Assemblages

Studies of consumers' demand for a specific product can help us adjust our production timeline with a careful, profitable timetable for likely sales. The data from the Halieis Houses is helpful when considering the presence of kraters in average houses of the Classical period. Kraters (fineware, but not necessarily calyx kraters or kraters with figured decoration) range from 1–21 in five houses at Halieis, accounting for 1–3% of all ceramics and when considered only within finewares and within pouring vessels, they account for 1–6% of all finewares and for 2–17% of all pouring vessels (Fig. 7).[41] The assemblage from the Classical Dema House in the Athenian countryside includes only two red-figured kraters.[42]

Bentz has undertaken another larger-scale analysis of domestic assemblages, while studying how the production capacity of the Kerameikos at Selinous met the local needs.[43] He conducted a large-scale comparative analysis of household assemblages from Dema, Vari, Halieis, Olynthos, and Himera to estimate averages of pithoi, of large, medium (amphorae, bowls), and small-sized (tableware) vessels, and of very small objects (loomweights, lamps), figurines, and rooftiles.

---

37 Denoyelle (1990).

38 For a comprehensive review of the scholarship on additional firing, see Walton *et al.* (2013); (2015); Luhl *et al.* (2014); Ciancetta *et al.* (2015a), (2015b), (2016); Saunders/Trentelman/Maish (2020); Jones (2021).

39 While admitting that we are entering slippery ground, it may be interesting to estimate the output of large vases during a painter's career. If Euphronios could produce ca. 15–20 kraters per year (if one assumes such a constant demand and considering a lower output during his early career vs. an increased output during his master phase and was active for 20 years, then his total output could have been 300–400 kraters; if 6 of them survive, then we have recovered about 1.5–2% of Euphronios total production. Roberto Paolini has produced ca. 50 kraters and similarly-sized vessels in 20 years (pers. comm. April 2024).

40 In the context of energetics we could consider pricemarks for ancient vases: Johnston (2006); or modern equivalents, such as the $225 price for an amphora of the Berlin Painter, suggested by Stansbury O'Donnell (2011) 162.

41 Ault (2005); Chapman (2022). The frequency of kraters in the domestic assemblages from the Halieis Houses: House 7: 21 kraters; House A: 1 krater; House C: 7 kraters; House D: 12 kraters; House E: 6 kraters.

42 Jones/Sackett/Graham (1962) 88. These include a bell krater and a column krater.

43 Bentz (2020).

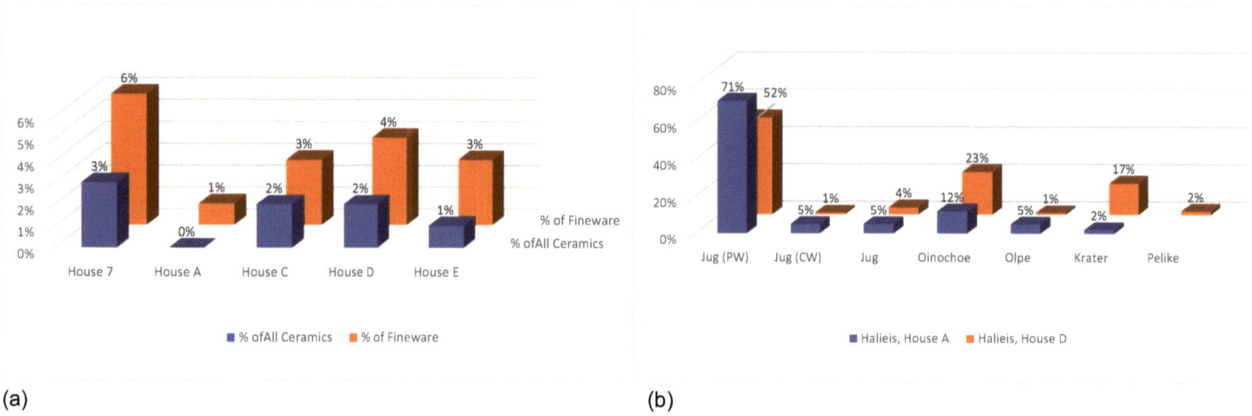

Fig. 7: (a) Percentages of all ceramics and finewares in Halieis Houses; **(b)** Comparison of percentages of pouring vessels in Halieis House A and D. Data from Ault (2006), Chapman (2022). © Eleni Hasaki.

For the Selinous population of ca. 2500 households he estimated annual needs of ca. 68000 ceramic items, and century needs of 6250000.[44]

When scaling up to a settlement's needs, it is useful to look again at how P. Acton attempts to estimate the needs of the Athenian population in clothing, shoes, basic furniture, and ceramics. He divides the estimated 250000 inhabitants of Classical Athens into 65000 households, either according to wealth (wealthy, moderately wealthy, of modest means, and poor) or according to the much broader categories of citizens vs. metics and slaves.[45] Whether it is 10000 shields per year,[46] or 16000 chairs, it is useful to consider the larger picture when discussing scale of output.[47]

For the sake of some general approximation, we can hypothesize that the wealthy and moderately wealthy had 3–5 kraters per household (some only slipped and some with figured decoration), while those of modest means and poor had 1–2 kraters, and most likely slipped and in smaller sizes than the Sarpedon krater (Tab. 3). With a replacement rate of 1–2 for the less wealthy, and about 2–4 kraters for the wealthier, one can estimate that the Athenian production of kraters in 100 years could amount to ca. 290000 kraters of various sizes and decorative techniques.

Tab. 3: Preliminary estimates of Athenian krater consumption in 5th century B.C. using demographic data from Acton (2014). © Eleni Hasaki.

	Level of Wealth				Total
	Wealthy	Moderately Wealthy	Modest Means	Poor	
N of Athenian Households	5000	10000	25000	25000	65000
Kraters per Household	3	2	2	1	
Peak Annual Demand/Use	15000	20000	50000	25000	130000
Replacement Rate/100 years	2	2	2	2	
N of Kraters purchased/100years	5	4	4	3	
N of Kraters purchased/100 years	75000	40000	100000	75000	290000

44 The replacement rates may be better refined as the breakage of 1 cup every decade is perhaps too low.
45 Acton estimates 30000 to 40000 households for citizens and 20000 to 30000 households for immigrants, freed slaves, and slaves.
46 Acton (2014) 142 provides a stimulating calculation of the production output of the shield workshop in Athens and the market demand. If one shield was needed every five years for male citizens every five years and one shield for male slaves and metics every ten years the peak demand in Classical Athens per year would be 10000 shields. If Lysia's well-manned shield workshop could produce 2000 shields a year, it would cover 20% of market share. Either these large armament industries were few with guaranteed sales (similar to how defense industries operate in modern countries), or Lysias' large-scale business remains an outlier and the annual demand of shields was met by the output of many several smaller, workshops.
47 Acton (2014) 302, table A1; 311–316. In another case, P. Acton (2014) 315 calculates the annual needs of basic furniture according to different socio-economic strata: the wealthy would need 500 items annually, the moderately wealthy 6000, those with modest means about 6500 and the poor 3100 chairs.

An average quantity of 290000 for a century translates into 2900 kraters per year produced by Athenian workshops which amounts to a definitely doable output of 40 kraters from ca. 75 workshops (in varying ratios of slipped to decorated and in various sizes) which could have distributed these labor-intensive vessels across several firings in a year. While this admittedly relies on guesswork, it could provide a preliminary reference point that balances the macro perspective of the ceramic needs of a polis and the micro-perspective of workshops' output. Workshops, of course, may have opted to increase their output to meet local or regional demands, targeting local clientele, while others may have chosen to specialize more in exports. At this point it is difficult to estimate what percentage of each Athenian workshop's output was earmarked for export, or if Athenian workshops considered the exported pottery their main source of revenue.[48]

Kraters were heavy, event-specific, and not prone to relocation and therefore the breakage risk of is reduced. Wealthy households which could afford to own several kraters, may have still opted to own only a couple at one time, thus reducing the risk of breakage, even when some kraters were rarely in use.

Turning to heavily used vessels, like water jars to fetch water from a well, K. Lynch in her study of a Late Archaic well in the Athenian Agora estimated that the household broke 4–6 water jars a year (ca. 100 estimated water jars in various fabrics over 25 years of the well's life).[49] Even with no wealth differentiation among households, we could estimate 65000 Athenian households would need 260000 water jars a year (if we take the lower range), almost the equivalent of the centennial output of kraters. Therefore, the ratio of water jars to kraters could be as extreme as 100:1.

These estimates could be further fine-tuned if we had any ancient evidence on uselife for ceramics. Some delicately formed vessels, like kylikes, must have had a short uselife, and they were routinely recycled as ostraca for ostracism.[50] Estimation of the uselife of different Greek ceramic shapes has not been undertaken to date. Ethnoarchaeological studies on domestic assemblages in the Philippines and elsewhere have helped archaeologists develop some reference data on how long shapes of certain height, weight, and function last in a household. From the village

Tab. 4: Select uselife and metric data on Dogon ceramics. After Shott (1996) 466, table 2.

Pots from Dogon	Max. Diam. (cm)	Height (cm)	Weight (kg)	Median Use Life
Cooking				
Dei juno	39.9	31.2	6,586	3.70
Pana dei	29.3	21.5	2,704	2.8
Lajiru dei	28.6	19.7	2,954	4.5
Joni	21.4	15.5	1.395	2
Storage				
Dei pogo	60.5	46.8	21,021	9.70
Dei dam	49.1	38.0	12,385	9.90
Dei no	38.9	29.2	4,845	4
Tegere	38.3	27.3	6,590	3
Ogono neu	22.8	17.8	1,827	7.5
Ogono seme	22.3	20.1	1,818	6.5
Ibonu ogono	13.5	9.8	0,568	3.5

48 The Beazley Archive Pottery Database (last access 08.04.2024) includes ca. 11600 kraters (of various types and in various states of preservation from a wide chronological range. If we make modest calculations that ca. 30 workshops in Athens produced each 20 kraters for annual export we reach a conservative figure of 60000 labor-intensive kraters produced in Athens for export in a century. We need to consider whether the potters' quarter in Athens included export-focused clusters of workshops, or and if we have indeed captured archaeologically a considerable percentage of the larger size vessels.

49 Lynch (2011) 152. No krater was found in that same well that otherwise had a rich variety of household items in plain, black gloss, and decorated vessels, as well as several types of drinking cups.

50 Brenne (2018).

of Dogon in Mali (Tab. 4), a cooking jar (*joni*) 15.5 cm high and 21 cm in maximum diameter, weighing 1.395 kg, has a median use life of 2 years, while a storage pot (*dei pogo*) of 46.8 cm height, 50.5 cm max. diameter, and weighing 21.021 kg, enjoys a median use life of 9.7 years.[51]

Conclusions

In this study, I combined iconographical evidence and ethnoarchaeological information from potters and vase-painters to reconstruct the energetics of potting and decorating the Sarpedon krater, one of the largest decorated vessels of the Classical period, and to shed light on the many skills that the ancient artisans had to master: the timing of each phase in potting and painting, the delicate handling of the vessel in various positions over 2 months, the optimal positioning while working on the vessels, deploying their toolkits, and achieving the right composition of slip for various stages. I contextualized the *chaîne opératoire* for the Sarpedon krater, a vessel that would have weighed at least 19 kgs in its first week of production. I highlighted the accuracy of ancient depictions that show specialists holding vases in similar positions. Studies such as this might be undertaken for other shapes, such as small cups or midsize jugs, to approximate a workshop's production. To estimate the annual productivity and the overall population of the Athenian potters' quarters, we need to keep several factors in balance: for example, time requirements to produce a vessel, workforce size, kiln size, spatial layout of workshop, typical frequencies and replacement rate of a shape in domestic assemblages, and attribution rates from surviving pots.

By considering in tandem supply and demand, we can rephrase our research questions on pottery production not as if it is a timed competition (How many pieces could the potter and painter produce?), but as a savvy business equilibrium of supply and demand (What level of production will bring revenue to the workshop? How many kraters were needed for Athenian consumption so as to ensure revenue without stockpiling?). Thus, we could better reconstruct the survival rate for each shape, instead of using a blanket average for all shapes of various sizes and use life spans. Moreover, we could more accurately gauge the market value of these vessels in their primary and subsequent contexts. If the large vases with figured decoration represent such a major labor investment, we need to reassess the annual productivity rates of these specialized workshops and reevaluate the value of these labor-intensive pots in the local and export markets. It is also important to consider whether diversified shape specialization in ancient workshops provided them with greater flexibility in the organization of production.

Bibliography

Abrams (1989): Elliot M. Abrams, "Architecture and Energy: An Evolutionary Perspective", in: *Archaeological Method and Theory* 1, 47–87.

Abrams/McCurdy (2019): Elliot M. Abrams and Leah McCurdy, "Massive Assumptions and Moundbuilders: The History, Methods, and Relevance of Architectural Energetics", in: McCurdy/Abrams (2019) 1–25.

Acton (2014): Peter Acton, *Poiesis: Manufacturing in Classical Athens*, Oxford.

Ault (2005): Bradley A. Ault, *The Excavations at Ancient Halieis Vol. 2: The Houses, The Organization and Use of Domestic Space*, Bloomington.

Balachandran (2019): Sanchita Balachandran, "Bringing Back the (Ancient) Bodies: The Potters' Sensory Experiences and the Firing of Red, Black and Purple Greek Vases", in: *Arts* 2. https//www.doi.org.10.3390/arts8020080.

Barker/Russell (2012): Simon Barker and Benjamin Russell, "Labour Figures for Roman Stone-Working: Pitfalls and Potential", in: Stefano Camporeale, Hélène Dessales and Antonio Pizzo (eds.), *Arqueología de la Construcción III. Los procesos constructivos en el mundo romano: la economía de las obras*, *Anejos de Archivo Español de Arqueologa* 64, Madrid/Merida, 83–94.

Bentz (1998): Martin Bentz, *Panathenäische Preisamphoren: Eine athenische Vasengattung und ihre Funktion vom 6.-4. Jahrhundert v. Chr.*, *Antike Kunst* Beiheft 18, Basel.

Bentz (2020): Martin Bentz, "Production and Consumption of Ceramics in Selinous: A Quantitative Approach", in: Hasaki/Bentz (2020) 81–96.

51 Shott (1996), (2022) with earlier bibliography. Use life or use expectancy is a topic pursued in New World Archaeology since the 1960s but it has not received much attention in the Mediterranean world.

Bolmarcich/Muskett (2017): Sarah Bolmarchich and Gina Muskett, "'Artists' Signatures on Archaic Greek Vases from Athens", in: Kristen Seaman and Peter Schultz (eds.), *Artists and Artistic Production in Ancient Greece*, Cambridge, 154–176.
Boswinkel (2021): Yannick Boswinkel, *Labouring with Large Stones: A Study into the Investment and Impact of Construction Projects on Mycenaean Communities in Late Bronze Age Greece*, Leiden.
Brenne (2018): Stefan Brenne, *Die Ostraka von Kerameikos*. Kerameikos 20, Wiesbaden.
Brinkman/Koch-Brinkman/Piening (2017): Vinzenz Brinkman, Ulrike Kock-Brinkman and Heinrich Piening, "Ancient Paints and Painting Techniques", in: Vinzenz Brinkman, Renne Dreyfus and Ulrike Koch-Brinkmann (eds.), *Gods in Color: Polychromy in the Ancient World*, San Francisco, 87–97.
Brysbaert *et al.* (2018): Ann Brysbaert, Victor Klinkenberg, Anna Gutierrez-Garcia and M. Irene Vikatou (eds.), *Constructing Monuments Perceiving Monumentality and the Economics of Building. Theoretical and Methodological Approaches to Built Environment*, Leiden.
Chapman (2022): Alice Chapman, *The Duties of a Household Manager. Women's Roles in the Creation and presentation of Social Identity in Classical/Late Classical Greece: An Interpretation of the Domestic and Mortuary Ceramic Assemblages at Halieis*, PhD Dissertation, State University of New York at Buffalo.
Chatzidimitriou (2005): Athena Chatzidimitriou, *Παραστάσεις εργαστηρίων και εμπορίου στην εικονογραφία των αρχαϊκών και κλασικών χρόνων*, Publications of *ArchDelt* 92, Athens.
Chiavara/Aloupi-Siotis (2016): Artemi Chiavara and Eleni Aloupi-Siotis, "The Story of a Soil that Became a Glaze: Chemical and Microscopic Fingerprints on the Attic Vases", in: *Journal of Archaeological Sceince: Reports* 7, 510–518.
Cianchetta *et al.* (2015a): Ilaria Cianchetta, Karen Trentelman, Jeff Maish, David Saunders, Brendan Foran, Marc Walton, Philippe Sciau, Tian Wang, Emeline Pouyet, Marine Cotte, Florian Meirer, Yijin Liu, Piero Pianetta and Apurva Mehta, "Evidence for an Unorthodox Firing Sequence Employed by the Berlin Painter: Deciphering Ancient Ceramic Firing Conditions through High Resolution Material Characterization and Replication", in: *Journal of Analytical Atomic Spectroscopy* 30, 666–676.
Ciancetta *et al.* (2015b): Ilaria Cianchetta, Jeff Maish, David Saunders, Marc Walton, Apurva Mehta, Brendan Foran and Karen Trentelman, "Investigating the Firing Protocol of Athenian Pottery Production: A Raman Study of Replicate and Ancient Sherds", in: *Journal of Raman Microscopy* 46, 996–1002.
Ciancetta *et al.* (2016): Ilaria Cianchetta, Karen Trentelman, Marc Walton, Jeffrey Maish, Apurva Mehta and Brendan Foran, "Reverse Engineering Ancient Greek Ceramics: Morphological and Spectral Characterization of Replicates", in: *Journal of American Ceramic Society* 99, 1792–1801.
Cuomo di Caprio (2017): Ninina Cuomo di Caprio, *Ceramics in Archaeology: From Prehistoric to Medieval Times in Europe and the Mediterranean. Ancient Craftsmanship and Modern Laboratory Techniques*, Rome.
DeLaine (1997): Janet DeLaine, *The Baths of Caracalla. A Study in the Design, Construction, and Economics of Large-Scale Building Projects in Imperial Rome*, *Journal of Roman Archaeology* Supplement 25, Portsmouth.
DeLaine (2015): Janet DeLaine, "The Pantheon Builders: Estimating Manpower for Construction", in: Tod A. Marder and Mark Wilson Jones (eds.), *The Pantheon: From Antiquity to the Present*, Cambridge, 160–192.
Denoyelle (1990): Martine Denoyelle, *Euphronios. Peintre à Athènes au VIe siècle avant J.C.*, Paris.
Devolder (2012): Maud Devolder, "Labour Costs and Neopalatial Architecture. A Study of the Buildings at Klimataria-Manares and Achladia and the Palace at Gournia", in: Diamantis Panagiotopoulos and Ute Günkel-Maschek (eds.), *Minoan Realities: Approaches to Images, Architecture, and Society in the Aegean Bronze Age*, Aegis 5, Leuven, 165–179.
Devolder (2013): Maud Devolder, *Construire en Crète minoenne. Une approche énergétique de l'architecture néopalatiale, Aegaeum* 35, Leuven.
Fitzsimons (2020): Rodney D. Fitzsimons, "Crossing Thresholds and Building States: Labor Investment, Tomb Construction, and Early State Formation in the Bronze Age Argolid", in: Hasaki/Bentz (2020) 17–38.
Jazwa (2020): Kyle Jazwa, "Comparing the Labor Investment and Production of Early and Late Bronze Age Ceramic Roofing Tiles in Mainland Greece", in: Hasaki/Bentz (2020) 3–15.
Jazwa (2024): Kyle Jazwa, *The Tiled-Roof Phenomenon in Early Helladic Greece: Evidence from Zygouries*, *Hesperia* Supplement 53, Princeton N.J.
Harper (2016): Charles R. Harper, *Laboring with the Economics of Mycenaean Architecture: Theories, Methods and Explorations of Mycenaean Architectural Production.* PhD Dissertation, University of Florida.
Hasaki (2011): Eleni Hasaki, "Crafting Spaces: Archaeological, Ethnographic, and Ethnoarchaeological Studies on Spatial Organization in Pottery Workshops in Greece and Tunisia", in: Mark Lawall and John Lund (eds.), *Pottery in the Archaeological Record: Greece and Beyond* (Acts of the International Colloquium held at the Danish and Canadian Institutes in Athens, June 20–22, 2008), Aarhus, 12–28.
Hasaki (2016): Eleni Hasaki, "Craft Apprenticeship in Ancient Greece. Reaching Beyond the Masters", in: Willeke Wendrich (ed.), *Archaeology and Apprenticeship: Body Knowledge, Identity and Communities of Practice*, revised edition, Tucson, 171–202.
Hasaki (2019): Eleni Hasaki, "Potters and their Wheels in Ancient Greece: Skills and Secrets in Communities of Practice", in: Mario Denti and Mathilde Villette (eds.), *Archéologie des espaces artisanaux. Fouiller et comprendre les gestes des potiers*, Rennes, 297–314.
Hasaki (2021): Eleni Hasaki, *Potters at Work in Ancient Corinth: Industry, Religion, and the Penteskouphia Pinakes*, *Hesperia* Supplement 51, Princeton N.J.
Hasaki (2024): Eleni Hasaki, "Handling Greek Vases", in: Nikos Dimakis and Eirene Dimitriadou (eds.), *Καλλίνικος: Honorary Volume for Professor Panos Valavanis*, Athens, 25–36.
Hasaki/Bentz (2020): Eleni Hasaki and Martin Bentz (eds.), *Reconstructing Scales of Production in the Ancient Greek World: Producers, Processes, Products, People* (Panel 3.4, Proceedings of the 19th International Congress of Classical Archaeology, vol. 8, Cologne/Bonn, May 22–26, 2018), Heidelberg, 81–96. https://doi.org/10.11588/propylaeum.639.c9978
Hasaki/DeLozier (2018): Eleni Hasaki and Rachel DeLozier, "Terracotta Statues from Ayia Irini, Kea: An Experimental Replication", in: Erika Angliker and John Tully (eds.), *Cycladic Archaeology and Research: New Approaches and Discoveries*, Oxford, 3–26.
Hurwit (2015): Jeffrey M. Hurwit, *Artists and Signatures in Ancient Greece*, Cambridge.
Johnston (2006): Alan Johnston, *Trademarks on Greek Vases: Addenda*, Warminster.

Jones, R. (2021). "The Decoration and Firing of Ancient Greek Pottery: A Review of Recent Investigations", in *Advances in Archaeomaterials* 2, 67–127.
Jones/Sackett/Graham (1962): John E. Jones, L. Hugh Sackett, Alexander J. Graham, "The Dema House in Attica", in: *Annual of the British School in Athens* 57, 75–114.
Korres (2001): Manolis Korres, *From Penteli to Parthenon*, Los Angeles/Malibu.
Kozelj (1988): Tony Kozelj, "Les carriers des époques grecque, romaine, et byzantine in: J. Clayton Fant (ed.), *Ancient Marble Quarrying and Trade* (Papers from a Colloquium Held at the Annual Meeting of the Archaeological Institute of America, San Antonio, Texas, December 1986), Oxford, 3–80.
Lancaster (2019): Jerrad Lancaster, "To House and to Defend: The Applications of Architectural Energetics to Southeast Archaic Greek Sicily", in: McCurdy/Abrams (2019) 95–113.
Lancaster (2021): Jerrad Lancaster, *Virtually Reconstructing the Past: Estimating Labour Costs through Digital Technologies*, Studia Archaeologica 247, Rome.
Landwehr (2010): Christa Landwehr, "The Baiae Casts and the Uniqueness of Roman Copies", in: Rune Frederiksen and Eckard Marchand (eds.), *Plaster Casts: Making, Collecting, and Displaying from Classical Antiquity to the Present*, Berlin, 35–46.
Lühl *et al.* (2014): Lars Lühl, Bernhard Hesse, Ioanna Mantouvalou, Max Wilke, Samnia Mahlkow, Eleni Aloupi-Sotis and Birgit Kanngiesser, "Confocal XANES and the Attic Black Glaze: The Three-Stage Firing Process through Modern Reproduction", in: *Analytical Chemistry* 86, 6924–6930.
Lynch (2011): Kathleen Lynch, *The Symposium in Context: Pottery from a Late Archaic House near the Athena Agora*, *Hesperia* Supplement 46, Princeton N.J.
McCurdy/Abrams (2019): Leah McCurdy and Elliot M. Abrams, *Architectural Energetics in Archaeology. Analytical Expansions and Global Explorations*, London.
Neth/Hasaki (2021): Brandon Neth and Eleni Hasaki, "The Potter's Wheel in Ancient Greece: Experimental Archaeology and Web Applications for Velocity", in: *Interdisciplinaria Archaeologica Natural Sciences in Antiquity* 2, 115–125.
Noble (1988): Joseph V. Noble, *The Techniques of Painted Attic Pottery*, 2nd revised edition, London.
Pakkanen *et al.* (2020): Jari Pakkanen, Ann Brysbaert, Daniel Turner and Yannick Boswinkel, "Efficient Three-dimensional Field Documentation Methods for Labour Cost Studies: Case Studies from Archaeological and Heritage Contexts", in: *Digital Applications in Archaeology and Cultural Heritage* 17. https://doi.org/10.1016/j.daach.2020.e00141.
Pallecchi (2008): Silvia Pallecchi, "Le fornaci romane di Albinia: Identificazione delle unità funzionali e ricostruzione delle line di produzione", in: Valeria Acconcia and Claudia Rizzitelli (eds.), *Materiali per Populonia* 7, Pisa, 323–338.
Rice (1987): Prudence M. Rice, *Pottery Analysis. A Sourcebook*, Chicago.
Russell (2014); Ben Russell, *The Economics of the Roman Stone Trade*, Oxford.
Russell/Leidwanger (2020): Ben Russell and Justin Leidwanger, "The Energetics of Lost Cargoes: A New Perspective on the Late Antique Marzamemi 2 Wreck," in: *Memoirs of the American Academy in Rome* 65, 194–260.
Sapirstein (2013): Philip Sapirstein, "Painters, Potters and the Scale of the Attic Vase-Painting Industry", in: *American Journal of Archaeology* 117, 493–510.
Sapirstein (2014): Philip Sapirstein, "Demographics and Productivity in the Ancient Athenian Pottery Industry", in: John H. Oakley (ed.), *Athenian Potters and Painters* III, Oxford. 175–186.
Sapirstein (2020): Philip Sapirstein, "Productivity of Athenian Vase-painters and Workshops," in: Hasaki/Bentz (2020) 81–96.
Sapirstein (2021): Philip Sapirstein, "Labour Organisation and Energetics of Early Archaic Architecture in Korinthos", in: Catherine Cooper (ed.), *New Approaches to Ancient Material Culture in the Greek and Roman World: 21st-century Methods and Classical Antiquity*, Leiden, 103–123.
Saunders/Trentelman/Maish (2020): David Saunders, Karen Trentelman and Jon Maish, "Collaborative Investigations into the Production of Athenian Pottery", in: Catherine Cooper (ed.), *New Approaches to Ancient Material Culture in the Greek and Roman World: 21st-century Methods and Classical Antiquity*, Leiden, 124–135.
Schreiber (1999): Toby Schreiber, *Athenian Vase Construction. A Potter's* Analysis, Malibu.
Shott (1996): Michael J. Shott, "Mortal Pots: On Use Life and Vessel Size in the Formation of Archaeological Assemblages", in: *American Antiquity* 61, 463–482.
Shott (2022): Michael J. Shott, "Inferring Use-Life Mean and Distribution: A Pottery Ethnoarchaeological Case Study from Michoacàn", in: *American Antiquity* 87, 794–815.
Spivey (2020): Nigel Spivey, *The Sarpedon Krater: The Life and Afterlife of a Greek Vase*, Chicago.
Stansbury O'Donnell (2011): Mark Stansbury O'Donnell, *Looking at Greek Art*, Cambridge.
Stissi (2002): Vladimir Stissi, *Pottery to the People: The Production, Distribution, and Consumption of Decorated Pottery in the Greek World in the Archaic Period (650–480 BC)*, PhD Dissertation, University of Amsterdam.
Stissi (2012): Vladimir Stissi, "Giving the Kerameikos a Context: Ancient Greek Potters' Quarters as Part of the Polis Space, Economy, and Society", in: Arianna Esposito and Giorgos Sanidas (eds.), *Quartiers artisanaux en Grèce ancienne: Une perspective méditerranéenne*, Villeneuve d'Ascq, 201–230.
Stissi (2016): Vladimir Stissi, "Minor Artisans, Major Impact?", in: Norbert Eschbach and Stefan Schmidt (eds.), *Töpfer-Maler-Werkstatt: Zuschreibung in der griechischen Vasenmalerei und die Organisation antiker Keramikproduction*, *CVA* Beiheft Deutschland 7, Munich, 47–53.
Stissi (2020): Vladimir Stissi, "From Counting Pots to Counting People: Assessing the Scale of Athenian Pottery Production and Its Impact on Workshop Staff", in: Hasaki/Bentz (2020) 97–108.
Verri/Opper/Lazzarini 2014: Giovanni Verri, Thorsten Opper, Lorenzo Lazzarini, "'In picturae modum variata circumlitio: The Reconstruction of a Roman Ideal Female Head (Treu Head)", in: Paolo Liverani and Ulderico Santamaria (eds.), *Diversamente bianco: la policromia della scultura romana*, Rome, 149–183.
Walton *et al.* (2013): Marc Walton, Karen Trentelman, Marvin Cummins, Guilia Poretti, Jeff Maish, David Saunders, Brendan Foran, Miles Brodie and Apurva Mehta, "Material Evidence for Multiple Firings of Ancient Athenian Red-Figure Pottery", in: *Journal of the American Ceramic Society* 96, 2031–2035.

Walton *et al.* (2015): Mark Walton, Karen Trentelman, Ilaria Cianchetta, Jeffrey Maish, David Saunders, Brendan Flora and Apurva Mehta, "Zn in Athenian Black Gloss Ceramic Slips: A Trace Element Marker for Fabrication Technology", in: *Journal of the American Ceramic Society* 98, 430–436.
Watson/Todeschini (2007): Peter Watson and Cecilia Todeschini, *The Medici Conspiracy: The Illicit Journey of Looted Antiquities. From Italy's Tomb Raiders to the World's Greatest Museums*, New York.
Wehgartner (1992): Irmas Wehgartner, *Euphronios und Seine Zeit*, Berlin.
Williams (2009): Dyfri Williams. "Picturing Potters and Painters", in: John H. Oakley and Olga Palagia (eds.), *Athenian Potters and Painters* II, Oxford, 306–317.

Philip Sapirstein

9 Reconstructing the Scale of Athenian Vase-Painting Workshops

Abstract: This paper builds upon recent quantitative research on the Athenian pottery industry which demonstrated that the number of vases attributed to a painter is strongly correlated to that individual's years of activity. It considers the broader question of workshop staffing in light of this new approach to painters. This is a more difficult problem since we possess fewer means of extracting individualized information about potters and their lesser-skilled co-workers. It reviews the evidence for studying potters – some of whom have been identified through *poietes*-inscriptions and shape analysis – alongside other methods for modelling workshop production, including ethnographic analogy, energetics, and economic analysis. A generalized model is developed and applied to the workshops around the potter Nikosthenes and the Penthesilea Painter, two of the larger operations of the Late Archaic and Classical periods. The Nikosthenes workshop appears to have employed a smaller group of artisans than previously believed, and the Penthesilean group, while certainly larger, is too ill-defined to determine the precise scale of operations.

Introduction

This paper extends my recent statistical investigations of the productivity of individual painters in 6th and 5th-century B.C. Athens to the structure of workshops. Moving from painters to workshops raises new problems, since we possess fewer sources of information about potters and workspaces than we do for painters. Their relationships to painters may nonetheless be examined through proxies such as inscriptions and potting styles, and broader generalizations are possible through ethnographic analogy. By focusing on two unusually well-defined groups of painters, the paper explores new methods for visualizing associations within workshops and estimating the size of their staff.

Approaches to Athenian Workshops

A substantial body of scholarship has examined Attic workshops producing figural vases. The monumental works of J. D. Beazley (*ABV, ARV², Paralipomena*) identify thousands of artisans, supplying an unusually specific resource for assessing the production of figural vases from the 6th and 5th centuries B.C. that continues to be expanded. His catalogues are organized by painting style and vessel form but generally avoid assigning workshops, a category reserved for a wide range of stylistic connections that were not specific to an individual painter.[1]

The ensuing scholarship on workshops integrates classifications by painting style and potting technique with the names of artisans written on many vases.[2] Our knowledge of painters is richer than for potters, who are more difficult to identify individually, and the locales where painting took place are often unknown. Excavations in Athens have encountered some workshops or kiln debris from figural pottery production, although the paucity of archaeological material published to date discourages generalization about their organization.[3] Several vases depict potters in action, providing fascinating albeit potentially fanciful vignettes showing the interactions among various artisans and clients.[4]

Our understanding of these corpora has been enriched by comparison to ethnographic accounts of pottery production. Potters still relying on premodern technologies were documented by ethnographers at many Mediterranean centres, and others now catering to the art market have revived some traditional methods.[5] Many aspects of Attic figural pottery,

1 Beazley (1944), (1986) did expand on his occasional comments about workshops in catalogues and earlier articles about individual painters.
2 Notably Webster (1972); Scheibler (1986); Stissi (2002) 123–144; Lüdorf (2010); Williams (2016), (2017); with citations to more of this voluminous body of scholarship.
3 Stissi (2002) 35–73; Sanidas (2013) 69–102, 194–197.
4 Beazley (1944) 1–21; Stissi (2002) 75–95; Williams (2016); Hasaki (2021).
5 Scheibler (1984); Hasaki (2011); Sapirstein (2013) 499–500.

https://doi.org/10.1515/9783111189635-009

however, are unparalleled in the recent past. Its potters were able to form shapes resembling metalwork that are unsustainable for most Mediterranean clays; the elaborate shapes and paints likewise exceed the technical sophistication of most wares studied by ethnographers.[6] The figural painting might be compared to majolica and other painted wares from recent centuries, but the analogy is limited by the technological divide between black gloss and lead-glazed pottery, the delicate linework of Attic painting, and the scarcity of detailed modern ethnographies.[7] Chinese porcelain producers and imitators established in Europe in the 18th and 19th centuries supply a better-documented, albeit technologically distant comparison.

Workshop Staffing

A central question raised here is where the vase-painting industry fell on the spectrum of production intensity. The high degree of technical sophistication rules out occasional or even specialized household-based production. The evident concentration in the Kerameikos of hundreds of producers, including migrants identifiable by foreign names, immediately suggests a "nucleated workshop industry" on Peacock's scale, which is toward the upper end of the spectrum in premodern Mediterranean ethnographies.[8] Such workshops are relatively small, employing one or two potters with a few assistants for preparing clay, fuel, and related lower-skilled tasks. In face of rising demand, the individual workshops in nucleated industries do not typically take on more staff, but rather workers from existing shops or immigrants establish new enterprises.

The largest premodern operations are "manufactories", where production stages are separated among larger crews in order to intensify output, such as by parallel production or improving efficiencies through specialization and managerial oversight.[9] Any vase-painting operation employing a dozen or more workers should qualify as such, although even the massive Roman production of Arretine and Gaulish sigillata may not have developed into true manufactories.[10] Limited working space, a lack of capital for horizontal integration (acquiring neighbouring workshops), and the costs of management might discourage the emergence of manufactories from within a nucleated pottery industry, but these structures had certainly developed in other sectors of the economy in Classical Athens.

An extensive body of scholarship has examined the scale of Attic pottery workshops. Two groups in particular have been identified as potential manufactories. The first is around Nikosthenes, whose name is signed on many vases from the latter 6th century B.C.[11] The workshop appears to have been one of the largest producing black-figured pottery, and it also participated in the experimentation with decorative techniques that culminated in the introduction of red-figured painting. The sheer number of associated hands at first encouraged scholars to restore a full-time staff of 10–20, dominated by painters alongside multiple potters and assistants.[12] Because many of the hands may not have been active very long, however, more recent studies have preferred to restore about eight workers total.[13]

The second workshop is named after the Penthesilea Painter, a group active more than four decades in the middle-5th century B.C.[14] Many of the painters decorated different sides of the same cup, an unusual practice implying a division of labour consistent with a manufactory. The Penthesileans have been less thoroughly studied than the Nikosthenic workshop, but they are usually assumed to have been the largest workshop making figural pottery in Classical Athens.[15]

6 Replication experiments have provided important insights into such methods: e.g., Schreiber (1999).

7 Examples include the 15th century manuscript by Piccolpasso and the study of faience producers in Fes: Bel (1918) 187–298; Lightbown/Caiger-Smith (1980).

8 Peacock (1982); Scheibler (1984); Stissi (2002) 170–176. Some Attic producers must have surpassed the threshold for Peacock's "individual workshop" scale of production at some time during the Archaic period; nucleated industries are located near favourable resources, engage in full-time production, and make highly standardized, recognizable products intended for export beyond the local market.

9 The concept was rooted in preindustrial Europe, where the largest 19th century potteries like Wedgwood fell short of a factory by employing fewer people, at smaller facilities, and not much relying on machine power: Peacock (1982) 43–46. Increases in staff numbers and workspaces in latter 18th century potteries signalled the emergence of manufactories from workshop-based industries.

10 Peacock (1982) 114–128; his case for pottery manufactories is questioned by Fülle (1997) and Wilson (2008).

11 Tosto (1999).

12 E.g., Webster (1972) 41; Immerwahr (1984) 345–347; Valavanis (1997) 86–87.

13 Scheibler (1984) 132–134; Tosto (1999) 193–200; Stissi (2002) 135–136; Osborne (2004) 91–92; Hasaki (2011) 27; see below for general literature on the workshop.

14 Beazley (1944) 111–112; Osborne (2004) 78–86; Williams (2017) 165–166.

15 Webster (1972) 15–17; Scheibler (1983) 116; (1986) 803.

If these two groups indeed employed the most artisans in the Kerameikos, then their sizes are critical to understanding the organization of Athenian pottery production in general. What is presently lacking is a consistent methodology by which to estimate how many painters they employed simultaneously. The current project derives such a methodology from my previous investigations of the productivity of painters.[16] Since that work focused exclusively on individuals, it must be revisited and expanded to accommodate the ambiguities introduced by workshops.

Although definitions in the foregoing literature varies, the "workshop" examined here is the physical space where potters, painters, and assistants prepared clay, shaped, decorated, and dried vessels.[17] This unit of analysis is useful in that it can be compared directly to excavated remains of ancient pottery workshops as well as ethnographic studies. However, connecting Beazley's vase-painters to a particular workspace is more challenging.

Attribution Rates of Painters

Although the great majority of figural Attic vases are unsigned, common elements in the figural painting link vases together into discrete clusters which J. D. Beazley identified as a Painter, Group, Class, and related subcategories. The most specific form of attribution, the Painter, implies the output of an individual. Due to ambiguities it can be unclear whether a cluster of related vases represents the entire surviving *oeuvre* of an individual artisan.[18] As a result, I will employ the term "hand" to refer to a group of works linked in the modern scholarship by painting style or signatures, and "artisan" when referring to the ancient individual who produced pottery in Athens.

The earlier research was inspired by R. M. Cook, who had suggested that if a typical hand were responsible for 3–4 extant vases per year, one could divide the total number of extant vases to estimate the population employed at painting in any given year.[19] R. M. Cook did not reveal exactly how he arrived at this estimate, but one method is to tally every vase connected with a well-known hand and divide by the approximate number of years spanned by those vases, resulting in a ratio that I define as the annual "attribution rate" of a hand. I reviewed the 36 hands with at least 150 attributions – which is enough for their career lengths to be estimated with reasonable precision – and found a remarkably narrow range of attribution rates close to 8.2 vases annually (Fig. 1).[20] Artisans may have worked at relatively consistent speeds during the 6th and 5th centuries B.C. since they were in competition to decorate and sell a fairly homogenous product.[21]

For comparison, I included additional hands with fewer known vases but reasonably well established chronologies.[22] Some were as productive annually as the prolific hands, but many others had markedly lower outputs despite having been active for decades (Fig. 1, hands below 150 attributions).

We might explain this falloff in various ways. For example, artisans lacking a distinctive and consistent style, or hands subjected to less study, might have incomplete lists of attributions. Some painters might have invested more time in each composition than their peers: massive showpieces like the François Vase certainly took Kleitias much longer to paint than the normal fare.[23] Rather than integrating measures of complexity for individual vases,[24] I pursued another possible explanation: painters who spent parts of their year at other tasks would reduce their attribution rate compared to those who painted full time.[25]

16 Sapirstein (2013), (2014).

17 See Hasaki (2011) 13–16; Sanidas (2013) 18–24. Workshops might also have fired vessels on site, but shared kilns outside the production area are also conceivable.

18 Vases "near" a Painter are possibly by another artisan, while those in the "manner", "circle", and similar are probably imitators. J. D. Beazley also left open the possibility that some painters represented different modes or periods of the same artisan: Sapirstein (2013) 494, 503–504, online 1–3. On potential merges and individual style: Pevnick (2016); Whitley (2018).

19 Cook (1959) 119.

20 On the methods, see Sapirstein (2013) online 2–9. The cut-off was at least 150 likely attributions published by 2011, which eliminates short-lived artisans whose career lengths cannot be determined with much confidence. Only Makron (Fig. 1) is an outlier with an exceptionally high output, perhaps due to collection practices.

21 Despite the individuality sought through Beazleyan attribution, it is difficult to distinguish the work of contemporary painters without careful scrutiny of minor variations in the painting.

22 Sapirstein (2013) figs. 2, 4.

23 *BAPD* 300000; compare to his small stand painted with a *gorgoneion* (*BAPD* 300735). Citations are to vase numbers in the Beazley Archive Pottery Database: https://www.carc.ox.ac.uk (last access 01.03.2024).

24 This approach would be welcome, but I did not attempt it for the nearly 13000 vessels treated in the study.

25 Sapirstein (2013) 497–501; building on remarks by Osborne (2004) 88–93.

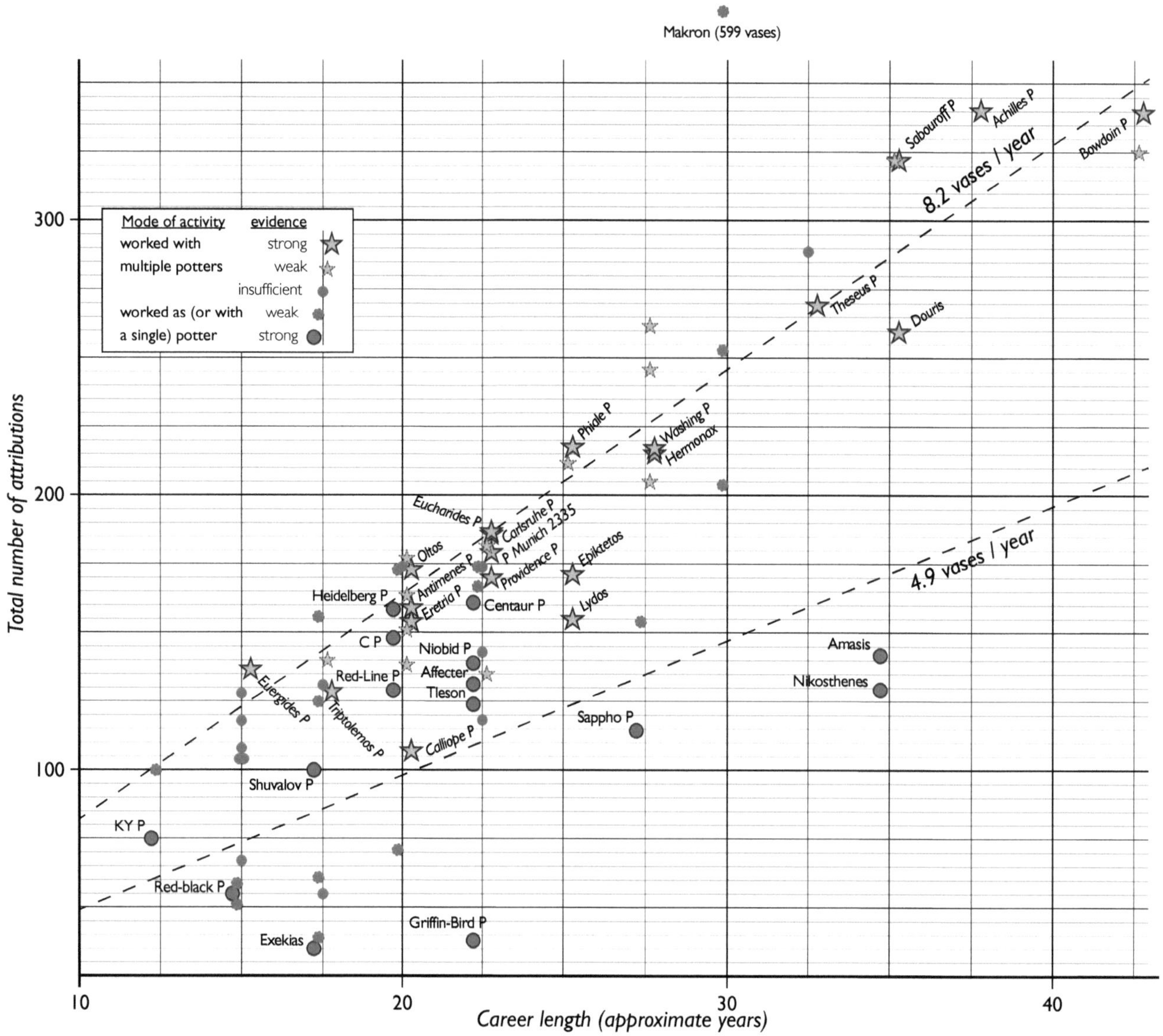

Fig. 1: Productivity of long-lived Attic hands and mode of activity. The approximate career length is plotted against the number of attributed vases. Hands associated with multiple potters are likely specialists (stars), hands correlated to a single potting style might have been potter-painters (outlined circles), and others are ambiguous (dots). The dashed lines plot the typical attribution rate for a hand operating as a specialist (8.2) and potter-painter (4.9). © Philip Sapirstein.

Modes of Activity

We can reframe the latter proposal by mode of activity. The hands with the highest outputs have similar productivities, as if approaching a maximum that could be achieved by a full-time "specialist painter". We would expect lesser productivity of an 'occasional painter', representing a potentially diverse cohort, such as apprentices who never engaged in full-time painting, or an assistant who painted a few vases when a specialist was unavailable. Their work would be difficult to identify through attribution, since they would have painted too infrequently to develop a distinctive style of their own, while borrowing heavily from original compositions by more experienced artisans. Some of this work may underlie the numerous attributions J. D. Beazley made to the "circle" of a Painter, or the many hands with only a few attributions.[26] Such artisans may have worked slowly relative to more experienced specialists, further depressing their productivity.

One mode of occasional activity, however, should be identifiable: the 'potter-painter', who both threw and painted vessels routinely while developing skills at both. We should expect such hands to have a lower attribution rate than a full-time specialist, since this rate tracks their painting output alone. Apprentices and others who failed to develop a recognizable body of work were already filtered out from the attribution-rate study due to their low outputs, but we might detect potter-painters with long careers among the moderately prolific hands. Comparing these two modes yields a postulate: attribution rates approaching 8.2 vases per year are typical of specialist painters, and lower rates are expected for potter-painters.

A hypothesis – whether these two modes in fact existed – can be tested with the availability of evidence independent of the attribution rate. Mobility is an important first consideration. Based on ethnographic analogy, we may predict that Athenian potters would have been relatively immobile compared to painters. On the one hand, potters require room to process clay, shape pots at the turntable, apply handles and decoration, and store the vessels before firing, all of which would tie them to a single workshop space.[27] On the other, painters increased the value of a vase through an investment of skilled labour; since they employed a lightweight toolkit consisting primarily of brushes, they would have had more freedom to rove about various workshops. We may predict that potter-painters would not only exhibit lower annual attribution rates, but also that their painting would be limited to vases they had thrown, or at least were shaped in the same workshop space.

The vases attributed to each hand in this study supply two independent indications for mode of activity: artisanal inscriptions, and the consistency of the potting. Many vases name their painter with ἔγραφσεν or similar, while others employ ἐποίησεν to denote its "maker" – which in its most literal reading would name the potter, and more generally the workshop.[28] Several specialist painters, such as Euphronios or Oltos, are immediately recognizable from inscriptions stating that they "painted" a vase, along with inscriptions naming various other individuals as the *poietes* who made them. Many other hands whose vases lack signatures can still be recognized as specialists from the diversity in the potting within their attributed vases, indicating that they too worked with multiple potters and workshops.

While roving painters are thus identifiable, the mode of activity is ambiguous when a hand is consistently associated with a single *poietes* or style of potting. Exekias, who signed with both verbs for making and drawing on some of his vases, identifies as a potter-painter.[29] Most other signatories who might been potter-painters identify only as *poietes*, as if painting were implicit in making the vase. A hand with multiple references to the same *poietes* might have belonged to a potter-painter, but it could also have been a painter who happened to have worked with a single potter for most of her/his life – a new mode of activity that might be dubbed the "sedentary specialist painter". A few who signed with ἔγραψεν can be distinguished from a separately named *poietes*, such as Makron, who seems to have worked with Hieron throughout his career. In the majority of cases where a hand is strongly correlated with a single class of potting, however, we know only the name of the *poietes*, and more often all the vases are anonymous. Such clusters could originate from either a potter-painter, or a stable collaboration of a painter with one potter.

Recalling that specialist painters would have been more mobile than potters, we should note that sedentary specialists would still face pressures to develop new collaborations from time to time. For example, imagine a specialist painter working for her parents who eventually outlived them, and subsequently painted vases thrown by another relative who took over potting. The workshop would have remained the same, although this hypothetical hand would still exhibit linkages with two different potters. In other words, the only sedentary painting specialists that would completely elude detection must have had shorter careers than their affiliated potter. We should expect, then, that most of the hands with consistent potting throughout their entire *oeuvre* were potter-painters.

26 J. D. Beazley identified many low-output hands. Of the 637 with at least two vases, more than 60% have no more than 10 works, whereas the top 60 most prolific hands (the top 6%) painted half of these vases: Sapirstein (2013) 503–505; (2020) 86–87, fig. 5.

27 Hasaki (2011); Sapirstein (2013) 500.

28 Beazley (1944) 25–36; Stissi (2002) 97–121; Hurwit (2015) 71–96; Williams (2016) 64. Some have argued that a shop-owner is indicated, although ποιεῖν typically indicates some physical agency in making the object.

29 Douris and Epiktetos, whose affiliations with multiple potters and signatures strongly indicate a specialization in painting, also occasionally signed as *poietes*: *BAPD* 200630; 205305; 205321. Dual-mode signatures are both unusual and concentrated on Acropolis votives: Sapirstein (2013) 500, n. 28.

Figure 1 plots the application of this model, including hands whose vase profiles have yet to be adequately assessed to assign to either mode of activity (pale dots), and those with some ambiguities (dashed outlines). With few exceptions, the hands who roved among multiple potters (stars) and thus were likely painting specialists also have high attribution rates. The hands with consistent potter-work (outlined circles) have more diverse productivities, though some high-output examples like the Heidelberg Painter could have been sedentary specialists. Most importantly, the latter type is exclusively found among the least-productive hands: attribution rates below five suggest a potter-painter or occasional painter.[30] This pattern holds across the graph and is reiterated if we focus on hands with similar career lengths: the starred hands have more attributed vases than their peers in each column (Fig. 1). Since such a distribution is extraordinarily unlikely to have come about by chance, we may conclude that diversity in potting is strongly correlated with high attribution rates.[31] The hypothesis should be accepted, while leaving it open to future refinements that might accommodate other modes of activity.

From Hands to Workshops

The painter-productivity study has important bearings on the analysis of Athenian painters and workshops. First, it confirms that Beazleyan attributions to the better-studied hands represent the productivity of actual ancient painters. The list of vases attributed to the hands in this study appear not only to comprise the surviving, published works of an individual artisan, but also are a statistically representative of that artisan's original output.[32] Thus we can assert that the ca. 35 vases attributed to Exekias over a two-decade career is meaningfully different than the 173 given to the Swing Painter over roughly the same period. The Swing Painter worked at near the maximum rate for specialists, whereas Exekias appears to have painted not just less hurriedly than his peers, but less frequently.

Second, potter-painters with relatively low attribution rates were the norm for much of the 6th century. Along the with increasing popularity of Athenian pottery on the export market, and the advent of new techniques – in particular red-figured painting – full-time painters became far more numerous toward the end of the century and are most frequently encountered throughout the 5th century.[33] As such, we should expect different workshop structures depending on the period. The early dominance of potter-painters already argues against an emergence in Archaic Athens of manufactories, whose divisions of labour are conceptually distant from cases like Exekias, who spent time away from the wheel to hand-paint his own vases. The subsequent rise of specialist painters is a sign of intensification, though it is insufficient to determine whether manufactories yet existed.[34]

Third, vases attributed to a workshop's painters are a valid indication of its actual productivity. However, the painters it employed would have variable productivities depending on whether they were specialist or occasional painters, something which must be factored into any quantitative analysis. Likewise, the potential mobility of specialist painters complicates the tallying of workshops, since we cannot assume the vases attributed to any roving hand were all made there. Only the fraction of a specialist hand's attributions which are connected directly by signatures or potting to the workshop should be counted.

My painter-productivity study focused on the most recognizable hands, but when examining workshops we must also include the impact of occasional painters. Such "trial" paintings would likely belong to various members of the shop,

30 The Calliope Painter has an unusually low productivity for a hand affiliated with more than one potter, but the attribution rate is still greater than five.

31 Sapirstein (2013) 501, note 31.

32 The preservation ratio, or how many vases have survived relative to the total output in Athens, is a separate problem and will be excluded here. On the validation of this method: Sapirstein (2013) online 7–9; (2020) 83–90.

33 Sapirstein (2013) 504–506, fig. 7.

34 Larger facilities may not have been necessary, since the total number of painters simultaneously employed in the Kerameikos probably never exceeded 100: Sapirstein (2013) 507–508, pl. 1. This describes only full-time positions that due to overturn would be occupied by many more individual artisans. Stissi (2016, 2020) has recently argued for a much larger industry in the thousands but bases the assertion on statistically invalid arguments: Sapirstein (2020) 84–90. Nonetheless, the total employment in the Kerameikos would have been moderately larger if we allow for more slow-paced work by inexperienced occasional painters, and these figures exclude what was surely a substantial population making plain black gloss and coarse wares.

such as the potter's relatives, or enslaved and hired assistants. Thus, we might expect that an operation dominated by one experienced painter would still generate some minor works by less-skilled, slower assistants.

We might predict similar variations in the potting. Because any potter-painter would leave the turntable unoccupied while painting, an associate might have been invited to throw pots during downtimes. An apprentice might practice throwing vases, yielding some good enough for the market, and a workshop with a sufficiently skilled assistant could produce a supplemental line of saleable vases whose potting would differ from that of the lead potter. In workshops centred on a full-time potter employing specialist painters, a large enough facility might install a second turntable, which could be open to internal staff as well as roving potters commissioned to throw a supplemental series of vessels.

Nikosthenes

The Nikosthenic workshop is a good case to explore these issues due to the thorough documentation by V. Tosto (1999).[35] The cluster is unusually well defined by the frequency with which the name Nikosthenes appears as *poietes*. Altogether 80 of his 149 signatures appear on Nikosthenic amphora, one of the distinctive forms that increase the workshop's recognizability. Most are black-figure, but 13 signed vases employ special techniques or red-figured painting. The latter belonged to specialists Oltos, Epiktetos, and the Nikosthenes Painter, all of whom also worked with the *poietes* Pamphaios. Through these linkages, Pamphaios has been tentatively identified as a junior potter and eventual successor to Nikosthenes.[36] Tosto separates a large number of entities within the signed vases, with at least 33 hands and 3 potters tied to the *poietes* (excluding Pamphaios), yet he arrives at a relatively small crew of about eight by assuming a high rate of turnover.[37]

In order to assess this quantification, we should visualize these relationships. The new *Social Networks of Athenian Potters* project has made promising advances by linking potters, painters, and other entities in a database for network analysis.[38] The centrality measures and linkages have already yielded new insights into the connections across the Kerameikos, although additional data are pertinent to examining interactions within a workshop.

One solution for visualizing the connections within the Nikosthenic group is presented (Fig. 2).[39] The links to the two main *poietes* are distinguished for those based on signatures (solid) or potting-work (dashed), and the number of links emphasized by varying line widths. It is immediately apparent that the most prominent hands like Lydos, Epiktetos, or the Epeleios Painter decorated very few vases in the workshop, and most of these hands have evidence of working elsewhere (starred names) through attributions with other *poietes*, painting on different styles of potter-work, or activity before or after Nikothenes' time (545–510 B.C.). Painter N is widely believed to be Nikosthenes himself, due to the tight correlation of this hand with the signatures. Other hands in the "Affiliated entities" on Fig. 2 might have been his work as well. Although rightly kept separate due to uncertainties, any vases from these lists which were actually painted by Nikosthenes would exaggerate the apparent complexity of his associations.[40]

The "unattributed" vases are problematic. Tosto distinguishes more than a dozen examples as *not* Painter N, based on differences in the figures or ornament. Some of these vases could belong to occasional painters drawn from the shop staff, but we also cannot assume that any of these isolated pieces represent the whole surviving output of an actual artisan. Each combination of these minor hands with only a single work to the same artisan would eliminate one of the 33 individual hands separated by V. Tosto within the Nikosthenic workshop. Another problem with the unattributed list is the presence of 21 vases with Nikosthenes' *poietes*-inscription but in plain black gloss, which obviously would have taken much less time to paint than those with figures.[41]

35 See the dissertation by Tafe (2022) for more recent literature, commentary on the workshop, and appraisal of its affiliated hands.

36 Immerwahr (1984); Tosto (1999) 193–195.

37 Tosto (1999) 198–200.

38 Hasaki/Cline (2020).

39 Sources are the catalogues in Tosto (1999) supplemented by *ABV*; Immerwahr (1984); *CVA* France 41 (Nassi Malagardis and Athena Tsingarida, *CVA* France 41 – *Louvre 27. Céramique attique archaïque, goblets "mastoids" à figures noires et rouges*, 103–104); Tafe (2022); shape attributions for Pamphaios are drawn from Bloesch (1940). Tafe counts the *poietes* Theozotos (*BAPD* 301977) in the workshop, although the style is closer to Amasis: Tonglet (2015). Including or excluding the two vessels in question has little impact on the workshop productivity as a whole.

40 On merges within the Group of Louvre F 166, see: Nassi Malagardis and Athena Tsingarida, *CVA* France 41 – *Louvre 27. Céramique attique archaïque, goblets "mastoids" à figures noires et rouges*, 47–49.

41 A typical composition on a figural vase has ten or more people as well as the floral borders: e.g., *BAPD* 302752. The unusual dual-*poietes* in-

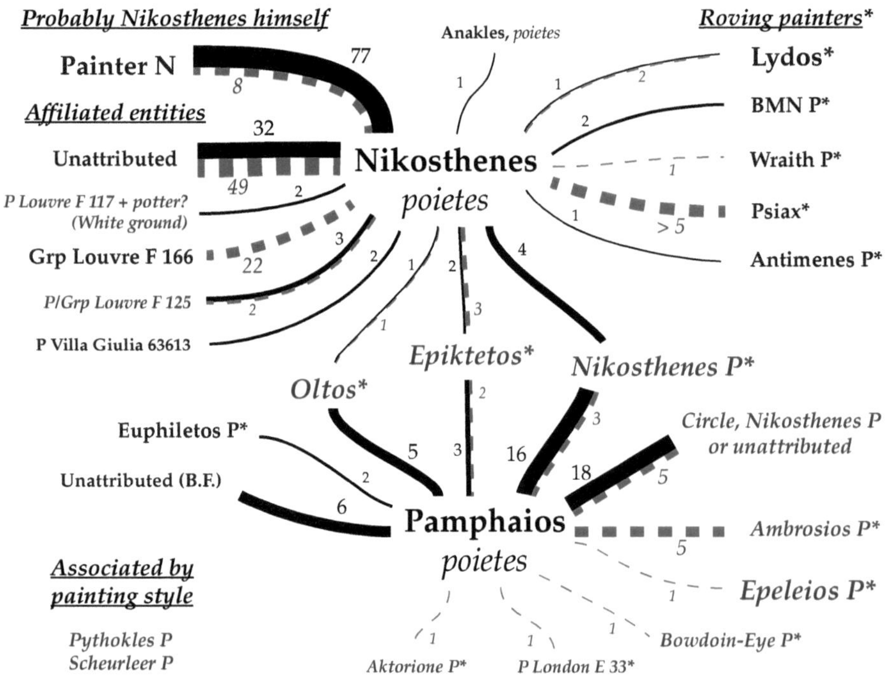

Fig. 2: Associations of *poietes* Nikosthenes and Pamphaios with hands based on inscriptions (solid lines) and potting (dashed lines). Line thickness corresponds to the number of links (written by the line) to these *poietes*. Black-figured hands are written in regular font face, red-figured in italics, and those with evidence for painting at other workshops are starred. © Philip Sapirstein.

Tosto's study of the workshop is framed around signed vases supplemented by potter-work. If we were to take the tally of 196 vases as the total surviving output and discount the undecorated cases, the shop attribution rate is only 5.0 vases per year. Nikosthenes (as Painter N) would have decorated about 75% of the material himself; while occupied at the wheel, he would have given the other 25% to shop staff or roving specialists to paint. Similarly, three vases attributed to two separate potters might have been made by an assistant while he painted.[42]

This is a minimalist approach excluding many other vases that probably originated in this workshop, but which happen to lack a Nikosthenic signature or identifiable potting style. A maximalist approach, expanding attributions to include more vases painted by the hands affiliated with the workshop is shown (Fig. 3a). Almost two-thirds of the resulting 1832 vases must have been painted elsewhere in the *kerameikos* than the Nikosthenic workshop, however, since they belong to prolific hands with slim connections to Nikosthenes or Pamphaios (Figs. 2; 3a).[43] After omitting 95 vases lacking figure decoration, which cannot have occupied much of a painter's time, the combined tally of figural vases that likely originated in the workshops of the two *poietes* is reduced to 657.[44]

Although it may be an overestimate, this larger count is probably closer to the reality, since many workshop vases might not preserve a signature or enough of the profile to be included in Tosto's potting analysis. Nikosthenes' first two decades see a steady output just over ten extant figural vases per year; with his share (as Painter N) of the total comprising less than half the production for the workshop (Fig. 3b: 545–525). This is consistent with the signs that he collaborated with older painters, notably Lydos and the BMN Painter, who arrived from other shops and may have dominated painting at first (unless their careers were somewhat longer than shown in Fig. 3b). Had he been unable to throw enough vases to

scriptions with Anakles (*BAPD* 301233) on a band cup, perhaps painted by an older potter-painter affiliated with the young Nikosthenes, is plain black gloss except for three tiny figures.

42 Tosto (1999) 11–12.

43 The hands connected with Nikosthenes have 1050 attributions, and those with Pamphaios have 782. The total number of attributions and the estimated fraction of each hand's output within the workshop is indicated in Fig. 3a. The counts are updated to 2011, following the methods described in Sapirstein (2013) online 2–6. Works near or in the manner of a hand are also tallied as potential workshop products by occasional painters.

44 That is, a total of 1080 figural are excluded. Most of this is due to hands with at least 50 vases that undoubtedly worked elsewhere: the Antimenes Painter, Epeleios Painter, Epiktetos, Euphiletos Painter, Lydos, Oltos, and Psiax. Only 173 of their 885 vases are likely to have been painted within the Nikosthenic workshop.

employ a full-time specialist, Nikosthenes might have allowed his associated painter to at times to pick up work elsewhere, and thus Nikosthenes would wish to maintain several relationships in case his preferred collaborator was unavailable when more vases became available to paint.[45] Productivity either rose modestly around 520 B.C., or else the contributions

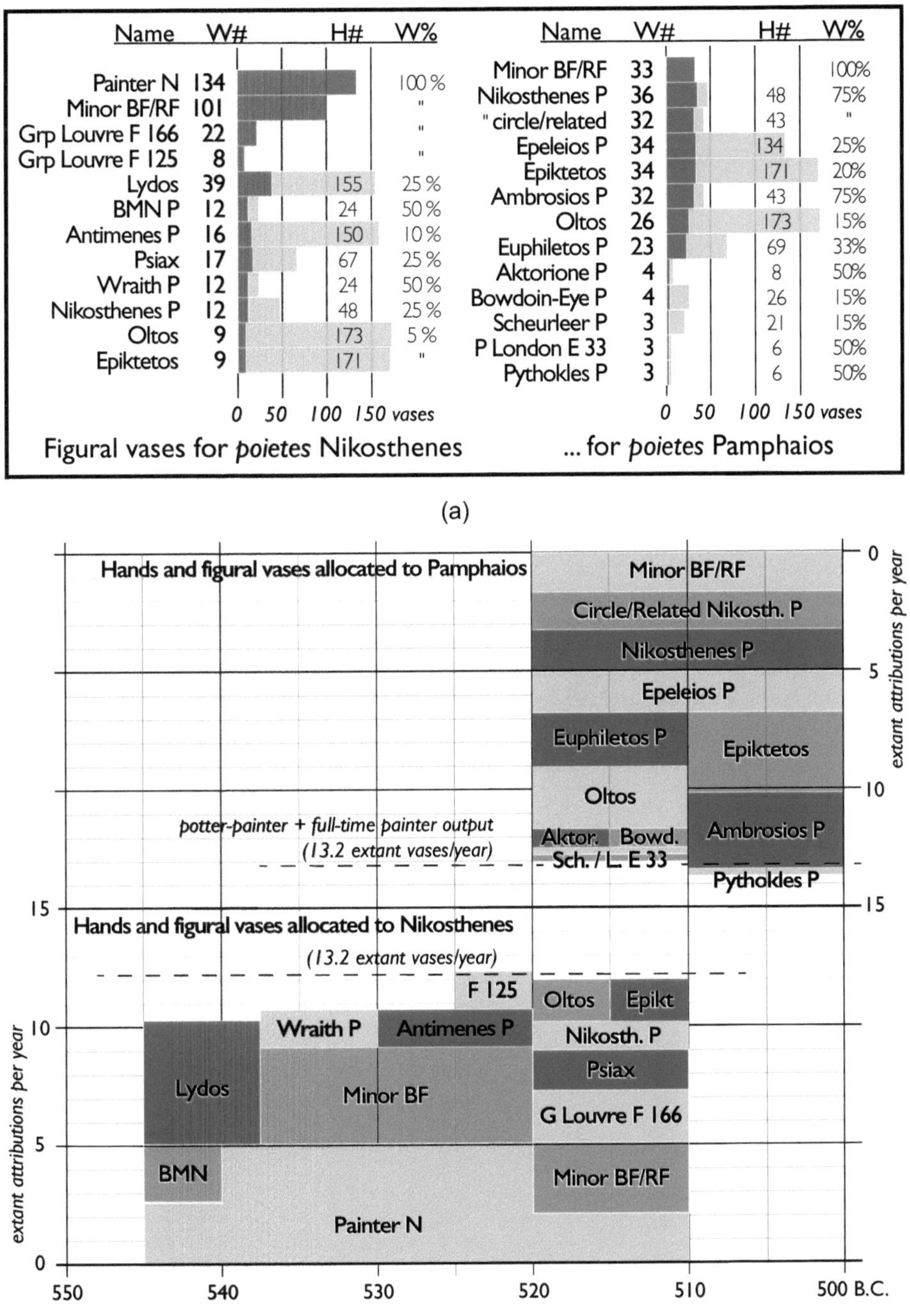

Fig. 3: **(a)** Vases allocated to the Nikosthenes and Pamphaios workshops by their associated hands. The number of vases within the workshop (W# / shaded bar) is derived from the total number of vases by each hand (H#) and the estimated percentage of the painter's career spent working with Nikosthenes or Pamphaios (W%); **(b)** Annual productivity in the workshops of Nikosthenes (below) and Pamphaios (above). Extant vases by each affiliated hand are plotted over its apparent dates of activity within the workshop, and the output of each hand stacked to represent the total extant vases per year for the two *poietes*. © Philip Sapirstein.

45 Fluid relationships of this kind were observed with the majolica painters at Fes, who often moved to a new workshops after completing a batch of pots on commission: Bel (1918) 188, note 1 200–201.

of roving red-figured painters like Epiktetos have been overestimated; if the former, Nikosthenes may have increased profits by spending more of his time potting and giving more of the decorative work to a series of full-time specialist painters (Fig. 3b: 525–515). Another possibility is that earlier in his career, Nikosthenes preferred to collaborate with occasional painters who worked slowly, and his later collaborations with red-figured specialists resulted in faster painting.

Pamphaios has a slightly higher productivity than the late Nikosthenes, though still less than the expected output of two specialist hands. Since his linkage to the older *poietes* is tenuous, and their combined output would abruptly double productivity and space requirements in ca. 520 B.C., he appears to have thrown pots in another workshop in the Kerameikos.

In sum, when we consider proxies for actual productivity, we find an even smaller workshop than envisaged in past studies. Nikosthenes and Pamphaios clearly cultivated relationships with many artisans over their decades-long careers, but they simply did not produce enough vases to occupy more than two painters at a time. Nikosthenes, a potter-painter himself, seems to have employed one assistant painter at times when he had available vases. Pamphaios, who is more likely to have been a full-time potter, supplied two painters at a time while cycling through many associates, who moved frequently among competitors' shops.

The Penthesileans

The specialists in red-figured cups grouped around the Penthesilea Painter will be treated more briefly. Influenced by the Early Classical Pistoxenos Painter, the Penthesilea Painter retired in the mid-440s B.C. and was succeeded by numerous associates.[46] Similar methods to those used to visualize the Nikosthenic workshop are applied here, although a central *poietes* is absent due to the absence of signatures and only limited study of the potting. More than 15 distinctive hands are present, over half of which shared the painting of the interior and exterior of a single cup (Fig. 4). With some deductions for painters known to have painted elsewhere, we still have ca. 1760 vases, of which ca. 1310 are dated to its apogee in 460–440 B.C. (Fig. 5). The scale of production is roughly ten times that of Pamphaios, equivalent to at least seven full-time specialist painters.

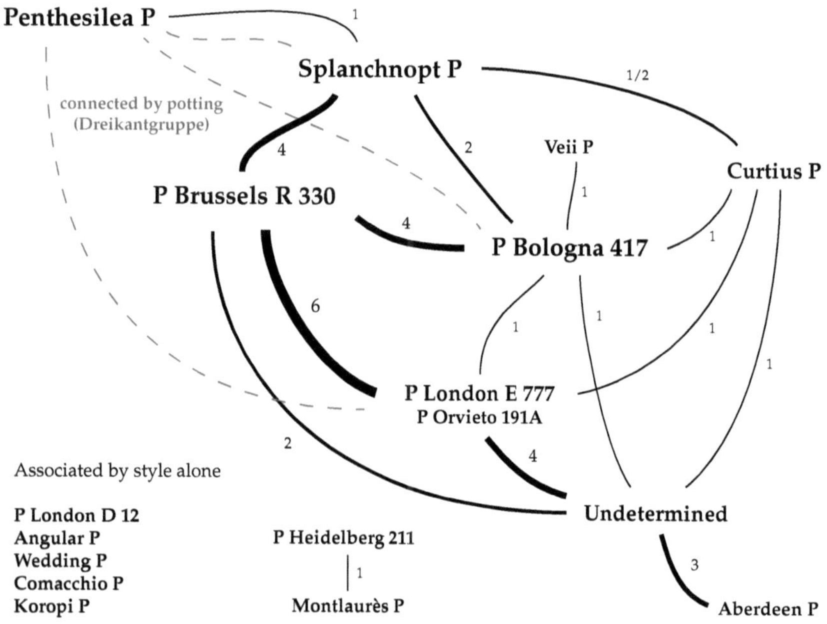

Fig. 4: Associations among the Penthesilean painters who shared the decoration of one vase (solid lines) or are linked by potting style (dashed). Conventions are similar to those of Fig. 2, except all hands are red-figured specialists, and no *poietes* is identified. © Philip Sapirstein.

46 Above, note 12, with Bloesch (1940) 103–107; Sapirstein (2013) online, entries for the Penthesilea, Splanchnopt, Veii, and Villa Giulia Painters, and the Painters of Bologna 417 and London D 12.

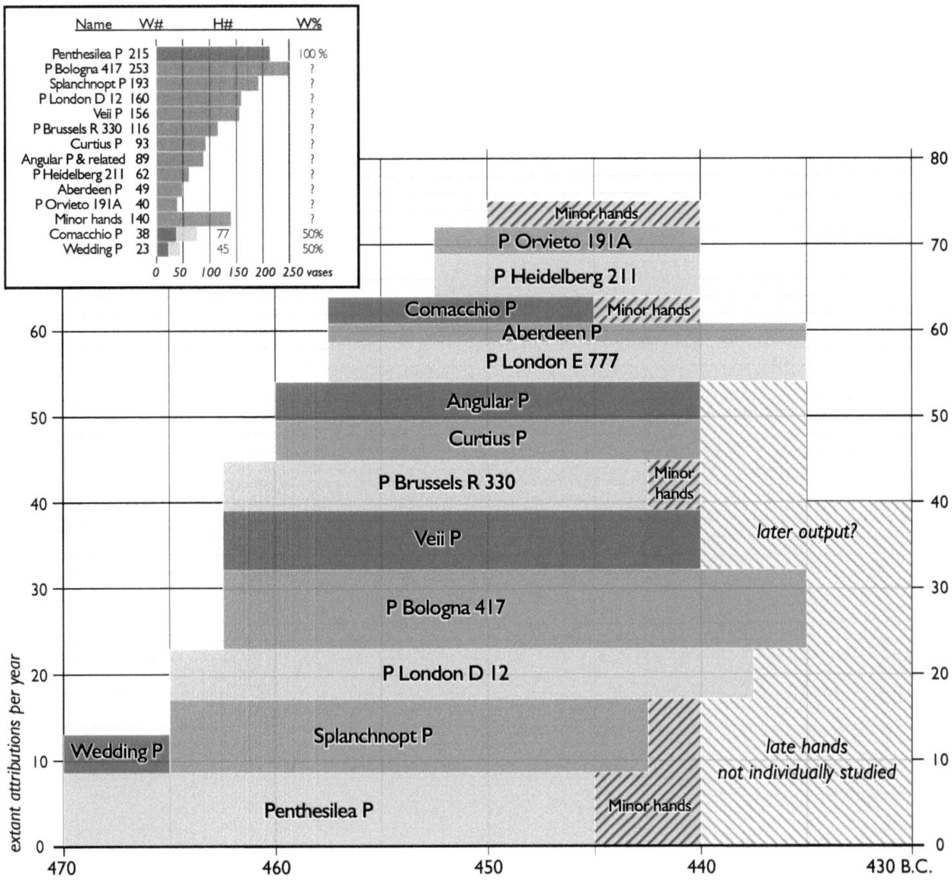

Fig. 5: Vases allocated to the Penthesilean painters. Conventions follow those for Fig. 3. At least the Comacchio and Wedding Painters worked elsewhere, as may have many others of the affiliated hands. Workshop activity after 440 B.C., which is poorly documented, has been estimated by assuming a gradual falloff in production. © Philip Sapirstein.

The impressive output could suggest a major intensification in Athens during the half century since Nikosthenes retired. However, the direct associations with a physical workspace are found on only a tiny fraction of the attributions to the Penthesilean hands. Fewer than 40 cups are shared, and the connection via potting are still fewer, leaving open the possibility that roving painters decorated much of this corpus in different workshops – which potentially could drastically reduce the estimated volume of production that took place in the same working space shown in the maximalist version (Fig. 5). We would need an analysis of the potting for a larger sample of cups dated between 460 and 440 B.C. to determine which scenario is the more likely.

Conclusions

A major finding is that the Nikosthenic workshop, considered one of the largest of its time, was smaller than expected, and probably separate from that of Pamphaios. Either potter might have had two assistants for preparing clay and turning the wheel, and perhaps a third full-time equivalent to prepare for firing and other odd jobs.[47] They appear to have employed only one painter at a time, even if many artisans passed through the shops over their decades of operation. The previous reconstructions of about eight workers in the Nikosthenic group are not far from the mark, but probably exaggerate an actual staffing of five or six artisans. On the ethnographic spectrum, the workshops in the Archaic Kerameikos would rep-

47 Scheibler (1983) 110–117, (1984) 133–134, (1986); Stissi (2002) 86–95.

resent a nucleated workshop industry. Family-centred operations with few paid employees would be dominant, and rising demand was probably met by the establishment of new workshops rather than internal intensification.

The analysis of the Penthesilean group shows how some operations were changing during the 5th century. If all the painters worked here for their entire careers, we would have a staff of about a dozen workers engaging in parallel production that had crossed the threshold to a small manufactory. The workshop where many of these painters exchanged cups surely did have a larger staff than that of Nikosthenes through its employment of full-time painting specialists, but the uncertainty over where each of the Penthesilean painters spent their whole careers leave open the possibility of more incremental changes.

I have focused on the Nikosthenic and Penthesilean operations because of the assumption that they were among the largest workshops of their respective eras. Future study of signatures and potting work in other groups may yield indications of other large operations, though it would without doubt also reveal that many smaller producers continued to be active throughout Archaic and Classical Athens. The persistence of the potter-painter model into the 5th century, such as the Shuvalov Painter (Fig. 1), signals that many small operations continued to thrive, especially those making larger vase shapes, and that intensification of the production of hand-painted vases was not sufficiently profitable to transform the industry radically.

Abbreviations

ABV *Attic Black-figure Vase-Painters*.
BAPD Beazley Archive Pottery Database, https://www.carc.ox.ac.uk/carc/pottery (last access 02.07.2024).
CVA *Corpus Vasorum Antiquorum*.

Bibliography

Beazley (1944): John Davidson Beazley, *Potter and Painter in Ancient Athens* (Proceedings of the British Academy 30), London.
Beazley (1986): John Davidson Beazley, *The Development of Attic Black-Figure*, Berkeley.
Bel (1918): Alfred Bel, *Les industries de la céramique à Fés*, Paris.
Bloesch (1940): Hansjörg Bloesch, *Formen Attischer Schalen von Exekias bis zum Ende des Strengen Stils*, Bern.
Cook (1959): Robert M. Cook, “Die Bedeutung der bemalten Keramik für den griechischen Handel”, in: *Jahrbuch des Deutschen Archäologischen Instituts* 74, 114–23.
Eschbach/Schmidt (2016): Norbert Eschbach and Stefan Schmidt (eds.), *Topfer-Maler-Werkstatt: Zuschreibungen in der griechischen Vasenmalerei und die Organisation antiker Keramikproduktion, CVA* Deutschland Beihefte 7, Munich.
Fülle (1997): Gunnar Fülle, “The Internal Organization of the Arretine Terra Sigillata Industry: Problems of Evidence and Interpretation”, in: *Journal of Roman Studies* 87, 111–155.
Hasaki (2011): Eleni Hasaki, “Crafting Spaces: Archaeological, Ethnographic, and Ethnoarchaeological Studies of Spatial Organization in Pottery Workshops in Greece and Tunisia”, in: Mark L. Lawall, John Lund (eds.), *Pottery in the Archaeological Record: Greece and Beyond* (Acts of the International Colloquium held at the Danish and Canadian Institutes in Athens, June 20–22, 2008), Aarhus, 12–28.
Hasaki (2021): Eleni Hasaki, *Potters at Work in Ancient Corinth: Industry, Religion, and the Penteskouphia Pinakes, Hesperia* Supplement 51, Princeton N.J.
Hasaki/Bentz (2020): Eleni Hasaki and Martin Bentz (eds.), *Reconstructing Scales of Production in the Ancient Greek World: Producers, Processes, Products, People* (Panel 3.4, Proceedings of the 19th International Congress of Classical Archaeology, vol. 8, Cologne/Bonn, May 22–26, 2018), Heidelberg https://doi.org/10.11588/propylaeum.639.c9978
Hasaki/Cline (2020): Eleni Hasaki and Diane H. Cline, “Social Network Analysis and Connoisseurship in the Study of Athenian Potters’ Communities”, in Hasaki/Bentz (2020) 59–80.
Hurwit (2015): Jeffrey Hurwit, *Artists and Signatures in Ancient Greece*, Cambridge.
Immerwahr (1984): Henry R. Immerwahr, “The Signatures of Pamphaios”, in: *American Journal of Archaeology* 88, 341–352.
Lightbown/Caiger-Smith (1980): Ronald W. Lightbown and Alan Caiger-Smith (trans. and eds.), *I Tre Libri dell’Arte del Vasaio by Cipriano Piccolpasso = The Three Books of the Potter’s Art: A Facsimile of the Manuscript in the Victoria and Albert Museum, London*, London.
Lüdorf (2010): Gundula Lüdorf, “Zur Größe athenischer Töpferwerkstätten des 6. und 5. Jhs. v. Chr. Familienunternehmen oder zentralisierte Massenproduktion?”, in: Hans Lohmann, Torsten Mattern (eds.), *Attika. Archäologie einer “zentralen” Kulturlandschaft* (Akten der Internationalen Tagung, Marburg, May 18–20, 2007), Wiesbaden, 151–159.
Osborne (2004): Robin Osborne, “Workshops and the Iconography and Distribution of Athenian Red-figure Pottery: A Case Study”, in: Simon Keay and Stephanie Moser (eds.), *Greek Art in View. Essays in Honour of Brian Sparks*, Oxford, 78–94.

Peacock (1982): D.P.S. Peacock, *Pottery in the Roman World: An Ethnoarchaeological Approach*, London.
Pevnick (2016): Seth D. Pevnick, "*Le style est l'homme même?* On Syriskan Attributions, Vase Shapes, and Scale of Decoration", in: Eschbach/Schmidt (2016) 36–46.
Sanidas (2013): Giorgos M. Sanidas, *La production artisanale en Grèce. Une approche spatiale et topographique à partir des exemples de l'Attique et du Péloponnèse du VIIe au 1er siècle avant J.-C.*, Lille.
Sapirstein (2013): Philip Sapirstein, "Painters, Potters, and the Scale of the Attic Vase-Painting Industry", in: *American Journal of Archaeology* 117, 493–510 (Online: 1–47). https://doi.org/10.3764/aja.117.4.0493.
Sapirstein (2014): Philip Sapirstein, "Demographics and Productivity in the Ancient Athenian Pottery Industry", in: John H. Oakley (ed.), *Athenian Potters and Painters* III, Oxford, 175–186.
Sapirstein (2020): Philip Sapirstein, "Productivity of Athenian Vase-painters and Workshops", in: Hasaki/Bentz (2020) 81–96.
Scheibler (1983): Ingeborg Scheibler, *Griechische Töpferkunst: Herstellung, Handel und Gebrauch der antiken Tongefässe*, Munich.
Scheibler (1984): Ingeborg Scheibler, "Zur mutmasslichen Größe attischer Töpfereien des 6. Jahrhunderts v. Chr.", in: Herman A.G. Brijder (ed.), *Ancient Greek and Related Pottery* (Proceedings of the International Vase Symposium, Amsterdam, April 12–15, 1984), Amsterdam, 130–134.
Scheibler (1986): Ingeborg Scheibler, "Formen der Zusammenarbeit in attischen Töpfereien des 6. und 5. Jahrhunderts v.Chr.", in: Siegfried Lauffer, Hansjörg Kalcyk, Brigitte Gullath and Andreas Graeber (eds.), *Studien zur alten Geschichte: Siegfried Lauffer zum 70. Geburtstag am 4. August 1981 dargebracht von Freunden, Kollegen und Schülern*, Rome, 787–804.
Schreiber (1999): Toby Schreiber, *Athenian Vase Construction: A Potter's Analysis*, Malibu.
Stissi (2002): Vladimir V. Stissi, *Pottery to the People: The Production, Distribution and Consumption of Decorated Pottery in the Greek World in the Archaic Period (650–480 BC)*, PhD Dissertation, University of Amsterdam. https://hdl.handle.net/11245/1.197192.
Stissi (2016): Vladimir V. Stissi, "Minor Artisans, Major Impact?", in: Eschbach/Schmidt (2016) 47–53.
Stissi (2020): Vladimir V. Stissi, "From Counting Pots to Counting People: Assessing the Scale of Athenian Pottery Production and Its Impact on Workshop Staff", in: Hasaki/Bentz (2020) 97–108.
Tafe (2022): Jennifer S. Tafe, *Nikosthenes: Innovation and Identity in Late Archaic Vase-Painting*, PhD Dissertation, Boston University. https://hdl.handle.net/2144/45323.
Tonglet (2015): Delphine M. Tonglet, "'Just-so stories': Plants and Animals on Two Early Attic Kyathoi by Theozotos", in: Claudia Lang-Auinger and Elisabeth Trinkl (eds.), *ΦΥΤΑ ΚΑΙ ΖΩΙΑ. Pflanzen und Tiere auf griechischen Vasen* (Akten des internationalen Symposiums an der Universität Graz, 26.–28. September 2013), *CVA* Österreich Beiheft 2, Vienna.
Tosto (1999): Vincent Tosto, *The Black-figure Pottery Signed ΝΙΚΟΣΘΕΝΕΣΕΠΟΙΕΣΕΝ*, Allard Pierson Series 11, Amsterdam.
Valavanis (1997): Panos Valavanis, "Βάκχιος, Κίττος και παναθηναϊκοί αμφορείς. Σκέψεις για τη δομή των αττικών κεραμικών εργαστηρίων του 4ου αι. π.Χ.", in: John H. Oakley, William D.E. Coulson and Olga Palagia (eds.), *Athenian Potters and Painters* I, Oxford, 85–95.
Webster (1972): Thomas B.L. Webster, *Potter and Patron in Classical Athens*, London.
Whitley (2018): James Whitley, "Style and Personhood: The Case of the Amasis Painter", in: *The Cambridge Classical Journal* 64, 178–203. https://doi.org/10.1017/S1750270518000088.
Williams (2016): Dyfri Williams, "Peopling Athenian *kerameia:* Beyond the Master Craftsmen", in: Eschbach/Schmidt (2016) 54–68.
Williams (2017): Dyfri Williams, "Beyond the Berlin Painter: Toward a Workshop View", in: Michael J. Padgett (ed.), *The Berlin Painter and His World. Athenian Vase-Painting in the Early Fifth Century B.C.*, New Haven, 144–187.
Wilson (2008): Andrew I. Wilson, "Large-Scale Manufacturing, Standardization, and Trade", in: John P. Oleson (ed.), *The Oxford Handbook of Engineering and Technology in the Classical World*, Oxford, 393–417.

Linda Adorno

10 Potters' Tools at the Kerameikos of Selinous

Abstract: The study of ceramicists-manufacturing tools provides invaluable information about the methodologies and production techniques of potters' workshops in the ancient world. Tools associated with ceramics production have rarely been attested in primary artisanal contexts, as these objects were often repurposed even when no longer intact. The results of excavations of Selinous' Kerameikos are thus exceptional, revealing in primary contexts a range of implements associated with various stages of production. The objects most frequently attested in work areas adjacent to the kilns are spacers and supports used for stacking ceramics within the kiln's combustion chamber. Along with the spacers, other tools include pumice millstones, pestles, smoothers, punches, and stencils for finishing and decorating pottery. This unusually high concentration of finds, rare within production contexts elsewhere, is likely due to the sudden abandonment of Selinous' after the Carthaginian sack in 409 B.C. These findings, drawn from the excavation of only a small portion of Selinous huge potters' quarter, help to rectify our previously incomplete knowledge of tools and stages of production in ancient Greek workshops.

The colony of Selinous, founded in 628 B.C. by Megara Hyblaea on the southwestern coast of Sicily,[1] played a prominent role in disseminating Greek culture in the West. The wealth manifested in Selinous regular street grid, characterized by a large Agora and lavish public and religious buildings, was the result of several factors, production being one of the most significant among them.[2] In fact, multiple factors contributed to Selinous economic prosperity.[3] The city's convenient location within a dense Mediterranean exchange network facilitated a quick expansion of the colony and its wide-ranging trade as early as the Archaic period. Through the creation of its manufacturing sector, the growth of Selinous local economy led to ceramic imports' cessation, at least for everyday objects.[4]

Not coincidentally, Selinous is one of the few Archaic period cities that developed a vast district exclusively dedicated to artisanal production, a manufacturing cluster that was considerably important for the urban economy.[5] The creation of this impressive potters' quarter, which would be intensively used for a long time, occurred contemporaneously with the original functional subdivision of the colony's urban area. Following D. Mertens's campaigns (1999–2002), H. Stümpel created a geophysical plan[6] indicating a large number of kilns extending for more than a kilometer along the Gorgo Cottone Valley (Fig. 1). For the first time, recent research by M. Bentz in this part of the city approached the topic of the craftsmanship in Selinous based on significant archaeological evidence. Excavation campaigns between 2010 and 2016 focused on *insula* S16/17 (Fig. 2). The area under examination dates back to the 5th century B.C., when the thriving city reached its maximum size, witnessing ambitious new building construction and an equally intense increase in manufacturing. There is also ample evidence of an earlier phase of structures partly integrated into these newer buildings.

A workshop approximately 85 meters long and 18 meters wide extends on four terraces covering an area of more than 1200 square meters.[7] Work rooms and areas devoted to different productive phases have been identified, including areas dedicated to clay working, with wells and settling tanks, and courtyards for drying and sorting the artifacts. It was a fully autonomous system whose kilns represent only the final moment of a production cycle marked by several phases, starting with the supply of raw materials and ending with the commercialization of the finished product.

Much has been written on the internal organization of installations for pottery production, especially in relation to their economy.[8] Nevertheless, there are some elements that remain elusive despite being central to ancient produc-

1 On the two foundation dates of Selinous, see especially: Marconi (2019); Robu (2014) 159–168; Parisi Presicce (2003) 264–270; De Angelis (2003) 124; Mertens (2003) 226–227 and 273; Villard/Vallet (1958).
2 On the topic of production, see the recent publication: Caminneci/Parello/Rizzo (2018); Atria *et al.* (2019).
3 De Angelis (2016) 137–212, 230–291.
4 On pottery production in Selinous, see Adorno (forthcoming).
5 Bentz (2018) 175.
6 Mertens (2003) 485–488.
7 Bentz *et al.* (2016) 59–62.
8 Recently Cucuzza *et al.* (2016); Caminneci/Parello/Rizzo (2018).

https://doi.org/10.1515/9783111189635-010

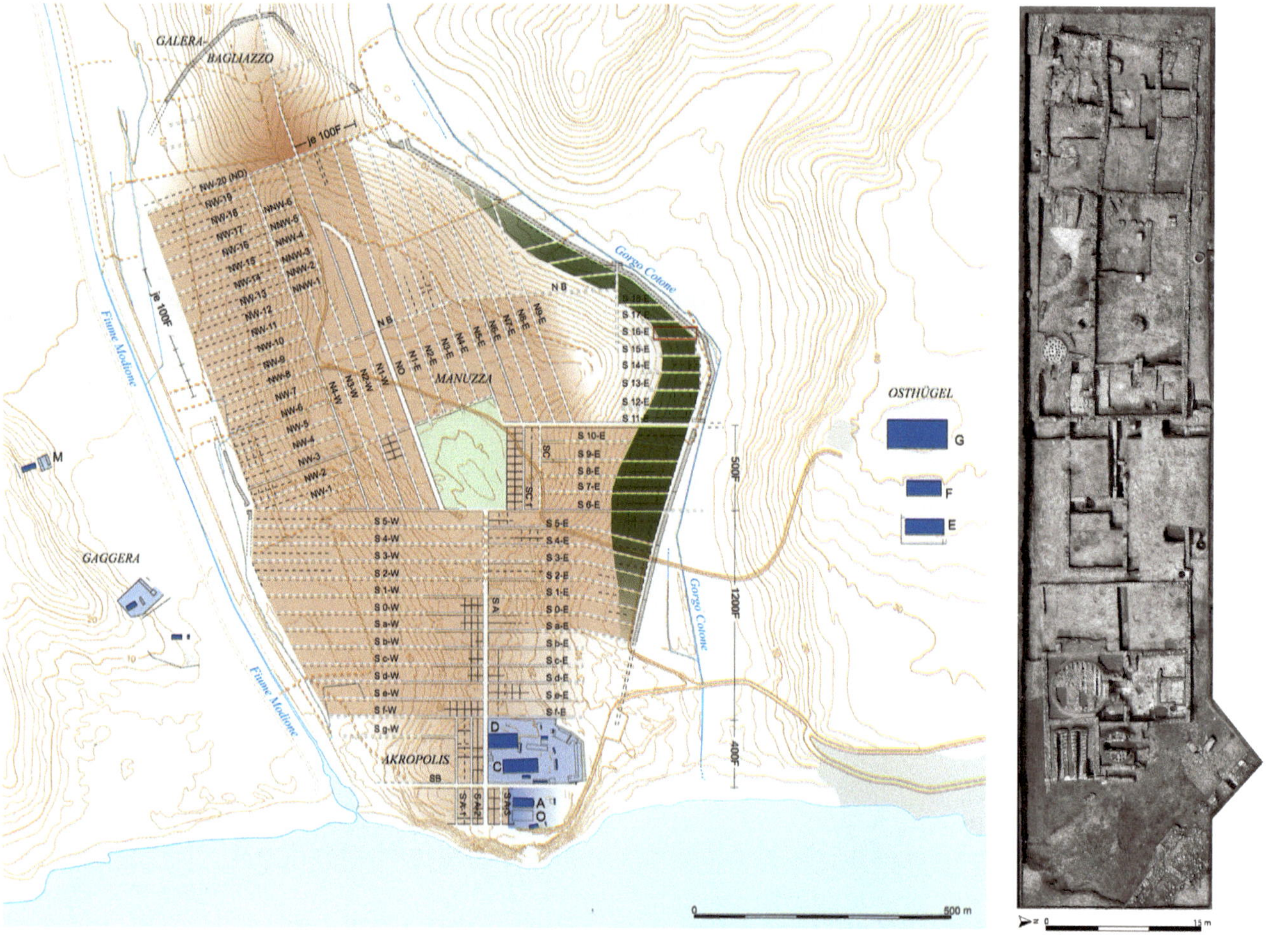

(a) (b)

Fig. 1: (a) Selinous, plan with indication of potters' quarter (Kerameikos) in green. After Mertens 2006, modified by M. Bentz. **(b)** Selinous, Kerameikos S16/17. After Bentz *et al.* 2016.

tion.[9] Such is the case with tools used to prepare pottery, a subject of great interest typologically, functionally, historically, and technologically, as they can provide insight into the organization of labor in ancient society. Despite their significance, the archaeological traces of these tools are not always easy to spot.

The excavation of Selinous Kerameikos thanks to its sudden abandonment after the Carthaginian sack in 409 B.C., brought to light an exceptional assemblage of tools associated with ceramics' preparation and firing.[10]

Kiln supports are the objects most represented; this is an artifact class frequently documented in Greece, Etruria, Magna Graecia, and Sicily starting from the 6th century B.C.[11] They were used by artisans to support and stabilize the vases stacked inside the kiln during firing. The variety of kiln supports' forms reflects the different ceramic shapes that had to be stacked. Finds from the excavated area amounted to about 1800 pieces. Five different types have been identified based on morphological features (Fig. 3):[12]

1. Wedge-shaped
2. Cylindrical
3. Bow-shaped

9 In this respect, I would like to thank the organizers of this timely conference for the emphasis given to the topic of manufacturing and craftmanship in the ancient world, and for inviting me to speak.

10 For a more detailed discussion on the topic, see the forthcoming monography on the excavation of potters' quarter.

11 E.g., Fusi (2020); Branca (2016); Cracolici (2003); Barra Bagnasco (1992).

12 Adorno (2018).

4. Ring-shaped
5. Heel-shaped

The majority of kiln supports found in the workshop were of Type 1, in the shape of a truncated pyramid and likely used for the separation of tiles or large storage containers in order to avoid any contact inside the kiln. The majority of Type 1 kiln supports have rectangular bases. Less common are square bases, and circular ones are rarer still. The analysis of kiln supports is of particular interest because of the symbols present on their surfaces (Fig. 4). These mainly take the form of letters engraved or imprinted on the raw material and can probably be interpreted as markers affixed during the manufacturing process. Among various Greek letters, H occurs on several supports. Remarkably, the punch for this marker has been found.[13] Other markers frequently engraved on the surface are crosses, circles, and bars.

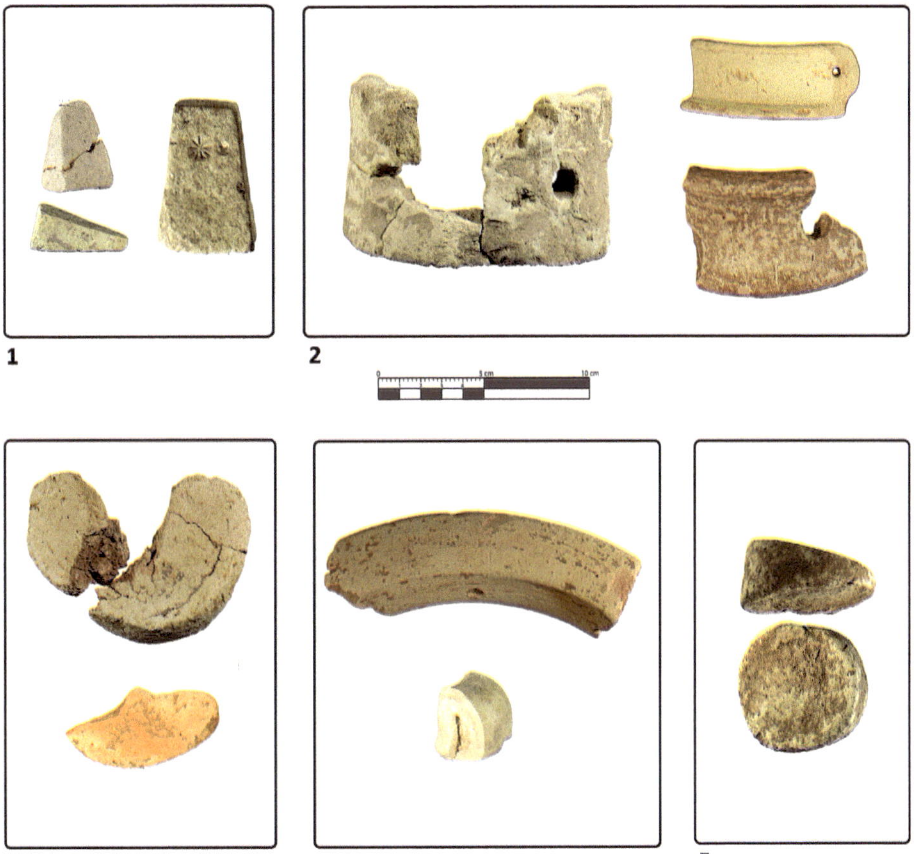

Fig. 3: Different types of kiln supports from the Kerameikos. Photos: Martin Bentz. Editing: Linda Adorno.

As was argued for kiln supports from Poseidonia,[14] it is likely that these symbols belonged to particular artisans and served as their marks in shared kilns. Such markers, therefore, as well as those present on tiles, may be useful to identify various ceramic workshops active in Selinous.[15] Type 2 kiln supports have small holes piercing their side walls to let out water vapor during the firing. In the smaller examples—10–14 cm in diameter—the side walls are concave. In the larger examples—25 cm in diameter—the side walls are convex. Type 3 is represented by only a few examples. It recalls crescent-shaped kiln supports from a 6th–5th century B.C. artisanal area in Taranto.[16] Type 4 kiln supports are circular with a

13 See *infra*.
14 Rizzo (2017).
15 See Conti (2018) for tile stamps.
16 Dell'Aglio (1996) 69, note 25.

diameter of more than 20 cm and a reduced height. Few examples have been found; they were probably used for artifacts of a particular size. The fifth and last type is called “heel-shaped” because of its peculiar circular shape with a triangular section. These kiln supports are less common and are found mainly in the oldest layers of the workshop in association with Archaic material. They are characterized by their hard ceramic fabric.

Other objects to be counted among the tools used by local potters are mills, basins of various shapes and sizes, mortars, pestles, smoothers, punches, and molds. Some large-scale mills stand out because they are made of volcanic stone, a material that cannot be found in the territory of Selinous but that is very common in the environs of Mount Etna, the Hyblaean Mountains, and the Aeolian Islands.[17]

Fig. 4: Selection of symbols on the kiln supports. Photos: Martin Bentz. Editing: Linda Adorno.

Pestles and smoothers represent a significant sample, with around twenty examples recovered. The hemispherical smoother (Fig. 5) is quite large and in almost all examples is characterized by a left-handed ergonomic handle. Such tools, whose diameter varies between 15 and 18 cm, could be used to prepare ceramic mixtures or polish the surfaces of large vases. A complete example can weigh up to around 1.4 kg.

The truncated cone–shaped smoother type, with its convex base and folded upper handle (Fig. 6), finds close comparisons in Taranto, where some of these artifacts, identified as “impastatoi fittili”, have been found together with the relevant molds.[18] River pebbles, small limestones, and marble cobbles could be used for the same purpose, as these are marked by clear traces of wear, suggestive of milling activities.

Some tools for marking and decorating artifacts before firing have been found inside the workshop, such as punches, small ceramic stamps, ceramic roller stamps, and small bone spatulas. The Greek letter H on the aforementioned punch appears on a number of artifacts, including tiles, bowls, kiln supports, and even a sarcophagus lid found close to the kilns (Fig. 7). This letter may be the identification mark of one of the workshops operating in the potters’ quarters of Selinous.

17 For the volcanic rock grinding stones from Selinous see Gluhak/Schwall (2015). An Aegean provenance is suspected for some of the analyzed samples.

18 Dell’Aglio (1996) 56–57, 70–71, figs. 37–40.

Fig. 5: Hemispherical smoother from Kerameikos. Photos: Martin Bentz. Editing: Linda Adorno.

Fig. 6: Other types of pestle. Photos: Martin Bentz. Editing: Linda Adorno.

Fig. 7: Selinous, terracotta punch and example of use. Photos: Martin Bentz. Editing: Linda Adorno.

Three small terracotta stamps have been found in the excavation of the Kerameikos: one is round, depicting the negative image of a crouched Silenus; another is rectangular, with its lower face featuring three spherical appendages enclosed by semicircles in relief; and the third is a cylindrical stamp with a decorative motif resembling a palmette petal or leaf, probably used to decorate louteria and large basin rims (Fig. 8a–c). The rather recent discovery of a stamp with a rosette motif (Fig. 9), found in a house of the 5th century B.C. on the northern hill of Selinous, was probably used for the decoration of larger vessels or tiles like the stamps mentioned above.[19]

Small bone spatulas conclude this brief overview of the tools found in Selinous (Fig. 10). Known examples from Selinous mainly come from the Agora, together with overfired pottery and fragments of a perforated clay floor, suggesting the existence of small production centers in that area as well.

19 The excavation on the northern hill was carried out by the German Archaeological Institute in May 2023. It is part of a project funded by the German Research Foundation to further investigate the urban layout and its transformation through time. For more information, see Jonasch *et al.* (2022).

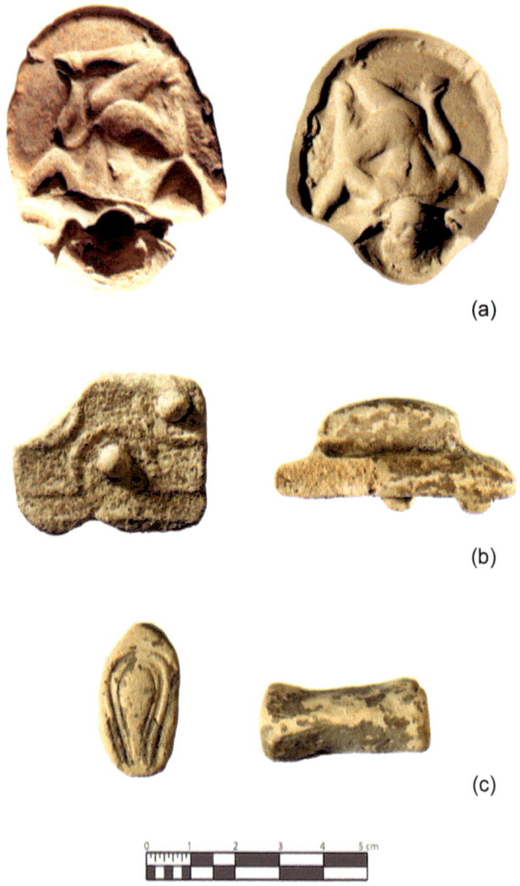

Fig. 8: (a-c) Selinous, terracotta stamps from the Kerameikos.

Fig. 10: Selinous, bone spatulas from Agora and Kerameikos. Photos: Linda Adorno.

Fig. 9: Selinous, northen hill house. Stamp with a rosette motif. Photo: Melanie Jonasch. Editing: Linda Adorno.

Finally, there are also many pieces of metal of uncertain function that may have been used as tools. Tools in perishable materials such as wood would also have been numerous. There is, however, no archaeological trace of such objects.

The numerous workshops yet to be investigated in Selinous likely preserve further tool types, crucial evidence for reconstructing the productive processes.

Bibliography

Adorno (2018): Linda Adorno, "Gli strumenti per la produzione ceramica del quartiere artigianale di Selinunte", in: Caminneci/Parello/Rizzo (2018) 223–226.

Adorno (forthcoming): Linda Adorno, *Selinus* VII. *La ceramica di produzione locale di Selinunte. Nascita ed evoluzione*, Rome, forthcoming.

Atria *et al.* (2019): Rosario Atria, Giuseppe L. Bonanno, Angelo Curti Giardina and Giovanna Titone (eds.), *Selinunte. Produzione ed economia di una colonia greca di frontiera* (Atti del Convegno internazionale, Selinunte, Baglio Florio, 15–16 Aprile 2016), *Sicilia Archeologica* 111, Trapani.

Barra Bagnasco (1992): Marcella Barra Bagnasco, "Sostegni", in: Marcella Barra Bagnasco (ed.), *Locri Epizefiri IV. Lo scavo di Marasà Sud. Il sacello tardo arcaico e "la casa dei leoni"*, Florence, 223–225.

Bentz *et al.* (2013): Martin Bentz, Linda Adorno, Jon Albers, Jan Marius Müller and Gabriel Zuchtriegel, "Das Handwerkerviertel von Selinunt. Die Werkstatt der Insula S16/17-E. Vorbericht zu den Kampagnen 2010–2012", in: *Römische Mitteilungen* 119, 69–98.

Bentz *et al.* (2016): Martin Bentz, Linda Adorno, Jon Albers, Annkatrin Benz, Axel Miß and Jan Marius Muller, "Das Handwerkerviertel von Selinunt – Vorbericht zu den archaischen Strukturen der Kampagnen 2010–2016", in: *Kölner und Bonner Archaeologica* 6, 59–74.

Bentz (2018): Martin Bentz, "Il ruolo della produzione ceramica nella città della Grecia classica", in: Caminneci/Parello/Rizzo (2018) 169–177.

Branca (2016): Alberto Branca, "La produzione artigianale a Siracusa: dalle materie prime ai sistemi di produzione", in: Cucuzza *et al.* (2016) 78–80.

Caminneci/Parello/Rizzo (2018): Valentina Caminneci, Maria Concetta Parello and Maria Serena Rizzo (eds.), *La città che produce. Archeologia della produzione negli spazi urbani* (Atti delle Giornate Gregoriane, X edizione, Agrigento, 10–11 Dicembre 2016), Bari.

Conti (2018): Maria Clara Conti, *I bolli su tegole e coppi a Selinunte*, *Biblioteca di Sicilia Antiqua* VII, Pisa-Rome.

Cracolici (2003): Vincenzo Cracolici, *I sostegni di fornace dal Kerameikos di Metaponto*, Bari.

Cucuzza *et al.* (2016): Nicola Cucuzza, Bianca Maria Giannattasio and Silvia Pallecchi (eds.), *Archeologia delle produzioni ceramiche nel mondo antico. Spazi, prodotti, strumenti e tecniche* (Atti del convegno, Genova, 1–2 Dicembre 2014), Rome.

De Angelis (2003): Franco De Angelis, *Megara Hyblaia and Selinous: The Development of Two Greek City-States in Archaic Sicily*, Oxford.

De Angelis (2016): Franco De Angelis, *Archaic and Classical Greek Sicily. A Social and Economic History*, Oxford.

Dell'Aglio (1996): Antonietta Dell'Aglio, "Taranto. L'argilla", in: Enzo Lippolis (ed.), *Greci in Occidente. Arte e artigianato in Magna Grecia* (Catalogo della mostra, Napoli, ex Convento di San Domenico, 29 Giugno 1996), Naples, 50–71.

Fusi (2020): Martina Fusi, "I distanziatori da fornace come indicatori di produzione. Nuovi dati per Populonia", in: *Fasti Online Documents and Research* 464, 1–20. https://www.fastionline.org/docs/FOLDER-it-2020-464.pdf (last access 27.03.2023).

Gluhak/Schwall (2015): Tatjana Gluhak and Christoph Schwall, "Provenance Analyses of the Volcanic Rock Grinding Stones from the Greek Colony Selinunte, Sicily (Italy) – Constraints and Possibilities", in: *Archaeometry* 57, 246–268. https://doi.org/10.1111/arcm.12086.

Jonasch *et al.* (2022): Melanie Jonasch, Linda Adorno, Aurelio Burgio, Antonio Di Maggio, Morten Harms, Marlen Schlöffel, Steffen Schneider and Dennis Wilken, "Forschungen in Selinunt, Teil 1. Ein neues Modell für die Stadt Selinunt. Die Feldarbeiten des Jahres 2021", in: *e-Forschungsberichte des Deutschen Archäologischen Instituts*, in: *e-Forschungsberichte* 1, 18–37. https://doi.org/10.34780/ftd4-fm24.

Marconi (2019): Clemente Marconi, "New Evidence for Early Greek Settlement on the Acropolis of Selinunte", in: Elena Partida and Barbara Schmidt-Dounas (eds.), *Listening to the Stones: Essays on Architecture and Function in Ancient Greek Sanctuaries in Honour of Richard Alan Tomlinson*, Oxford, 252–261.

Mertens (2003): Dieter Mertens (ed.), *Selinus I: Die Stadt und ihre Mauern*, Mainz.

Mertens (2006): Dieter Mertens, *Città e monumenti dei Greci d'Occidente*, Rome.

Parisi Presicce (2003): Claudio Parisi Presicce, "Selinunte dalla scelta del sito alle prime fasi di vita", in: *Römische Mitteilungen* 110, 263–284.

Rizzo (2017): Maria Luigia Rizzo, "Iscrizioni sui sostegni di fornace da Poseidonia: un contributo alle attestazioni dal mondo greco e magno-greco", in: Stefano De Caro, Fausto Longo, Michele Scafuro and Antonia Serritella (eds.), *Percorsi. Scritti di archeologia di e per Angela Pontrandolfo*, Salerno, 107–118.

Robu (2014): Adrian Robu, *Mégare et les établissements mégariens de Sicile, de la Propontide et du Pont-Euxin. Histoire et institution*, Bern.

Villard/Vallet (1958): François Villard and Georges Vallet, "La date de fondation de Sélinonte: les données archéologiques", in: *Bulletin de Correspondance Hellénique* 82, 16–26.

Laetitia Cavassa, Guilhem Chapelin, Alberto De Bonis, Chiara Germinario, Celestino Grifa, Bastien Lemaire, Emilie Mannocci, Fabrice Paul and Marie Pawlowicz

11 Pottery Production in Pompeii, Step by Step: From Excavations to Experimental Archaeology

Abstract: Between 2012 and 2017, two potters' workshops were excavated that had been active in 79 A.D. on the day when Vesuvius erupted. The first, discovered in 1838, lies outside the city walls in the necropolis sector in front of the Herculaneum Gate. The second, discovered in 1958, is within the city, installed within a domus near the Nocera Gate.

Recent research has focused on the extensive excavations of these two establishments to clarify the production, chronology, and organization of each workshop. The information collected on these two workshops, coupled with archaeometric analysis, made it possible to reconstruct the operating sequence used by the Pompeian potters at these sites in 79 A.D., from processing of raw material to product distribution. The experimental approach concentrated on the potter's wheels. Five have been identified in Pompeii, four in the Herculaneum Gate workshop and one in the Nocera Gate workshop. Combining excavated material, iconographical depictions, and comparative evidence from other sites, we reconstructed one potter's wheel to better understand its mechanism and operation. We will present the various stages of this experiment: from the details of the structure to the results obtained, both successes and failures. We will conclude with a virtual rendering of one workshop, placing each element in its location and offering a reconstruction of one kiln and the potter's wheel.

Introduction

Scholarship has often covered the pottery production in Pompeii. We have many sources available, but the kilns and the archaeological structures related to this craft remained fundamental objects of study. Although we now have many indications from several periods concerning this activity in Pompeii, only three workshops active in 79 A.D. are known, and only two of these have been excavated: one inside the town (Reg. I, ins. 20, 3) and the other outside the city walls, located at nos. 28–30 along the Via dei Sepolcri, in the area of the Herculaneum Gate necropolis (Fig. 1). The third was outside the city in a part of the territory called *Pagus Felix Suburbanus*. This workshop has not been entirely uncovered. Only the vault of the kiln was documented and studied but information on its production is limited.[1]

The findings from the first two workshops' excavations, which were carried out between 2012 and 2017, document the entire *chaîne opératoire* of this craft, from the extraction of raw materials to the distribution of finished products, at the time of the destruction of the town in 79 A.D.[2] The archaeological evidence, combined with experiments and the work of real potters, have helped us gain a better understanding of how certain steps of this work functioned, in particular those relating to potter's wheels, and to propose reconstructions based on Pompeian excavated remains. Our project over the last few years has also focused on showcasing the remains of these workshops, by offering a 3D reconstruction and a virtual tour using augmented-reality headsets, providing a real immersion experience in a Pompeian potters' workshop.

1 Jacobelli/Antico Gallina/Di Maio (2020).

2 These campaigns of excavations and study of the potters' workshop (under multi-year excavation concessions issued by the Ministero Beni Culturali) were funded by the MEAE as part of the 'Southern Italy' archaeological mission, by the École française de Rome, the Centre Jean Bérard, the Centre Camille Jullian, the Arpamed endowment fund and private funds (CMD2 and Neptunia): Cavassa/Lemaire/Piffeteau (2013); Cavassa *et al.* (2014), (2015); Cavassa/Lacombe/Lemaire (2015); Cavassa/Lemaire (2019).

Note: Numbers preceded by SB (structure bâtie) refer to built structures and those with FR (four) refer to kilns.

https://doi.org/10.1515/9783111189635-011

The Herculaneum Gate Workshop

The studied area is situated in the Herculaneum Gate zone, just outside the settlement's walls in the northwest of the city.[3] From an organizational point of view, this is a mixed sector, combining mausolea funerary enclosures with opulent villas (e.g., Villa delle Colonne a mosaico, Villa of Diomedes, Villa of Cicero) as well as an area dedicated to trade and crafts. The latter is a complex marked by the presence of a portico about 85 m long, under which fifteen shops were set up. The penultimate, no. 29, which was identified as a potter's workshop because of the presence of a kiln, was the first stage of our study (Fig. 1).

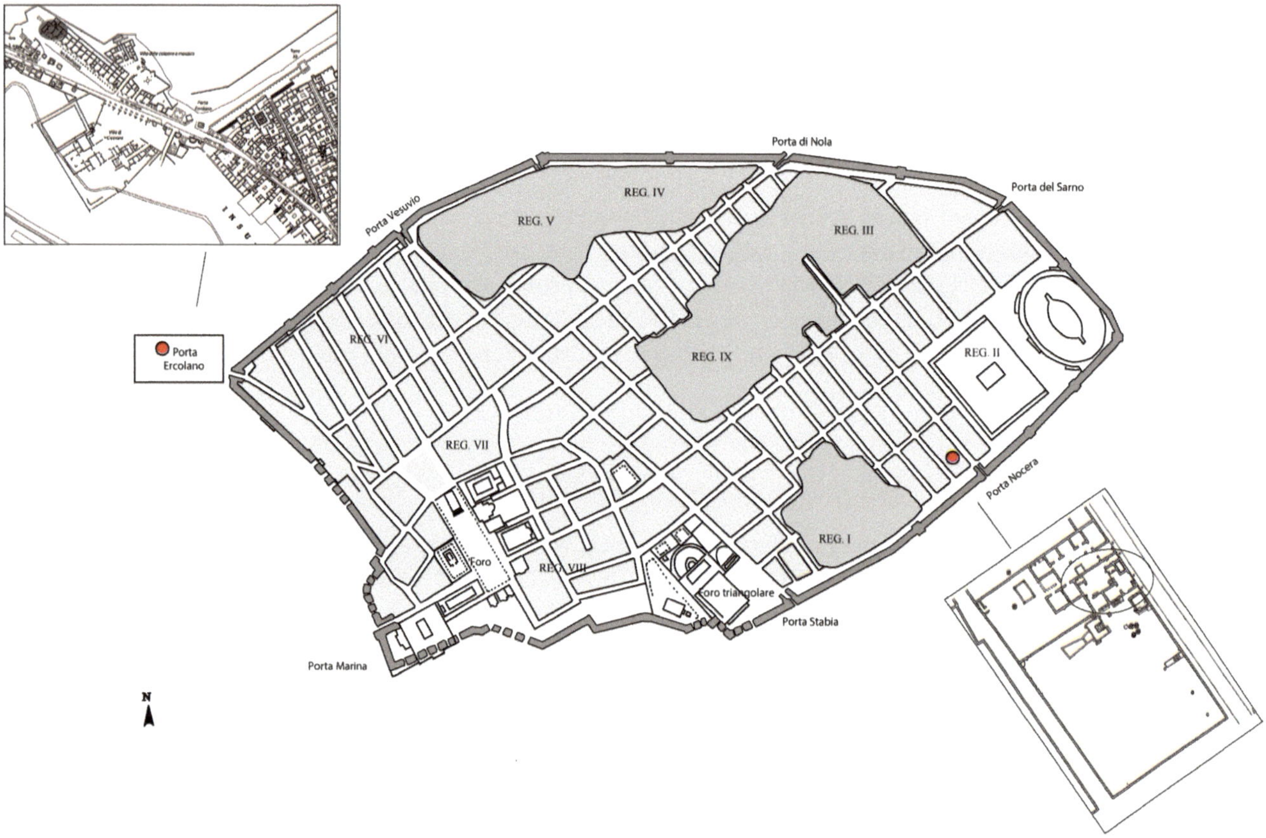

Fig. 1: General map of Pompeii with the localization of the two workshops: **(a)** Herculaneum Gate potters' workshop (Van der Poel 1984); **(b)** Nocera Gate potters' workshop located in Reg. I, ins. 20, 2–3 (Cerulli Irelli 1977, 69, fig. 1). Editing: Laetitia Cavassa, with the permission of the Italian Ministry of Culture – Archaeological Park of Pompeii. Forbidding on further reproduction or duplication by any means.

This workshop was uncovered between 15th October and 11th November 1838, and a few pages from the excavation diaries mention its discovery, between 15th and 21st October:

> in seguito di detta bottega si è disterrato un forno di terracotta, in dove si sono rinvenuti trentaquattro pignattini ad un manico. . . .22 ottobre—Si seguita sempre a travagliare nella strada dei Sepolcri, e propriamente in seguito della cosi voluta pubblica locanda, ove nell'ultima bottega si è palesata una fornace per cuocere gli oggetti di creta, ed in un angolo della medesima si sono raccolte trenta pignatte di diversa forma e grandezza, e tra esse una col manico lungo.[4]

3 The Herculaneum Gate workshop was also part of a wider, multi-year programme conduct by the École française de Rome (2012–2016) focusing on the development of the district.: https://www.efrome.it/it/la-ricerca/programmi/dettagli-programmi/organisation-gestion-et-transformations-dune-zone-suburbaine-le-secteur-de-la-porte-dherculanum-a-pompei-entre-espace-funeraire-et-commercial (last access 01.03.2024). See also: Zanella (2017); Zanella *et al.* (2016), (2017).

4 Translated by author: ". . . after this workshop, a kiln was unearthed, in which thirty-four one-handled pignattini (potteries) were found [. . .] 22 October – We continue our excavation in the street of the Sepolcri, and specifically after the so called public inn, where in the last workshop a

These notes provide information only on the presence of the kiln and of the vases that were discovered. These vases are called *pignatte* or *pignattini*, a very common name during the 18th century that can designate several vase shapes. In the Apulian region, for example, it is used to designate a pitcher used for preserving or serving liquid products;[5] in Sicily, it is a dish used to cook food in. Therefore, these terms cannot help us understand the production of the ancient workshop. Moreover, the sources do not agree on the total number of objects discovered: one source mentions "trenta pignatte"[6]; another "34 petites marmites de terre cuite dont une munie d'un long manche",[7]; a third one "382 vasetti di terra cotta della medesima forma"[8]; a fourth one "34 marmitte e altri vasi";[9] and a fifth one mentions "92 pignatte et pignattini".[10] These sources provide varying information on the workshop production, but in most of them these *pignattini* seem to refer to cooking pots.

There are also a few sources that cite the workshop itself. In 1855 E. Breton published a travel guide for visitors in which he mentions a pottery workshop. His description of the kiln is quite detailed, including information on dimensions and building techniques. Breton also mentions a second kiln:

> Le four qui se trouve dans l'arrière-boutique, est d'une construction très-remarquable. C'est un fourneau à réverbère, bâti en pierres et en briques, formant un massif carré de 2m,60 sur 2m,40, et 0m,90 de hauteur. La voûte du fourneau est plate est percée de petits trous pour laisser entrer la flamme dans le four, qui se trouve au-dessus. Celui-ci avait à l'intérieur 1m,25 en tous sens et 1m,80 de hauteur au milieu de sa voûte. Cette voûte, qui est la partie la plus singulière de sa construction, est formée de vases de terre cuite emboîtés les uns dans les autres, comme on le fit dans le VIe siècle pour la fameuse coupole de Saint-Vital de Ravenne. Des ouvertures ménagées dans les parois du four et garnies de tuyaux de terre cuite, permettaient de modérer la chaleur à volonté. Dans une pièce à gauche, et communiquant également avec la boutique, se trouve un four plus petit que l'on a point entièrement dégagé, mais dont la construction ne présente rien de remarquable.[11]

This piece of information was repeated a few years later by G. Fiorelli: "no. 29: L'ultima di tali botteghe richiude un forno a riverbero per la cottura delle stoviglie, ed ha accosto altra cella con un secondo forno non del tutto scoperto".[12] Later, in 1879, a few cleanings were made by L. Fulvio during which he unearthed "382 vasetti di terracotta della medesima forma".[13] He also published a detailed description of the kiln and provided a beautiful and quite realistic representation of it.[14]

In 1983, a German team carried out a series of studies on the construction and assemblages of all these shops. A few plans and pictures of the kiln were published, but it was not excavated.[15]

Finally, in 2009, J. Th. Peña and M. McCallum published an article reviewing the issue of production establishments in Pompeii. Based upon the visible structures and the maps of the buildings, they put forward several hypotheses on the repartition of the activities inside a workshop, such as possible positions of the potter's wheels, and fuel storage.[16] In short, the potter's workshop has been known since 1838, thanks to the presence of the kiln, but there was not much data available for its study. Besides, its production was known only through the references to *pignattini*.

---

kiln for firing clay objects was revealed, and in one of its corners thirty pignatte of different shapes and sizes were found, and among them one with a long handle": Fiorelli (1863) 361–362.

5 Cuomo di Caprio (2007) 537–538.

6 Fiorelli (1863). Translated by author: "thirty pignatte (potteries)".

7 Breton (1855) 234–235. Translated by author: "34 small terracotta cooking pots, one with a long stick-handle".

8 *Notizie degli scavi di Antichità* (1879) 207. Translated by author: "382 small terracotta pots of the same shape".

9 Annecchino (1977a) 106. Translated by author: "34 cooking pots and other vessels".

10 Kockel/Weber (1983) 79.

11 Breton (1855). Translated by author: "The kiln in the back room is a remarkable construction. It is a reverberatory furnace, built of stone and brick, forming a solid square measuring 2.60 m by 2.40 m and 0.90 m high. The vault of the kiln is flat and pierced with small holes to let the flame into the kiln. Inside, the kiln was 1.25 meters in all directions and 1.80 meters high in the middle of the vault. This vault, which is the most unusual part of its construction, is made of interlocking terracotta vases, as was done in the 6th century A.D. for the famous dome of Saint Vital in Ravenna. Openings in the walls of the kiln, lined with terracotta pipes, allowed the heat to be moderated at will. In a room to the left, also communicating with the shop, there is a smaller kiln that has not been fully excavated, but whose construction is unremarkable".

12 Fiorelli (1875) 416. Translated by author: "The last one of these shops contain a reverberatory kiln for burning potteries, and in an adjoining room there is a second kiln not completely excavated".

13 See *supra*, note 8.

14 Fulvio (1879) 280–281 and pl. II, 1.

15 Kockel/Weber (1983) 77–79, abb. 8–11 and taf. 32, n. 3.

16 Peña/McCallum (2009) 72–76.

Our excavation and study aimed to review and enrich these incomplete data. The objectives were many, including the detailed study of the kiln itself, the use period of the workshop, its production, the identification of potting activities (e.g., traces of the wheel and presence of settling tanks). The present state of our research allows us to establish that, in 79 A.D., the potter's workshop extended over at least three workshops: Workshops nos. 28, 29, and 30. We have also been able to reconstruct part of the potters' *chaîne opératoire* at the end of the 1st century A.D. in Pompeii.

Organization of the Workshop: Step by Step

The potter's entire workshop studied constitutes the western end of the long portico. Workshops 28 and 29 were built following the same plan (a first, almost square room giving access to a second, smaller room), whereas Workshop 30 has a different layout due to a later reorganization. Several stages of production activity have been identified in the three shops (Fig. 2), and excavations have shown that the potter's workshop was in operation when Mount Vesuvius erupted.[17]

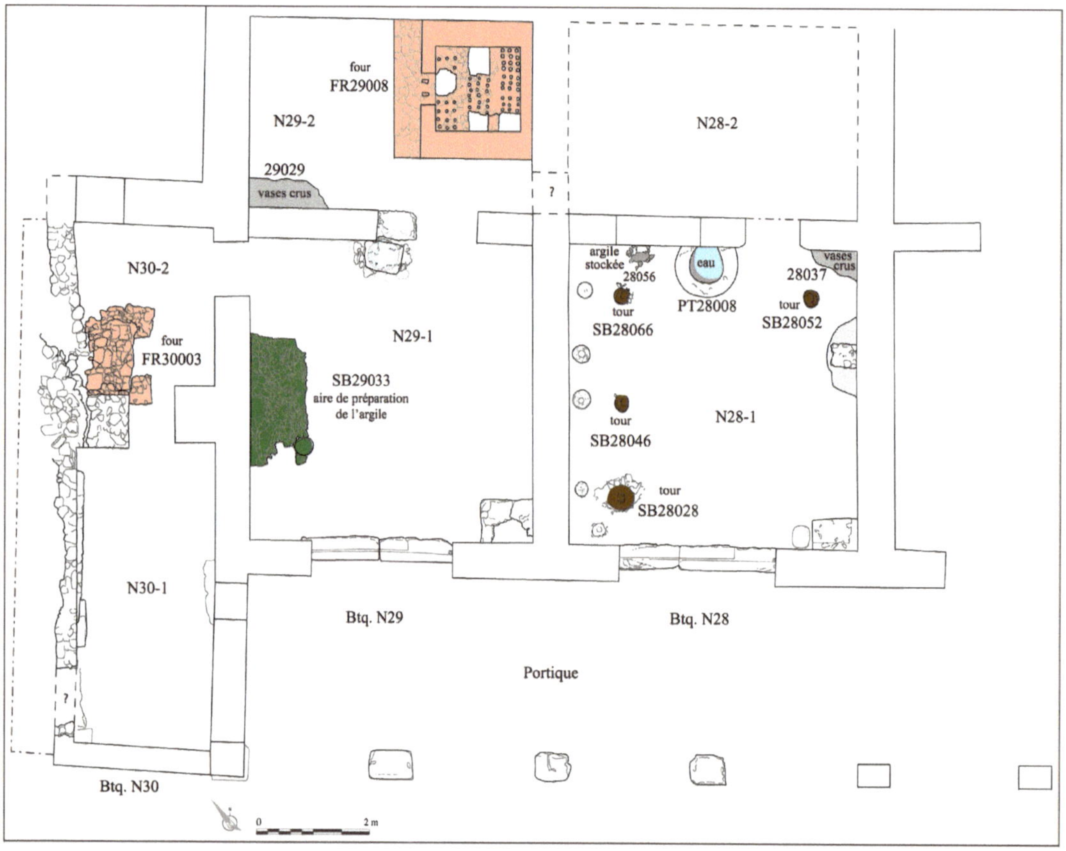

Fig. 2: General map of the Herculaneum Gate workshop during 79 A.D. Drawing: Guilhem Chapelin, J.-A. Delorme, Bastien Lemaire and John Marc Piffeteau, with the permission of the Italian Ministry of Culture – Archaeological Park of Pompeii. Forbidding on further reproduction or duplication by any means.

First step: access to the raw materials, water and clay. In rooms 28–1 and 29–1, two wells (or cisterns) were discovered. The one in room 29 was blocked up when the volcano erupted and therefore was placed out of order. However, the one in workshop 28 had been in a working state because it was filled with eruption materials. As for the clay gathering areas, against the northwestern wall of room 29–1, we uncovered a rectangular structure (SB29033) that is 0.98 m wide, 2.32 m

17 The workshop was in operation for several years during the 1st century A.D. and underwent various phases of occupation, but here we will only present the data relating to the last phase in 79 A.D.

long, and made of tile fragments that had been cut and arranged on the floor over a leveling backfill. The structure has a slight dip toward the southern corner, where there is the bottom of a cut Italic amphora in which rough sand was discovered, varying in color from black to gray. This sand looked like the volcanic temper found in the clay of the ceramics produced in the workshop.[18] This structure may be where the clay was prepared.

Second step: shaping of the vases. Room 28–1 is thought to have been a work room. Its preservation as the time of the discovery enabled us to determine the different work areas. Although our 19th-century predecessors have most certainly removed a large number of objects, the entire area of the excavated room was partly covered with lapilli.

Four structures were discovered and identified as potter's wheels. A first kind of wheel is represented by only one of them (SB28028), located in the western corner of the room next to the shop's threshold. This wheel is shaped like a circular depression measuring 46 cm in diameter. Its northeastern half is marked by tile fragments and a few basalt elements refined by mortar and smoothed on the inside. The bottom of this depression 25 to 28 cm deep, is flat, where fragments of raw vases were discovered, the results of turning waste. In the center of this depression, there was a rectangular cavity (14 cm long on each side and 46 cm deep) that used to hold the wooden axle of the wheel. These specifics allow us to say that the wheel had a fixed axle whose upper part was mobile, made out of the wheelhead and of its rotation system.

There are three wheels of a second kind: two of them (SB28046 and SB28066) are located northeast of the previous wheel, and one (SB28052) is in the eastern corner of the room. The disc is made of the bellies of cut Italic amphorae, with internal diameters between 23 and 30 cm and between 20 and 26 cm high (Fig. 3). The bottom of the first wheel's depression presented a cavity of a rectangular or irregular section, depending on the wheels, which varied in size between 12 and 18 cm on the sides and 37 to 40 cm deep where the axle used to stand. These turning structures also had fixed axles.

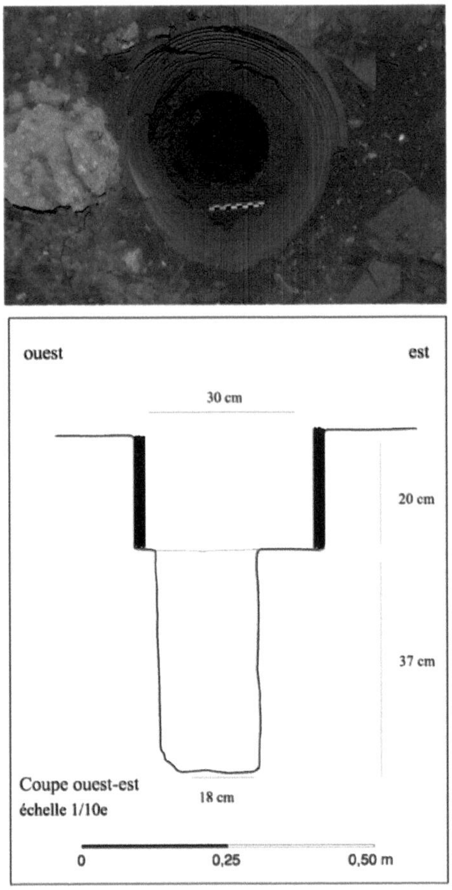

Fig. 3: Picture and section of a Pompiean potter's wheel in the Herculaneum Gate workshop. Photo and drawing: Laetitia Cavassa and Bastien Lemaire with the permission of the Italian Ministry of Culture – Archaeological Park of Pompeii. Forbidding on further reproduction or duplication by any means.

18 Minero-petrographic and chemical analysis of this sand and of a series of ceramic samples were made by the research teams of the Università di Studi del Sannio and Università degli Studi di Napoli Federico II: Grifa et *al.* (2021a), (2021b). Preliminary observations are reported in the third part of this article.

Each of the three wheels in the northwestern half of room 28–1 was found with amphora bottoms set in the floor next to them. Two of these wheels are pierced, but the third is not. Such layout suggests they served as containers for water or clay at the potter's immediate disposal. Once the vases were finished, turned, covered with slip and decorated, they were left to dry. We have uncovered at least two drying places.

Third step: the drying phase. Areas for drying were indicated in the eastern corner of room 28–1, just next to wheel (SB29052), through fragments from unfired vases covered with slip and decorated. The same is true of the western corner of room 29–2, where more unfired vases, decorated and covered with slip, were discovered close to kiln FR29008. In this second zone, vases were laid out and aligned on the bare floor. When Vesuvius erupted, they were ready to be put into the kiln to be fired.

Fourth step: firing. Once prepared, the vases were then fired. The workshop had at least two kilns active in 79 A.D. The first kiln (FR29008), located in room 29–2, is rectangular, 2.48 m long (north–west / south–east), and 2.41 m wide (north–east / south–west). It is vertical,[19] belonging to N. Cuomo di Caprio's type IIb: "fornace con camera di combustione a pianta quadrata a corridoio centrale."[20] It stands in the eastern corner of the room; its external facings are made of small tuff and basalt blocks bound to gray mortar. The internal facings of the combustion chamber, of the furnace chamber, and of their access are made of *tegulae* cut in such a way as to make the most of the refractory characteristics of this material. The kiln escharарests upon two retaining arches. The cover of the kiln is no longer preserved, but the modern written sources tell us that the superior part ended in a vault built with vases piled up one inside another.[21]

A second kiln (FR30003) was also active during the last phase of the workshop. Smaller in size, it is located in the western corner of room 30–2. It was very substantially damaged, most likely during October 1943's bombings.[22] Only the inferior part of the combustion chamber remains, 1.24 m wide (north–west / south–east) and 1.73 m long (north–east / south–west). This kiln is mainly made of cut tegulae, except for the external facings made with tuff blocks and reused elements such as Dressel 2/4 Italic amphora fragments as well as two-handled pots. No traces of coatings have been observed on the walls; the back of the kiln is made of several overturned and cut tegulae. We were able to distinguish the remains of a retaining arch that used to hold the hearth plate.

In room 30–1, no structures dating back to the last phase of the workshop have been discovered. Despite the destruction of the west wall, probably caused by the bombing raids of 1943, it was possible to uncover a doorway leading to the area located just to the west of the building and most likely to the street below.

The Workshop's Production

As noted previously, we had very little information on the workshop's production. Only a few references about several vases discovered in the shops were available to us: "pignatte di diversa forma e grandezza, e tra esse una col manico lungo." As we were unable to locate any materials uncovered in 1838, this brief reference remained our only clue.

From the first days of the excavation, in room 29–2 a few dozen centimeters west of kiln FR29008, we uncovered a level of lapilli dating from the Vesuvius eruption, intact and untouched by the 19th–century excavations. This small heap, direct evidence of the catastrophe, stood in the western corner of the room, at the foot of wall MR29006, which separates rooms 29–1 and 29–2. It contained about ten vases laid out on the bare floor that were unfired and were shaped, turned, decorated with guilloche, and covered with slip. Their preservation enabled us to determine that the workshop was producing cups with a thin coating of slip. More specifically, those discovered at the drying stage were cups of forms Mayet XXb[23] and Ricci I/111,[24] cups with a globular belly, a flat bottom, and a small, flared rim. They feature a vertical handle (Fig. 4). This extraordinary discovery enabled us to identify securely the production and to picture the *pignattini* mentioned in excavation diaries and to argue beyond any doubt that the workshop was indeed in full working mode when the volcano

19 In opposition to some kinds of kilns for which the heating area and part of the structure are half-buried.
20 Cuomo di Caprio (1971–1972) 429–434.
21 Breton (1855).
22 Garcia y Garcia (2006) 163.
23 Mayet (1975).
24 Ricci (1985).

erupted. These vases had most likely been stored in this room's corner, probably awaiting their firing in the kiln. This first discovery was reinforced in 2014 when other fragments of unfired vases, a pile of vases not as well preserved, were discovered in the eastern corner of room 28–1. However, the same cup shape can easily be recognized. Many fragments of decorated body areas, of rims, and of handles confirm this identification. These discoveries further confirm that the workshop was active when the volcano erupted. The excavation warehouses contain many examples of these. A. Carandini listed 176 of them in the storehouses of the Naples Archaeological Museum, divided according to three size units.[25] Finally, we were also able to clarify an element mentioned in the excavation diaries: the presence of a vase with a "long stick-handle." It is a particular vase that belongs to the group of cups of this precise type, but its handle is shaped like a stick, not like a loop. Indeed, we have uncovered several examples of these stick-handles. A whole cup unearthed in Pompeii and stored in Naples, previously published by A. Carandini, represents this type of production.[26]

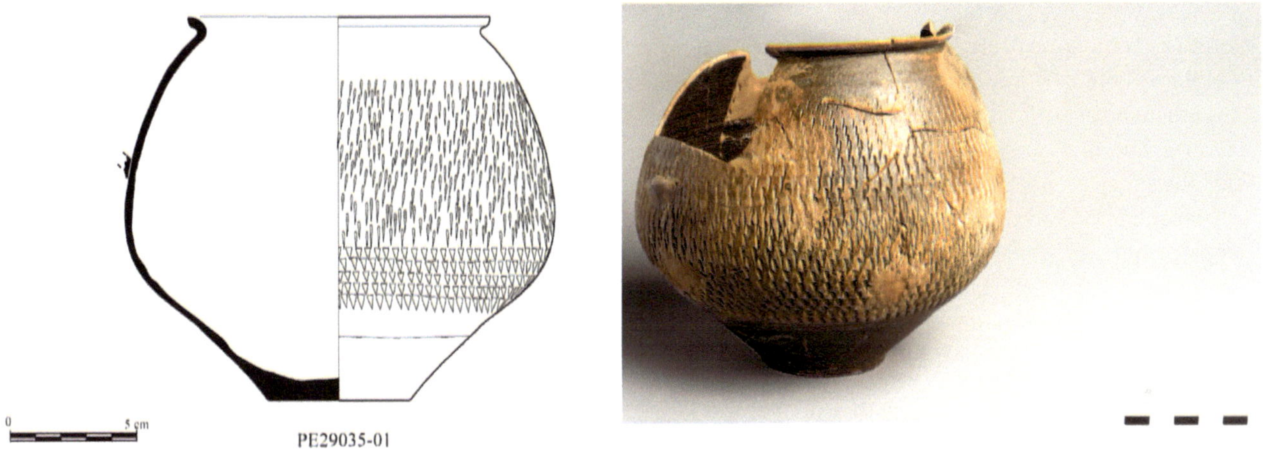

Fig. 4: Thin-walled pottery produce in the Herculaneum Gate workshop in 79 A.D. Photo and drawing: Laetitia Cavassa and Christine Durand. With the permission of the Italian Ministry of Culture – Archaeological Park of Pompeii. Forbidding on further reproduction or duplication by any means.

The Nocera Gate Workshop

This second workshop studied is located in the southeastern part of the town, near the Nocera Gate, in Reg. I, *insula* 20, 2–3. It was excavated between December 1958 and February 1959 by Amedeo Maiuri. The workshop was immediately identified thanks to the discovery of two kilns, and the production likewise with the discovery of 61 ceramic oil lamps and 24 plaster molds in one kiln, and 123 small vases known as *fritilli* whose function is unclear, possibly pottery for making vaults, cup to play thimbles, as shown on wall paintings, or pots for holding pigments? Further excavations were carried out in 1973.[27]

The workshop was set up in a domus after the 62/63 A.D. earthquakes. Several commercial and craft structures were established in houses after the disaster, probably because of the massive influx of labor needed to rebuild the city.[28] At the time of the eruption, the workshop was mainly organized around two rooms in which all the activities were concentrated: rooms 1 and 4 (Fig. 5). Room 4 was the workshop's main room, the epicenter of production. It is accessed directly from the street to the east through door 3, via a corridor. This room measures 7.25 m from east to west and 4.95 m from north to south (i.e., almost 36 m^2). During new excavations in 2016 and 2017,[29] we completed our knowledge of this workshop, and, as for the Herculaneum Gate workshop, we are now able to reconstruct the entire *chaîne opératoire* of the Pompeian potters.

25 Carandini (1977) 25–31. According to A. Carandini, 2 vases belong to the small size unit (8 cm high), 143 belong to the medium size unit (9 cm), and 31 to the big size unit (12 cm).
26 Naples, Museo Archeologico Nazionale, inv. no. 16899.
27 Cerulli Irelli (1977); Annecchino (1977b).
28 Andreau (1979) 43.
29 Cavassa *et al.* (2017), (2018).

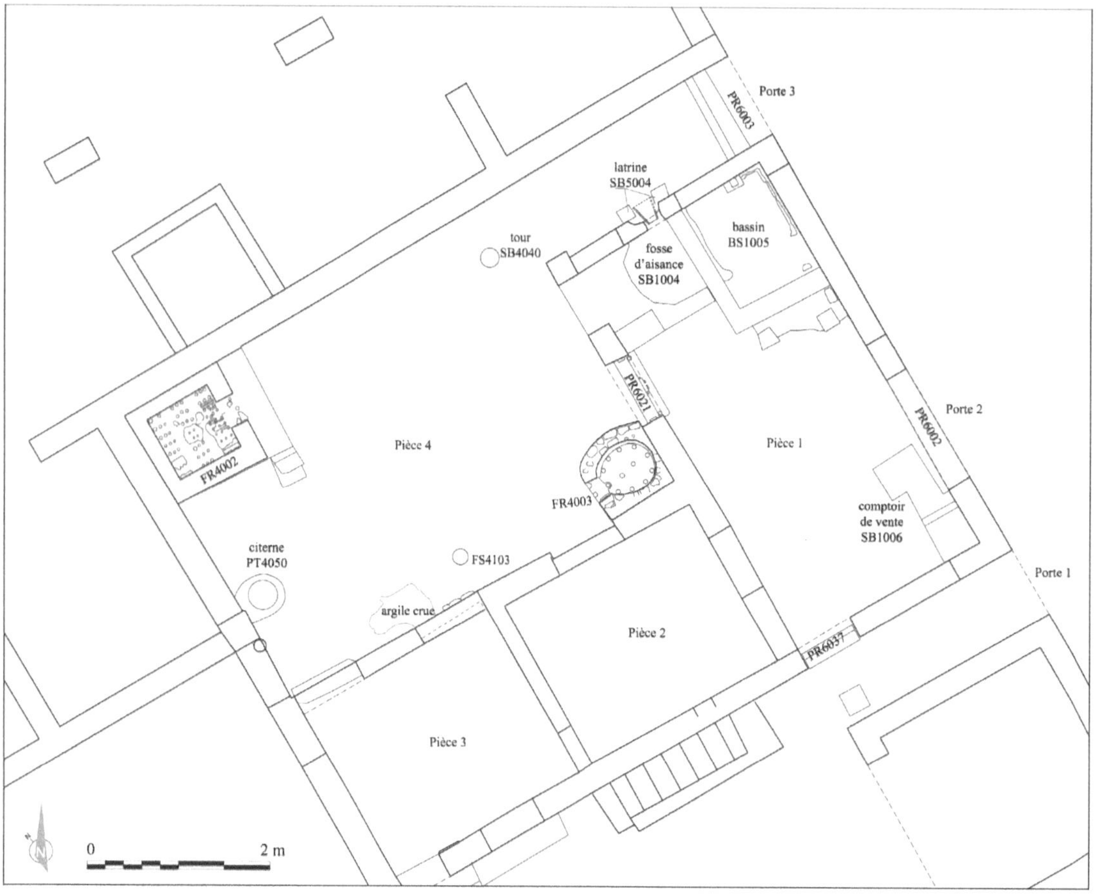

Fig. 5: General map of the Nocera Gate workshop during 79 A.D. Drawing: Guilhem Chapelin with the permission of the Italian Ministry of Culture – Archaeological Park of Pompeii. Forbidding on further reproduction or duplication by any means.

First step: the raw material. In the first room, there is a quadrangular basin measuring 1.16 m wide and almost 2 m long, covered with a hydraulic plaster. This structure seems to be directly linked to the clay-cleaning phase, despite the absence of a water supply and drainage system. Possibly the clay was filled using buckets, especially as there is a cistern fed by a pipe made from reused amphorae and installed in a wall in room 4.

On the other hand, there was no space in this workshop dedicated to clay working or preparation, but we did uncover a reserve of clay (nearly 1 m long), raw and ready for use, in the kiln room.

Second step: shaping of the vases. This workshop was known for making oil lamps (using bivalve molds), as well as small wheel thrown vases. Although the entire workshop was excavated at the end of 1959, one detail seems to have escaped the excavators: the location of the potter's wheel. It was uncovered in room 4 during the 2016 excavations. Based on the same model as the potter's wheels in the Herculaneum Gate workshop, the wheel consists of a circular cavity dug into the ground and held in place by a carved Vesuvian Italic amphora; a negative (16 and 12 cm x 30 cm) of the wooden stake that held the hoe was found in its center. Two elements that were essential in understanding how this structure functioned were also uncovered: a pebble almost 3.5 cm long, marked by a circular cavity in its center, and an iron nail 5 cm long, the lower part of which still bears traces of wood. The discovery of this fifth wheel structure in Pompeii completes the data uncovered in the Pompeian workshops and confirmed that a wheel with a fixed-axle was always used.

Third step: firing. The third documented phase in this workshop is the firing. As in the previous one, two kilns were in operation in this building at the time of the eruption. The first kiln is located in the western corner of the room. It is a vertical rectangular kiln (of type IIb, the same type as the Herculaneum Gate kiln) measuring 2.09 m by 1.79 m and standing almost 2.48 m high. The firing chamber also features a permanent vault, as shown by the photographs and sketches made at the time of its discovery, and also by the impressions of the vases used for the construction of the kiln.[30] The second kiln is circular and more modest in size (0.88 m in diameter and 1.38 m high). It is located in the southeastern corner of the room. It belongs to N. Cuomo di Caprio's type Ia, a kiln with a central pillar.[31] It is made of cut tiles and similarly to the other kiln, parts of its walls have been restored.

Fourth step: production. The last step illustrated here is the diffusion of the production. In this workshop, in room 1, there is an L-shaped masonry counter facing the street, an ideal position for selling products directly. It is 1.95 m long and 0.90 m wide. Although we have no clear clues, we can speculate that the lamps and small vases may have been sold from this structure.

Regarding the production, oil lamps and little vases were produced in this workshop. The lamps are common for this period in Pompeii, with various decorative themes, including divinities, animals or figures, such as winged lovers. The only peculiarity is the production of lamps with two opposed nozzles and a central handle.[32] No lamp bears the signature of a potter.

The Pottery Production in Pompeii

This research, focusing on Pompeian potters' workshops, has helped us understand the organization of the pottery production in this city in 79 A.D. We know two workshops, operating in different times and with different lines of production one for thin-walled potteries, and the other for lamps and little vases. All the findings, as part of an interdisciplinary approach, allow us to propose a reconstruction of the *chaîne opératoire*.

Various factors have been taken into consideration, and now we can better appreciate the different aspects, such as the origin of the clays, the operation of the wheels, the construction of the Pompeian kilns, and, finally, we can propose a way of showcasing all the data.

The first point is the clay. In most cases, workshops are set up close to deposits of raw materials, to clay quarries, to a water source, to fuel resources, or along communication routes for the supply or sale of products. In the case of Pompeii, it appears there were no clay quarries in the vicinity of the workshops. The discovery of unfired vases and clay reserves are important elements in understanding the supply networks. Thanks to the presence of characteristic and identifiable microfossils in the raw clay,[33] the archaeometric analyses for the material belonging to the last period of the workshops (79 A.D.) carried out on this material have identified the exact quarries from where the clays were extracted.[34] These quarries are located in the area of present-day Salerno, around thirty kilometers from Pompeii,[35] suggesting that clay was transported from its extraction site to the workshops.

Regarding the wheels, the elements found during the excavations allow us to understand and suggest a method of operation. The five examples known in Pompeii are all low potter's wheels, operating on a fixed wooden axle set into the ground. The upper part of the wheel is rotated by means of a nail driven into the stake, which acts as a pivot and on which the wheel rests. The wheel is rotated by means of a clamp (in this case a pebble) previously fixed to the underside of the wheel. The wheel was operated by hand or by means of a stick, as can be seen in the two Pompeian frescoes depicting potters at work. To back this up, we replicated these wheels to understand how they work. Using the elements at our disposal, namely a circular cavity dug in the ground, the imprint of a wooden stake, a nail, a pebble, and wood remains, we carried out some tests (Fig. 6). The first wheel, made of pine, was 58 cm in diameter, 6 cm thick, and weighed 7 kg. The tests showed that the wheel works; the rotation takes place between the nail and the pebble, but the speed of rotation is fairly low. We therefore carried out a second test using oak, with a wheel 58 cm in diameter and 8 cm thick, but heavier (11 kg). The results were much more encouraging and enabled us to achieve a satisfactory rotation speed. New experiments will help us have a better understanding of the entire process.

30 *PPM II*, 1967, note 1.
31 Cuomo di Caprio (1971–1972) 411–413.
32 Cerulli Irelli (1977); Cavassa (2020).
33 Grifa *et al.* (2021a), (2021b).
34 The Herculaneum Gate workshop has differents period of activities, and the analysis shows different results for the other periods. Grifa *et al.* (2021a), (2021b).
35 De Bonis *et al.* (2013) 3, fig. 1.

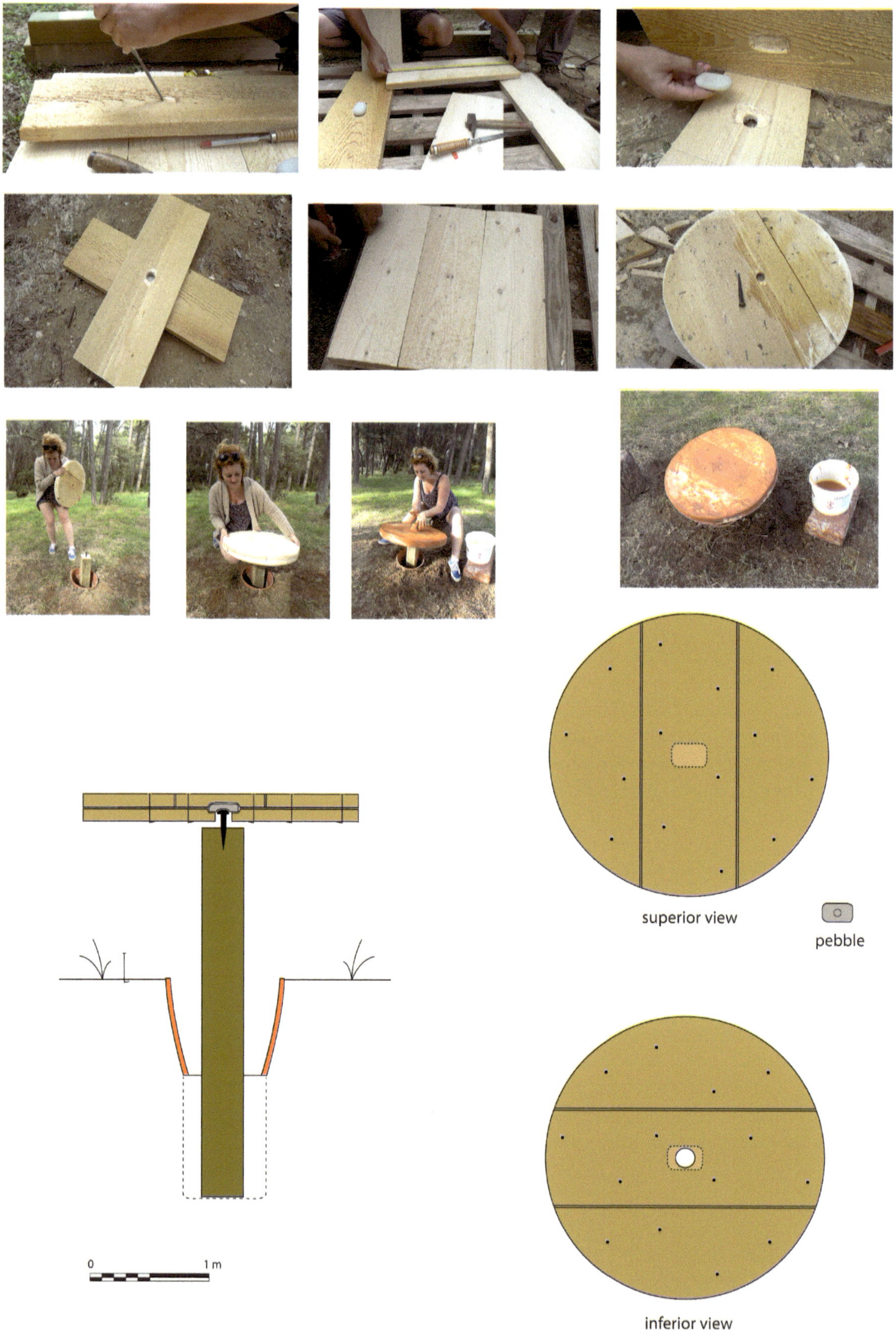

Fig. 6: Different steps of the reconstruction of a Pompeian wheel. Photos and drawings: Guilhelm Chapelin and Bastien Lemaire with the permission of the Italian Ministry of Culture – Archaeological Park of Pompeii. Forbidding on further reproduction or duplication by any means.

The last point is the construction of Pompeian kilns, and in particular the technique for the vault of the upper part of the kiln. With the exception of the circular kiln in the Nocera Gate workshop, all other kilns, including the unexcavated kiln unearthed on the outskirts of the city, had the distinctive feature of a permanently vaulted combustion chamber made of interlocking vases, traditionally known as amphorulae cusatae.[36] This technique was not unique to Pompeii, but it is overall rare: only around thirty sites are known to have used the same construction technique in antiquity, using specific vaulted vessels produced for this specific purpose.[37] In the case of the Pompeian kilns, the vessels used in construction are common ceramic pots produced in Pompeii that were reused in a context different than their primary purpose: Schoene-Mau 1 type pots for the kilns in the Herculaneum Gate workshop and Di Giovanni type 2211 ollae, produced locally for the kilns in Nocera Gate.

All the material uncovered during the excavation of the two workshops provides us with an accurate picture of the work of Pompeian potters at the time of the eruption of Vesuvius in 79 A.D. These scientific results have also been used to promote the findings among both academic and general audiences, by producing a 3D reconstruction of the Nocera Gate workshop (Fig. 7) and offering a virtual tour of the establishment during which all the elements come to life.[38]

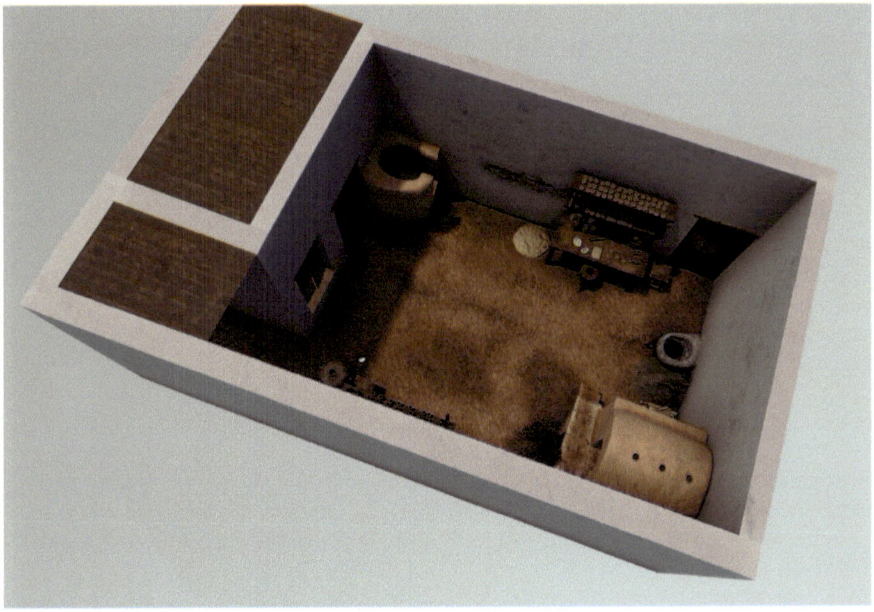

Fig. 7: 3D reconstruction of the Nocera Gate workshop. Rendering: Fabrice Paul and P. Poveda with the permission of the Italian Ministry of Culture – Archaeological Park of Pompeii. Forbidding on further reproduction or duplication by any means.

Conclusions

The results of our long-term research have helped us reconstruct the entire organization process of the Pompeian potters during the last years of life in the Roman city of Pompeii. These two workshops offer a fairly accurate picture of how this craft worked, from the arrival of the raw materials to the distribution of the finished products. These discoveries have also helped us to bring to light the presence of potter's wheels, which had previously been illustrated by frescoes. It is also important to note that these two structures produced basic products, such as lamps, and drinking cups, intended for the local market.

36 Promis (1875) 252.
37 Lancaster (2015).
38 Mannocci/Pawlowicz/Cavassa (2022).

Abbreviation

PPM II *Pompei: pitture e mosaici, vol. II. Regio I: parte seconda*, Rome, 1990.

Bibliography

Andreau (1979): Jean Andreau, "Il terremoto del 62", in: Fausto Zevi (ed.), *Pompei* 79, Naples, 40–44.
Annecchino (1977a): Maria Annecchino, "Suppellettile fittile da cucina di Pompei", in: *L'instrumentum domesticum di Ercolano e Pompei nella prima età imperiale*, Quaderni di cultura materiale, 1, Rome, 105–120.
Annecchino (1977b): Maria Annecchino, "Fritillus. Un piccolo vaso di terracotta", in: *Cronache pompeiane, III*, Rivista dell'associazione internazionale "Amici di Pompei", 198–213.
Breton (1855): Ernest Breton, *Pompeia décrite et dessinée suivie d'une Notice sur Herculanum*, seconde édition, Paris.
Carandini (1977): Andrea Carandini, "La ceramica a pareti sottili di Pompei e nel museo nazionale di Pompei", in: *L'instrumentum domesticum di Ercolano e Pompei nella prima età imperiale*, Quaderni di cultura materiale 1, Rome, 25–31.
Cavassa (2020): Laetitia Cavassa, "Pompei, città figulinaria nel 79 d.C.", in: Massimo Osanna and Luana Toniolo (eds.), *Fecisti cretaria. Dal frammento al contesto: studi sul vasellame ceramico del territorio vesuviano*, Studi e ricerche del parco archeologico di Pompei 40, Rome, 2020, 23–30. https://shs.hal.science/halshs-01478833.
Cavassa (2024): Laetitia Cavassa, *La production céramique à Pompéi jusqu'en 79 après J.-C.*, Archéologie de l'artisanat antique 12, Naples.
Cavassa/Lacombe/Lemaire (2015): Laetitia Cavassa, Aline Lacombe and Bastien Lemaire, "Une production de gobelets à paroi fine à Pompéi en 79 de notre ère", in: *SFECAG – Société Française d'Etude de la Céramique Antique en Gaule, Actes du Congrès de Nyon*, Marseille, 285–292.
Cavassa/Lemaire (2019): Laetitia Cavassa and Bastien Lemaire, "Un atelier de potier en activité en 79 de notre ère à Pompéi", in: Mario Denti and Mathilde Villette (eds.), *Archéologie des espaces artisanaux. Fouiller et comprendre les gestes de potiers* (Actes du colloque international de Rennes, 27–28 novembre 2014, Lattes, MAM Hors Série n°9, 191–202. https://shs.hal.science/halshs-03012866.
Cavassa/Lemaire/Piffeteau (2013): Laetitia Cavassa, Bastien Lemaire, John-Marc Piffeteau and Pompéi, "Via dei sepolcri, boutique NE, n. 29: l'atelier de potier", in: *Chronique des activités archéologiques de l'École française de Rome*. https://doi.org/10.4000/cefr.881.
Cavassa *et al.* (2014): Laetitia Cavassa, Bastien Lemaire, Guilhem Chapelin, Aline Lacombe, John-Marc Piffeteau and Giuseppina Stelo, "Pompéi. L'atelier de potier de la via dei Sepolcri, 29", in: *Chronique des activités archéologiques de l'École française de Rome*. https://doi.org/10.4000/cefr.1139.
Cavassa *et al.* (2015): Laetitia Cavassa, Bastien Lemaire, Guilhem Chapelin and Aline Lacombe, "Pompéi. L'atelier de potier de la *via dei Sepolcri*, 28–30", in: *Chronique des activités archéologiques de l'École française de Rome*. https://doi.org/10.4000/cefr.1279.
Cavassa *et al.* (2017): Laetitia Cavassa, Bastien Lemaire, Antoine Boisson, Guilhem Chapelin, Giulia Ciucci, Christine Durand, Jean-Marie Gassin, Giovanni Festa, Marcella Leone, Agnès Oboussier, Fabrice Paul, Marie Pawlowicz, Rémi Rêve, Carla Rosa, Giuseppina Stelo and Lorenzo Vollaro, "Recherches sur la production céramique à Pompéi: l'atelier des lampes à huile (*Reg.* I, *Ins.* 20, 2–3)", in: *Chronique des activités archéologiques de l'École française de Rome*. https://doi.org/10.4000/cefr.1761.
Cavassa *et al.* (2018): Laetitia Cavassa, Sylvie Coubray, Emilie Mannocci, Agnès Oboussier, Marie Pawlowicz, Carla Rosa and Marcello Turci, "Recherches sur la production céramique à Pompéi: l'atelier des lampes à huile (*Reg.* I, *Ins.* 20, 2–3). Campagne 2017", in: *Chronique des activités archéologiques de l'École française de Rome*. https://doi.org/10.4000/cefr.2230.
Cerulli Irelli (1977): Giuseppina Cerulli Irelli, "Officina di lucerne fittili a Pompei", in: *L'instrumentum domesticum di Ercolano e Pompei nella prima età imperiale*, Quaderni di cultura materiale 1, Rome, 53–72.
Cuomo di Caprio (1971–1972): Ninina Cuomo di Caprio, "Proposta di classificazione delle fornaci per ceramica e laterizi nell'area italiana dalla preistoria a tutta l'epoca romana", in: *Sibrium* 11, 429–434.
Cuomo di Caprio (2007): Ninina Cuomo di Caprio, *Ceramica in archeologia 2. Antiche tecniche di lavorazione e moderni metodi di indagine*, Rome.
De Bonis *et al.* (2013): Alberto De Bonis, Celestino Grifa, Giuseppe Cultrone, Pantaleone De Vita, Alessio Langella and Vincenzo Morra, "Raw Materials for Archaeological Pottery from the Campania Region of Italy: a Petrophysical Characterization", in: *Geoarchaeology: An International Journal* 28, 478–503.
Fiorelli (1863): Giuseppe Fiorelli, *Pompeianarum antiquitatum historia* II, Naples.
Fiorelli (1875): Giuseppe Fiorelli, *Descrizione di Pompei*, Naples.
Fulvio (1879): Luigi Fulvio, "Delle fornaci e dei forni pompeiani", in: *Pompei e la regione sotterrata dal vesuvio nell'anno LXXIX*, Memorie e notizie pubblicate dall'ufficio tecnico degli scavi delle province meridionali, Naples, 273–291.
Garcia Y Garcia (2006): Laurentino Garcia y Garcia, *Danni di guerra a Pompei. Una dolorosa vicenda quasi dimenticata. Con numerose notizie sul 'Museo Pompeiano' distrutto nel 1943*, Studi della Soprintendenza Archeologica di Pompei 15, Rome.
Grifa *et al.* (2021a): Celestino Grifa, Chiara Germinario, Alberto De Bonis, Laetitia Cavassa, Francesco Izzo, Mariano Mercurio, Alessio Langella, Ioanna Kakoulli, Christian Fischer, Diana Barra, Giuseppe Aiello, Gianluca Soricelli, Christopher R. Vyhnal and Vincenzo Morra, "A Pottery Workshop in Pompeii Unveils New Insights in the Roman Ceramics Crafting Tradition and Raw Materials Trade", in: *Journal of Archaeological Science* 126, 105305. https://doi.org/10.1016/j.jas.2020.105305.

Grifa *et al.* (2021b): Celestino Grifa, Chiara Germinario, Alberto De Bonis, Laetitia Cavassa, Francesco Izzo, Mariano Mercurio, Alessio Langella, Ioanna Kakoulli, Christian Fischer, Diana Barra, Giuseppe Aiello, Gianluca Soricelli, Christopher R. Vyhnal and Vincenzo Morra, "Archaeometric Data from the *Via dei Sepolcri* Ceramic Workshop in Pompeii (Southern Italy)", in: *Data in Brief*, 24, 106706. https://doi.org/10.1016/j.dib.2020.106706.

Jacobelli/Antico Gallina/Di Maio (2020): Luciana Jacobelli, Maria Vittoria Antico Gallina, Giovanni Di Maio, "Pompei suburbio Nord-Ovest: Nuovi importanti indizi di un quartiere industriale", in: *Riflessioni su dati di scavo, tecniche edilizie, ambiente e paesaggio*, *Archeologia dell'Architettura* XXV, 193–208.

Kockel/Weber (1983): Valentin Kockel and Berthold F. Weber, "Die Villa delle colonne a mosaico in Pompeji", in: *Römischen Mitteilungen* 90, 51–89.

Lancaster (2015): Lynne C. Lancaster, *Innovative Vaulting in the Architecture of the Roman Empire. 1st to 4th Centuries CE*, New York.

Mannocci/Pawlowicz/Cavassa (2022): Emilie Mannocci, Marie Pawlowicz, Laetitia Cavassa, "'Être potier à Pompéi' ou comment faire vivre l'expérimentation archéologique au grand public", in: *Bulletin de l'Apera*, 91–106.

Mayet (1975): Françoise Mayet, *Les céramiques à parois fines dans la péninsule ibérique*, Paris.

Peña/McCallum (2009): J. Theodore Peña and Myles McCallum, "The Production and Distribution of Pottery at Pompeii: A Review of the Evidence. Part 1, Production", in: *American Journal of Archaeology* 113, 57–79.

Promis (1875): Carlo Promis, *Vocaboli latini di architettura posteriori a Vitruvio*, Turin.

Ricci (1985): Andreina Ricci, "Ceramica a pareti sottili", in: *Enciclopedia dell'Arte Antica,* vol. II, Rome, 233–357.

Van der Poel (1984): Halsted B. Van der Poel, *Corpus Topographicum Pompeianum: The Rica Maps of Pompeii*, Austin, 1984.

Zanella (2017): Sandra Zanella, "Progetto: Porta Ercolano 2012–2016", in: *Rivista di Studi Pompeiani* 28, 131–136.

Zanella *et al.* (2016): Sandra Zanella, Laetitia Cavassa, Nicolas Laubry, Nicolas Monteix, Bastien Lemaire, "Pompéi, *Porta Ercolano*: organisation, gestion et transformations d'une zone suburbaine", in: *Chronique des activités archéologiques de l'École française de Rome*. https://doi.org/10.4000/cefr.1581.

Zanella *et al.* (2017): Sandra Zanella, Laetitia Cavassa, Nicolas Laubry, Nicolas Monteix, Guilhem Chapelin, Arnaud Coutelas, Audrey Delvigne Ryrko, Mélanie Errera, Léa Gerardin, Bastien Lemaire, Raphael Macario, Florien Ortis, Vincenzo Pellegrino and Géraldine Sachau-Carcel, "Pompéi, Porta Ercolano : organisation, gestion et transformations d'une zone suburbaine", in: *Chronique des activités archéologiques de l'École française de Rome*. https://doi.org/10.4000/cefr.1676.

Ludovica Xavier de Silva

12 Republican Kilns at Pollentia-Urbs Salvia: An Interdisciplinary Approach to the Study of a Local Ceramic Production

Abstract: The paper gives an overview of the "pilot action" developed for the Archaeological Park of Pollentia-Urbs Salvia within the European TRANSFER project, which had among its focuses the use of ICTs to enhance the cultural heritage. The outcome of this project was the development of a Virtual Reality experience concerning the organisation and functioning of a Roman ceramic workshop. The discovery of a group of three kilns dated to the 3rd–2nd century B.C. led to the study of the workshop organisation and pottery production of the site. At the same time, the obliteration of these structures due to their poor state of conservation highlighted the necessity of visualising and communicating the ceramic workshop differently. The interdisciplinary approach used for developing the VR experience, involving 3D modelling and printing, experimental archaeology, video making, and public archaeology, has eventually provided the chance to approach both research and its communication from different perspectives.

In 2017, the enhancement works at the so-called Republican Building of Pollentia-Urbs Salvia provided an opportunity to resume the excavation of the area, which started in 2000, when the existence of a productive area was first documented. Three pottery kilns have been brought to light (Fig. 1a). Excavation data suggest that two of these kilns, named A and B, must have worked together.[1] The third kiln—Kiln C—is slightly smaller than the other two, and its foundation in front of the praefurnium of Kiln B suggests that the workshop must have had a second phase, in which at least one of the two initial kilns was not working anymore. In the latest phase of the workshop, a forge was built on the remains of Kiln C, suggesting a conversion of the productive area to producing metal tools.[2]

Located south of the later forum, the workshop must have been part of a larger artisanal area, since in the 2000s some remains of another kiln were found nearby, slightly to the east.[3] Here we will focus exclusively on kilns A and B, as they belong to the phase of the workshop that we decided to work on. Both kilns have a south-to-north orientation and belong to Cuomo di Caprio's Type Ia. They both have a pseudocircular combustion chamber with an elliptical pillar positioned slightly to the left of center. The combustion chamber of Kiln A has a diameter of approximately 1.77 m. Kiln B's diameter is about 1.60 m. Both kilns are partially dug in the ground, and the rest is built with fragments of bricks mixed with clay; the thickness of the structure is approximately 50–60 cm. Excavation data suggest that the length of the built *praefurnia* must have been around 1.3–1.5 m, with a continuation of the uncovered corridor up to 2.3 m. The area in front of the kilns is lower than the walking surface (SU 126) that covers the entire area where the kilns are built.[4]

Belonging to the same phase of the kilns is a posthole (its evidence is visible on the east side of the excavation area; Fig. 1b). Its remains have been interpreted as part of a covering structure, possibly a shed that may have had approximately the dimensions of the later Republican Building, whose foundations could have covered the traces of the other pillars.[5]

The study of the pottery found both in the layers that filled the destroyed kilns and in the ones above, connected to the foundation of the Republican Building, led to the workshop's activity being dated between the end of the 3rd and the 2nd century B.C.[6]

---

1 Cuomo di Caprio (2007) 512–513.
2 Perna *et al.* (2021).
3 Perna (2006) 71, note 30.
4 Perna *et al.* (2021) 379–381.
5 Unpublished excavation data coming from the excavation diary and Abaco's excavation report for Soprintendenza Belle Arti e Paesaggio delle Marche.
6 Perna *et al.* (2021) 382–392.

https://doi.org/10.1515/9783111189635-012

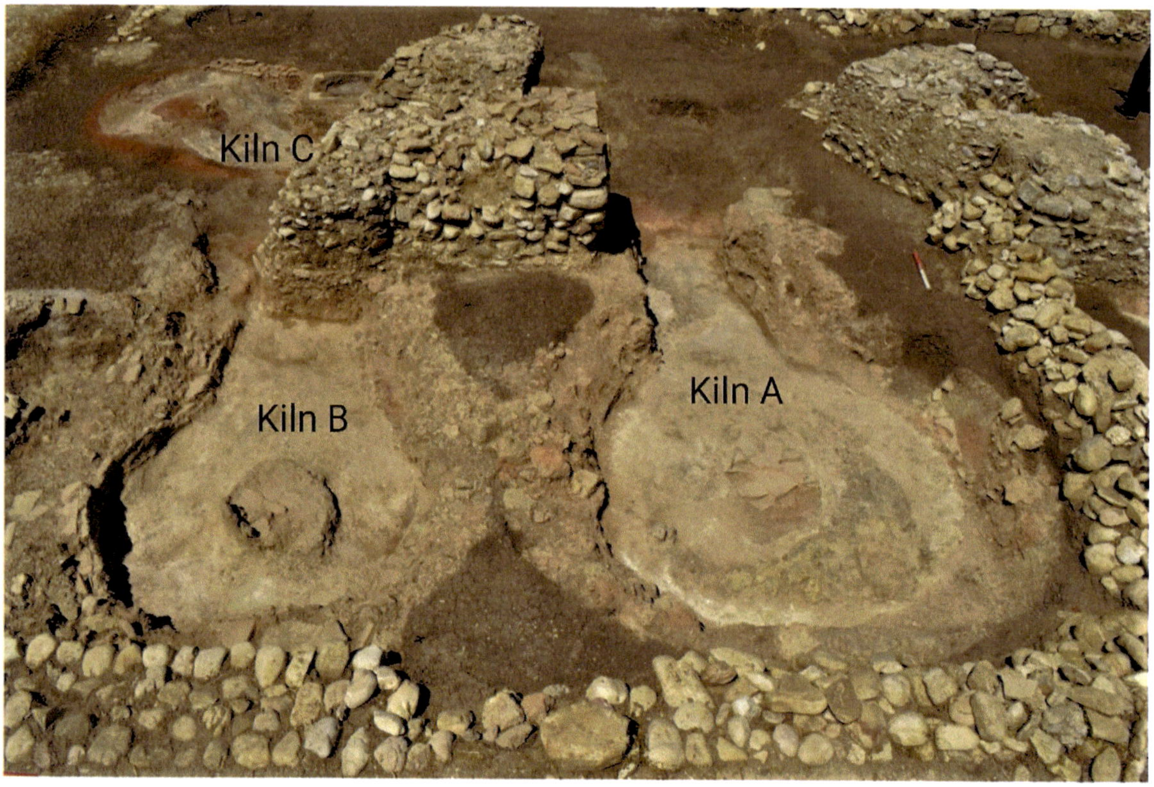

(a)

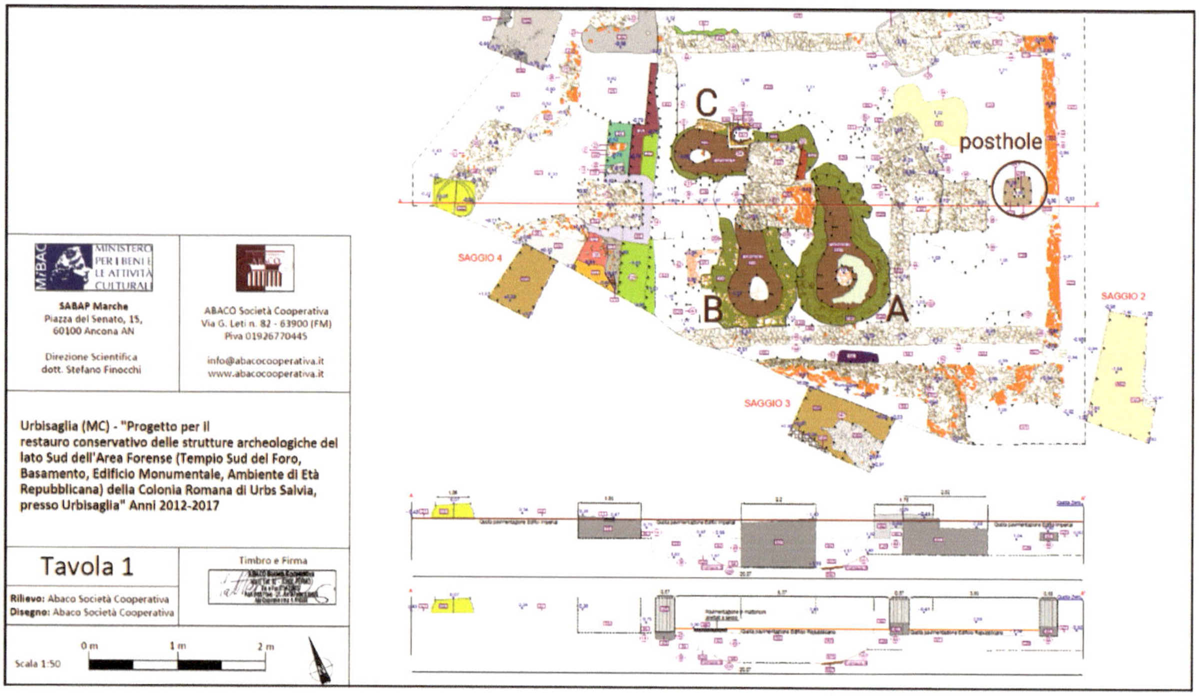

(b)

Fig. 1: (a) The kilns at the end of the excavation carried out by Abaco soc.coop. in 2017 (Photo: Matteo Tadolti, Abaco soc. coop.); **(b)** Plan of the excavation (from the excavation report of Abaco soc.coop.). © University of Macerata.

A preliminary catalogue of the documented shapes, together with a macroscopic examination of clay fabrics, has led to the identification of a local production primarily dedicated to coarse wares. This conclusion was also confirmed by wasters and flawed pottery found in the same context.

Because of their poor conservation, the kilns were covered during the enhancement works at the Republican Building. Thus, the remains of the artisanal area belonging to the first phase of Pollentia-Urbs Salvia are no longer visible in the archaeological park.

From 2019 to 2023, the Archaeological Park of Urbisaglia was part of the Adrion TRANSFER project, whose objective was to produce a common sustainable management model for the Adriatic archaeological areas.[7] One of the project's foci was using ICTs (Information and Communication Technologies) to manage the parks and enhance their visits.[8] In this context, the research group guided by the University of Macerata chose to develop a virtual reality (VR) experience for the Republican kilns of Urbisaglia. The choice was seen as an opportunity to show an archaeological asset no longer visible, thus giving the park a chance to offer an otherwise impossible experience. Moreover, the artisanal area is pertinent and important evidence of the phase before the foundation of the colony Pollentia, then named Urbs Salvia in the Imperial period. The kilns were probably part of a larger complex of artisanal activities connected with the organization of a *conciliabulum*.[9]

Following the first of the Sevilla Principles,[10] the project started with building an interdisciplinary "archaeological and digital work chain" that included not only the actors and institutions involved in the management, research, and protection of the park but also different professionals and companies, starting with archaeologists and computer engineers and also including experts in experimental archaeology, communication, and 3D reconstruction and printings, potters, video makers, sound producers, speakers, and translators.

I will mainly focus on the archaeological research and some of the most interesting interdisciplinary interactions for the reconstruction of the Pollentia-Urbs Salvia artisanal area. As noted above, the excavations have been limited to a small portion of the area. Therefore, we have archaeological data only about the foundation of the kilns and the remains of a covering structure. On the other hand, limited evidence points to a larger artisanal area, most of which remains unexcavated.[11]

Nevertheless, for a virtual reconstruction project, it is necessary to include the hypothetical anastylosis of the archaeological remains in a plausible environment. This procedure is crucial both for the research aspect of contextualization and for developing both an authentic experience in an early Roman workshop, and a compelling story about the process of making pottery.

Establishing an accurate and interdisciplinary workflow was essential to validate each step from an archaeological and communication/educational point of view. Input from experimental archaeology was crucial to develop a first draft for the story of the VR experience. This first step helped the team establish the research parameters on which to focus in order to produce an accurate environment. At a later stage, experimental archaeology research was also used to define the scenes to film, some characteristics of the environment, and the appropriate gestures and details of the artisanal work to be shown in the 360° video of the artisans working at the kilns.

After establishing these premises, the main focus was the archaeological research—the interpretation of the remains and the finds and the reconstruction of the artisanal area. Concerning the kilns and their function, documented experimental kiln firings[12] and recent studies on structure and function of kilns have been crucial to developing a hypothetical reconstruction of the kilns.[13] The experience derived from experimental archaeology and local potters' artisanal work have better clarified specific features of the kilns.[14] Debate concerning the organization of artisanal workshops and their

7 For a brief overview of the TRANSFER project and its objectives, see Perna (2023) 17–21.

8 For a brief overview of the methodological approach and the objectives of the project with reference to ICTs, see Perna (2023) 47–49, 58, 60, 80–81. For a synthesis of the Pilot Action developed for the Archaeological Park of Urbisaglia, see Perna (2023) 158.

9 Perna *et al.* (2021).

10 *The Sevilla Principles* (2017).

11 Perna (2006) 71.

12 See, for instance: Greff (1991).

13 For an overview of the subject: Cuomo di Caprio (2007) 483–542. For the study of the ancient representations of the kilns, see Hasaki (2021).

14 One of these issues has been how to shape and close the opening in the firing chamber.

potential definition as planned quarters is still ongoing. Nevertheless, some characteristics of the organization of pottery-production activities are common to most of the better-known and published artisanal areas as well as to modern workshops that still work in a traditional way.[15] Among these characteristics, access to a road seemed to be crucial for spatial and functional organization. Artisanal activities needed to be supplied with raw materials and fuel, to move products, and to be easily accessible to people who worked there or potential buyers in cases of small-scale commerce.[16] Recent extensive studies on artisanal workshops have shown complex systems of covered and open spaces where raw materials were stored and worked, and pots were made, fired, and stored.[17] The research and the need to represent all the phases of pottery production in the VR experience led the team to include the kilns and their shed (Fig. 2c–d) in an environment composed of a potter's workshop (Fig. 2a–b), where the clay was worked and the pots were made, and a storage room (Fig. 2e), where fired pots could have been not only stored but also sold, given the proximity to the road (Fig. 2).[18]

The area of the later forum of Urbs Salvia has been represented as a scattered productive settlement where houses alternated with sheds (visible in Fig. 2d). This representation was based on recent finds in the forum area, where the excavations of 2022 identified traces of artisanal structures.

The virtual environment created for the Republican artisanal workshop of Pollentia-Urbs Salvia is therefore the result of the combination of excavation records, the hypothesis of reconstructions based on the remains, and the interpolations based on the study of the organization of artisanal areas in antiquity. The level of reliability of each 3D object and its components is visible in the proxy version of the reconstruction (Fig. 3a), which is therefore a research tool and an essential source for the visualization of the research stages.[19] To achieve this result, all the information and data mentioned above have been uploaded to Blender, an open-access software, in an extended matrix that carries the sources connected to each reconstruction stage and makes the reliability levels distinguishable by different colors.[20]

The final stage of the project was filming a 360° video showing artisans working at the kilns. The VR experience was divided into six scenes. The first scene is presents the workshop and informs the user about the reconstruction. In the second, the viewer is offered the opportunity to explore the building where pots were made and is told about the first stages of pottery production. The next three scenes are 360° videos set in the kilns' shed with artisans at work. These videos show the phase of stacking the pottery in the kiln (Fig. 3b), the firing process (Fig. 3c), and the collection of the vases from the cooled down kiln. In the final scene of the experience, users have the chance to interact with the pots in the storage and sales building (Fig. 3d).[21]

15 Cuomo di Caprio (2007) 512–513 and Sparkes (1991) 8–13. For general studies on artisanal areas and their organization: Denti/Villette 2019; Esposito/Sanidas 2012; Hasaki (2011).

16 On the importance of roads for the organization of artisanal areas: Helmann (2012).

17 For some specific references: Bergamini (2007); Desbat (2000).

18 The area is close to a diverticulum of Salaria Gallica. Moreover, unpublished data suggest the existence of a flat, empty space immediately north of the kilns. This combination has led us to hypothesize a spatial organisation of the workshop influenced by the road network.

19 For more information about the workflow and the possibility to work with levels of reliability, see Ferdani *et al.* (2020).

20 The work has been carried out, under our coordination, by Mirco D'Alessio (Università Politecnica delle Marche).

21 The pots shown throughout the entire experience are some peculiar and distinctive shapes associated with the kilns' production of Pollentia-Urbs Salvia. They have been reconstructed virtually but also reproduced as props both in clay and as 3D prints.

Fig. 2: The virtual environment of the artisanal workshop at Pollentia-Urbs Salvia. Select shots: **(a, b)** The potter's workshop; **(c, d)** The kilns, **(e)** The storage-sales building. All the pictures have been captured before the final detailing process. © University of Macerata.

Fig. 3: (a) Proxy version of the virtual environment in Blender (Courtesy: Mirco d'Alessio); **(b)** Artisans loading the kiln – a frame of the scene filmed "in green screen"; **(c)** Male artisan firing the kiln – a frame of the scene filmed "in green screen"; **(d)** A frame of the user's interaction with the pots in the storage and selling building. © University of Macerata.

Abbreviation

The Sevilla Principles (2017) *The Sevilla Principles. International Principles of Virtual Archaeology* (ratified by the 19th ICOMOS General Assembly in New Delhi, December 2017).

Bibliography

Bergamini (2007): Margherita Bergamini, *Scoppieto I. Il territorio e i materiali*, Florence.

Cuomo di Caprio (2007): Ninina Cuomo di Caprio, *La ceramica in archeologia 2*, Rome.

Denti/Villette (2019): Mario Denti and Mathilde Villette (eds.), *Archéologie des espaces artisanaux. Fouiller et comprendre les gestes des potiers*, Monographies d'Archéologie Méditerranéenne 9, Lattes.

Desbat (2000): Armand Desbat, *L'atelier de potiers antique de la rue du Chapeau Rouge à Vaise (Lyon 9e)*, Lyon.

Esposito/Sanidas (2012): Arianna Esposito and Giorgos M. Sanidas (eds.), *Quartiers artisanaux en Grèce ancienne: Une perspective méditerranéenne*, Villeneuve d'Ascq.

Ferdani *et al.* (2020): Daniele Ferdani, Bruno Fanini, Maria Claudia Piccioli, Fabiana Carboni and Paolo Vigliarolo, "3D reconstruction and Validation of Historical Background for Immersive VR Applications and Games: The Case Study of the Forum of Augustus in Rome", in: *Journal of Cultural Heritage* 43, 129–143. https://doi.org/10.1016/j.culher.2019.12.004.

Greff (1991): Gilles Greff, "Les potiers du site Gallo-Romain de Bliesbruck (Moselle): reconstruction expérimentale de fours et production de céramiques", in: *Archéologie expérimentale. Tome 1. Le feu: métal et céramique* (Actes du Colloque international "Expérimentation en archéologie: bilan et perspectives, Beaune, 6–9 Avril 1988), Paris, 214–226.

Hasaki (2011): Eleni Hasaki, "Crafting Spaces: Archaeological, Ethnographic and Ethnoarchaeological Studies on Spatial Organization in Pottery Workshops in Greece and Tunisia", in: Mark L. Lawall and John Lund (eds.), *Pottery in the Archaeological Record: Greece and Beyond* (Acts of the International Colloquium held at the Danish and Canadian Institutes in Athens, June 20–22, 2008), Aarhus, 12–28.

Hasaki (2021): Eleni Hasaki, *Potters at Work in Ancient Corinth: Industry, Religion, and the Penteskouphia Pinakes*, *Hesperia* Supplement 51, Princeton N.J.

Hellmann (2012): Marie-Christine Helmann, "Quartiers ou rues? La notion de quartier économique spécialisé dans le monde grec: comparaison des données textuelles et archéologiques", in: Arianna Esposito and Giorgos M. Sanidas (eds.), *Quartiers artisanaux en Grèce ancienne. Une perspective méditerranéenne*, Villeneuve d'Ascq, 23–37.

Perna (2006): Roberto Perna, Urbs Salvia. *Forma e urbanistica*, Città dell'Italia antica 7, Rome.

Perna (2023): Roberto Perna (ed.), *Common Sustainable Governance Model for Archaeological Parks*, Macerata.

Perna *et al.* (2021): Roberto Perna, Giorgio Postrioti, Marzia Giuliodori, Matteo Tadolti, Valeria Tubaldi and Ludovica Xavier de Silva, "Fornaci e produzioni ceramiche a Pollentia-Urbs Salvia in età repubblicana", in: Daniela Rigato, Manuela Mongardi and Mattia Vitelli Casella (eds.), *Produzioni artigianali in area adriatica: manufatti, ateliers e attori (III sec. a.C. – V sec. d.C.)*, *AdriAtlas 4*, Bordeaux-Montaigne, 21–37.

Sparkes (1991): Brian A. Sparkes, *Greek Pottery. An Introduction*, Manchester/New York.

Section III: **Workshop Recipies and Craft Production Systems: Technologies Magnified**

Diego Elia, Patrizia Davit, Alessandro Re and Monica Gulmini

13 Magnific Magnification at Locri Epizephyrii: An Insight into the Surface of Western Red-figured Vases

Abstract: Recent studies on the red-figured pottery found at Locri Epizephyrii have led to an in-depth knowledge of the local production and the characteristics of the Locrian workshops (the Locri Group, the Kneeling Eros Painter, the Painter of the Pyxis RC 5089) between 390 and 340 B.C. In recent years, attention has been focused on the surface of these figured vases, with the aim of developing suitable procedures to support the archaeologist's naked-eye examination in the detection of micromorphological features that can shed light on the techniques, manufacturing practices, gestures and tools used by ancient artisans. A project has been launched to develop the methodological approach by examining a selection of figured vases and fragments belonging to the Locrian production using several instrumental techniques, namely reflectance transformation imaging, optical microscopy, scanning electron microscopy coupled with energy dispersive spectrometry and confocal optical profilometry. Results have shown that this approach makes it possible to obtain a more detailed picture of the characteristics of the *miltos* and black gloss, the stratigraphy of the contour and relief-lines, the strategy in the preliminary sketch and in the drawing sequence, and the use of overpainted decorations. The analytical approach presented has great potential in the observation and study of the traces left on the surfaces by the coating and decoration processes and it has shed further light on some of the techniques and procedures used by the Locrian painters.

Introduction

In the last two decades, the studies on red-figured pottery at Locri Epizephyrii have allowed for an in-depth knowledge of some western workshops and their activity, which was especially aimed at satisfying the demand of the city and its territory during a short period, namely between 390 and 340 B.C.

These studies, based on a rich and detailed body of data, proved that an important part of the activity of the Locri Group took place at Locri itself,[1] after a previous Sicilian phase.[2] This is shown by the concentration in the city of a considerable amount of finds with specific, common characteristics: more than 90% of the vases and fragments of the Group found in southern Calabria come from Locri itself, and this production dominates the circulation in the area as a kind of monopoly for about half a century. In fact, even though the Sicilian roots are evident in style and composition, Locrian workshops show strict affinities in drawing and iconography, revealing that they belong to a common productive environment.

Archaeometric studies, moreover, proved extremely important in confirming this hypothesis put forward on the basis of style and distribution analysis. In fact, the similarity of the clay composition of figured and black glaze pottery, which makes up a reference group for Locrian origin, testifies to the local production of many red-figured samples.[3]

The systematic study of the vases and, above all, of the hundreds of fragments from various contexts in the residential area and the necropoleis allows us to determine how the workshops – and in some cases the activity of individual craftsmen – were organized at Locri in order to respond to particular requests from the local market, producing specific shapes and iconographies.

In the 'Locrian phase' the morphological repertoire shows different features compared to other contemporary outputs in southern Italy: besides the predominant production of kraters – mainly bell-kraters, although calyx and volute examples are also present –, lekanai and skyphoi are well attested, as well as various shapes that are rather rare in western production, such as kylikes, barrel-vases, shallow askoi, and miniature vases. The iconographic repertoire also shows adaptations to specific themes particularly appreciated in relation to local religious tradition and ritual practices.[4]

1 Elia (2005), (2010) 148–160, (2014); Barresi/Elia (2017).

2 Barresi (2013), (2018).

3 Mirti *et al.* (2004a).

4 For the relationship between Locrian red-figured production and local market demands: Elia (2019), (2024). For an overview of Locrian craftsmanship between the 6th and 4th centuries B.C., see Elia/Meirano (2022).

https://doi.org/10.1515/9783111189635-013

Thus, at the same time as with the transfer from Sicily, a flourishing local craftsmanship developed at Locri, with the involvement of numerous artisans.

The iconographic and stylistic analysis, based on the composition of the scenes, the drawing of the figures and the details, as well as the identification of specific modes of rendering, has made it possible to identify several painters who worked closely at Locri: the Locri Painter – let us say the beginner – and others, more or less influenced by him, such as the Painter of the Kylix 105400, the Kneeling Eros Painter, the Painter of the Pyxis RC 5089, and possibly other craftsmen not yet identified, whose products are gathered within the Locri Group. The stylistic analysis has allowed for the identification of the drawing rendering of some details for each individual 'hand': the schemes used for male, female and winged figures, as well as for heads, eyes, hands, clothes. In this way, even small fragments can be traced back to well-known patterns and schemes.[5]

The beginning of this figurative production represents an extremely significant episode in the history of Locrian craftsmanship: it comes after an interval of several decades during which the circulation of red-figured pottery was essentially exhausted, and seems to be determined by the same needs that had sustained a substantial import of Attic vases until the decade from 430 to 420 B.C., as shown by its wide diffusion mainly in necropoleis and sanctuaries, but also in the inhabited area.[6]

Concerning the technological aspects, since the first identification of the Locri Group, A. D. Trendall reported for many vessels the presence of "a red wash" and of a "black glaze. . . often very inferior in quality".[7] These remarks have been substantially confirmed in light of the hundreds of new attributions subsequently proposed.[8]

In the last few years, we have started a project in order to develop the analysis of the surfaces of the red-figured vases of the Locrian production.[9] This ongoing project is aimed at enriching the information we have already obtained on the thickness of the coatings and the firing temperatures,[10] and at developing methods and testing instruments already used in previous research.[11] In particular, the goal is to support the archaeologist naked-eye examination of the surfaces in order to enable the detection of micro-morphological features and gain insights into the techniques, the manufacturing practices, the gestures and the tools used by ancient craftsmen. We therefore developed a procedure involving selected instrumental techniques.

Techniques and Methods for the Examination of the Surface

Instrumental techniques were selected starting with those that allowed access to some macro aspects and then moving on to those that would allow the detection of micro features.

Reflectance Transformation Imaging (RTI) is a photographic technique that captures high quality digital images of the surface by varying the direction of illumination. Computational post-processing generates new images using any direction of light in the hemisphere around the camera location. By virtually moving the light across the surface of red-figured vases, we found that details barely visible or not visible at all under 'normal' (i.e. diffuse) lighting conditions can be revealed.[12] Then, optical microscopy (OM) allowed us to get closer to some specific details, with magnifications ranging

5 About the characteristics of the production of specific craftsmen, see: Elia (2010) 148–160. For the Painter of the Pyxis RC 5089, identified by A. D. Trendall in relation to the eponymous vessel only (*LCS* 220, no. 113) and initially considered as a minor craftsman due to his preference for more serial subjects: see also the analysis in Spigo (2000) 40–41.

6 Elia (2010) 140, note 116.

7 *LCS* 74. The same characteristics are also common to the other figured productions recognized as local at Locri.

8 Elia (2010) 113: these technological characteristics constitute useful elements in attributing to local production even small fragments, for which iconographic and stylistic details are necessarily limited. A. D. Trendall's initial observations were based solely on the analysis of a small number of vases (nine in *LCS*, eighteen in *LCS III*) and fragments (four in *LCS*).

9 It is worth noticing that there is still some discussion about how to refer to the black layer that develops on the surface of a clay body covered with a (suitable) clay slip and fired in a single firing through three (oxidizing-reducing-oxidizing) stages of the kiln atmosphere. Here we will follow the recommendation of R.E. Jones (1980, 2021) and use the term black gloss throughout the text.

10 Mirti *et al.* (2004b).

11 For a first example of the application of RTI: Bovero *et al.* (2012); for the use of microscopy: Elia *et al.* (forthcoming).

12 For the application of RTI in the analysis of red-figured production: Artal Isbrand/Klausmeyer (2013); Saunders et *al.* (2017); Balachandran (2018); see also Serino in this volume. For the RTI processing used in this contribution, the authors would like to thank Alessandro Bovero, expert in computational imaging applied to cultural heritage. An average of about 50 shots were taken for each fragment at different lighting angles. A Nikon d800E DSLR equipped with AF-S Micro Nikkor 60 mm lens and a Nikon SB-910 flash was used.

from 6× to 60×. Observations were made in visible light, with different geometries of illumination. This made it possible to capture the different shades of color and further improve the visibility of the morphological features on the surface. A stereomicroscope was used in the laboratory[13] and a portable digital microscope in the field.[14]

Scanning Electron Microscopy (SEM) allowed an even closer look. Fragments several centimeters in size can be examined as they are, although invasive procedures, such as the preparation of polished cross-sections, can be carried out. Images are obtained by using an electron beam and so the highly magnified images (up to 10.000×) are in grayscale, with no color information retained. They can be obtained by detecting secondary electrons, which capture the surface morphology, or by backscattered electrons. In the latter case, the gray level is related to the average atomic number, and therefore it provides a qualitative indication of the distribution of different chemical elements on the surface. Unfortunately, SEM does not provide any information if products for conservation or restoration purposes have been applied on the surface. SEM can be coupled with an X-ray detector (EDS), which makes it possible to obtain the quantitative elemental composition.

Finally, confocal optical profilometry was used to obtain 2D and 3D morphological maps (in grayscale and in false colors) and profile measurements.[15] The latter make it possible to measure the depth of the grooves and the height of the ridges of overlapping material, which are important traces of the painter's hand.

Miltos

As mentioned above, since the first studies, these productions have been associated with the characteristic presence of an evident *miltos* layer, spread on the ceramic body in order to obtain a deeper red on the reserved surfaces. In the bibliography on Western red-figured productions it is not always clear whether this coating of red ochre was applied before the slip and on the entire surface of the ceramic shapes, or whether it was applied at an advanced stage of the technological process, only on the surfaces of the vase that were not intended to be covered by the black gloss.[16]

In the Locrian productions, the *miltos* layer, besides being richer in red tone than the body itself, is characterized by a very vivid color. It is particularly visible in the uncovered areas, and its presence is also more or less recognizable on surfaces characterized by the partial or complete detachment of the black gloss. The examination under OM of the micrometric cracks of the black coating, which allow a glimpse of the underlying surface, makes it possible to detect the evident red tone of the *miltos* over the paler clay body. A representative example is the network of cracks that developed on the black surface of the fragmentary lid of an oversized lekane (Fig. 1a).[17] Each crack reveals the red coating underneath, which is clearly distinguishable from the lighter ceramic body (Fig. 1c). As the cracks are only about 100 µm wide, the *miltos* is not visible to the naked-eye. Similarly, along the edges of the right fracture of the shard, the microscope shows the red layer under the black coating (Fig. 1b). The relatively frequent occurrence of such observations on vases and fragments confirms the hypothesis that in many cases the surface was completely covered with *miltos* before the figured scenes were painted, and thus before the slip was applied; *miltos* is also often visible on the inner surfaces of kraters (Fig. 1e).[18]

---

13 Leica MZ 9.5 equipped with a Leica MC 190 digital camera (10 megapixel sensor) managed by the LAS X software.

14 Dino-Lite AMZ795MZT (5 megapixel sensor).

15 Optical profilometry is a well-established approach to extract topographic information from a surface, and it has many applications in archaeology: Carcagni *et al.* (2005); Giumlia Mair *et al.* (2010). Artal-Isbrand/Klausmeyer (2013) report the use of this technique for the examination of red-figured vases. For the optical profilometry acquisition and processing used in this contribution, the authors would like to thank Marco Cavagnero, ITT Motion Technologies (Barge, Cuneo). The employed instrumentation is a Nanovea PS50 with a scanned area up to 5 cm × 5 cm and a maximum high range of 300 µm.

16 Gadaleta (2012) 118–119; based on current knowledge, the two different procedures do not appear to be exclusive to individual painters/workshops: however, it is unclear on what basis these observations were made. *Miltos* was used especially by South Italian workshops because of the paler clay body that characterizes many Western productions: Cuomo di Caprio (2007) 309. On issues related to *miltos* see: Photos-Jones *et al.* (1997); Schreiber (1999) 48–52; Cuomo di Caprio (2017) 285–286; Photos-Jones *et al.* (2018); Jones (2021) 106–107.

17 Reggio Calabria, Museo Archeologico Nazionale, without inv. no.: Barresi/Elia (2017) fig. 1.

18 This technological sequence was found on many large vessels, such as kraters (e.g., Elia (2010), Appendice I, CC6; Elia (2018) fig. 6b: Locri, Museo Archeologico Nazionale, inv. no. 75221; Locri, Museo Archeologico Nazionale, without inv. no.; Reggio Calabria, Museo Archeologico Nazionale, without inv. no.) and oversized lekanai lids (e.g., Barresi/Elia (2017): 29, fig. 1), but also on smaller shapes, such as lids of small and medium-sized lekanai (e.g., Elia (2010), Appendice I, Le1, Le10). In a number of cases no traces of *miltos* remain on the surfaces originally covered with black gloss: this is explained by the fact that such checks can usually be made only in the parts where the glossy black has peeled off, leaving

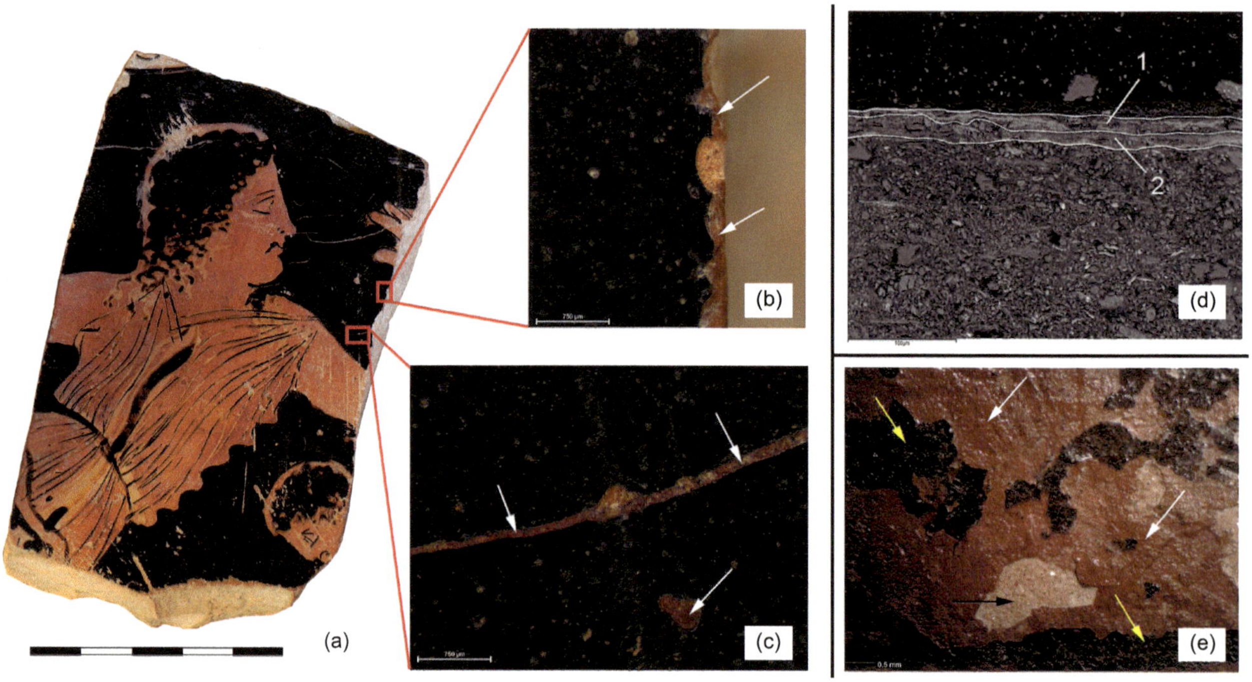

Fig. 1: Lekane lid fragment, (without inv. no; after Barresi/Elia (2017) fig. 1): **(a)** Macroscopic image, the two red frames illustrate the positions of the details highlighted by the corresponding OM images; **(b)** and **(c)**, OM images, the white arrows point to the *miltos* layer beneath the black surface.
Bell-krater fragment (Reggio Calabria, Museo Archeologico Nazionale, inv. no. 105434): **(d)** SEM image (backscattered electrons) of a cross-section image showing the black gloss (layer 1) on top of the *miltos* layer (layer 2); **(e)** OM image, the black gloss layer is indicated by the yellow arrows, the *miltos* layer by the white arrows and the body emerging from underneath the *miltos* layer by the black arrow.

The stratigraphy of the *miltos* and of the black gloss, and their micro-morphological characteristics, are particularly evident in the SEM image (Fig. 1d) obtained from a sample detached from a fragment of a bell-krater and prepared as a cross section.[19] The picture shows the layer of the black gloss (layer 1) on the *miltos* (layer 2) laying on the porous ceramic body visible in the bottom part. The higher vitrification of the black gloss with respect to the *miltos* is visible in the picture; on the other hand, although the *miltos* and the body share a similar morphology, the *miltos* appears lighter than the body in the grayscale of the backscattered electron image. This feature is related to the higher content of iron, which differentiates the *miltos* from the body of the vase.

Even if the procedures and tools used for the application of *miltos* are rarely recognizable, thin and regular parallel lines are sometimes visible on the surfaces: the recurring pattern of horizontal traces testifies that the red ochre layer was applied by turning the vase on the potter's wheel.[20] A different pattern can be observed in vertical details, such as a ribbon handle likely belonging to a pelike,[21] on which the brushstroke follows the longitudinal development of the morphological element, which is in fact the easiest and most effective way to apply the *miltos* in this case (Fig. 2a–c).

---

the underlying surface visible; in these cases it cannot be ruled out that the black slip has torn off any underlying *miltos* coating, removing it completely. The same phenomenon occurs when sections of the inner details of the figures, made with painted lines, come off, leaving the precise imprint on the *miltos*.

19 Reggio Calabria, Museo Archeologico Nazionale, inv. no. 105434. Technological analysis on this vase was started during conservation treatments: Gobbato *et al.* (2019); Elia *et al.* (forthcoming).

20 Locri, Museo Archeologico Nazionale, inv. no. 75221. For this technique see also Noble (1988) 60.

21 Locri, Museo Archeologico Nazionale, inv. no. 1954/641-98888.

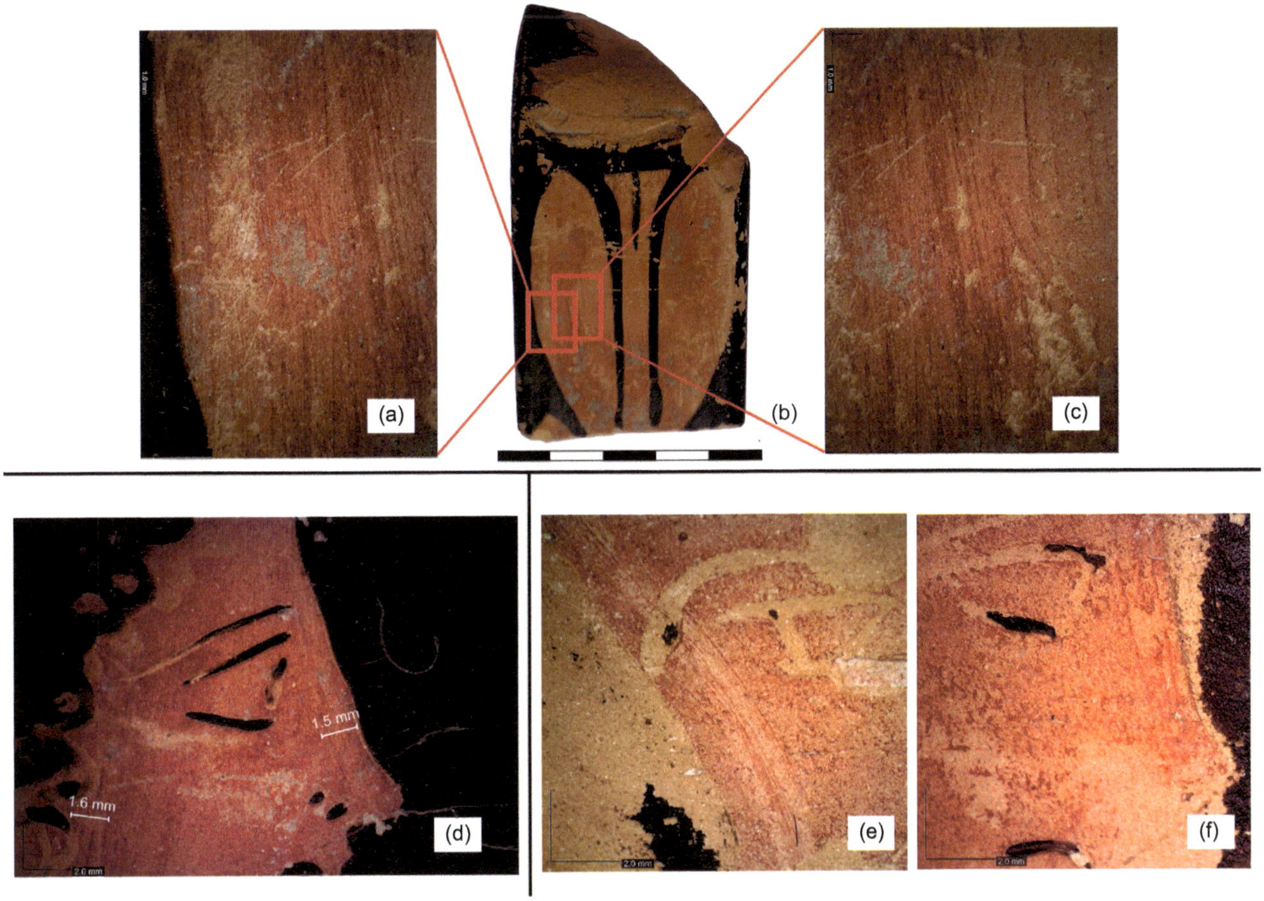

Fig. 2: Pelike ribbon handle (Locri, Museo Archeologico Nazionale, inv. no. 1954/641-98888): **(a)** and **(c)** OM images illustrating the brushstrokes for the *miltos* application in the two areas of the fragment highlighted by the red squares in the **(b)** macroscopic image.

Lekane lid fragment (Reggio Calabria, Museo Archeologico Nazionale, without inv. no; after Barresi/Elia (2017) fig. 1): **(d)** OM image showing the brushstroke width on the nose and on the cheek of the female figure on the left of the fragment.

Bell-krater fragment (Reggio Calabria, Museo Archeologico Nazionale, inv. no. 105434): **(e)** and **(f)** OM images showing the morphological characteristics of the brushstrokes.

The use of a brush is also particularly evident in the rendering of specific details, as highlighted by both the OM and RTI images, probably with the aim of correcting inaccuracies and increasing the saturation of the red color. This can be seen in several noses, such as those of the figures depicted on the lekane lid (Fig. 2d) and of other figures on a bell-krater (Fig. 2e–f), where brushstrokes about 1.5 mm wide are visible.[22] The figure on the lekane lid shows similar strokes also on the cheek.

Preliminary Sketch

Among the vases produced at Locri, only a very small number bear residual traces of the preliminary sketch: this would indicate that this operation should not normally leave permanent traces.[23]

Thanks to the complementary application of various techniques (OM, RTI, Profilometry: Fig. 3a–d), these fortunate cases have made it possible to gather a large amount of information on the use and manner of drawing the preliminary sketch in Locrian production, also in relation to the different artisans.[24] For example, on the bell-krater attributed to the

22 See *supra*, note 19.

23 For a review of the proposals put forward concerning tools and techniques: Gadaleta (2012) 120.

24 The preliminary sketch has been investigated in detail for Attic red-figured production: Corbett (1965); Noble (1988); Boss (1997); Balachandran (2018). For the preliminary sketch in Western red-figured production: Gadaleta (2012) 120–121; see also Serino in this volume.

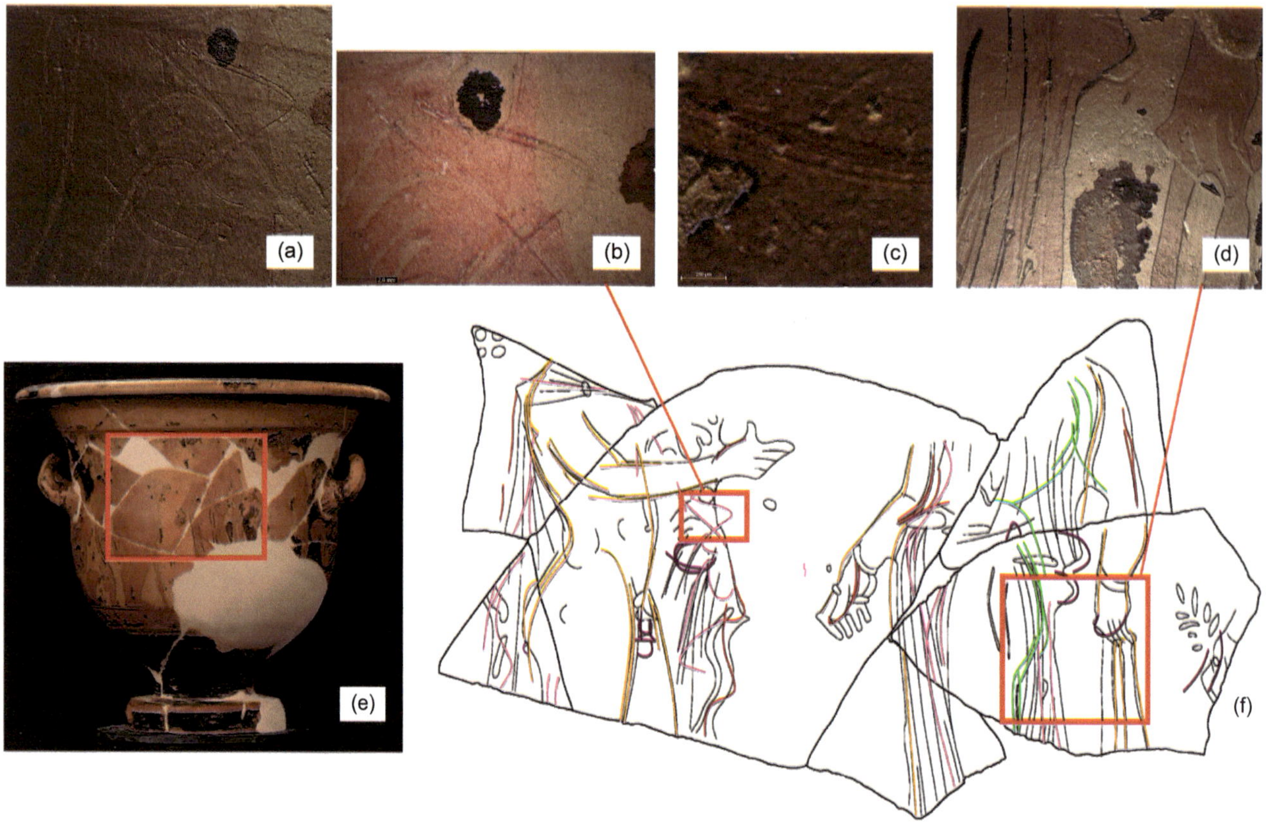

Fig. 3: Bell-krater attributed to the Locri Painter (Reggio Calabria, Museo Archeologico Nazionale, inv. no. 105434): **(a)–(d)** RTI and **(b)–(c)** OM images highlighting details of the preliminary sketch; **(e)** Macroscopic image, the red frame corresponds to the area sketched in the **(f)** drawing where the reconstruction of the drawing sequence is illustrated: elaboration from Gobbato (2016–2017). The red frames in **(f)** illustrate the positions of the details magnified in **(b)** and **(d)**.

Locri Painter (Fig. 3e)[25] a structured strategy in the use of the different lines of the sketch is recognizable, as in the part in which two figures – male and female – face each other (Fig. 3f).[26]

In fact, it is noticeable that some of the lines very precisely indicate some essential elements of the profile of the subjects (yellow lines) – and then are carefully reproduced in black slip – in order to place the subjects themselves in the space: for example, the lines that define the curve of the back and the buttocks, the belly and the legs of the male figure. These lines are sometimes doubled, almost in an attempt to achieve a more effective volumetric rendering.

On the contrary a few short, discontinuous lines (red lines) are used to mark the position and profile of some details, such as the *himation* hanging from the shoulders and under the left hand; similar remarks can be made for the woman. In this figure, however, the part covered by the *chiton* is drawn in a peculiar way: the painter has indicated the contours of the body (green lines), even if this profile has obviously disappeared in the final version, wrapped up in the voluminous folds of the garment. This is a clear trick adopted to give the painter a reference for the correct balance and proportions of the subject.

A less accurate sketch with a few selected lines is commonly used in order to obtain various elements, for example, in the representation of the hands of the female figure. In the right one, the palm is precisely indicated (again in red); instead, in the left one, the painter has made only an approximate mark (in purple). In both cases, the definition of the fingers is left to the successive black lines.

A confused 'bundle' of lines, even more approximate (again in purple), is made to suggest other details, such as the left hand of the male figure wrapped in the *himation*.

25 Reggio Calabria, Museo Archeologico Nazionale, inv. no. 105434.
26 The figure on the left has been identified as Apollo *Daphnephoros*, while the female figure on the right as a worshiper: Elia (2010) 144–145.

A final observation is suggested by the preliminary sketch of the dense folds of the garments in both figures (pink lines): some lines are faithfully reproduced in black slip and serve as an orientation to the closest ones (for example, in the upper part of the *himation*), while others are completely ignored in the final version.

Pending the collection of the entire *corpus* of evidence, we can recognize in this use of the preliminary sketch a peculiar trait of the Locri Painter: in fact, the same sketching strategy is attested both on other large vases produced at Locri and on the hydria from Syracuse belonging in the Sicilian phase.[27] Thus, at this early stage of the research, it seems possible to support the hypothesis that the types of details identified – the characteristics of the sketch, the choice and the pressure exerted by the tools – can be idiosyncratic to the activities of craftsmen working in Locrian workshops.

In comparison to this elaborate preparation, other painters working in the same productive environment – perhaps in the same workshop – and dealing mainly with smaller vases probably adopted different strategies. It is rare to find the traces of preliminary sketches on their vases and, even when they are visible, we can count only a few sketch lines, as in a fragment (Fig. 4a)[28] belonging to a barrel vase that can be attributed to the activity of a minor painter of the Locri Group, a pupil of the Locri Painter, probably the Painter of the Kylix RC 105400.

Here the sketch is used only for the profile of the face and the bust of the male figure, and then it is completely disregarded, since the face was made further to the right (Fig. 4b). The combination of profilometry (Fig. 4c–d) with the other techniques makes it possible to precisely define the shape of the marks left on the surface and their dimensions, useful elements for hypothesizing the type of instrument used. Sometimes the dimensions of the grooves are very small, as in the case of a fragment of a large open shape (Fig. 4e–h),[29] where the width is only 200 μm. The measurement of the depth repeated on the same groove leads to interesting results: in this curved line, the depth changes from 13 to 4 μm. The data seem to support the recognition of a single gesture, directed from upper left to the lower right, where the groove ends (Fig. 4f).

In other cases, the grooves are larger: on the barrel vase fragment (Fig. 4c–d), their width reaches 400 μm, more or less the same on the bell-krater we examined at the beginning (Fig. 3c). In the last cases, a rigid tool with a flat tip was used, leaving deep marks of constant width and sharp edges.[30]

Another interesting element is the sporadic appearance of a purple-reddish material in the grooves.

If the *miltos* was applied after that the preliminary sketches were traced, then the red traces could be the consequence of the accumulation of the *miltos* in the grooves. Nevertheless, we cannot ignore the fact that in the Locrian products the color tone of the material in the grooves – at least in some cases – differs significantly from that of the *miltos*. In addition, SEM images with backscattered electrons would indicate a higher mean atomic number for this material compared to that of the *miltos* (Fig. 5).

On the other hand, if the preparatory sketch was made after the application of the *miltos*, then we have to assume that the painter had to use another pigment to highlight the sketches on the red surface, and that this pigment was richer in iron than the *miltos* itself. Natural ochres may show an extremely wide range of colors, from yellow to orange, to red and to violet.[31] Recent research has demonstrated that hematite is turned purple by varying the dimension of the grains through annealing. The violet hue does not develop from natural red ochres, but is obtained by heating (annealing) high-purity hematite or other iron ores.[32] The use of such a pigment would explain both the characteristics observed under the OM (the violet color) and the high mean atomic number observed under the SEM in those grooves where the violet pigment is preserved.

---

27 Siracusa, Museo Archeologico Regionale "Paolo Orsi", inv. no. 35187. For the Locrian phase, see for example the preliminary sketch clearly visible on the bell-krater from grave 1119 (Elia 2010, Appendice I, CC5); for the hydria from Syracuse: Elia (2010) Appendice I, Hy1. We are grateful to Marco Serino for providing an RTI of a part of the figured scene in which the preliminary sketch is also clearly visible.

28 Reggio Calabria, Museo Archeologico Nazionale, without inv. no.: Barresi/Elia (2017) fig. 9.

29 Reggio Calabria, Museo Archeologico Nazionale, without inv. no.

30 This particular shape of the tool tip, possibly a small spatula or a rigid chisel brush, is also recognizable in the traces visible in other cases (*e.g.*, see the vases mentioned in note 27).

31 On the colors of ochres, see Elias *et al.* (2006).

32 The research reported by Castagnotto *et al.* (2021) involved the production of iron-based purple pigments from various natural and artificial raw materials.

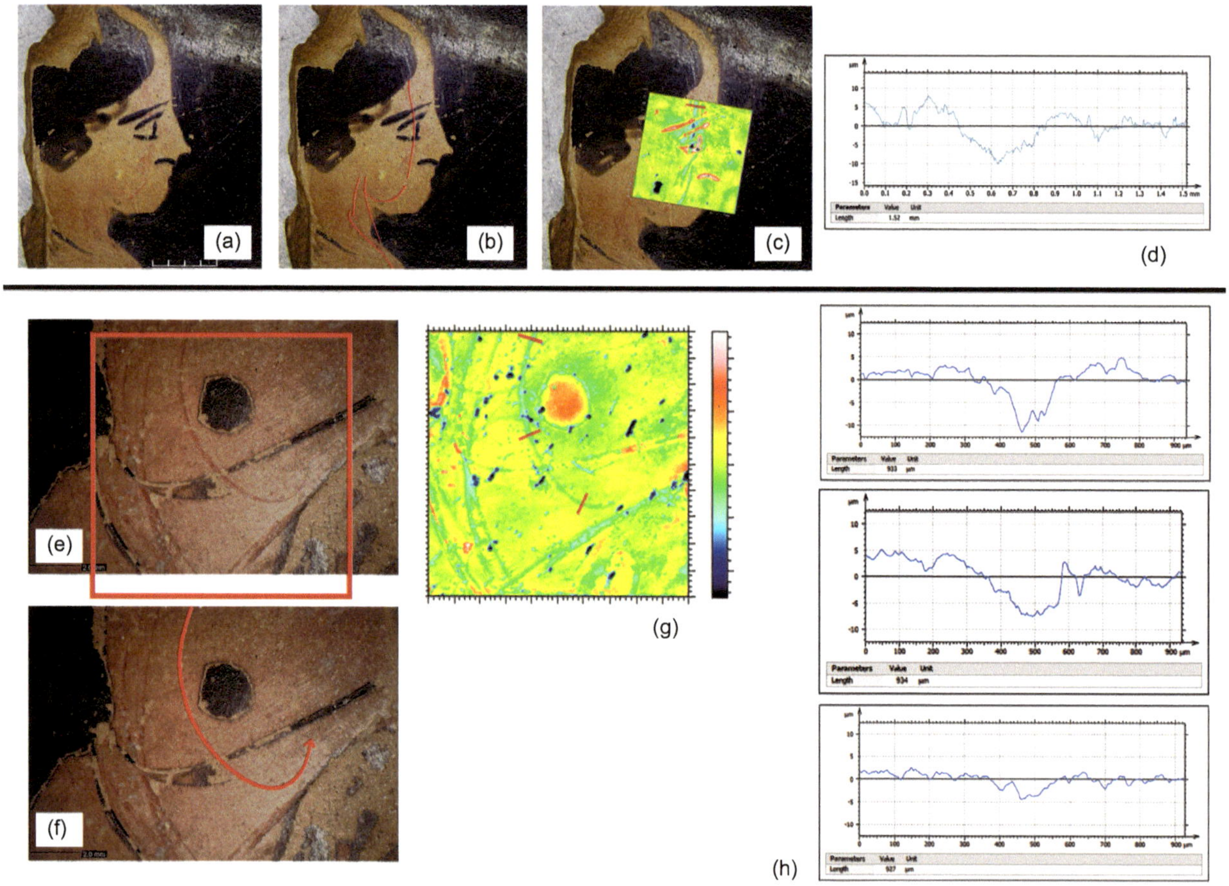

Fig. 4: Barrel vase fragment (Reggio Calabria, Museo Archeologico Nazionale, without inv. no; after Barresi/Elia (2017) fig. 9): **(a)** Macroscopic image; **(b)** Macroscopic image, in red the face profile sketch; **(c)** Color map of the data obtained by surface profilometry, color code as in **(g)**; **(d)** Depth profile extracted from profilometry data at the red straight line in **(c)**.

Large open shape fragment (Reggio Calabria, Museo Archeologico Nazionale, without inv. no.): **(e)** OM image, the red frame corresponds to the area examined by profilometry; **(f)** OM image, the red arrow highlights the direction of the sketch gesture; **(g)** Color map of the data obtained by surface profilometry; **(h)** Depth profiles extracted from profilometry data at the three red straight lines (from top to bottom) in **(g)**.

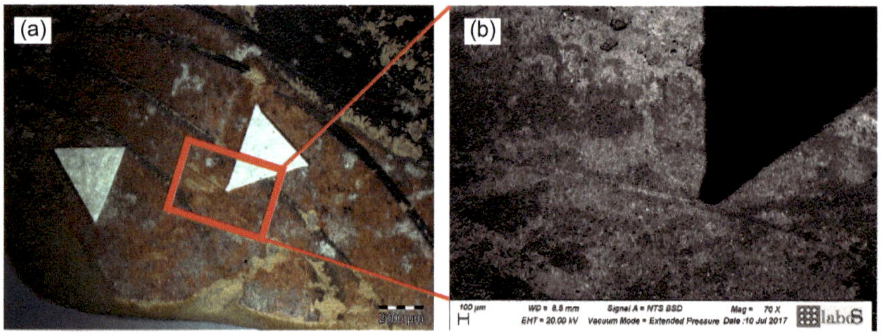

Fig. 5: Krater fragment (Reggio Calabria, Museo Archeologico Nazionale, without inv. no.): **(a)** OM image. The red line frames the area visible in the SEM image (backscattered electrons) shown on **(b)**; The paper tape markers appearing as white triangles under the optical microscope indicate the details to be examined in the electron microscope.

These observations, along with the low frequency of traces of the preliminary sketches, suggest that the preliminary drawing could have been traced with a pigment, and the sporadic traces could be an unintended consequence of using a tool to make the traces on the still soft surface.[33]

Features of the Black Gloss, Contour and Relief-lines

The low quality of the black gloss has been noticed since the first studies on the figured materials from Locri Epizephirii.[34] Lack of homogeneity in the color, with reddish and brownish areas, cracks and detachments of the surface finishing are common among the figured shards from the site (Fig. 6). Similar features characterize the monochrome black gloss ceramics at Locri between the second half of the 4th and 3rd centuries B.C.[35] As for these wares, although EDS analysis does not highlight any significant compositional difference between good- and poor-quality glosses, in the latter case the morphology of the black finishing investigated under SEM shows that it is thin and irregularly distributed.[36]

Contour line/Relief-line

In the Locrian red-figured pottery, the poor homogeneity of the black gloss allows the contour line to be particularly evident on many vases, and thus its width is easily measured under OM. The width of the contour lines is consistent in a significant number of artifacts, with values concentrating around 2.5–3.5 mm, and with rare examples below (1.6–1.8 mm) or above (up to 4 mm) this range (Fig. 7a–d).[37] This suggests that a tool of similar size was used to trace the contour lines both for the large figures painted on large vases (such as kraters, lids of large lekanai, large closed forms) and for the smaller subjects painted on small forms (such as skyphoi, lekane lids, and barrel-vases).

In a few cases, the contour line was examined by profilometry. It turned out that its thickness is barely distinguishable from the black background, even in those cases where it is evident from a macroscopic examination. A significant example is reported in Fig. 7 (e–h), where RTI reveals the contour line used here by the painter to reshape the nose of a female figure.

In numerous cases, the black background does not appear to follow the contour line precisely. A detailed examination of the RTI and OM images shows that, in some instances, the slip was applied inaccurately and overlaps the contour line.

Previous studies performed mostly on Attic productions have suggested which morphological features, detected under optical magnification, may reveal the order in which the contour line and the relief-line were traced by the painters.[38] The close examination of samples from the Locrian productions, and the detection of specific features, allowed us to hypothesize the order in which the painter used the contour line and the relief-lines (both pulled and laid ones) to trace the internal details (Fig. 8). The contour line usually appears to have been traced after the figures had already been outlined, with the inner details drawn with the relief-line before the background was applied. However, in some cases – such as the

---

33 Also in Gadaleta (2012) 120, the use of "pigmenti" is suggested.

34 See *supra,* note 7.

35 For a description of the evolution of the black gloss in Locrian production see Ercolin *et al.* (2023); see also Ercolin *et al.* in this volume.

36 The discussion on composition and morphology is given in Ercolin *et al.* (2023).

37 Measurements were obtained from seven items of large-medium size at Locri, Museo Archeologico Nazionale: 1) inv. no. 75220, fragment of krater: 2.7–3.4 mm; 2) inv. no. 1952/187, fragment of krater: 3–4 mm; 3) inv. no. 1955/49B, fragment of krater: 2,2 mm; 4) inv. no. 461, lekane lid: 3–3,7 mm; 5) inv. no. 1952/115, fragment of a small krater: 2–2,1 mm; 6) inv. no. 1952/656, fragment of a small krater: 3,9 mm; 7) inv. no. 1954/639, fragment of a medium closed shape: 2,7–2,8 mm; additional examples are in Reggio Calabria, Museo Archeologico Nazionale: 1) without inv. no., fragment of lid of oversized lekane: 2,5–3 mm; 2) without inv. no. fragment of krater: 3,4–3,6 mm. In addition, measurements were obtained from three small shapes at Locri, Museo Archeologico Nazionale: 1) without inv. no. fragment of barrel-vase: 2,2 mm; 2) inv. no. 1975/275A, fragment of lid of lekane: 2,2 mm; 3) inv. no. 1954/375, fragment of skyphos: 3,5 mm; a fourth small vase is at Reggio Calabria, Museo Archeologico Nazionale: inv. no. 8072, lekane lid: 1,7–2,5 mm.

38 The features that can suggest the sequence of drawing and painting are discussed by Artal-Isbrand/Klausmeyer (2013). In their work, the authors argue that the relief-lines might be pulled or laid on the surface and show that, for Attic productions, the contour line always overlapped the relief-lines. They also suggested that Western figured vases may show more varied sequences, although only a very small number of artifacts from different contexts was considered in their study.

Fig. 6: Krater fragments: **(a–b)**, Reggio Calabria, Museo Archeologico Nazionale, without inv. nos.; **(c)** Locri, Museo Archeologico Nazionale, inv. no 75221; **(d)** Locri, Museo Archeologico Nazionale, inv. no. 75220; some examples of the typical features observed on the black gloss: **(a)** Detachment of the surface finishing; **(b)** Presence of brownish areas; **(c–d)** Uneven color of the surface finishing.

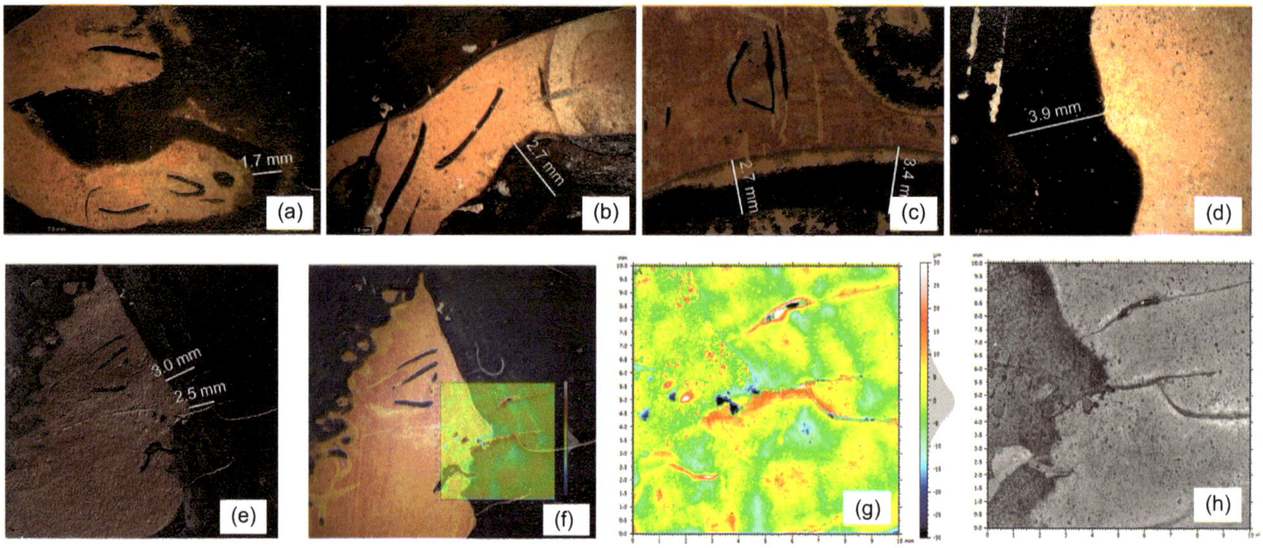

Fig. 7: (a) Lekane lid (Reggio Calabria, Museo Archeologico Nazionale, inv. no. 8072); **(b)** Medium close shape fragment (Locri, Museo Archeologico Nazionale, inv. no. 1954/639); **(c)** Bell-krater (Reggio Calabria, Museo Archeologico Nazionale, without inv. no.; after Barresi/Elia (2017) fig. 5a); **(d)** Lekane lid (Locri, Museo Archeologico Nazionale, inv. no. 1952/656-79665): examples of contour lines with different widths.

Lekane lid fragment (Reggio Calabria, Museo Archeologico Nazionale, without inv. no.; after Barresi/Elia (2017) fig. 1): **(e)** RTI image with measurement of the widths of the contour line; **(f)** OM image, the color map of profilometric data highlights the nose detail; **(g)** Color map of profilometry data; **(h)** Grayscale profilometric image of the nose detail.

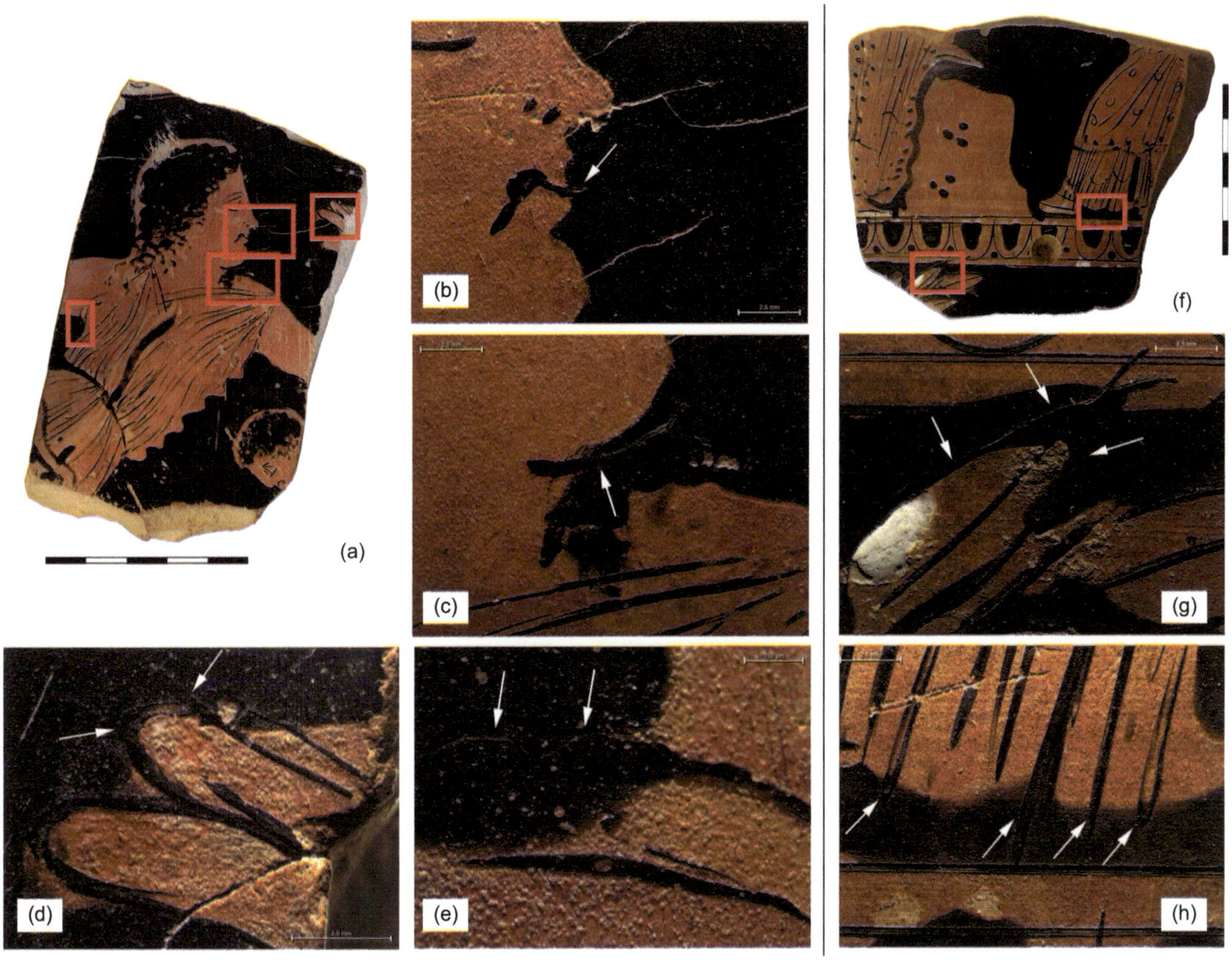

Fig. 8: Lekane lid fragment (Reggio Calabria, Museo Archeologico Nazionale, without inv. no.; after Barresi/Elia (2017) fig. 1): **(a)** Macroscopic image, the four red frames illustrate the positions of the details highlighted by the OM images: in **(b)** the line for the mouth, and in **(c)** the line for the chin are traced over the contour line; on the contrary, in **(d)** the lines for the fingers and in **(e)** the lines for the fold edge of the garment near the right armpit are traced before the contour line of the female figure on the left (the picture is rotated 90 degrees clockwise).

Calyx krater fragment (Locri, Museo Archeologico Nazionale, inv. no. 75221): **(f)** Macroscopic image, the two red frames illustrate the positions of the details highlighted by the corresponding OM images: in **(g)** the lines for the leaves at the bottom left of the fragment and in **(h)** the lines for the *chiton* folds of the figure on the right area traced after the contour lines.

fragment of the lip of an oversized lekane (Fig. 8a) – a more varied strategy can be discerned: while the above-mentioned sequence is often recognizable (e.g., in the fingers or along the edge of the garment: Fig. 8d–e), some details seem to have been traced as final touches to the vase. This is evidenced by a few short strokes of the relief-line superimposed on the contour line and used to render the mouth and the profile of the chin (Fig. 8b–c), probably in order to give these features greater prominence. A similar solution can also be seen, for example, on a fragment of a calyx krater (Fig. 8,f) where the leaves have been retouched (Fig. 8g): the relief-line is traced after the contour line and the background slip to sharpen and enhance the profile; in this fragment, the same sequence is also recognizable in the rendering of the *chiton* folds, drawn above the contour line (Fig. 8h).

Preliminary examination suggests that it is also possible to identify different technical approaches among the painters operating at Locri: while the Locri Painter (and, more generally, the Locri Group craftsmen) appears to employ frequently a thin contour line to outline the figures, as shown by the profilometric results, the Painter of the Pyxis RC 5089 extensively uses a thick relief-line to stress, mask or modify the figures with a very thin brush stroke, becoming in some situations a distinctive personal sign for the rendering of the anatomic details.

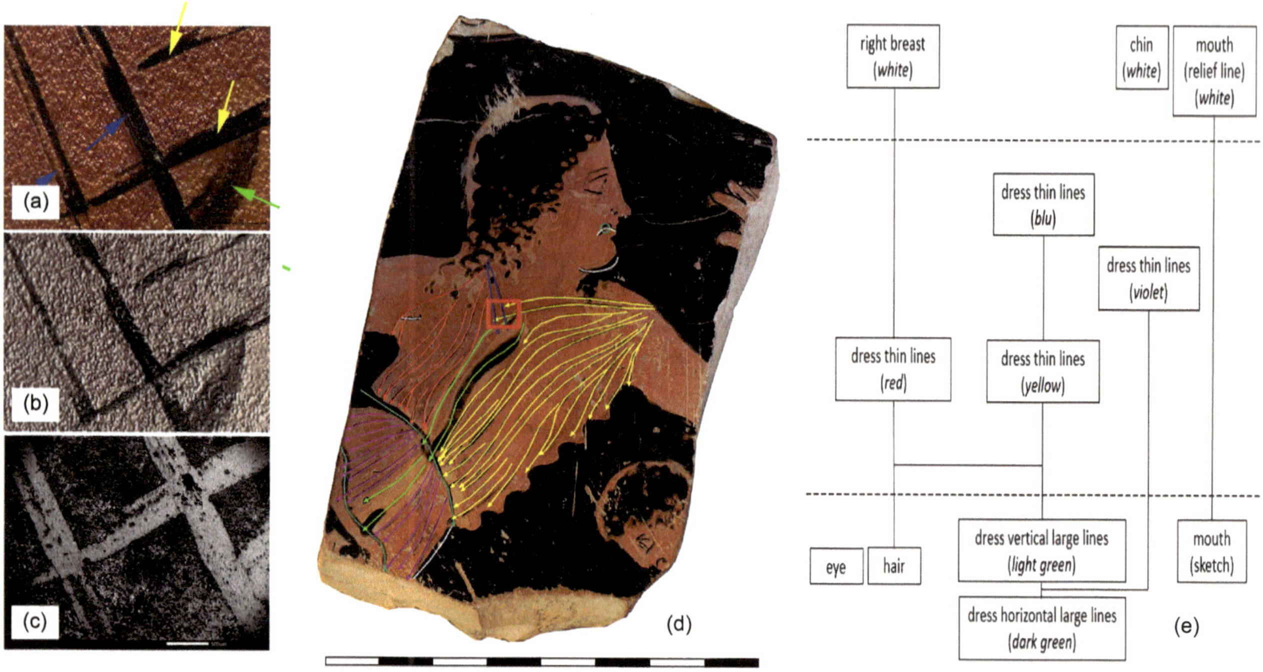

Fig. 9: Bell krater (Reggio Calabria, Museo Archeologico Nazionale, inv. no. 105434): **(a–c)** OM, RTI and BS-SEM images with the same area corresponding to the red frame in **(d)**, with the drawing sequence: firstly the vertical band on the chest (green arrow), then the thin lines for the folds of the *chiton* (yellow arrows), finally the fold of the neckline (blue arrows); **(d)** Complete drawing sequence; **(e)** Stratigraphic diagram of the drawing sequence.

Drawing Sequence

To better understand the operational sequence in the work of the Locrian painters, we have associated different technical approaches in order to reconstruct the steps in the painting of the internal details (Fig. 9a–c). The aim is to trace the stratigraphic sequence on the basis of the physical relationship between the individual brushstrokes: a significant example is provided by the female figure depicted on the oversized lekane lid by a painter within the Locri Group, as reported in Fig. 9d.

The first step consisted of the rendering of the hair and tracing the long curls with the diluted slip (average width 0.7–1.1 mm), perhaps preceded only by the drawing of the eye; at this stage, the depiction of the mouth was only represented by a small spot of diluted slip, just to mark its position (dark green).

At a later moment, but also to be traced back to the first step, the belt (width from 0.7 mm on the right edge to 2.3 mm on the left edge) and the lower edge of the hem of the garment were added, both in diluted slip again (dark green lines); then, the craftsman painted the vertical bands on the chest and the belly (light green lines).

At this point, the craftsman began painting the thin lines for the folds of the *chiton* (mean width 0.3–0.5 mm) from top to bottom, and – likely – starting from the left: first, the right side of the figure (red lines), then the left side (yellow lines),[39] then, the folds of the neckline, always directing the brushstrokes from top to bottom (blue lines) and the thin lines under the belt in the same direction (violet lines).

At this step the contour line was probably drawn, as we described above (Fig. 8d–e). Finally, with a broad brush, some thick slip strokes were applied (white lines): a short stroke on the breast (perhaps an error), the retouching of some details of the face, such as the mouth and the profile of the chin (Fig. 8c–b), and the lower edge of the dress (on the right).

It is thus possible to trace back the sequence of painters' gestures reflected on the surfaces of the vases in a kind of stratigraphic diagram (Fig. 9e) and, in a broader framework, to verify the extent to which, in a production context as

39 There is no certain overlap between the group of lines in the right part of the *chiton* (in yellow) and those in the left part (in red), but the succession in the two groups seems to indicate this sequence.

limited and well-known as the Locrian one, specific practices attributable to individual painters can be identified. Obviously, many variables must be taken into account, such as the scheme of the figures and their position on the surface, the shape of the vase and, therefore, the inclination and/or curvature of the walls.

Overpainted Decorations

The final touches after spreading the black slip were the application of the so-called overpainted decorations. In the Locrian production – and in that of the Locri Group in particular – these seem limited to a few details. The surface of some vases and fragments, however, keeps the traces of a higher number of interventions, which should have had the aim of improving the chromatic rendering of the decorations. Most of them are not visible anymore, due to their complete detachment from the underlying figured and ornamental motifs.[40] In most cases, however, the analytical microscopic approach allowed us to increase the possibility of capturing the remaining traces of the lost paintings and to conjure up a more accurate idea of the original decorations.

This is the case, for example, with the lekane lid, from which numerous overpainted retouches have completely disappeared. However, the OM and RTI images make it possible to recognize significant elements of the original polychromy and iconography (Fig. 10a–b) such as the vertical sequence of round decorative elements on the robe and the vegetal crown on the female head (Fig. 10c–d). Here, SEM images (Fig. 10g) taken in backscattered mode show that the composition of the overpainted materials was different from that of the underlying black gloss, possibly poorer in iron, as the black gloss under the (lost) decorations appears darker in the images. While some of the traces can be at least partly guessed with the naked-eye, in other cases the OM images are indispensable, as in the case of a fragment of a calyx krater, where the microscope reveals an ivy shoot – otherwise completely invisible – decorating the black-painted background (Fig. 10e–f).

Furthermore, the examinations under the microscope allowed us to recognize differences in the overpainting technique among Locrian productions. In fact, in the vases attributed to the Locri Group, the overpainted decoration is frequently superimposed (partially or completely) onto a previously applied slip layer; in contrast, this approach does not seem to be documented in the other local productions, such as those of the Kneeling Eros Group and the Painter of the Pyxis RC 5089. Systematic analysis will make it possible to assess whether these are specific traits of individual artisans.

Finally, SEM-EDX allowed us to tentatively determine the elemental composition from a small residue of overpainting on a fragment of a bell krater (Fig. 11a–b). Despite the poor accuracy that is expected for quantitative data obtained under low-vacuum conditions, some parallels – as well as some differences – emerge when comparing the data with those relating to materials and techniques used by Apulian artisans.[41] The small residue of a lost overpaint appears as having a white background, with more yellowish areas in some parts; a red grain is also visible. This micro-morphology, already documented in Apulian vases, suggests that more than one color may have been obtained by layering different materials on the white touches. In the case of the Apulian vases, the high alkali content of the white paint would explain the greater adhesion to the underlying surface. On the other hand, the most evident compositional characteristic of the overpainting we studied on the Locrian fragment (Fig. 11c) is the low concentration of alkalis, which would explain why these details are usually lost. Moreover, the analyses highlight the calcareous nature of the Locrian overpaintings, in contrast to the Apulian ones. Similar to what has been observed in the Apulian red-figured vases, the color of the overpainted decorations is mostly determined by iron compounds. In fact, iron is present in increasing concentrations from the white to the yellow parts – and its concentration is definitely very high in the red particle – as highlighted by the different grayscales of the corresponding areas in the backscattered image. The overall morphological and compositional data could be explained by the skillful use of a diluted wash of the black slip spread over the white overpaint, to add further depth to the overpainted details. One sample is not enough to draw any firm conclusion, but the preliminary information obtained here can guide the focus of future studies.

---

40 Detached overpaintings can leave more or less faint traces, such as opaque spots on the shine gloss, imprints on the *miltos*, or unevenness in the decay of the surfaces, which can sometimes be seen with the naked-eye. This subject is discussed in detail in relation to a krater from the Lucifero necropolis: Elia (2010) 113. For this krater, the reconstruction of the lost overpainted elements was crucial for the interpretation of gestures and figures.

41 Overpaintings have been investigated throughout by means of a combination of analytical techniques in Apulian red-figured pottery: see Mangone *et al.* (2009) and Giannossa *et al.* (2022) for results.

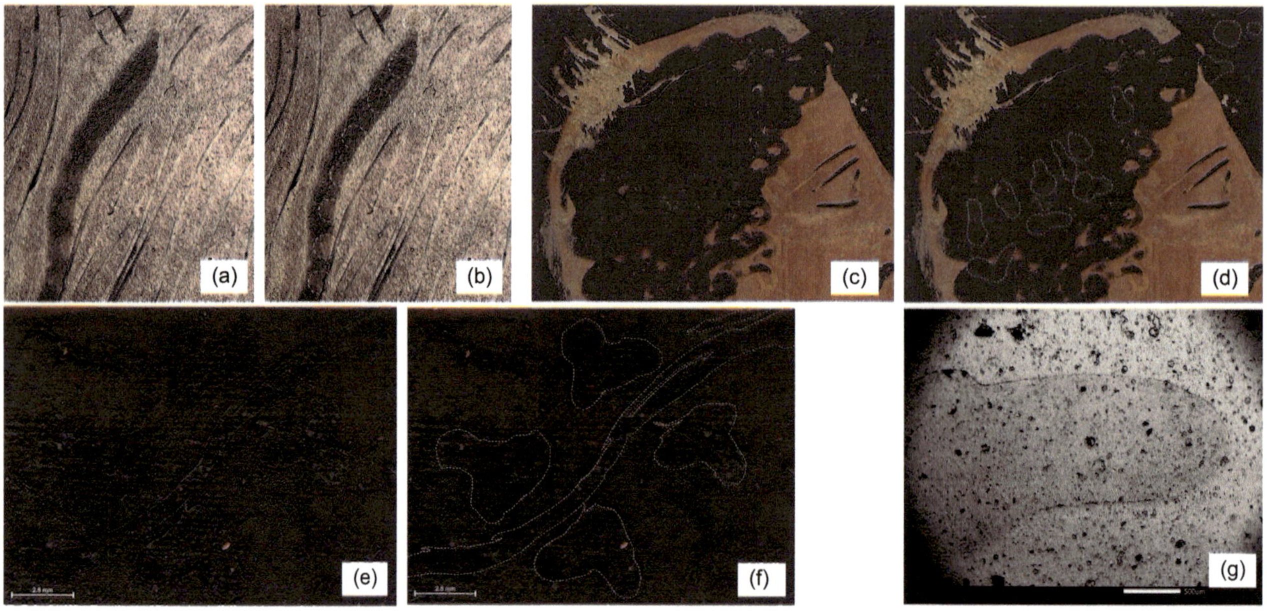

Fig. 10: Lekane lid fragment (Reggio Calabria, Museo Archeologico Nazionale, without inv. no.; after Barresi/Elia (2017) fig. 1): **(a)** RTI image of the robe of the female figure on the left of the fragment; **(b)** The corresponding reconstruction of the vertical sequence of the detached round overpainted decorative elements on the robe; **(c)** RTI image of the head of the female figure on the left of the fragment; **(d)** the corresponding reconstruction of the detached overpainted details of the vegetal crown and **(g)** BS-SEM image of one of the details of the vegetal crown.

Calyx krater fragment (Locri, Museo Archeologico Nazionale, inv. no. 75221): **(e)** OM image and **(f)** the corresponding reconstruction of the detached overpainted ivy shoot decoration.

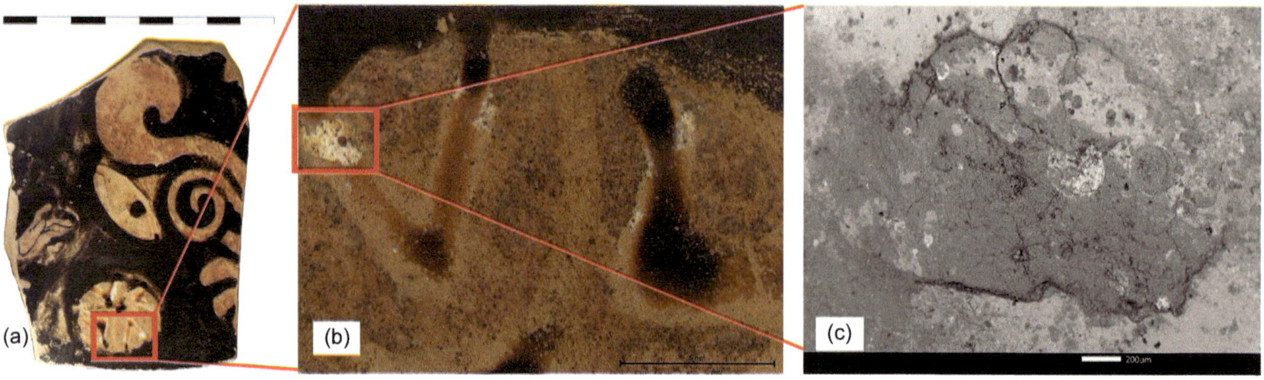

Fig. 11: Bell krater fragment (Locri, Museo Archeologico Nazionale, without inv. no.): **(a)** Macroscopic image, the red frame illustrates the position of the area magnified in the **(b)** OM image, where the red frame highlights the detail magnified in the **(c)** BS-SEM image.

Conclusions

This project is currently at its preliminary stages, but the results obtained so far have clearly demonstrated the potential of the analytical approach presented here for the observation and study of the traces left on the surfaces of red-figured vases found at Locri Epizephyrii by the coating and decoration processes, shedding further light on some of the techniques and procedures used by the Locrian painters.

The working hypotheses formulated so far have been based on a combination of methods and instruments, some of which can only be used in a laboratory. For the continuation of the research, we envisage the extensive and systematic use of portable instruments, in particular the digital microscope. Increasing the number of the samples examined – for example, by scrutinizing a large number of significant objects conserved at the National Archaeological Museums of

Reggio Calabria and Locri – may allow the identification of other clues that will be useful for a more detailed interpretation of the traces of the painter's gestures, in order to verify and enrich the hypothesis developed so far.

All this information can significantly contribute to the reconstruction of the world of the workshops in this city: a more detailed and in-depth understanding of the production units also enriches our knowledge about the interactions that took place between craft and society in Locri Epizephyrii.

Abbreviations

LCS Arthur Dale Trendall, 1967: *The Red-figured Vases of Lucania, Campania and Sicily*, Oxford.

LCS III Arthur Dale Trendall, 1983: *The Red-figured Vases of Lucania, Campania and Sicily – Third Supplement (consolidated)*, BICS, Supplement 41, Oxford.

Bibliography

Artal Isbrand/Klausmeyer (2013): Paula Artal Isbrand and Philip Klausmeyer, "Evaluation of the Relief Line and the Contour Line on Greek Red-figure Vases Using Reflectance Transformation Imaging and Three-dimensional Laser Scanning Confocal Microscopy", in: *Studies in Conservation* 58, 338–359.

Balachandran (2018): Sanchita Balachandran, "Uncovering Ancient Preparatory Drawings on Greek Ceramics", in: *The Iris*. http://blogs.getty.edu/iris/uncovering-ancient-preparatory-drawings-on-greek-ceramics (last access 25.04.2024).

Barresi (2013): Sebastiano Barresi, "Sicilian Red-figure Vase Painting", in: Claire Lyons, Michael Bennett and Clemente Marconi (eds.), with the collaboration of Alexandra Sofroniew, *Sicily. Art and Invention between Greece and Rome* (Exhibition Catalogue, J. Paul Getty Museum, Los Angeles, 2013), Los Angeles, 210–218.

Barresi (2018): Sebastiano Barresi, "Il Gruppo di Locri in Sicilia: analisi ed elementi di riflessione", in Martine Denoyelle, Claude Pouzadoux and Francesca Silvestrelli (eds.), *Mobilità dei pittori e identità delle produzioni*, Cahier du Centre Jean Bérard 25, Naples, 41–63.

Barresi/Elia (2017): Sebastiano Barresi and Diego Elia, "Locri Epizefiri: nuove testimonianze relative alla ceramica figurata di produzione locale", in: *Orizzonti. Rassegna di archeologia* XVIII, 27–42.

Boss (1997): Martin Boss, "Preliminary Sketches on Attic Red-Figured Vases of the Early Fifth Century B.C.", in: John H. Oakley, William D.E. Coulson and Olga Palagia (eds.), *Athenian Potters and Painters* I, Oxford, 345–351.

Bovero *et al.* (2012): Alessandro Bovero, Marco Demmelbauer, Diego Elia, Maria Gargano, Tommaso Poli and Carla Scilabra, "Studio e intervento di restauro su un cratere apulo a figure rosse", in: *Lo Stato dell'arte 10* (Atti del X Congresso Nazionale del Gruppo Italiano dell'IGIIC International Institute for Conservation, Roma, Accademia Nazionale di San Luca, 22–24 novembre 2012), Rome, 181–188.

Carcagni *et al.* (2005): Pierluigi Carcagni, Claudia Daffara, Raffaella Fontana, Maria Chiara Gambino, Maria Mastroianni, Cinazia Mazzotta, Enrica Pampaloni and Luca Pezzati, "Optical Micro-Profilometry for Archaeology", in: *Optical Methods for Arts and Archaeology* 5857, 118–128.

Castagnotto *et al.* (2021): Elena Castagnotto, Federico Locardi, Sawssen Slimani, Davide Peddis, Laura Gaggero and Maurizio Ferretti, "Characterization of the Caput Mortuum Purple Hematite Pigment and Synthesis of a Modern Analogue", in: *Dyes and Pigments* 185, 108881.

Corbett (1965): Peter Edgard Corbett, *Preliminary Sketch in Greek Vase-Painting*, in: *The Journal of Hellenic Studies* 85, 16–28

Cuomo di Caprio (2007): Ninina Cuomo di Caprio, *Ceramica in archeologia 2. Antiche tecniche di lavorazione e moderni metodi di indagine*, Rome.

Cuomo di Caprio (2017): Ninina Cuomo di Caprio, *Ceramics in Archaeology. From Prehistoric to Medieval Times in Europe and the Mediterranean: Ancient Craftsmanship and Modern Laboratory Techniques*, Rome.

Elia (2005): Diego Elia, "La diffusione della ceramica figurata a Locri Epizefiri nella prima metà del IV secolo: problemi di stile, produzione e cronologia", in: Martine Denoyelle, Enzo Lippolis, Marina Mazzei and Claude Pouzadoux (eds.), *La céramique apulienne. Bilan et perspectives* (Actes de la Table Ronde, École française de Rome – Centre Jean Bérard, Naples, 30 Novembre-2 Décembre 2000), Naples, 155–162.

Elia (2010): Diego Elia, *Locri Epizefiri VI.* Nelle case di Ade. *La Necropoli in Contrada Lucifero, Nuovi Documenti*, Alessandria.

Elia (2014): Diego Elia, "Local Production of Red-figure Pottery at Locri Epizephyrii: A Synthesis on the Last Decade of Studies", in: Stine Schierup and Victoria Sabetai (eds.), *The Regional Production of Red-figure Pottery: Greece, Magna Graecia and Etruria*, Gösta Enbom Monograph 4, Aarhus, 279–290.

Elia (2018): Diego Elia, "Il Gruppo di Locri in Calabria meridionale: sviluppo di una tradizione siceliota", in: Martine Denoyelle, Claude Pouzadoux and Francesca Silvestrelli (eds.), *Mobilità dei pittori e identità delle produzioni*, Cahier du Centre Jean Bérard 25, Naples, 77–95.

Elia (2019): Diego Elia, "Produzione, distribuzione e consumo da Locri a Caulonia nella prima metà del IV sec. a.C.", in: *Produzioni e committenze in Magna Grecia* (Atti del LV Convegno di Studi sulla Magna Grecia, Taranto, 24–27 settembre 2015), Taranto, 553–567.

Elia (2024): Diego Elia, "Locri Epizefiri: per una contestualizzazione della ceramica figurata di produzione occidentale. L'abitato e l'area esterna alle mura tra "agorà emporica" e Adonion", in: *Thiasos. Rivista di Archeologia e Architettura antica* 13, 83–102.

Elia/Meirano (2022): Diego Elia, Valeria Meirano, "Produzioni artigianali, contesti e forme di consumo a Locri Epizefiri. Appunti e riflessioni, tra periodo arcaico ed età classica", in: *Orizzonti. Rassegna di archeologia* XXIII, 39–66.

Elia *et al.* (forthcoming): Diego Elia, Patrizia Davit, Marco Demmelbauer, Silvia Gobbato and Monica Gulmini, "Le produzioni di ceramica figurata dell'Italia meridionale: problemi e prospettive di ricerca", in: Alfonsina Benincasa, Eliana Mugione and Claude Pouzadoux (eds.), *Le produzioni di ceramica figurata dell'Italia meridionale: problemi e prospettive di ricerca*, forthcoming.

Elias *et al.* (2006): Mady Elias, Céline Chartier, G. Prévot, Hélène Garay and Colette Vignaud, "The Colour of Ochres Explained by Their Composition", in: *Materials Science and Engineering: B*, 127, 70–80.

Ercolin *et al.* (2023): Elisa Ercolin, Diego Elia, Patrizia Davit and Monica Gulmini, "Oltre il visibile. Un approccio interdisciplinare per la ricostruzione della tecnologia di produzione della ceramica a vernice nera da Locri Epizefiri", in: Chiara Andreatta, Marco Argentina, Greta Boldorini, Federica Bosio, Eliana Bridi, Chiara Casarin, Andrea Chiocca, Francesca Daniele, Mirko Fecchio, Ylenia Saretta, Federica Stella Mosimann, Gianantonio Urbani and Romina Zanon (eds.), *Svelare l'invisibile. Tracce nascoste di storie, opere e contesti* (Atti del Convegno internazionale, Padova, 17–18 Novembre 2022), Padua, 25–34.

Gadaleta (2012): Giuseppina Gadaleta, "Tecniche", in: Luigi Todisco (ed.), *La Ceramica a Figure Rosse della Magna Grecia e della Sicilia*, Vol. II. Rome, 111–128.

Giannossa *et al.* (2022): Lorena Carla Giannossa, Italo Maria Muntoni, Tiziana Forleo, Salvatore Patete, Claude Pouzadoux, Rocco Laviano and Annarosa Mangone, "Artisanal Experiences in Arpi: Colours and Raw Materials of Apulian Red Figure Production", in: *Journal of Archaeological Science: Reports* 45, 103618. https://doi.org/10.1016/j.jasrep.2022.103618.

Giumlia-Mair *et al.* (2010): Alessandra Giumlia-Mair, Cindy Albertson, Giovanni Boschian, Gianna Giachi, Paola Iacomussi, Pasquino Pallecchi, Giuseppe Rossi, Aaron N. Shugar and Susan Stock, "Surface Characterisation Techniques in the Study and Conservation of Art and Archaeological Artefacts: A Review", in: *Materials Technology* 25, 245–261.

Gobbato (2016–2017): Silvia Gobbato, *Un cratere a figure rosse del Pittore di Locri da Locri Epizefiri: studio della tecnica esecutiva, delle problematiche conservative e restauro*, Tesi Magistrale in Conservazione e restauro dei Beni Culturali, University of Turin.

Gobbato *et al.* (2019): Silvia Gobbato, Marco Demmelbauer, Diego Elia and Monica Gulmini, "Studio e restauro di un cratere a figure rosse del Pittore di Locri", in: *Faenza. Bollettino del Museo Internazionale delle ceramiche in Faenza* 1–2, 83–90.

Jones (1986): Richard E. Jones, *Greek and Cypriot Pottery: A Review of Scientific Studies*, Fitch Laboratory Occasional Paper 1, Athens.

Jones (2021): Richard E. Jones, "The Decoration and Firing of Ancient Greek Pottery: A Review of Recent Investigations", in: *Advances in Archaeomaterials* 2, 67–127.

Mangone *et al.* (2004): Annarosa Mangone, Lorena Carla Giannossa, Giuseppe Colafemmina, Rocco Laviano and Angela Traini, "Use of Various Spectroscopy Techniques to Investigate Raw Materials and Define Processes in the Overpainting of Apulian Red Figured Pottery (4th Century BC) from Southern Italy", in: *Microchemical Journal* 92, 97–102.

Mirti *et al.* (2004a): Piero Mirti, Monica Gulmini, Marco Pace and Diego Elia, "The Provenance of Red Figure Vases from Locri Epizephirii (Southern Italy): New Evidence by Chemical Analysis", in: *Archaeometry* 46, 183–200.

Mirti *et al.* (2004b): Piero Mirti, Monica Gulmini, Alessandra Perardi, Patrizia Davit and Diego Elia, "Technology of Production of Red Figure Pottery from Attic and Southern Italian Workshops", in: *Analytical and Bioanalytical Chemistry* 380, 712–718.

Noble (1988): Joseph Veach Noble, *The Techniques of Painted Attic Pottery*, 2nd revised edition, New York.

Photos-Jones *et al.* (1997): Effie Photos-Jones, Arbory McNulty Cottier, Allan J. Hall and Lina Mendoni, "Kean Miltos: The Well-Known Iron Oxides of Antiquity", in: *The Annual of the British School at Athens* 92, 359–371.

Photos-Jones *et al.* (2018): Effie Photos-Jones, Charles W. Knapp, Danae Venieri, Georgios E. Christidis, Christine Elgy, Eugenia, Iosifina Gouniaki and Nafsica C. Andriopoulou, "Greco-Roman Mineral (Litho)therapeutics and Their Relationship to Their Microbiome: The Case of the Red Pigment Miltos", in: *Journal of Archaeological Science: Reports* 22, 179–192.

Saunders *et al.* (2017): David Saunders, Richard Collmann and Ann Borda, "Reflectance Transformation Imaging and ImageJ: Comparing Imaging Methodologies for Cultural Heritage Artefacts", in: *Proceedings of EVA-Electronic Visualisation and the Arts* (London, 11–13 July 2017). https://doi.org/10.14236/ewic/EVA2017.71.

Schreiber (1999): Toby Schreiber, *Athenian Vase Construction: A Potter's Analysis*, Malibu.

Spigo (2000): Umberto Spigo, "Fra Lipára e Lokroi Epizephýrioi. Il cratere a campana del Gruppo di Locri nella tomba 1155 di c/da Diana e alcune considerazioni sui rapporti fra ceramografi presenti a Lipari e nella Calabria meridionale nel secondo e terzo venticinquennio del IV secolo a.C.", in: Umberto Spigo and Maria Clara Martinelli (eds.), *Nuovi studi di archeologia eoliana*, Quaderni del Museo Archeologico Regionale Eoliano "Luigi Bernabò Brea" 3, 29–54.

Annarosa Mangone, Lorena Carla Giannossa, Fabrizio Mastrorocco and Italo Maria Muntoni

14 Examining the Red and Black Surfaces of Apulian Red-figured Pottery: Raw Materials and Technological Features

Abstract: The red and black surfaces of Apulian red-figured pottery from significant sites located in various geographic-cultural areas in Apulia were investigated in order to test the hypothesis that local pottery production grew progressively independent of Attic manufacturing technology, its original inspiration. Inductively Coupled Plasma Mass Spectrometry (ICP-MS), Scanning Electron Microscopy (SEM) with Energy Dispersive X-ray Spectroscopy (EDS), Powder X-ray Diffraction (PXRD), and Raman Spectroscopy were used to obtain a complete chemical and mineralogical characterization of the surfaces. The archaeometric results indicate that despite being very similar externally, Apulian and Attic surfaces are in fact different, having been made with locally available raw materials and via similar but not identical manufacture. In particular, the finest fraction of *terra rossa*, very common throughout Apulia, was probably the raw material for black gloss. As for the red surfaces, various methods were employed to redden the whitish/grayish ceramic surface of Late Apulian vases, whose color is due to the coarse clay fraction used in producing them. Refined from the same clays as the body, a clayey engobe could have been applied to the body, or else a clayey layer enriched with iron oxides or a wash with ferruginous water. These are the expedients identified so far. It remains unclear why a clay coarse fraction was used, though typically discarded by Attic potters. It is unclear whether this choice was mechanically or structurally motivated, a more appropriate material being needed for the larger vases, characteristic of Late Apulian production, or if it was economically motivated, enabling significant savings in raw materials as compared with Attic red-figured pottery while still achieving visual effects like those of traditional technology. All the results support the hypothesis that at least since the 4th century B.C. Apulian production is distinct in character and completely autonomous from Attic traditions of production. The results also suggest that the potters in Magna Graecia profoundly understood pottery craftsmanship, as is evident in their deliberately selecting raw materials and their adopting suitable technologies for the specific needs of producing each vase.

Introduction

This chapter considers the surfaces of Apulian red-figured pottery as part of a broader research project. That project aims to identify where archaeological artifacts were crafted and to investigate the production technologies of their creation.[1]

There were five specific archaeometric goals:

- Broaden our knowledge of the manufacturing technology of this ceramic class in southern Italy.
- Assess the gradual divergence of this technology from Attic production, initially drawing inspiration thence but developing its own distinct character: this, to strengthen the conclusion that Apulian pottery production evolved independently from the Attic, encompassing distinctive manufacturing technologies.
- Determine whether different technologies were used at different sites.
- Obtain objective determinants of finds, essential for tracing the mobility of artifacts, artisans, and ancient knowledge.
- The ultimate and most ambitious goal: comparison of archaeometric results with those from a stylistic textual and contextual classification may allow us to arrive at better definitions of "local works," "imported works," "works in imitation of" and therefore make possible explaining the chronology, rate of production, and the mechanism of the Hellenization process of non-Greek centers in southern Italy.

Archaeometric investigations into red-figured pottery have primarily concentrated on Attic production, often examining vases in museums or private collections with the aim of reconstructing the technologies of making black gloss.[2]

1 Mangone *et al.* (2008), (2009a), (2009b), (2013); Bitetto *et al.* (2016); Giannossa *et al.* (2016), (2017a), (2017b), (2019), (2020), (2021), (2022); Eramo/Mangone (2019); Domenech-Carbo *et al.* (2022a), (2022b); Forleo *et al.* (2022).

2 Jones (2021); Saunders *et al.* (2020); Solard *et al.* (2023) and references therein.

https://doi.org/10.1515/9783111189635-014

A clay selected for specific characteristics or else the finer fraction obtained by elutriating the same clay as the ceramic body are the raw materials hypothesized to produce the Attic black gloss. Adding compounds to the clay that can improve its technological characteristics has also been hypothesized.

Because of the strong compositional diversity between the black gloss and the ceramic body and because of the results of experimental archaeology, scholars concur that the hypothesis that the same clay as the ceramic body was used (except in specific and sporadic instances) should be dismissed. Instead, they argue that using a clay rich in minerals, low in mica and calcite, and subjected to a thorough purification could provide the optimal raw material for the gloss.

It is still under debate why there is more zinc (Zn) in the black gloss than in the body in some sherds of Athenian red and black-figured pottery, one Corinthian black-figured sherd dating from the 6th to the 5th century B.C., and one Etruscan red-figured sherd from the 4th century B.C. According to some scholars this difference is due to the treatment of the clayey raw material with vitriol to induce flocculation and promote calcium carbonate ($CaCO_3$) removal. For others, the difference is due to the use of a clay different from that used for the body and extracted from clay deposits whose composition is characterized by high zinc (Zn) contents.[3] Similarly under debate is the cause of the negative cerium (Ce) anomaly in the rare-earth element profile evidenced in them. Such depletion could be associated with anoxic weathering environments but may also have been caused by chemical treatment of the clay during its refining, although according to R. Jones other factors may also have contributed to this anomaly.[4]

Regarding the manufacturing, consensus exists that after the clay slip was applied, the vessel underwent a single firing. This firing involved three phases—oxidation, reduction, oxidation—with a maximum firing temperature reaching 890°–950°C. During the reduction, hematite (Fe_2O_3) was transformed predominantly into magnetite (Fe_3O_4) as well as ferrous oxide (FeO) and hercynite ($FeAl_2O_4$).

Very few studies have concerned the production of southern Italy and particularly Apulia,[5] where intense production is documented from the third quarter of the 5th century B.C. to the end of the 4th, since Apulian potters have been supposed to have uncritically acquired their knowledge from Attic manufacturing, Apulian pottery having hitherto been wrongly considered a mere imitation of the Attic. Differences in manufacturing and/or raw materials cannot, however, be excluded a priori.

This chapter summarizes the results of red and black surface analyses of Apulian red-figured pottery dating to the 4th century B.C. from significant sites of different geographic-cultural areas of Apulia: Peucetia, Daunia, and Messapia.[6] Objects in museums or private collections were omitted from analysis because of seldom adequately documented restorations they had previously undergone, especially the older ones. Frequently, vases underwent reconstruction prior to acquisition (with pieces assembled from different vessels or missing parts integrated, intended to enhance the vases' commercial value) and repainting using creative materials and methods.[7] Additionally, crucial details such as provenance and excavation context had often been compromised or lost.

A thorough archaeological analysis was conducted on all finds, and readers are directed to the cited papers for comprehensive archaeological details. All archaeometric data were obtained using integrated analytical techniques that provided compositional and structural information (composition, microstructure, profile at various depths, distinct phases, inclusions, and segregations) for comprehensive characterization.

Black Gloss

All Apulian black gloss shares distinctive characteristics, including extremely compact structure—about 20 μm thick, exhibiting high sintering with no visible clay structure. Its chemical composition is marked by higher quantities of aluminum (Al), iron (Fe), and potassium (K), alongside less calcium (Ca) and silicon (Si) as compared with the ceramic body, with the ratio of aluminum to silicon (Al/Si) increasing from 0.33 to 0.63 and a 90% increase in the amount of iron (Fe) and

---

3 Chaviara/Aloupi-Siotis (2016); Aloupi-Siotis (2020).

4 Jones (2021).

5 Mangone *et al.* (2008), (2009a), (2009b), (2013); Giannossa *et al.* (2016), (2017a), (2017b), (2019), (2020), (2021), (2022); Bitetto *et al.* (2016); Eramo/Mangone (2019); Domenech-Carbo *et al.* (2022a), (2022b); Forleo *et al.* (2022).

6 Mangone *et al.* (2008), (2009a), (2009b), (2013); Giannossa *et al.* (2017a), (2019), (2020), (2021), (2022).

7 Giannossa *et al.* (2016), (2017b).

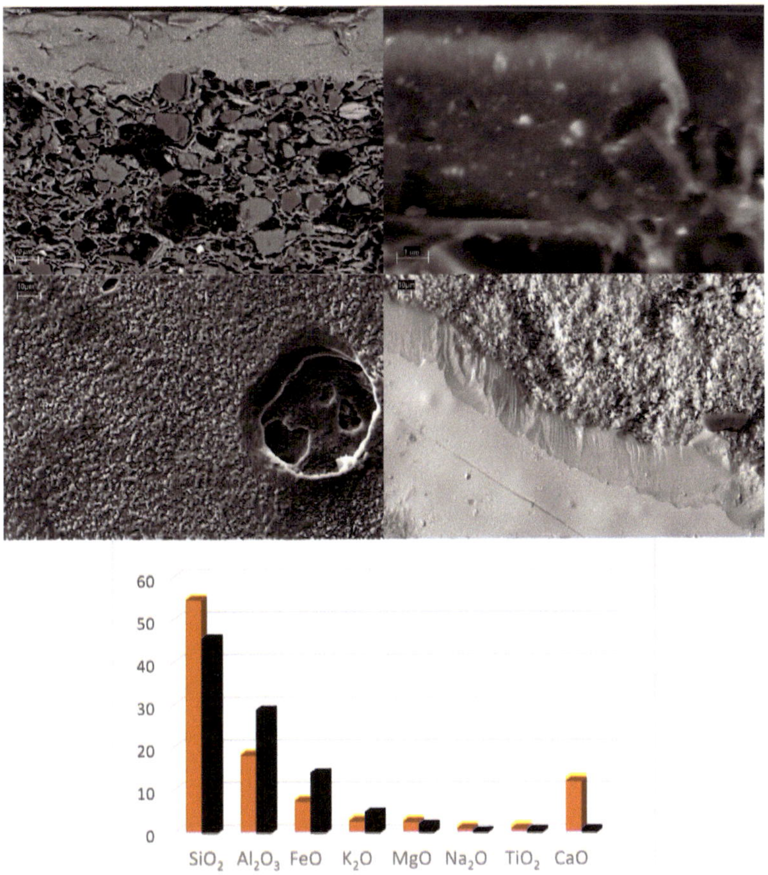

Fig. 1: From top to bottom clockwise: SEM-BSE photomicrographs of E29 and M15 cross sections, Ar3 and Ar5 surfaces, showing the structure of the black gloss and microprobe analysis of black gloss and ceramic body (mean values). Fe oxides particles (smaller brighter dots) and Fe-Ti oxides (larger brighter dots) are visible inside the M15 black gloss. © Mangone/Giannossa/Mastrorocco/Muntoni.

a 93% decrease in the amount of calcium (Ca) (Fig. 1). The amount of zinc (Zn) in the gloss and in the body is comparable, and the pattern of rare-earth elements shows a negative anomaly of cerium (Ce).

The notable variations in composition rule out the hypothesis that the suspension employed for the black gloss could have originated from refining the same clay as the ceramic body. Instead, these differences endorse the hypothesis that a distinct clay underwent refinement to produce the black gloss.

Quartz, iron oxides (mostly magnetite, but also wüstite, maghemite, and hematite), feldspars, pyroxenes, more rarely hercynite, ilmenite, and sometimes corundum are the mineralogical phases of black gloss. The distribution of iron oxides varies across the thickness, with a predominant presence of magnetite in the outer areas and hematite (α-Fe_2O_3) and maghemite (γ-Fe_2O_3) gradually appearing moving toward the ceramic body (Fig. 2).[8]
Based on the minerals present and their association and spatial arrangement within the black gloss some assumptions can be made about raw materials and manufacturing technology:

- The formation of corundum suggests the use of aluminum (Al)-rich minerals (i.e., kaolinite and illite) as raw materials.[9]
- The presence of both hercynite and magnetite makes it possible to assume that the firing reached a maximum temperature of about 900°C.[10] This temperature was also confirmed through High Resolution Magic Angle Spinning Nuclear Magnetic Resonance (CP-MAS NMR), utilizing the temperature-dependent chemical-shift values of silicon

8 Forleo *et al.* (2022).
9 See previous note.
10 Maritan *et al.* (2006).

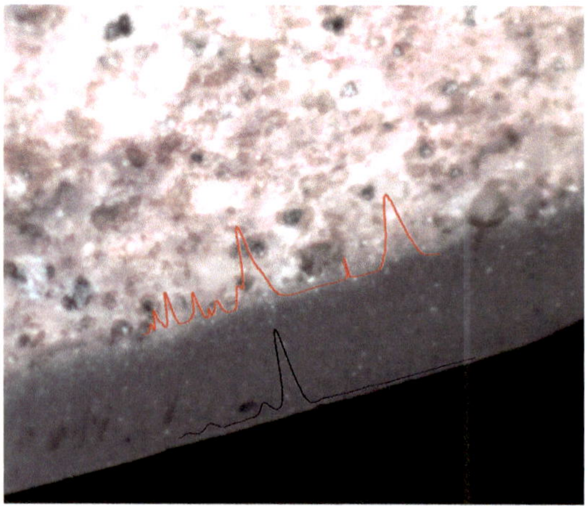

Fig. 2: MO image of the transversal section of the A6 sample, displaying the black gloss on the ceramic body. Raman baseline-subtracted representative spectra of the outermost layer (black), indicating the exclusive presence of magnetite and the innermost layer (red) revealing the simultaneous presence of magnetite, hematite, and maghemite. © Mangone/Giannossa/Mastrorocco/Muntoni.

(^{29}Si) and aluminum (^{27}Al), observed in kaolinite within the white pigment beneath the black gloss layer in a fragment from Monte Sannace (Gioia del Colle, BA).[11]

- The presence of both hematite and maghemite in the innermost area of the gloss layer suggests that magnetite was reoxidized during the last step of the firing, implicitly confirming the three-step (ox-red-ox) firing.
- The initial particle size of magnetite in the gloss was below 1 micron.[12] Maghemite is, indeed, like a chemical intermediate in the oxidation process of magnetite to hematite when the starting magnetite has an average size of less than 1 µm. This size, as measured by SEM for iron oxide in the gloss, aligns with the findings shown in Fig. 1.

The possibility that organic materials (i.e., tannins, wood ash) were added to the suspension to reduce the slip's melting point and enhance its sintering properties should also be ruled out.[13] Raman analyses revealed the presence of amorphous carbon on all the black and red surfaces, and SEM investigations did not detect any residues or vacuoles within the gloss with plant-ash morphology that would suggest the use of organic material.[14]

The difference between Apulian and Attic gloss is also evident in the results of Principal Components Analysis (PCA). Through the application of multivariate statistical analysis on compositional data, it is indeed possible to extract information related to the origin of objects/raw materials,[15] the manufacturing process,[16] and classification within an archaeological category,[17] based on groupings that arise from items according to their chemical composition.

Eleven chemical parameters (Mg, Al, Si, K, Fe, Cr, Ni, Co, Zn, Rb, Sr) were selected via a statistical criterion such as assessing the means and variances of different variables before applying PCA and eliminating a few unnecessary parameters that only complicate a good classification.[18] The software package Minitab® was applied on compositional and standardized data to perform the multivariate statistical treatment.

Figure 3 shows the diagram of the scores and loadings of black gloss of the Apulian and Attic gloss[19] in the subspace of the first three principal components. Two markedly distinct groups can be identified. In the same figure the 95% isoprobability ellipsoids, whose surfaces define the boundaries of clusters A and B, are also shown.

Cluster A groups all Apulian finds except FT38; cluster B groups all Attic finds. They differ along PC1 primarily because of loadings of the aluminum (Al), silicon (Si), cobalt (Co), strontium (Sr), rubidium (Rb), and nickel (Ni) parameters.

11 Mangone *et al.* (2009b).
12 Modesto Lopez *et al.* (2009).
13 Goffer/Winefordner/Dovichi (2008).
14 Forleo *et al.* (2022).
15 Giannotta *et al.* (2006); Giannossa *et al.* (2017b).
16 Mangone *et al.* (2009a), (2009c), (2013); Eramo *et al.* (2014); Bitetto *et al.* (2016).
17 Mangone *et al.* (2009d).
18 Massart *et al.* (1988).
19 Walton *et al.* (2015).

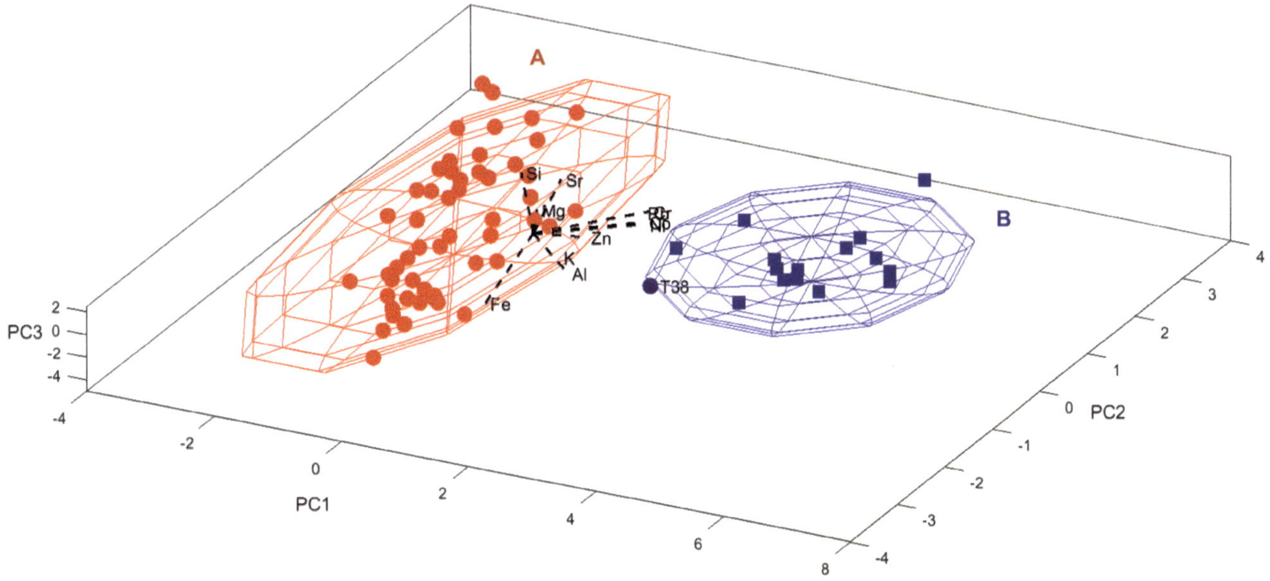

Fig. 3: Scores and loadings diagram for the first three principal components related to Apulian and Attic – Walton *et al.* (2015) red-figured black gloss. The accounted variance is 75% of the total one. © Mangone/Giannossa/Mastrorocco/Muntoni.

Cluster A scores in particular are distinguished by negative PC1 values because of lesser quantities of aluminum (Al), nickel (Ni), cobalt (Co), and rubidium (Rb), whereas cluster B scores are distinguished by positive PC1 values because of higher silicon (Si) and strontium (Sr) contents. The enlargement of cluster A along negative PC3 is mostly owing to iron and magnesium loadings.

Sample T38 and Attic artifacts merge into the same macrogroup, highlighting a marked compositional similarity. Despite being excavated in Taranto, archaeological investigations suggested that sample T38 was imported. This hypothesis has been substantiated through our analysis, indicating that its ceramic body has a greater chemical compositional and mineral-petrographic similarity to Attic samples than to Apulian ones.[20] The distinctiveness of T38 is further underscored by the zinc content in its gloss, which is significantly higher than in its body—approximately ~ 1,000 ppm—and unusually elevated as compared with the contemporaneous black glosses of Apulian samples (~ 400 ppm).

All the analytical results indicate that the primary material used for Apulian black glosses is the finest fraction (< 2 μm) of *terra rossa*, which is rich in kaolinite and illite, as well as iron (Fe) and aluminum (Al) oxides. This fine fraction is easily obtained through elutriation. *Terre rosse* are continental sedimentary layers prevalent throughout Apulia.[21] They are predominantly composed of partially crystalline iron (Fe) and aluminum (Al) oxides and hydroxides, clay minerals such as illite and kaolinite, and traces of quartz, feldspars, micas, pyroxenes, and other minerals.

The identification of *terra rossa* in an Apulian red-figured vase (Ar43) retrieved from a tomb in Arpi (Fig. 4) is not *per se* conclusive evidence for *terra rossa* as a raw material. However, it does strongly imply the presence of *terra rossa* in the immediate vicinity of the tomb. Interesting insights into raw materials and production have emerged from two unfired vases with female-figure appliques. These vases were discovered alongside red-figured vases at the entrance of a grotticella tomb located along the Celone River (Foggia) and dated to the second half of the 4th century B.C.[22] The raw material in its natural state, straightforward manufacturing, and the absence of firing renders these well-suited vessels for recognizing raw materials. Analysis revealed that the vessels were crafted from local clays near Arpi, confirming that pottery was produced in the city. An iron-based clay with manganese added for the black hue was what colored the surfaces.

20 Giannossa *et al.* (2017a).
21 Dell'Anna (1967); Dell'Anna/Laviano (1991).
22 Giannossa *et al.* (2019).

Fig. 4: *Terra rossa* within a vase (RCE 2068) recovered from tomb 4/2004 in Arpi (lef), SEM-BSE photomicrograph of *terra rossa* (right, upper) and its microprobe analysis (right, lower). © Mangone/Giannossa/Mastrorocco/Muntoni.

Red Surfaces

Many Late Apulian vases exhibit a noticeable contrast between the whitish/grayish ceramic body and the dark-red surface. Archaeometric findings mostly reveal an engobe over the ceramic body (Fig. 5). This engobe is distinguished from the ceramic body by its more compact and finer paste, characterized by a higher matrix and lower silt. The thickness varies between 40 and 200 µm, reaching 1 mm in some samples from Taranto. The minerals in the engobe and ceramic body are the same, but their quantitative ratios differ: the body contains more quartz, a primary component of the coarse fraction, and feldspars, whereas the engobe has more micas (biotites and muscovites) and clay minerals.

Compositional analysis of the two layers reveals an ascending trend in the ratios of aluminium to silicon (Al/Si) and of potassium to calcium (K/Ca), along with increased iron (Fe), moving from the ceramic body to the engobe. Notably, when the ceramic body is markedly calcium (Ca)-rich, the clay fraction of the engobe exhibits more calcium (Ca). These findings suggest the same raw material for both the engobe and the ceramic body, with a refinement specifically applied to the engobe. So far as we know, a clayey engobe for a red surface has not been documented for Attic production.

In the samples from Arpi, no engobe was detected. However, various methods imparted a reddish hue to the surfaces. Some have a fine silty-grained clay layer enriched with micas, quartz, and alkali feldspar, reinforced with iron oxides on the ceramic body, as depicted in Fig. 6. Others exhibit concentrated iron-oxide aggregates solely on the surface (Fig. 6). In both cases, the size of iron oxides on the surfaces exceeds that in the bodies, as indicated by the center-position difference

Fig. 5: Upper: MO (left) and SEM-BSE (right) photomicrographs of the cross sections of T200A. Lower: SEM-BSE photomicrograph of fragments 39N (lower), showing the engobe layer on the ceramic body. © Mangone/Giannossa/Mastrorocco/Muntoni.

between the 660 cm^{-1} and 225 cm^{-1} Raman bands of hematite—a parameter sensitive to grain size.[23] These findings suggest a clay layer enriched with iron oxides for the former and ferruginous water (*miltos*) for the latter on the ceramic body. The *miltos* wash before the black gloss resulted in the black gloss's flaking off (Fig. 6). Unlike the sieved clay binder used in other instances between the ceramic body and the gloss, *miltos* probably did not exhibit the same behavior. The nonclay nature of *miltos* may have contributed to the flaking off of the black surface. Its properties may also have hindered the retention of gloss adhesiveness, making it less resistant to degradation during burial.

Conclusions

Although Apulian and Attic red-figured surfaces (both red and black) may exhibit a very similar external appearance, archaeometric analyses have demonstrated their inherent differences. These distinctions arise from the use of locally available raw materials and similar though not identical manufacture. The disparities in the raw materials highlighted for Apulian black gloss in comparison with the Attic offer objective criteria for distinguishing regional production. Particularly in late productions, these differences strongly argue against dried Attic black glaze being imported from Greece, a hypothesis entertained by some scholars,[24] despite high costs and commercial impediments.

As for the black gloss, one may speculate that the finest fraction from abundant raw materials in Apulia, namely *terre rosse*, was used without adding any organic material to the clay. Furthermore, the use of an iron (Fe)-based clay with manganese (Mn) added for the black, to produce red and black colors on unfired vases discovered in a tomb at Arpi alongside red-figured ones, indicates that the practice of using manganese (Mn) and iron (Fe) oxides to achieve both red and black hues on the same vessel persisted among potters in Magna Graecia during the 4th century B.C. This practice also suggests that the intentional adoption of the Attic process was aimed at creating more sophisticated pottery, such as red-figured vases.

---

23 Modesto Lopez *et al.* (2009).

24 Noble (1960).

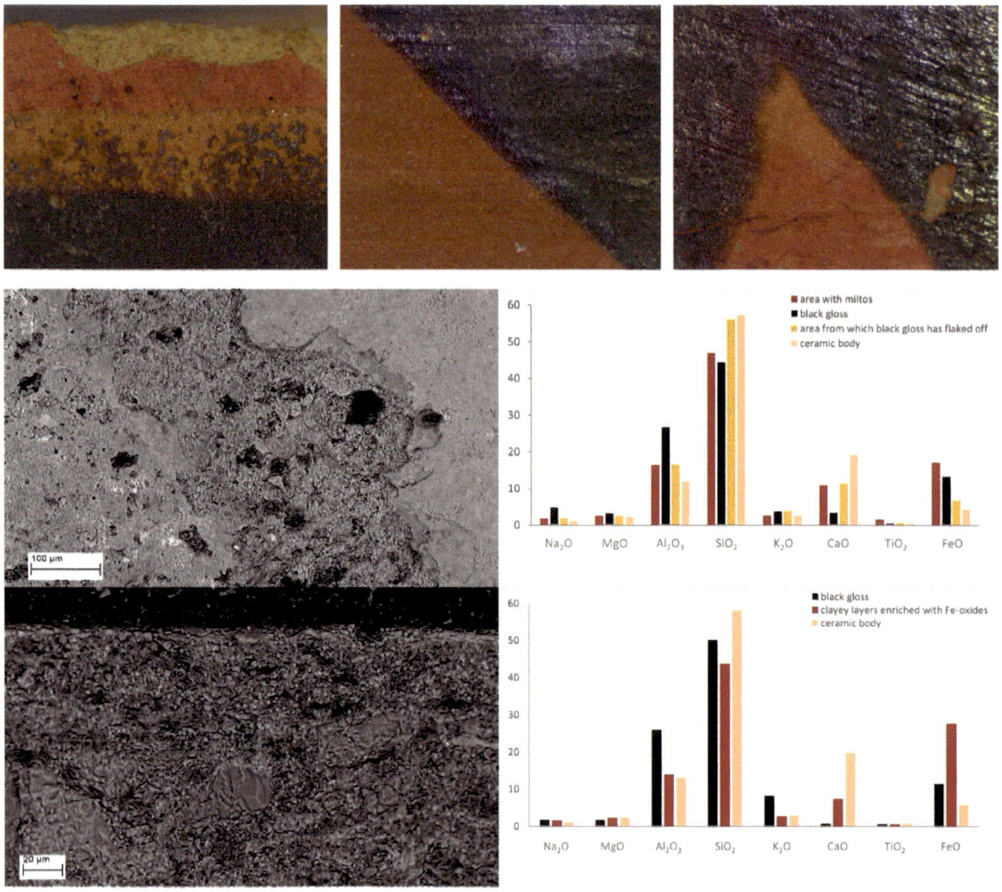

Fig. 6: Upper: MO photomicrographs depicting the surfaces of fragments with ceramic bodies covered by *miltos* (left), clayey layers enriched with Fe-oxides (intermediate) and clayey engobe (right). Intermediate: SEM-BSE photomicrograph showing the surface of fragment ArB (left), chemical composition of the different surface areas (right). Lower: SEM-BSE photomicrograph of the cross-sections of fragment ArA (left), chemical composition of the ceramic body and the various overlapped layers on it (right). © Mangone/Giannossa/Mastrorocco/Muntoni.

Apulian potters employed various techniques to replicate the red surface of Attic vases, which proved challenging to achieve using the coarse fraction of local clay for the body, given its high calcium (Ca) and low iron (Fe) content. Their methods included using a clay engobe, a clay layer enriched with iron (Fe) oxides, and the use of *miltos* (ferruginous water). To the best of the authors' knowledge, Attic potters, on the other hand, utilized only an ochre wash to enhance the red color of the ceramic body. Using the coarse clay fraction typically discarded by Attic potters may have originated from the need for a material with mechanical-structural characteristics better suited for creating the monumental vases characteristic of Late Apulian production. Alternatively, it may have been driven by economic considerations associated with more cost-effective productions. This approach allowed for the conservation of significant quantities of raw materials compared with traditional Attic technology while achieving similar visual effects. The multistep firing (ox-red-ox) itself was, however, similar, reaching a maximum temperature of about 900° C.

In addition to widening knowledge about technological-productive aspects of Apulian red-figured pottery, the results obtained validate the hypothesis that at least from the 4th century B.C. on it assumed distinctive characteristics and a complete independence from Attic production, and besides as regards the raw materials and manufacturing employed. They also suggest that the potters working in Magna Graecia at that time possessed a thorough understanding of pottery craftsmanship. They demonstrated a deliberate selection of raw materials, opting for the most appropriate technological processes tailored to their vases, considering factors such as market value, significance, and intended purpose of the pottery. Perhaps familiarity rather than real and proper continuity with Attic pottery traditions explains the analytical results.

Bibliography

Aloupi-Siotis (2020): Eleni Aloupi-Siotis, Ceramic Technology: "How to Characterize Black Fe-Based Glass-Ceramic Coatings", in: *Archaeological and Anthropological Sciences* 12, 191–205.

Bitetto *et al.* (2016): Alessandro Bitetto, Annarosa Mangone, Rosa Maria Mininni and Lorena Carla Giannossa, "A Nonlinear Principal Component Analysis to Study Archeometric Data", in: *Journal of Chemometrics* 30, 405–415.

Chaviara/Aloupi-Siotis (2016): Artemi Chaviara and Eleni Aloupi-Siotis, "The Story of a Soil that Became a Glaze: Chemical and Microscopic Fingerprints on the Attic Vases", in: *Journal of Archaeological Science: Reports* 7, 510–518.

Dell'Anna (1967): Luigi Dell'Anna, "Ricerche su alcune terre rosse della regione pugliese", in: *Periodico di Mineralogia* 36, 539–592.

Dell'Anna/Laviano (1991): Luigi Dell'Anna and Rocco Laviano, "Mineralogical and Chemical Classification of Pleistocene Clays from the Lucanian Basin (Southern Italy) for the Use in the Italian Tile Industry", in: *Applied Clay Science* 6, 233–243.

Doménech-Carbó *et al.* (2022a): Antonio Doménech-Carbó, Michele Giannuzzi, Annarosa Mangone and Lorena Carla Giannossa, "Electrochemical Methods to Discriminate Technology and Provenance of Apulian Red-figured Pottery I: VIMP and SECM", in: *Archaeometry* 64, 1325–1339. https://doi.org/10.1111/arcm.12785.

Doménech-Carbó *et al.* (2022b): Antonio Doménech-Carbó, Michele Giannuzzi, Annarosa Mangone and Lorena Carla Giannossa, "Electrochemical Methods to Discriminate Technology and Provenance of Apulian Red-figured Pottery II: EIS", in: *Archaeometry* 64, 1124–1137. https://doi.org/10.1111/arcm.12774.

Eramo *et al.* (2014): Giacomo Eramo, Lorena Carla Giannossa, Anita Rocco, Annarosa Mangone, Fabio Sossio Graziano and Rocco Laviano, "Oil Lamps from the Catacombs of Canosa (Apulia)", in: *Archaeometry* 56, 375–391.

Eramo/Mangone (2019): Giacomo Eramo and Annarosa Mangone, "Archaeometry of Ceramic Materials", in: Luigi Sabbatini and Inez Dorothé van der Werf (eds.), *Chemical Analysis in Cultural Heritage*, Berlin-Boston, 331–355.

Forleo *et al.* (2022): Tiziana Forleo, Lorena Carla Giannossa, Rocco Laviano and Annarosa Mangone, "Exploring the Raw Materials and Technological Practice to Obtain Red and Black Surfaces of Apulian Red Figure Pottery by Raman and SEM-EDS Investigations", in: *Journal of Raman Spectroscopy*, 53, 810. https://doi.org/10.1002/jrs.6307.

Giannossa *et al.* (2016): Lorena Carla Giannossa, Rocco Laviano, Fabrizio Mastrorocco, Gabriele Giannelli, Italo Maria Muntoni and Annarosa Mangone, "A Pottery Jigsaw Puzzle: Distinguish True and False Pieces in Two Apulian Red Figured Vases by a Poli-Technique Action Plan", in: *Applied Physics A: Materials Science and Processing* 122, 68–77.

Giannossa *et al.* (2017a): Lorena Carla Giannossa, Rosa Maria Mininni, Rocco Laviano, Fabrizio Mastrorocco, Maria Cristina Caggiani and Annarosa Mangone, "An Archaeometric Approach to Gain Knowledge on Technology and Provenance of Apulian Red Figured Pottery from Taranto", in: *Archaeological and Anthropological Sciences* 9, 1125–1135.

Giannossa *et al.* (2017b): Lorena Carla Giannossa, Federica Giacobello, Italo Maria Muntoni, Rocco Laviano, Maria Cristina Caggiani, Fabrizio Mastrorocco, Giovanni Lagioia and Annarosa Mangone, "An Analytical Techniques Pool to Hit the Target. A Comprehensive Examination on an Apulian Red Figured Pottery Collection", in: *Microchemical Journal* 134, 354–361.

Giannossa *et al.* (2019): Lorena Carla Giannossa, Italo Maria Muntoni, Rocco Laviano and Annarosa Mangone, "Contribution of Mineralogical and Analytical Techniques to Investigate Provenance and Technologies of Hellenistic Pottery from Arpi (Southern Italy)", in: *Journal of Archaeological Science: Reports* 24, 729–737. https://doi.org/10.1016/j.jasrep.2019.02.011.

Giannossa *et al.* (2020): Lorena Carla Giannossa, Italo Maria Muntoni, Rocco Laviano and Annarosa Mangone, "Building a Step-by-Step Result in Archaeometry. Raw Materials, Provenance and Production Technology of Apulian Red Figure Pottery", in: *Journal of Cultural Heritage* 43, 242–248. https://doi.org/10.1016/j.culher.2019.11.002.

Giannossa *et al.* (2021): Lorena Carla Giannossa, Tiziana Forleo and Annarosa Mangone, "The Distinctive Role of Metal Elements in Archaeometry. The Case of Apulian Red Figure Pottery", in: *Applied Sciences* 11, 3073–3089. https://doi.org/10.3390/app11073073.

Giannossa *et al.* (2022): Lorena Carla Giannossa, Italo Maria Muntoni, Tiziana Forleo, Salvatore Patete, Claude Pouzadoux, Rocco Laviano and Annarosa Mangone, "Artisanal Experiences in Arpi: Colours and Raw Materials of Apulian Red Figure Production", in: *Journal of Archaeological Science: Reports* 45, 103618. https://doi.org/10.1016/j.jasrep.2022.103618.

Giannotta *et al.* (2006): Cosimo Giannotta, Caterina Laganara, Rocco Laviano, Annarosa Mangone and Angela Traini, "Medieval Islamic Type Pottery from Siponto (Italy): an Integrated Physical-Chemical and Mineralogical Investigation", in: *X-Ray Spectrometry* 35, 338–346. https://doi.org/10.1002/xrs.915.

Goffer/Winefordner/Dovichi (2008): Zvi Goffer, James D. Winefordner and Norman J. Dovichi, *Archaeological Chemistry*, 2nd edition, New York.

Jones (2021): Richard Jones, "The Decoration and Firing of Ancient Greek Pottery: A Review of Recent Investigations", in: *Advances in Archaeomaterial* 2, 67–127. https://doi.org/10.1016/j.aia.2021.07.002.

Mangone *et al.* (2008): Annarosa Mangone, Lorena Carla Giannossa, Angela Ciancio, Rocco Laviano and Angela Traini, "Technological Features of Apulian Red Figured Pottery", in: *Journal of Archaeological Science* 35, 1533–1541.

Mangone *et al.* (2009a): Annarosa Mangone, Lorena Carla Giannossa, Giacomo Eramo, Rocco Laviano, Custode Silvio Fioriello, Viriana Redavid and Angela Traini, "Technological Features of Gnathia Pottery", in: *X-Ray Spectrometry* 38, 386–393.

Mangone *et al.* (2009b): Annarosa Mangone, Lorena Carla Giannossa, Giuseppe Colafemmina, Rocco Laviano and Angela Traini, "Use of Various Spectroscopy Techniques to Investigate Raw Materials and Define Processes in the Overpainting of Apulian Red Figured Pottery (4th century B.C.) from Southern Italy", in: *Microchemical Journal* 92, 97–102.

Mangone *et al.* (2009c): Annarosa Mangone, Lorena Carla Giannossa, Caterina Laganara, Rocco Laviano and Angela Traini, "Manufacturing Expedients in Medieval Ceramics in Apulia", in: *Journal of Cultural Heritage* 10, 134–143.

Mangone *et al.* (2009d): Annarosa Mangone, Lorena Carla Giannossa, Rocco Laviano, Custode Silvio Fioriello and Angela Traini, "Investigations by Various Analytical Techniques to the Correct Classification of Archaeological Finds and Delineation of Technological Features. Late Roman Lamps from Egnatia: From Imports to Local Production", in: *Microchemical Journal* 91, 214–221.

Mangone *et al.* (2013): Annarosa Mangone, Maria Cristina Caggiani, Lorena Carla Giannossa, Giacomo Eramo, Viriana Redavid and Rocco Laviano, "Diversified Production of Red Figured Pottery in Apulia (Southern Italy) in the Late Period", in: *Journal of Cultural Heritage* 14, 82–88.

Maritan *et al.* (2006): Lara Maritan, Luca Nodari, Claudio Mazzoli, Alessandro Milano and Umberto Russo, "Influence of Firing Conditions on Ceramic Products: Experimental Study on Clay Rich in Organic Matter", in: *Applied Clay Science* 31, 1–15.

Massart *et al.* (1988): Désiré Luc Massart, Bernerd G.M. Vandeginste, Stanley N. Deming, Yvette Michotte and Leonard Kaufman, *Principal Components and Factor Analysis, Chemometrics: a Textbook*, Amsterdam.

Modesto Lopez *et al.* (2009): Luis B. Modesto Lopez, Jill Dill Pasteris and Pratim Biswas, "Sensitivity of Micro-Raman Spectrum to Crystallite Size of Electrospray-Deposited and Post-Annealed Films of Iron-Oxide Nanoparticle Suspensions", in: *Applied Spectroscopy* 63, 627–635.

Noble (1960): Joseph Veach Noble, "The Technique of Attic Vase Painting", in: *American Journal of Archaeology* 64, 307–313.

Saunders *et al.* (2020): David Saunders, Karen Trentelman and Maish Jeffrey, "Chapter 5 – Collaborative Investigations into the Production of Athenian Pottery", in: Catherine Cooper (ed.), *New Approaches to Ancient Material Culture in the Greek and Roman World: 21st-Century Methods and Classical Antiquity*, Leiden, 124–135.

Solard *et al.* (2023): Baptiste Solard, Silvia Amicone, Eleni Aloupi-Siotis, Lars Heinze, Fede Berti, Claudia Lambrugo and Christoph Berthold, "Back to Black: a Mineralogical and Chemical Characterisation of Atticising Fourth Century BCE Black Gloss Ware", in: *Archaeological and Anthropological Sciences* 15, 135–157. https://doi.org/10.1007/s12520-023-01822-4.

Walton *et al.* (2015): Marc Walton, Karen Trentelman, Ilaria Cianchetta, Jeff Maish, David Saunders, Brendan Foran and Mehta Apurva, "Zn in Athenian Black Gloss Ceramic Slips: A Trace Element Marker for Fabrication Technology", in: *Journal of the American Ceramic Society* 98, 430–436.

Mattia D’Acri

15 Reconstructing Pottery Production in Archaic Rome and Latium: Integrating Traditional Approach and Archaeometry

Abstract: This chapter presents preliminary results from an interdisciplinary study investigating the intricate relationship between pottery production and social complexity in Archaic Rome and Latium. Employing traditional typological study and Neutron Activation Analysis (NAA) on excavated ceramics from sites like Sant’Omobono, Regia, Forum of Caesar, Gabii, and San Giovenale, the research explores technological choices and standardization within and across different wares. NAA, a nuclear technique providing comprehensive multielement analyses, reveals distinct chemical compositions, challenging established classifications. Principal Component Analysis (PCA) identifies significant elemental drivers, unveiling unexpected patterns and outliers. Groups with shared chemical imprints and manufacturing processes emerge, shedding light on local production and trade networks. The study’s innovative approach, combining archaeological and chemical analyses, promises a deeper understanding of pottery production in Archaic Rome and Latium, challenging existing paradigms and opening avenues for future research.

Introduction

This chapter presents the preliminary results of an interdisciplinary study that forms part of a broader PhD dissertation investigating the relationship between pottery production and social complexity in Archaic Rome and Latium. The work is based on the integration between traditional typological study and geochemistry, using recently excavated assemblages from relevant sites in Rome (Sant’Omobono, Regia, and Forum of Caesar) and Gabii (Rome’s closest neighbor), in addition to San Giovenale (as a comparative case study from Etruria).[1] Having conducted a classification of the vessels according to their morphology,[2] I have employed Neutron Activation Analysis (NAA), a nuclear technique in which a sample is bombarded with neutrons to produce radioactive isotopes with distinct decay patterns in order to characterize the ceramic paste. This method is ideal to perform bulk qualitative and quantitative multielement analyses of major, minor, and trace elements.[3] Combining the two techniques gives us a more precise idea of the variability and the level of standardization within and across the different wares, significantly enhancing the standardization of the ceramic record. Thus, this research strategy provides a more consistent framework to explain certain technological parameters such as potter’s skills.[4]

Even though archaeometry has so far played a minor role in previous studies in Latium for the Archaic period,[5] the available data allow cross-referencing and a broader analysis of regional and extraregional sites. The results of this methodology in the research-proposed question established chronologies,[6] with implications for our understanding of technology, development through time, impact within the society, and diffusion of products.

1 I express my thanks to the Soprintendenza Speciale Archeologia, Belle Arti e Paesaggio di Roma (especially Dr. Rocco Bochicchio and Chiara Andreotti), the Soprintendenza Archeologia Belle Arti Paesaggio Etruria Meridionale (Dr. Barbara Barbaro), the Parco Archeologico del Colosseo (Dr. Roberta Alteri), and the Soprintendenza di Roma Capitale (Dr. Monica Ceci) for granting the necessary permits. Furthermore, I would also like to thank Prof. Nicola Terrenato and Prof. Paolo Brocato, as directors of the Sant’Omobono and Regia projects, Dr. Jan Kindberg Jacobsen, Prof. Rubina Raja, and Dr. Claudio Parisi Presicce the directors of the *Caesar’s Forum Project*, Dr. Fredrik Tobin of the Swedish Institute in Rome, who helped me to get access to the material from San Giovenale, and finally, my advisor, Dr. Marcello Mogetta, for his constant help and support.

2 First results in D’Acri (2016); D’Acri/De Luca (2016), (forthcoming); D’Acri *et al.* (2018), (forthcoming).

3 Glascock (1992); Riehle *et al.* (2023).

4 Santacreu (2014) 183–184.

5 E.g., Ammerman *et al.* (2008); Revello Lami (2017). On the other hand, archaeometric analysis has a long tradition in other Mediterranean regions and periods, and this research aims to fill the gap in the scholarship; e.g., Riehle *et al.* (2023), with further references.

6 E.g., Carafa (1995), (2017).

https://doi.org/10.1515/9783111189635-015

Methods and Goals

A total of 209 samples were selected for analysis at the University of Missouri Research Reactor (MURR). The traditional approach to identify comparable shapes and types across sites has been used as a basis for selecting samples. Rome, Gabii, and San Giovenale share the same material culture in the period examined, consisting of Impasto Bruno, Impasto Rosso, Impasto Rosso-Bruno, Impasto Chiaro-Sabbioso, Finewares, and Bucchero.[7] All these pottery classes are represented in the study, with the ultimate objective of examining whether similarities in shapes or series are confirmed by geochemical analysis and of identifying shared or different technological choices. As correctly argued by D. A. Santacreu, the traditional typology is indeed a procedure that creates broad categories of artifacts related to specific spatial, chronological, or cultural frameworks which are useful for comparison. However, its use should not be limited to purely taxonomic or classificatory aims. Moving the debate beyond exclusively functional concerns, the combined analysis of types and fabrics is an essential tool to document technological choices in response to particular ways of organizing production. Therefore, correlations can be detected between the typological variability of the vessels and their manufacture. It is possible to determine, for instance, whether certain fabrics are related only to certain pottery types and to evaluate the correspondence between marginal types and fabrics that could indicate different technological choices.[8]

Neutron Activation Analysis

Neutron activation analysis of ceramics at MURR is a unique procedure that consists of two irradiations and three gamma counts to determine the concentrations of 33 elements.[9] This is more comprehensive than most other NAA labs, which use a single irradiation and gamma count. The first irradiation is short (5 seconds) and produces gamma spectra for nine short-lived elements: aluminum (Al), barium (Ba), calcium (Ca), dysprosium (Dy), potassium (K), manganese (Mn), sodium (Na), titanium (Ti), and vanadium (V). The second irradiation is long (24 hours) and produces gamma spectra for seven medium half-life elements: arsenic (As), lanthanum (La), lutetium (Lu), neodymium (Nd), samarium (Sm), uranium (U), and ytterbium (Yb). After a decay period, samples are counted twice: once for 1,800 seconds to measure seven medium half-life elements, and again for 8,500 seconds to measure 17 long half-life elements: cerium (Ce), cobalt (Co), chromium (Cr), cesium (Cs), europium (Eu), iron (Fe), hafnium (Hf), nickel (Ni), rubidium (Rb), antimony (Sb), scandium (Sc), strontium (Sr), tantalum (Ta), terbium (Tb), thorium (Th), zinc (Zn), and zirconium (Zr). Statistical analysis performed on the elemental concentration data identifies distinct groups of samples.[10] The goal is to identify groups with similar elemental compositions, which can indicate shared origins or manufacturing processes. In addition to PCA and Mahalanobis Distance and Group Membership Probabilities, I also applied other pattern-recognition techniques to the multivariate chemical data to interpret the compositional data. This procedure allowed me to identify more subtle patterns and relationships in the data.

Results

A principal component analysis (PCA) was undertaken in order to understand the elements most significantly driving variation within the sample (Fig. 1; Tab. 1). Elements such as arsenic, caesium, and thorium were found to be significant drivers of variation within the dataset.

7 In addition, there are also 6 samples in Opaque Red Ware, an Early Republican pottery, included for a side project with the results published in D'Acri (2023).

8 Santacreu (2014) 184.

9 Glascock (1992). This phase of the work was funded by the Archaeometry Lab at the University of Missouri Research Reactor (MURR) through NSF Grant BCS-2208558.

10 The software GAUSS Run-time Module was used for data processing.

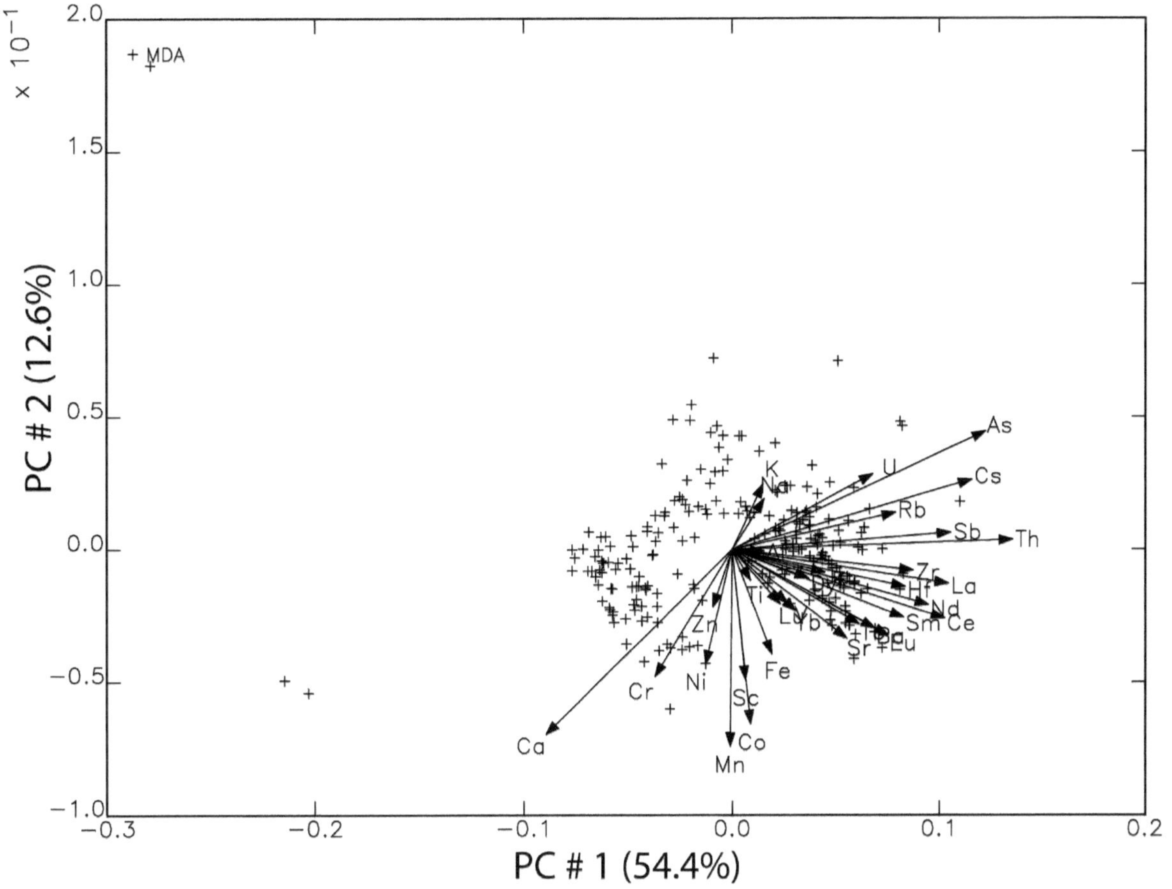

Fig. 1: Biplot of the first two principal components calculated for the current dataset with labeled vectors demonstrating the loading of individual elements. © Mattia D'Acri.

Overall, most samples clustered together, but some examples were different. Therefore, considering the PCA, the archaeological information, and the Mahalanobis distance, I assessed and interpreted the possible internal subgroups and outliers (Fig. 2). The results show an interesting pattern among the groups, with some unexpected evidence and a few outliers (Tabs. 2 and 3). In particular, Group 2 has a distinctive chemical composition related to only a specific ware (Impasto Chiaro-Sabbioso) attested both in Rome and in Gabii. The only shape documented is the basin, as is typical of Impasto Chiaro-Sabbioso production. Group 3 is particularly interesting, because it encompasses all the Fineware and the Opaque Red Ware samples, regardless of the shape, site, and date (Fig. 3a–b). The chronology ranges from the 7th century B.C. for some types of Fineware to the 5th–4th centuries B.C. with the Opaque Red Ware. In addition, this group includes two Impasto Chiaro-Sabbioso basins smaller in size than previous ones.

Group 4 is similar, because most of the Bucchero samples from Latium are in it. They are mostly open shapes, such as bowls. Oddly enough, one Impasto Chiaro-Sabbioso sample from San Giovenale and one Impasto Rosso-Bruno bowl from Caesar's Forum belong to the group as well. Group 5 gathers features the largest number of Impasto-production samples in Rome and Gabii, together with some examples from San Giovenale. In particular, it is worth noticing that 12 out of 19 Impasto Rosso samples belong to the group, including all the typical ollae with decoration in the internal part of the lip. Group 6 is of interest, with an unexpected relationship between Gabii and San Giovenale material, leaving Rome attested with only three samples out of twenty-five. Also, in this case, Impasto predominates. However, it is significant that there are also 4 out of 29 Bucchero samples. Group 7 may be called the San Giovenale Group, because regardless of the pottery class, the samples come from that Etruscan site. The chemical imprint is strong, perhaps suggesting that all the pots were made with local clay. Group 8 includes only two Impasto samples from all the Latin sites; it has two peculiarities. The first is the higher number of Impasto Bruno samples, and the second is the Impasto Rosso forms attested, including large basins. Finally, the outliers are, unsurprisingly, mainly Finewares rather than Impasti, which may suggest importations from other regions or geological areas.

Tab. 1: Elemental loadings for principal component axes 1-6 based on the variance-covariance matrix. Values in bold explain the greatest amount of variation within each component. Over 88% of the cumulative variance in the dataset is explained by the first 6 principal components. © Mattia D'Acri.

Variable	*Average*	*Magnitude*	*PC1*	*PC2*	*PC3*	*PC4*	*PC5*	*PC6*
Na	6607.752	1	0.039093	0.100693	−0.0809	0.01278	−0.12707	**0.349796**
Al	89368.04	1	0.038964	−0.00269	−0.07508	0.018647	−0.00988	−0.09383
K	24525.86	1	0.037693	0.130312	−0.02884	0.208041	0.108926	−0.12366
Ca	29338.49	1	−0.22776	−0.36901	**0.45603**	**0.540431**	0.133491	0.185003
Sc	15.63969	1	0.015935	−0.25921	−0.03619	−0.04711	0.143918	0.015918
Ti	4558.804	1	0.022282	−0.06089	−0.07023	0.042525	−0.01851	−0.04191
V	152.0247	1	0.06766	−0.10689	0.10071	−0.00581	−0.09665	−0.13489
Cr	118.5529	1	−0.09521	−0.25336	0.031834	−0.2891	0.224307	0.027605
Mn	950.0633	1	−0.00249	−0.39218	0.067125	0.054975	−0.40403	−0.50808
Fe	48027.16	1	0.048762	−0.2086	−0.02785	−0.10218	0.044602	0.001402
Co	20.89424	1	0.022308	−0.34786	0.10051	−0.23436	−0.08563	−0.23799
Ni	63.60513	1	−0.03263	−0.22915	0.094904	−0.50066	**0.405024**	0.131815
Zn	100.255	1	−0.02341	−0.11475	−0.04188	−0.0253	**0.25457**	−0.16296
As	19.60118	1	**0.311667**	**0.236529**	**0.749304**	−0.31104	−0.17504	0.037645
Rb	193.8035	1	0.200527	0.07372	0.026015	0.151651	0.180043	−0.19971
Sr	364.7463	1	0.140652	−0.17727	**0.254638**	**0.257142**	0.071012	**0.322743**
Zr	215.6708	1	**0.22233**	−0.04134	−0.09227	−0.05949	0.002894	0.180073
Sb	1.180184	1	**0.268812**	0.033593	−0.05713	−0.04276	−0.0023	−0.02484
Cs	16.89592	1	**0.294998**	0.139717	0.074464	0.065473	0.218702	−0.29404
Ba	669.9755	1	0.175175	−0.15761	−0.11058	−0.05459	−0.02812	**0.228875**
La	63.30737	1	**0.26602**	−0.06773	−0.09748	0.08183	−0.05369	0.044578
Ce	129.2561	1	**0.261565**	−0.13763	−0.05538	0.03619	−0.14954	−0.04491
Nd	52.94327	1	**0.241536**	−0.1099	−0.05326	0.079874	−0.01785	0.04866
Sm	10.24087	1	0.210996	−0.13585	−0.05299	0.057997	−0.00857	0.041628
Eu	1.951764	1	0.191564	−0.17142	−0.04462	0.023356	−0.03536	0.077782
Tb	1.095846	1	0.156013	−0.14894	−0.04922	0.0318	0.054278	0.048983
Dy	6.053216	1	0.093216	−0.05702	−0.03076	0.070395	−0.09309	−0.03361
Yb	2.897182	1	0.077487	−0.12251	−0.09223	0.017093	0.102765	0.013387
Lu	0.401633	1	0.058908	−0.10819	−0.10542	0.012912	0.117848	−0.01011
Hf	6.812423	1	0.212463	−0.07413	−0.13077	−0.00853	−0.04299	0.194159
Ta	1.343852	1	0.11356	−0.0451	−0.09799	0.085865	0.132412	−0.00875
Th	27.14647	1	**0.345119**	0.020099	−0.0298	0.026354	−0.05684	0.038274
U	4.416906	1	0.172776	0.150515	0.090397	0.159552	**0.519287**	−0.25511
Eigenvalues:			0.615384	0.142207	0.094837	0.067377	0.053772	0.026488
% variation:			54.38631	12.56797	8.381454	5.954614	4.752276	2.340921

Discussion

Archaeologically the results can be read in two ways: looking at the specific pottery classes or the contexts. The evidence that specific groups (especially Groups 2–4) are associated with particular pottery classes regardless of their provenance may have several explanations. One may speculate that potters and workshops shared the same recipe for certain types of pottery, handed down generation after generation, or that the vessels were traded from a single production center. That justifies how, regarding Group 3 (Fig. 3a-b), the 7th-century B.C. Fineware sherds have the same chemical composition as the 5th-century B.C. Opaque Red Ware. As far as the various Impasto pottery productions are concerned, however, Group 2 shows Impasto Chiaro-Sabbioso with a completely different pattern, and the considerations suggested for the previous group may be adopted. Instead, Impasto Bruno, Impasto Rosso, and Rosso-Bruno seem to share the same chemical composition, and any difference should be found in surface treatment and technology adopted by the potters, such as the use of the wheel, the polishing of the surfaces, or the different firing processes. Otherwise, from the 8th to the 6th

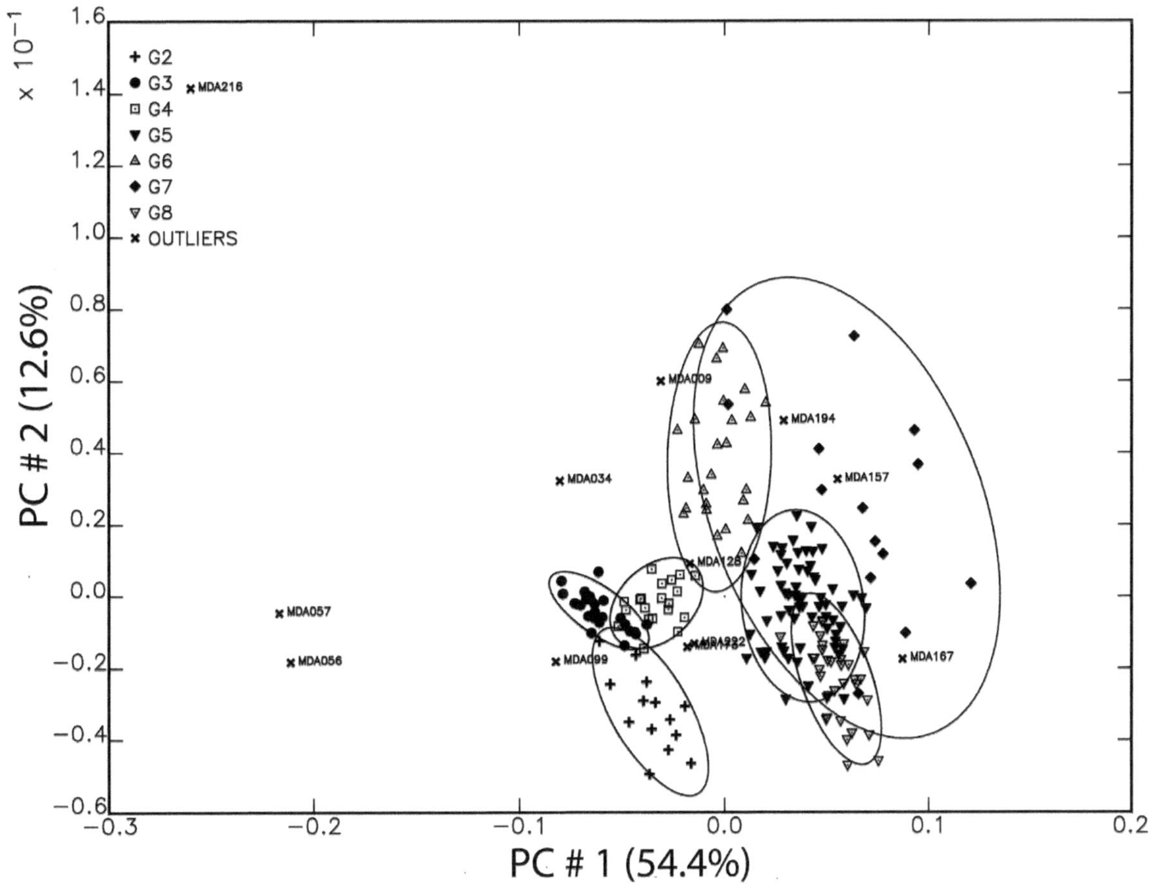

Fig. 2: Biplot of the first two principal components calculated for the current dataset with internal subgroups and samples labeled for the outliers. Ellipses represent 90% confidence interval of group or subgroup membership. This plot describes 67% of the cumulative variation in the dataset © Mattia D'Acri.

Tab. 2: Sample distribution by pottery classes. © Mattia D'Acri.

	Fineware	Bucchero	Opaque Red Ware	Impasto Chiaro-Sabbioso	Impasto Rosso	Impasto Rosso-Bruno	Impasto Bruno	Total
Group 2	–	–	–	14	–	–	–	14 (6.69%)
Group 3	18	–	6	2	–	–	–	26 (12.44%)
Group 4	–	18	–	1	–	1	–	20 (9.57%)
Group 5	–	1	–	–	12	34	18	65 (31.10%)
Group 6	–	4	–	–	–	16	5	25 (11.96%)
Group 7	–	2	–	1	1	8	4	16 (7.66%)
Group 8	–	–	–	–	6	12	15	33 (15.79%)
Outliers	3	4	–	1	–	1	1	10 (4.79%)
	21	29	6	19	19	72	43	209

century B.C. the vessels share the same clay types of temper. A final note on Group 8 is that it may represent the samples with a coarser clay, with more added temper, because of the presence of vessels with bigger dimensions, and therefore a less depurated clay.

Moving on to the sites, the plot in Figure 2 shows that the ellipse for Group 7 is much wider compared with the others, implying that San Giovenale pottery had a more heterogeneous chemical composition, perhaps resulting from a different clay supply or by its treatment. On the other hand, Groups 2–6 and 8 are narrow and chemically strong, suggesting diverse

Tab. 3: Sample distribution by sites. © Mattia D'Acri.

	Sant'Omobono	Regia	Caesar's Forum	Gabii	San Giovenale	Total
Group 2	5	3	–	6	–	14 (6.69%)
Group 3	11	2	1	9	3	26 (12.44%)
Group 4	7	7	4	1	1	20 (9.57%)
Group 5	22	13	17	11	2	65 (31.10%)
Group 6	1	–	2	8	14	25 (11.96%)
Group 7	–	–	–	–	16	16 (7.66%)
Group 8	10	7	4	12	–	33 (15.79%)
Outliers	1	1	2	4	2	10 (4.79%)
	57	33	30	51	38	209

chaînes opératoires. In addition, Rome and Gabii seem to be unsurprisingly related. The relationship between Gabii and San Giovenale that emerges in Group 6 is unexpected, deserving further study.

Finally, on a broader level, there are clear connections within and among the sites investigated, suggested already by the traditional approach and confirmed by the chemical analysis. Generally, the Impasto wares were locally produced (with a few exceptions), and both Finewares and Bucchero seem to share a specific workshop spread in the contexts investigated.

Conclusions and Future Perspectives

With this chapter, I presented the preliminary results of the ongoing study, with a few general observations and suggestions that may be further clarified in a future study. The dataset will be expanded soon with seven additional raw clay samples from Rome and thirty-four pottery samples coming from the Archaic building on the Quirinal, located at the Ex-Istituto Geologico.[11] Integration with XRD and thin sections will be applied to at least one pottery class, specifically Impasto Chiaro-Sabbioso, to better understand and interpret the NAA results. Nevertheless, this chapter highlights the potential of the study, sharing a first glimpse of the results that may unpack pottery production in Archaic Rome and Latium.

11 I had the opportunity to include the new site thanks to a new research project led by Nicola Terrenato (University of Michigan), John Hopkins (New York University), and Mirella Serlorenzi (Soprintendenza Speciale Archeologia, Belle Arti e Paesaggio di Roma).

Fig. 3a: Group 3: 18/21 Fineware samples (remaining 3/21 are outliers) © Mattia D'Acri.

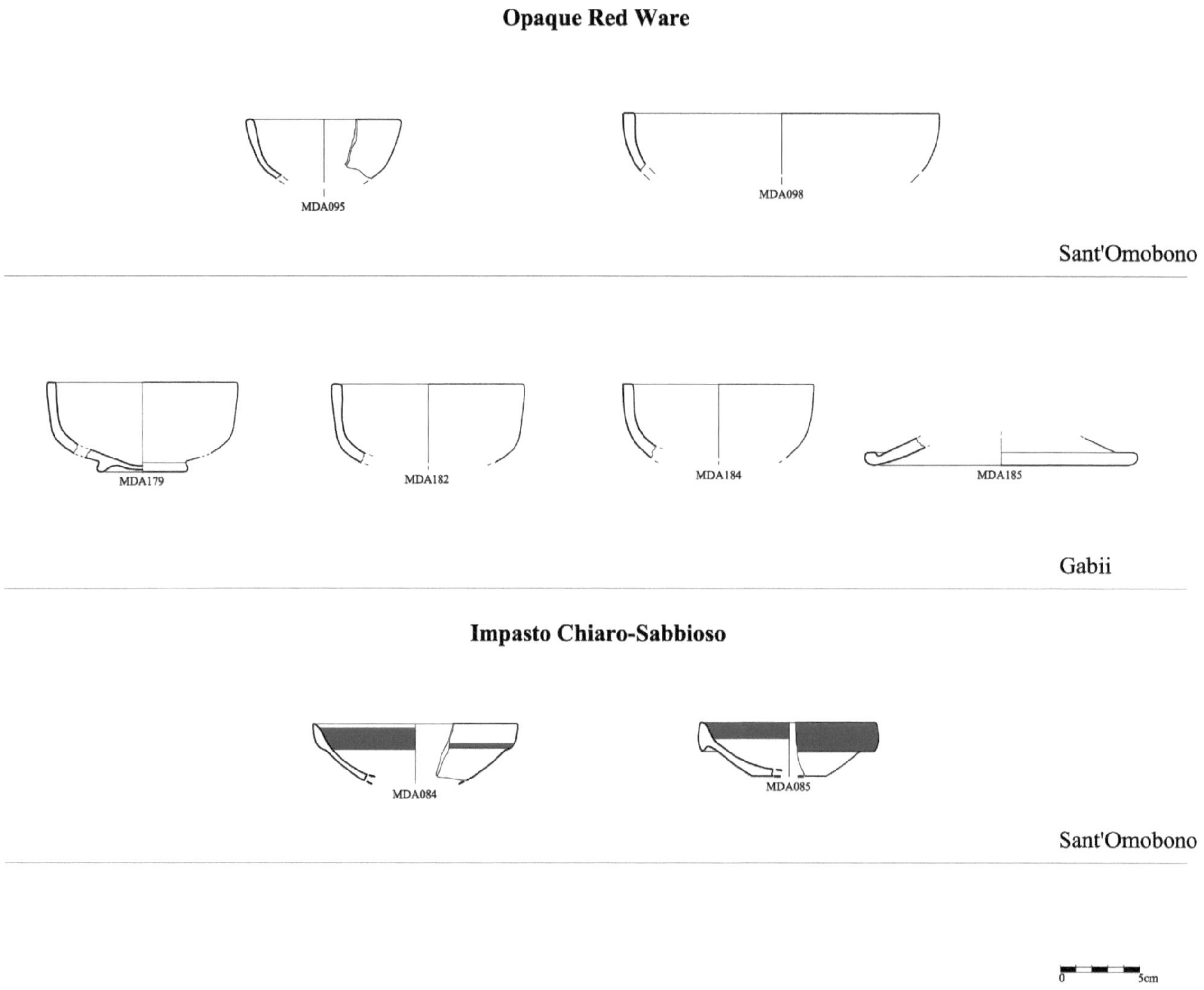

Fig. 3b: Group 3: 6/6 Opaque Red Ware samples and 2/19 Impasto Chiaro-Sabbioso samples © Mattia D'Acri.

Bibliography

Ammerman *et al.* (2008): Albert J. Ammermann, Ioannis Iliopoulos, Federica Bondioli, Dunia Filippi, Jill Hilditct, Alessandra Manfredini, Licio Pennisi and Nancy A. Winter, "The Clay Beds in the Velabrum and the Earliest Tiles in Rome", in: *Journal of Roman Archaeology* 21, 7–30.

Carafa (1995): Paolo Carafa, *Officine ceramiche di età regia*, Rome.

Carafa (2017): Paolo Carafa, "Artigiani e officine di età regia a Roma", in: *Scienze dell'Antichità* 23.2, 5–19.

D'Acri (2016): Mattia D'Acri, *Gli scavi dell'abitato di Gabii: l'Area D*, Tesi di Specializzazione, Università degli Studi della Basilicata.

D'Acri (2023): Mattia D'Acri, "Notes on the Opaque Red Ware at Gabii", in: Laura Banducci and Mattia D'Acri (eds.), *Gabii Through Its Artifacts*, Oxford, 10–17.

D'Acri *et al.* (forthcoming): Mattia D'Acri, Carmelo Colelli and Jan Kindberg Jacobsen, "Le ceramiche d'impasto", in: Rubina Raja, Jan Kindberg Jacobsen, Sine Grove Saxkjær and Claudio Parisi Presicce (eds.), *Foro di Cesare. I materiali ceramici dallo scavo del 1998–2000*, forthcoming.

D'Acri/De Luca (2016): Mattia D'Acri and Luca De Luca, "Il settore 4 della Regia: i reperti ceramici", in: Paolo Brocato and Nicola Terrenato (eds.), *Nuovi studi sulla Regia di Roma*, Cosenza, 71–93.

D'Acri/De Luca (forthcoming): Mattia D'Acri and Luca De Luca, "Le ceramiche arcaiche dai recenti scavi nell'area archeologica di S. Omobono (Settore D-10)", in: Paolo Brocato, Nicola Terrenato and Monica Ceci (eds.), *Ricerche nell'area dei templi di Fortuna e Mater Matuta III*, forthcoming.

D'Acri *et al.* (2018): Mattia D'Acri, Luca De Luca, Margherita Perri, Geraldine Pizzitutti and Carlo Regoli, "Appendice 1. I reperti ceramici", in: Paolo Brocato, Nicola Terrenato and Monica Ceci (eds.), *Ricerche nell'area dei templi di Fortuna e Mater Matuta II*, Arcavacata di Rende, 37–117.

Glascock (1992): Michael D. Glascock, "Characterization of Archaeological Ceramics at MURR by Neutron Activation Analysis and Multivariate Statistics", in: Hector Neff (ed.) *Chemical Characterization of Ceramic Pastes in Archaeology*, Madison, 11–26.

Revello Lami (2017): Martina Revello Lami, “Evidenze dirette, indirette, o circostanziali? Topografia e archeometria della produzione ceramica a Satricum durante il periodo arcaico”, in: *Scienze dell'Antichità* 23.2, 389–412.
Riehle *et al.* (2023): Kai Riehle, Erich Kistler, Birgit Öhlinger, Christian Heitz, David Ben-Shlomo, Reinhard Jung, Hans Mommsen, Johannes H. Sterba, Stefanos Gimatzidis, Alexander Fantalkin, Susanne Prillwitz, Anno Hein, Leonhard Geissler, Gunnar Lehmann, Jan Kindberg Jacobsen, Richard Posamentir and Udo Schlotzhauer, “Neutron activation analysis in Mediterranean Archaeology: current applications and future perspectives”, in: *Archaeological and Anthropological Sciences* 15, 25.
Santacreu (2014): Daniel Albero Santacreu, *Materiality, Techniques and Society in Pottery Production. The Technological Study of Archaeological Ceramics through Paste Analysis*, Warsaw/Berlin.

Elisa Ercolin, Diego Elia, Patrizia Davit and Monica Gulmini

16 Black-glazed Pottery from Locri Epizephyrii: New Evidence from the Technological Study of the Local Hellenistic Production

Abstract: For the study of black-glazed pottery from the recent excavations of the University of Turin in Locri Epizephyrii (Casino Macrì area) we developed an interdisciplinary approach to present a diachronic reconstruction of the local production technology. Our goal was to follow the manufacturing process, from the "visible" aspects to the "invisible" ones. The study was based on the typological classification of more than 3000 sherds dated between the Late Archaic and the Hellenistic periods. The macroscopic study was supported by optical microscopy (OM) and Scanning Electron Microscopy (SEM) analysis in order to establish a preliminary classification of clay bodies and define production sequences and artisanal gestures through the detection of surface processing traces. Thanks to archaeometric investigations – both chemical (SEM-EDX) and mineralogical (XRD and micro-XRD) – we were able to investigate compositional and micro-morphological aspects of clay bodies and coatings, as well as the general conditions of temperature and atmosphere during the firing process. In particular, a systematic study of the thermal behavior of the iron oxides was used as a marker to determine the success of the last steps of firing, and to reconstruct the positioning of vessels with not-reduced/re-oxidized coatings inside the kiln.

Introduction

The black-glazed pottery at Locri Epizephyrii has been studied for at least fifty years within the long-term fieldwork conducted by the University of Turin at the site. After the first analysis conducted in 1977,[1] the local production was fully captured through the typological classification published in 1989 and 1992, based on the full exploration of the artisanal quarter of Centocamere[2] and the so-called "Adonion-Casa dei Leoni" complex located in Marasà Sud.[3] Additional data on the local ceramic repertoire came from the studies of the Lucifero-Parapezza necropolis published by D. Elia.[4] Besides the morphological analyses, provenance and archaeometric investigations focusing on the provenance and technology for both black-glazed and red-figured fine wares confirmed the presence of local workshops from the Late Archaic to the Hellenistic periods.[5]

The recent excavations in the Casino Macrì area have improved our knowledge of this ceramic production and allowed us to develop an interdisciplinary approach for the diachronic reconstruction of the manufacturing processes and craft organization.[6] This research was carried out at the University of Turin within the framework of a Ph.D. project focusing on black-glazed pottery from the end of the 6th to the 2nd centuries B.C.[7] The typological classification of more than 3000 sherds was coupled with extensive archaeometric analyses of their technological processes. The archaeometric study took place at the University of Turin (Chemistry and Earth Sciences departments) and at the scientific laboratories of the "Centro di Conservazione e Restauro La Venaria Reale". Macroscopic analysis and optical microscopy allowed a preliminary classification of clay bodies and highlighted production sequences and gestures through the detection of pro-

1 Bacci (1977); Sabbione (1977).
2 Bitti (1989); Buzzi Auxilia (1989); Olivero Ferrero (1989); Preacco (1989).
3 Preacco Ancona (1992); Rubinich (1992).
4 Elia (2010) 198–227, (2019b).
5 Mirti *et al.* (1995), (1996), (2004a), (2004b); see also Elia *et al.* in this volume.
6 The excavation in these archaeological areas, located in the central part of the ancient polis, were started in 2002 by M. Barra Bagnasco and continued between 2010 and 2019 under the direction of D. Elia and V. Meirano: Elia (2019a); Elia/Meirano (2016), (2017), (2020); Sudano (2021).
7 The project is entitled *Tecnologia di produzione della ceramica a vernice nera in Calabria meridionale (fine VI-inizi II secolo a.C.). Studio dei contesti artigianali e analisi archeometriche*, under the supervision of D. Elia (Department of Historical Studies) and with the collaboration of M. Gulmini and P. Davit (Department of Chemistry). Preliminary results of the study are in Ercolin (forthcoming) and in Ercolin *et al.* (2023).

Note: Authorship credits: Introduction and conclusion by Elisa Ercolin. Technological Characterization of the Local Hellenistic Pottery by all authors.

https://doi.org/10.1515/9783111189635-016

cessing traces on the surfaces of the pots.[8] Through micro-morphological and compositional investigations we were able to develop a comprehensive chemical and mineralogical characterization of the local pottery, and to investigate whether – and to which extent – the organization of the ceramic production underwent any systematic changes between the late 6th and the early 2nd century B.C.

Technological Characterization of the Local Hellenistic Pottery

The diachronic analysis of the shapes and their technological features revealed noticeable transformations in the local ceramic production during the Hellenistic period. From the last decades of the 4th century B.C., Locrian workshops, in addition to the usual imitations of Attic prototypes (such as skyphoi, bolsal "light wall" cup-skyphoi, small bowls with echinus wall or with convex-concave profile) started developing an articulated repertoire of cup-kantharoi, plates with outturned rims, overpainted mastoi and basins, which reproduced the morphological features and decorative patterns of Apulian, Lucanian, and Sicilian ceramics.

In this complex heterogeneity, the 3rd century B.C. progressively marked a new technological horizon characterized by a generalized morphological, dimensional and proportional standardization. Elements of continuity are only represented in this phase by the most recent variants of small bowls with echinus wall, cup-kantharoi and plates, whose production seems to have continued in Locri until the first decades of the 2nd century B.C., similarly to other sites in southern Calabria and eastern Sicily (Fig. 1).

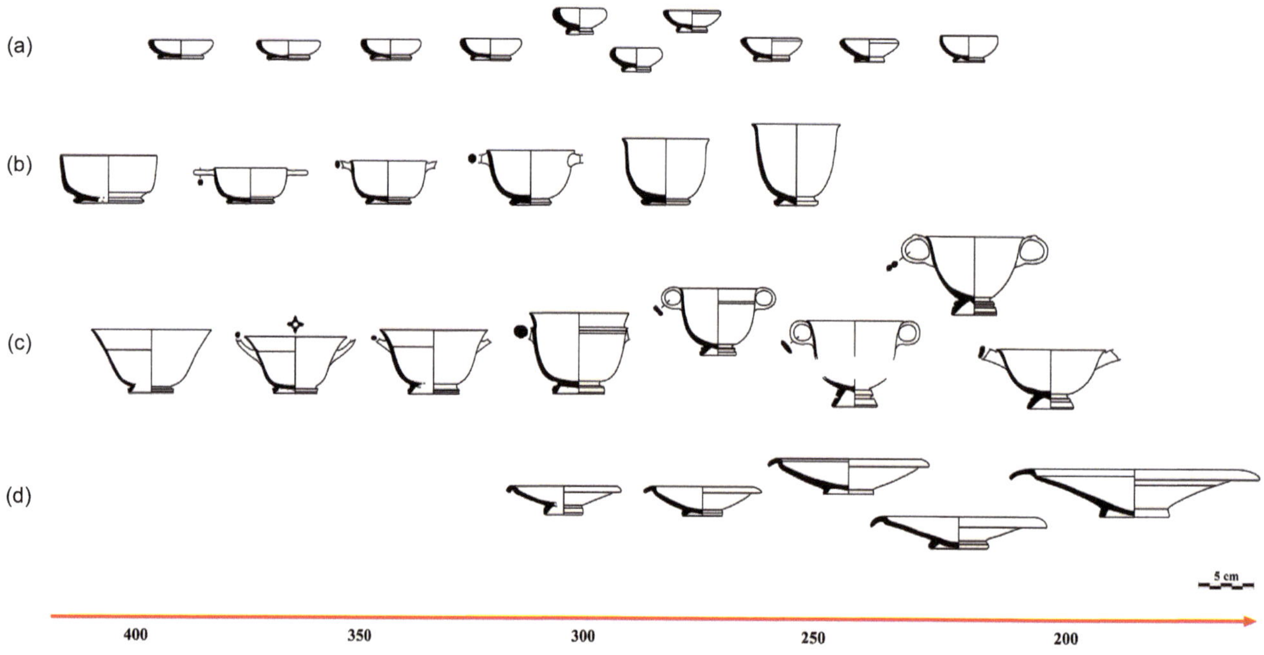

Fig. 1: Locri Epizephyrii, black-glazed pottery. Morphological development of select shapes between the Classical and the Hellenistic periods: **(a)** Small bowls with echinus wall; **(b)** Bolsal; **(c)** Cup-skyphoi; **(d)** Cup-kantharoi, and plates with outturned rims. Drawings: Elisa Ercolin.

During the Early Hellenistic period, potters focused more on the detailed finishing of the pots. They accomplished this by adding stamped decorations and minute morphological details, such as strips, grooves, pods and molded feet (Fig. 2a–b), and by making the surface remarkably compact, regular and smooth, as is clearly documented in the imaging from Scan-

8 Optical microscopy analysis was carried out with a DinoLite portable microscope (AM 311 3T, equipped with 1.3 megapixel sensor and integrated LED lighting system, images acquired with DinoCapture 2.0 software) and a Leica MZ95 stereomicroscope, equipped with optical fibres and Leica MC190 HD camera (images acquired with LAS X software).

ning Electron Microscopy backscattered electrons (Fig. 2c).[9] On the other hand, shapes produced between the first half of the 3rd and the beginning of the 2nd century B.C. lack partially or completely final surface treatments, which are mostly limited to the molding and application of rims, handles, or feet. In particular, both Optical and Scanning Electron Microscopy analysis revealed clay residues, porosities and stripes on both coated and reserved areas: in some cases, the shape and orientation of these traces suggest the use of damp sponges (Fig. 2d) or "scraper" tools (Fig. 2e) during the wheel forming of the vessel, which potters may have used either to make the walls uniform in thickness or to define the foot by cutting away excess clay along the profile of the vessel.[10]

Specific preferences in the manufacturing sequences were also recognized for the application of the gloss coating. Since the final decades of the 4th century B.C., the technique of immersion is clearly documented by the presence of fingerprints, drippings and splashes on the external surfaces, suggesting decreased attention to final cleaning treatments (Fig. 2f). Furthermore, the use of brushes is gradually reduced during the 3rd century B.C. and limited to the creation of simple decorative bands on the foot of bolsal and cup-kantharoi. In several pots, one can see a differentiation of reduction between inner and outer surfaces, or surfaces with iridescent, partially reduced and fully not-reduced/re-oxidized coatings, probably caused by the ineffective use of staking rings or by the direct superposition of the vases inside the kiln.

This complex survey of technological procedures was further augmented by archaeometric investigations. While the compositional analyses of the ceramic bodies and the coatings did not highlight any significant variations in clay working processes compared to those of the Classical period, it seems that substantial changes had affected the management of kiln firings during the 3rd century B.C. A systematic study of the thermal behavior of the iron oxides through X-Ray Powder Diffraction (XRPD) on the ceramic bodies and micro-X-Ray Diffraction (micro-XRD) on the slips showed a diminished attention towards the regulation of the firing process itself, especially in terms of monitoring the duration and general conditions of temperature and atmosphere during the last step of firing and the cooling phase.[11] In particular, we studied the behavior of select minerals to determine the sequences of the firing process: hematite, which develops in ceramics fired above 900°C under oxidizing conditions; magnetite and hercynite, the main iron compounds when the ceramics go through reduction conditions between 950 and 1050°C; and maghemite, which forms when reduced iron compounds are re-oxidized at temperatures below 800°C.[12]

Interesting data emerged from the mineralogical analysis of the coatings with different degrees of reduction between inner and outer surfaces, or with completely not-reduced/re-oxidized stages. The presence, absence or low concentration of iron oxides provided new data about the effectiveness and success of the reduction phase in the firing, as well as the positioning of the vessels inside the kiln. For example, in the case of one patera, maghemite was not detected in the inner red slip, suggesting that the hot gasses did not evenly cover the entire surface of the vase during the reduction phase, probably due to ineffective stacking (Fig. 3a). On the other hand, the presence of maghemite in the orange slips of a cup-kantharos may indicate that the reduction step was too short or insufficiently effective, so that a widespread re-oxidation occurred during the cooling phase (Fig. 3b).

Equally significant for the morphological study of gloss coatings is the SEM imaging. Although the firing reached maximum temperatures constantly ranging between 900 and 1000°C, only limited areas of the gloss coatings are vitrified (Fig. 3, in yellow), and they often show micro-cracks (Fig. 3, in red), possibly resulting from inadequate conditions during the reduction and cooling phases, which amplified the differences in thermal expansion between bodies and slips.

9 The images were acquired through a scanning electron microscope equipped with both secondary and backscattered electrons detector and coupled to an Energy-Dispersive X-ray (EDX) microprobe. Images were acquired in special acquisition mode which combines the high spatial resolution of secondary electrons to the – qualitative – compositional information of the backscattered ones. EDX chemical composition was obtained for both bodies and coatings.

10 Morphological comparisons for these traces were found in the experimental tests by T. Schreiber and in the micro-morphological analysis on prehistoric productions by C. Ionescu and V. Hoeck: Schreiber (1999) 39–41, figs. 4.1, 4.2; Ionescu/Hoeck (2020) 5–9, figs. 4, 6.

11 Our warm thanks to Roberto Giustetto and Giulia Berruto (University of Turin, Department of Heart Sciences), who performed the diffractograms for the XRPD and micro-XRD analysis.

12 Gliozzo (2020) 14–15.

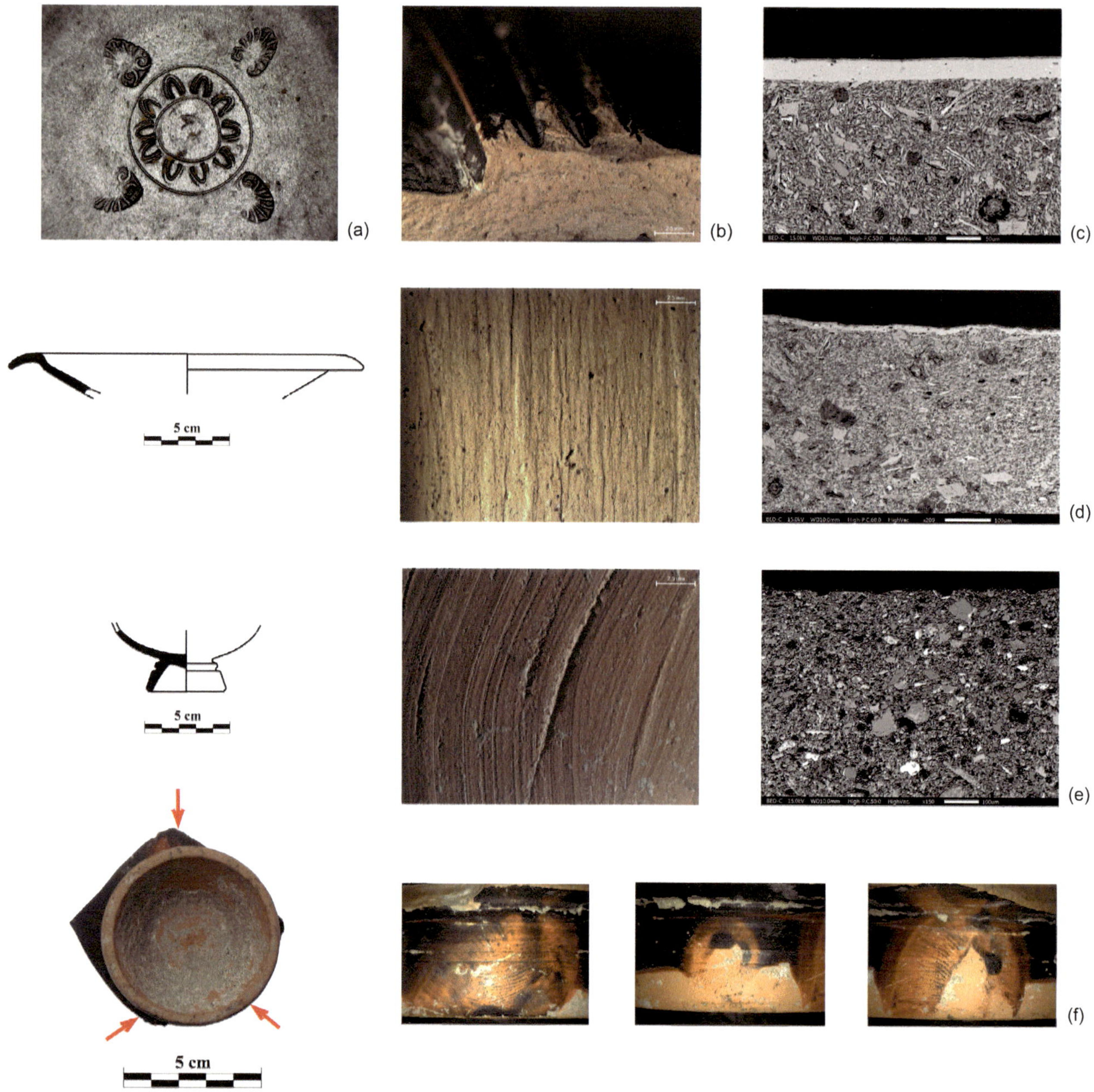

Fig. 2: Locri Epizephyrii, black-glazed pottery: **(a)** Cup-skyphos, 4th century B.C., OM image (6.3X); **(b)** Cup-skyphos, late 4th century B.C., OM image (6.3X); **(c)** Bolsal, beginning of the 4th century B.C., SEM backscattered electrons image (300X); **(d)** Plate with outturned rim, mid-3rd/beginning of the 2nd century B.C., OM image (6.3X, left) and SEM backscattered electrons image (200X, right); **(e)** Cup-kantharos, 3rd century B.C., OM images (6.3X, left) and SEM backscattered electrons image (150X, right); **(f)** Plate with outturned rim, second half of the 3rd century B.C., OM images (6.3X). Drawings: Elisa Ercolin. Photos: Elisa Ercolin, Patrizia Davit and Monica Gulmini.

Conclusions

Studying morphological repertoires and craft procedures together provides a reliable framework for understanding the cultural, social and economic elements of ceramic production. Far from representing a qualitative "deterioration" of the *savoir-faire* reached between the 5th and 4th centuries B.C., the technological elements highlighted in the local Hellenistic pottery of Locri Epizephyrii suggest a new artisanal framework as well as a widespread economic discontinuity, which probably would have forced local workshops to reorganize their *chaîne opératoire* and develop different production protocols.

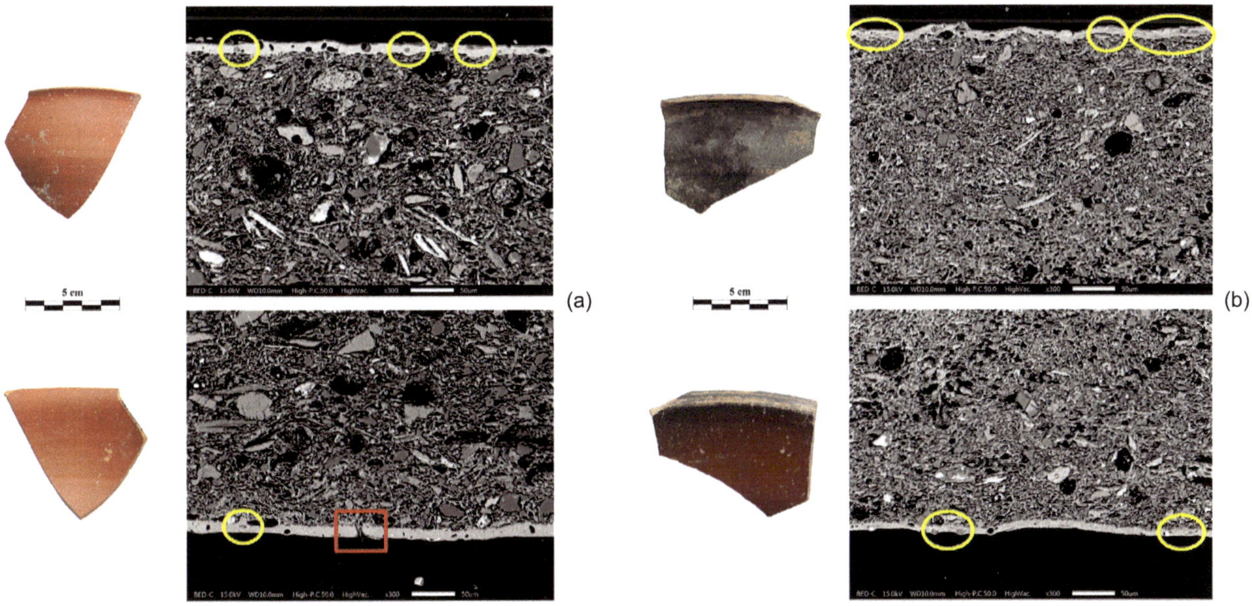

Fig. 3: SEM backscattered electrons images 300X: **(a)** Plate with rilled rim, 3rd century B.C.; **(b)** Cup-kantharos, 3rd century B.C. Photos: Elisa Ercolin, Patrizia Davit and Monica Gulmini.

In these new circumstances, procedures connected with the shaping, surface finishing, and distribution of the coatings reveal more rapid and approximate gestures, suggesting that potters no longer prioritized the accurate finishing of their vases. Instead, a diachronic study of the pottery technology reveals that significant changes took place during the 3rd century B.C. and potters redirected their attention to a better management of the firing process in order to simplify kiln loading, with standardization of shapes and sizes in the pottery repertoire facilitating stacking. Moreover, the potters had successfully regulated the general conditions during the reducing and cooling phases in the previous decades and were able to guarantee successful chemical-mineralogical transformations into highly reduced and sintered gloss coatings.

Abbreviation

Locri Epizefiri II — Marcella Barra Bagnasco (ed.), *Locri Epizefiri, II. Gli isolati I2 e I3 dell'area di Centocamere*, Florence.

Bibliography

Bacci (1977): Giovanna Bacci, "Ceramica a vernice nera del VI e della prima metà del V secolo", in: *Locri Epizefiri, I. Ricerche nella zona di Centocamere*, Florence, 89–92.

Bitti (1989): Maria Cristina Bitti, "Vasellame di II strato", in: *Locri Epizefiri II*, 141–193.

Buzzi Auxilia (1989): Maria Consolata Buzzi Auxilia, "Vasellame di II strato", in: *Locri Epizefiri II*, 107–141.

Elia (2010): Diego Elia, *Locri Epizefiri VI.* Nelle case di Ade. *La necropoli in contrada Lucifero. Nuovi documenti*, Alessandria.

Elia (2019a): Diego Elia, "La gestione delle acque a Locri Epizefiri. Criticità e soluzioni, dalle origini alla conquista romana", in: Sophie Bouffier, Oscar Belvedere and Stefano Vassallo (eds.), *Gérer l'eau en Méditerranée au premier millénaire avant J.-C.*, Aix-en-Provence, 87–108.

Elia (2019b): Diego Elia, "Produzione, distribuzione e consumo da Locri a Caulonia nella prima metà del IV secolo a.C.", in: *Produzioni e committenze in* Magna Grecia (Atti del LV Convegno di Studi sulla Magna Grecia, Taranto, 24–27 settembre 2015), Taranto, 553–567.

Elia/Meirano (2016): Diego Elia and Valeria Meirano, "Il sacro e l'acqua a Locri Epizefiri: osservazioni alla luce delle recenti scoperte", in: Alfonsina Russo Tagliente and Francesca Guarneri (eds.), *Santuari mediterranei tra Oriente e Occidente. Interazioni e contatti culturali* (Atti del Convegno Internazionale, Civitavecchia-Roma, 18–22 giugno 2014), Rome, 419–434.

Elia/Meirano (2017): Diego Elia and Valeria Meirano, "Locri Epizefiri: al cuore della città antica. Vecchi problemi e nuove scoperte dalla fondazione all'età romana", in: Angela Pontrandolfo and Michele Scafuro (eds.), *Dialoghi sull'archeologia della Magna Grecia e del Mediterraneo* (Atti del I Convegno Internazionale di Studi, Paestum, 7–9 settembre 2016), Paestum, 265–274.

Elia/Meirano (2020): Diego Elia and Valeria Meirano, "Locri Epizefiri, in città e oltre. A cinquant'anni dall'avvio delle ricerche UNITO", in: Diego Elia (ed.), *Chiedi alla terra. Scavi e ricerche archeologiche del Dipartimento di Studi Storici dell'Università di Torino*, Turin, 144–159.

Ercolin (forthcoming): Elisa Ercolin, "Ceramica a vernice nera da Locri Epizefiri: nuove prospettive di ricerca dallo studio di reperti inediti dall'abitato", in: Monica Salvadori, Monica Baggio and Luca Zamparo (eds.), *Forma e Immagine. Vasi, cultura e società tra Grecia e Magna Grecia*, Padova, forthcoming.

Ercolin *et al.* (2023): Elisa Ercolin, Diego Elia, Patrizia Davit and Monica Gulmini, "Oltre il visibile. Un approccio interdisciplinare per la ricostruzione della tecnologia di produzione della ceramica a vernice nera da Locri Epizefiri", in: Chiara Andreatta, Marco Argentina, Greta Boldorini, Federica Bosio, Eliana Bridi, Chiara Casarin, Andrea Chiocca, Francesca Daniele, Mirko Fecchio, Ylenia Saretta, Federica Stella Mosimann, Gianantonio Urbani and Romina Zanon (eds.), *Svelare l'invisibile. Tracce nascoste di storie, opere e contesti*, Padova, 25–34.

Gliozzo (2020): Elisabetta Gliozzo, "Ceramic Technology. How to Reconstruct the Firing Process", in: *Archaeological and Anthropological Sciences* XII, 1–35.

Ionescu/Hoeck (2020): Corina Ionescu and Volker Hoeck, "Ceramic Technology. How to Investigate Surface Finishing", in: *Archaeological and Anthropological Sciences* XII, 1–16.

Mirti *et al.* (1995): Piero Mirti, Antonella Casoli, Marcella Barra Bagnasco and Maria Cristina Preacco Ancona, "Fine Ware from Locri Epizephirii: A Provenance Study by Inductively Coupled Plasma Emission Spectroscopy", in: *Archaeometry* 37, 41–51.

Mirti *et al.* (1996): Piero Mirti, Antonella Casoli and Lino Calzetti, "Technology of Production of Fine Pottery Excavated on a Western Greek Site Investigated by Scanning Electron Microscopy Coupled with Energy-Dispersive X-ray Detection", in: *X-Ray Spectrometry* 25, 103–109.

Mirti *et al.* (2004a): Piero Mirti, Monica Gulmini, Marco Pace and Diego Elia, "The Provenance of Red Figure Vases from Locri Epizephirii (Southern Italy): New Evidence by Chemical Analysis", in: *Archaeometry* 46, 183–200.

Mirti *et al.* (2004b): Piero Mirti, Monica Gulmini, Alessandra Perardi, Patrizia Davit and Diego Elia, "Technology of Production of Red Figure Pottery from Attic and Southern Italian Workshops", in: *Analytical and Bioanalytical Chemistry* 380, 712–718.

Olivero Ferrero (1989): Gabriella Olivero Ferrero, "Il vasellame di IV e III strato", in: *Locri Epizefiri II*, 71–106.

Preacco (1989): Maria Cristina Preacco, "Il vasellame del I strato", in: *Locri Epizefiri II*, 195–255.

Preacco Ancona (1992): Maria Cristina Preacco Ancona, "La ceramica fine dalla Casa dei Leoni", in: *Locri Epizefiri II*, 113–171.

Rubinich (1992): Marina Rubinich, "La ceramica fine della frequentazione arcaica e del sacello", in: *Locri Epizefiri II*, 85–111.

Sabbione (1977): Claudio Sabbione, "La ceramica a vernice nera dalla metà del V al III secolo", in: *Locri Epizefiri I. Ricerche nella zona di Centocamere*, Florence, 97–127.

Schreiber (1999): Toby Schreiber, *Athenian Vase Construction. A Potter's Analysis*, Malibu.

Sudano (2021): Fabrizio Sudano, "Attività della Soprintendenza Archeologia Belle Arti e Paesaggio per la città metropolitana di Reggio Calabria e la provincia di Vibo Valentia (2016–2017)", in: *Gli altri Achei. Kaulonia e Terina, contesti e nuovi apporti* (Atti del LVII Convegno di Studi sulla Magna Grecia, Taranto, 28–30 settembre 2017), Taranto, 1061–1095.

Dario Giuliano, Giuseppe Montana and Elisa Chiara Portale

17 Roof Tiles with the ΣΩΣΗΝΟΣ Stamp: Calabrian Production in Sicily

Abstract: Ever since the 19th century, roof tiles have been found in Sicily and southern Italy that were stamped with the name ΣΩΣΗΝΟΣ. The first discovery occurred at Solunto, and this was then followed by further examples identified at Solunto, as well as Lipari, Messina, Caulonia, and especially Reggio Calabria, where the greatest quantity has been found. Given their area of distribution, it has been claimed that the workshop producing these roof tiles was located either in Sicily or Reggio Calabria. In order to resolve this issue, a study of these artifacts was conducted, employing petrographic analysis. The results obtained have conclusively identified their place of production as the extreme tip of Calabria.

Introduction

This paper is part of the research activity that the Department of "Culture e Società" of the University of Palermo has been carrying out at the Solunto site. Since 2021, following an agreement with the Archaeological Park of Himera, Solunto and Monte Iato, archaeological exploration has taken place over two distinct areas.[1] The first is the area of the northern baths, located in the NE corner of the Agora; the second is the sacred area, situated NW of the theatre, in which several buildings are located. The sacred function of this area had already been hypothesised in the 19th century following the discovery of two valuable sculptures, one of Zeus the other Astarte enthroned.[2]

During the initial research phase, while examining the account recorded by the first excavator, Dotto de' Dauli, we found information regarding the presence of a "greca iscrizione, che si osserva in molte tegole di terra cotta e ben grandi, che forse coprivan la cella del tempio, dice (ΣοΣΗΝοΣ) Sosinos".[3] Although he hypothesized that these pan tiles belonged to a temple roof, we can plausibly assume that they actually belonged to the portico located behind the sacred building (the so-called "edificio sacro a due navate" that indeed would be uncovered over a century later), which de' Dauli had interpreted as the temple's postico. The interest in the roof tiles bearing this stamp lies in the fact that if they had been used in the portico, they would have provided us with a terminus ad quem for its construction. Furthermore, from viewing the fragments of roof tiles on display in the Solunto antiquarium, it was possible to ascertain that the "fabric" these pan tiles were produced was not typical of local production, but rather more foreign. Therefore, in order to resolve this issue, and thus better identify their area of production, a study was undertaken with the intention of re-examining the roof tiles with this stamp and selecting some for petrographic analysis.

The ΣΩΣΗΝΟΣ Stamp

A few years after de' Dauli's publication, Avolio reported the discovery of "una tegola [. . .] di recente cavata dall'antica Solanto col nome greco Sosinos",[4] which must plausibly have been one of those mentioned by de' Dauli.[5] Both scholars gave the name Σωσίνος instead of Σωσήνος, a misreading probably due to the letters *H* and *N* being written in ligature.

1 For a preliminary account of the research conducted at Solunto during the 2021 campaign, see Portale *et al.* (2021).

2 Recently Portale (2019) 245–249; and related bibliography.

3 Dotto de' Dauli (1825) 290. The author incorrectly reports the *omicron* between the first two *sigma* instead of the *omega*.

4 Avolio (1829) 31.

5 This is probably the pan tile of which Moretti says "traces have been lost for a century and a half": Moretti (1984) 322, note 4. The stamp is currently on display (Palermo, Antonino Salinas Archaeological Museum. inv. no. 65818).

Note: Authorship credits: Introduction by Dario Giuliano and Elisa Chiara Portale; The ΣΩΣΗΝΟΣ Stamp and Conclusions by Dario Giuliano; Petrographic Analyses by Giuseppe Montana. The authors wish to express their thanks to Dr. Caterina Greco, Director of the Antonio Salinas Archaeological Museum of Palermo, and to Dr. Costanza Polizzi of the same institution; to the Superintendent of Messina and architect Mirella Vinci and to Dr. Maria Ravesi of the same institution; to Stefano Zangara, architect and former director of the Archaeological Park of Himera, Solunto and Monte Iato; to the President of the Cultural Foundation Mandralisca Dr. Vincenzo Garbo, and to the Director of the Civic Museum 'B. Romano' of Termini Imerese Dr. Fabio lo Bono, for allowing this study.

https://doi.org/10.1515/9783111189635-017

In the two hundred years since the first report, pan tiles bearing this stamp have been found in other Sicilian sites (Tab. 1, Fig. 1). One, found in Lipari[6] in the 19th century during research carried out on the island by Baron Mandralisca, is currently on display at the Mandralisca Museum in Cefalù. In the same museum, this stamp is seen in two other fragments of pan tiles of uncertain provenance,[7] though probably unearthed as part of Baron Mandralisca's research activities.[8] From Solunto come two other stamps found during research carried out by C. Greco;[9] another, also of unknown provenance, is on display at the Museo Civico di Termini Imerese;[10] and, the last find comes from Messina, brought to light during the excavation of the rural settlement near Via Catania, where it was re-used in the construction of a wall.[11]

Tab. 1: Synoptic table of the roof tiles with ΣΩΣΗΝΟΣ stamp. © Dario Giuliano.

ΣΩΣΗΝΟΣ stamps found in Sicily							
Thin Section	**Current location**	**Inv. No.**	**Discovery**	**Context**	**Tile**	***IG/SEG***	**Bibliography**
	Palermo, Antonino Salinas Archaeological Museum	65818	Solunto	Sanctuary	Pan-tile, right cutaway	*CIG* 5577b = *IG* XIV 2395, 6b	Dotto dè Dauli (1825) 290; Avolio (1829) tav. I, no. 6
	Solunto, Antiquarium	M 1534	Solunto		Pan-tile, left cutaway		Garozzo (2003) 573, no. 12
	Solunto, Antiquarium	SA 152	Solunto		Pan-tile, left cutaway		Unpublished
SOS 1	Cefalù, Mandralisca Museum	1184	Probably Lipari		Pan-tile, right cutaway		Cavalier/Brugnone (1986) 240, note 97; Garozzo (2003) 617, no. 54
SOS 2	Cefalù, Mandralisca Museum	1548	Lipari	Necropolis	Pan-tile, right cutaway	*IG* XIV 2395, 6a = *SEG* XXXIV 957 (1984); *SEG* XXXVII 765 (1987)	Moretti (1984); Cavalier/Brugnone (1986) 239, no. 73; Garozzo (2003) 617, no. 53
SOS 3	Cefalù, Mandralisca Museum	1607	Probably Lipari		Upper surface of pan tile		Cavalier/Brugnone (1986) 240, note 97
SOS 4	Termini Imerese, Civic Museum	H.I. 849	Unknown		Cover tile?		Brugnone (1974); Moretti (1984)
SOS 5	Messina, Superintendence	9570	Messina	Rural settlement	Pan-tile, left cutaway	*SEG* LII, 924 (2002)	Bonanno (1997/98) 406; Bonanno (2001) 206, VCT/11

Tab. 2: Synoptic table of the ΣΩΣΗΝΟΣ stamp found at Reggio Calabria. © Dario Giuliano.

***ΣΩΣΗΝΟΣ* stamps found at Reggio Calabria. After D'Amore (2019), 80. All stamps are currently located in the Archaeological Museum of Reggio Calabria**			
Inv. No.	**Discovery**	**D'Amore definition of the tile**	**Tile**
125C	Reggio Calabria	Fragment	Pan tile?
125C	Reggio Calabria	Roof tile	Pan tile?
31741	Reggio Calabria	Big roof tile	Pan tile
31768	Reggio Calabria	Big roof tile	Pan tile
31769	Reggio Calabria	Big roof tile	Pan tile, right cutaway
31770	Reggio Calabria	Big roof tile	Pan tile
31771	Reggio Calabria	Big roof tile	Pan tile
31780	Reggio Calabria	Big roof tile	Pan tile, left cutaway

Outside Sicily, a large quantity of pan tiles with this stamp comes from Reggio Calabria, all found in necropoleis. Two of these were found during the excavation in the Pentimeli area, where two variants of the stamp are present in the roof tiles: one shows the ligature between the letters H and N, while the other does not;[12] three further attestations come from tombs excavated in an area in the current district of S. Caterina;[13] the last one comes from a tomb excavated in 1942 in the

6 Inv. no. 1548, Cavalier/Brugnone (1986) 239–240.
7 Inv. nos. 1184 and 1607. Cavalier/Brugnone (1986) 240.
8 Crisà (2009) 149.
9 Garozzo (2003) 557. Currently on display (Palermo, Antiquarium of Solunto: inv. no. M1534; SA 152).
10 Brugnone (1974) 252.
11 Bonanno (1997–1998) 405, (2001) 197.
12 Di Lorenzo (1884) 282.
13 Orsi (1922) 184–185.

area where the Archaeological Museum was built.[14] Recently, L. D'Amore published a study presenting eight occurrences of stamp, probably including the roof tiles discussed above (Tab. 2). Besides Reggio, there is one example from Caulonia.[15]

The stamp is normally embossed in the cutaway of pan tiles from the so-called "hybrid" system related roofing systems (Fig. 2).[16] The "hybrid" style consists of rectangular pan tiles with semi-circular covers. The pan tiles have large, flat surface bordered along the long sides by flanges; in the front corners there are two cutaways useful for the overlapping system. More rarely, the stamp is applied on the flat surface, as in the case of a roof tile from Cefalù[17] and one from Termini Imerese (Fig. 2).

Scholars agree in interpreting the name Σωσήνος as that of the manufacturer and owner of the workshop that produced these pan tiles. For some time, it was thought that Σωσήνος could be compared to the nominative Σωσίνος, of which there are various attestations in the Greek world.[18] In 1984, Moretti proposed the convincing interpretation that *Σωσήνος* would be the genitive of the name Σωσήν, i.e. one of the anthroponyms "diffusi nella zona di influenza corinzia sulla sponda orientale dell'Adriatico",[19] advancing the hypothesis that it either originated in the area colonised by Corinthians in Sicily or with an immigrant from the eastern shore of the Adriatic who was "proprietario di una fabbrica non localizzabile".[20]

On the basis of palaeographic characteristics, the stamp has been dated by several scholars to the 3rd–2nd century B.C.[21]. Simonetti dates the Caulonia example to the 3rd century B.C.[22] Moretti proposes a date of the 2nd century B.C. for the fragment exposed at the Termini Imerese Museum.[23] Our study allowed us to recognize the use of three different types of stamps, which will be the subject of a further publication.

Overall, roof tiles with the ΣΩΣΗΝΟΣ stamp have been found exclusively on the Tyrrhenian coasts of Sicily and Calabria (Fig. 2). Muscolino therefore put forward the hypothesis that the workshop must have been located in one of these two regions.[24] Initially, the Calabrian examples were also included in this study; however, due to work being carried out in the storerooms of the Archaeological Museum of Reggio Calabria, it was not possible to access them in time for publication.[25] The study therefore focuses on the eight Sicilian examples, five of which (SOS 1-SOS 5; Figs. 2 and 3) were sampled for petrographic analysis. The microphotographs of thin sections of these artifacts were compared with the vast fabric collection in the FACEM database.[26]

The Petrographic Analyses

The petrographic analyses on thin sections (0.03 mm thick; Fig. 3) of the five ceramic samples were carried out by using optical microscopy under transmitted polarized light. The aim was to characterize the ceramic pastes in detail and reveal the possible production area. It is well known that this method of investigation allows one to define the mineralogical composition and various textural parameters of ceramic samples, relative both to the aplastic inclusions and to the groundmass. By recognizing the constituents of the sandy skeleton (monomineralic granules and lithic fragments), while evaluating their size, morphology and relative frequency, it is possible to characterize the clayey raw material used for the ceramic paste, acquire information on the production processes and narrow down or even locate the area of origin.[27] For this particular analysis, the observations were conducted using a Leica DM-SLP model polarizing microscope equipped

14 Galli (1942) 186.
15 Simonetti (2001) 432–433.
16 This definition was adopted by Akerström (1966) 198; see also Wikander (1988) 214.
17 Inv. no. 1607.
18 Galli (1942) 186, note 1.
19 Moretti (1984) 322.
20 Moretti (1984) 323.
21 Recently Garozzo (2011) 668–669, and related bibliography.
22 Simonetti (2001) 432.
23 Moretti (1984) 322.
24 Lentini-Muscolino (2013) 280.
25 I would like to thank the director of the Archaeological Museum of Reggio Calabria, Dr. Carmelo Malacrino, for his availability.
26 See also Gassner in this volume.
27 Tite (2008).

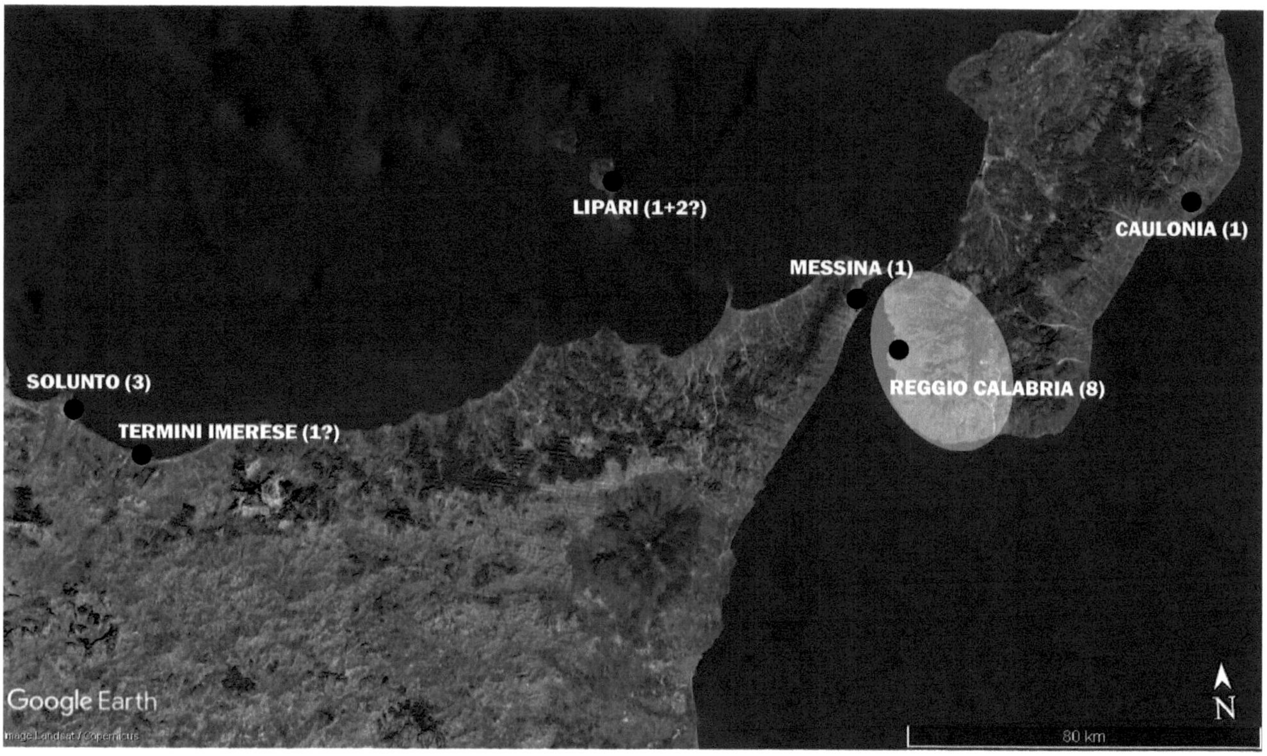

Fig. 1: Distribution map of pan tiles with ΣΩΣΗΝΟΣ stamp. The circle indicates the production area identified following petrographic analysis of samples. Map: Google Earth. Editing: Dario Giuliano.

with a digital camera (Leica IM100 Image Manager software). Appropriate comparative tables were used to evaluate the relative abundance of aplastic granules.[28]

The examined pan tiles showed multiple shared compositional and textural characteristics. They were characterized by a medium-high, aplastic inclusion frequency (25-30%), with granules mainly falling into the medium sand-size class (0.5–0.25 mm). The association of monocrystalline quartz, polycrystalline quartz, K-feldspar (also microcline), plagioclase, biotite, garnet (rare) and lamellae of mica (silt-sized) characterize all the samples analyzed. Among the lithic fragments, medium-high grade metamorphites (mica schists and gneisses) and sandstones are common. The calcareous component is also well represented and microfossils, more or less transformed by the firing process, could be recognized.

These results, on the basis of the specific minero-petrographic markers, indicate the Southern Calabrian-Peloritan Arch as a zone of production, being fully compatible with the lithologies in this area. In particular, the stringent comparison with the ceramic fabrics collected within the FACEM database allowed us to assign this production to the Reggio Calabria area (Fig. 3).

Conclusions

Thanks to petrographic analysis, we have now established with certainty the origin of the pan tiles bearing the ΣΩΣΗΝΟΣ stamp, placing the workshop that produced them in the Reggio Calabria area (Fig. 2). These data reveal the great economic dynamism of Reggio Calabria during the 3rd century B.C. From this city various products were exported,[29] including roof tiles, and, for this class of material, in recent years more and more data seem to indicate that it was traded on a larger scale than hitherto assumed. In fact, whereas for the Roman period this phenomenon is more evident due to the presence

28 Matthew *et al.* (1991).
29 Spadea (1993) 350.

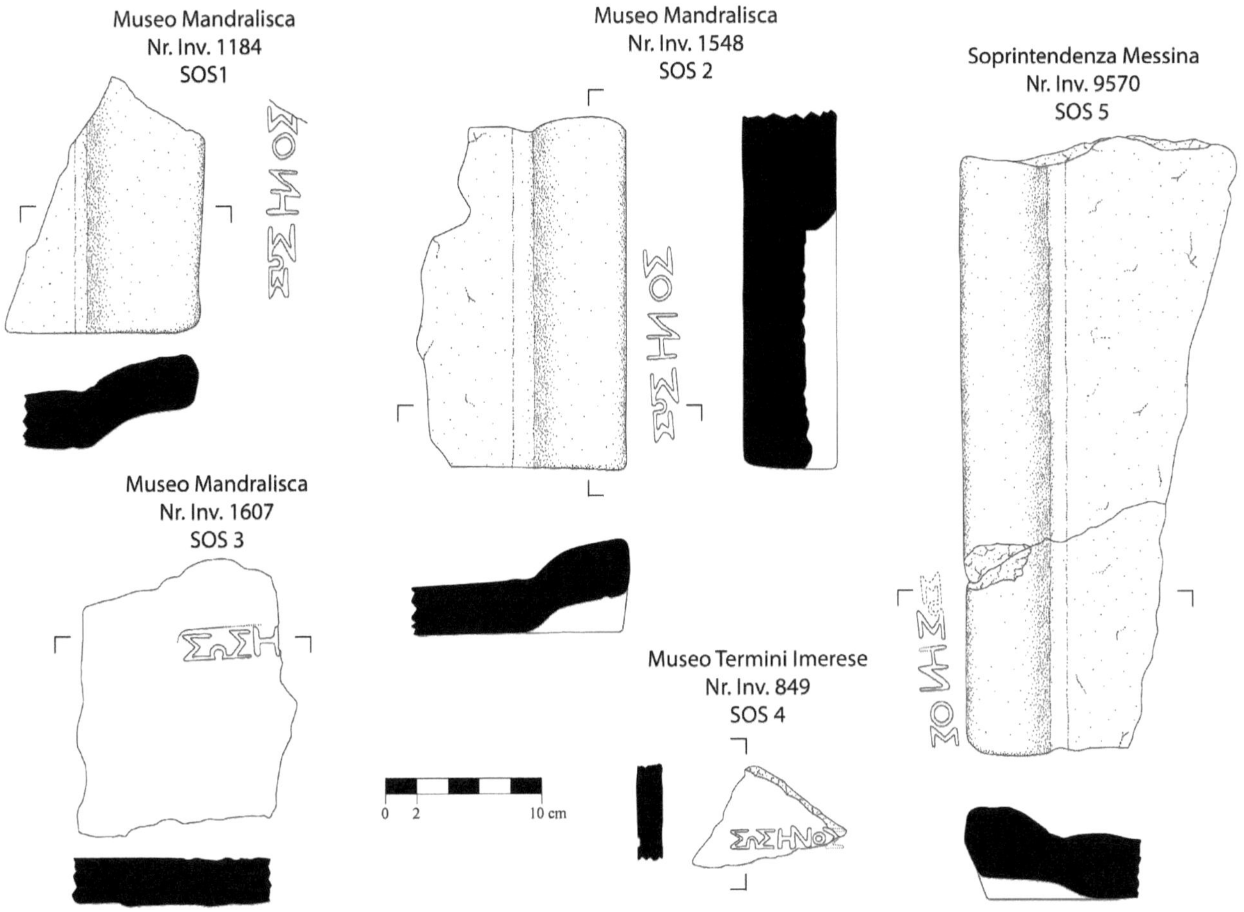

Fig. 2: Drawings of the roof tiles from which the *fabric* samples has been taken. Drawings: Dario Giuliano.

of numerous stamps on roof tiles, as well as the discovery of several shipwrecks with cargoes of roof tiles,[30] fewer clues are available for the Classical and Hellenistic periods. Apart from the famous (and for a long time isolated) case of the *thesauros* of Gela in Olympia,[31] we have additional data from Greece,[32] Velia,[33] Lipari,[34] Himera[35] and also from western Sicily in the sanctuary of Contrada Mango in Segesta.[36]

Returning to the pan tiles with the ΣΩΣΗΝΟΣ stamp, the study conducted would allow us to reconstruct the output of an original, foreign producer who, having settled in Reggio, set up his workshop producing roof tiles. The particularity of the products of Σωσήν is their presence not only in the city of Reggio Calabria where we assume they were produced, but also outside this Chalcidian colony, in several Sicilian sites and in Caulonia. One explanation could lie in the search for new markets to which to sell their products, since the *ergasterion* of Σωσήν seems to have been established in a city whose internal market was 'dominated' by another private producer, Μέμνων, whose roof tiles are abundantly documented in Reggio Calabria;[37] however, they do not seem to have been attested elsewhere, based on the existing finds.

30 Shepherd (2007) 69–72.

31 Although in this case we are not dealing with trade but rather with a state commission, produced in Gela and transported to Olympia. Most recently: Ferrara (2017).

32 De Domenico (2015) 98.

33 Gassner (2003) 157.

34 Cavalier/Brugnone (1986).

35 Epifanio Vanni (1993) 39.

36 De Cesare/Giuliano (forthcoming).

37 For the roof tiles with the stamp ΜΕΜΝΟΝΟΣ and its variants, see D'Amore (2019) for a list of 28 pan tiles plus other fragments of unclear identification.

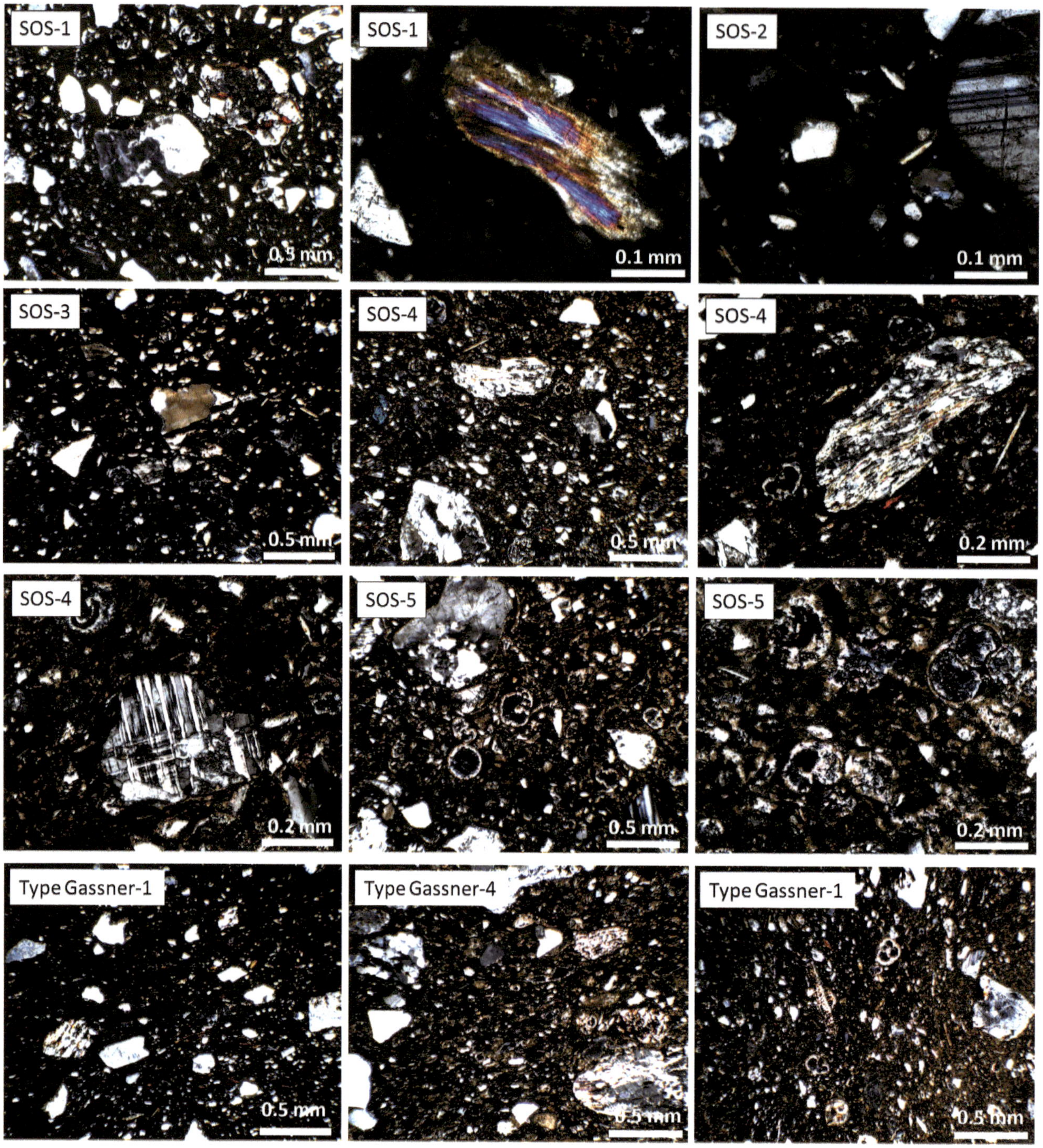

Fig. 3: Comparison of photomicrographs of the thin section SOS 1-SOS 5 with fabrics of Calabrian amphorae.

Bibliography

Akerström (1966): Ake Akerström, *Die Architektonischen Terrakotten Kleinasiens*, Lund.

Avolio (1829): Francesco di Paola Avolio, *Delle antiche fatture di argilla che si ritrovano in Sicilia*, Palermo.

Bonanno (1997–1998): Carmela Bonanno, "Messina. L'insediamento rurale suburbano nei pressi della dia per Catania", in: *Kokalos. Studi di Storia antica pubblicati dall'Università di Palermo* 43–44, 401–421.

Bonanno (2001): Carmela Bonanno, "L'insediamento rurale suburbano nei pressi della via per Catania", in: Giovanni Maria Bacci and Gabriella Tigano (eds.), *Da Zancle a Messina. II**, 195–213.

Brugnone (1974): Antonietta Brugnone, "Iscrizioni greche del Museo Civico di Termini Imerese", in: *Kokalos. Studi di Storia antica pubblicati dall'Università di Palermo* 20, 218–264.

Cavalier/Brugnone (1986): Madeleine Cavalier-Antonietta Brugnone, "I bolli delle tegole della necropoli di Lipari", in: *Kokalos. Studi di Storia antica pubblicati dall'Università di Palermo* 32, 181–282.

Crisà (2009): Antonino Crisà, "Lettera di Enrico Pirajno di Mandralisca a Karl von Estorff con notizie di scavi e ricerche numismatiche a Lipari ed acquisti antiquari a Tindari", in: *LANX. Rivista della Scuola di Specializzazione in Beni Archeologici dell'Università degli Studi di Milano* 4, 146–155.

D'Amore (2019): Lucia D'Amore, *Bolli figulini greci da Reggio Calabria e dal suo territorio*, Reggio Calabria.

de Cesare/Giuliano (forthcoming): "Elementi di copertura e decorazioni architettoniche", in: Monica de Cesare (ed.) *Segesta, Santuario di Contrada Mango. Materiali e contesti degli scavi Tusa*, forthcoming.

De Domenico (2015): Carlo De Domenico, *Lateres Signati Graeci I. Athene et Attica*, Athens-Paestum.

Di Lorenzo (1884): A. M. di Lorenzo, "Reggio di Calabria. Note del can. prof. Fr. M.ª di Lorenzo vice-direttore del Museo Provinciale, sopra le nuove scoperte di antichità avvenute nel territorio reggino", in: *Notizie degli Scavi di Antichità*, 28–87.

Dotto de' Dauli (1825): Girolamo Dotto de' Dauli, "Sopra alcuni frammenti di scultura, e sopra gli avanzi di talune fabbriche scoperte nell'antica città di Solunto", in: *Giornale di Scienze, Lettere ed Arti per la Sicilia* 11, 273–291.

Epifanio Vanni (1993): Elena Epifanio Vanni, "Antefisse di tipo campano a Himera", in: Eva Rystdt, Charlotte Wikander and Örjan Wikander (eds.), *Deliciae Fictiles I. Proceedings of the International Conference on Central Italic Architectural Terracottas*, 39–43.

Ferrara 2017: Bianca Ferrara, "Il *thesauros* di Gela a Olimpia: l'affermazione di un'identità ellenica", in: Rosalba Panvini (ed.), *Migrazioni e commerci in Sicilia. Modelli del passato come paradigma del presente*, Palermo, 185–204.

Galli (1942): Edoardo Galli, "Reggio di Calabria. Necropoli ellenistica scoperta nell'area del Museo", in: *Notizie degli Scavi di Antichità*, 166–234.

Garozzo (2003): Bruno Garozzo, "Nuovi dati sull'instrumentum domesticum bollato – anfore e laterizi – dal palermitano", in: Alessandro Coretti (ed.), *Quarte giornate Internazionali di Studi sull'area Elima* (Erice, 1–4 dicembre 2000), Pisa, 557–683.

Garozzo (2011): Bruno Garozzo, *Bolli su anfore e laterizi in Sicilia (Agrigento, Palermo, Trapani)*, Pisa.

Gassner (2003): Verena Gassner, *Materielle Kultur und kulturelle Identität in Elea in spätarchaisch-frühklassischer Zeit. Untersuchungen zur Gefäß- und Baukeramik aus der Unterstadt (Grabungen 1987–1994)*, Archäologische Forschungen 8. Velia-Studien 2, Vienna.

Lentini/Muscolino (2013): Maria Costanza Lentini and Francesco Muscolino, "Fornaci e produzioni di anfore e laterizi tra Naxos e Taormina (III-I secolo a.C.) e rapporti con le aree tirreniche", in: Gloria Olcese (ed.), *Immensa Aequora Workshop. Ricerche archeologiche, archeometriche e informatiche per la costruzione dell'economia de dei commerci nel bacino occidentale del Mediterraneo (metà IV sec. a.C.- I sec. d.C.)* (Atti del Convegno, Roma, 24–26 gennaio 2011), Rome, 275–285.

Matthew *et al.* (1991): A.J. Matthew, A.J. Woods and C. Oliver, "Spot Before the Eyes: New Comparison Charts for Visual Percentage Estimation in Archaeological Material", in: Andrew Middleton and Ian Freestone (eds.), *Recent Developments in Ceramic Petrology*, British Museum Occasional Paper 81, London, 211–263.

Moretti (1984): Luigi Moretti, "Epigraphica", in: *Rivista di Filologia e di Istruzione Classica* CXII, 314–327.

Portale (2019): Elisa Chiara Portale, "Scultura ellenistica e paesaggio urbano: i casi di Tindari e Solunto", in: Monika Trümper, Gianfranco Adornato and Lappi Thomas (eds.), *Cityscapes of Hellenistic Sicily*, Rome, 239–262.

Portale *et al.* (2021): Elisa Chiara Portale, Gilberto Montali, Massimo Limoncelli, Giovanni Polizzi, Leonarda Fazio and Dario Giuliano, "Nuove ricerche a Solunto (2021)", in: *Mare Internum. Archeologia e culture del Mediterraneo* 13, 115–154.

Orsi (1922): Paolo Orsi, *Reggio Calabria. Scoperte negli anni dal 1911 al 1921*, in: *Notizie degli Scavi di Antichità*, 151–186.

Shepherd (2007): Elizabeth J. Shepherd, "Considerazioni sulla tipologia e diffusione dei laterizi da copertura nell'Italia tardo-repubblicana", in: *Bullettino della Commissione Archeologica Comunale di Roma* 108, 55–88.

Simonetti (2001): Maddalena Simonetti, *I bolli di Caulonia*, in: Maria Cecilia Parra (ed.), *Kaulonia, Caulonia, Stilida (e oltre). Nuovi contributi storici, archeologici, topografici*, in: *Annali Scuola Normale Superiore di Pisa, Classe di Lettere e Filosofia, Quaderni* 11–12, Pisa, 417–464.

Spadea (1993): Roberto Spadea, "Produzioni ellenistiche sullo stretto", in: *Lo stretto, crocevia di culture* (Atti del XXVI Convegno di Studi sulla Magna Grecia, Taranto-Reggio Calabria, 9–14 ottobre 1986), Taranto, 337–360.

Tite (2008): Mike S. Tite, "Ceramic Production, Provenance and Use—A Review", in: *Archaeometry* 50, 216–231.

Wikander (1988): Örjan Wikander, "Ancient Roof-Tiles — Use and Function", in: *Opuscula Atheniensia* 17, 203–216.

Gloria Olcese, Andrea Razza and Domenico Michele Surace

18 Preliminary Observations on the Interdisciplinary Project of Roman Amphorae from the Territory of Ostia

Abstract: This chapter presents preliminary data about amphorae from the territory of Ostia, already investigated within the *Immensa Aequora* project (www.immensaaequora.org [last access 02.07.2024]).

The new research aims to reconstruct settlement and production patterns of rural farms and *villae* during the Republican and Imperial periods, considering both major and minor sites; in particular, Villa A in Dragoncello (Acilia, Rome), currently under investigation. The study employs fieldwork, excavations, surveys, and typological and archaeometric analyses, with a focus on amphorae as indicators of trade and food distribution. The findings from Ostia reveal several amphora types of different origins, showcasing shifts in imports over time. Chemical-mineralogical analyses can help to identify their origins, and gas chromatography–mass spectrometry offers insights into their original contents. The data highlight changing trends in food circulation and indicate a prevalence of Italian wine during the Republican period and a predominance of provincial imports, especially from Hispania, in the Imperial age.

The Ongoing Project in the Territory of Ostia

This chapter provides preliminary data, already covered in previous publications, regarding amphorae from some contexts of Ostia and its territory (Fig. 1). It is one of the branches of the ongoing *Immensa Aequora* project, which intends to study this area of great historical importance and informative potential.[1] The main foci of the new research are the excavation of Villa A in Dragoncello (Acilia, Rome) and the investigation of the ancient agricultural landscape.[2]

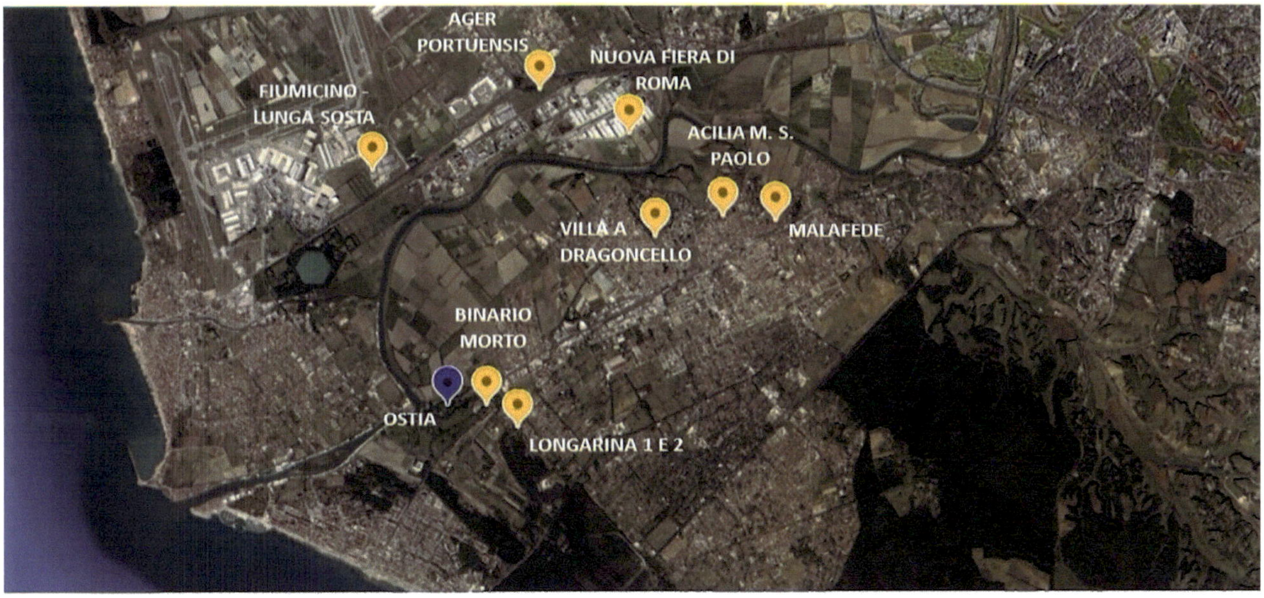

Fig. 1: The sites of *Ager Ostiensis* and *Ager Portuensis* considered in the project. © Immensa Aequora Project.

1 The research is coordinated by G. Olcese, in collaboration with the Superintendencies in charge of the area of Ostia, that we thank. For our previous studies on the amphorae of the territorys see e.g., Olcese/Coletti (2016); Olcese (2018); Olcese *et al.* (2017), (2019), (2023); Rizzo *et al.* (2020). For data relating to the amphorae forming part of the shipwrecks' load, see Olcese (2021).
2 Olcese *et al.* (2018), (2021), (forthcoming), with bibliography. The data on the amphorae and the ceramics from the territory of Dragoncello, which are not presented in this work, will be published in a specific contribution.

https://doi.org/10.1515/9783111189635-018

There are currently numerous studies of amphorae—among the most important indicators of food preservation and distribution (of wine and oil in particular)—coming from different parts of Italy and the Mediterranean in the Late Republican and Early Imperial periods; production in Ostia remains unknown, though it is an important area for redistribution of goods.[3] The new data, obtained with archaeological and archaeometric methods (above all, mineropetrographic analyses), allow us to expand our understanding of ceramics in the territory over the centuries.[4]

The study took into consideration a number of sites in the Ager Portuensis dating back to the Republican era (Fiumicino, Nuova Fiera di Roma, Lunga Sosta, Casal Bernocchi, Acilia–Monti S. Paolo).[5] Moreover, our research has also recently been focusing on various aspects of the Ager Ostiensis: different contexts in which amphorae were used for land reclamation enabled us to chart a chronological sequence between the end of the Republic era and the start of the Augustan (Longarina 2),[6] and the early years of the Augustan era (Longarina 1)[7] and the late Augustan years (Binario Morto).[8] This research allowed us to obtain new information regarding the presence of these amphorae in the Augustan era, one of the lesser-known periods in this area.

The Preliminary Data from the Archaeological and Archaeometric Study of the Amphorae

In the Republican-era sites of Ostia and the Ager Portuensis, amphorae from the end of the 4th century B.C. up until the 3rd century B.C. are documented (see Fig. 3 for the quantifications). Most of them are Graeco-Italic amphorae (types van der Mersch IV, V–V/VI, VI) of Tyrrhenian origin; Punic amphorae from North Africa and areas under Carthaginian influence are fewer in number. In addition, a few Dressel 1 amphorae dating from the 2nd century B.C. are recorded, made in the Tyrrhenian region of Italy, as well as a single example of a so-called tubular Punic amphora.

At Longarina 2 we find a net prevalence of Italian wine containers dating back to the last two centuries of the Republican era (in particular, Lamboglia 2 and Dr. 1).[9] A review of the amphorae found at Longarina 1 (Fig. 2) determined[10] a prevalence of Campanian Dr. 2–4, in particular with Vesuvius fabrics, and of Baetic Dr. 7–11 for transporting fish sauce. Baetic Haltern 70 are also numerous, used for wine derivatives, as well as vinary Dr. 6A. These determine the significance of Adriatic imports to Ostia during the same period. Other amphorae can be identified as Africane Antiche and their production can be traced back to North Africa. These make up the majority of oil amphorae dating back to that period.[11] Baetic oil amphorae (Oberaden 83/Ovoid 7) are rather scarce in number. A few amphorae with different origins (from Tarraconensis, North African, Gallic, and Eastern regions) were also identified, as well as a few smaller flat-bottomed amphorae (lagynoi): mineralogical analyses are currently being carried out on these samples in order to establish whether they could have possibly been produced in Campania.

Amphorae from Binario Morto (Fig. 3) reveal jn the late Augustan era more Baetic Haltern 70 as compared with those found at Longarina 1 and a large increase in Dr. 2–4 produced in Hispania Tarraconensis. This site also provides a notable presence of Dr. 7–11, which along with the appearance of other new types (Beltran IIA and IIB) demonstrates how imports into the area from the Iberian Peninsula continued. Last, there are also numerous Baetic Dr. 20 examples.

3 Hesnard (1980) and (2001); Rizzo (2018); Rizzo *et al.* (2020).
4 Olcese/Thierrin Michael (2009); Olcese/Coletti (2016). For the research within the *Immensa Aequora* project (*www.immensaaequora.org): Atlante* 2011–2012; *Immensa Aequora Workshop* 2013, part I.
5 Olcese/Coletti (2016); Olcese (2018).
6 D'Alessandro/Pannuzi (2016).
7 Hesnard (1980); Olcese *et al.* (2023).
8 Olcese *et al.* (2017).
9 D'Alessandro/Pannuzi (2016); Contino *et al.* (2022).
10 Hesnard (1980). For the full review of Longarina 1 context, see Olcese *et al.* (2023), in which updated data on the documented amphora types and stamps are presented, along with new photographic documentation and preliminary macroscopic analyses of the amphorae fabrics. See also Olcese *et al.* (2019), an extended abstract of our presentation on this topic presented during the Sesto seminario ostiense (Roma-Ostia, 10–11 aprile 2019) and uploded to the website Academia.edu since the autumn of 2019.
11 Contino *et al.* (2019).

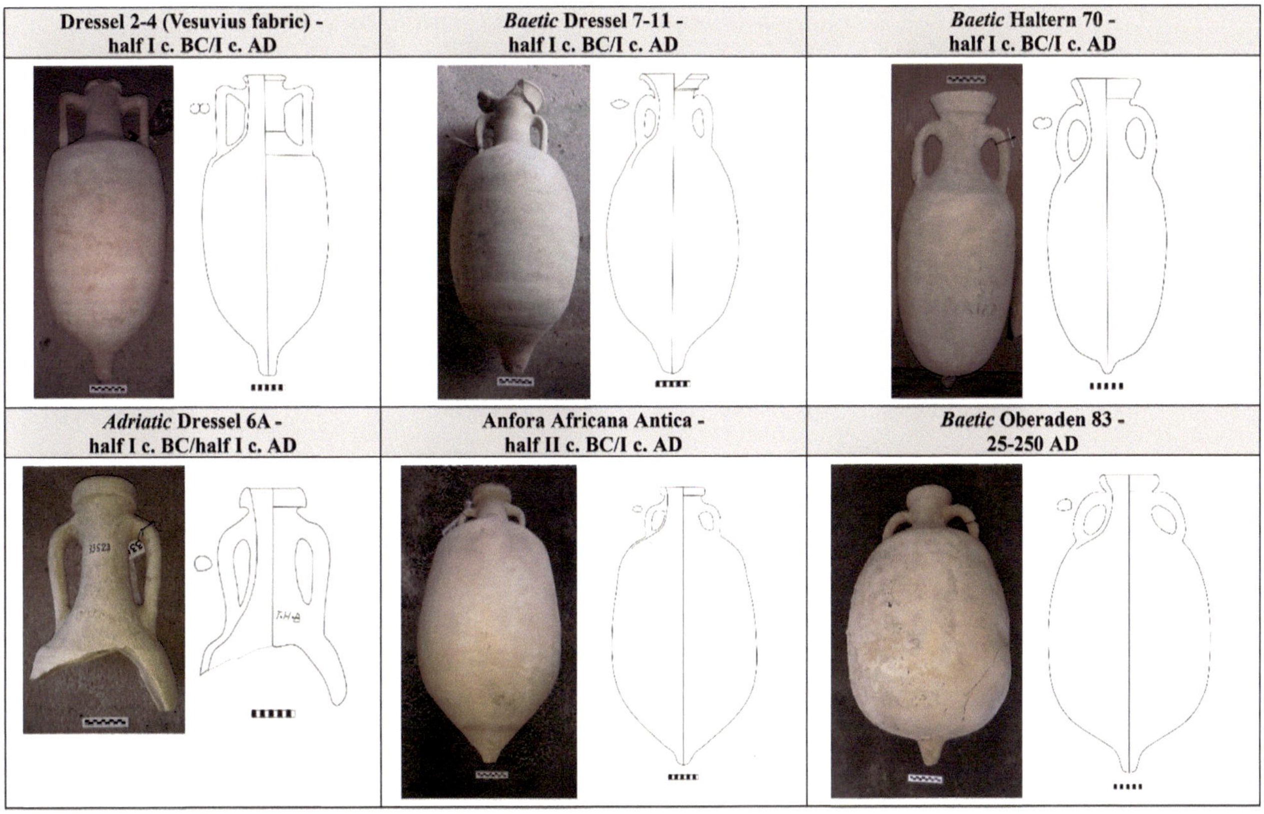

Fig. 2: Table of the most documented amphora types in the Longarina 1. After Olcese *et al.* (2023), pl. 4–6.

Thanks to a comparison with reference groups of the *Immensa Aequora* project, analysis (XRF and polarized microscopy) of some Graeco-Italic amphorae found in Ostia and the Ager Portuensis allowed us in some cases to identify the origin as areas around Minturno and Mondragone, whereas most examples resembled those originating from a few unidentified geological areas in Lazio.[12]

The chemical-mineralogical analysis carried out on eighteen examples of different amphorae from Binario Morto allowed us to identify three archaeometric groups (Fig. 4).[13] These can be compared with reference data of the Baetic areas of the Sierra Morena (Haltern 70, Dr. 2–4, and Dr. 20), of the Guadalete Delta (Dr. 7–11 and Dr. 9, Haltern 70 and Dr. 20), and of the Adriatic coast of the Italian Peninsula (Dr. 6A and Forlimpopoli amphora). Production traceable to the Vesuvius area (Dr. 2–4), to the central-northern Tarraconensis coast (Dr. 2–4), and to western Sicily (Dr. 21–22) was also identified. Further analyses of the amphorae from Longarina 1 are still being carried out.

In order to determine the original contents of the amphorae and the nature of these foodstuffs, relevant data were also obtained in collaboration with N. Garnier through a series of gas chromatography-–mass spectrometry (GC–MS) analyses applied to the organic residues from some samples from Binario Morto.[14] It was possible to verify the actual contents of a Dr. 2–4 as red wine and to note the reuse of a Dr. 7–11 (traditionally used for transporting fish sauce) since it retained traces of grape derivatives. GC–MS analysis of amphorae from Longarina 1 is ongoing, and results will be published in specific contributions. Laboratory data obtained in the project provide us with information regarding the circulation of food in Ostia between the Late Republican and Augustan eras: some known trends were confirmed, but it was also possible to detail them.[15] During the Late Republican phase, we observe a marked prevalence of Italian-produced wine (mainly from

12 Olcese/Thierrin-Michael (2009); Olcese/Coletti (2016).

13 Olcese *et al.* (2017). Analyses in collaboration with ERAAUB (with M.À. Cau, L. Fantuzzi, E. Tsantini).

14 Olcese *et al.* (2023) 100–103. For another application, see Garnier/Olcese (2021).

15 For the first successful archaeological attempts, see Hesnard (1980), (2001); more recently, Rizzo (2018). For a preliminary synthesis of the situation of food imports in Ostia during the Augustan period, see Olcese *et al.* (2023) 99–104.

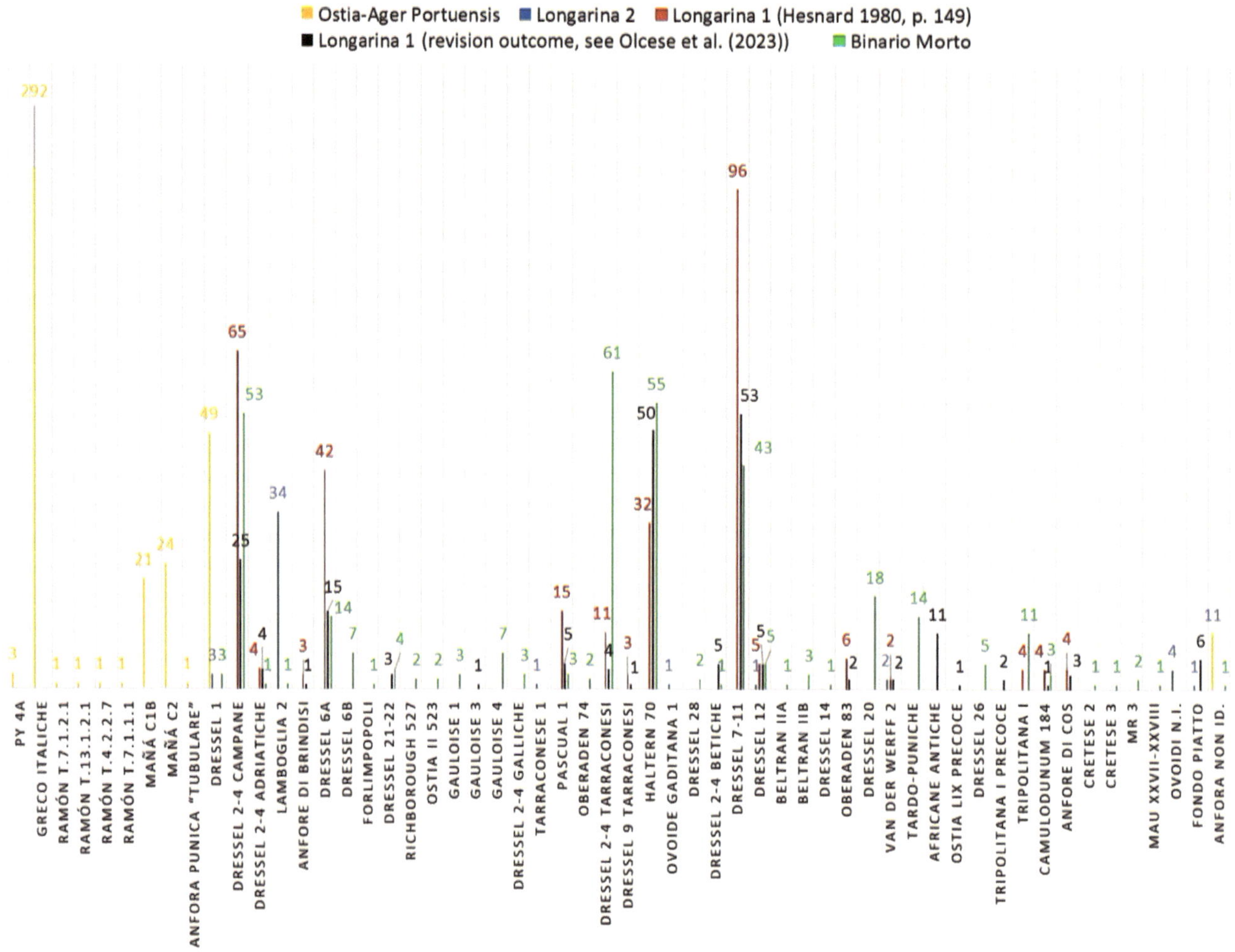

Fig. 3: Histogram of the amphora types documented in the different sites of investigation within the ongoing project: the quantification of the different amphora types is useful to highlight overlaps and differences between the different context. Data from previous publications by the authors related to amphorae in the territory. © Immensa Aequora Project.

Fig. 4: Identified archaeometric groups due to chemical-mineralogical analysis carried out on 18 examples of amphorae from Binario Morto. After Olcese *et al.* (2017) tav. 2, 219.

Tyrrhenia), above all at Ager Portuensis. We also witness a gradual increase in provincial imports from the early Augustan era onward until those from Hispania become predominant, as at Binario Morto.

A predominance of oil imports from North Africa emerges right from the early years of the Augustan era, with an increase in those from Baetica in later years: these data anticipate a trend already known for the eras following the Antonine age, when the crisis of Baetic products began.[16] Moreover, the small quantity of oil amphorae from Italy attests the low incidence of oil produced in this area. Imports of fish sauces in the Augustan age seem to come mainly from Baetica, increasing over what was found for this territory during the Late Republican period.

Ongoing research on Villa A of Dragoncello will add useful data not only for the study of amphorae from that context and the territory around but also to provide other important information concerning the agricultural landscape, the settlement system, and alimentation during the Republican and Imperial ages.

Abbreviations

Atlante (2011–2012)	Gloria Olcese, *Atlante dei siti di produzione ceramica (Toscana, Lazio, Campania e Sicilia)*, Rome.
Immensa Aequora Workshop (2013)	Gloria Olcese (ed.), *Immensa Aequora Workshop Ricerche archeologiche, archeometriche e informatiche per la ricostruzione dell'economia e dei commerci nel bacino occidentale del Mediterraneo (meta IV sec. a.C.-I sec. d.C.)* (Atti del Convegno, Roma 24–26 gennaio 2011), Rome.

Bibliography

Contino *et al.* (2019): Alessia Contino, Lucilla D'Alessandro, Guillermo Pascual Berlanga, Albert Ribera i Lacomba and Claudio Capelli, "Ovoid African and Hispanic amphorae in Italy. Some exemples from Ostia and Pompeii", in: Enrique García Vargas, Rui Roberto de Almeida, Horacio González Cesteros and Antonio Sáez Romero (eds.), *The Ovoid amphorae in the Central and Western Mediterranean*, Oxford, 237–273.

Contino *et al.* (2022): Alessia Contino, Lucilla D'Alessandro, Alessandro D'Alessio, Horacio Gonzdlez Cesteros and Simona Pannuzi, "La Longarina 2 (Ostia). Revisione dei dati provenienti dai contenitori da trasporto. Rapporto Preliminare", in: *Fasti online* 540. https://www.fastionline.org/docs/FOLDER-it-2022-540.pdf (last access 03.08.2024).

D'Alessandro/Pannuzi (2016): Lucilla D'Alessandro and Simona Pannuzi, "Le anfore dello scavo di Longarina 2 ad Ostia antica (RM)", in: Ramon Járrega Domínguez and Piero Berni Millet (eds.), *Amphorae ex Hispania: paisajes de producción y consumo* (III Congreso Internacional SECAH, Tarragona, 10–13 de diciembre de 2014), Tarragona, 530–537.

Garnier/Olcese (2021): Nicolas Garnier and Gloria Olcese, "The Contents of Ancient Graeco-Italic Amphorae. First Analyses on the Amphorae of the Filicudi F and Secca di Capistello Wrecks (Aeolian Islands, Sicily)", in: Dario Bernal Casasola, Michel Bonifay, Alessandra Pecci and Victoria Leitch (eds.), *Roman Amphora Contents. Reflecting on the Maritime Trade of Foodstuffs in Antiquity. In Honour of Miguel Beltrán Lloris* (Proceedings of the Roman Amphora Contents International Interactive Conference – RACIIC, Cadiz, 5–7 October 2015), Oxford, 141–148.

González Cesteros *et al.* (2024): Horacio González Cesteros, Enrique García Vargas, Iván González Tobar, Piero Berni Millet and Rui Roberto de Almeida, "Before the Dressel 20: Pottery Workshops and Olive Oil Amphorae of the Guadalquivir Valley between the Late Republic and Augustan-Tiberian Time", in: *Journal of Roman Archaeology* 37, 111–147. https://www.cambridge.org/core/journals/journal-of-roman-archaeology/article/before-the-dressel-20-pottery-workshops-and-olive-oil-amphorae-of-the-guadalquivir-valley-between-the-late-republic-and-augustantiberian-times/EF523D160396F97D6AE1FF988FD3B87B (last access 23.08.2024).

Hesnard (1980): Antoinette Hesnard, "Un dépôt augustéen d'amphores à la Longarina (Ostie)", in: John Haughton D'Arms and Edward Christian Kopff (eds.), *The Seaborne Commerce of Ancient Rome*, Rome, 141–156.

Hesnard (2001): Antoinette Hesnard, "L'approvisionnement alimentaire", in: *La ville de Rome sous le Haut-Empire. Nouvelles connaissances, nouvelles réflexions*, Toulouse, 285–302.

16 Rizzo (2018) 237; Rizzo/Moreno Megías (2019); see also, to some extent, González Cesteros *et al.* (2024): of this paper we only became aware while our article was in press. In their work, colleagues rightly lament the scarcity of specific data on the import of these amphorae into the Ostia area, as well as the lack of drawings and quantifications. However, the existing data published in Olcese *et al.* (2023) 102–103 are not taken into account, perhaps due to a coincidence of publication dates. Our article, already online in Academia.edu since 2022, before its final publication, summarises our previously published work in some contexts in the Ostia area (Binario Morto, Longarina 1) and anticipates some results on recent work in progress within the *Immensa Aequora* project on imported amphorae and their contents. Among them are also some data (albeit preliminary) on the import of Baetican oil amphorae into Ostia in the Augustan age.

Olcese (2018): Gloria Olcese, "Il Laboratorio per lo studio delle ceramiche e dei commerci. Un'esperienza didattica e di ricerca e i dati sulle ceramiche del territorio ostiense", in: Claire De Ruyt, Thomas Morard and Françoise Van Haeperen (eds.), *Ostia Antica. Nouvelles études et recherches sur les quartiers occidentaux de la cité* (Actes du colloque international, Rome-Ostia Antica, 22–24 Septembre 2014), Rome, 99–111.

Olcese (2021): Gloria Olcese, *Relitti dall'Italia tirrenica nel Mediterraneo occidentale (fine IV secolo a.C. – I d.C.). Archeologia e archeometria delle anfore*, Rome.

Olcese *et al.* (2017): Gloria Olcese, Miguel Ángel Cau Ontiveros, Leandro Fantuzzi, Andrea Razza, Domenico Michele Surace and Eva Tsantini, "Le anfore del contesto della ruota idraulica di Ostia Antica: archeologia e archeometria", in: *Archeologia Classica* 68, 197–224.

Olcese *et al.* (2018): Gloria Olcese, Xabier González Muro, Angelo Pellegrino, Andrea Razza, and Domenico Michele Surace, "La villa A di Dragoncello (Acilia). Alcuni dati dello scavo e dallo studio dei reperti", in: *Mélanges de l'École française de Rome* 130, 325–340.

Olcese *et al.* (2019): Gloria Olcese, Andrea Razza and Domenico Michele Surace, "La cultura materiale del territorio di Ostia nella prima età imperiale: Le ceramiche dai contesti della Longarina e del cosiddetto Binario Morto" (abstract lungo), in: *Atti del Sesto seminario ostiense*, (Roma-Ostia, 10–11 aprile 2019), 2019.

Olcese *et al.* (2021): Gloria Olcese, Xabier González Muro, Michele Di Filippo, Angelo Pellegrino, Andrea Razza, and Domenico Michele Surace, "Prospezioni geofisiche e scavo della villa A di Dragoncello (Acilia, RM): la campagna del 2019", in: *Fasti online* 518. https://www.fastionline.org/docs/FOLDER-it-2021-518.pdf (last access 01.03.2024).

Olcese *et al.* (2023): Gloria Olcese, Andrea Razza and Domenico Michele Surace, "Un progetto multidisciplinare sulla cultura materiale del territorio di Ostia. Alcuni nuovi dati sulle anfore dei contesti augustei della Longarina e del Binario Morto", in: Maria Letizia Caldelli, Nicolas Laubry and Fausto Zevi (eds.), *Ostia e Portus dalla Repubblica alla Tarda Antichità*, Rome, 85–128.

Olcese *et al.* (forthcoming): Gloria Olcese, Angelo Pellegrino, Xabier González Muro and Andrea Razza, "Lo scavo della villa A e le ricognizioni nel territorio di Dragoncello (Acilia, Roma): un progetto multidisciplinare in corso", in: *Ostia prima del II secolo d.C.* (Atti dell'Ottavo Seminario Ostiense, Roma, 21–23 marzo 2024).

Olcese/Coletti (2016): Gloria Olcese and Caterina Coletti, *Ceramiche da contesti repubblicani del territorio di Ostia*, Rome.

Olcese/Thierrin-Michael (2009): Gloria Olcese and Gisela Thierrin-Michael, "Graeco-italic Amphorae in the Region of Ostia: Archaeology and Archaeometry", in: Katalin Biró, Veronika Szilágyi and Attila Kreiter (eds.), *Vessels: Inside and Outside* (Proceedings of EMAC '07 – 9th *European Meeting on Ancient Ceramics*, Budapest, 24–27 October 2007), Budapest, 159–164.

Rizzo (2018): Giorgio Rizzo, "Ostia, le anfore e i commerci mediterranei. Un bilancio preliminare", in: *Archeologia Classica* 69, 223–266.

Rizzo *et al.* (2020): Giorgio Rizzo, Andrea Razza, Domenico Michele Surace and Violeta Moreno Megías, "Importazioni ispaniche di generi alimentari trasportati in anfora ad Ostia nella prima età imperiale: nuovi dati" in: Jean Pierre Brun, Nicolas Garnier, and Gloria Olcese (eds.), *A. Making Wine in Western-Mediterranean B. Production and Trade of Amphorae: Some New Data from Italy* (Proceedings of the 19th International Congress of Classical Archaeology, Vol. 9), Heidelberg, 143–158.

Rizzo/Moreno Megías (2019): G. Rizzo, V. Moreno Megías, "Roma e le prime importazioni di alimenti ispanici trasportati in anfora", in: *Journal of Roman Archaeology 32*, 23–38.

Barbara Borgers and Francesca Diosono

19 A True Melting Pot: The Production of Cooking Ware at Fregellae, Southern Lazio (Italy), between the 4th and the 2nd Centuries B.C.

Abstract: The Latin colony of Fregellae was one of the largest in the Middle and Late Republican periods and a cornerstone of Rome's political dominion in southern Latium. In this important city lived many people from other territories, especially Italians such as Samnites and Peligni but also Carthaginians as well as people probably from Greece and Magna Graecia. The multiculturalism of Fregellae is also reflected in its domestic contexts, as shown in this chapter, focusing on the production and use of cooking ware. The results indicate that cooking ware was produced locally, incorporating typical forms of several culinary cultures rather than from the Roman-Latin tradition only. However, it was also this influx of new inhabitants that led to the destruction of the city, following from their revolt against Rome to request the extension of citizenship rights.

Introduction

Fregellae was founded as a Latin colony in 328 B.C. in hostile territory disputed between the Samnites and the Volscians. The town fulfilled its obligations to Rome during the Second Punic War, but it was besieged and destroyed in 125 B.C. by the Romans because of its citizens' claim to the rights for Roman citizenship. Uncontaminated by later settlement, Fregellae is considered a unicum for the study of Roman archaeology dated between the Middle and the Late Republican eras.

Located in a fertile and well-connected territory, with a wealthy local aristocracy proud of its military and political alliance with Rome, Fregellae was a large and flourishing city; situated in Latium Adiectum in a strategic position along the Via Latina, trade occurred both by land and by water along the route connecting Rome to Capua and from the Sacco and Liri River valleys to the port at Minturnae on the Tyrrhenian coast. Imported goods arrived from neighboring areas as well as from Sicily, the Punic area, and the Aegean Sea.[1]

Studies have led to the hypothesis that most of the ceramic objects found in Fregellae were produced locally, based on the following arguments: the area is rich in good-quality clay; coarse ware, ceramic building material, and *opus doliare* seem to display a similar composition; a pottery kiln and pottery waste, dated between the 3rd and 2nd centuries B.C., were found during archaeological surveys and excavations.[2] This chapter examines the composition of coarse cooking ware dated to the Middle and Late Republican eras, with the aims of testing the hypothesis and understanding the town's position within local production networks at that time.

Background

The University of Perugia has carried out excavations at Fregellae since 1978, under the direction of F. Coarelli. Several public buildings have been found, including three temples, the Forum, the Comitium, the Curia, and the public baths (Fig. 1).[3] In addition, two areas with domestic properties were excavated in the town;[4] in one of these areas, located east of

1 Diosono (2008); Diosono *et al.* (2019).
2 Crawford/Keppie/Vercnocke (1985) 94.
3 Coarelli 1986; Coarelli/Monti (1998); Tsiolis (2013); Battaglini/Coarelli/Diosono (2019); Diosono (2023).
4 Battaglini/Diosono (2010), Battaglini/Braconi (2019).

Note: We would like to thank the Austrian Science Fund (through the FWF project T-1085 G) for their financial support of this research, and the members of the organizing committee, including Diego Elia, Eleni Hasaki, and Marco Serino, for the opportunity to present our research at the stimulating conference entitled "Technology, Crafting and Artisanal Networks in the Greek and Roman World: Interdisciplinary Approaches to the Study of Ceramics." We also thank the two anonymous reviewers for their suggestions on a previous version of this article, and the team of pottery specialists from Fregellae and Dr. Carlo Molle from SABAP Frosinone e Latina for the permission to analyze the ceramic samples in thin-section petrography.

https://doi.org/10.1515/9783111189635-019

the Forum along the Decumanus I (i.e., Opi 2 on Fig. 1), a total of 18 houses have been excavated, highlighting two different construction phases as well as the context of destruction, dated to 125 B.C. They are the context of study of this chapter.[5]

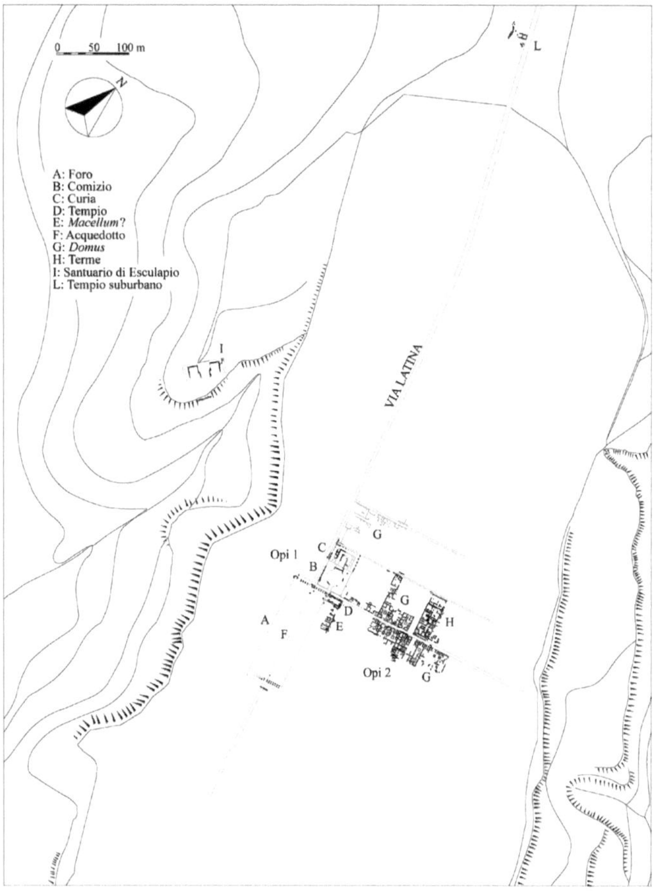

Fig. 1: General plan of the excavations of the University of Perugia (1978–2004) at Fregellae. After Battaglini/Coarelli/Diosono 2019, fig. 1.

The study of the archaeological objects found in these houses has shown that, at the time of its destruction, Fregellae was a multicultural city where the descendants of Latin colonists lived together with those of Punic hostages confined there and with numerous Samnite and Peligni immigrants seeking better living conditions.[6] This is indicated by the typomorphological study of shapes of coarse cooking ware: some shapes, including pots (ollae), casseroles, saucepans, and testi,[7] are part of the traditional kitchenware used in the Roman-Latin world (Fig. 2, 1–10), whereas other shapes, including kakkabai, lopades, and chytrai, were typical of cookery in Magna Graecia and southern Italy (Fig. 2, 11–21). Other forms are specific to the Punic world (Fig. 2, 22–24), but this is not surprising given that two hundred Punic aristocratic families were present in Fregellae from 202 B.C., hostages after the battle of Zama, who asked to reside there and not in Rome;[8] the presence of people of Carthaginian origin is also demonstrated by the large number of Punic amphorae found in the excavations of the city.

This chapter examines the composition of cooking ware from four houses along the Decumanus I in detail,[9] with the objectives to reconstruct production technology on the one hand and aspects of use on the other.

5 Diosono *et al.* (forthcoming).
6 Diosono (2019); Diosono *et al.* (2019); Diosono (2023), (forthcoming).
7 On the use of the term testum to indicate the large lid with flaps with which bread was made, traditionally called clibanus in current scientific literature, see Braconi (forthcoming).
8 Nep. *Annib.* 7.2–3.
9 See also Borgers (forthcoming).

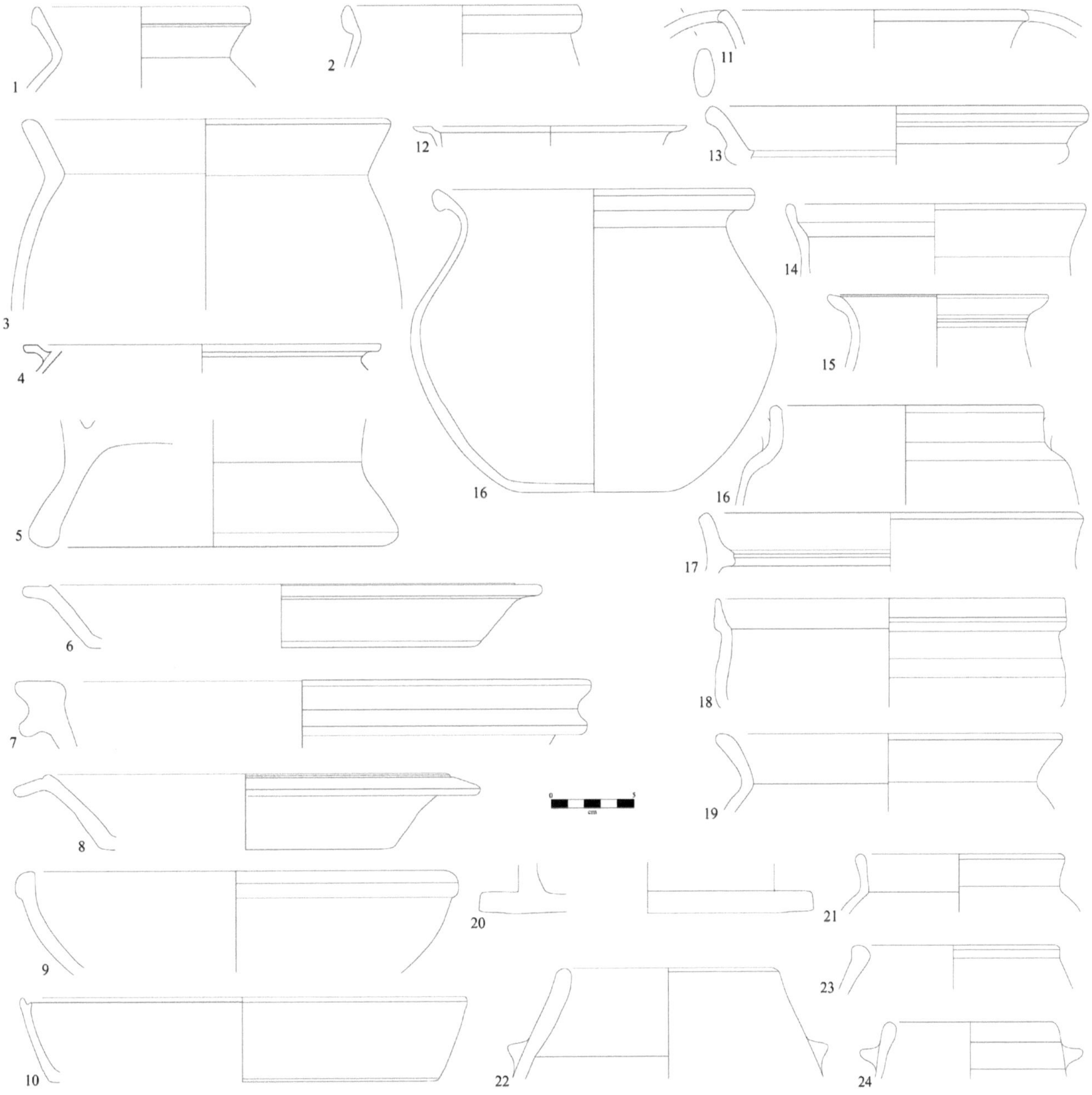

Fig. 2: Select cooking ware forms from the domus of Fregellae. After Diosono (forthcoming).

Method

A total of 36 samples of coarse cooking ware were selected from four houses along the Decumanus I, including Domus I.2, I.7, I.9, and I.13 (Tab. 1). To reconstruct aspects of production technology and use, cooking ware sherds were first examined in hand sample (i.e., macroscopic analysis) and then in thin-section petrography (i.e., microscopic analysis). More specifically:

Tab. 1: A total of 36 cooking ware samples analyzed from Fregellae: with reference to *domus*, shape, abundance of coarse inclusions, firing temperature, surface, and core colour, as well as use traces observed on the external surface of the cooking ware analyzed.

CrtNr	Sample Nr	Domus	Shape	Abun	Firing	Munsell		Use Traces
				dance	Temp	Surface	Core	
1	FREG2	7	Olla (Diosono *et al.* 2019, fig. 4, type 5)	Very	Opaque	Reddish	Gray	Yes
2	FREG3	7	Olla (Diosono *et al.* 2019, fig. 4, type 5)	Moderate	↓	Reddish		/
1	FREG11	7	Olla (Diosono 2019, fig. 2, type 12)	Moderate	↑	Reddish	Gray	
4	FREG13	7	Olla (Diosono 2019, fig. 2, type 8)	Moderate	↓	Reddish		Yes
5	FREG15	7	Olla (Diosono 2019, fig. 2, type 11)	Very	Opaque	Reddish	Gray	Yes
6	FREG16	7	Lopas (Diosono 2019, fig. 2, type 19)	Moderate	Opaque	Reddish		Yes
7	FREG19	7	Lid	Moderate	Opaque	Gray	Reddish	Yes
8	FREG28	7	Lid	Very	↑	Reddish	Gray	Yes
9	FREG31	7	Testum	Moderate	↓	Reddish		Yes
10	FREG32	7	Lid	Very	↓	Reddish		Yes
11	FREG33	7	Chytra	Moderate	↓	Reddish		No
12	FREG34	7	Olla (Diosono *et al.* 2019, fig. 4. type 1)	Very	↑	Reddish	Gray	No
13	FREG35	7	Lopas (Diosono 2019, fig. 2, type 19/21)	Moderate	↑	Reddish		No
14	FREG37	7	Lopas (Diosono 2019, fig. 2, type 20)	Moderate	Opaque	Reddish	Gray	Yes
15	FREG38	7	Olla (Diosono 2019, fig. 2, type 3)	Moderate	↓	Reddish		No
16	FREG39	7	Lid	Very	↑	Reddish		No
17	FREG40	7	Olla (Diosono *et al.* 2019, fig. 4, type 5)	Moderate	Opaque	Reddish		Yes
18	FREG41	7	Testum	Very	↓	Reddish		Yes
19	FREG42	7	Olla (Diosono *et al.* 2019, fig. 4, type 5)	Very	Opaque	Reddish	Gray	Yes
20	FREG43	13	Testum	Very	↓	Reddish	Gray	Yes
21	FREG44	13	Olla (Diosono 2019, fig. 2, type 3)	Moderate	↑	Reddish	Gray	No
22	FREG45	13	Olla (Diosono *et al.* 2019, fig. 4, type 1)	Moderate	↓	Reddish		Yes
23	FREG46	13	Mortarium	Moderate	↓	Reddish		Yes
24	FREG47	13	Lopas	Moderate	Opaque	Reddish		Yes
25	FREG48	9	Olla (Diosono *et al.* 2019, fig. 4, type 1)	Moderate	Opaque	Reddish		Yes
26	FREG49	9	Pan (Diosono 2019, fig. 2, type 24)	Moderate	↑	Reddish	Gray	No
27	FREG50	9	Olla (Diosono 2019, fig. 2, type 3)	Moderate	↑	Reddish		No
28	FREG52	9	Olla (Diosono et al. 2019, fig. 4, type 8)	Moderate	Opaque	Reddish		Yes
29	FREG53	9	Olla (Diosono 2019, fig. 2, type 7)	Moderate	Opaque	Reddish		Yes
30	FREG54	9	Olla (Diosono *et al.* 2019, fig. 4, type 5)	Very	Opaque	Reddish		Yes
31	FREG61	2	Testum	Very	Opaque	Reddish		Yes
32	FREG62	2	Lopas (Diosono 2019, fig. 2, type 22)	Moderate	↓	Reddish		No
33	FREG63	2	Chytra	Very	↑	Reddish	Gray	No
34	FREG64	2	Olla (Diosono 2019, fig. 2, type 4)	Very	↑	Reddish		Yes
35	FREG70	7	Testum	Moderate	↓	Reddish		No
36	FREG71	7	Olla (Diosono 2019, fig. 2, type 5)	Moderate	Opaque	Reddish	Gray	No

– The first stage of analysis was conducted at the macroscopic scale, to identify broad groups based on inclusion variability and abundance. This was complemented by observation of the interior base and color on vessel surface and fresh breaks of the pottery at the Department of Classical Archaeology, University of Vienna, to identify basic information on forming techniques[10] and firing conditions,[11] respectively. In addition, examination of accretion (e.g., black

10 Cuomo di Caprio (2017).
11 Orton/Tyers/Vince (1993).

discoloration on the ceramic surface consisting of deposited carbon) permitted reconstructing the use of the cooking ware.[12]

- Second, all samples were prepared as standard ceramic thin sections and studied with a Leica DM4500 P polarizing-light microscope at the Department of Lithospheric Research, University of Vienna. The clayey matrix, inclusions, and voids of the ceramic thin section were examined in order to identify aspects of the technology, including paste recipes (e.g., intentionally added temper) and firing temperature.[13]

Results

Macroscopic Analysis

Based on the type and abundance of coarse inclusions, as well as the surface and core color of the sherds, all 36 samples belong to the same macrogroup (Tab. 1). The samples display predominantly coarse light-colored and black inclusions, varying in size between 250 µm and 400 µm. The coarse inclusions comprise between 20% and 30% of the clay body (Fig. 3a). Although there seems to be some variability in their abundance – some samples display very abundant coarse inclusions (ca. 25–30%), whereas others display moderately abundant coarse inclusions (ca. 20–25%) – they all seem to have been deliberately added as temper.

(a) (b)

Fig. 3: (a) Macroscopic fabric with reddish surface and core colour, and coarse light-coloured and black inclusions; **(b)** Microscopic fabric with coarse quartz, K-feldspar and clinopyroxene, as well as some carbonate inclusions, as seen in cooking ware from Fregellae.

The interior bases of the cooking ware display spiral rhythmic grooves, indicating that they were thrown on the fast wheel. Most of the samples (N = 22) display a reddish color, resulting from a well-controlled oxidizing firing process. The remaining samples (N = 14) display a sandwich structure – that is, with a reddish-yellow color on the surface and a gray color in the core – suggesting that the firing process was not long enough for oxygen to fully penetrate into the clay body.[14] No surface treatment, in the form of a slip or burnishing, has been identified on the cooking ware.

Evidence for accretion has been found on some of the cooking vessels examined; this is witnessed by a grayish color on the interior rims of the cooking wares examined and their associated lids, following from their use and contact with fire. These traces occur most often on the ovoid jars with flared opening known as ollae[15] (Fig. 2).

12 Banducci (2014).
13 Quinn (2013).
14 Orton/Tyers/Vince (1993).
15 Banducci (2014).

Microscopic Analysis

The compositional analysis indicates that all 36 samples belong to the same petrographic group (Tab. 1). They are defined by coarse subangular (light-colored) quartz and potassium feldspar (250–400 μm), with common (black) clinopyroxene (300–400 μm) and some carbonate inclusions (300–400 μm). The latter seem to consist of numerous (white) micritic grains, which have also been found as pore fill. However, some carbonate fragments show cleavage, indicating that they are primary inclusions. Rare, rounded iron-rich inclusions and angular sedimentary rock fragments, including quartzite, siltstone, and chert (150–400 μm), have also been found (Fig. 3b). The coarse inclusions comprise between 20% and 30% of the clay body and seem to be the result of intentionally added temper.

The clayey matrix of all sections comprises numerous fine quartz and iron-rich inclusions (15–20 μm) as well as common to rare microcline and muscovite mica inclusions (100 μm). Some voids are the result of organic material that has been fired out. About a third of the samples have been fired at a high temperature; another third display an opaque clay matrix, and the final third have been fired at a comparatively low temperature.

Discussion and Conclusions

The results of the macroscopic and microscopic analyses of the 36 samples of cooking ware from Fregellae will be used to reconstruct the framework of their production and use in the city in the Middle and Late Republican eras.

The cooking ware studied seems to have been produced with similar raw materials and technology. More specifically, potters produced cooking ware with a ferrous or iron-rich clay (as suggested by its reddish color) and tempered it with coarse inclusions (consisting of light-colored quartz and potassium (K)-feldspar, and black clinopyroxene). The pottery seems to have been thrown on the fast wheel and was fired in a well-controlled oxidizing atmosphere (i.e., a vertical-up-draft kiln).

Differences exist in the typomorphology of the cooking ware. More specifically, some shapes find their origin in the Roman-Latin world (e.g., ollae, pans, testi), whereas others are typical for culinary practices in Magna Graecia (e.g., chytrai, lopades) both of Greeks and Italics; finally, others still, characterized by the recessed rim and small lateral grip handles, are typical of the Punic world. In particular, ollae display evidence for accretion, attesting their use in cookery. Ollae are often associated with lids, suggesting that they were used for long, slow boiling of foodstuffs.[16] It should also be noted that these jars tend to display an opaque clay body or matrix, which may be explained by their frequency or longevity of use, or by temperature.

Cooking ware belonging to these three different cultural traditions – of the Roman-Latin world, southern Italy, and the Punic world – was found in the excavated houses along the Decumanus I of Fregellae. This is taken to suggest one of two things: either people with different culinary traditions lived together in these domus, or the inhabitants of these houses had developed a hybrid cuisine, consisting of cooking practices prevalent in all three culinary traditions.[17] In fact, the shape has no aesthetic value but serves to obtain different results in cooking: different rims, different widths, and different heights guarantee less or more evaporation, for example, or a different distribution of heat on the food being cooked.

Although only a small assemblage of coarse cooking ware was studied here, future application of the method to a large assemblage may provide a richer understanding of ancient cooking methods and foodways, exploring the importance of food behavior as a marker to express group affiliation.

16 Banducci (2014).
17 Diosono (2019), (forthcoming).

Bibliography

Banducci (2014): Laura M. Banducci, “Function and Use of Roman Pottery: A Quantitative Method for Assessing Use-Wear”, in: *Journal of Mediterranean Archaeology* 27, 187–210.

Battaglini/Braconi (2019): Giovanna Battaglini and Paolo Braconi, “Dalla tegola al mattone. Laterizi sperimentali a *Fregellae*”, in: Jacopo Bonetto, Evelyne Bukowiecki and Rita Volpe (eds.), *Alle origini del laterizio romano. Nascita e diffusione del mattone cotto nel Mediterraneo tra IV e I secolo a.C. Atti del Convegno internazionale di studio* (Padova, 26–28 Aprile 2016), Rome, 495–506.

Battaglini/Coarelli/Diosono (2019): Giovanna Battaglini, Filippo Coarelli and Francesca Diosono (eds.), *Fregellae, Il tempio del Foro e il tempio suburbano sulla via Latina*. Accademia dei Lincei, Rome.

Battaglini/Diosono (2010): Giovanna Battaglini and Francesca Diosono, “Le *domus* di *Fregellae*: case aristocratiche di ambito coloniale”, in: Martin Bentz and Christoph Reusser (eds.), *Etruskisch-italische und römisch-republikanische Häuser* (Kolloquium, Bonn, 23–25 January 2009), Wiesbaden, 217–231.

Borgers (forthcoming): Barbara Borgers, “Caratterizzazione della ceramica da cucina di età medio e tardo repubblicana dalla Domus di *Fregellae*”, in: Diosono *et al.* (forthcoming), forthcoming.

Braconi (forthcoming): Paolo Braconi, “The Syriac Bread of Hellenistic Rome and the *Clibanus*: a Problem of Method”, in: *Proceedings of the 6th IARPotHP Conference. New theories and Methods for Old Pottery: Innovating Perspectives on Hellenistic Ceramics (*Catania - Syracuse, October 9–12, 2023), forthcoming.

Coarelli (1986): Filippo Coarelli (ed.), *Fregellae II. Il santuario di Esculapio*, Rome.

Coarelli/Monti (1998): Filippo Coarelli and Pier Giorgio Monti (eds.), *Fregellae I, Le fonti, la storia, il territorio*, Rome.

Crawford/Keppie/Vercnocke (1985): Michael H. Crawford, Lawrence Keppie and Michael Vercnocke, “Excavations at *Fregellae*, 1978–84. An Interim Report on the Work of the British Team, 2”, in: *Papers British School at Rome* 53, 72–96.

Cuomo di Caprio (2017): Ninina Cuomo di Caprio, *Ceramics in Archaeology. From Prehistoric to Medieval Times in Europe and the Mediterranean: Ancient Craftsmanship and Modern Laboratory Techniques*, Rome.

Diosono (2008): Francesca Diosono, “Materiali e modelli orientali nella colonia latina di Fregellae (Ceprano/Arce, FR)”, in: *Rei Cretariae Romanae Fautorom Acta* 40, 393–396.

Diosono *et al.* (2019): Francesca Diosono, Alessandra Caselli, Simona Consigli, Marzia de Minicis, Vanessa Forcatura, David Lanzi, Sofia Sepiacci, Silvia Staiano and Nicol Tiburzi, “Living in *Fregellae*: Pottery from the *domus*”, in: Annette Peignard-Giros (ed.), *Daily Life in a Cosmopolitan World. Pottery and Culture during the Hellenistic Period* (Proceedings of the 2nd Conference of IARPotHP, Lyon, November 5–8, 2015), Vienna, 551–562.

Diosono (2019): Francesca Diosono, “Siamo quello che mangiamo. Pratiche alimentari e identità dalle domus di *Fregellae*”, in: Marina Cipriani, Emanuele Greco, Angela Pontrandolfo and Michele Scafuro (eds.), *Dialoghi sull’Archeologia della Magna Grecia e del Mediterraneo* (Atti del III Convegno Internazionale di Studi, Paestum, 16–18 novembre 2018), Salerno, vol. I, 139–146.

Diosono (2023): Francesca Diosono, “The Archaeology of *Fregellae*: An Update”, in: Alessandro Launaro (ed.), *Roman Urbanism in Italy: Recent Discoveries and New Directions*, Oxford.

Diosono (forthcoming): Francesca Diosono, “Ceramica da cucina”, in: Diosono *et al.* (forthcoming), forthcoming.

Diosono *et al.* (forthcoming): Francesca Diosono, Giovanna Battaglini, Paolo Braconi, Filippo Coarelli and Manuel Moreno-Alcaide (eds.), *Fregellae. Le Domus del quartiere ad Est del Foro*, forthcoming.

Orton/Tyers/Vince (1993): Clive Orton, Paul Tyers and Alan Vince, *Pottery in Archaeology*, London.

Quinn (2013): Patrick S. Quinn, *Ceramic Petrography: The Interpretation of Archaeological Pottery and Related Artefacts in Thin Section*, Oxford.

Tsiolis (2013): Vassilis Tsiolis, “The Baths at Fregellae and the Transition from Balaneion to Balneum”, in: Sandra K. Lucore and Monika Trümper (eds.), *Greek Baths and Bathing Culture. New Discoveries and Approaches*, Leuven, 89–111.

Section IV: **From Invisible to Visible: Artisanal Gestures, Ontologies, and Ceramic Networks**

Marco Serino

20 From Micro to Macro and Vice Versa: Technology Studies and Network Analysis on Red-figured Vase Production between Sicily and Campania

Abstract: This chapter aims to present some of the main results from the A.G.A.T.H.O.C.L.E.S. project. With a highly interdisciplinary approach, some figurative traditions active between the end of the 5th and the first half of the 4th century B.C. in Sicily (Himera and the area of Syracuse) and along the Tyrrhenian coast (primarily Lipari and Campania) have been analyzed in depth. Technological features were explored through the support of computational imaging. Data come from multiscale perspectives. By using a microscale and a midscale perspective, it is possible to detect technological details related to the ancient painters' gestures (e.g., images coming from Reflectance Transformation Imaging – RTI), especially concerning idiosyncratic ways of drawing preliminary sketches and the differing sequences of applying relief-lines and added colors. Moreover, the combination of these data with those from the experimental archaeological sessions was crucial to better test various technological aspects, creating analogies with ancient technological procedures, tools, and gestures. Finally, using a macroscale perspective through the application of Social-Network Analysis (SNA) allows us to reconstruct a wider historical overview of these artisanal networks, expressed with nodes and edges. This network approach, based on the raw Big Data from the seminal volumes by A. D. Trendall and A. Cambitoglou, helps us to visualize in new and innovative ways the — sometimes elusive — possible collaborations and relationships between different South Italian workshops and to explore interesting links suggestive of movements and migrations of artists between Sicily and Magna Graecia.

Introduction

Catalogues, related supplements, and handbooks published by A. D. Trendall and A. Cambitoglou until 1989 have provided us with the most important and consistent systematization of red-figured vases made in southern Italy.[1] In the last three decades, this systematic framework has been partially amended. This shared—albeit piecemeal—effort to update, integrate, and even overturn, in some cases, aspects of the framework, has produced an emerging picture for South Italian red-figured industries. A number of scholars have recently revised single specific productions and suggested a new general overview for some regional areas, combining different and innovative methodologies. In particular, Sicily and its supposed local workshops[2] have received particular attention by several scholars.[3] Thus, the time is now ripe to argue that every workshop operating on the island during the last decades of the 5th century was born and developed independently and autonomously, making artifacts aimed at satisfying local, subregional markets. However, an even more intriguing aspect is worth mentioning here: most of these workshops seem to have had additional experiences elsewhere, before or later on.[4] It is precisely this 'elsewhere' and the strong propensity for mobility among certain South Italian artisans that constitutes one of the main topics of the A.G.A.T.H.O.C.L.E.S. project, namely investigating the workshops' dynamics of apprenticeship and phenomena of artisanal mobility.[5]

1 *LCS* and its supplements (*LCS I*; *LCS II*; *LCS III*); *RVAp* and its supplements (*RVAp I*; *RVAp II*), Trendall (1987), (1989), (1992).

2 Unfortunately, there is no archaeological evidence regarding a local production of red-figured pottery in Sicily since no archaeologically attested potter's quarter or single workshops have returned fragments of this specific production. For a general overview of the archaeological finds related to kilns in Sicily see Fischer Hansen (2000).

3 Some of the most important post-Trendall general overviews can be found in: Denoyelle/Iozzo (2009) 97–136, 165–170; see also the chapter by G. Gadaleta, in Todisco (2012) 93–98. In particular, related to Sicilian workshops (and their relation to the Campanian area): Spigo (1996), (2001); Barresi (2002), (2013), (2018); de Cesare (2009); Madella (2010); Denoyelle (2011); Elia (2012). For a brief overview of the last two decades of studies, see Soleti (2012) 66–71.

4 For an overview of vase-distribution within Magna Graecia and Sicily, also with comparisons with the Attic imports in the second half of the 5th century B.C., see Serino (2014), (2019) 17–26.

5 Within the framework of the A.G.A.T.H.O.C.L.E.S project, to date the first and the second generations of Sicilian red-figured workshops have received particular attention. For further details about the EU-funded project, see the Introduction in this volume. For a general overview of this

https://doi.org/10.1515/9783111189635-020

Mapping the Movements of South Italian Red-figured Vase Painters: between Sicily, Lipari, Campania and Paestum

Concerning mobility, from his earliest papers, A. D. Trendall recognized significant stylistic affinities between some of the Sicilian and the first Campanian and Paestan ateliers.[6] He suggested a strong link between the Campanian "schools" and some Sicilian workshops. In particular, he referred to the vases produced by the Prado-Fienga Group and the Revel Painters. These painters provided the inspiration for one of the first Campanian workshops, that of the so-called Cassandra Painter, which is still considered today to belong to the Capuan School.[7]

Even earlier than Campanian workshops—as is now unanimously accepted, following A. D. Trendall—the first Paestan painters acquired their artisanal skills from the early 4th century Sicilian red-figured workshops and, in particular, from the Dirce Group and, again, from the Prado-Fienga Group.[8] The first unquestionably recognizable Paestan painters are Assteas and Python, which are the real names of the two artisans as attested by their signatures on some of their vases.

The strongest links between these two regional areas (Sicily, Paestum) are to be seen in the so-called Painter of Louvre K236, the Painter of Naples 2074, and the Painter of Louvre K240.[9] Several vases belonging to these figurative traditions were found on the island of Lipari. According to A. D. Trendall and, more recently, M. Denoyelle, the vases attributed to the Painter of Louvre K240 are very close to the work of Assteas in both subject and treatment.[10] A. Hughes even stated that Louvre K240 was Assteas himself and that he started to put his signature on his vases—*Asteas egraphe*—only when he arrived in Paestum.[11] M. Denoyelle stated that the transmission of the Sicilian figurative tradition to Paestum was achieved directly and naturally and that we can consider this an indirect clue about the Sicilian origin of Assteas and Python themselves.[12] While accepting the argument of M. Denoyelle, greater investigation of the (social) network of these itinerant artisans requires focusing on another (social) actor. In this regard, it is worth mentioning the so-called Painter of the Geneva Orestes, who seems to be a close contemporary of Assteas in the earlier stages of his career or perhaps his master, and whose work provides a strong connecting link to the Sicilian workshops.[13]

A very significant stylistic comparison between the Dirce Painter and the early Paestan figurative traditions can be made by looking at side A of two kraters by the Dirce Painter (Syracuse, Museo Archeologico Regionale "Paolo Orsi", inv. nos. 36332 and 36334) and a fragment of a neck amphora (New York, Metropolitan Museum, inv. no. 1985.74), attributed to the Painter of Geneva Orestes (Fig. 1a–b, f). Here a strong link between the two supposed ateliers is evident from various perspectives: for

renewed interest in mobility in the ancient world (not only from an archaeological perspective), see Horden/Purcell (2000) 342–400; Rouillard (2007), (2009), (2010); Jockey (2009); Archibald (2011) 53–55; Isayev (2017). Mobility in South Italian red-figured production was also recently (2018) explored in the volume published by the Centre Jean Bérard – *Mobilità dei pittori e identità delle produzioni* (2018) – and in Robinson (2022).

6 Since Trendall (1960).

7 According to Trendall, it is possible to recognize at least three different "fabrics" named "Capua I", "Capua II" and Cumae. Again, according to Trendall, all of these "schools" have their origin in Sicily in the early 4th century, especially the Prado-Fienga Group, the Revel Painter and the Eros and the Hare Painter: Trendall (1989) 157–163.

8 In the Paestan area there is an early phase of local experimentation between the late 5th and the early 4th centuries B.C. that led to the production of some rather peculiar artifacts made with a peculiar technique probably imported by Etruscan craftsmen, and made with black glaze as the background and figures overpainted in red. Among these vases, there are some that can be stylistically compared to the Etruscan artifacts attributed by J.D. Beazley to the so-called Argonaut Group: Beazley (1976) 33–36, pls. VIII, 3–4 and IX, 1.

9 Painters of Louvre K240 and K236 probably never worked in Paestum; however, they are considered as "forerunners" of the Paestan stylistic language. With regard to the Painter of Louvre K236, it should be noted that the eponymous krater was initially placed by Trendall within the late phase of the Prado-Fienga Group: *LCS III*, 110 and Trendall (1987) 34–37. Vases by the Painter of Louvre K236 were found in Sicily, Calabria and Campania, but none were from Paestum. Among these groups which can be considered 'borderline' between Campania and Sicily, only a lekanis attributed to the Painter of Naples 2074 comes from Santa Venera, Paestum.

10 Denoyelle (2011) 32–34. See also Trendall (1987) 44 and (1989) 198. Vases by the Painter of Louvre K240 come from Lipari (3), Gela, Syracuse and from Campania. See also Todisco (2012), vol. I, 338–340.About the likely migrations of artisans coming from various parts of Magna Graecia towards Campania, and in particular to the Paestan area, it is worth mentioning the Sydney Painter, probably with Lucanian origin (*LCS*, 127–129; Trendall (1987) 364–367) and the Thyrsus Painter (with Apulian origin; *RVAp* I, 274–283 and especially 280–281 for the ones found in Paestum).

11 Hughes (2003).

12 Denoyelle (2011) 12.

13 About the Painter of the Geneva Orestes, Trendall argued that this painter produced "[. . .] the first vases which may with reasonable certainty be regarded as true Paestan". The vessels attributed to this Painter are currently 4. According to M. Denoyelle – Denoyelle (2011) 29, note 25 – the artifacts from the tomb with *oinochoe* 601 should be further investigated because several vases in this funerary deposition seem to belong to the same artisanal hand (see also *Poseidonia e i Lucani*, 152–155). For the Painter of the Geneva Orestes see also Todisco (2012), vol. I, 381–382.

example, it is possible to recognize how cupid's bow and the faces of the central female figures were drawn, especially the nose, and the philtrum (with a sort of ring under the nose). Such similarities in these very detailed drawing traits look very peculiar, especially if we consider that they are assumed to be by two different artisans. Moreover, the iconography on inv. no. 36334 is the same of that on the MET's fragment: it is related to a *kanephoros*,[14] and the way of drawing, especially the *kanoun* and its patterns, reveals close connections between the two artists. An attempt of overlapping all the three patterns related to the face of one of the main characters on the scene can show very significant affinities in drawings and in the ponderation of a few, but crucial lines to create the face and its internal details (Fig. 1c–e, g–h).

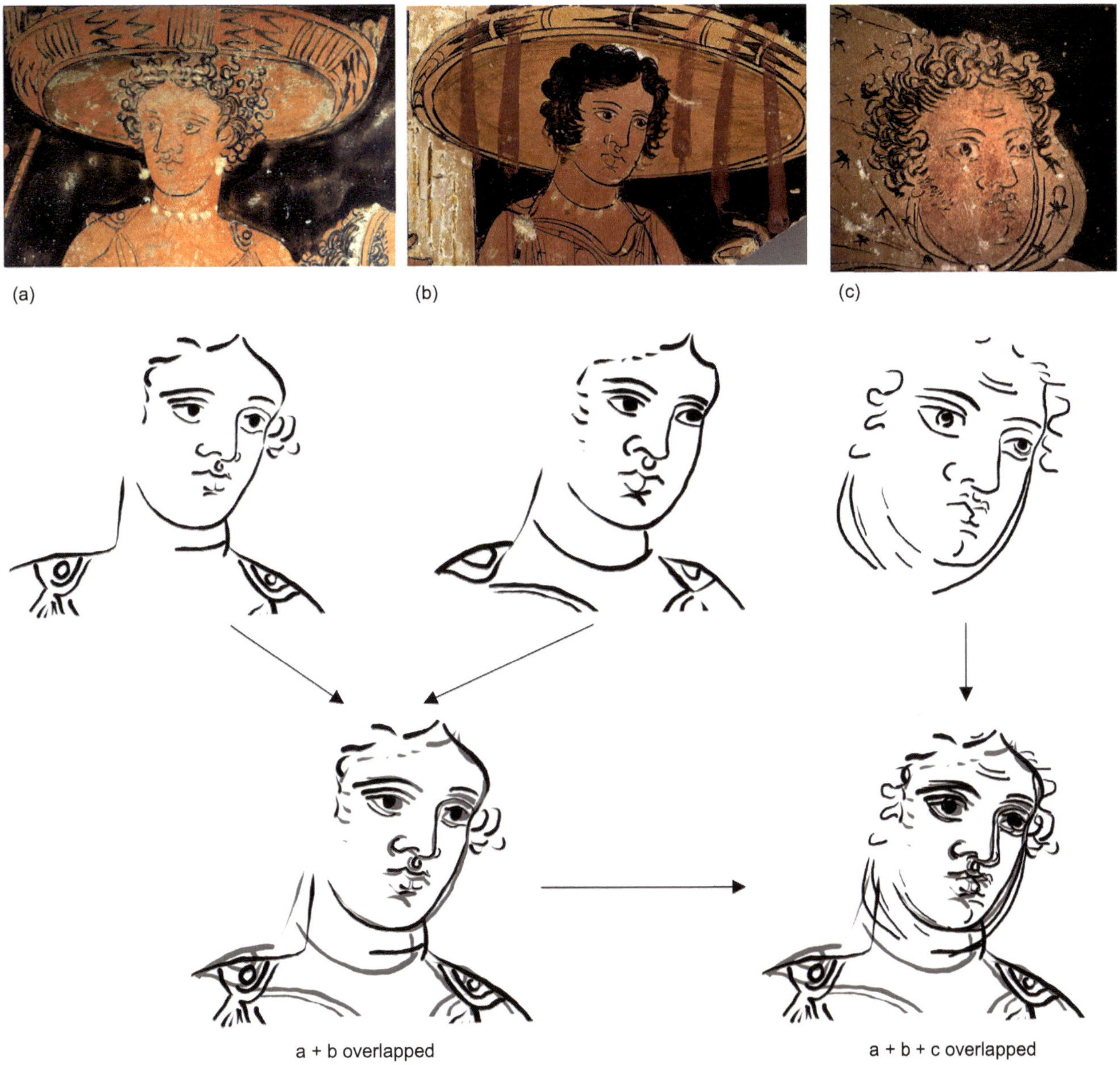

Fig. 1: Dirce Painter and the Painter of Geneva Orestes, some stylistic comparisons: (**a**) Detail of side A from calyx krater (Syracuse, Museo Archeologico Regionale "Paolo Orsi", inv. no. 36334), Dirce Painter (permission by Assessorato Regionale Siciliano dei Beni Culturali e dell'Identità Siciliana) - drawing of the relief-lines applied by the Painter to create the face and its details; (**b**) Fragment of a neck amphora (New York, Metropolitan Museum, inv. no. 1985.74), Painter of Geneva Orestes (© New York, Metropolitan Museum of Art) - drawing of the relief-lines applied by the Painter to create the face and its details; (**c**) Detail of side A from calyx krater (Syracuse, Museo Archeologico Regionale "Paolo Orsi", inv. no. 36332), Dirce Painter (permission by Assessorato Regionale Siciliano dei Beni Culturali e dell'Identità Siciliana) - drawing (with 20° of rotation) of the relief-lines applied by the Painter to create the face and its details; (a+b) and (a+b+c) Overlapping of drawings. Editing: Marco Serino.

14 Young women bearing a basket on her head, taking part in religious processions.

As M. Denoyelle argued before, with the Painter of the Geneva Orestes, it is possible to appreciate the first high-quality local figured pottery in Paestum, deriving from an authentic Greek artisanal tradition. Thus, the presence of this painter can be considered a strong clue that allows us to consider Paestum as an urban center ready to host the red-figured technology and its related artisanal skills, replacing the previous local attempts produced with a vase technique likely imported from Etruria.[15] Therefore, it was around 390 B.C. when the Paestan red-figured production started in a solid and systematic way. Important connections between Sicily and Magna Graecia have been recognized not only around the first decades of the 4th century B.C. but also within the first generation of Sicilian Painters (420–400 B.C.). Concerning the artisanal displacements, it is worthwhile mentioning two early probably Sicilian workshops: that of the Himera Painter, whose training, however, may have taken place in Apulia, and that of the Locri Painter, who started in Sicily and then moved to the Ionian coast, as demonstrated in studies by S. Barresi and D. Elia (Fig. 2).[16] Another interesting recent development—in terms of mobility and early connections between Magna Graecia and Sicily—comes from a recent reappraisal by I. McPhee regarding the so-called Spinelli Painter and Dundee Painter and their likely connections with the Chequer Painter and the Eros and the Hare Painter.[17] As A. Pontrandolfo already stated in 1996, a possible explanation of these networking dynamics is that the supposedly Sicilian Chequer Painter may really have worked in Campania for a certain period, before or after his experience in Sicily, perhaps in the same workshop as the Spinelli Painter.[18]

Likewise, it is possible to appreciate similar dynamics in another early Sicilian workshop, that of the Santapaola Painter. Most of his vases were found in the east of Sicily, some fragments in Himera, and some vases come from tombs on the island of Lipari. In a recent volume of the *CVA* series (Italy 78), a kylix—probably found at Suessula, in Campania—shows very close stylistic connections with the early Sicilian Santapaola Painter.[19] Some rather emblematic stylistic comparisons between the exterior of this kylix (Fig. 3a–b) and various vases already attributed to the workshop of the Santapaola Painter can better explain the argument: see, for example, the face of the female figure on the main side, both the mantle figures on side B of a krater in Lipari (Museo Archeologico Regionale "Luigi Bernabò Brea", inv. no. 11839; cf. Fig. 3c for a detail of the mantle figure on the right), the face of the male figure on the main side of the krater in Catania (Museo Civico Castello Ursino), and again the faces on skyphoi fragments in Syracuse (Museo Archeologico Regionale "Paolo Orsi", inv. no. 56961; Fig. 3d–e).[20] Also, the tondo of the kylix (Fig. 3f) deserves particular attention because it was

15 Cf. *supra*, notes 8–13.

16 Barresi (2013), (2018); Elia (2010), (2014).

17 McPhee (2018). The Spinelli Painter is regarded by J. D. Beazley as an Attic craftsman and appears in his list with only 2 vases: an oinochoe and a kylix from the Spinelli collection of the Archaeological Museum of Naples, both likely from Suessula. I. McPhee has recently attributed another 2 vases (kylikes) to the Spinelli Painter: one is now at the Ackland Art Museum of Chapel Hill, University of North Carolina (USA), probably coming from Capua. While, the other one is traditionally attributed to the Chequer Painter. However, I. McPhee, through an in-depth stylistic reappraisal of this figurative tradition, proposes placing them within the Campanian figurative tradition of the late 5th century B.C. and very close to the Spinelli Painter. Some of the comparisons proposed by I. McPhee between Spinelli and Scacchiera are very convincing indeed, and I believe that today the kylix traditionally attributed to the early Sicilian Chequer Painter should be more correctly placed, as argued by I. McPhee, within the Spinelli Painter's figurative tradition. About the likely late fifth century Campanian red-figured workshops, see also McPhee (2015), in relation to the so-called Dundee Painter.

18 Pontrandolfo (1996) 35–38. *Contra*, Barresi (2002) 69. In Serino (2017) 123–127, a systematic reappraisal of the Chequer Painter's vases makes it possible to observe a significant internal stylistic and qualitative heterogeneity. Regarding style, iconography, context, and morphology, it is possible to recognize different modalities of decorative interventions on these products. There are high-quality vases, close to the Spinelli Painter and most of them come from Campania. Among other details, these vases have constantly three overpainted colors: white, yellow, and red. A second group encompasses all the sides B of the Campania vases. They seem to be made systematically by a sole artisan who applied a very standardized scheme, even in the detail of mantle men facing each other and aryballoi suspended in the middle of the scene. The third group has all the Sicilian vases, characterized by different morphology than the previous groups. Among these vases, some evident drawing comparisons associated with poor figurative skills can be made with the Spinelli Painter and the Campanian products by the Chequer Painter. Moreover, these artifacts reveal a much more limited use of overpainting colors than Campanian artifacts and only the overpainted white seems to appear here. Overpainting is a technological and productive aspect that should not be underestimated in attempting to define, even through technological-productive details, this complex craft network. The three groups proposed do not necessarily indicate three different hands/painters. It is a matter of grasping the association of a specific drawing tradition with forms and repertoires well anchored to specific geographic areas and thus identifying trends that document the connection between Campania and Sicily. On this topic, see also Serino (forthcoming a).

19 It was attributed by M. Borriello to the Meleager Painter, an Attic craftsman active in the last decades of the 5th century B.C. (*CVA* Italy 78 – Maria Rosaria Borriello, *Napoli, Museo Nazionale, Collezione Spinelli* 3: 82–83, pl. 80, inv. no. 164407).

20 Cf. Serino (2019) tab. on p. 197, nos. CK29-Sa9, CK26-Sa3 and Sk11-Sa12.

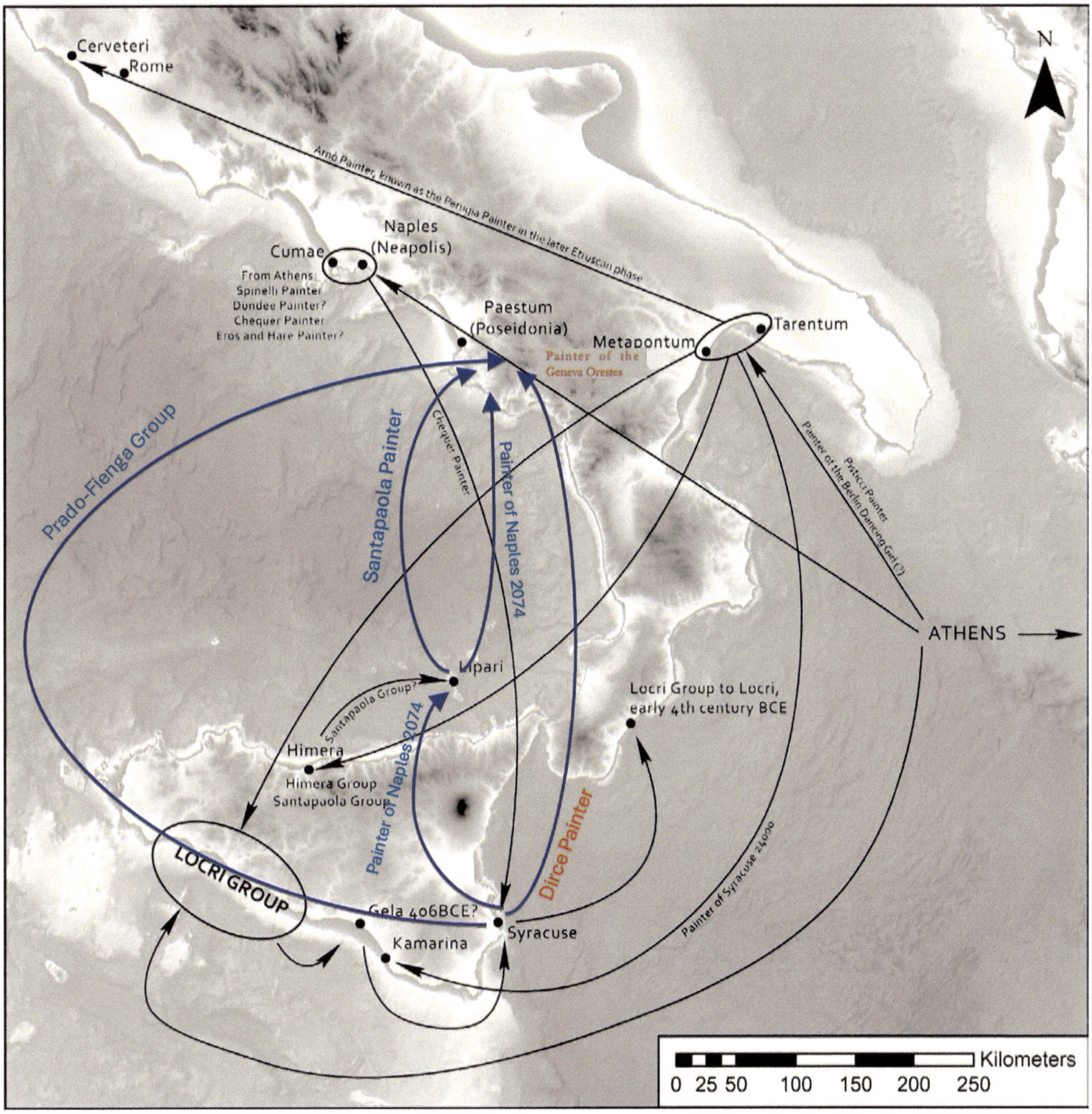

Fig. 2: Movements of early South Italian Painters in Southern Italy and Sicily 440–390 B.C. Map: After Robinson (2022) fig. 5.1. Editing: Marco Serino.

probably decorated by an artisan very close to the Paestan figurative tradition (Fig. 3g–h),[21] which would mean that the Sicilian Santapaola Painter and a Paestan painter worked together on the same vase. We also know that the kylix was one of the favorite shapes during these decades in Campania and that this shape is totally absent in the early Sicilian red-figured vase tradition. Consequently, the presence of the Santapaola Painter on a kylix in Campania together with another artisan cannot be considered a mere coincidence. And this combination should open up other big questions related to supposedly sedentary potters on one hand (who made specific shapes in specific regions, such as the kylix in Campania),

21 This kylix was already published with this new attribution in Serino (2022) 96–98, and (forthcoming a) where various cross-references to the Tyrrhenian figurative tradition were already mentioned. For the comparisons with the Painter of the Geneva Orestes see the Nolan amphora at the Musée d'Art et d'Histoire, inv. no. HR29. For some close stylistic affinities with some mantle figures and their lower borders of *himatia*, see also the Paestan artifacts made by Assteas, such as the neck amphora at the Museo Arqueológico Nacional de Madrid, inv. no. 11235.

and purportedly itinerant painters on the other hand (who decorated new shapes when they moved to another region), starting from at least the last two decades of the 5th century B.C.

Fig. 3: (a, b, f) Kylix (Naples, Museo Archeologico Nazionale, inv. no. 164407), Santapaola Painter (exterior) and Painter of the Geneva Orestes (tondo) (?). After Borriello (2015) pl. 80; **(c)** Mantle figure on side B of the krater (Lipari, Museo Archeologico Regionale "Luigi Bernabò Brea", inv. no. 11839), Santapaola Painter (permission by Assessorato Regionale Siciliano dei Beni Culturali e dell'Identità Siciliana); **(d–e)** Fragments of skyphos (Syracuse, Museo Archeologico Regionale "Paolo Orsi", inv. no. 56961), Santapaola Painter (permission by Assessorato Regionale Siciliano dei Beni Culturali e dell'Identità Siciliana); **(g)** Neck amphora (Geneva, Musée d'Art et d'Histoire, inv. no. HR29), Painter of Geneva Orestes. After Denoyelle (2011) fig. 8; **(h)** Neck amphora (Madrid, Museo Arqueológico Nacional, inv. no. 11235), Assteas. After *RVP*, pl. 69e.

All the examples mentioned above, together with other case studies, invite us to take a new look at our theoretical model in relation to the organization of these early workshops. It is very important to consider that in Magna Graecia and Sicily the beginning of the new dynamics of the red-figured industry related to the first generations of painters highly specialized in red-figured productions must surely have involved close contacts with existing industries and established local workshops, with their own facilities, kilns, and workers already active on site.[22]

One of the few contexts that can provide us with some answers on this subject is certainly the area of Metapontum, which has been examined in depth and very well by F. Silvestrelli.[23] Of course, Metapontum is an important model, but not necessarily the only model universally valid also for other regional areas and other chronological frameworks that can evince diversified dynamics in terms of market, buyers, vase choices, social organization, production systems, interregional connections, and so forth. Concerning the Pisticci Painter (around 440 B.C.), the revolution in the repertory of vases brought about by his arrival—especially in terms of local black gloss pottery—can be explained with the presence/arrival of different new artisans, both painters and potters, resulting from a migration of entire workshops from another place (probably Greece, probably Athens).

After about one generation, starting from 425–420 B.C., various local regional workshops would have been stabilized. Several painters must have been trained in the Magna Graecia workshops (and not in Athens), and thus market demand would have required a greater local presence of red-figured painters. South Italian painters could move and relocate to other local workshops, already active, without revolutionizing the local vase repertoire (because they are only painters). They needed (only) to adapt their drawing skills, personal patterns and decorations, and cultural background to local trends and preferences (especially with respect to shapes), as in the case of the kylix with the Santapaola Painter's intervention.

Accordingly, the mobility of an entire workshop against the mobility of a single painter is an issue that must be addressed case by case, context by context. For all these reasons, craftspeople involved in the making of the vase shape and vase image need to be considered not only as two different artisans but as two different social actors related to different social dynamics of mobility.[24]

Computational Imaging and Experimental Archaeology: Investigating Technological Features and Crafting Procedures in South Italian Red-figured Workshops

Mobility is also a matter of the displacement of knowledge. Some innovative diagnostic techniques can open a new pathway to distinguish the dynamics and the combinations of different painters and potters within a workshop. To better define specific vase productions and recognize both the similarities and the differences in technological procedures within the network of the first and the second generations of Sicilian workshops, new analyses coming from the field of computational imaging are becoming increasingly more crucial.[25] Thanks to a portable digital microscope and to RTI (Reflectance Transformation Imaging),[26] we can recognize many details hidden to the naked eye: fingerprints, the way of applying *miltos* on the vase, the sequence of the different coatings, marks left by the different brushes; or we can observe from a

22 Mannino (1996), (2008). On the introduction of specialists (mainly painters) of red-figured production into local workshops already active *in loco* during the last decades of the 5th century B.C., see Silvestrelli (2018), (2019) (for Metapontum); Elia (2001), (2010) 221–227, and (2019) 554–558 (for Locri Epizephyrii); Fontannaz (2014) 81–90; dell'Aglio/Masiello (2019) (for Tarentum).

23 Silvestrelli (2018), (2019).

24 Preliminary considerations about this theoretical model can be found in Serino (2022), (forthcoming a).

25 These data will be processed again in the next step of the A.G.A.T.H.O.C.L.E.S. project thanks to archaeometric analysis in collaboration with the Departments of Chemistry of both the University of Turin and the University of Bari. For previous technology studies applied to red-figured pottery production see: Seiterle (1976); Noble (1988); *TonArt*, 28–34.

26 For the RTI and photogrammetry training and for the 3D concept of the tool presented I would like to thank Alessandro Bovero, a professional photographer and an expert in computational imaging applied to cultural heritage.

microscale and a midscale point of view the thickness of black gloss and how it was applied on the vase surface, as well as the overpainting and the preliminary sketches.[27]

These technological details naturally lead us to some important questions: Do these features allow us to distinguish one workshop from another? Is it possible to find some specific trends that help us to claim that different technological procedures reflect different workshops? Consequently, would these transfers of technological procedures suggest likely migrations of specific artisans?

First of all, technological features alone cannot say anything. However, if we combine these new data in a general framework, and if we ponder them together with other qualitative information such as style, iconography, morphology, and distribution, as well as archaeometric data, we will be increasingly able to follow ancient artisanal movements, because we will be able to recognize a combination of the same artisanal, figurative, and technological practices belonging to a specific artisan or, at least, to a workshop or school.

Given the interesting and significant (likely) dynamics of mobility between Sicily and Campania around the early 4th century B.C., some vases produced by these workshops have been examined as a test case.

Thanks to RTI on a calyx krater very close in style to the Scoglitti Group[28] the overpainted white and red can be observed in terms of the tools used (dimension and tips), the way the painters moved the brushes, and the sequence of different lines and coatings applied on the various figures depicted, especially in relation to the relief-lines (Fig. 4a).[29] For example, if it is true that for the overpainted white the black lines lie under the white coating, it is totally the opposite for the overpainted red, where the black relief-line creates an outstanding effect resembling chiaroscuro. This is a very interesting situation in terms of technological procedures, because it is important to consider the application of black-glazed relief-lines, and then the drying time (this is a dense coating) and the efficiency of the reduction phase during the firing (in order to allow the coating to change from red to black) and then, within this very systematic sequence of artisanal gestures, the overpainted layer applied on these relief-lines in order to create a fold effect for all the *chitoniskoi* depicted on the krater (Fig. 4b). All these technological features have been tested with dedicated sessions of experimental archaeology:[30] the relief-lines under the red overpainting took the reduction and became black without any problems, despite the coating on top.

Diagnostic techniques using digital tools allow us to also find (at least partial) fingerprints on these vases. In spite of the expected results, fewer fingerprints have been found on the group of samples analyzed so far, but they can still be considered useful and important data because this in-depth survey on fingerprints can tell us that these artifacts usually received considerable attention during the workflow within the workshop. Although fingerprints seem to be very rare in this category of artifacts, sometimes we are able to observe some very peculiar traces that, as in the case of a krater in Zurich (Archäologische Sammlung der Universität, inv. no. 4991) and attributed to the early Sicilian Himera Painter, may be interpreted as signs of an apprentice working inside the workshop. The presence of an apprentice, with his clumsy handling of this vase, could explain such traces as these directly on the coating, because the smudges on the glaze are on various parts of the vase (Fig. 4c).[31]

RTI allows us to observe other interesting features in relation to the tools used for decorating the various parts of the vases. Sometimes we can see tool traces invisible to the naked eye, and we can even measure them with a digital microscope, recognizing different dimensions and different types in relation to the brushes as well as to other tools used in dif-

27 *TonArt*; Boss (1997); Corbett (1965); Noble (1988); Todisco (2012) vol. II, 120–121. For RTI see Artal Isbrand/Klausmeyer (2013); Saunders/Collmann/Borda (2017). Recently, also Balachandran (2018). See also Elia *et al.* in this volume.

28 Syracuse, Museo Archeologico Regionale "Paolo Orsi", inv. no. 55484.

29 In this regard, it is also worth mentioning the attempts to reconstruct the ancient Painter's tools by Sanchita Balachandran, with very interesting and stimulating results obtained thanks to the study of some Attic vases at the John Hopkins Archaeological Museum. See the online Iris Foundation Awards Lecture entitled "Marked in Clay: Intedisciplinary Methods to Re-imagine Ancient Greek Potters at Work" (April 27, 2021): https://www.bgc.bard.edu/research-forum/articles/608/marked-in-clay-interdisciplinary-methods (last access 01.03.2024).

30 Thanks to a strategic collaboration with Roberto Paolini – Pithos Ancient Reproductions (https://www.pithosriproduzioniceramiche.it [last access 01.03.2024]), see note *infra*. On the firing process related to red-figured vases, see also Balachandran (2019). Concerning experimental sessions of firing with an ancient kiln replica see Wissinger/Kahn (2008) and Hasaki (2021) 246–264. This latter experiment is related to the *AIA Tucson Greek Kiln Project*, directed by Eleni Hasaki at the University of Arizona: https://aiatucson.arizona.edu/kiln-construction (last access 01.03.2024).

31 Zurich, Archäologische Sammlung der Universität, inv. no. 4991. See also *LCS*, 35, no. 25j, tab. VIII,1; *LCS III*, 97, no. 40; New York market, Sotheby's, December 8, sale 7572, 102–103, no. 130, figs. at p. 102–103 (Sotheby's 2000); Serino (2016) 48–49, fig. 17; (2019) no. CK6-Hi6. Thanks to Ian McPhee for informing me about the current location of this calyx krater, after rather intricate collecting episodes.

Fig. 4: Technological details. **(a–b)** Calyx krater (Syracuse, Museo Archeologico Regionale "Paolo Orsi", inv. no. 55484), Prado-Fienga Group (permission by Assessorato Regionale Siciliano dei Beni Culturali e dell'Identità Siciliana); **(c)** Calyx krater (Zurich, Archäologische Sammlung der Universität, inv. no. 4991), Himera Painter (permission by Archäologische Sammlung der Universität); **(d)** Calyx krater stored in Lipari, Museo Archeologico Regionale "Luigi Bernabò Brea", inv. no. 168F, Painter of Naples 2074 (permission by Assessorato Regionale Siciliano dei Beni Culturali e dell'Identità Siciliana) **(e–f)** Experimental archaeology: sequence of gestures between preliminary sketch and *miltos*. Photo: Marco Serino. © A.G.A.T.H.O.C.L.E.S. project.

ferent specific practices. For example, the preliminary sketch is an aspect related to an internal workshop's technological procedures that can help scholars to recognize different figurative (and technological) traditions within the social network of ancient red-figured painters in southern Italy.

Experimental archaeology can also be combined with these analytical observations, testing various brushes and observing what kind of tools could have been used for every single gesture that created preliminary sketches.[32] Focusing on the second generation of Sicilian painters and their likely migratory movements toward Campania and Paestum, in previous studies we consistently find a number of workshops involved in these dynamics. Thus, the question is again: What can technology studies provide to better define these productions? One possible answer is related to the different ways of making preliminary sketches. For example, taking into account a vase made by the so-called Painter of Naples 2074 (Museo Archeologico Regionale "Luigi Bernabò Brea", inv. no. 168F, from Lipari), the preliminary sketch also involves in some way the use of *miltos*.[33] There is a thin line that is a deeper red compared with the rest of the surrounded area, and this is an imperceptible groove used for the preliminary sketches.[34] It is evident how it was made before the application of the *miltos*; an important confirmation of this hypothesis comes from the same effect that was observed in the experimental sessions. Experimental archaeological attempts have tried to accurately replicate this sequence of gestures related to the preliminary sketch and its relationship with the use of *miltos*. It is particularly evident how, following this sequence of applications, part of the *miltos* applied as a second step (after the preliminary sketch) tends to accumulate within the grooves made by a pointed tool, which is exactly the same effect we can observe on some ancient artifacts (Fig. 4d–f).

We can also observe other very different ways of making preliminary sketches, with a much thinner pointed tool and with a constant repetition of very thin engravings to create the anatomical parts, as in most of the products of the Prado-Fienga Group.[35] Within the productions analyzed, on vases by the Santapaola Painter we can find the same way of making preliminary sketches as on works by the Prado-Fienga Group.[36] Conversely, it is possible to recognize other procedures of making sketches (like the ones on the Painter of Naples 2074's vases) on the artifacts attributed to the Dirce Painter.[37]

Thanks to the combination of RTI and the digital microscope,[38] it is also possible to take a look at the tool's width and, above all, measure it. In the case of preliminary sketches on vases by the Dirce Painter and the Painter of Naples of 2074, we find measurements between 1.1 and 1.4 mm. Other contemporary workshops have completely different measures: the Prado-Fienga Group has groove widths between 0.30 and 0.38 mm, and the Santapaola Painter's measure between 0.28 and 0.32 mm. Such differences can be explained only by different tools used to make these preliminary sketches.

Briefly, this specific technological evidence is a further indication of a strong relationship between the Dirce Painter and the Painter of Naples 2074. In addition, it indicates a certain distance between these latter artisans on the one hand and the Santapaola Painter and especially the Prado-Fienga Group on the other.

Diagnostic analyses on South Italian red-figured pottery need to consider other general but decisive features, mostly belonging only to this regional vase production.

---

32 Sessions of experimental archaeology have been carried out in collaboration with Roberto Paolini (Pithos Ancient Reproductions, Cerveteri) at the LTT (Laboratory of Traditional Technology) of the University of Arizona (Tucson, USA) and at the Laboratory of the "Officina della Ricerca", Department of Historical Studies, University of Turin. Many thanks to Eleni Hasaki (LTT co-director) for her support and coordination in all the management of these scientific sessions. Moreover, I am very grateful to Cynthia Jones, the resident potter at the LTT in Tucson, for her availability and for giving us the opportunity to use her kilns to reproduce the three stage firing (oxidation-reduction-reoxidation). Last, but not least, I have to thank all the UA students who helped us during the experimental activities in the Lab and especially during all dissemination moments within the campus.

33 Lipari, Museo Archeologico Regionale "Luigi Bernabò Brea", inv. no. 168F. *Miltos* is a ferruginous red ocher that artisans used in red-figured production to make the parts of the vase that were not covered by black gloss appear deeper red. Since the South Italian clay body was usually paler than the original Attic one, *miltos* was used especially in the South Italian workshops. On issues related to *miltos* see: Richter (1923) 53–59, 96–98; Cuomo di Caprio (2017) 285–286; Photos-Jones *et al.* (1997); Schreiber (1999) 48–52; Todisco (2012), vol. 2, 118–119. See also Elia *et al.* in this volume.

34 See Lipari, Museo Archeologico Regionale "Luigi Bernabò Brea", inv. no. 168F.

35 See Syracuse, Museo Archeologico Regionale "Paolo Orsi", inv. no. 47102, and Lipari, Museo Archeologico Regionale "Luigi Bernabò Brea", inv. nos. 6814, 9671, 15138.

36 See Lipari, Museo Archeologico Regionale "Luigi Bernabò Brea", inv. no. 11839.

37 For the Dirce Painter's vases, see: Syracuse, Museo Archeologico Regionale "Paolo Orsi", inv. nos. 36319, 36332, 36334.

38 Dino-Lite Edge Digital Microscope Model AM7915MZTL (Long Working Distance).

Fig. 5: Experimental archaeology: chromatic issues with raw materials. The case-study of the calyx krater: Syracuse, Museo Archeologico Regionale "Paolo Orsi", inv. no. 47102. Photo and digital image: Marco Serino. © A.G.A.T.H.O.C.L.E.S. project.

The pivotal element of all these issues is *miltos*. The use of *miltos* must be considered in relation to the sequence of gestures related to the coatings, and especially to the color they have when they are raw. Experimental archaeology is again crucial to argue considerations on technology: raw *miltos* (rich in iron oxides) can only be red; raw black gloss can be red, but not only red. In-depth analytical observation, thanks to experimental archaeology, combined with technology studies, highlighted how the red raw black gloss might be a huge problem for an artisan who wanted to make fine details (Fig. 4f, Fig. 5, and Fig. 6 option 1), especially with relief-lines inside the silhouettes of the figures in the scene (e.g., clothing, eyes, muscles). In light of this evidence, we tried to recreate different solutions during the experimental sessions with different possible combinations of chromaticism. In this regard, an extraordinary result is related to the possibility of making the black gloss from both red clay and gray clay. Only in the second case did the artisan have enough chromatic contrast between *miltos* and the relief-line to draw details without any problems (Fig. 5 and Fig. 6, option 2). Red raw black gloss can come from a different clay than the clay body. Gray raw black gloss can come from the same clay as the clay body, and vice versa.

Thus, a possible solution for these chromatic issues is that when ancient painters used *miltos*, they needed to apply a gray coating as black gloss, and the real chromatic contrast was gray background, red figures and red secondary patterns, gray details (relief-lines) inside the figures with *miltos*, thus already red when they were still raw. This hypothesis can also explain why in many South Italian vases we can see so many problems related to the coatings and the black gloss (e.g., poor quality, signs of brushes visible to the naked eye, poor attention to the background, especially close to the handles): if the artisan was not able to see a clear contrast between the clay body and the background (gray on gray), he could not have been sure whether he had applied the coating correctly. Only after the firing would the black gloss (now black) have been clearly visible. At that point, however, it would have been too late for any corrections (Fig. 5).

There are also other two possibilities, very close each other: a background with red clay (for the future black gloss) and a relief-line with gray clay (for the future black gloss) or with red clay but mixed with some organic material that alters the color of the original red clay to something more grayish (e.g., charcoal?) that allowed the artisan to see the different colors between the *miltos* and the drawing details while also allowing him to avoid problems with the background (Fig. 6, option 3). In this way, it could have been checked better during its raw stage on the vase.[39] All these scenarios must be checked, of course, context by context, and we need to consider how important the dynamics of local supply were in relation to the raw materials. Some workshops could have worked with different clays, so they could prepare two different recipes for the clay body and the coatings. Some other workshops probably prepared the clay body and black gloss from the same clay.[40]

	Option 1	Option 2	Option 3
Background, contour line	Raw red / fired "black" gloss	Raw gray / fired "black" gloss	Raw red / fired "black" gloss
Relief-line			Raw gray / fired "black" gloss
Unpainted surface	Miltos	Miltos	Miltos

Fig. 6: Experimental archaeology and the color of clay. Raw materials and their related chromatic contrasts during the sequence of gestures and paintings. Photo and digital editing: Marco Serino. © A.G.A.T.H.O.C.L.E.S. project.

39 For more details on these experimental archaeological sessions see Serino (forthcoming b).

40 Archaeometric analyses on the same samples can be very useful to understand if there could be differences between the black gloss used for the background and the one used for the "internal" relief-lines.

The Social Network Analysis of South Italian Red-figured Vases

Moving from a micro- to a macroscale perspective, the social network analysis (SNA) of South Italian vase painters is another innovative and interesting approach to explore and visualize the tangled and elusive dynamics of mobility, temporary transfers, and collaborations.

Inspired by the SNA developed by E. Hasaki and D. Harris Cline in connection with the industry of Attic figured pottery,[41] this attempt was made to integrate the data obtained by the microscale analyses outlined above and to provide an unusual perspective related to the whole network of artisans mentioned by A. D. Trendall and. A. Cambitoglou within their master works.

SNA is a toolbox related to digital humanities that can support the study of relationships between entities, and in this case between ancient artisans. A sociogram was created, converting written text and vocabularies in ties. From more than 3000 pages and 11 volumes, the social vocabulary and the qualitative terms used by Trendall and Cambitoglou were considered, and an undirected network model was applied, meaning that links between entities were symmetrical.[42]
Since we have masters and pupils within Trendall and Cambitoglou's catalogues, it may seem obvious to apply a directed network:[43] master to pupil, or vice versa. On the contrary, a symmetrical network model allows us to eliminate some apriorisms and preconceptions, and we can see only the larger nodes with greater ties and those that are in a strategic tie position, without giving them a direction. Brutally translated: we are trying to make Trendall's mind map work in a different way. We are using the same information that Trendall had, but we connect the information differently: not necessarily hierarchically but horizontally, and according to an idea of a whole network. It will be the qualitative observations and studies, such as the stylistic, iconographic, and morphological context analyses, as well as technological features like the ones mentioned above, that will suggest a hierarchical order and a diachronic development of the various workshops and painters.

Taking this into account, the new sociogram maps artisanal relationships within the South Italian red-figured workshops, providing an innovative panoramic view never seen before for this artisanal production (Fig. 7, on the left). This structural approach for the analysis of the social phenomena is based on a very intuitive consideration: the borders of action by a social actor within a system of relations depend on the position he or she occupies and, at the same time, on the position that others occupy. Some social actors will be in particularly favorable positions and, for this reason, will play a leading role within the dynamics we are studying.[44]

Starting from the vertices sheet, we will be able to clearly identify the most central nodes and also to specify what their prominence consists of, whether it is prestige, activity, brokerage capacity, or a combination thereof. So, this new way of visualizing connections through Trendall's and Cambitoglou's raw data set allows us to better visualize how some workshops have played a central role in the development of certain figurative traditions, or how influential they may have been as brokers between one production and another.

Furthermore, using the metrics tool included in NodeXL,[45] it was also possible to measure this centrality for every single Painter and Group. In general, having high levels of degree means being highly involved within the dynamics of interaction in the sociogram. High levels of betweenness give nodes a high power of brokerage.

In relation to the workshop groups examined here, it is possible to consider the metrics of Assteas, the Dirce Painter, Cassandra Painter, Chequer Painter, and the Painter of Naples 2074. Among the most interesting data are the different results provided by different metrics such as degree and betweenness: the values of these two rankings do not appear—as usually happens—proportional to each other, especially for the Dirce, the Chequer, and the Revel Groups. This dispro-

---

41 Namely the SNAP project. The main research goal of the SNAP project was to map and analyze the structure of the communities of Athenian potters and painters who worked in the Kerameikos during the period from 600 to 400 B.C. For further details see Hasaki/Cline (2020); Harris Cline/Hasaki (2023). I thank Eleni Hasaki for her assistance in developing the sociogram for the South Italian painters.

42 For an overview of Trendall's and Cambitoglou's works, see *supra*, note 1.

43 A different approach, especially in relation to the SNA developed for the Attic vases by D. Harris Cline and E. Hasaki (see *supra*, note 41).

44 Some important works on Network theories and their relationship with the ancient material culture, and on Social Network Analysis are: Knappett 2011; Kadushin 2012; Hodder/Mol 2016; Collar 2020.

45 NodeXL Pro is a software developed by the Social Media Research Foundation (California, USA) and it is an add-in for Microsoft Excel that supports social network and content analysis. It can calculate overall network metrics (for example, density, modularity), basic vertex metrics (degree, indegree, outdegree), advanced vertex metrics (for example, betweenness, closeness, eigenvector, pagerank), and group vertices by cluster or attributes. In addition to metrics, it allows you to visualize network graphs, choosing from various layout algorithms.

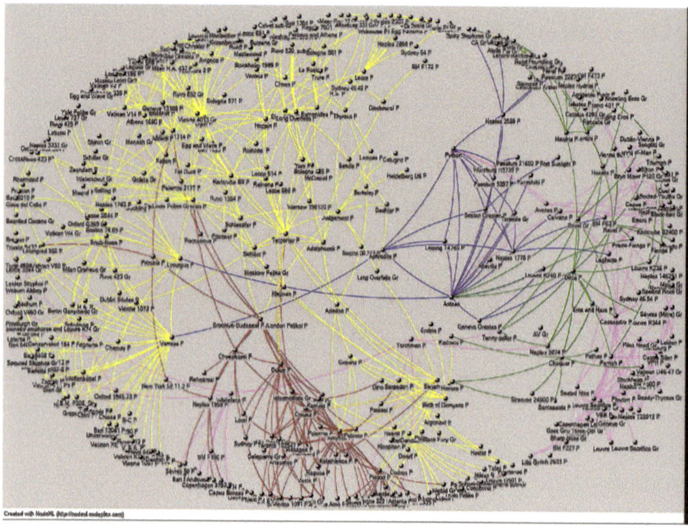

	A	B	C	D	E
1	Label	X	Y	Degree	Betweenness Centrality
2	Asteas	6.172,9	7.041,7	19	33725,592
3	Lycurgus P	2.433,8	5.304,5	15	33708,343
4	Iiupersis P	2.791,4	6.215,2	24	25228,433
5	Dirce P	7.733,0	3.641,0	13	16963,305
6	Dijon P	2.422,4	7.939,7	40	16263,515
7	Tarporley P	3.852,8	6.290,3	19	13187,192
8	Varrese P	3.463,8	4.516,7	28	12007,591
9	Snub-Nose P	1.977,7	4.601,4	12	10582,508
10	Brooklyn-Budapest P /London Pelikai P	2.807,6	4.383,1	13	8734,621
11	Birth of Dionysos P	4.653,0	3.931,5	10	8031,322
12	Cassandra P	8.659,6	3.711,2	16	7816,278
13	Parrish P	9.130,8	3.519,5	15	7198,388
14	Sisyphus P	4.621,2	3.332,9	16	6779,335
15	Amykos P	3.721,1	2.182,3	27	6519,519
16	Chequer P	7.693,6	2.217,4	8	6211,197
17	Naples 2074 P	7.031,4	5.178,6	6	6188,681
18	Revel Gr	8.272,9	4.824,6	7	6169,035
19	Torchmen P	5.008,6	4.617,3	5	5639,000
20	Kadmos P	6.357,3	2.908,7	2	5239,397
21	Manfria P	9.193,4	6.369,4	6	5058,446
22	Hoppin P	4.021,7	8.166,9	14	4872,539
23	Whiteface P	1.587,6	4.241,3	8	4478,000
24	Primato P	1.848,3	5.813,2	5	4371,214
25	Aphrodite P	5.643,1	5.600,8	6	4186,010
26	Athens 1714 P	2.840,4	8.141,6	16	4173,168
27	Schulman P	3.195,0	2.246,6	11	4009,108
28	Long Overfalls P	4.313,5	8.729,9	13	3427,380

Fig. 7: Social Network Analysis of South Italian red-figured painters/workshops. © A.G.A.T.H.O.C.L.E.S. project.

portion could indicate the importance of these workshops not as central actors but rather as bridges or brokers between different workshops located in different areas (Fig. 7, on the right).

To sum up: what is worth underlining is that these data fit well with the technological information we collected, especially as related to the preliminary sketches we have seen before. Both SNA and technology studies provide, in fact, new information on the Prada-Fienga Group, which seems to stay on its own path and does not intersect with the Painter of Naples 2074.[46] On the other hand, this latter painter would seem to have strong connections with both the Dirce Painter and the Painter of the Geneva Orestes. Although this is just the first step towards revising the networks of Sicilian, early Campanian, and early Paestan productions from new perspectives, the preliminary results are promising in helping us develop a new framework for understanding the artisanal mobility between Sicily and Magna Graecia during the end of the 5th and the beginning of the 4th century B.C.

New Ways of Exploring—and Visualizing—the Red-figured Workshops and the Social Network of Ancient Artisans

Within the framework of the A.G.A.T.H.O.C.L.E.S. project, a preliminary concept of a 3D tool was developed in order to interrogate all the data throughout a unique interface (Fig. 8).[47]

This 3D tool is based on a photogrammetric campaign, and its main goal is to provide a user-friendly interface for various types of users—from the everyday museum visitor to the specialist scholar.

You can explore the entire vase in 3D, with zoom options. While exploring the vase itself, it is possible to interrogate the system and obtain different levels of information: from the morphological features and vases' profiles—which constitute important details to better understand the technological procedures used by potters in a specific workshop—to information related to the RTI campaign in regard to the painters' gestures and technological solutions.

Potters and painters—queried on the same artifact—can help to develop and follow a renewed theoretical model, where the workshop organization (and not only the "Painter of") is the core of the whole analysis and research case study. Moreover, data related to the upcoming archaeometric analysis of the samples collected will be available soon, and they will be integrated in the tool as well.

46 Other technological information will soon be available from the archaeometric analysis on samples related to the same workshops: SEM, XRF, Laser Ablation are some of the methodologies applied on these artifacts.

47 For the RTI and photogrammetry training, and for the 3D concept of the tool presented, I would like to thank Alessandro Bovero.

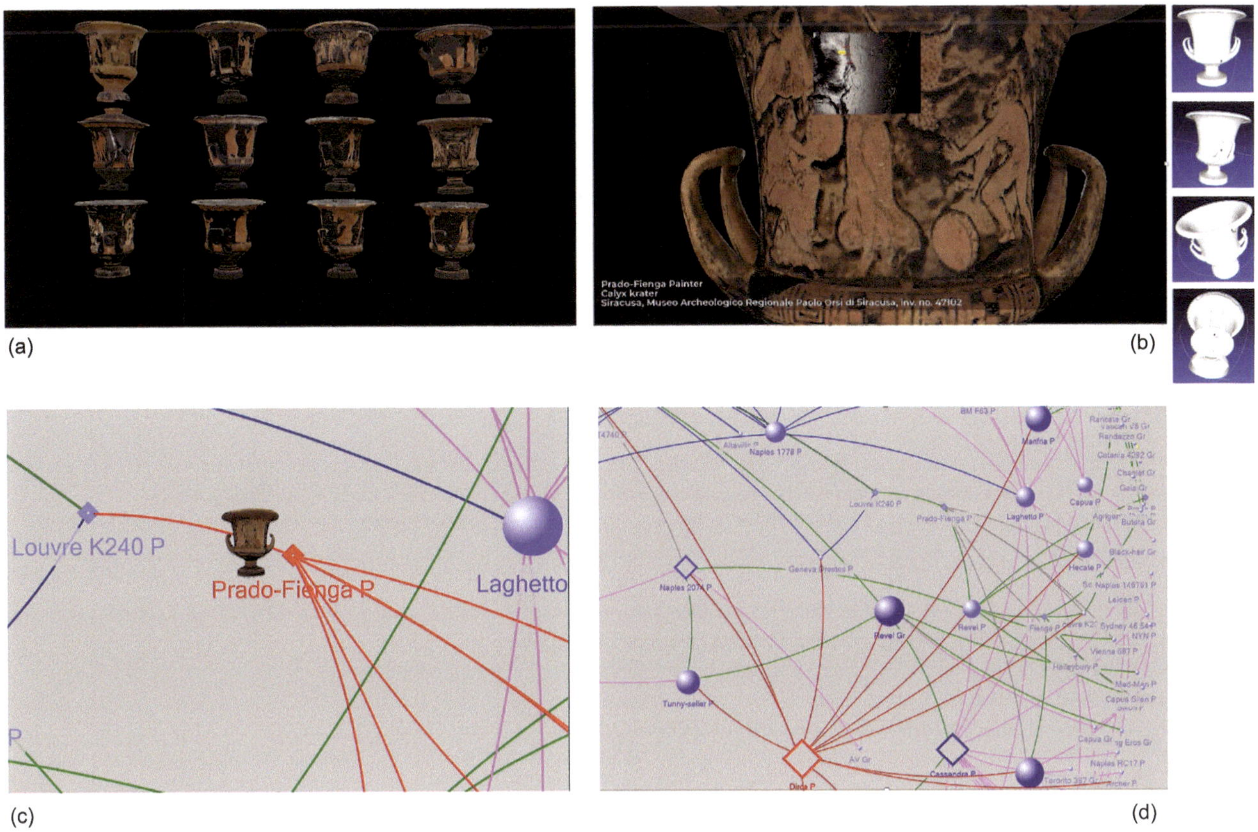

Fig. 8: 3D tool developed in the A.G.A.T.H.O.C.L.E.S. project: **(a)** Tool interface; **(b)** Exploring RTI and photogrammetric results; **(c)** Browsing in SNA; **(d)** Visualizing connections in SNA. Tool Developer: Alessandro Bovero. Editing: Marco Serino. © A.G.A.T.H.O.C.L.E.S. project.

Last, but not least, you can search for information related to the artifact's position within the ancient community of practice thanks to the SNA visualization.[48]

Briefly, this is a preliminary concept developed and aimed at providing a user-friendly tool that allows everyone to explore the world of South Italian red-figured vases, from their micro technological features to a macro overview of all the artisanal networks that the vase belongs to. From micro to macro and vice versa: because this is a reversible tool, it is also possible to explore the whole overview first, and then go inside and explore specific micro features, following the questions that have been inspired by the network perspective provided by the SNA.

Abbreviations

CVA	*Corpus Vasorum Antiquorum*.
LCS	Arthur Dale Trendall, 1967: *The Red-figured Vases of Lucania, Campania and Sicily*, Oxford.
LCS I	Arthur Dale Trendall, 1970: *The Red-figured Vases of Lucania, Campania and Sicily – First Supplement*, BICS, Supplement 26, Oxford.
LCS II	Arthur Dale Trendall, 1973: *The Red-figured Vases of Lucania, Campania and Sicily – Second Supplement*, BICS, Supplement 31, Oxford.
LCS III	Arthur Dale Trendall, 1983: *The Red-figured Vases of Lucania, Campania and Sicily – Third Supplement (consolidated)*, BICS, Supplement 41, Oxford.

48 Which is a totally different way of visualizing data and, in particular, links between painters/workshops – as it is possible to see in Trendall (1989) tabs. at 270–271 – since SNA provides more links and creates a real network of possible contacts between ateliers. For the definition of 'communities of practice', see Lave/Wenger (1991); Wenger (1996).

Poseidonia e i Lucani	Marina Cipriani and Fausto Longo (eds.), *I Greci in Occidente. Poseidonia e i Lucani*, Exhibition Catalogue (Paestum 1996), Naples.
RVP	Arthur Dale Trendall, 1987: *The Red-Figured Vases of Paestum*, British School at Rome, Rome.
RVAp	Arthur Dale Trendall and Alexander Cambitoglou, 1978: *The Red-Figured Vases of Apulia. Early and Middle Apulian, Vol. I*, Oxford.
RVAp I	Arthur Dale Trendall and Alexander Cambitoglou, 1982: *The Red-Figured Vases of Apulia. Late Apulian, Vol.* II, Oxford.
RVAp II	Arthur Dale Trendall and Alexander Cambitoglou, 1991: *The Red-Figured Vases of Apulia, Supplement II, BICS*, Supplement 60, London.
Sotheby's 2000	*Antiquities and Islamic Works of Art*, Sale Catalogue Sotheby's, New York.
TonArt	Martin Bentz, Geominy and Jan Marius Müller (eds.), 2010: *TonArt: Virtuosität antiker Töpfertechnik*, Petersberg.

Bibliography

Archibald (2011): Zosia H. Archibald, "Mobility and Innovation in Hellenistic Economies: The Causes and Consequences of Human Traffic", in: Zosia H. Archibald, John K. Davies and Vincent Gabrielsen (eds.), *Economies of Hellenistic Societies, Third to First Centuries BC*, Oxford, 42–65.

Artal Isbrand/Klausmeyer (2013): Paula Artal Isbrand and Philip Klausmeyer, "Evaluation of the Relief Line and the Contour Line on Greek Red-figure Vases Using Reflectance Transformation Imaging and Three-dimensional Laser Scanning Confocal Microscopy", in: *Studies in Conservation* 58, 338–359.

Balachandran (2018): Sanchita Balachandran, "Uncovering Ancient Preparatory Drawings on Greek Ceramics", *The Iris*. http://blogs.getty.edu/iris/uncovering-ancient-preparatory-drawings-on-greek-ceramics (last access 01.03.2024).

Balachandran (2019): Sanchita Balachandran, "Bringing Back the (Ancient) Bodies: The Potters' Sensory Experiences and the Firing of Red, Black and Purple Greek Vases", in: *Arts* 8: 70.

Barresi (2002): Sebastiano Barresi, "Teseo, Sini, Melicerte ed il Pittore della Scacchiera", in: *Ostraka* 11, 55–79.

Barresi (2013): Sebastiano Barresi, "Sicilian Red-figure Vase Painting", in: Claire Lyons, Michael Bennett and Clemente Marconi (eds.), with the collaboration of Alexandra Sofroniew, *Sicily. Art and Invention between Greece and Rome* (Exhibition Catalogue, J. Paul Getty Museum, Los Angeles, 2013), Los Angeles, 210–218.

Barresi (2018): Sebastiano Barresi, "Il Gruppo di Locri in Sicilia: analisi ed elementi di riflessione", in: Martine Denoyelle, Claude Pouzadoux and Francesca Silvestrelli (eds.), *Mobilità dei pittori e identità delle produzioni*, Cahier du Centre Jean Bérard 25, Naples, 41–63.

Beazley (1976): John Davidson Beazley, *Etruscan Vase Painting*, reprint, New York.

Boss (1997): Martin Boss, "Preliminary Sketches on Attic Red-Figured Vases of the Early Fifth Century B.C.", in: John H. Oakley, William D.E. Coulson and Olga Palagia (eds.), *Athenian Potters and Painters* I, Oxford, 345–351.

Collar (2020): Anna Collar, "Networks, Connectivity, and Material Culture", in: Catherine Cooper (ed.), *New Approaches to Ancient Material Culture in the Greek and Roman World: 21st-century Methods and Classical Antiquity*, Leiden, 47–62.

Corbett (1965): Peter Edgar Corbett, "Preliminary Sketch in Greek Vase-Painting", in: *Journal of Hellenic Studies* 85, 16–28.

Cuomo di Caprio (2017): Ninina Cuomo di Caprio, *Ceramics in Archaeology. From Prehistoric to Medieval Times in Europe and the Mediterranean: Ancient Craftsmanship and Modern Laboratory Techniques*, Rome.

de Cesare (2009): Monica de Cesare, "Il Pittore della Scacchiera e la nascita della ceramica figurata siceliota", in: Carmine Ampolo (ed.), *Immagine e immagini della Sicilia e di altre isole del Mediterraneo antico*, vol. I, Pisa, 277–294.

Denoyelle/Iozzo (2009): Martine Denoyelle and Mario Iozzo, *La céramique grecque d'Italie méridionale et de Sicile. Productions coloniales et apparentées du viiie au iiie siècle av. J.-C.*, Paris.

Denoyelle (2011): Martine Denoyelle, *La céramique grecque de Paestum. La collection du musée du Louvre*, Paris.

Dell'Aglio/Masiello (2019): Antonietta Dell'Aglio and Laura Masiello, "Produzioni artigianali e consumi", in: *Produzioni e committenze in Magna Grecia* (Atti del LV Convegno di Studi sulla Magna Grecia, Taranto, 24–27 settembre 2015), Taranto, 617–661.

Elia (2001): Diego Elia, *La Necropoli in Contrada Lucifero: Nuove Ricerche sulle Aree Funerarie di Locri Epizefiri tra V e IV Secolo a.C. (Campagne di Scavo Degli Anni 1913–1914)*. PhD Dissertation, Università di Messina.

Elia (2010): Diego Elia, *Locri Epizefiri VI.* Nelle case di Ade. *La Necropoli in Contrada Lucifero, Nuovi Documenti*, Alessandria.

Elia (2012): Diego Elia, "Birth and Development of Red-figured Pottery between Sicily and South Calabria", in: Dimitris Paleothodoros (ed.), *The Contexts of Painted Pottery in the Ancient Mediterranean World (Seventh-Fourth Centuries BCE)*, BAR International Series 2364, Oxford, 101–116.

Elia (2014): Diego Elia, "Local Production of Red-figure Pottery at Locri Epizephyrii: A Synthesis on the Last Decade of Studies", in: Victoria Sabetai and Stine Schierup (eds.), *The Regional Production of Red-figure Pottery: Greece, Magna Graecia and Etruria*, Gösta Enbom Monograph 4, Aarhus, 279–290.

Elia (2019): Diego Elia, "Produzione, distribuzione e consumo da Locri a Caulonia nella prima metà del IV sec. a.C.", in: *Produzioni e committenze in Magna Grecia* (Atti del LV Convegno di Studi sulla Magna Grecia, Taranto, 24–27 settembre 2015), Taranto, 553–567.

Fischer Hansen (2000): Tobias Fischer Hansen, "Ergasteria in the Western Greek World", in: Pernille Flensted-Jensen, Thomas Heine Nielsen and Lene Rubinstein (eds.), *Polis and Politics. Studies in Ancient Greek History, presented to Mogens Herman Hansen on his Sixtieth Birthday* (Conference Proceedings, Copenhagen, 2000), Copenhagen, 91–120.

Fontannaz (2014): Didier Fontannaz, "Production and Functions of Apulian Red-Figure Pottery in Taras: New Contexts and Problems of Interpretation", in: Thomas H. Carpenter, Kathleen Lynch and Ted Robinson (eds.), *The Italic People of Ancient Apulia. New Evidence from Pottery for Workshops, Markets, and Customs*, New York, 71–95.

Harris Cline/Hasaki (2023): Diane Harris Cline and Eleni Hasaki, "Assortative Mixing in the Social Networks of Athenian Potters and the Search for Communities of Practice", in: *Journal of Historical Network Research* 8, 21–55.

Hasaki (2021): Eleni Hasaki, *Potters at Work in Ancient Corinth: Industry, Religion, and the Penteskouphia Pinakes*, Hesperia Supplement 51, Princeton N.J.

Hasaki/Cline (2020): Eleni Hasaki and Diane H. Cline, "Social Network Analysis and Connoisseurship in the Study of Athenian Potters' Communities", in: Eleni Hasaki and Martin Bentz (eds.), *Reconstructing Scales of Production in the Ancient Greek World: Producers, Processes, Products, People* (Panel 3.4, Proceedings of the 19th International Congress of Classical Archaeology, vol. 8, Cologne/Bonn, May 22–26, 2018), Heidelberg, 59–80. https://doi.org/10.11588/propylaeum.639.c9977.

Hodder/Mol (2016): Ian Hodder and Angus Mol, "Network Analysis and Entanglement", in: *Journal of Archaeological Method and Theory* 23, 1066–1094. https://doi.org/10.1007/s10816-015-9259-6.

Horden/Purcell (2000): Peregrine Horden and Nicholas Purcell, *The Corrupting Sea. A Study of Mediterranean History*, Oxford.

Hughes (2003): Alan Hughes, "Comedy in Paestan Vase-Painting", in: *Oxford Journal of Archaeology* 22, 281–301.

Isayev (2017): Elena Isayev, *Migration, Mobility and Place in Ancient Italy*, Cambridge.

Jockey (2009): Philippe Jockey, "D'une cité l'autre. Brèves réflexions sur la mobilité des artisans de la pierre dans l'antiquité Classique", in: Claudia Moatti, Wolfgang Kaiser and Christophe Pébarthe (eds.), *Le monde de l'itinérance en Méditerranée de l'antiquité à l'époque moderne. Procédures de contrôle et d'identification* (Table-rondes, Madrid, 2004–Istanbul, 2005), Bordeaux, 139–159.

Kadushin (2012): Charles Kadushin, *Understanding Social Networks: Theories, Concepts, and Findings*, New York.

Knappett (2011): Carl Knappett, *An Archaeology of Interaction: Network Perspectives on Material Culture and Society*, Oxford.

Lave/Wenger (1991): Jean Lave and Etienne Wenger, *Situated Learning: Legitimate Peripheral Participation*, Cambridge.

Lippolis (2018): Enzo Lippolis, "La mobilità del ceramografo dalla formazione alla produzione. Problemi generali e un caso di studio: il Pittore di Dario e il suo ambiente artigianale", in: *Archeologia Classica* 59, 73–112.

Madella (2010): Paolo Madella, "Nuove osservazioni su un cratere del pittore della Scacchiera del Museo Paolo Orsi", in: Filippo Giudice and Rosalba Panvini (eds.), *Il greco, il barbaro e la ceramica attica* I, 77–92.

Mannino (1996): Katia Mannino, "Gli *ateliers* attici e la nascita della produzione figurata", in: Enzo Lippolis (ed.), *I Greci in Occidente. Arte e artigianato in Magna Grecia* (Catalogo della mostra, Taranto, 1996), Naples, 363–370.

Mannino (2008): Katia Mannino, "Dalle importazioni attiche alle produzioni italiote: la documentazione dell'area apulo-lucana", in: *Atene e la Magna Grecia dall'età arcaica all'ellenismo* (Atti del XLVII Convegno di Studi sulla Magna Grecia, Taranto, 27–30 settembre 2007), Taranto, 425–443.

McPhee (2015): Ian McPhee, "A Red-figure Bell-Krater in Dundee: A Campanian Vase-Painter of the Late 5th Century?", in: *Numismatica e Antichità Classiche – Quaderni Ticinesi* 44, 33–49.

McPhee (2018): Ian McPhee, "The Ackland Cup, the Chequer Painter and the Spinelli Painter", in: Maria Bernabò Brea, Massimo Cultraro, Michel Gras, Maria Clara Martinelli, Claude Pouzadoux and Umberto Spigo (eds.), *A Madeleine Cavalier*, Collection du Centre Jean Bérard, Naples, 295–303.

Noble (1988): Joseph Veach Noble, *The Techniques of Painted Attic Pottery*, 2nd revised edition, New York.

Photos-Jones *et al.* (1997): Effie Photos-Jones, Arbory McNulty Cottier, Allan J. Hall and Lina Mendoni, "Kean Miltos: The Well-Known Iron Oxides of Antiquity", in: *Annual of the British School at Athens* 92, 359–371.

Pontrandolfo (1996): Angela Pontrandolfo, "Le prime esperienze dei ceramografi sicelioti e le altre officine tirreniche", in: *I Vasi Attici ed altre Ceramiche Coeve in Sicilia, Vol. II* (Atti del Convegno Internazionale, Catania, Camarina, Gela, Vittoria, 1990), *Cronache di Archeologia* 30, Catania, 35–49.

Richter (1923): Gisela Richter, *The Craft of Athenian Pottery*, New Haven.

Robinson (2022): Ted Robinson, "Potters and Mobility in Southern Italy (500-300 BCE)", in: Jeremy Armstrong and Sheira Cohen (eds.), *Production, Trade, and Connectivity in Pre-Roman Italy*, London/New York, 82–98.

Rouillard (2007): Pierre Rouillard (ed.), with the collaboration of Catherine Perlès and Emmanuel Grimaud, *Mobilités, immobilismes. L'emprunt et son refus*, Colloques de la Maison René Ginouvès. Archéologie et ethnologie 3, Paris.

Rouillard (2009): Pierre Rouillard (ed.), *Portraits de migrants, portraits de colons*, 1, Colloques de la Maison René-Ginouvès. Archéologie et ethnologie 5, Paris.

Rouillard (2010): Pierre Rouillard (ed.), *Portraits de migrants, portraits de colons*, 2, Colloques de la Maison René-Ginouvès. Archéologie et ethnologie 6, Paris.

Saunders/Collmann/Borda (2017): D. Saunders, Richard Collmann and Ann Borda, "Reflectance Transformation Imaging and ImageJ: Comparing Imaging Methodologies for Cultural Heritage Artefacts", in: *Proceedings of EVA-Electronic Visualisation and the Arts* (London, 11–13 July 2017). https://doi.org/10.14236/ewic/EVA2017.71.

Schreiber (1999): Toby Schreiber, *Athenian Vase Construction: A Potter's Analysis*, Malibu.

Seiterle (1976): Gérard Seiterle, "Die Zeichentechnik in der rotfigurigen Vasenmalerei: Das Rätsel der Relieflinien", in: *Antike Welt* 7, 2–10.

Serino (2014): Marco Serino, "The Beginnings of Sicilian Red-figured Pottery and its Relationship with Early South Italian Productions: A Reappraisal through the Case-Study of the Himera Painter Workshop", in: Stine Schierup and Victoria Sabetai (eds.), *The Regional Production of Red-figure Pottery: Greece, Magna Graecia and Etruria*, Gösta Enbom Monographs 4, Aarhus, 247–267.

Serino (2016): Marco Serino, "Indizi di ritualità nuziale in abitato. Una proposta di ricontestualizzazione iconografica per i prodotti della bottega del Pittore di Himera", in: *Orizzonti. Rassegna di Archeologia* XVII, 35–52.

Serino (2017): Marco Serino, "Le prime botteghe di ceramica a figure rosse in Sicilia. Stato dell'arte e nuove prospettive di ricerca", in: Elvia Giudice and Giada Giudice (eds.), *Studi Miscellanei di Ceramografia Greca* III, Catania, 121–175.

Serino (2019): Marco Serino, *La Bottega del Pittore di Himera e le altre Officine Protosiceliote. Stile, Iconografia, Contesti, Cronologia*, Fecit te, Rome.

Serino (2022): Marco Serino, "Some New Perspectives on the Study of Craftspeople's Mobility in the Red-figure Pottery Production of Magna Graecia and Sicily", in: Chiara Maria Mauro, Diego Chapinal-Heras and Miriam Valdés Guía (eds.), *People on the Move Across the Greek World*, Sevilla, 91–107.

Serino (forthcoming a): Marco Serino, "Forma e immagine nelle officine protosiceliote. Una rilettura degli "inizi" delle prime produzioni a figure rosse attraverso possibili indizi di mobilità artigianale tra Campania e Sicilia", in: Monica Salvadori, Monica Baggio and Luca Zamparo (eds.), *Forma e Immagine* (Atti del Convegno internazionale, Castelfranco Veneto, 4–6 dicembre 2019), Padua, forthcoming.

Serino (forthcoming b): Marco Serino, "Per un'archeologia del gesto artigianale. Aspetti tecnologici e network analysis. Organizzazione interna, flussi di lavoro e trasferimento di *know-how* nelle prime botteghe di ceramica a figure rosse tra Sicilia, Paestum e Campania", in: Eukene Bilbao Zubiri and Alexandra Attia (eds.), *L'artisanat de l'argile en Italie Méridionale (VIIIe – IIIe s. av. J.-C.)* (Colloque International du Centre Jean Bérard – École française de Rome – Scuola Superiore Meridionale, Naples, 23–24 marzo 2023), forthcoming.

Silvestrelli (2018): Francesca Silvestrelli, "Il repertorio morfologico delle ceramiche fini di Metaponto. Caratteristiche e trasformazioni nel V e nel IV secolo a.C.", in: Claude Pouzadoux, Martine Denoyelle and Francesca Silvestrelli (eds.), *Mobilità dei pittori e identità delle produzioni*, Cahier du Centre Jean Bérard 25, Naples, 135–160.

Silvestrelli (2019): Francesca Silvestrelli, "La *polis* di Metaponto e il suo territorio dalle prime esperienze alla metà del IV sec. a.C.", in: *Produzioni e committenze in Magna Grecia* (Atti del LV Convegno di Studi sulla Magna Grecia, Taranto, 24–27 settembre 2015), Taranto, 583–601.

Soleti (2012): Vita Maria Soleti, "Storia degli studi", in: Todisco (2012) 37–71.

Spigo (1996): Umberto Spigo, "Il problema degli influssi della pittura vascolare attica nella ceramica a figure rosse siceliota", in: *I Vasi attici ed altre ceramiche coeve in Sicilia*, II (Atti Convegno Internazionale Catania, Camarina, Gela, Vittoria, 1990), *Cronache di Archeologia e di storia dell'arte* 30, Catania, 51–65.

Spigo (2001): Umberto Spigo, "Brevi considerazioni sui caratteri figurativi delle officine di ceramica siceliota della prima metà del IV sec. a.C. ed alcuni nuovi dati", in: *La Sicilia dei due Dionisî* (Atti della settimana di studio, Agrigento, 24–29 febbraio 1999), Rome, 265–292.

Todisco (2012): Luigi Todisco (ed.), *La Ceramica a Figure Rosse della Magna Grecia e della Sicilia*, Vols. I–III. Rome.

Trendall (1960): Arthur Dale Trendall, "The Cassandra Painter and his Circle", in: *Jahrbuch der Berliner Museen* 2, 7–33.

Trendall (1989): Arthur Dale Trendall, *Red Figure Vase of South Italy and Sicily. A Handbook*, London.

Trendall (1992): Arthur Dale Trendall, "New Vases by the Chequer Painter", in: Heide Froning, Tonio Hölscher and Harald Mielsch (eds.), *Kotinos. Festschrift für Erika Simon*, Mainz am Rhein, 301–305.

Wenger (1996): Etienne Wenger, "How We Learn. Communities of Practice: The Social Fabric of a Learning Organization", in: *The Healthcare Forum Journal* 39, 20–26.

Wissinger/Kahn (2008): John Wissinger and Lisa Kahn, "Re-creating and Firing a Greek Kiln", in: Kenneth D.S. Lapatin (ed.), *Papers on Special Techniques in Athenian Vases*, Los Angeles, 129–138.

Elisa Chiara Portale and Gabriella Chirco

21 The Craft of Centuripe Vases: Techniques, Colors, Patterns

Abstract: Polychrome Centuripe ware is a specialty of the Hellenistic workshops of the ancient city of Kentoripai, in inland Sicily. It has caught scholarly attention since the beginning of the last century because of its formal and iconographic features, but it was not until the early 1970s that Paul Deussen, in his PhD dissertation, ventured to investigate workshop organization on the basis of a systematic reappraisal of morphological, decorative, and stylistic features. He recognized four different workshops with their own painters, whom he named after favorite subjects or decorations (Tendrill Altar Master, Gorgon Portrait Master, Medusa D-Veil Painter, Dionysos Painter, Dolphin Coroplast, Imitation Coroplast). While highlighting some objective characteristics, this general picture was nevertheless weakened by some interpretative fallacies and the lack of in-depth study of the technical features. Moreover, the state of conservation of the paintings (which were often barely legible or altered by modern restorations) further complicates their study. An interdisciplinary research project, on selected Centuripe vases kept in the archaeological museums of Syracuse and Palermo, sheds new light on technological and chromatic aspects of these vases through morphological and iconographic study closely coupled with archaeometric and computational imaging analyses. On the basis of this interdisciplinary approach, some groupings have been reconsidered in order to penetrate deeper into the production logic.

Introduction

Since the earliest discoveries, the Centuripe vases have posed exegetical problems that are difficult to resolve because of the perishable nature of the decorations, the lack of scientifically investigated contexts, and the nature of the production, which was evidently a niche phenomenon linked to a specific clientele, given the small output (around 130 known vases and 30 clipei) and the tiny number of tombs with these objects in Centuripe alone.[1] Another problem, and sometimes an obstacle to correct evaluation, has been the intervention of clandestine excavators and restorers who have also polluted the evidence by putting on the market partial or total forgeries of the artifacts.[2] In recent decades, the range of interdisciplinary research has widened, combining historical-artistic and technical studies with the analysis of materials and their application methods, thanks to the emergence of noninvasive, nondestructive, and portable analytical techniques and documentation and imaging programs that allow for closer examination.[3] Specifically, our research, which is still in progress, has primarily focused on identifying discriminators between originals and fakes, and on reconstructing the original appearance of the artifacts.[4] At the same time, however, the in-depth study of techniques, colors, morphological, iconographic, and stylistic aspects allows us to delve deeper into the question of production and to test the criteria proposed so far for the chronological classification of artifacts and for attribution to different workshops. As part of a general review of the documentation, provenance data, and groupings that can be made on a contextual, formal, and stylistic basis, a series of targeted analyses were carried out on some representative artifacts held in the two main archaeological museums in Sicily (respectively at Palermo and Syracuse), with a view to comparing these data with those acquired from artifacts stored both in Sicilian museums and in foreign collections, for a better understanding of the Centuripe vases.

Centuripe Workshops-Painters-Techniques

This paper aims to present an overall perspective rather than the still-provisional results of the research. It focuses on four high-quality vases that came to the Antonino Salinas Museum in Palermo in the first two decades of the 1900s (for which

1 For a first comprehensive survey: Libertini (1926). See recently: Portale (2011), with previous bibliography; Baas (2016); Biondi/Romano (2018); Portale (2021); Hodgson (2021).
2 Jost (1997); Barbanera (2003); Biondi (2014); Portale (2022).
3 Portale *et al.* (2017); Biondi/Romano (2018).
4 Portale *et al.* (2019); Chirco *et al.* (2021).

https://doi.org/10.1515/9783111189635-021

a comprehensive investigation was carried out),[5] and on fifteen pieces from the largest collection of Centuripe vases preserved in the Paolo Orsi Museum in Syracuse (still under examination). Only a few of the artifacts in the Syracuse museum come from controlled excavations; most of them were acquired in the second quarter of the 20th century, close to the first wave of imitations and/or falsifications but before the further plundering of the necropolis of Centuripe, which occurred up until the early 1990s.[6] As a consequence of the looting, some valuable artifacts were placed on the European and American antiquities markets, completely decontextualized (and sometimes problematic in authenticity) and were included in the systematic corpora of Paul Deussen[7] and Ulrike Wintermeyer[8] at the beginning of the 1970s.

The selected sample includes the main shapes – pyxides, lekanides, lebetes gamikoi, clipei and bottles – and specimens both with black backgrounds (a tiny minority) and with pink backgrounds: Tab. 1 lists the vessels we will discuss in this article.

Tab. 1: Centuripe Vases discussed in this study. (Abbreviations: Boston: Museum of Fine Arts, Boston; CT: University Archaeological Museum, Catania; NY: Metropolitan Museum of Art, New York City; PA: Antonio Salinas Museum, Palermo; SR: Paolo Orsi Museum, Syracuse).

Vases Concordance of catalogue numbers-Wintermeyer (W.) and Deussen (D.)			**Classification by P. Deussen**				**Background colour**
Inv. No.	**W. No.**	**D. No.**	**Workshop**	**Type**	**Shape**	**Painter**	
PA 2231	K21	57	1	lekanis	4	Medusa D-Veil Master	pink
SR 32408	K12	49	1	lekanis	4	Tendrill Altar Master	black
PA 2236	K20	100	1	lekanis	5	Tendrill Altar Master	pink
PA 2237	K82	71	1	lebes	4	Tendrill Altar Master	pink
SR 41943	K35	10	2	pyxis	2	Gorgon Portrait Master	black
SR 49051 bis	K44	25	2	pyxis	3	Gorgon Portrait Master	pink
49052	K49	77	2	pyxis	4	Gorgon Portrait Master	pink
SR no inv. no.	K5	86	2	clipeus	1	Gorgon Portrait Master	–
SR 43170	K92	83	2	clipeus	1	Gorgon Portrait Master	–
SR 35946	K56	29	2	pyxis	3	Not identified	pink
SR 32408	K51	35	3	pyxis	3	Dionysos Painter	pink
PA 2235	K57	81	4	pyxis	4	Dolphin Coroplast	pink
NY 30.11.4	K11	50	1	lekanis	4	Not identified	pink
Boston 1970.479	K34	10A	2	pyxis	2	Gorgon Portrait Master	black
NY 29.131.2	K50	34	4	pyxis	3	Dionysos Painter	pink
CT 05/191	K52	27	2	pyxis	3	Gorgon Portrait Master	pink
NY 53.11.5	K76	72	1	lebes	4	Medusa D-Veil Master	pink
CT private collection	K77	73	1	lebes	4	Medusa D-Veil Master	pink
NY 24.97.120	K90	69	1	finial lebes	–	Medusa D-Veil Master	pink
SR 40773	K93	163	2	clipeus	1	–	–

The first investigations of the Palermo vases, carried out using FT-IR (Fourier Transform Infrared Spectroscopy), revealed a technical fact relevant to distinguishing genuine work from forgeries. As a matter of fact, the preparation layer of the paintings, white in color, consists of the monohydrated form of calcium sulphate, bassanite, which is formed by firing gypsum at less than 130°. Therefore, a gypsum layer was spread on the already-fired vase, and, with further firing at low temperature, stabilized as bassanite to serve as a base for the tempera painting. As a result, the actual presence of gypsum (which has not undergone the second firing at low temperature), in only a few places on the vase, identifies modern retouches, which are further proven by stylistic or pigment examination.[9]

5 For a first assessment, see Chirco *et al.* (2021).

6 See Portale (2022).

7 Deussen (1970). Despite its importance, this comprehensive study (PhD thesis, published 1988) has been far less influential than the short paper by the same author on the "nuptial" interpretation of the vases: Deussen (1973).

8 Wintermeyer (1975).

9 Portale *et al.* (2017); Portale *et al.* (2019); Chirco *et al.* (2021).

Additionally, the originality of the vases is also confirmed by the presence of Egyptian blue and cinnabar identified by the VIL (Visible-induced Luminescence) and XRF (X-ray Fluorescence Spectroscopy). The first pigment, in fact, was not used after the Early Byzantine period; its presence allowed Giacomo Biondi and Francesco Romano to confirm the authenticity of a Centuripe pyxis that had been considered a forgery, seized by the police and deposited at the Manifattura Tabacchi in Catania.[10] As for cinnabar, although it was used until modern times, it was generally replaced with other red pigments because of its high cost, and banned because of its toxicity. The value of the original artifacts is confirmed by the gold leaf used on the plastic decorations, applied with the bole technique; in retouched areas, modern restorers/forgers have imitated this with much-cheaper yellow ochre.[11] The other pigments present, derived from iron oxides, are the more common ochres or earths. Particularly for the pink ground vases UV (Ultraviolet) and FT-IR revealed the use of an organic dye, madder or alizarin, which was popular in the Hellenistic age. Furthermore, FT-IR spectroscopy identified the binder used to apply the pigments: it is an egg protein (tempera technique).[12]

In addition to questions of authentication and the identification of pigments and binders, the investigation of materials and the detailed examination of painting technique, forms, and decorations can contribute to an understanding of production aspects and workshops: left undefined by Wintermeyer and other researchers,[13] this subject was instead explored in depth by Deussen, though on a basis that was not always solid.[14] The German-American researcher divided the corpus of artifacts into four main groups, four different workshops with six painters-coroplasts[15] that in his opinion would have operated over a few decades (early-mid-3rd century B.C.) but with a clear evolution (Fig. 1). Leaving aside questions of the chronology, framing, and exegesis of the Centuripe artifacts, which have already been dealt with elsewhere,[16] we will attempt to shed more light on production through the artifacts we examined, which are mainly representative of the hypothetical Workshops 1 and 2 and the painters named Tendrill Altar Master and Gorgon Portrait Master according to Deussen's classification. The scientific study of the paintings and especially the results from imaging techniques made it possible to identify shared similarities and distinct differences. Although the results from our analysis support the presence of various more or less specialized painters, it is difficult to categorize and assign them to different workshops according to Deussen's classification. Instead, it is more plausible to assume that the painters worked within a single workshop and cooperated/succeeded each other over the course of approximately three generations.

The main differences in technique concern the sequence of layers and the texture of the painting. Within the group attributed to Workshop 1, a development in the morphology of the vases was observed by Deussen, particularly in the lekanides, which are one of its main products, classified as shapes 3, 4, and 5, with increasingly articulated plastic decorations. However, it is probably more appropriate to call them variants, not so far apart in time, according to the chronology of the few known contexts of discovery.[17] Contextual data do not allow us to separate the so-called Workshops 1 and 3 because of the probable association of the eponymous pyxis of the Dionysos Painter **K51**,[18] identified by Deussen as the main artist of Workshop 3, with the lekanis **K12** in our sample (see below), which was attributed instead by him to Workshop 1, and to a relatively early phase, because of its black background and its simpler decoration, marked by racemes and rosettes with traces of blue.

As for the other lekanides we examined,[19] **K21** of the same shape (4), pink-ground, is distinguished by a more elaborate plant motif where two types of rosettes and small erotes are inserted between the racemes; while the lekanis **K20**, less

---

10 Biondi/Romano (2018).

11 Portale *et al.* (2017); Chirco *et al.* (2021).

12 Chirco *et al.* (2021).

13 Wintermeyer (1975) 210, 212, 220–221, 225. Lastly, Hodgson (2021) 215 merely reports the few stylistic groupings made by this scholar, with some dubious additions.

14 Deussen (1970).

15 Deussen (1970) 175–195 for the painters.

16 Portale (2011), (2014), (2021).

17 Hereafter we will identify each vase by the entry in Wintermeyer's corpus: Wintermeyer (1975). The tomb (25) that yielded the "oldest" *lekanis* K1 (shape 3 early developed, mold A Io-Isis) dates at least to the last quarter of the 3rd century, even after the contexts in which the shape (or rather the plastic decoration) which was considered more "evolved" appeared (tomb 31, shape 4 fully developed: mold D, assigned to the "Medusa-D Veil Master"), and not far from funerary contexts including the latter (tomb 33, final of shape 4, mold C Aphrodite): see Portale (2011), (2021).

18 Deussen (1970) 297, no. 35; Wintermeyer (1975) 225–226, no. K51, figs. 53–54; Portale (2011) 170; Biondi/Romano (2018) 113, fig. 15; Portale (2021) 111–112, figs. 25–26.

19 For the first analyses, see Chirco *et al.* (2019).

Fig. 1: Groupings of Centuripine vases according to Deussen (1970).

Fig. 2: Evolution of the plastic decoration of racemes and Medusa heads, on lekanides (from left: K12, K21, K20).

Fig. 3: Left: detail of the lekanis' lid K12 with the paint superimposed directly on the ceramic ware. Right: detail of the lekanis' lid K20 with the paint superimposed on the preparation layer.

contoured in profile and more developed in height (shape 5), also enriches the peopled scrolls with small swan figurines, in addition to rosettes, erotes, and long whorls, giving more liveliness to the racemes (Fig. 2).

It should be noted that the lekanis **K21** is distinguished by Deussen and attributed to the Medusa D-Veil Painter on the basis of the mold of the Medusa head placed at the center of the vegetal frieze; this painter owes his name, alongside the Medusa mold ("D"), to a veiled figure recurring in the painted scenes of ritual processions on the typical lebetes gamikoi attributed to him (we will return later to these paintings).[20]

In contrast, the above-mentioned lekanides **K12** and **K20** would be assigned to another craftsman, reputed to be his master and the leader of Workshop 1, the Tendrill Altar Master, named for the rendering of the vegetal girals and the predilection for ritual scenes painted on the lids of lekanides, representing the bulk of his production.[21] If such ritual subject matter—but with different details—does indeed unite our lekanides **K12** and **K20** the same cannot be said for the mold of Medusa's head, originally painted white (probably functioning as a basis of other, now-vanished, colors) with gilded hair. As for secondary decorations, lekanis **K12** shows a painted-leaf ornament on the basin, while the others bear a corolla of relief acanthus leaves applied to the base and two painted bands on the body, black and white, which in the case of lekanis **K20** show a running-dog motif. Lekanides **K20** and **K21** also have a *"denti di lupo"* pattern (visible only under UV light) painted on the pedestal on a white background, which is also found on the foot of the lebes gamikos **K82**,[22] also attributed to the Tendrill Altar Master. However, for the latter Deussen considers pertinent the finial in the form of a miniature lebes

20 For this iconography, see Trendall (1955); Wintermeyer (1975) 148–150, 173–174; Green (1990); Portale (2014) 292–294.
21 Most recently, Portale (forthcoming).
22 Deussen (1970) 299, no. 81; Wintermeyer (1975) 235, no. K82, fig. 64. See Portale *et al.* (2017) 82–97, pls. 20–23; Chirco *et al.* (2019).

with a painted female bust and a similar secondary decoration, possibly found together and now in the Metropolitan Museum,[23] attributed by him to the pupil of the Tendrill Altar Master, the Medusa D-Veil Painter.

Indeed, analytical study of the paintings of the three vessels in our sample assigned to the Tendrill Altar Master (**K12; K20; K82**) reveals common elements but also differences between the first one (with a black background) and the others. On all artifacts we can note the overlapping of the paint on the background, either black or pink, but in the first case (**K12**) the black color is overpainted directly on the ceramic body, while in the other two (**K20; K82**) a white preparation layer is interposed between the ceramic body and the pink background on the decorated part of the vessel (the front half; Fig. 3). On the lid of **K12** the paint is fluid and thin, as is evident from macrophotographs and UV observation. Consistent with other black-ground vases, there is no preparation layer, and the black background layer (carbon black), also executed in tempera, is in direct contact with the pottery surface. The figures are painted onto the black ground, and the color fields for details or larger areas (e.g., *himatia*) are in turn superimposed on the main paint layers. A technical difference can be seen in the use of pink pigment, whereby the painter exploits the transparency of alizarin to recreate a different color rendering on the *himation* of the woman on the left of the central ovoid object, on the "plume" and the band of the same object, and on the "Doric" frame along the lower edge of the lid. Alizarin is superimposed on a white background that acts as a base and enlivens the tone, while for the pink hues on the female figure on the right, the dye is applied directly on the black background, taking on a less vivid tone. The simultaneous use of the two techniques with alizarin denotes the artist's knowledge of materials and his painterly skills.

On lekanis **K20** and lebes gamikos **K82**, both pink-ground, XRF spectroscopy and the application of the SmArt_Scan program reveal a homogenous preparation layer made of calcium sulphate (bassanite) onto which the pink alizarin background was applied. Macrophotographs and UV-light recordings make it possible to observe the stratigraphic succession, showing that here (unlike the black-ground vase) the silhouette for figures was reserved in painting the pink background, and the figures were painted directly onto the white preparation. The painted details (relating to cloaks, hair, etc.) are in any case added after the application of the pink color on the ground, as some overlapping in the marginal parts of the figures reveals. There are no traces of preparatory drawing, but the sequence of execution suggests the existence of a preliminary plan and shows the painter's skill. The details, such as hair, umbrella, drapery, or other objects, are added as a final step, totally or partly overlapping the backgrounds below. On the pink-ground vases, the pictorial palette is broader: iron-oxide pigments (ochres or earths); the alizarin dye; Egyptian blue and cinnabar, found only on lebes gamikos **K82**. In most cases, the pigments are applied pure, but there are also areas where the painter uses mixtures or overlays. For example, cinnabar is mixed with red ochre, probably for a more plastic effect. Egyptian blue, also on the lebes gamikos, is used as a glaze over the pink of the shoulder of the first figure, to create a purplish-mauve tone. On the lid of the black-ground lekanis **K12**, several blue areas with a vivid tone are clearly visible, while on lekanis **K20** the Egyptian blue is highlighted by VIL both on white areas and on areas barely discernible to the naked eye (Fig. 4). Egyptian blue is also used in a mixture with white, as in the case with pyxis **K57** attributed instead to the Dolphin Coroplast,[24] and thus to Workshop 4. On lekanis **K20**, in fact, the VIL shows fluorescence on two adjacent white areas, those of the chiton and the *himation* of the female figure standing behind the mound like element around which the ritual action revolves:[25] the chiton, most probably, originally looked bright white (as a result of mixing white with Egyptian blue), and the *himation* blue. This diversified use of pigment, to indicate different color tones on two adjacent surfaces, shows the competence and pictorial skill of the craftsman (Fig. 4, bottom left).

As these preliminary results indicate, verifying the relationship of vases attributed to the same master or workshop, or comparing them with artifacts attributed to a different workshop, reveals to be useful. For the first point, one can consider the famous pink-ground lekanis **K11** (shape 4) in the Metropolitan Museum in New York, attributed by Deussen to the same painter (Tendrilll Altar Master) and the same Workshop (1) as the group above, which was analyzed by Richter in the 1930s.[26] The plastic decoration of the New York vase is of the more elaborate variant, similar to that of our lekanis **K20**, although the outline of the body is clearly profiled and less evolved. Here, too, the silhouette of the figures is reserved on the pink background, and the details overlap the main colored fields; however, unlike ours, in the New York vase **K11**

23 Libertini (1926) 166, n. 34, pl. LX; Deussen (1970) 286, no. 69, pl. XXIX; Wintermeyer (1975) 237, no. K90: lastly, Portale (2021) 112, fig. 28.

24 Wintermeyer (1975) 228, no. K57, fig. 26. See previously Portale *et al.* (2017); Chirco *et al.* (2019).

25 Portale (forthcoming); Chirco (forthcoming).

26 Richter (1932); Deussen (1970) 277–278, no. 50; Wintermeyer 1975, 211–212, no. K11, fig. 5; Portale (forthcoming).

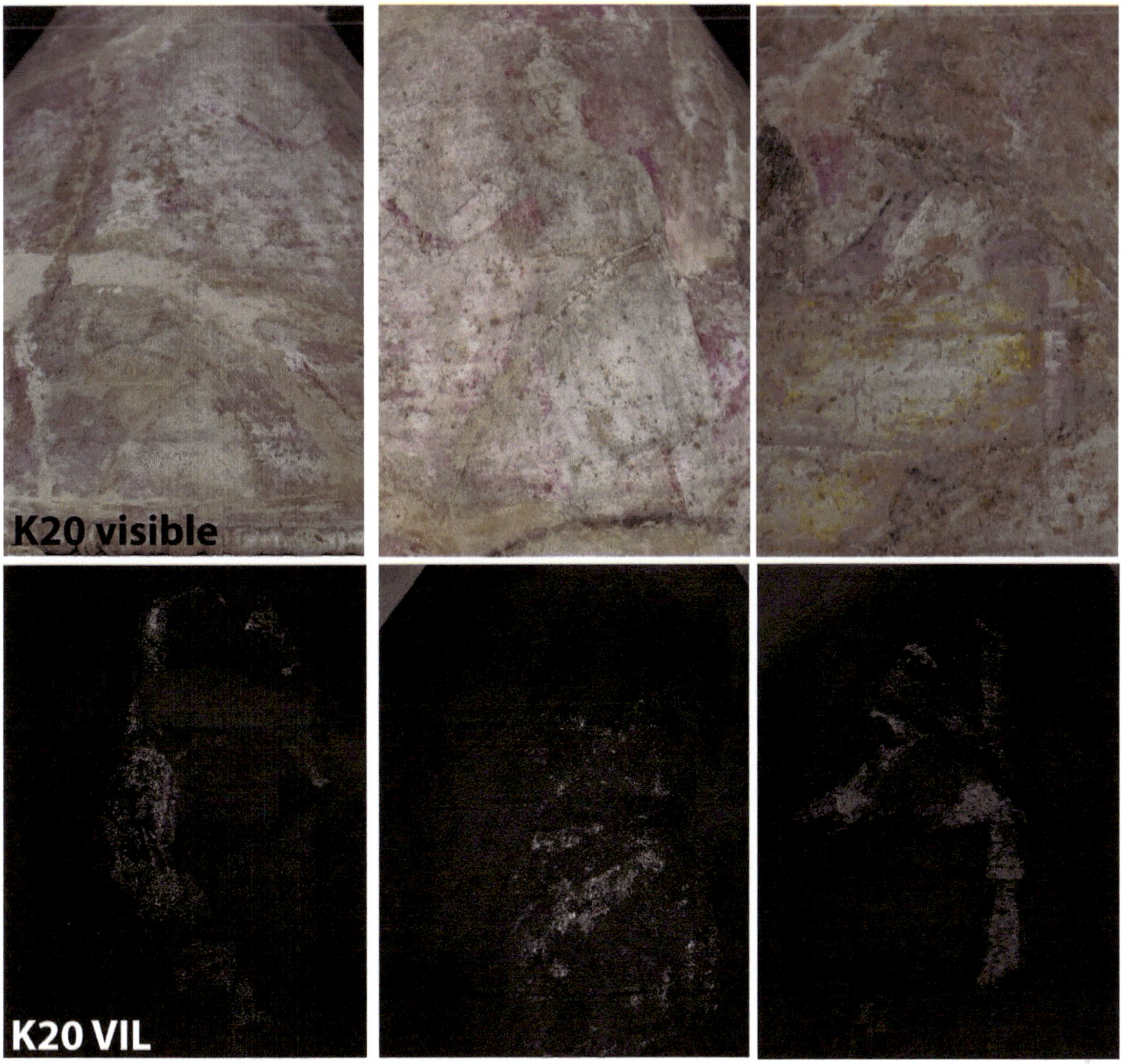

Fig. 4: Details of areas affected with Egyptian blue on the cover of the lekanis K20. Top: Under visible light. Bottom: With VIL.

the pink background is applied directly onto the ceramic body, while a white layer under the colored fields is found only in association with the figures.

This different process (white preparatory layer only under the figures) is rather reminiscent of what was found in a pink-ground pyxis, **K52**,[27] which Deussen assigned to the Gorgon Portrait Master of Workshop 2, as well as in the new pyxis in Catania Museum,[28] which can also be linked to the group with an abduction scene painted on the body, attributed to the same painter. In contrast, the black-ground pyxis **K35**,[29] also attributed to the Gorgon Portrait Master by Deussen, differs from that painter's modus operandi in that the figures are superimposed *on* the black ground, making attribution uncertain.

As far as the color palette is concerned, the New York lekanis **K11** is the only one – apart from the two monumental Palermo vases examined by us (**K82**; **K57**)[30] – on which cinnabar has been identified, although it is not specified where and how: the only bright red element visible in the painted scene is the flame/object on the altar representing the center of the ritual performed by the protagonists. Cinnabar, combined with red ochre, is used on lebes gamikos **K82**, to give a

---

27 Deussen (1970) 288, no. 27, pl. XXV; Wintermeyer (1975) 226, no. K52; Pappalardo *et al.* (2014).

28 Biondi/Romano (2018).

29 Deussen (1970) 287, no. 10; Wintermeyer (1975) 219, no. K35.

30 Portale *et al.* (2017); Chirco *et al.* (2019). See previously: Libertini (1926) 162–163, nos. 30–31, pls. XLV,6; LIII; Deussen (1970) 299, 279, nos. 81, 71; Wintermeyer (1975) 228, 235, nos. K57, K82, figs. 26, 64.

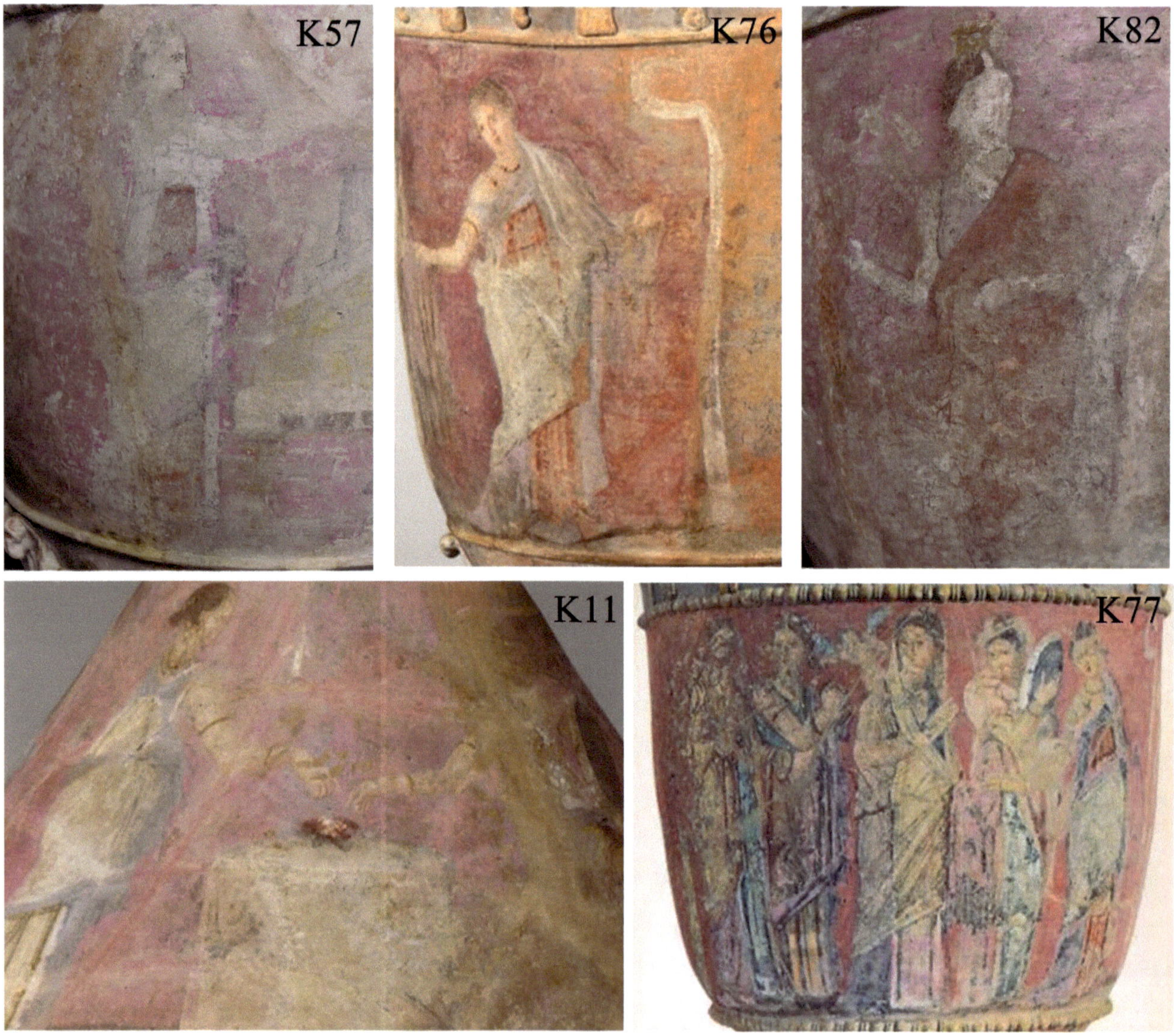

Fig. 5: Details of the Centuripine vases where cinnabar red is used for the stoles, the *himation* and the fire on the altar. Top row, from left: pyxis K57, lebes K76, lebes K82.; Lower row, from left: lekanis K11 and lebes K77 (unknown pigments).

bright red tone to the cloak of the first female figure, characterized by a polos and an arrangement of drapery recurring for Kore-Persephone and her priestesses: the highly selective use of cinnabar could support the idea that this pigment conferred a particular symbolic value referring to the sacred sphere, like a Dionysian-chthonian overtone.[31] On the other hand, in comparison with the monumental pyxis **K57**, which Deussen assumed belonged to the same funerary context,[32] but which he attributed to a different workshop (4) and craftsman (the Dolphin Coroplast), there are no significant similarities in the lid's plastic decoration: this could validate Deussen's attribution to a different workshop. However, in terms of painting, lekanis **K20** and lebes gamikos **K82**, attributed to the Workshop 1, and pyxis **K57**, attributed to the Workshop 4, share the use of thin, fluid color and its method of application. The pigments are similar and applied in both pure form or mixtures or as glazes; even cinnabar, as mentioned, is used in pyxis **K57**, for the vertical band distinguishing the dress of the two women on either side of the main figure (Fig. 5). It is likely that it represents a ritual garment, as confirmed by female figures wearing this dress in the typical procession scenes on lebetes gamikoi, in particular the splendid examples

31 Portale *et al.* (2017) 94, 96; Portale (2021) 112–114.
32 Deussen (1970) 261–262, 279, no. 81.

K76 and **K77** (belonging to a private collection),[33] although the pigments of these two vases have not been yet analyzed. **K76** and **K77** are assigned by Deussen to the Medusa D-Veil Master, named, as said, for the veiled main figure of these choral scenes.[34] The painter, already mentioned in association with lekanis **K20** in Palermo, is considered to be a pupil of the Tendrill Altar Master in Workshop 1.

The monumental pyxis **K57** is attributed by Deussen to the Dolphin Coroplast of Workshop 4 because of the relief frieze with sea animals on the lid, marked by ichthyomorphic and anguiform figures,[35] and dolphins (recently identified through UV images) painted on either side of a seahorse on the lid itself. It should be noted, however, that similarly painted figures appear on vases assigned to Workshop 2, and that similarly-rendered relief figures of marine creatures decorate two pyxides of the same shape that appeared on the antiquities market in the 1980s. The first one is in Boston (Museum of Fine Arts, inv. no. 1983.558) and shows on the body a scene of "homage to the bride" very similar to that of Palermo pyxis **K57**, but in association with a female portrait on its lid;[36] the same combination of painted "portrait" and marine frieze in relief on the lid appears on the second pyxis, now in Kyoto (Kitazonocho Greek and Roman Museum).[37] This latter represents on the body the abduction scene recurrent on the pyxides that Deussen assigns to the Gorgon Portrait Master of Workshop 2, but in a stylistically inferior version. The portrait that gives the painter in question his conventional name is the female bust painted on pyxis lids in connection with a main scene of abduction or homage to the bride on the body of the vase; the same subject is represented in isolation on the painted shields/plates (clipei).[38]

To Workshop 2, and in particular to the Gorgon Portrait Master, Deussen assigns the pyxides **K35**, **K44** and **K49** – the last two marked by the above-mentioned portrait on the lid – and the clipei **K5** and **K92**.[39] As far as the plastic decoration is concerned, there are no special similarities between these three pyxides except for some general traits such as the shape of the handles (tripartite and with two relief acanthus leaves applied to the attachment) or the foot (low and relatively simple, flared at the bottom). Significant similarities in the pictorial technique were, instead, found in the pink-ground pyxides **K44** (showing a "homage to the bride") (Fig. 6) and **K49** (with a scene of rape), where the pink ground color, superimposed on a white preparatory layer, covers the entire decorated surface (except the back, that is always unadorned), and the contours of the figures and objects 'overlay' this pink background. This contrasts with the procedure typically used for figures which are spared from background, found in the large vases in Palermo referring to both Workshop 1 (lekanides **K20** and **K21**; lebes gamikos **K82**) and Workshop 4 (pyxis **K57**).

In contrast, the black-ground pyxis **K35**, also assigned to the Gorgon Portrait Master and to Workshop 2, has, as mentioned, a white layer overpainted on the black ground only under the female figures (Fig. 6). Finally, in the pyxis **K52**, attributed to the same painter and workshop and with a similar iconography, the pink background of the figural scene was applied directly on the ceramic body, but sparing the figures that overlay a first white background. One can observe the same procedure in this pink-ground lekanis in New York (**K11**), which has been attributed, however, to a different painter (Tendrill Altar Master) and a different Workshop (Workshop 1).

The painting on the three above-mentioned vessels attributed to the Gorgon Portrait Master (**K35**; **K44**; **K49**) is full-bodied and thick but the brushstrokes are not very fluid and consistent. In contrast to the Tendrill Altar Master, the Gorgon Portrait Master does not reserve the outlines from the background but superimposes the colored fields on it, regardless of the color and transparency of the pigment. Moreover, the colored fields are superimposed on each other, as in the *himation* of the bride on pyxis **K44** or the arms of the female character on pyxis **K49**. Although the diagnostic study of these vases is still incomplete, the use of color mixtures or glazes as employed by the Tendrill Altar Master has not been yet observed. The paint palette includes the traditional earth colors and Egyptian blue. The craftsman also uses red on larger fields, as on the lid of pyxis **K35** (red ochre) and for the *himation* of the female figure on the left on the body of pyxis **K44** (pigment

33 Trendall (1955); Deussen (1970) 282, nos. 72–73; Wintermeyer (1975) 232–234, nos. K76 and K77.

34 Deussen (1970) 139; Wintermeyer (1975) 148–149, 173; lastly, Portale (2021) 82, fig. 4, with other references.

35 Deussen (1970) 116–117.

36 Padgett (1993) 222–223, no. 152; Portale (2011) 170–171, 174, 179; Baas (2016) 90, 92, fig. 2; the Boston vase is paralleled with **K57** in Portale *et al.* (2017) 83–84, fig. 10.

37 Simon (1989): at that time the vase was in temporary display the University of Wurzburg. Portale (2011) 184, pl. XIb; Portale (2014) 292–294, fig. 10; Baas (2016) 92, 100; Portale (2022) 222–223, fig. 6.

38 Portale (2011) 173, 175–177, figs. 65a, 66b, 67b; Portale (2014) 292, 295, 297–298; for the debate about the authenticity of the "portrait shields", see Barbanera (2003); Biondi (2014), with earlier references; lastly, with new data, Portale (2022) 267–268, fig. 2.

39 Deussen (1970) 287, no. 10, 181–182, no. 25A, 290, no. 77; Wintermeyer (1975) 219, 224–225, 227, nos. K35, K44, K49. For the plates/shields, Wintermeyer (1975) 210, 237, nos. K5, K92; Deussen (1970) 291, no. 83.

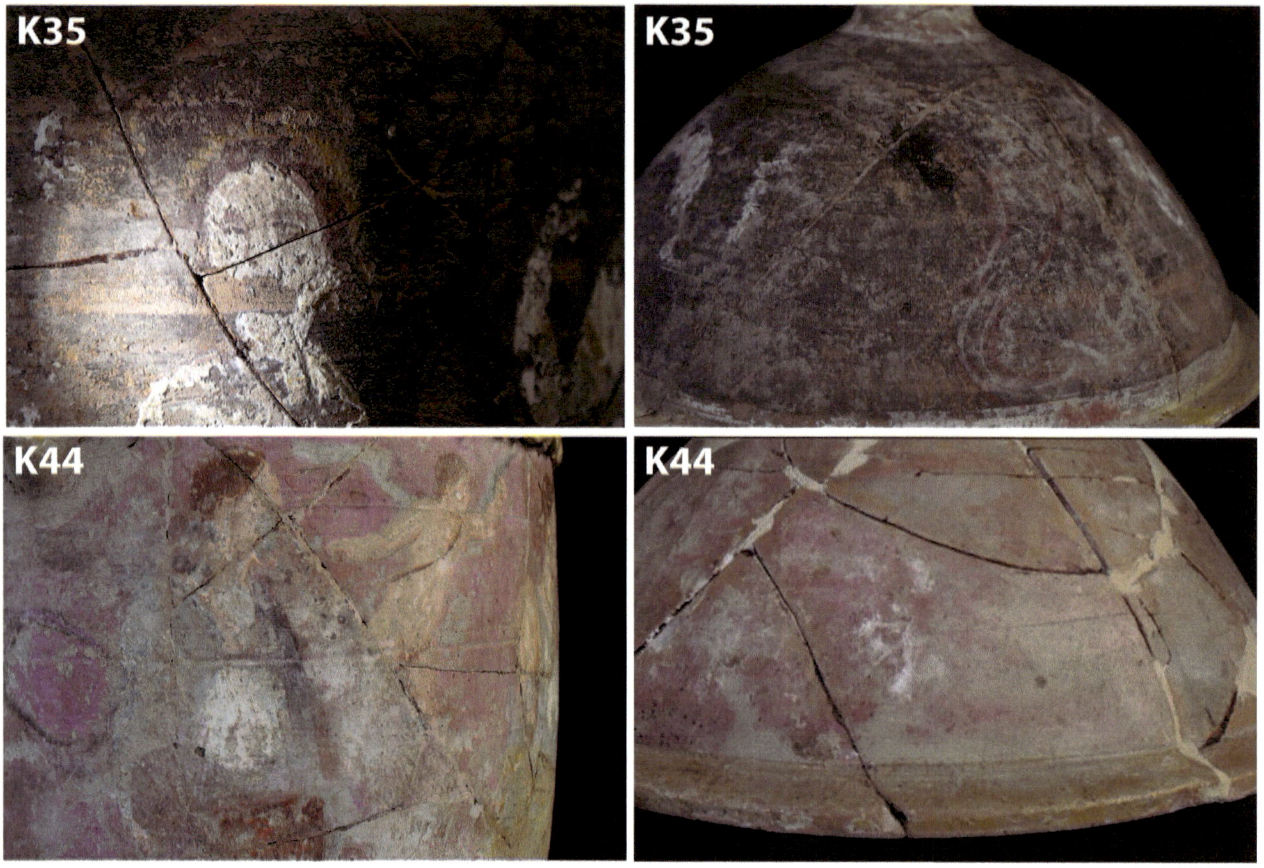

Fig. 6: Details of the pyxides K35 and K44 attributed to the Gorgon Portrait Master. K35: interposition of a white preparation layer between the black background and the color layer; K44: superimposition of color on the pink background.

not examined). The painter is especially skilled for the pictorial rendering of faces—and particularly of the 'portraits' that characterize (Fig. 1) some of the painted lids and clipei attributed to him. Dashes, shading, and highlighting add plastic prominence to the faces, which are generally arranged in a three-quarter view. Clipei **K5** (found in a regular excavation) and **K92** also reveal the same technical characteristics, despite their poor state of preservation.[40]

The last piece of our selection attributed by Deussen to Workshop 2 is pyxis **K56** (Fig. 1).[41] It differs in technique and design from all previously mentioned vases as the pink background is directly upon the ceramic body and the figures are painted on a white preparatory layer, although the painting is full-bodied and the faces are elegantly defined, as with the pyxides assigned to the Gorgon Portrait Master. The dress of the second figure on the right appears to be made with blue veils on the pink, a trait more similar to the manner of the Tendrill Altar Master.

40 The former shows the superimposition of the surviving fragments of paint film on the background, while in the latter, the thickness of the brushstrokes for both the main strokes and the details can be clearly observed from the grazing light images. The same can be said for the clipeus **K93**, from the same workshop but not attributed to a painter; the central figure is characterized by a robe made with slightly more fluid brushstrokes, while the arms and face show a more full-bodied painting made, as with the pyxis **K35**, on a whitish base. The execution technique of this clipeus, therefore, can be likened to that of the Gorgon Portrait Master even if the full-length iconography adopted remains unparalleled. A comparison of the three Syracusan clipei of our sample with a clipeus in the museum of Centuripe, which is certainly authentic – see Portale (2022) 268, fig. 2 – and with two others in the Archaeological Museum of Naples, which are proven to be unequivocally forgeries – see Biondi (2014)–, highlights substantial technical differences. In the Centuripe tondo, the "portrait" seems to be spared from the background, and the main outline of the face is overlaid with facial details. In the infamous Neapolitan "tondi", on the other hand, the differences are, or rather were, not only from a technical but also from a conservation point of view: the modern paintings were in excellent condition, with the face and hair added on the ceramic body, and the clothes superimposed on the bright pink background; but, following a mold attack, the pigments disappeared completely, ingloriously putting an end to the lively controversy over the claimed authenticity of these "Roman" style portraits; Barbanera (2003).
41 Deussen (1970) 293, no. 29; Wintermeyer (1975) 227–228, no. K56, figs. 57–58.

The analysis presented above is still partial and could be further refined by a more systematic study of a larger sample, following all the links to the remaining evidence. However, our observations can indeed support the identification of a few different painters' hands, each artist with his own approach to color and composition. In general, each craftsman seems to prefer an iconography related to a monumental shape of the (circumscribed) repertoire of the Centuripe class, but this does not imply exclusivity. For example, a monumental vase such the Kyoto pyxis is decorated by a modest painter taking up schemes (both for the abduction scene and the portrait bust) that were developed by a more gifted artist (the Gorgon Portrait Master of Deussen's classification). Actually, even the more refined vases with a similar abduction scene show at least two different technical solutions, depending on whether the figures are depicted on a reserved background (as in the new vase in Catania), or overpainted on the colored ground (as in **K49**).[42] A distinction must also be made for the second standard iconographical scheme employed by the Gorgon Portrait Master, namely the "homage to the bride". This scene at first presents a formula more closely related to the technique employed in the late red-figured vessels with overpainting of the figures on a black background (**K35**; and the similar pyxis **K34**),[43] with an additional white background applied only for the figures. This white background for the figures, whose silhouette is, however, spared on the pink ground, is found in the pyxis in Catania, reminding one of the two different solutions mentioned above.

Moreover, some vases that appeared in pairs on the antiquarian market (belonging most probably to the same funerary context) and show a clear kinship – as noted by Wintermeyer – pose the problem of the link between the two Workshops 1 and 2 and/or the associated painters as identified by Deussen.[44] It may seem difficult to disassociate these objects and to define them in Deussen's terms (as products of different painters and workshops), also because of the additional links created by plastic forms and decorations.

Against the model of a proliferation of workshops, if one were to follow the articulation proposed by Deussen, other contextual associations are to be noticed, that would imply the co-existence of different painters and different workshops in the same set of vases and even in the same tomb: a difficult fact to understand, this, in the case of such particular objects, which were most likely created ad hoc for the burial ceremony.[45]

Conclusions

In view of the intersections and associations, as well as the links between painting technique, style, and schemes, a more plausible key to understanding the production of Centuripe vases, in our opinion, is to recognize a specialized workshop, operating for no more than three generations, from the third quarter of the 3rd century to mid- 2nd century B.C., with a close cooperation of potters, coroplasts and painters. Some of them are more talented, such as (alongside the ones examined above) the Dionysos Painter, who seem to superimpose the paint directly on the background,[46] and belong to the most inventive phase of this factory. Other painters, decidedly less gifted (such as the creator of the Kyoto pyxis), work in the final phase of activity of the workshop, in which a "baroque" increase in ornamentation is no longer matched by a high pictorial quality of the painting. The Dolphin Coroplast (similarly to the Imitation Coroplast, who is not considered here) could be a workshop figure in charge of plastic decoration. Other workshop figures (or perhaps the same artisan as the one in charge of plastic decoration) may have been responsible for other elements of phytomorphic and pseudo-architectural ornamentation, made either with a mold or freehand, allowing for different combinations and sub-aggregations. He was not, however, an autonomous artist, at least not the creator of the paintings of the monumental vessels with plastic

---

42 See above.

43 Deussen (1970) 287, respectively nos. 10a and 10; Wintermeyer (1975) 218–219, nos. K34; Padgett (1993) 222, no. 151, and K35, fig. 10.

44 Portale (2011) 172–173, with references.

45 Contrarily to Deussen's explanation, the vases cannot be intended as true wedding gifts, held in the marital house by Centuripan women and finally buried with them, after a long span of time: Deussen (1970), (1973). Funerary connotations of the plastic and painted decorations, technical features precluding a proper usage as containers, extreme delicacy of paintings and bodies (assembled from several pieces), and find-contexts, all militate against the thesis that such vases were employed before/out their funerary (or funerary-votive) use. Hence, it might be hard to hypothesize that the customers bought the vases for the same funerary set from different workshops. For a comprehensive, contextual explanation of Centuripe vases, see Portale (2014), (2021), (forthcoming).

46 See the two famous Centuripe pyxides depicting Dionysos, respectively in the museum of Syracuse – Deussen (1970) 297, no. 35; Wintermeyer (1975) 225–226, no. K51, figs. 53–53; Biondi/Romano (2018) 113 – and New York – Richter (1930) 196–198, figs. 9–10; Deussen (1970) 282, no. 34; Wintermeyer (1975) 225, no. K50. On these special vases and their contextual associations, see Portale (2021) 111–115.

decorations of marine subjects and similar morphology (like the pyxides in Palermo **K57**, Boston, Kyoto), as these latter exhibit different pictorial traits.

Of course, our investigation needs to continue, and the proposed working hypothesis needs to be tested in light of other considerations, especially the relationship between the coroplastic workshops (which were quite lively in Centuripe, but equally have been affected by the plagues of clandestine excavations and counterfeits)[47] and the ceramic workshops which are also attested on the site, but not yet sufficiently investigated.

In any case, from the kaleidoscope of detailed observations (from the micro level) to the systemic reconsideration (to the macro level), it is certain that Centuripe vases, which Libertini described as "a strange and unexpected phenomenon" of Sicel material culture,[48] will take on an increasingly concrete and reliable appearance.

Bibliography

Baas (2016): Philipp Baas, "Zu den Vasenbildern aus Centuripe", in: *Jahreshefte des Österreichischen Archäologischen Instituts in Wien* 85, 83–106.

Barbanera (2003): Marcello Barbanera, "False impressioni. La polemica sui 'Tondi di Centuripe' tra Giulio Emanuele Rizzo e Carlo Albizzati", in: *Bollettino d'Arte* 125–126, 79–98.

Biondi (2014): Giacomo Biondi, "Autentici falsi. La Collezione Libertini tra pitture ritoccate, falsificazioni e protagonisti non ufficiali dell'archeologia di Centuripe", in: Giacomo Biondi, Graziella Buscemi Felici and Edoardo Tortorici (eds.), *Il Museo di Archeologia dell'Università di Catania. Collezione Libertini*, Acireale/Rome, 51–84.

Biondi/Romano (2018): Giacomo Biondi and Francesco Paolo Romano, "Un nuovo vaso policromo e riflessioni sull'iconografia delle scene sui vasi di Centuripe", in: *BABesch. Annual Paper on Mediterranean Archaeology* 93, 105–129.

Chirco (forthcoming): Gabriella Chirco, "Appendice. Indagini diagnostiche su due *lekanides* centuripine", in: Monica Salvadori, Monica Baggio and Luca Zamparo (eds.), *Forma e Immagine* (Atti del Convegno internazionale, Castelfranco Veneto, 4–6 dicembre 2019), Padua, 187–192.

Chirco *et al.* (2021): Gabriella Chirco, Elisa Chiara Portale, Eugenio Caponetti, Vincenzo Renda and Delia Francesca Chillura Martino, "Investigation on Four Centuripe Vases (late 3rd- 2nd cent. B.C.) by Portable X-ray Fluorescence and Total Reflectance-FTIR", in: *Journal of Cultural Heritage* 48, 326–335.

Deussen (1970): Paul Wolfgang Deussen, *The Polychromatic Ceramics of Centuripe*, PhD Dissertation, Princeton University.

Deussen (1973): Paul Wolfgang Deussen, "The Nuptial Theme of Centuripe Vases", in: *Opuscula Romana* 9, 125–133.

Green (1990): J. Richard Green, "The Iconography of Marriage and Death in Ancient Italy: A Funerary Vase from Centuripe", in: *North Carolina Museum of Art Bulletin* 14, 50–58.

Hodgson (2021): I. Hodgson, *Polychrome hellenistische Mattfarbenkeramik: Funde aus Italien und Griechenland*, Bibliotheca Academica 3, Baden-Baden.

Jost (1997): Mirjam Jost, "Eine Centuripe-Pyxis in der Antikensammlung Bern-Untersuchungen zu einem 'Problemkind'", in: *Hefte des Archäologischen Seminars der Universität Bern* 16, 27–40.

Libertini (1926): Guido Libertini, *Centuripe*, Catania.

Padgett (1993): J. Michael Padgett (ed.), *Vase-Painting in Italy. Red-figure and Related Works in the Museum of Fine Arts*, Boston.

Pappalardo *et al.* (2014): Lighea Pappalardo, Francesco Paolo Romano and Francesca Rizzo, "Caratterizzazione non-distruttiva ed in-situ della composizione di pigmenti su ceramica ellenistica centuripina appartenente alla Collezione Libertini: uso combinato degli spettrometri portatili PIXE-alfa e XRD dell'IBAM/CNR – LNS/INFN di Catania", in: Giacomo Biondi, Graziella Buscemi Felici and Edoardo Tortorici (eds.), *Il Museo di Archeologia dell'Università di Catania. Collezione Libertini*, Acireale–Rome, 223–228.

Portale (2011): Elisa Chiara Portale, "Un 'fenomeno strano e inatteso': riflessioni sulla ceramica di Centuripe", in: Gioachino Francesco La Torre and Mario Torelli (eds.), *Pittura ellenistica in Italia e in Sicilia: linguaggi e tradizioni* (Atti del Convegno di studi, Messina, 24–25 settembre 2009), Rome, 157–182.

Portale (2014): Elisa Chiara Portale, "Musica e danza nell'iconografia funeraria centuripina", in: Angela Bellia (ed.), *Musica, culti e riti nell'Occidente greco*, Telestes 1, Pisa-Rome, 287–303.

Portale (2021): Elisa Chiara Portale, "Dioniso a Centuripe: iconografia 'teatrale' e *imagerie* dionisiaca in contesto funerario", in: *Ocnus* 29, 87–118.

Portale (2022): Elisa Chiara Portale, "Old Controversies and New Data: Assessing Again the Polychrome Vases from Centuripe", in: Monica Salvadori, Elisa Bernard, Luca Zamparo and Monica Baggio (eds.), *Beyond Forgery. Collecting, Authentication and Protection of Cultural Heritage*, Antenor Quaderni 52, Padua, 265–279.

Portale (forthcoming): Elisa Chiara Portale, "Forma e immagine. Il caso delle *lekanides* centuripine", in: Monica Salvadori, Monica Baggio and Luca Zamparo (eds.), *Forma e Immagine* (Atti del Convegno internazionale, Castelfranco Veneto, 4–6 dicembre 2019), Padua, 165–186, forthcoming.

Portale *et al.* (2017): Elisa Chiara Portale, Delia Chillura Martino, Maria Luisa Saladino, Eugenio Caponetti and Gabriella Chirco, "I "vasi di Centuripe": per un approccio integrato", in: Elisa Chiara Portale and Giusj Galioto (eds.), *Scienza e archeologia. Un efficace connubio per la divulgazione della cultura scientifica*, Pisa, 75–99.

47 Biondi (2014). For a parallel reading of funerary terracottas and vases, see Portale (2021).

48 Libertini (1926).

Portale *et al.* (2019): Elisa Chiara Portale, Delia Chillura Martino, Eugenio Caponetti, Maria Luisa Saladino, Francesca Spatafora and Maria Costanza Polizzi, "Indagine composizionale sulle pitture di due vasi di Centuripe", in: Monica Salvadori, Francesca Fagioli and Clelia Sbrolli (eds.), *Nuovi dati per la conoscenza della pittura antica. Atti del I Colloquio AIRPA* (Aquileia, 16–17 giugno 2017), Rome, 295–300.
Richter (1930): G.M.A. Richter, "Polychrome Vases from Centuripe in the Metropolitan Museum", in: *Metropolitan Museum Studies* 2, 187–205.
Richter (1932): Gisela M.A. Richter, "A Polychrome Vase from Centuripe", in: *Metropolitan Museum Studies* 4, 45–54.
Simon (1989): Erika Simon, *Menander in Centuripe*, Wissenschaftliche Gesellschaft an der Johann Wolfgang Goethe-Universität Frankfurt am Main, Sitzungsberichte 25.2, Stuttgart.
Trendall (1955): Arthur Dale Trendall, "A New Polychrome Vase from Centuripe", in: *The Metropolitan Museum of Art Bulletin* 13, 161–166.
Wintermeyer (1975): Ulrike Wintermeyer, "Die polychrome Reliefkeramik aus Centuripe", in: *Jahrbuch des Deutschen Archäologischen Instituts* 90, 136–241.

Tyler Jo Smith, Ethan Gruber and Nicholas A. Harokopos

22 *Kerameikos.org* and Digital Accessibility for Ancient Greek Vases

Abstract: Launched in 2013, *Kerameikos.org* is an international collaborative project whose primary focus is defining the concepts used within the study of ancient Greek figure-decorated pottery following the tenets of Linked Open Date (LOD). This paper presents the background and on-going work of *Kerameikos.org* as it relates to the issue of digital accessibility. We explain the reasons *Kerameikos.org* complements existing resources, especially given its potential for cross-querying across the Semantic Web; the ways it is bringing new prospects and greater possibilities to the teaching and studying of ancient Greek vases; and how it invites changing perceptions of them in the modern world. The paper also addresses our commitment to using FAIR principles through the implementation of LOD, and the issues around accessibility of Greek vases in Greece and in other multilingual environments. Finally, we anticipate the impact of *Kerameikos.org* on the museums, libraries, and cultural institutions of Greece and in other countries where Classical Archaeology plays a fundamental role.

Introduction

Launched in 2013, *Kerameikos.org* (http://kerameikos.org [last access 01.03.2024]) is an international collaborative project whose primary focus is defining the concepts used within the study of ancient Greek figure-decorated pottery (or 'vases') following the tenets of Linked Open Date (LOD). With its ontologies and controlled vocabulary formulated, *Kerameikos.org* presents the opportunity for sharing relevant data across disparate systems (e.g., Getty Art & Architecture Thesaurus, Beazley Archive Pottery Database, Wikipedia). In this paper, we present the background and on-going work of *Kerameikos.org* as it relates to the issue of digital accessibility. It is our conviction that this web-based enterprise neither replaces others nor renders them obsolete. Rather, in the spirit of LOD, *Kerameikos.org* complements existing resources, especially given its potential for cross-querying across the Semantic Web.[1] At the same time, it brings new prospects and greater possibilities to the teaching and studying of ancient Greek vases, and invites changing perceptions of them in the modern world. In order to demonstrate these points, we present: first, some key background to the project and the needs we hope it can address as an openly accessible tool (Smith); second, the ways we are using FAIR principles through the implementation of LOD and some envisioned next steps (E. Gruber); finally, the perspective of our Greek editor-translator on issues of the accessibility of Greek vases in Greece and in other multilingual environments (N. A. Harokopos). Our remarks combined and our brief conclusions are a testament to the collective nature of the project and the inherent need for the multiple voices and diverse skill sets that underpin a joint Digital Humanities project such as this one.

Digital Access and Greek Vases

The study of ancient Greek vases has evolved notably over the past few decades and access to them has never been greater. Following the death of Sir J. D. Beazley in 1970, the Beazley Archive was established at Oxford University in an effort to organize the photographs, notes, and drawings of the 20th century's most influential scholar of Athenian vases.[2] Although

1 For Linked Data and the Semantic Web see Bizer/Heath/Berners-Lee (2009), which includes the set of 'principles' established by T. Berners-Lee for publishing data on the Web. Cf. Mayer (2015) 5 ("a ranking system for all datasets online"); and Candela *et al.* (2022) 21–22.
2 On Beazley's life and contribution, see Rouet (2002).

Acknowledgments: In addition to the editors of this volume, the authors wish to thank the following institutions and individuals for their assistance in the preparation of this paper and for their support of the *Kerameikos.org* project in general: National Endowment for the Humanities; Classical Art Research Centre and Beazley Archive, Oxford; Institute for Advanced Technology in the Humanities, University of Virginia; Interdisciplinary Archaeology Program, University of Virginia; and Abigail Bradford, Giada Giudice, Renee Gondek, and Worthy Martin.

https://doi.org/10.1515/9783111189635-022

vase specialists had already become accustomed to using Beazley's famous 'lists' of black-figured and red-figured painters as a research tool, the advent of the paper archive, and its eventual manifestation as a searchable online database – a process which took years to achieve - resulted in a new level of accessibility. Initially, a person in need of information stored in the database would be required to request a paper print-out from the Archive using specific 'search terms'. Thus, if one needed a complete list of all Athenian black-figured cups depicting dance, all Tyrrhenian Group vessels identified as showing Amazons, or all red-figured amphorae found on the Acropolis, such lists could be provided and, in turn, saved researchers a tremendous amount of time spent tracking down examples and images for further study. In order to see those examples and images, however, it was still (up until the late 1990s) necessary to travel to Oxford and to consult the Archive's holdings in person. With the growth of the World Wide Web and the creation of the Beazley Archive Pottery Database (*BAPD*), over time a new level of potential and accessibility was again reached for our discipline. Under the leadership of D.C. Kurtz, the *BAPD* united for the first time in history the contents of Beazley's lists, subsequent published data on individual vases, and the legacy data housed at the Archive including thousands of Beazley's own photographs.[3] Now housed under the umbrella of the Classical Art Research Centre at Oxford (and with an updated URL to reflect that change), the *BAPD* continues to grow, containing (according to their website) more than 130000 individual object records and 250000 images.[4]

It is important to remember, however, that the *BAPD* was never and still is not the only tool available to individuals interested in learning more about or conducting researching on Greek vases. For a start, it bears reminding that Greek vases, in a conceptual sense as figure decorated pottery, occupy an unusual space between art and archaeology - that is to say, they have been appreciated through the lens of aesthetics and connoisseurship as well as being artifacts that derive from ancient sites and contexts.[5] While many beautifully made objects that have been subjected to detailed study of their shapes or iconography can be classified as the *better* or museum-quality pieces (and are indeed on display in museums), a great many others produced in Athens and elsewhere are published in site reports resulting from excavations and are in a *poorer*, often fragmentary, state of preservation. As a result, scholars of Greek vase-painting, be they concerned with artistic personalities or iconographic themes, often need to cast the net as widely as possible in an effort to find as many examples or comparisons as possible, while being increasingly aware of the limitations of poorly provenanced, missing, or de-contextualized pieces.[6] Furthermore, traditional presentations of Greek vases in museum collections, such as those appearing in volumes of the *Corpus Vasorum Antiquorum* (*CVA*), are not limited to intact examples nor are they entirely devoid of archaeological context.[7] And, regardless of where a particular vase or sherd is ultimately published – a *CVA* volume, a museum catalogue, or a site report – the same basic information about an individual vessel must still be gleaned about it. Where was it made? How was it made? When was it made? What is the form? Who made it? In attempting to answer these questions, whether about a specific vase or a group of them associated with a common technique, shape, region, or artist, immediate problems arise. Inconsistencies abound between disparate systems, vocabularies, and languages, with terms such as 'style', 'technique', 'ware', and 'fabric' being used differently and/or interchangeably by museums, excavators, and published scholarship in general. Furthermore, the problem of data sharing for Classical Archaeology continues given the resistance of some stakeholders, or the restrictions placed on them, to make their finds (including figure-decorated Greek pottery) freely available or digitally accessible.[8]

This brief background is both necessary and important to understanding the advent of *Kerameikos.org*, a collaborative international project which aims to define the intellectual concepts of ancient Greek figure-decorated pottery, or 'vases', in accordance with the principles of LOD.[9] As an instrumental component of the Semantic Web, LOD at its core aims to link

3 Smith (2005) details the transformation from paper archive to online database.

4 The website, which was for decades www.beazley.ox.ac.uk (last access 01.03.2024) has more recently been embedded into that of the Classical Art Research Centre as www.carc.ox.uk/carc/pottery (last access 01.03.2024); the statistics mentioned derive from the latter.

5 The inherent interdisciplinarity of the sub-field situates it ideally into the Digital Humanities; see Nurmikko-Fuller (2024) 2–5.

6 Issues of context and audience have become among the most relevant to scholars currently publishing in the field. For both Greek and non-Greek contexts see e.g., Alexandridou (2011); Paleothodoros (2012); and Bundrick (2019).

7 The recent *CVA* for Ruvo is based entirely on material excavated from the site's necropoleis in the 19th century; *CVA* Italy 86 – Elvia Giudice and Giada Giudice, *Corpus Vasorum Antiquorum Italia 86. Ruvo di Puglia. Museo Nazionale Jatta. 6. Ceramica Attica a Figure Rosse e a Figure Nere, 1*, Rome.

8 However, some projects are now using Open Context (https://opencontext.org [last access 01.03.2024]) to publish primary research data from archaeological fieldwork, as well as geospatial data and 3D models. See further Kansa/Kansa (2021).

9 Gruber/Smith (2014); Gondek/Gruber/Smith (2019).

published online data deriving from different machine-readable sources. In recent years it has seen increased use and a certain amount of success in particular within the so-called GLAM community (Galleries, Libraries, Archives, Museums), where its implementation is easily justifiable.[10] Historically, in such settings and including the *BAPD* itself, metadata was open but not linked, creating what's known as a "data silo" (i.e., isolated, maintained, and accessed internally) replete with "tombstone data" (i.e., "static, factual information about an object").[11] To remedy this problem, a primary aim of the *Kerameikos.org* project has been to create streamlined definitions for key concepts in vase-painting, including pottery shapes, painting techniques, time periods, and persons (i.e., painters and potters), that are in-turn machine-readable, discoverable, and interoperable. This work has been undertaken by a cultural heritage LOD specialist (E. Gruber) in conjunction with vase-painting specialists (T. J. Smith, R. Gondek) and translators (e.g., N. A. Harokopos), and a team of editors, many of whom are graduate students in Classical Archaeology.[12] In our effort to maintain quality in creating definitions (i.e., controlled vocabulary using an ontology), and in order to address the challenges of individual and cultural bias, our editors work in teams whereby every definition undergoes a series of editorial stages prior to its publication.[13] It is our hope that *Kerameikos.org*, as a result, will serve as a fundamental (human-readable) resource for Classical Archaeology students, instructors, and researchers, as well as for those based across the GLAM sector with an interest in ancient Greek vases and ceramics in general.

With a basic understanding of how LOD is applicable to Greek vases, there are several other questions we should ask and considerations to be made about *Kerameikos.org*. The first question is what is digital accessibility and why is it important for Greek vases in general?[14] Digital access, as the *BAPD* has shown us, allows knowledge (i.e., published research, museum objects, scientific findings) to be freely available and free of charge (i.e., rather than behind a paywall). This brings to the forefront issues of equity and impact (who has access), as well as efficiency (ease of access).[15] Because LOD "is about surmounting differences" and "overcoming silos",[16] *Kerameikos.org* democratizes Greek vases as data both by merging information across systems and cultural sectors and, importantly, by leveling the playing field between museum-quality objects and excavation pottery.[17] Secondly, and as a result of this process, can we in turn use less jargon and be more inclusive? The answer here is, yes, we can. One need not know in advance what a skyphos is, or understand who Euphronios was, in order to begin researching Greek vases. Rather, the controlled vocabulary provided by *Kerameikos.org* and the LOD format it supports lead the user to simple, yet accurate definitions of these terms as well as to examples of them in museum collections, other cultural heritage intuitions, or elsewhere on the Web (e.g., 3D models, maps). By default, this again lends better access to vases and grows the potential audience(s) for them. Finally, what can *Kerameikos.org* provide that no one else has? Apart from what has been articulated so far, we might list the following benefits of the site:

- it is collaborative at every level and credits the labor of all contributors (e.g., technicians, subject specialists, editors)[18]
- it is a dynamic platform with the ability to forge new alliances, incorporate new data, and increase bridge building between systems
- it provides the best in 'peer review', having endless potential for feedback
- it makes archaeometric links (e.g., to websites and articles) and connections between other resources and collections
- it has the prospect for quantitative analysis, allowing for greater creativity, and thus generating new knowledge
- its content is defined and translated by humans rather than machines
- it brings Classical Studies further into the Digital Humanities sphere

---

10 Yoose/Perkins (2013) 209; Mayer (2015); Candela *et al.* (2022) 21–22.
11 For these terms, see Mayer (2015) 5–6; Candela *et al.* (2022) 22.
12 See Smith/Bradford/Gruber (forthcoming), where our workflow and method of crediting labor are summarized.
13 Cf. Mayer (2015) 6; Brown (2022).
14 Cf. Fleming *et al.* (2021).
15 On different forms and standards of Open Access, see Fleming *et al.* (2021) 12–13.
16 Brown (2022) 2.
17 Schmidt/Thiery/Trognitz (2022) encourage the creation and re-use of LOD in archaeology. See also Bellia/Sabba (2017).
18 Gervits/Orcutt (2016) 223; and see *supra*, note 12.

It is with this background and these perspectives in mind that we were able to envision the need for *Kerameikos.org* in the past. As we shall see in the following section, it remains a priority to ensure that its ethics as a web-based platform and as a digital tool align both now and in the future with a sustainable vision for LOD as it applies to archaeological data and the Digital Humanities.[19]

Making Greek Vases FAIR

As scientific data have become more open through FAIR principles in recent years, so too have archaeological and humanities datasets.[20] *Kerameikos.org*, as a Digital Humanities project focused on data related to Classical Archaeology, is no exception, having applied FAIR principles since its inception in 2013. FAIR – *Findable, Accessible, Interoperable, Reusable* – principles have emerged partly as a product of the change in philosophy regarding the sharing of research data, but also as a requirement of publicly funded projects.[21] Funding bodies desire the broadest possible impact for their financial investment by making research data and the resulting syntheses openly accessible and reusable by third parties. With this in mind, it is useful to consider how *Kerameikos.org* functions and its potential for altering the landscape of Greek figure-decorated pottery studies through the implementation of LOD.

For a start, the third tenet of FAIR, interoperability, was the primary aim of *Kerameikos.org* when it was initially conceived. At that time, E. Gruber, a software developer trained in Classical Archaeology and employed at the American Numismatic Society, was already working on *Nomisma.org* (https://nomisma.org [last access 01.03.2024]), an international, collaborative project to define the intellectual concepts of numismatics according to LOD principles.[22] By 2013, *Nomisma.org* was no longer a prototype. Rather, it had evolved into the core technical infrastructure underlying Online Coins of the Roman Empire (OCRE; http://numismatics.org/ocre [last access 01.03.2024]), which remains the American Numismatic Society's most heavily-used digital project. The creation of *Kerameikos.org* was thus inspired by the success of *Nomisma.org* with a nearly identical software architecture and similar purpose: to facilitate the interoperability between disparate databases of figure-decorated Greek pottery, whether published by museums or archaeological excavations, and to answer new types of research questions. It was within this framework that E. Gruber approached T. J. Smith, a Classical Archaeologist specializing in ancient Greek vases and having vast prior experience with the Beazley Archive (as both a student volunteer and as a researcher), with an idea that would launch the existing digital access to Greek vases into the next generation. Because pottery and coins are not the same, many issues unique to Greek vases and their presentation had to be considered from the outset, leading to the creation of a Scientific Committee composed of pottery scholars, museum personnel, and data specialists, and thus to the formal establishment of *Kerameikos.org* (Fig. 1).[23]

When we question the 'why' of *Kerameikos.org*, the most succinct explanation for its creators is language. Take, for example, Exekias, an Athenian 'artist' (painter and potter) active during the Archaic period, the 6th century B.C.[24] If one searches for this name in an English-language database, such as the *BAPD* or the British Museum (https://www.britishmuseum.org/research [last access 01.03.2024]), the list of artists' names, including this one, would be expectedly in English. However, in the database of the Musée du Louvre (https://collections.louvre.fr [last access 01.03.2024]) or the Bibliothèque national de France (https://data.bnf.fr [last access 01.03.2024]), this same person is referred to as "Exékias," or becomes "Εξηκίας" in Greek, "エクセキアス" in Japanese, and so forth. Regardless of the translation or transliteration of the name into various languages or writing systems, scholars of Athenian vases fundamentally agree on the concept of "Exekias". Therefore, a common, unique identifier can be applied to this concept which enables systems authored in different languages to share the same vocabulary. This, in short, is the fundamental building block of interoperability and reusability.[25]

19 On LOD and Digital Humanities, see Antopolsky (2022); Nurmikko-Fuller (2024).

20 See *supra*, note 8; Scheffler *et al.* (2022), materials science; and Pawlicka-Deger (2021) 40–41, open research data.

21 Wilkinson *et al.* (2016); Jacobson *et al.* (2020); Schmidt *et al.* (2022) 343.

22 *Nomisma.org* was created in 2010 at the American Numismatic Society by Andrew Meadows and Sebastian Heath, and Ethan Gruber took over software development in 2011; see https://nomisma.org/about (last access 01.03.2024).

23 Gruber/Smith (2014).

24 Moignard (2015).

25 Kansa/Kansa (2022).

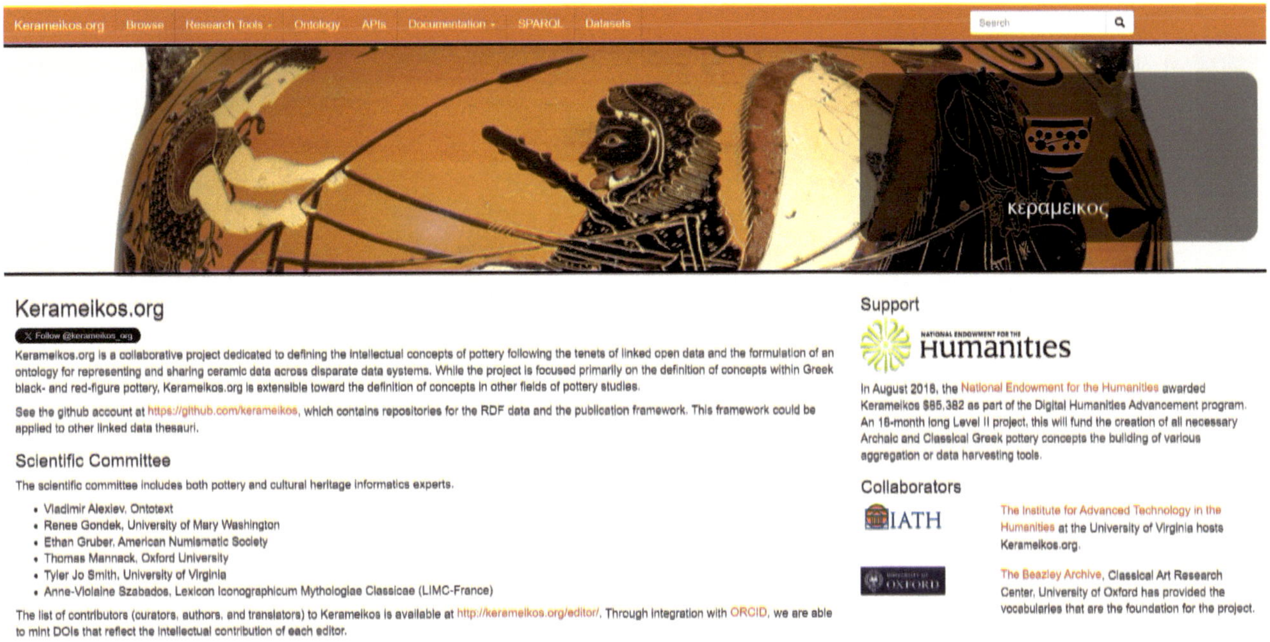

Fig. 1: Detail of website homepage for *Kerameikos.org* (last access 01.03.2024).

According to LOD, this unambiguous, stable ID is in the form of a Uniform Resource Identifier (URI) in the Hypertext Transfer Protocol (HTTP), plainly put: a URL that anyone anywhere can access in their web browser.

In *Kerameikos.org*, the intellectual concept of Exekias is represented by http://kerameikos.org/id/exekias (last access 01.03.2024) (Fig. 2). Any user accessing this URI – whether a seasoned scholar of the subject or a complete novice to it – will receive HTML, which will be rendered into a user-friendly web page in a browser. This page contains information about Exekias: preferred labels for this entity in various languages, a well-researched definition in English (with translations into other languages gradually being added), the role that Exekias plays with respect to ceramics (as both a painter and potter), and a list of URIs that also define the concept of Exekias in other information systems, such as the Getty Union List of Artist Names (ULAN),[26] Lexicon of Greek Personal Names,[27] Gemeinsame Normdatei,[28] and Wikidata.org, among others. This concordance is particularly important as it enables, for example, databases that conform to the Getty thesauri to map to German ones that implement their own international authority file, and so on. *Kerameikos.org* therefore serves as a bridge between these vocabulary systems, making it possible to aggregate data regardless of the native language of the database. Allowing these systems to 'speak' to each other is more than a simple conversation starter or a matter of convenience, however. It presents the opportunity for users to access digital information (i.e., scholarship, objects, perspectives) about Exekias that they may never have encountered otherwise – a step which leads to creativity and originality in research and teaching, not to mention to greater community building. The page for Exekias, and other pottery concepts defined by *Kerameikos.org*, may also display a map showing the distribution of objects related to the concept (where the vases were produced and found, based on available information) and a list of related objects from datasets that have been contributed to *Kerameikos.org*.

What is not apparent to a casual visitor to the URI, http://kerameikos.org/id/exekias (last access 01.03.2024), is the underlying machine-readable data, which is an integral tenet of LOD architecture. Although the default response to requesting a URI in *Kerameikos.org* is an HTML web page, it is possible to request alternative formats, particularly Resource Description Framework (RDF). RDF is a standardized, structured data format pioneered by T. Berners-Lee, the inventor of the World Wide Web, and maintained by the international World Wide Web Consortium (W3C).[29] RDF is

26 See https://www.getty.edu/research/tools/vocabularies/ulan (last access 01.03.2024), based at the Getty Research Institute.
27 See https://www.lgpn.ox.ac.uk (last access 01.03.2024), based at Oxford University.
28 See https://www.dnb.de/DE/Professionell/Standardisierung/GND/gnd_node.html (last access 01.03.2024), based at the Deutsche Nationalbibliothek.
29 See https://www.w3.org/DesignIssues/LinkedData.html (last access 01.03.2024); also, Berthoud/Hartley (2021); and *supra*, note 1.

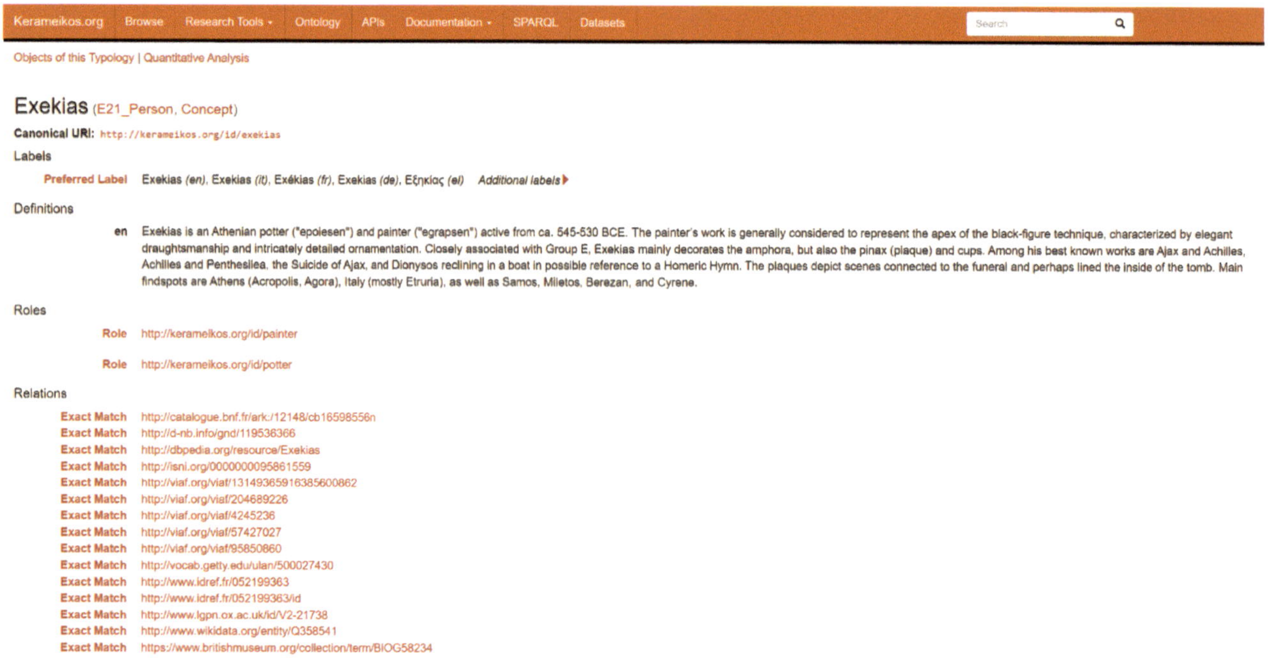

Fig. 2: *Kerameikos.org* search result for the artist "Exekias" (last access 01.03.2024).

modeled by an ontology, which gives semantic meaning to the data, enhancing its reusability. *Kerameikos.org* implements several ontologies to represent data in various contexts. The intellectual concepts are primarily represented in the Simple Knowledge Organization System (SKOS: https://www.w3.org/TR/skos-reference [last access 01.03.2024]), which is used to encode labels and definitions in different languages, as well as the lists of related URIs in external thesauri. 'Artist' is not the only category of concept defined in *Kerameikos.org*: there are production places, time periods, materials, vessel shapes (Fig. 3), decorative techniques, museum collections, to mention a few. Furthermore, these categories are classified by the CIDOC-CRM ontology (https://cidoc-crm.org [last access 01.03.2024]), developed by the International Council of Museums and considered the standard for use amongst the GLAM community and for other cultural institutions.[30]

Essentially, a software developer can extract data for reuse directly from the *Kerameikos.org* platform through Application Programming Interfaces (APIs) that conform to protocols standardized by the W3C in a data model that also meets well-known and documented specifications.[31] For example, an information system that links its geographic entities for production places to *Kerameikos.org* URIs, e.g., the city of Corinth (http://kerameikos.org/id/corinth [last access 01.03.2024]), can extract geographic coordinates from the RDF in order to generate a map showing the location of Corinth (Fig. 4). Another major use is the extraction of preferred labels in alternative languages in order to build a multilingual user interface. This capability has not yet been built by platforms that rely on the technical infrastructure provided by *Kerameikos.org*; but its capability is illustrated by digital numismatic projects that are built upon *Nomisma.org*, such as OCRE, currently available in nearly twenty languages.[32] An analysis of web traffic to OCRE illustrates a correlation between available translations and the country of origin of users: Bulgaria, Turkey, Poland, and Ukraine are routinely among the top-ten in visitors. A large-scale ceramics platform, such as the BAPD, could actually make their data more readily accessible to a far broader range of audiences, especially those based in different countries, by relying on open data and tools that *Kerameikos.org* provides. At present, we are aware that both the Österreichisches Archäologisches Institut and the Sicily project of Sicilian inscriptions based at Oxford University (https://isicily.org [last access 01.03.2024]) have begun to integrate *Kerameikos.org* URIs into their databases, but much work remains to done before fully internationalizing access to Greek vase data.

30 Tzitzikas *et al.* (2022); and see Rodríguez (2021).

31 See https://www.w3.org/api (last access 01.03.2024).

32 See http://numismatics.org/ocre (last access 01.03.2024); and see above.

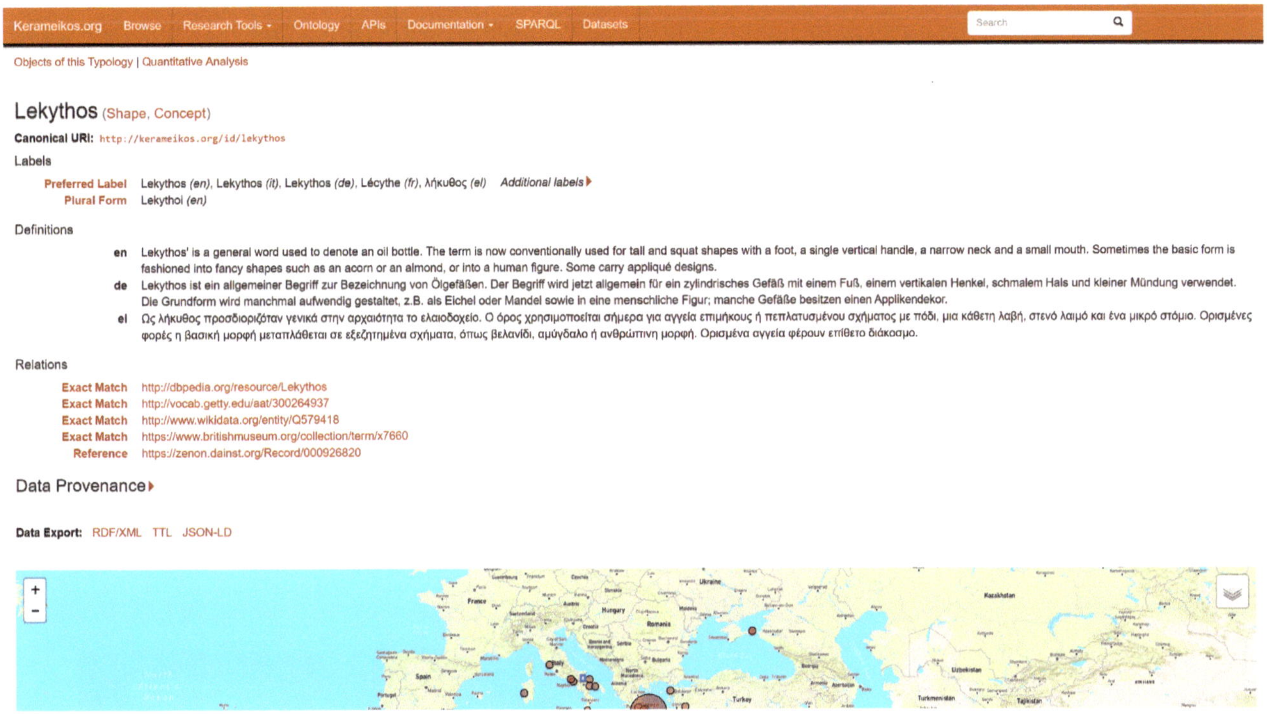

Fig. 3: *Kerameikos.org* search result for the shape “lekythos” (last access 01.03.2024).

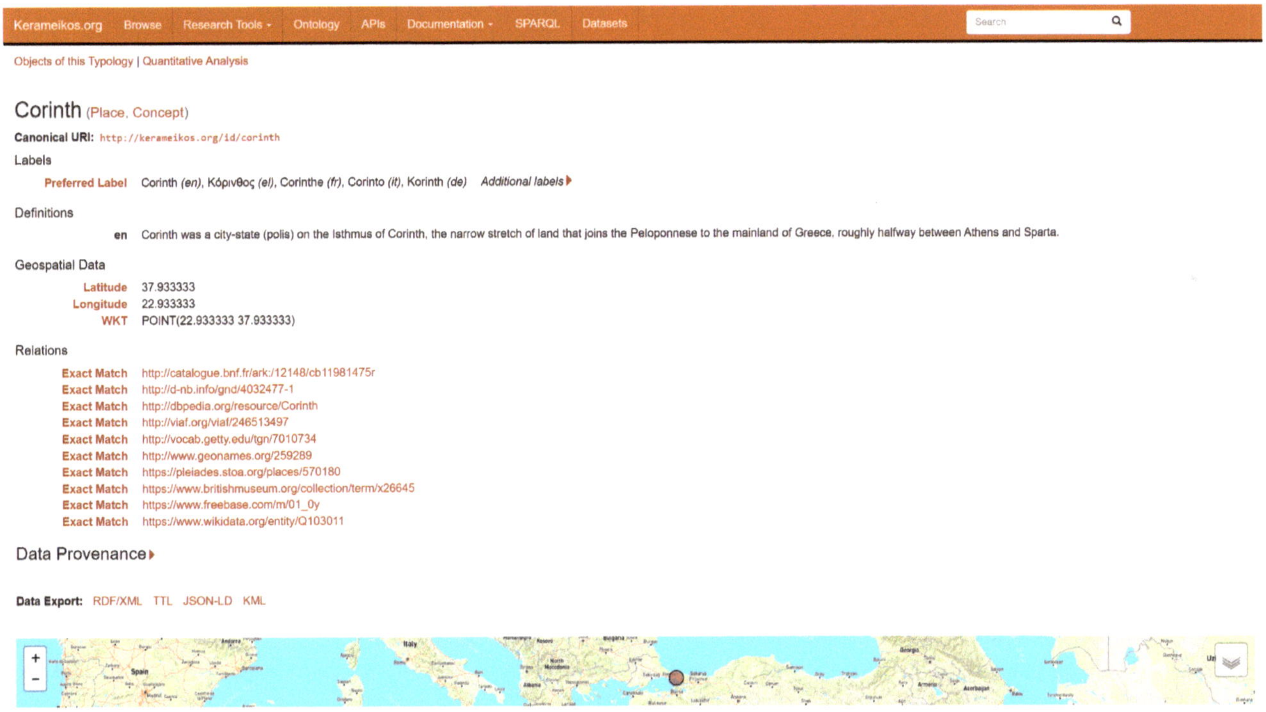

Fig. 4: *Kerameikos.org* search result for the city of “Corinth” (last access 01.03.2024).

The primary function of *Kerameikos.org* is a LOD thesaurus of pottery concepts that includes well-researched definitions and which enhances the platform as a resource for its users, among them students, scholars, museums, and the public. Students may be directed to the site by their course syllabi, while casual enthusiasts will likely find the site in increasing numbers through search indexes. Although the algorithms developed by search engines, such as Google and DuckDuckGo, are opaque, certain *Kerameikos.org* URIs are beginning to appear on the first page of search results for more specialized terms, such as the Agrigento Painter. In addition to multilingual labels and definitions for *Kerameikos.org* concepts, each search leading to one such *Kerameikos.org* web page includes additional research context, if available: maps, lists of related objects, and a basic interface for generating charts illustrating statistical distributions (e.g., shapes decorated by a particular painter; Fig. 5).These features are derived from vase data aggregated into the *Kerameikos.org* LOD database, which itself conforms to the W3C SPARQL query syntax and protocol for RDF.[33] Over time, as more projects and institutions link to *Kerameikos.org*, more general terms should appear higher in search results, thus completing the Findable and Accessible tenets of the FAIR paradigm.

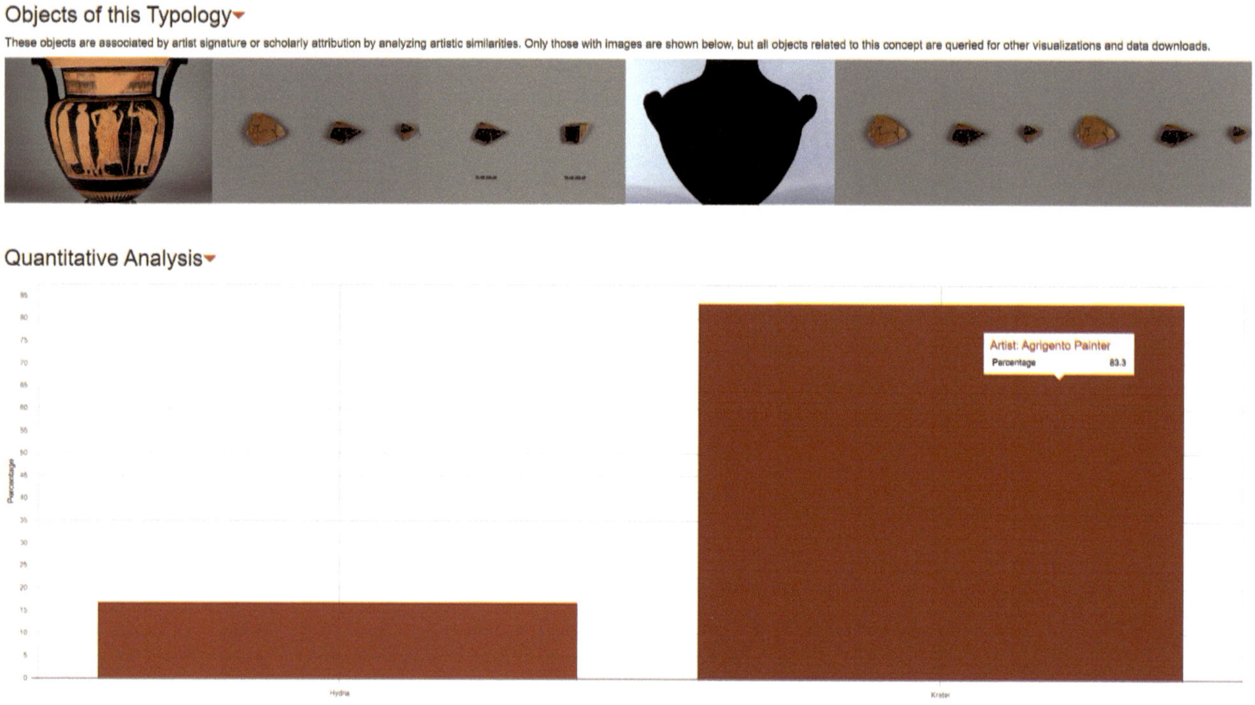

Fig. 5: *Kerameikos.org* search result for shape distribution of the "Agrigento Painter" (last access 01.03.2024).

Expanding Reach and Reducing Bias

In its earliest phase, the *Kerameikos.org* prototype included a number of Greek vases from several museum collections, which had been linked together through the common vocabulary system. Among them were several hundred vases from the British Museum and several dozen from the J. Paul Getty Museum, which link to one or more high-resolution photographs. Because each museum collection publishes data conforming to its own thesaurus, *Kerameikos.org* LOD became the bridge linking these disparate datasets together. But a common vocabulary system is only the first step in any large-scale aggregation of databases, and a common data model is also necessary. In *Kerameikos.org*, the vase data from each museum collection is mapped into RDF following the Linked Art profile of CIDOC-CRM (https://linked.art [last access 01.03.2024]), which is a community-driven project to define a shared model for the exchange of museum data. As more museums and archaeological excavation databases begin to adopt the Linked Art API specifications, *Kerameikos.org* will be able to harvest these objects into its SPARQL endpoint and make them available to any individual or institution interested in ancient Greek vases.

Importantly, the more data integrated into the *Kerameikos.org* system, the more accurate and complete the maps and statistical analyses will become. At present, the distribution of materials in *Kerameikos.org* is heavily biased toward the

33 See https://www.w3.org/TR/sparql11-query (last access 01.03.2024); see also Zou (2018).

thousands of artifacts published by the two museums we first partnered with (the British Museum and the Getty). And while there are a handful of other contributors to *Kerameikos.org* so far (e.g., Fitzwilliam Museum, Cambridge), there are many thousands of Archaic and Classical Athenian vases that live outside the LOD ecosystem on the web. In the short term, a primary goal of *Kerameikos.org* is to continue to grow our alliances with relevant databases and collections in an effort not only to discover and share Greek vase information with our users, but also to provide digital access that is accurate and information that is transferable. In the long term, however, we look ahead to our next major challenge. The technical principles established by the project are extensible into other types of archaeological ceramics both from the Mediterranean (e.g., Corinthian, South Italian, Roman) and elsewhere (e.g., Indigenous North America, China, Pre-Columbian Maya), and the infrastructure has been proven (through *Nomisma.org*, specifically) to be scalable into the millions of objects.[34] The long-term sustainability of the project is dependent upon growing the community of contributors and users: not only reaching those exclusively invested in the study of ancient Greek figure-decorated vessels, but rather by expanding our global reach beyond the time period and culture currently reflected in the system.

From Theory to Praxis: A Greek Perspective

The efforts to reach a larger audience for ancient Greek vases, as outlined above, are already yielding results. *Kerameikos.org* has begun to reach non-English audiences, not least in part because its content is being translated in German, Modern Greek, and other languages. Thus, the perspective given here arises from personal involvement in the project (as translator and editor) as well as from teaching experience and scholarly interests (as a specialist in Athenian vase-painting). From the beginning, the idea and scope of *Kerameikos.org* seemed exciting for a Greek audience, not only because of the features of the project and its aspiration to offer a broad interconnected database of ancient Greek pottery, but also because it would be made available in other languages, including Greek. This excitement has had nothing to do with notions of national pride or the requirements of the Greek University curriculum. As we have entered the era of digital education and research, the Humanities fields are no exception. More and more undergraduate and graduate students are turning to the Internet to find easily accessible resources when it comes to studying for a course, writing an essay, or researching a thesis, be it for Classical Archaeology or another discipline. The same is true, of course, for scholars, and not just young ones, who by necessity are becoming better acquainted with the necessary tools and technology.[35]

For many years now, students in Greece, both undergraduate and graduate, have been asking for online resources for ancient Greek figure-decorated pottery. We routinely provide them with an extensive bibliography, as we have no real shortage of relevant handbooks and books on the subject (many of them translations into Greek from English), and also with some major online resources by foreign universities and institutions: the *BAPD*; WebAtlas of Ceramic Greek Kilns in Ancient Greece;[36] the major collections of ancient Greek vases, such as the British Museum and the Louvre already mentioned, as well as the Metropolitan Museum of Art and the Museum of Fine Arts Boston; and the digital collection of the American School of Classical Studies at Athens, which includes excavation materials, among them the Greek vases from the Athenian Agora.[37] But when it comes to Greek-born websites, there exists thus far no single major online resource for ancient Greek vases of the type being provided by *Kerameikos.org*. There are, however, a limited number of freely acces-

34 For the potential to expand into the realm of vase iconography, see Schmidt *et al.* (2022) 351–353. On cross- referencing with other categories and collections, see Mayer (2015) 3 ("a global picture of cultural holdings").

35 See Candela *et al.* (2022) 21–22, who describe LOD as the "bridge between Digital Humanities practitioners and less technically proficient researchers". See also Schmidt *et al.* (2022) 334.

36 https://atlasgreekkilns.arizona.edu (last access 01.03.2024), directed by E. Hasaki at the University of Arizona.

37 See https://www.metmuseum.org/art/collection; https://www.mfa.org/collections; https://ascsa.net/research?v=default (all last access 01.03.2024).

sible sites which present some basic aspects of vases in general,[38] their techniques,[39] not to mention specialized repositories,[40] relevant research projects, or multimedia presentations. In addition, within the broader field of ceramic studies, there are some webpages and online resources dedicated to traditional Greek pottery in the modern era, the techniques of which have become an object of study and comparison with ancient ceramic production.[41] This trend has encouraged some interesting experimental archaeology workshops and educational programs with the cooperation of living ceramic artists.[42] Apart from the above, there are also some webpages that offer general, encyclopedic information about ancient Greek pottery and vase-painting, such as the "Hellenic Culture" website hosted by the Panhellenic School Network, which includes a catalogue of vase shapes.[43]

Despite this impressive variety of current digital resources available in Greece, *Kerameikos.org* fills an important gap for Greek students and scholars alike. This is not to suggest that Greek students or scholars are hindered by languages other than their own in accessing foreign websites. But what should be stressed is the importance of having the various categories and major aspects of Athenian vases digitally accessible in a single location, with concise authoritative multilingual texts, and all the necessary tools to highlight connections between the categories that provide access to multi-faceted data. It is possible to illustrate this point using the example of the first category translated into Greek for the *Kerameikos.org* project: vase shapes. This extensive category includes a multitude of forms, names, and variations, sometimes subtle ones, between them. Having the corpus of well-known shapes easily available in Greek, and supplemented by the LOD resources of *Kerameikos.org*, is already proving valuable to many students, as evidenced by the feedback being received in classes and seminars. The website is a good starting point for Greek undergraduate students approaching the subject of ancient Greek vases for the first time, as it helps to acquaint them with a large corpus of information and with often confusing or inconsistent terminology. It has also been a useful resource for graduate students, who have found the shapes category particularly helpful in bettering their understanding of the various decorated vessels they encounter in their research, and in improving their ability to describe and write about them correctly, both with confidence and clarity.

Important in this direction are the project's multilingual interfaces which allow users to access aspects and intellectual concepts of Athenian black- and red-figured pottery in different languages so that they may easily compare them and see how they are rendered in those languages. This approach will undoubtedly further enhance the ability of Greek students, and perhaps even more advanced scholars, to apply these concepts accurately in Modern Greek. Returning to the category of shapes, one might encounter some difficulties or find inconsistencies in the rendering of certain shape-related concepts in Greek, both in essays and theses, especially when relying on non-Greek bibliography.[44] As a result, we sincerely hope that visitors to *Kerameikos.org*, whatever their language or country of origin, will use the site to become more familiar with the inter-lingual approach necessary to and underlying the field of Greek vase studies.

Finally, in the last two decades, Greek universities have seen an encouraging upswing in the number of master's and doctoral theses on topics related to ancient Greek pottery, including vases with figure decoration, an area that had been

---

38 E.g., the summary presentation of the pottery collection at the National Archaeological Museum of Athens: https://www.namuseum.gr/en/permanent_exhibition/syllogi-aggeion-kai-mikrotechnias (last access 1.03.2024, in Greek and in English). The website of the Archaeological Museum of Thessaloniki offers a brief description of its pottery and minor arts collection: https://www.amth.gr/sylloges-keramikis-metallotehnias-kai-mikrotehnias (last access 01.03.2024, in Greek), and sometimes vases appear as "Exhibits of the Month", e.g., a late 5th century B.C. Attic red-figured pyxis in July 2022: https://www.amth.gr/en/exhibitions/exhibit-of-the-month/2207 (last access 01.03.2024, in Greek and in English).

39 E.g., the concise introduction to production methods on the website of the Museum of Cycladic Art: https://cycladic.gr/en/page/archaia-technologia (last access 01.03.2024). See also the website of ATTIC VASES / PRECIOUS EARTH, a collaborative project focused on pottery technology through the use of innovative techniques: http://www.atticvases.gr (last access 01.03.2024).

40 E.g., the *AtticPOT* project examines the Athenian presence in ancient Thrace through the diffusion of Attic painted pottery; its repository includes published vases that have been found on sites in ancient Thrace, as well as relevant bibliography for each example: https://atticpot.athenarc.gr/index.php/en (last access 01.03.2024, Greek and English).

41 E.g., the website of the Centre for The Study of Modern Greek Pottery: http://potterymuseum.gr/reduction/?lang=en (last access 01.03.2024, Greek and English, with video).

42 See e.g., the announcement of the experimental workshop on Archaic pottery in Macedonia co-organized by the Archaeological Museum of Thessaloniki in 2016: https://www.amth.gr/news/peiramatiko-ergastirio-gia-tin-keramiki-stin-arhaiki-makedonia (last access 01.03.2024, in Greek); and a similar event for children hosted by the Stavros Niarchos Foundation Cultural Center in 2019: "Ceramics Lab: From the Tile Maker to the Potter" (https://www.snfcc.org/en/events/ceramics-lab-tile-maker-potter/4913 [last access 01.03.2024].

43 See http://users.sch.gr/ipap/Ellinikos%20Politismos/AR/ar.ag/aggeia.htm (last access 01.03.2024).

44 A common issue encountered in student writing is inconsistency when it comes to terms such as "flaring lips", or "tapering body", or "echinus foot".

somewhat sidelined as an independent field of study in previous decades.[45] Moving away from traditional art historical methods and iconographical approaches to ancient Greek vases, a new generation of scholars are generating theses (both completed and currently in progress) covering an impressive range of topics, including archaeological and social context, production and distribution, and reception.[46] This trend is underpinned by a number of specialized seminars on different aspects of ancient Greek vases offered to undergraduate and graduate students in various departments of Greek universities. We are confident that *Kerameikos.org*, which takes a similarly broad approach, will contribute positively to this momentum in the study of ancient Greek vase-painting, and pottery studies in general, both in Greek universities and in our ongoing and future scholarship. We may also anticipate its impact on the museums, libraries, and other cultural institutions of Greece, as in other countries, where Classical Archaeology plays such a fundamental role.

Conclusions: Greek Vases for All

Accessibility for Greek vases, be it digital or otherwise, begins with visibility. As we have tried to show in the three parts of this paper, there are many challenges involved in increasing digital accessibility for ancient Greek vases: justifying the need to do so, recognizing and implementing FAIR principles, and building a multilingual, more inclusive community of stakeholders. Unlike other fields of study and including some STEM fields, advancements in the study of ancient Greek pottery, both figural and non-figural, are not especially fast-moving.[47] Furthermore, there are still far too many contributions to the discipline – most replete with illustrated examples - that remain invisible online, with chapters in edited volumes, museum catalogues, and conferences often not listed in citation data.[48] This problem is only further exaggerated by the disciplinary silos we inhabit and the ways that information about Greek vases is labelled and shared. But the issues we face as practitioners of Classical Archaeology or specialists in Greek art are not simply to do with research and research assessments. Collectively, we should be better invested in making our seemingly esoteric sub-discipline available and comprehensible to a wider audience, while still maximizing its potential as research data.[49] The groundwork that was laid a generation ago by the Beazley Archive enables a digital project such as ours to continue building on that success, given that the quality of our dataset of Athenian examples, while not perfect, is still largely intact and reliable.[50] The human factor remains, however, at the forefront of our minds. With every concept definition we create for our portal – be it a painter, a shape, or a technique - we are aware of making the best choices for the data we have (e.g., what to put in and what to leave out; how much detail to include; how to avoid bias).[51] Thus, by harnessing the potential of Linked Open Data and adhering to FAIR principles, the *Kerameikos.org* project is attempting to place Greek vases more readily into the public eye and more open to global connection and engagement.[52] We envision a future for Greek vases, and for pottery studies in general, that is both linked and open; and one where student or specialist, amateur or enthusiast, will be able to access reliable information that adheres to the ethics and protocols that underlie best practices in the Digital Humanities and to the rigors of archaeology as a data-driven discipline.[53]

---

45 This is based on the data provided by the National Archive of PhD Theses (https://www.didaktorika.gr [last access 01.03.2024]) and a search (available in Greek or in English) using the terms "pottery", or "pottery" and "excavation context", or "pottery and "iconography" (a brief summary of each thesis in English is included).

46 These results are available using the search terms "pottery" and "excavation context" on the website of the National Archive of PhD Theses; see previous note.

47 Cf. Gervits/Orcutt (2016) 221.

48 See Gervits/Orcutt (2016) 227–229, an appendix of tools for citation analysis and metrics. The full picture is not captured for Greek pottery by Research Gate, Web of Knowledge, Scopus, and other search engines because the subject rests between disciplines, thus only exaggerating the issue of discovery. As a result, Classical archaeologists, including Greek vase scholars, increasingly depend on Academia.edu, despite certain restrictions, in order to share their research. The same problem exists for citation analysis in various disciplines; Fleming *et al.* (2021) 15–17.

49 Cf. Pawlicka-Deger (2021).

50 Rula/Maurino/Batini (2016) 87.

51 Cf. Candela *et al.* (2022) 22; Brown (2022) 10–11.

52 See contributions of relevance in Golub/Lui (2021).

53 Mayer (2015) 5; Schmidt/Thiery/Trognitz (2022) 342.

Abbreviation

CVA *Corpus Vasorum Antiquorum*.

Bibliography

Alexandridou (2011): Alexandra Alexandridou, *The Early Black-Figured Pottery of Attika in Context (c. 630–570 BCE)*, Leiden/Boston.

Antopolsky (2022): Alexander B. Antopolsky, "Linked Open Data in the Digital Humanities (Review of Publications)", in: *Scientific and Technical Information Processing* 49, 119–126.

Bellia/Sabba (2017): Angela Bellia and Fiammetta Sabba, "Communication Patterns for Archaeological Research Data: The Case of the Getty Vocabularies as Linked Open Data", in: *AIB studi. Rivista di biblioteconomia e scienze dell'informazione* 57, 124–127.

Berthoud/Hartley (2021): Heidy Berthoud and Jeannie Hartley, "Practical Approaches to Linked Data", in: *The Serials Librarian* 80, 19–25.

Bizer/Heath/Berners-Lee (2009): Christian Bizer, Tom Heath and Tim Berners-Lee, "Linked Data: The Story so Far", in: *International Journal on Semantic Web Information Systems* 5, 1–22.

Brown (2022): Susan Brown, "Same Difference: Identity and Diversity in Linked Open Cultural Data", in: *International Journal of Humanities and Arts Computing* 16, 1–16.

Bundrick (2019): Sheramy D. Bundrick, *Athens, Etruria, and the Many Lives of Greek Figured Pottery*, Madison.

Candela *et al.* (2022): Gustavo Candela, Pilar Escobar, Rafael C. Carrasco and Manuel Marco-Such, "Evaluating the Quality of Linked Open Data in Digital Libraries", in: *Journal of Information Science* 48, 21–43.

Fleming *et al.* (2021): Jessie I. Fleming, Sarah E. Wilson, Sara A. Hart, William J. Therrien and Bryan G. Cook, "Open Accessibility in Education Research: Enhancing the Credibility, Equity, Impact, and Efficiency of Research", in: *Educational Psychologist* 56, 110–121.

Gervits/Orcutt (2016): Maya Gervits and Rose Orcutt, "Analysis and Tenure Metrics in Art, Architecture, and Design-Related Disciplines", in: *Art Documentation* 35, 218–229.

Golub/Liu (2021): Koraljka Golub and Ying-Hsang Liu (eds.), *Information and Knowledge Organisation in Digital Humanities: Global Perspectives*, London.

Gondek/Gruber/Smith (2019): Renee M. Gondek, Ethan Gruber and Tyler Jo Smith, "Kerameikos.org: Linked Open Data for Greek Pottery" (poster), in: *Check Object Integrity: Computer Applications and Quantitative Methods in Archaeology* (CAA Annual Meeting, Krakow, 23–27 April 2019). https://doi.org/10.5281/zenodo.3243539.

Gruber/Smith (2014): Ethan Gruber and Tyler Jo Smith, "Linked Open Greek Pottery", in: François Giligny *et al.* (eds.), *21st Century Archaeology: Concepts, Methods and Tools* (Proceedings of the 42nd Annual Conference on Computer Applications and Quantitative Methods in Archaeology), Oxford, 205–214.

Jacobson *et al.* (2020): Annika Jacobsen, Ricardo de Miranda Azevedo, Nick Juty, Dominique Batista, Simon Coles, Ronald Cornet, Mélanie Courtot, Mercè Crosas, Michel Dumontier, Chris T. Evelo, Carole Goble, Giancarlo Guizzardi, Karsten Kryger Hansen, Ali Hasnain, Kristina Hettne, Jaap Heringa, Rob W.W. Hooft, Melanie Imming, Keith G. Jeffery, Rajaram Kaliyaperumal, Martijn G. Kersloot, Christine R. Kirkpatrick, Tobias Kuhn, Ignasi Labastida, Barbara Magagna, Peter McQuilton, Natalie Meyers, Annalisa Montesanti, Mirjam van Reisen, Philippe Rocca-Serra, Robert Pergl, Susanna-Assunta Sansone, Luiz Olavo Bonino da Silva Santos, Juliane Schneider, George Strawn, Mark Thompson, Andra Waagmeester, Tobias Weigel, Mark D. Wilkinson, Egon L. Willighagen, Peter Wittenburg, Marco Roos, Barend Mons and Erik Schultes, "FAIR Principles: Interpretations and Implementation Considerations", in: *Data Intelligence* 2, 10–29. https://doi.org/10.1162/dint_r_00024.

Kansa/Kansa (2021): Eric C. Kansa and Sarah W. Kansa, "Digital Data and Data Literacy in Archaeology Now and in the New Decade", in: *Advances in Archaeological Practice* 9, 81–85. https://doi.org/10.1017/aap.2020.55.

Kansa/Kansa (2022): Eric C. Kansa and Sarah W. Kansa, "Promoting Data Quality and Reuse in Archaeology Through Collaborative Identifier Practices", in: *Proceedings of the National Academy of Sciences of the United States of America* 119, e2109313118. https://doi.org/10.1073/pnas.2109313118.

Mayer (2015): Allana Mayer, "Linked Open Data for Artistic and Cultural Resources", in: *Art Documentation* 34, 2–14.

Moignard (2015): Elizabeth Moignard, *Master of Attic Black-Figure Painting: The Art and Legacy of Exekias*, London/New York.

Nurmikko-Fuller (2024): Terhi Nurmikko-Fuller, *Linked Data for Digital Humanities*, London/New York.

Paleothodoros (2012): Dimitris Paleothodoros (ed.), *The Contexts of Painted Pottery in the Ancient Mediterranean World (Seventh–Fourth Centuries BCE)*, Oxford.

Pawlicka-Deger (2021): Urszula Pawlicka-Deger, "Digital Humanities and a New Research Culture: Between Promoting and Practicing Open Research Data", in: Shane Hawkins (ed.), *Access and Control in Digital Humanities*, London/New York, 40–57.

Rodríguez (2021): John Roberto Rodríguez, "Computational Ontologies for Accessing, Controlling, and Disseminating Knowledge in the Cultural Heritage Sector: A Case Study", in: Shane Hawkins (ed.), *Access and Control in Digital Humanities*, London/New York, 61–77.

Rouet (2002): Philippe Rouet, *Approaches to the Study of Attic Vases: Beazley and Pottier*, Oxford.

Rula/Maurino/Batini (2016): Anisa Rula, Andrea Maurino and Carlo Batini, "Data Quality Issues in Linked Open Data", in: Carlo Batini and Monica Scannapieco (eds.), *Data and Information Quality: Dimensions, Principles and Techniques*, London, 87–112.

Scheffler *et al.* (2022): Matthias Scheffler, Martin Aeschlimann, Martin Albrecht, Tristan Bereau, Hans-Joachim Burgartz, Claudia Felser, Mark Greiner, Axel Groß, Christoph T. Koch, Kurt Kremer, Wolfgang E. Nagel, Markus Scheidgen, Christof Wöll and Claudia Draxl, "FAIR Data Enabling New Horizons for Materials Research", in: *Nature* 604, 635–642. https://doi.org/10.1038/s41586-022-04501-x.

Schmidt/Thiery/Trognitz (2022): Sophie Charlotte Schmidt, Florian Thiery and Martina Trognitz, "Practices of Linked Open Data in Archaeology and their Realisation in Wikidata", in: *Digital* 2, 333–364. https://doi.org/10.3390/digital2030019.

Smith (2005): Tyler Jo Smith, "The Beazley Archive: Inside and Out", in: *Art Documentation* 24, 22–25.

Smith/Bradford/Gruber (forthcoming): Tyler Jo Smith, Abigail Bradford and Ethan Gruber, "Linked Open Data Vocabularies and Recognizing Intellectual Contributions via ORCID: the Example of Kerameikos.org", in: Phaedon Kyriakidis, Antonia Agapiou and Georgios Leventis (eds.), *Proceedings of the 48th CAA Conference* (CAA 2021 "Digital Crossroads", Limassol, 14–18 June 2021), Tübingen, forthcoming.

Tzitzikas *et al.* (2022): Yannis Tzitzikas, Michalis Mountantonakis, Pavlos Fafalios and Yannis Marketakis, "CIDOC-CRM and Machine Learning: A Survey and Future Research", in: *Heritage* 5, 1612–1636. https://doi.org/10.3390/heritage5030084.

Wilkinson *et al.* (2016): Mark D. Wilkinson, Michel Dumontier, IJsbrand Jan Aalbersberg, Gabrielle Appleton, Myles Axton, Arie Baak, Niklas Blomberg, Jan-Willem Boiten, Luiz Bonino da Silva Santos, Philip E. Bourne, Jildau Bouwman, Anthony J. Brookes, Tim Clark, Mercè Crosas, Ingrid Dillo, Olivier Dumon, Scott Edmunds, Chris T. Evelo, Richard Finkers, Alejandra Gonzales-Beltran, Alasdair J.G. Gray, Paul Groth, Crole Globe, Jeffrey S. Grethe, Jaap Heringa, Peter A.C.'t Hoen, Rob Hooft, Tobias Kuhn, Ruben Kok, Joost Kok, Scott J. Lusher, Maryann E. Martone, Albert Mons, Abel L. Packer, Bengt Persson, Philippe Rocca-Serra, Marco Roos, Rene van Schaik, Susanna-Assunta Sansone, Erik Schultes, Thierry Sengstag, Ted Slater, George Strawn, Morris A. Swertz, Mark Thompson, Johan van der Lei, Erik van Mulligen, Jan Velterop, Andra Waagmeester, Peter Wittenburg, Katherine Wolstencroft, Jun Zhao and Barend Mons, "The FAIR Guiding Principles for Scientific Data Management and Stewardship", in: *Scientific Data* 3 (160018) https://doi.org/10.1038/sdata.2016.18.

Yoose/Perkins (2013): Becky Yoose and Jody Perkins, "The Linked Open Data Landscape in Libraries and Beyond", in: *Journal of Library Metadata* 13, 197–211.

Zou (2018): Lei Zou, "SPARQL", in: Ling Liu and M. Tamer Özsu (eds.), *Encyclopedia of Database Systems*, 2nd edition, New York, 3554–3558.

Verena Gassner

23 New Perspectives in the Study of Western Greek Amphorae

Abstract: This contribution gives an overview of the study of Western Greek amphorae in the last 20 years – one involving serious challenges because morphologically identical vessel types were used over large geographical areas, including southern France, southern Italy, and Sicily. In particular, this scholarship has shown how the method of fabric classification made it possible to distinguish a series of production sites on the Ionian and Tyrrhenian coasts of Italy and Sicily. These have been made accessible online by a web-based information system called FACEM (Fabrics of the Central Mediterranean), first presented in 2011 and currently in its 8th release.

Introduction

In the last two decades the progress of interdisciplinary research in archaeology as well as the rapid development of Digital Humanities led to major changes in our approaches to the study of ancient pottery. This contribution presents the impressive developments in the study of Western Greek amphorae that became possible by an interdisciplinary approach, in particular by the combination of archaeometric analysis with the concept of fabrics. The widespread dissemination of the results of this research was supported by a web-based database (FACEM – www.facem.at [last access 02.07.2024]) that allowed users to engage in analyses of distribution patterns and exchange modalities.[1] With the term "Western Greek amphorae," we refer to a large class of amphorae, dating from the 6th to the 4th centuries B.C., which were produced in most of the Greek cities in southern Italy and Sicily, but also in Greek settlements in the western Mediterranean, such as Massalia and Ampurias.[2] The form of the amphorae is therefore characteristic for a large production area and not only for one city.[3] At the beginning, in the early 6th century B.C., we find vessels with a globular-bellied body and a cylindrical neck; the rim is rounded and outward-thickened, followed by an offset band around the top of the neck. In the late 6th and the early 5th centuries B.C., the shape of the body became more bulbous and the rim remained thickened and projecting outwards, but the stepped collar under the rim instead became a groove in the lower part of the rim. The final stage, from the middle of the 5th to the 4th centuries B.C., is characterized by slender ovoid vessel forms with strongly swollen necks and almond-shaped rims, the groove having developed into a small ridge in the uppermost part of the neck.

That we can understand these different shapes as steps of a continuous development of one class was the fundamental contribution of J.-Ch. Sourisseau at the annually Congress of Taranto in 2009 because, before they were perceived as different "types", defined for example as "Corinthian B amphorae", "amphores ioniennes", "amphores ionio-massaliètes" or "pseudo-chiote."[4] Sourisseau proposed five general Forms (Form 1α-5) as representative for the chronological development of Western Greek amphorae from the early 6th to the 4th century B.C. (Fig. 1).[5] For the later 5th and the 4th centuries B.C., the typology of Sourisseau overlapped with the already existing typology of Ch. Vandermersch, which concentrated on amphorae from the 4th to the 2nd centuries B.C., but already started in the later 5th century B.C.[6] For example, Form Sourisseau 4 corresponds to Vandermersch MGS II, form Sourisseau 5 to Vandermersch MGS I. As the typological sequence of J.-Ch. Sourisseau is mainly based on the general form of the vessel, which is often difficult to identify when only fragments are preserved, it was supplemented by a typology of rim forms by the present author (Fig. 2).[7] As research currently

1 My warmest thanks for the invitation to present these results at Torino go to the organizers of the congress: Diego Elia, Eleni Hasaki and Marco Serino.

2 This paper will concentrate on the situation in Southern Italy and Sicily. For the important production at Massalia, see Bertucchi (1992).

3 For the concept 'one amphorae type – one city' of the 1970s and 1980s, see Gras (1987). For the definition of the term 'production region', see Gassner (2003) 207–208, fig 104; Lawall (2010) with the previous bibliography.

4 For the history of research, often discussed in the last years, see Gassner (2003) 173–219; Savelli (2009); Sacchetti (2012) 32–48; Gassner (2015).

5 Sourisseau (2011), see in particular fig. 6 for an overview of the forms.

6 Vandermersch (1994). For the difficult history of research of the MGS-typology, see Cibecchini/Capelli (2013); Gassner (2024) ch. V.2.1.

7 Gassner (2003) 173–219; Gassner/Sauer (2015); Gassner (2024).

https://doi.org/10.1515/9783111189635-023

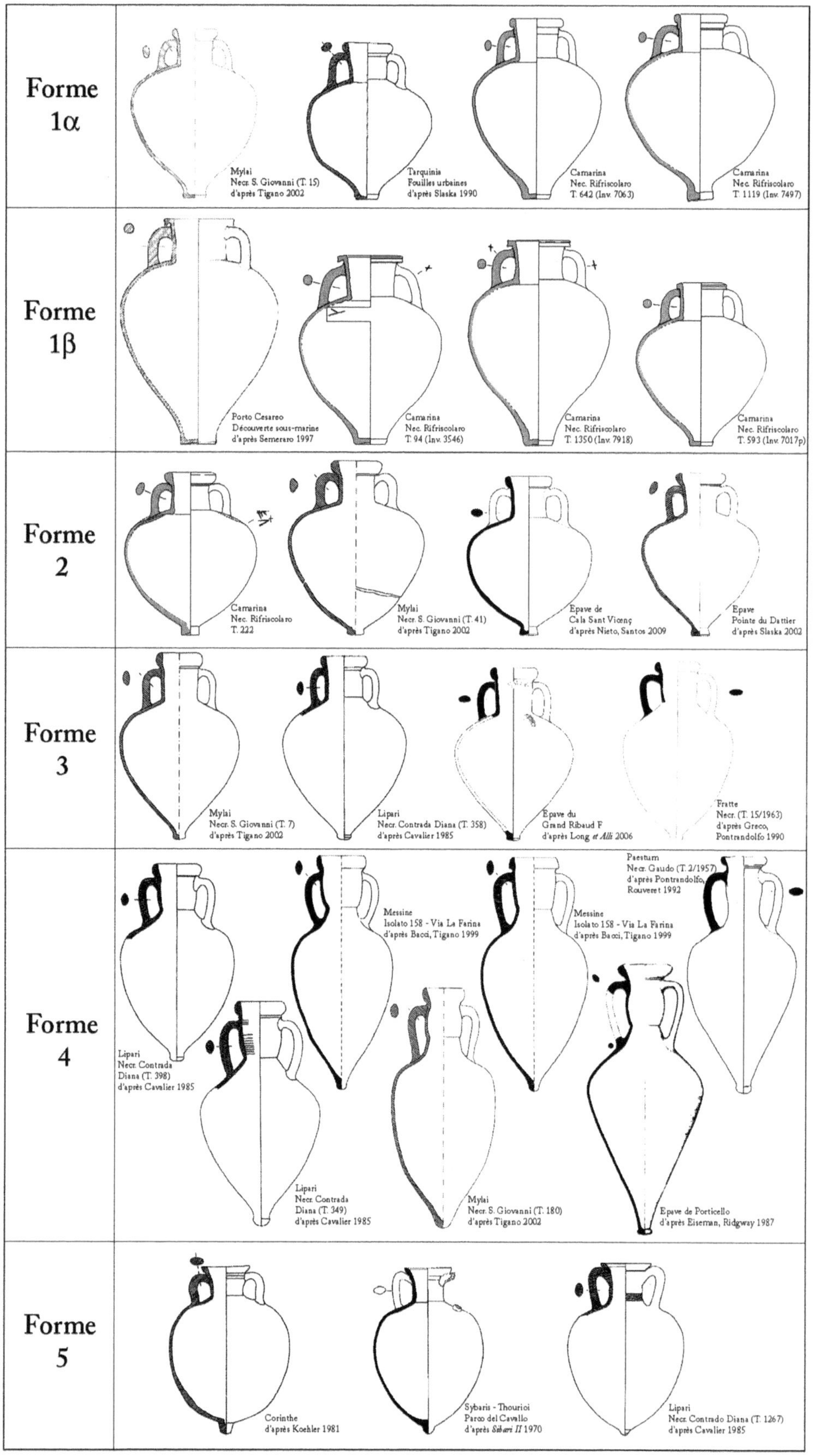

Fig. 1: Amphorae forms Sourisseau 1–5. After Sourisseau (2011) 176, fig. 6.

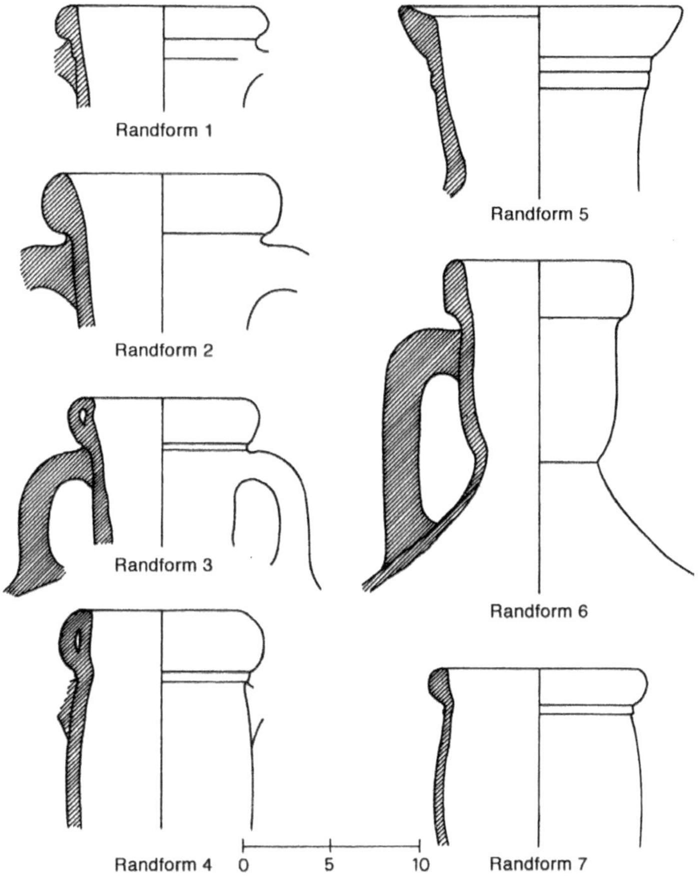

Fig. 2: Rim types Gassner 1–7. After Gassner (2003) 181, fig. 91.

stands, we are aware of the fact that some production centers in the region of the Ionic-Adriatic Sea also used another form type of amphora, subsumed under the term "Corinthian A" and now addressed by J.-Ch. Sourisseau as his Form 1β. This problem will be discussed at the end of the paper.[8]

While Sourriseau's typology of vessel forms allows us to follow the chronological development of Western Greek amphorae, the identification of their provenance remains very difficult, as in most cases minor differences in the morphology of vessels are not sufficient to attribute them to specific production centers or areas. Thus, from the 1980s onwards, archaeometric research gained increasing importance in studies of transport amphorae.[9] Archaeometric analysis, however, are restricted to the analyzed samples and, as they are expensive, cannot be extended to large assemblages of pottery. To face this problem, when we started the study of pottery at Elea/Velia in the 1990s we adopted the concept of pottery fabric, developed by British archaeologists, as a tool for classifying finds according to the composition and treatment of the clays used for their production.[10] This method of fabric description was used for the classification of all pottery finds and was also of enormous help during sampling activities for archaeometric analysis. Decisive for our development of fabric classification was the close collaboration with the geologist R. Sauer, who analyzed a large number of samples from various sites in southern Italy, including clay samples from various surveys which allowed the identification of numerous production sites, or at least general areas of production, there.[11]

8 The contribution does not discuss, however, the early production of amphorae in Pithekoussai, see Buchner/Ridgway (1993); Sourisseau (2011) 149–173.

9 E.g., Whitbread (1995) 255–246 for "Corinthian B" amphorae; in Italy we find various research groups e. g. at Milano (G. Olcese), Genova (C. Cappelli), Catania (G. Barone), Palermo (G. Montana).

10 For the concept of fabric, see https://facem.at/project/about.php#what-is-a-fabric (last access 23.05.2023).

11 E.g., Gassner *et al.* (2003), (2014); Gassner/Sauer (2015). These analyses were made possible by a project of the Austrian Science Fund (P-20597-G02, 2008-2010), that was followed by others, see https://facem.at/project/impressum.php (last access 23.05.2023). Important results were also achieved by a project of the Centre Jean Bérard, Naples, and the dissertation of G. Abbas (*Identification pétro-minéralogique des productions*

When we faced the difficult question of which kind of publication would be appropriate for our results, it soon became clear that opting for a printed book could not guarantee that the necessary quantity and high quality of images, indispensable for comparing and identifying fabrics, would be available. In accordance with the quick development of Digital Humanities, we therefore decided to create a web-based information system that would meet the criteria of open-access policy and also offer the possibility to grow continually.[12] The website FACEM (www.facem.at [last access 02.07.2024]) offers archaeologists the opportunity to compare their fabrics to a great number of images of fabrics attributed to identified production sites or regions. It went online in 2011 and to date has had eight releases presenting various classes of ceramics from modern Spain, Italy, Malta, Tunisia, Albania, Greece and Turkey.[13] Over the last few years, our research has extended the chronological and geographical frame considerably, and for this reason as well as the rapid changes in digital methods we plan a basic relaunch of the website for 2024/2025.[14]

The Production Sites of Western Greek Amphorae in Southern Italy

A production center of Western Greek amphorae was identified first at Massalia and thus in the periphery of the main distribution area of this class.[15] In Magna Graecia, Locri remained the only known production site for a long time; here, workshops for Western Greek and Greco-Italic amphorae have been found.[16] In the last decennium it was possible to identify further production areas by the rigorous application of fabric classification, combined with archaeometric research.[17]

Region of Sybaris/Kroton

J.-Ch. Sourisseau suggested that the production of Western Greek amphorae started at Sybaris. His hypothesis is based on examples of the form Sourisseau 1α, found in the necropolis of Rifriscolaro at Camarina together with Middle-Corinthian and Late-Corinthian pottery, dated to the first quarter of the 6th century B.C.[18] A similar chronology is attested for finds from Massalia and Megara Hyblaea.[19] However, at present no archaeometric analyses of the respective amphorae have been published that would confirm their provenance from Sybaris. Clay analyses as well as analyses of amphorae conducted by Sauer confirm a possible production of these amphorae in the region of Sybaris; however, similar clays also occur in the areas of Kroton and Caulonia.[20] Therefore, we cannot exclude that form Sourisseau 1α was also produced at Kroton, where C. Sabbione had already assumed a local production of amphorae,[21] or in a hitherto unknown production center in southern Italy or Sicily.[22]

---

céramiques anciennes. Application aux amphores de Grande Grèce, Thèse de Doctorat, Université Montpellier II, Montpellier 1999) which however remained unpublished.

12 For similar data bases, concentrated on transport amphorae, see for example that of Roman Amphorae of the University of Southampton: https://archaeologydataservice.ac.uk/archives/view/amphora_ahrb_2005 or the database of amphorae from Alexandria: http://amphoralex.org (both last access 23.05.2023). The database *Immensa Aequora* (https://www.immensaaequora.org/it/content/database [last access 23.05.2023]) was still under construction when submitting the paper in May 2023. For a database of chemical, archaeological, petrographic and mineralogical data of prehistoric ceramics from the Aegean and wider Eastern Mediterranean region, see https://ceradat.net (last access 23.05.2023).

13 For its history, see https://facem.at/project/about.php (last access 23.05.2023).

14 The database will keep the name FACEM, which then will stand for *Fabrics of Ancient Ceramics in the Mediterranean*. The chronological frame will start in the 7th century B.C. and continue to the 7th century A.D. See also the contributions of B. Borgers and C. Hasenzagl on FACEM in https://www.socminpet.it/344/plinius.html (forthcoming).

15 Bertucchi (1992); Gassner (2003) 204–206; Sourisseau (2011) 213. This is the reason why the class was called 'Ionio-Massaliote' for a long time.

16 Barra Bagnasco (1992); Gassner (2003) 189–191.

17 It is not possible to reproduce here all images of the fabrics mentioned in this paper, which are however found easily on https://facem.at/search (last access 23.05.2023).

18 Sourisseau (2011) 204–206 (production Group 1).

19 Sourisseau (2011) 177–189, fig. 7–14.

20 Sauer/Gassner (2008); see also Gassner (2003) 186–188.

21 Sabbione (1986) 260. According to his description the amphora could be form Sourisseau 1α.

22 See also Sourisseau (2011). An unknown production centre for form Sourisseau 1α was also identified among the finds from Pantelleria (impasto 21), Bechtold (2013a) 498–499, pl. 92,11.

Based on the current state of research, we have various fabrics that can be attributed to the east coast of Calabria between the Gulf of Taranto and the Gulf of Squillace.[23] Certainly, the region was producing a surplus of wine, as is demonstrated by exported amphorae of form Sourisseau 1α found, e.g., at Butrint in Albania[24] or in the shipwreck of Cala Sant Vicenç near Mallorca.[25] The archaeological context of the latter, dated to 520-510/500 B.C., is of particular importance, as its cargo consisted of amphorae from the region of Sybaris/Kroton belonging to form Sourisseau 1α (fabric CAL -A-6) and form Sourisseau 2 (fabric CAL-A-8) (Fig. 3).[26] If we assume Sybaris as production center for both fabrics (CAL-A-6 and CAL-A-8), then this would mean that the shift from form Sourisseau 1α to form 2 started before the destruction of Sybaris in 510 B.C. If the fabrics belong to different centers, they must have been active at the same time as well. The lifetime of form Sourisseau 2 is assumed from 540/530 B.C. until 480 B.C., according to archaeological contexts at Massalia, Megara Hyblaea and Athens.[27]

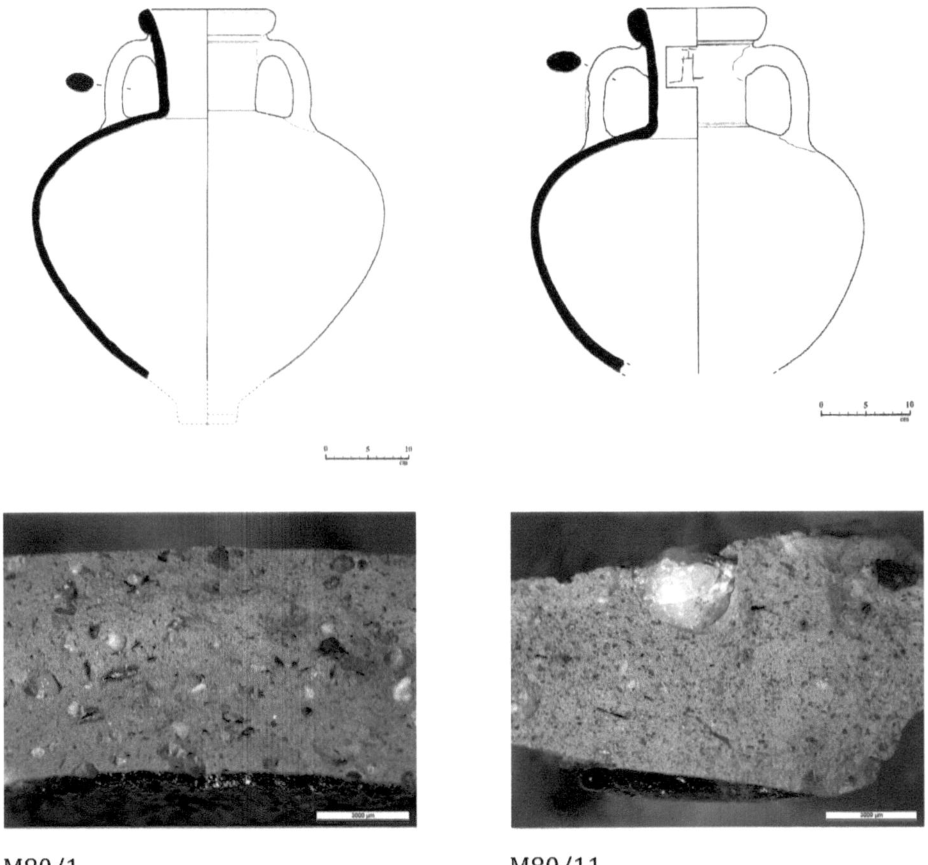

Fig. 3: Production of Sybaris/Kroton, samples from Cala Sant Vicenç (Mallorca): CAL-A-6 (FACEM – http://facem.at/m-80-1); CAL-A-8 (FACEM – http://facem.at/m-80-11).

23 Gassner (2003) 186–188, (2011a) 4-5 (CAL-A-6 to CAL-A-9).
24 Gassner (2011a) 4, note 25.
25 Nieto/Santos (2008); see in particular 125–152 (Greek amphorae) and 316–319 (conclusions); for the archaeometric analyses, see Sauer/Gassner (2008).
26 Sauer/Gassner (2008); see also Gassner (2011a).
27 Sourisseau (2011) 189–191.

Southern Calabria: from Rhegion to Locri

Similar, though clearly distinguishable, fabrics have been identified in the area from Rhegion to Locri. The fabrics CAL-A-4 and CAL-A-5 have been defined by Late Roman amphorae (type Keay 52), produced in the workshop area of Pellaro, località Fiumarella,[28] but have also been identified for amphorae with rim type Gassner 3 (CAL-A-5) and with later amphorae of the form MGS III.[29] Therefore, the production of Western Greek amphorae can be assumed from the 5th century B.C. until the Hellenistic period.

Both fabrics are very similar to a fabric identified at Locri (LOC-A-1), where the discovery of kilns confirmed the local production of the forms Sourisseau 2–4.[30] Unfortunately, the samples available for FACEM were few and certainly do not represent the entire spectrum of the site's/region's production.

Southern Calabria on the Tyrrhenian Coast

Another important and evidently early production region, characterized by a pale brown to yellow fabric, was localized in southern Calabria (CAL-A-1 to CAL-A-3).[31] Sauer suggested the area of Capo Vaticano on the Tyrrhenian coast as the most likely region of provenance, but he could not exclude the area to the south, between Capo Vaticano and Rhegion, or to the north as far as Hipponion/Vibo Valentia, since the clay samples collected during a survey in 1995 were few.[32] Sourisseau assumed that this production, his production 2, started later than that of Sybaris in the years between 580/570 B.C.[33] Evidently, the production continued to the later 6th and the 5th centuries B.C., with vessels of form Sourisseau 2 and 3. In the course of the 5th and the 4th centuries B.C., the amphorae developed a more slender shape (form Sourisseau 4). Characteristic of the later vessels of this production is an elongated rim without ridge (rim type Gassner 6) and a baggy bulge in the lower part of the neck.[34]

Amphorae of this production area were exported mainly in the 6th and the early 5th centuries B.C., as is demonstrated by finds at various sites along the Tyrrhenian coast and in the Bay of Naples area, Sicily and also North Africa.[35]

Poseidonia and Elea

The productions of Poseidonia and Elea show a very similar morphological development, but can be distinguished quite well by the fabric, as the clays used at Poseidonia are calcareous, while the area of Elea is dominated by non-calcareous raw materials.[36] In this wider region we also supposed a further production (NW-LUC-A), which could not be located precisely until now. From the characteristics of the natural landscape we would assume that Poseidonia represented the

28 Gassner (2011a). For Keay 52, see the Southampton Database https://doi.org/10.5284/1028192.

29 See https://facem.at/cal-a-4 and https://facem.at/cal-a-5 (both last access 23.05.2023).

30 Gassner (2011c). The kilns of Locri were published by Barra Bagnasco (1992), (1995); Barra Bagnasco *et al.* (2001); see also Sourisseau (2011) 207–208, (2018) 27–28, figs. 2.3.1-2.3.2. The fragment Barra Bagnasco (1992) 210, n. 184, pl. LX could be interpreted as rim type Gassner 1 = form Sourisseau 1α, but from the publication it is not clear if it belongs to the local production.

31 This production region can be equated with Sourisseau's production Group 2, while his production Group 3 until now has not been identified by us, see Sourisseau (2011) 206-209 (Group 2), 209 (Group 3). FACEM – http://facem.at/cal-a-1 to FACEM – http://facem.at/cal-a-3 (last access 23.05.2023).

32 In Gassner (2003) 191–193 this production was still attributed to the region of Rhegion; see now Gassner (2011a) 1–3.

33 Sourisseau (2011) 206–209.

34 Gassner (2003) 179–182, fig. 91, (2024).

35 Sourisseau (2011) 207 (Massalia); Gassner (2003) 191–193 (Poseidonia and Elea); Di Sandro (1986) 34–35, sg 80, pl. 7 = https://facem.at/m-37-1 (Pithekoussai) (last access 23.05.2023). As examples in FACEM, see the fabric CAL-A-1 (https://facem.at/cal-a-1 [last access 23.05.2023]) with Gassner rim-type 1: M45/4 (Sybaris), M92/55 (Carthage); with Gassner rim 2: M154/36 (Selinunte); with Gassner rim type 3: M6/13; M6/110 (Elea); M37/5, M37/7 (Pithekoussai); M55/1 (Rhegion); M65/5 (Caulonia); M92/58 (Carthage); M119/157; M119/61 (Pantelleria), M149/38; M149/39 (Jerba); with Gassner rim type 6: M6/59 (Elea); M 92/57 (Carthage); M149/41 (Jerba) (last access 23.05.2023). For Jerba, see Bechtold (2013b). It is clear that these samples from FACEM do not reflect all export areas and so are in no way representative.

36 Gassner *et al.* (2003); Sauer (2015).

most important production center for wine or oil, or both, in this region, but, even so, it still is not easy to reconstruct the development of the local amphorae production.[37] We depend mostly on the analysis of exported vessels, in particular from Elea, where imports from Poseidonia played an important role in the 5th century B.C. We found samples of Poseidoniate amphorae with rim types Gassner 3 and 4 (form Sourisseau 3) in contexts of the 1st half of the 5th century B.C.[38] At Naples, the most ancient vessels belong to form Sourisseau 2 and stem from contexts of the late 6th and early 5th centuries B.C., while samples of form Sourisseau 3 were attributed to the beginning of the 5th century B.C.[39] One fragment of rim type Gassner 3 from Pithekoussai (Scarico Gosetti) could be attributed to the 5th century B.C., while most of the identified fragments belong to form Sourisseau 4 (rim type Gassner 7). B. Bechtold also identified many imports from Poseidonia, dated to the 4th century B.C. (rim type Gassner 7) at North African sites.[40]

At Elea we can reconstruct the local production from the 5th to the 2nd century B.C. on the basis of more than 200 individuals from the early settlement in the Lower town.[41] The most ancient amphorae belong to rim types Gassner 3 and 4 (form Sourisseau 3) and appear in small numbers in contexts of the 2nd quarter of the 5th century B.C. This period seems to represent the beginning of the local production, as local amphorae are rare among the repertoire of the early settlement on the acropolis. During the later 5th and the 4th centuries B.C. we see the development of vessels of form Sourisseau 4 with rim type Gassner 7. Evidently, the 4th century B.C. was a flourishing period for the amphorae production of Elea, as Bechtold identified samples from Elea in various sites in Sicily and North Africa.[42]

Among the finds from Elea we were also able to identify amphorae from a production center that was subsumed under the general denomination NW-LUC-A (North-Western Lucania), as archaeometric analysis by Sauer suggested a location in this region.[43] In the repertoire of Elea, this production was very representative in the 5th century B.C. (form Sourisseau 3 = rim types Gassner 3 and 4), but became less frequent from the 4th century B.C. onwards.[44]

Bay of Naples

For a long period, research on the amphorae production in the Bay of Naples was concentrated on Greco-Italic amphorae for which production centers were confirmed by archaeometric analysis at Ischia, Cumae and Naples.[45] The identification of earlier amphora production centers in this area from the late-6th century B.C. to the beginning of the 4th century B.C. depends largely on finds from Piazza Nicola Amore at Naples.[46] At the present, the most ancient known samples belong to form Sourisseau 2 (rim type Gassner 3), found at Naples in a context of the 1st half of the 5th century B.C.[47] More frequent are amphorae of form Sourisseau 4 (= MGS II), belonging to the late 5th or early 4th centuries B.C.[48]

The archaeometric analysis, conducted by Sauer, made it possible to distinguish a production within the area of Naples (BNAP-A-1 to BNAP-A-3; Fig. 4) which started in the 1st half of the 5th century B.C.[49] A second group (BNAP-A-8 to BNAP-11)

37 Gassner (2003) 197–201 with the previous bibliography; Giacco (2017); De Bonis *et al.* (2020); Gassner (2024). For the fabrics, see http://facem.at/pae-a-1 to http://facem.at/pae-a-4 (last access 23.05.2023).

38 Gassner (2003) 197–201, fig. 100. At the present state of research imports from Poseidonia seem largely absent from the early materials on the acropolis of Elea.

39 Gassner/Scoppetta (2014) 124, cat. 2; cat. 8, fig. 1 (Sourisseau form 2), cat. 11-13, fig. 2 (Sourisseau form 3)

40 Bechtold (2018) 3–5, fig. 3.3, 8.2–3.

41 Gassner (2003) 195–197; Gassner/Sauer (2015). For the further development of the local amphora production, see Gassner (2024). For the fabrics, see http://facem.at/vel-a-1 to http://facem.at/vel-a-5 (both last access 23.05.2023).

42 Bechtold (2018) 5–6, 14–18, see fig. 3, 4; 7; see also for later amphorae fig. 3, 5–6; 8; fig. 8,1; Bechtold *et al.* (2018) 25–27, fig. 4.

43 Gassner (2003) 201–202, (2011d). Their morphological similarities to Calabrian amphorae might also be taken as indication that their production centre should be sought in the south of Elea, but here further investigations are necessary. For the fabrics, see http://facem.at/nw-luc-a-1 and http://facem.at/nw-luc-a-1 – http://facem.at/nw-luc-a-2 (last access 23.05.2023).

44 Gassner (2024).

45 Olcese (2010), in particular 185–230, (2012) 357–359; Olcese *et al.* (2013). For the excavations at Piazza Nicola Amore, see Febbraro/Giampaola (2012).

46 Savelli (2006); see also Gassner/Scoppetta (2014); Gassner (2015).

47 Gassner/Scoppetta (2014) cat. 5, fig. 1; 3.

48 Gassner/Scoppetta (2014) cat. 14–15, fig. 2.

49 Gassner/Scoppetta (2014) cat. 5, fig. 1; 3. For the archaeometric analyses, see Gassner/Sauer (2016). The production Group BNap-A-6 to BNap-A-7 have only been identified with Greco-Italic amphorae.

represents one or several production centers in the wider surroundings of Naples, documented mainly by Greco-Italic or Roman amphorae.[50] Only BNap-A-9 was already used for amphorae of Gassner rim type 3, belonging to Sourisseau form 3, which were found at Elea in contexts from the first half to the last quarter of the 5th century B.C.[51]

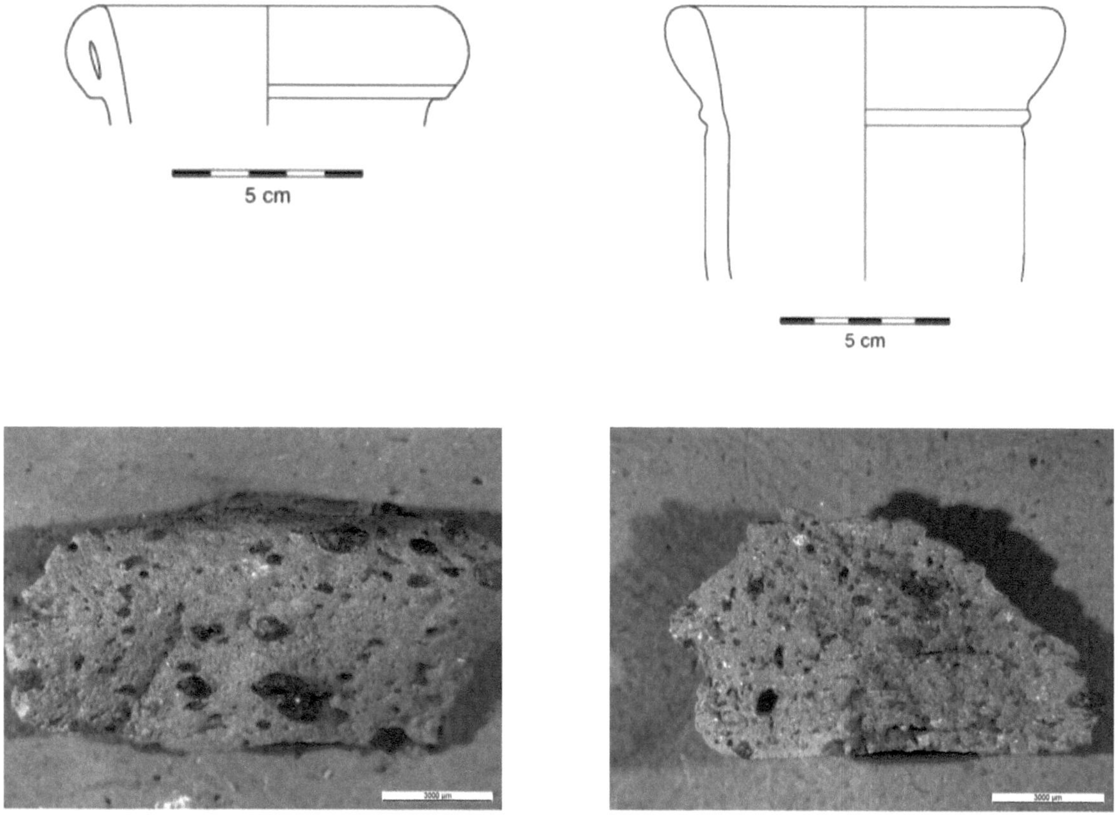

M137/1 M137/3

Fig. 4: Production of Naples, samples from Piazza Nicola Amore: BNap-A-1 (FACEM – http://facem.at/bnap-a-1); BNap-A-3 (FACEM – http://facem.at/bnap-a-3).

The Production Sites of Western Greek Amphorae in Sicily

In 2011, Sourisseau located his production Group 4 in southeastern Sicily because it was frequently identified with amphorae found in this area.[52] The most ancient vessels belonged to form Sourisseau 2 and were found in tombs of the 2nd half of the 6th century B.C. at Camarina, thus giving a reliable starting point for these workshops. Twelve years later and due to the groundbreaking work of B. Bechtold, including her studies of transport amphorae from the necropolis of Himera, we now have much better insights into the situation on the southern and northern coasts of Sicily, though we still are not able to follow the entire chronological development of shapes in all known productions.[53] The amphorae production of

50 Gassner/Sauer (2016).

51 https://facem.at/bnap-a-9 (M6/7); https://facem.at/m-6-120 (last access 23.5.2023); see also Gassner (2003) 202–203, fig. 103. Also the production Group CAMP-A-1 to CAMP-A-5 is only attested with Greco-Italic amphorae, see Gassner/Sauer (2016).

52 Sourisseau (2011) 209–210.

53 These studies are conducted in close collaboration with S. Vassallo and – for the archaeometric research – G. Montana; the project was funded by the Austrian Science Fund (P 25046-G19; P 30030-G25). See https://facem.at/project/papers.php s. v. Sicily for the papers by the team of B. Bechtold (last access 23.05.2023).

the eastern coast of Sicily included in Sourisseau's Group 4, with important Greek cities like Syracuse, Catania or Naxos, remained vague until now.[54]

Southern Coast of Sicily

The production of transport amphorae on the Southern coast of Sicily was confirmed for Gela by analysis conducted by G. Spagnolo and G. Barone in 2012. The production started in the Late Archaic period with amphorae of form Sourisseau 2 and continued until the later 4th century B.C. (form MGS IV).[55]

Recent research by Bechtold also identified the local production of transport amphorae at Akragas (AKR-A-1), where kilns have been found in the area outside Porta V.[56] The production started with vessels of form Sourisseau 2 (rim type Gassner 3) dated to the Late Archaic period, and continued in the 5th and the 4th centuries B.C. with forms Sourisseau 3 and 4 (rim type Gassner 7) (Fig. 5). Bechtold identified numerous exported amphorae of this important production in various sites in Sicily (Selinous, Segesta, Himera) as well as on Pantelleria and Malta.[57]

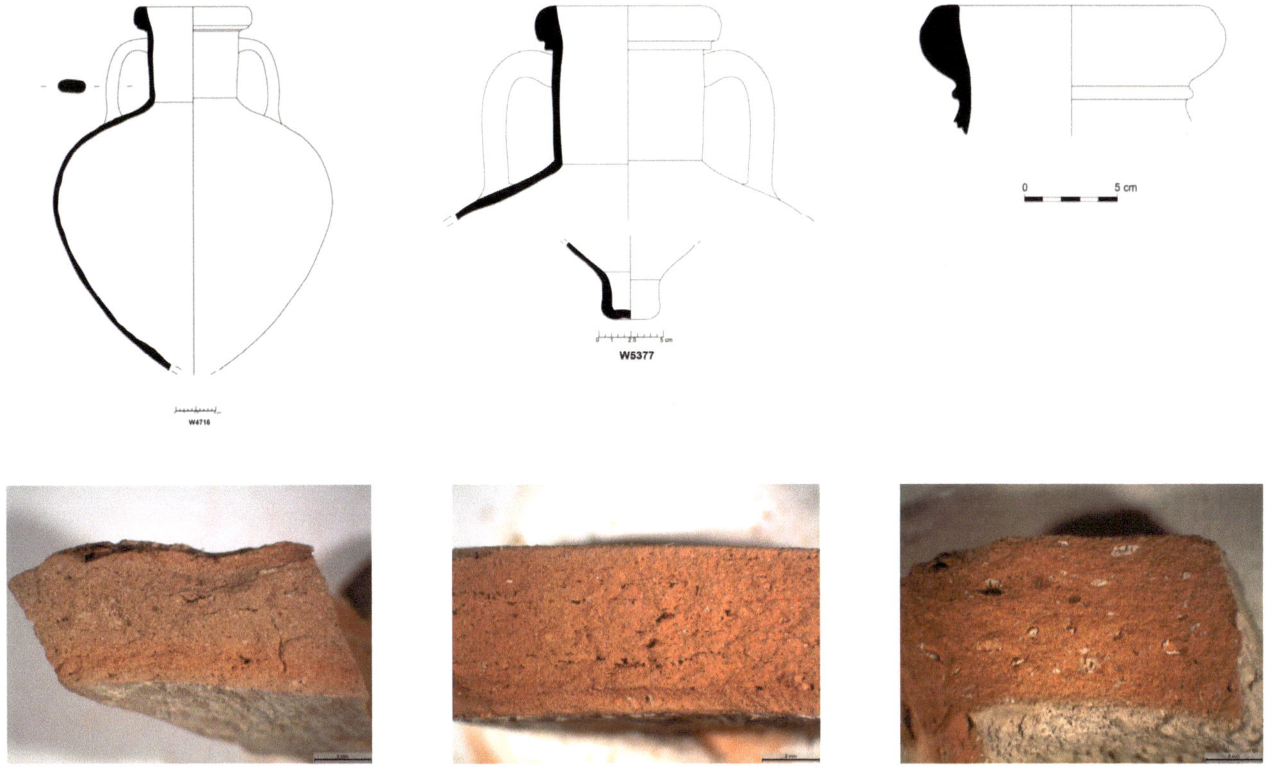

M179/168 M179/232 M208/1

Fig. 5: Production of Akragas, samples from Himera (M179) and Akragas (M208): FACEM – http://facem.at/m-179-168; FACEM – http://facem.at/m-179-232; FACEM – http://facem.at/m-208-1.

Further workshops existed at Selinous, where the production of Punic transport amphorae was attested by finds from the kiln area on the acropolis, in block FF1North. Based on current research, fragments of form Sourisseau 4 (rim type Gassner

54 For the state of research regarding the area of Catania, see the overview in Bechtold *et al.* (2020).

55 Barone *et al.* (2012); Spagnolo (2018).

56 Bechtold (2020a); see also Barone *et al.* (2003) 45; Barone *et al.* (2004).

57 Bechtold (2020a), fig. 5.

7) (SEL-A-4) also confirm the local production of Western Greek amphorae during the pre-409 B.C. phase of Selinous. B. Bechtold also did not exclude an earlier start of the production (rim types Gassner 2 and 3).[58]

Northern Coast of Sicily

On the northern coast of Sicily the production of Western Greek amphorae was identified at various sites. The situation at Messina and its surroundings remains difficult to understand until now, though the excavators assumed a local production of amphorae of form Sourisseau 4, to which G. Barone attributed two amphorae from Mylai.[59]

The recent studies of Bechtold identified the local production of Western Greek amphorae of Himera.[60] The most ancient vessels belong to form Sourisseau 1α and stem from contexts dating to the 3rd quarter of the 6th century B.C.[61] In the Late Archaic period they were followed by form Sourisseau 2. The latest examples belong to form Sourisseau 4 (rim type Gassner 7) and date to the period before the destruction of Himera in 409 B.C. Until now, amphorae from this production were identified only at Himera.[62]

Western Greek amphorae were also produced in Panormos or Solus (or both), where an important production of Punic amphorae is attested as well.[63] The typological sequence started in the Late Archaic period with samples of rim types Gassner 2 and 3, for which Bechtold assumed mainly a regional exchange, including indigenous communities in the area. While vessels belonging to the 2nd and 3rd quarters of the 5th century B.C. are still missing, amphorae of rim type Gassner 7 are present. The production probably continued into the later 4th century B.C. with fragments of rim types Gassner 8 and 9, which constitute the link to the class of Greco-Italic amphorae. It is interesting to note that in Sicily amphorae of form Sourisseau 4 of the later 5th and the 1st half of the 4th centuries B.C. often display rim type Gassner 7 and therefore follow (or were followed by?) the Campanian tradition, while amphorae form Sourisseau 4 from the south Calabrian area were characterized by rim type Gassner 6.[64]

Greek Amphorae from the Ionic Adriatic Region

Though the term “Western Greek amphorae” suggests that these amphorae represent the only class of Greek amphorae produced in the central and western Mediterranean, this does not correspond to ancient realities. In particular, on the Ionian coast of Southern Italy we find a very complex situation which reflects the long-lasting and tight cultural and economic contacts with the Albanian and northwestern Greek coasts.[65] We therefore subsume amphorae from these areas under the term of “Ionic-Adriatic,” and are aware of the fact that the situation is still difficult to understand.[66]

Many sites in this region, for example Corcyra, started their production of amphorae in the 6th century B.C. with vessels resembling the early shape of Western Greek amphorae, which are now identified as form Sourisseau 1α, but were originally named “Corinthian B”.[67] C. G. Koehler described the various chronological steps in the development of

---

58 Bechtold (2020b) with the earlier bibliography; for the archaeometric analyses, see Montana *et al.* (2018). Fabrics SEL-A-1 to SEL-A-3 belong to Punic amphorae.

59 Bacci/Tigano (2001) 29, fig. 10,17–19; see also Barone *et al.* (2005) 26; Barone *et al.* (2011) 90; 98–99, see also Bechtold *et al.* (2019) 10.

60 Bechtold *et al.* (2019); Bechtold (2020c) with earlier bibliography; see also Bechtold/Vassallo (forthcoming); for the archaeometric analyses, see Montana *et al.* (2020).

61 Bechtold *et al.* (2019) 5–6.

62 One exception is the rim fragment (M154/158), found at Selinous, see Bechtold (2020c) 8, fig. 4.

63 Bechtold (2015) 26–43; Bechtold/Schmidt (2015); Montana/Randazzo (2015). The clays of these two sites could not be distinguished by archaeometric analyses.

64 See also Bechtold *et al.* (2019) 9–10.

65 Gassner (2015).

66 Gassner (2011b).

67 Kourkoumelis (1990); Preka-Alexandri (1992); Gassner (2011b); Sacchetti (2012) 39–43. The term ‘Corinthian’ results from the – not correct, as we know now – assumption that the class was produced at Corinth.

these so-called "Corinthian B" amphorae, but did not propose an explicit typological classification for them.[68] However, it became clear that from the 5th century B.C. onwards the shape of these amphorae developed in a different way from that described by Sourisseau for Western Greek amphorae (forms Sourisseau 2–4).[69]

In the Ionic-Adriatic region we find also amphorae denominated "Corinthian A". The term goes back to C. G. Koehler as well and was given to globular-bellied vessels with a horizontal, projecting rim that were produced at Corinth.[70] From the 2nd quarter of the 5th century B.C. onwards, the body of these amphorae became more elongated and C. G. Koehler denominated the type "Corinthian A'"[71] At a rather early period of research it became clear that amphorae of this shape were not only produced at Corinth, but also at various sites of the Illyrian-Epirote coast and the Ionian coast of Italy, often Corinthian foundations; they were, however, referred to as "imitations."[72] It was once again J.-Ch. Sourisseau who suggested giving up the concept of "imitations" and also the misleading and often incorrect name "Corinthian A" amphorae and to replace it with "Western Greek amphorae form 1β" for vessels produced in Southern Italy or Sicily.[73] This type, however, refers only to vessels of the 6th century B.C., while we have again to revert to the publication of C. G. Koehler for the later development.[74] Evidently, the present situation is not satisfying, as the new terminology refers only to the Archaic period (form Sourisseau 1β) or constitutes a simple addition to typologies of amphorae from southern Italy and Sicily (form Sourisseau 5; rim type Gassner 5). As a coherent typology for amphorae of the Ionic-Adriatic region remains a desideratum, in the meantime I would prefer to change simply the denomination "Corinthian A" to "Koehler A" in order to avoid probably short-lived new types, but also to correct the often incorrect provenance indication.[75]

Based mainly on archaeometric analysis of amphorae from sites in Albania such as Apollonia and Butrint, we currently are able to distinguish three regions of production which, however, cannot be located unambiguously, since the regions in question show raw materials with very similar characteristics.[76] Most important seems the region of Corfu/Butrint, where both amphorae of the type "Corinthian B"/Western Greek and of the type Koehler A had been produced.[77] Archaeometric analysis distinguished a second production region, located in the central Albanian region as well.[78] It produced amphorae of the type Koehler A and continued its activities until the 4th century B.C. A third production region is attested for the Hellenistic period.[79] The existence of workshops, producing amphorae of the type Koehler A in southern Italy and Sicily, remains a big question mark, as comprehensive archaeometric analysis is still missing. F. Sacchetti called attention to the fact that amphorae of the type Koehler A and in particular Koehler A' are very frequent at sites like Siris, Metapontum or Incoronata.[80] Evidently we cannot exclude a local production of amphorae of these types in southern Italy, but only further archaeometric analysis will make it possible to evaluate their importance.[81]

68 Koehler (1979) 33–49; see now also Sacchetti (2012) 32–39; Sourisseau (2018) 28–29, fig. 2.3.6.

69 The first stage of these amphorae in the 5th century B.C. was labelled form 5 by Sourisseau (2011) 193–195, MGS I by Vandermersch (1994) 61–65 and rim type 5 by Gassner (2003) 180–182, fig. 91, but this does not help to understand the typo-chronological development of the class.

70 Koehler (1979) 9–32; Whitbread (1995) 255–346.

71 Koehler (1981).

72 Sacchetti (2012) 24, notes 63–64, with the previous bibliography to 'imitations'; see also Gassner (2015).

73 Sourisseau (2006), (2011) 186–188, figs. 13–14.

74 For a detailed description of the situation, see Sacchetti (2012) 15–32.

75 Gassner (2024) chap. V.3. and V.4.1. Unfortunately, form Sourisseau 1α now corresponds to Koehler's type "Corinthian B", while his form 1β is equated with Koehler's form "Corinthian A", creating thus further confusion.

76 The archaeometric analyses were conducted by R. Sauer on the material from the excavations of K. Neeft and R. Docter, but unfortunately this important volume was never published. They were supplemented by studies of Greco-Italic amphorae from Apollonia from the excavations of B. Lahi, H. von Hesberg and M. Fiedler. See Gassner (2011b) and Gassner (2015) with earlier bibliography.

77 ION-ADR-A-1 to ION-ADR-A-4, Gassner (2011b). For the workshops of Corfu, see Kourkoumelis (1990) 42–47; Preka-Alexandri (1992) 41–52.

78 ION-ADR-A-5 to ION-ADR-A-6, Gassner (2011b).

79 ION-ADR-A-7 to ION-ADR-A-11, Gassner (2011b).

80 Sacchetti (2012) 24–32; see also Sourisseau (2018) 29; for Metapontum, see Morter/Leonard (1998).

81 During our sampling campaign in the 1990s it was not possible to have access to finds from this region.

Conclusions

As has become clear from this short synthesis, the conditions and dynamics of the production of Greek amphorae in the central and western Mediterranean are complex and not yet fully understood. However, in the last 20 years, research on the topic made enormous progress and was able to bring order to the puzzling variety of "types"; we also succeeded in identifying a conspicuous number of production sites. This was only made possible by both the publication of a relevant number of transport amphorae from securely dated archaeological contexts and a steady intensification of interdisciplinary cooperation. Only the growing integration of archaeometric analysis into the study of pottery allowed the definition of production sites and production regions. Of similar importance was the introduction of the concept of fabric, which made it possible to extend the classification according to the provenance of finds to large assemblages of amphorae.

As a result of these efforts, we can now identify the local production of transport amphorae in many, though not all, important Greek settlements in southern Italy and Sicily and can follow their typo-chronological evolution. It also has become clear that the thesis that every *apoikia* adopted the amphora shape of its mother-city resulted from a colonialist approach and does not relate to ancient realities. In fact, the shape of Western Greek amphorae was not characteristic for one city, but for a rather large area which ranged from the Epirote coast to southern Italy and Sicily and the western Mediterranean. While starting in the first half of the 6th century B.C. from a very similar shape, denominated form Sourisseau 1α, the various production regions developed diverse morphological languages from the 5th century B.C. onwards. In the area of the Ionian Sea amphorae from the 5th century B.C. show form Sourisseau 5 (= MGS I) while simultaneously in Sicily and on the Tyrrhenian side of southern Italy the development continues with vessels of form Sourisseau 4 and rim type Gassner 7. The production on the Tyrrhenian coast of Calabria was characterized by amphorae of form Sourisseau 4, but rims of type Gassner 6.

For future research on Western Greek amphorae, in particular for studies of ancient exchange activities, it seems necessary to extend the method of fabric classification – with due caution and with control by archaeometric analysis – to large assemblages of transport amphorae. Only in this way and with the use of databases such as FACEM, will it be possible to determine the correct provenance of all samples, since our studies have given clear evidence that the morphological shape of an amphora often is not sufficient for determining its provenance. For example, so-called "Corinthian A" amphorae could have been produced at Corinth, at sites on the Albanian-Epirote coast or on the Ionian coast of southern Italy, and so on. Only studies also comprising the fabric of pottery items will allow us to get a reliable picture of the economic strength of certain regions and of both long-distance and regional trade, as has been demonstrated by the study of amphorae from Himera. Equally important will be the further reconstruction of local series of transport amphorae that will only be possible when vessels from chronologically reliable contexts are also classified according to the fabric. This is true not only of the class of Western Greek amphorae, but also of amphorae from the Ionic-Adriatic region ("Corinthian A" and "Corinthian B"), for which a reliable chronological typology, maybe under the label Koehler A and Koehler B, should be developed. Only the deepening of studies on early Koehler A amphorae will make it possible to find out if amphorae of form Sourisseau 1α from the early 6th century B.C. are, in fact, the most ancient Greek amphorae produced in southern Italy and Sicily, or if we also have to take into consideration Koehler A amphorae, which could be earlier, since they are already dated to the 7th century B.C. at Corinth.[82]

Bibliography

Bacci/Tigano (2001): Giovanna Maria Bacci and Gabriella Tigano (eds.), *Da Zancle a Messina. Un percorso archeologico attraverso gli scavi* 2, Messina.

Barone *et al.* (2003): Germana Barone, Catia Branca, Vincenza Crupi, Salvatore Ioppolo, Domencio Majolino, Giusy Puglisi, Grazia Spagnolo and Gabriella Tigano, "Archaeometric Analyses on Ceramics from Sicilian Greek Colonies: A Contribution to the Knowledge of Messina, Gela and Agrigento Production", in: *Periodico di Mineralogia* 73, 43–56.

Barone *et al.* (2004): Germana Barone, Vincenza Crupi, S. Galli and Domenico Majolino and P. Migliardo and Grazia Spagnolo, "Mineralogical-Petrographic and Spectroscopic Investigations on Coarse Pottery and Transport Amphorae from Agrigento", in: *Mediterranean Archaeology and Archaeometry* 4, 47–59.

82 Sourisseau (2006).

Barone *et al.* (2005): Germana Barone, Salvatore Ioppolo, Domenico Majolino, Caterina Branca, Lidia Sannino, Grazia Spagnolo and Gabriella Tigano, "Archaeometric Analyses on Pottery from Archaeological Excavations in Messina (Sicily, Italy) from the Greek Archaic to the Medieval Age", in: *Periodico di Mineralogia* 74.1, 11–41.

Barone *et al.* (2011): Germana Barone, Cristina Maria Belfiore, Paolo Mazzoleni, Grazia Spagnolo, Gabriella Tigano and Maurizio Triscari, "Le anfore commerciali dell'Occidente greco in età arcaica e classica: problemi di localizzazione delle fabbriche", in: Sabrina Gualtieri, Elisabetta Starnini and Roberto Cabella (eds.), *La ceramica e il mare* (Atti della XII Giornata di Archeometria della Ceramica, Genova, 2008), Rome, 87–102.

Barone *et al.* (2012): Germana Barone, Paolo Mazzoleni, Grazia Spagnolo and Erica Aquilia, "The Transport Amphorae of Gela: A Multidisciplinary Study on Provenance and Technological Aspects", in: *Journal of Applied Science and Computetions* 39, 11–22.

Barra Bagnasco (1992): Marcella Barra Bagnasco (ed.), *Lo scavo di Marasà Sud. Il sacello tardo arcaico e la «casa dei leoni»*. Locri Epizefiri IV, Florence.

Barra Bagnasco (1995): Marcella Barra Bagnasco, "Anfore locresi: documenti a favore di una produzione locale tra VI e IV sec. a. C.", in: Màrius Vendrell-Saz, Trinitat Pradell, Judit Molera and Maite Garcia (eds.), *Estudis sobre ceràmica antigua* (Actes del simposi sobre ceràmica antigua), Barcelona, 77–81.

Barra Bagnasco *et al.* (2001): Marcella Barra Bagnasco, Antonella Casoli, Giacomo Chiari, Roberto Compagnoni, Patrizia Davit and Piero Mirti, "Mineralogical and Chemical Composition of Transport Amphorae Excavated at Locri Epizephiri (Southern Italy)", in: *Journal of Cultural Heritage* 2.3, 229–239.

Bechtold (2013a): Babette Bechtold, "Le anfore da trasporto da Cossyra: un'analisi diacronica (VIII sec. a.C. – VI sec. d.C.) attraverso lo studio del materiale della ricognizione", in: Martina Almonte (ed.), *Cossyra II. Ricognizione topografica. Storia di un paesaggio mediterraneo*, Tübinger Archäologische Forschungen 11, Rahden/Westfalen, 409–517.

Bechtold (2013b): Babette Bechtold, "Distribution Patterns of Western Greek and Punic Sardinian Amphorae in the Carthaginian Sphere of Influence (6th–3rd century BCE)", in: *Carthage Studies* 7, Ghent, 43–119.

Bechtold (2015): Babette Bechtold, "Western Greek Amphorae Produced at Panormos and/or Solus", in: *FACEM* (version December/06/2020). http://www.facem.at/project-papers.php (last access 23.05.2023).

Bechtold (2018): Babette Bechtold, "La distribuzione della produzione anforica di Poseidonia/Paestum (V-I sec. a.C.) nell'area di influenza punica (Sicilia, Tunisia, Malta): una revisione dei dati editi e prospettive di ricerca", in: *FACEM* (version December/06/2018). http://www.facem.at/project-papers.php (last access 23.05.2023).

Bechtold (2020a): Babette Bechtold, "First Observations on Western Greek Amphorae Produced at Akragas", in: *FACEM* (version December/06/2020). http://www.facem.at/project-papers.php (last access 23.05.2023).

Bechtold (2020b): Babette Bechtold, "First Evidences for Western Greek Amphorae Produced at Selinus", in: *FACEM* (version December/06/2020). http://www.facem.at/project-papers.php (last access 23.05.2023).

Bechtold (2020c): Babette Bechtold, "Prime osservazioni sulla produzione di anfore greco-occidentali in alcune città della Sicilia: il contributo degli scavi di Himera", in: Monica Chiovaro and Riccardo Sapia (eds.), *Studi in onore di Stefano Vassallo*, Palermo, 135–143.

Bechtold *et al.* (2018): Babette Bechtold, Giuseppe Montana and Lucia Randazzo, "Campanian Wine for Punic Sicily: Petrographic and Archaeological Studies of Graeco-Italic Amphorae from Palermo", in: *Mediterranean Archaeology and Archaeometry* 18, 11–33. https://doi.org/10.5281/zenodo.1297404.

Bechtold *et al.* (2019): Babette Bechtold, Stefano Vassallo and Flavio Ferlito, "La produzione delle anfore greco-occidentali di Himera: uno studio sulla loro identificazione", in *Notiziario Archeologico della Soprintendenza di Palermo* 51, 121.

Bechtold *et al.* (2020): Babette Bechtold, Laura Maniscalco, Maria Turco and Flavio Ferlito, "Studi di provenienza su anfore da trasporto rinvenute in alcuni siti indigeni della provincia di Catania (VI-IV sec. a.C.): primi dati", in: *FACEM* (version December/06/2020). http://www.facem.at/project-papers.php (last access 23.05.2023).

Bechtold/Schmidt (2015): Babette Bechtold and Karin Schmidt, "Amphorae Fabrics of Solus: Evidences for Local Production and Export", in: *FACEM* (version June/06/2015). http://www.facem.at/project-papers.php (last access 23.05.2023).

Bechtold/Vassallo (forthcoming): Babette Bechtold and Stefano Vassallo, *Le anfore greco-occidentali dalle necropoli di Himera (VI-V sec. a.C.): produzioni e circolazione*, Bulletin Antieke Beschaving-Supplements, forthcoming.

Bertucchi (1992): Guy Bertucchi, *Les amphores et le vin de Marseille: VIe s. avant J.-C.-IIe s. après J.-C.*, in: *Revue Archéoloique de Narboinnese*, Supplément 25, Paris.

Buchner/Ridgway (1993): Giorgio Buchner and David Ridgway, *Pithekoussai 1*, Accademia Nazionale dei Lincei, Monumenti antichi 4, Rome.

Cibecchini/Capelli (2013): Franca Cibecchini and Claudio Capelli, "Nuovi dati archeologici e archeometrici sulle anfore greco-italiche: i relitti di III secolo del Mediterraneo occidentale e la possibilità di una nuova classificazione", in: Fabienne Olmer (ed.), *Itinéraires des vins romains en Gaule (IIIe-Ier siècle av. J. C.)* (Confrontations de faciès, Lattes, 30 janvier – 2 février 2007), Lattes, 423–452.

De Bonis *et al.* (2020): Alberto De Bonis, Verena Gassner, Theodoros Naftlos, Maria Luigia Rizzo, Roman Sauer, Antonia Serritella, Stefano Vassallo and Babette Bechtold, "5th-Century BC Himera and the Campanian Connection: Petrographic and Archaeological Studies on Western Greek Amphorae from Poseidonia and Elea Unearthed in the Necropolis of Himera", in: *Minerals* 10, 227. https://doi.org/10.3390/min10030227.

Di Sandro (1986): Norma Di Sandro, *Le anfore arcaiche dallo scarico Gosetti, Pithecusa*. Cahiers du Centre Jean Bérard 12, Naples.

Febbraro/Giampaola (2012): Stefania Febbraro and Daniela Giampaola, "Ceramiche comuni e vernici nere dal quartiere artigianale di Piazza Nicola Amore a Napoli", in: *FACEM* (version 06/12/2012). http://www.facem.at/project-papers.php (last access 23.05.2023).

Gassner *et al.* (2003): Verena Gassner, Giovanna Greco and Roman Sauer, "Analisi archeometriche a Velia: ceramiche arcaiche e laterizi", in: Giovanna Greco (ed.), *Elea – Velia. Le nuove ricerche* (Atti del convegno di studi, Napoli, 14 dicembre 2001), Quaderni del Centro Studi Magna Grecia 1, Pozzuoli, 199–205.

Gassner *et al.* (2014): Verena Gassner, Roman Sauer and Maria Trapichler, "Pottery Production of Velia", in: Giovanna Greco and Luigi Cicala (eds.), *Archeometry. Comparing Experiences*, Quaderni del Centro Studi Magna Grecia 19, Naples, 191–269.

Gassner (2003): Verena Gassner, *Materielle Kultur und kulturelle Identität in Elea in spätarchaisch-frühklassischer Zeit. Untersuchungen zur Gefäß- und Baukeramik aus der Unterstadt (Grabungen 1987–1994)*. Archäologische Forschungen 8, Velia-Studien 2, Vienna.

Gassner (2011a): Verena Gassner, "Amphorae Production in Calabria", in: *FACEM* (version 06/06/2011). http://www.facem.at/project-papers.php (last access 23.05.2023).

Gassner (2011b): Verena Gassner, "Amphorae Production of the Ionic-Adriatic Region", in: *FACEM* (version 06/06/2011). http://www.facem.at/project-papers.php (last access 23.05.2023).

Gassner (2011c): Verena Gassner, "Amphorae Production of Locri", in: *FACEM* (version 06/06/2011). http://www.facem.at/project-papers.php (last access 23.05.2023).

Gassner (2011d): Verena Gassner, "Unidentified Amphorae Production in NW-Lucania", in: *FACEM* (version 06/06/2011). http://www.facem.at/project-papers.php (last access 23.05.2023).

Gassner (2015): Verena Gassner, "Le anfore Greco-occidentali: riconsiderando la loro evoluzione e l'identificazione dei centri di produzioni", in: Réjane Roure (ed.), *Contacts et acculturations en Méditerranée Occidentale. Hommages à Michel Bats* (Actes du colloque de Hyères, 15–18 septembre 2011), Études Massaliètes 12, Aix-en-Provence, 345–356.

Gassner (2024): Verena Gassner, "Griechische und punische Transportamphoren", in: Verena Gassner and Maria Trapichler (eds.), *Von Hyele zu Velia. Die Stadtmauern im urbanistischen Kontext. Die Funde. Die österreichischen Forschungen in der West- und Unterstadt (1974, 1990–199 und 1997–2001)*, Velia-Studien IV/2, 171–254, Vienna.

Gassner/Sauer (2015): Verena Gassner and Roman Sauer, "Transport Amphorae from Velia", in: *FACEM* (version 06/06/2015). http://www.facem.at/project-papers.php (last access 23.05.2023).

Gassner/Sauer (2016): Verena Gassner and Roman Sauer, "Fabrics of Western Greek Amphorae from Campania and from the Bay of Naples", in: *FACEM* (version 16/12/2016). http://www.facem.at/project-papers.php (last version 23.05.2023).

Gassner/Scoppetta (2014): Verena Gassner and Elda Scoppetta, "Le anfore greco-occidentali dello scavo Piazza Nicola Amore a Neapolis", in: Giovanna Greco and Luigi Cicala (eds.), *Archeometry. Comparing Experiences*, Quaderni del Centro Studi Magna Grecia 19, Naples, 111–124.

Giacco (2017): Marialucia Giacco, "Le anfore dal santuario di Hera alla foce del Sele: dati preliminari", in: *Dialoghi sull'Archeologia della Magna Grecia e del Mediterraneo* I.1, 2017, 1121–1126.

Gras (1987): Michel Gras, "Marseille, la bataille d'Alalia et Delphes", in: *Dialogues d'Histoire Ancienne* 13, 161–181.

Koehler (1979): Carolyn G. Koehler, *Corinth A and B Transport Amphoras*, Dissertation, Princeton University.

Koehler (1981): Carolyn G. Koehler, "Corinthian Developments in the Study of Trade in the Fifth Century", in: *Hesperia* 50, 449–458.

Kourkoumelis (1990): Dimitri Kourkoumelis, "Corcyrean Amphoras", in: *ENALIA* Annual 2, 42–47.

Lawall (2010): Mark L. Lawall, "Imitative Amphoras in the Greek World", in: *Münsterische Beiträge zur antiken Handelsgeschichte* 28, 45–88.

Montana *et al.* (2018): Giuseppe Montana, Babette Bechtold, Anna Maria Polito and Lucia Randazzo, "Selinunte (Sicily) and its Productive Context: the Clayey Raw Materials Applied in a Long-lived Ceramic Production (Seventh to Third Century B.C.E.)", in: *Archaeological and Anthropological Sciences* 10, 657–673. https://doi.org/10.1007/s12520-016-0381-5.

Montana *et al.* (2020): Giuseppe Montana, Lucia Randazzo and Babette Bechtold, "The Beginning of Western Greek Amphorae Production in Western Sicily: Archaeometric and Archaeological Studies on 6th-5th Centuries BCE Amphorae Manufactured at Himera", in: *Minerals* 10, 1–19.

Montana/Randazzo (2015): Giuseppe Montana and Lucia Randazzo, "Le ricerche archeometriche: la caratterizzazione delle produzioni di anfore punico-siciliane", in: Babette Bechtold (ed.), *Le produzioni di anfore puniche della Sicilia occidentale (VII-III/II sec. a.C.)*, Carthage Studies 9.1, 18–46.

Morter/Leonard (1998): Jon Morter and John R. Leonard, "Storage Amphorae", in: Joseph Coleman Carter (ed.), *The Chora of Metaponto. The Necropoleis*, Vol. 2, Austin, 731–755.

Nieto/Santos (2008): Xavier Nieto and Marta Santos, *El vaixell grec arcaic de Cala Sant Vicenç*. Monografies del CASC 7, Salamanca.

Olcese (2010): Gloria Olcese, *Le anfore greco-italiche antiche di Ischia: archeologia e archeometria. Artigianato ed economia nel Golfo di Napoli*, Immensa aequora 1, Rome.

Olcese (2012): Gloria Olcese, *Atlante dei siti di produzione ceramica (Toscana, Lazio, Campania e Sicilia): con le tabelle dei principali relitti del Mediterraneo occidentale con carichi dall'Italia centro meridionale; IV secolo a.C. – I secolo d.C.*, Immensa aequora 2, Rome.

Olcese *et al.* (2013): Gloria Olcese, Stefania Giunta and Ioannis Iliopoulos, "Ceramics Production in the Gulf of Naples and Northern Campania. Part II. Archaeometric Reference Collection of Ceramics of Some Important Production Sites in Campania: Ischia, Naples, Sorrento, Capua and Cales", in: Gloria Olcese (ed.), *Immensa aequora workshop: ricerche archeologiche, archeometriche e informatiche per la ricostruzione dell'economia e dei commerci nel bacino occidentale del Mediterraneo, metà IV sec. a.C. – I sec. d.C.* (Atti del convegno, Roma, 24–26 gennaio 2011), *Immensa aequora* 3, Rome, 50–73.

Preka-Alexandri (1992): Kalliope Preka-Alexandri, "A Ceramic Workshop in Figareto, Corfu", in: Francine Blondé and Jean Yves Perrault (eds.), *Les Ateliers de potiers dans le monde grec aux époques géometrique, archaïque et classique* (Actes de la Table Ronde, Athènes 1987), *Bulletin de Correspondance Hellénique* Supplément 23, Athens, 41–52.

Sabbione (1986): Claudio Sabbione, "L'artigianato artistico", in: *Crotone* (Atti XXIII Convegno di Studi sulla Magna Grecia, Taranto, 7–10 ottobre 1983), Taranto, 245–302.

Sacchetti (2012): Federica Sacchetti, *Les amphores grecques dans le Nord de l'Italie. Échanges commerciaux entre les Appennins et les Alpes aux époques archaïque et classique*, Bibliothèque d'Archéologie Méditerranéenne et Africaine 10, Aix-en-Provence.

Sauer (2015): Roman Sauer, "Pottery Production at Velia: Ceramic Raw Materials and Archaeometric Analyses", in: *FACEM* (version 06/06/2015). http://www.facem.at/project-papers.php (last access 23.5.2023).

Sauer/Gassner (2008): Verena Gassner and Roman Sauer, "The Shipwreck of Cala Sant Vicenç. Thin Section and Heavy Mineral Analyses of Amphorae Samples", in: Nieto/Santos (2008) 355–371.

Savelli (2006): Sveva Savelli, "Le anfore da trasporto", in: Mariassunta Cuozzo, Bruno D'Agostino, Francesca Fratta and Valentina Malpede (eds.), *Cuma. Le fortificazioni 2. I materiali dai terrapieni arcaici*, in: *Annali dell'Istituto universitario orientale di Napoli, Sezione di archeologia e storia* 16, Naples, 103–202.

Savelli (2009): Sveva Savelli, "Anfore greco-occidentali in Magna Grecia: un aggiornamento sul tipo 'corinzio B arcaico' e 'ionico-massaliota'", in: Francesco Camia and Santo Privitera (eds.), *Obeloi, contatti, scambi e valori nel Mediterraneo antico. Studi offerti a Nicola Parise*, Tekmeria 11, Paestum, 105–130.

Sourisseau (2006): Jean Christophe Sourisseau, "Les amphores commerciales de la nécropole de Rifriscolaro à Camarine. Remarques préliminaires sur les productions corinthiennes de type A", in: Paola Pelagatti, Giovanni Di Stefano and Lucilla de Lachenal (eds.), *Camarina: 2600 anni dopo la fondazione: nuovi studi sulla città e sul territorio* (Atti del convegno internazionale, Ragusa, 7 dicembre 2002 / 7–9 aprile 2003), Rome, 129–147.

Sourisseau (2011): Jean Christophe Sourisseau, "La diffusion des vins grecs d'Occident du VIIIe au IVe s. av. J.-C., sources écrites et documents archéologiques", in: *La vigna di Dioniso. Vite, vino e culti in Magna Grecia* (Atti del XLIX Convegno di Studi sulla Magna Grecia, Taranto, 24–28 settembre 2009), Taranto, 145–252.

Sourisseau (2018): Jean Christophe Sourisseau, "Amphores de transport des Ve-IVe s. av. J.-C.", in: Henry Tréziny (ed.), *Mégara Hyblaea. 7, La ville classique, hellénistique et romaine*, 1/7, Rome, 27–31.

Spagnolo (2018): Grazia Spagnolo, "La produzione di anfore da trasporto a Gela", in: Maria Bernabò Brea, Michel Gras, Maria Martinelli, Claude Pouzadoux, Umberto Spigo and Madeleine Cavalier (eds.), *A Madeleine Cavalier*, Collection du Centre Jean Bérard 49, Naples, 281–294.

Vandermersch (1994): Christian Vandermersch, *Vin et amphores de Grande Grèce et de Sicile IVe–IIIe s. avant J.-C.*, Centre Jean Bérard 1, Naples.

Whitbread (1995): Ian K. Whitbread, *Greek Transport Amphorae: A Petrological and Archaeological Study*, Fitch Laboratory Occasional Paper 4, Athens.

Allard W. Mees

24 Economic Principles in the Roman Economy: Modelling Mass Ceramic Production and Trade

Abstract: The production and marketing of terra sigillata (Samian) in the Roman Empire yielded a high level of standardized mass production and full coverage of all existing markets. This industry featured such economic practices as standardization, labor division, directional trade and differential marketing, which were driven by underlying dynamic market developments. The product cycles involved were based on knowledge transfer, risk management, investment capital and the achievement of a critical mass. These elements can be described by mathematical models, which are influenced by nonlinear dynamic systems. These distinguishable economic principles within the ancient economy are helpful for understanding the ongoing developments in earlier mass production centers, such as the black- and red-figured pottery industries in Athens.

Introduction

The rise of innovative mass production of ceramics in the Late Hellenistic and Roman economy was accompanied by legal innovations and made intrinsic economic principles visible, which has continued to be a function of economic developments ever since.[1] Although it is still a chicken-or-egg question whether these legal innovations were the cause or rather the result of the new legal developments in the Roman economy, the economic development itself was based on preconditions that were man-made as well. Long-held assumptions that legal changes occurred in response to new economic realities still lack corroborative evidence.[2] The core logic in Roman times was that security of property rights and integrity of contract underpin, respectively, investment and trade, which in turn fuel economic growth and development. However, property rights and contracts depend on institutions, which themselves depend on coalitions of interests. Formal institutions are important, but, particularly in less developed areas, informal institutional arrangements play a significant part as well.[3] In addition, dependence on underlying indigenous legal traditions was a widespread phenomenon, at least until the 3rd century A.D.[4]

As a core thesis of this contribution, archaeologically detectable economic phenomena – such as innovation cycles, critical mass, directional trade and differential marketing – should not be considered as intentionally man-made structures, but as belonging to generic properties of any developed economy, whose visibility is proportional to the economic scale involved. This is not limited to the written historic period only: entrepreneurship has been present even in Neolithic communities.[5]

At the conference held in Turin in 2022, many distribution maps were presented illustrating the mass export of black- and red-figured Greek pottery vessels to Italy. While interesting as a basic visualization of the extant data, it is not entirely clear whether the thin distribution of exported Greek pottery in modern Turkey is a function of the current state of research or reflects a historical phenomenon requiring further study. Moreover, the marketing principles behind this distribution remain largely unknown. Even more complicating is the fact that most attributions of these Greek decorated vessels to individual potters heavily depend on non-reproducible, vague stylistic criteria, which may be less useful for economic considerations because they were primarily developed for chronological classifications.[6] Hence these stylistic attributions prevent any verifiable research related to marketing questions.

New research, some not possible in the past, is making certain types of information readily available. For example, we are now fortunate to have a recently released online database comprising a quarter of a million precisely recorded potters' stamps on Samian pottery. These vessels were produced from the 1st century B.C. until the 3rd century A.D. and

1 Hasaki/Bentz (2020).
2 Verboven/Erdkamp (2022a) 22.
3 Haggard/MacIntyre/Tiede (2008).
4 Verboven/Erdkamp (2022a) 21.
5 Lidow (2022).
6 Arrington (2017).

https://doi.org/10.1515/9783111189635-024

distributed throughout the Roman Empire. With the help of this granular information, we can study the main economical market principles in antiquity. Since there were no clearly defined concepts about economic principles, much of what we know is derived from archaeological evidence of economically-oriented activities such as building harbors and roads, which were supposed to improve prosperity in a general manner. Alleviation of individual private poverty generally was not thought to be a state affair, hence the rarity of legal measures undertaken by state authorities in antiquity. However, commerce in all goods that were not essential to the state (and, following from that, the army) was left entirely to the free market, although this concept, as such, was not really defined in antiquity.[7]

However, there was no general belief that it was best to leave things fully to the natural workings of the market. The activities of landowners and merchants were limited by municipal and imperial rulings that were based on the conviction that the avarice of traders and landowners needed to be counterbalanced by (legal) regulations serving common interests.[8] Obviously, the absence of positive evidence regarding state involvement in luxury goods such as fine ware pottery did not mean that negative evidence is inconclusive. But it is worth noting that neither Egyptian military papyri and ostraca, nor inscriptions, nor juristic texts, nor the Vindolanda and London tablets feature even a single word attesting to any interest from the government in the ceramics trade.[9] Research on these topics based on ancient legal texts is difficult, because in Roman times there was no "commercial law" in the strict sense.[10] Since laws were intended to address particular unsolved situations, they were not focused on a subject such as "commercial law."[11]

Most of the arguments that are possible from the writings of Classical authors or legal texts have already been made, for the simple reason that hardly any new ancient literary or legal texts have become available since the 19th century. However, the more recent archaeological finds of tablets and papyri with commercial contracts have broadened modern research considerably.[12]

The large amount of data which became available in the last two decades through the emergence of online databases raises serious doubts about the validity of certain conclusions of prominent historians such as M. Finley, who in the 1960–80s assumed that the Roman economy was primitive, lacking technological innovation and missing any interdependent, long-distance trade.[13] First doubts about this approach were expressed by Kevin Green back in 2000.[14] Notably, for example. the CEIPAC database (ceipac.ub.edu) as represented in a searchable form in the invaluable resource of Roman Open Data (https://romanopendata.eu [last access 02.07.2024]) and the Samian Research databases (https://www.rgzm.de/samian [last access 02.07.2024]) are now providing hundreds of thousands of new pieces of evidence.[15] Therefore, I have adopted a data-driven, phenomenological approach in order to assess the intrinsic economic principles involved, foregrounding the archaeological evidence and deriving theories only from what is available archaeologically.

Product Cycles

Since the beginning of humanity, man-made objects would appear and then disappear after a certain amount of time. None of these new objects came into existence in a "big bang," but rather their appearance always started with small quantities being produced and eventually, after a period of maximum output, they would vanish. Only in rare cases would a market presence stop abruptly due to external reasons, such as a war or an earthquake.

As an example, the export phase of the southern Gaulish production center La Graufesenque – from the viewpoint of pure quantification – is marked by a typical bell shape (Fig. 1a). This bell-shaped pattern can be observed for virtually all Samian production sites as well as for other Roman bulk ware products, such as amphorae.[16]

7 Temin (2001); Reddé/Mees (2022) 69.
8 Verboven/Erdkamp (2022a) 12.
9 Fink (1971); Bowman/Thomas (1994); Bowman/Thomas/Tomlin (2010).
10 Gialdroni (2022) 255.
11 Gialdroni (2022) 256.
12 Mataix Ferrándiz (2022) 156.
13 Finley (1987).
14 Greene (2000) 32.
15 Temin (2001).
16 Franconi/Brughmans *et al.* (2023).

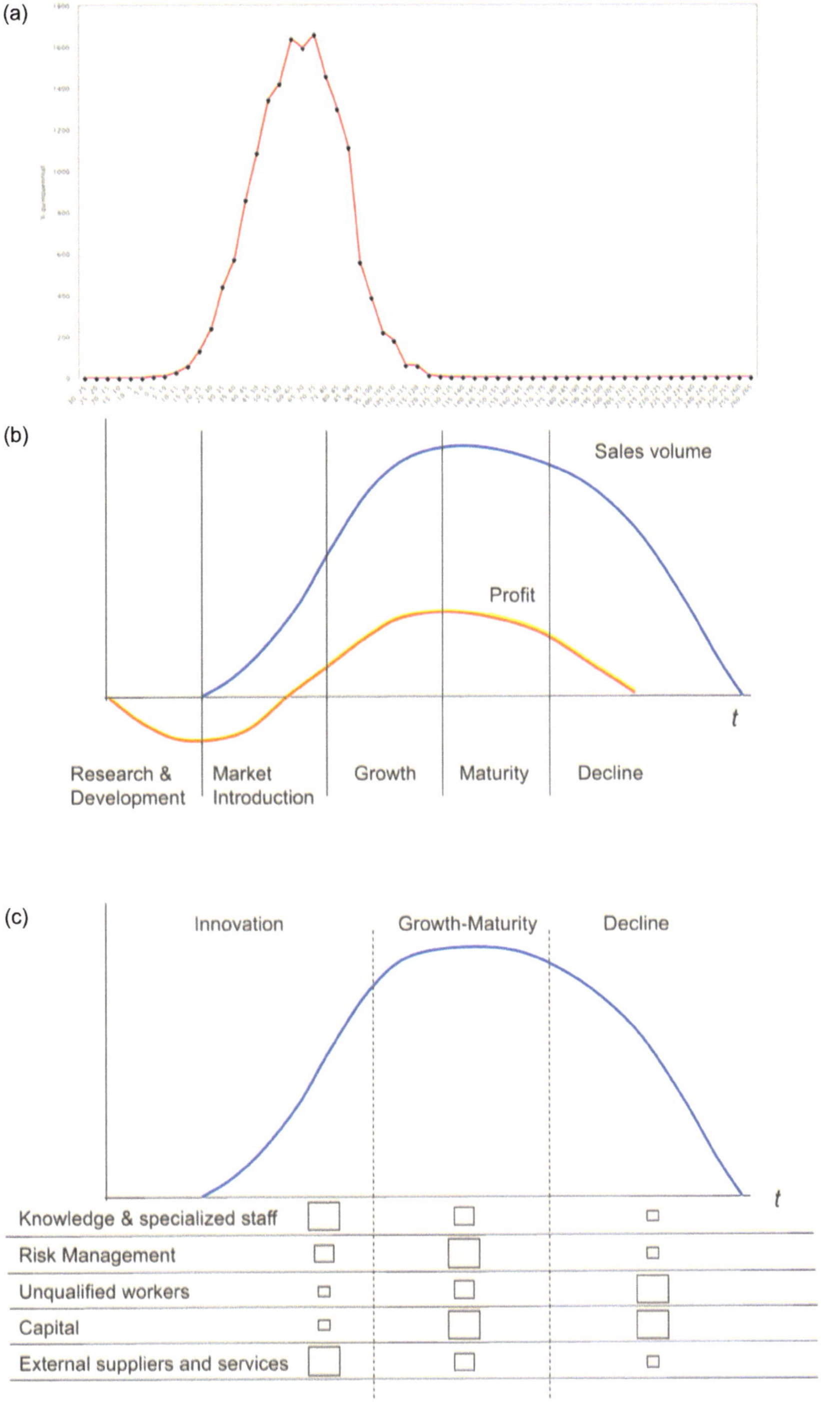

Fig. 1: (a) Quinquennial dating curve of the La Graufesenque exports. Graph generated at Samian Research (after https://www.rgzm.de/samian [last access 05.03.2024]); **(b)** Schematic presentation of the development of profit and sales volume (after https://diyinvestor.de/der-produkt-lebenszyklus-kurz-und-buendig [last access 05.03.2024]); **(c)** The importance of five economic location factors on a product cycle. After Haas/Neumair (2015) 58.

It is unclear whether the rise in production at this center around 10–20 A.D. was in some way connected to the sudden disappearance of the Samian potteries during the same period in Lyon, which on its own might have happened in the context of the defeat of governor Varus and his legions in Germania in 9 A.D. The causes of the end of the major La Graufesenque production center remain a matter of debate. The rise of terra sigillata production centers in the central and the north-eastern parts of Gaul around A.D. 100 is only one of the possible reasons for the disappearance of La Graufesenque wares from the markets (Fig. 1a).

In its simplest form, the development of the relation between profit and sales volume can be depicted in a basic diagram (Fig. 1b). It goes without saying that not only the research and development phase, but also any production start, is accompanied by an investment phase, resulting in negative profits for that early product cycle period (Fig. 1b). Although any overly detailed subdivisions of product cycles will be arbitrary, there are some major economic phases discernible: research and development, market introduction, growth, maturity and decline (Fig. 1b). However, when we look at the influences of economic location factors behind this, the general development of the sales volume can be reduced to three stages in which each of these factors plays a gradually different role (Fig. 1c). In the case of the Gaulish Samian industry, the following location factors are archaeologically quite tangible: the knowledge and availability of specialized staff, management, the availability of unskilled workers, available capital and, finally, the external suppliers and services.

Modifying these models for the "real world of archaeology" is, of course, a transformation from an easily describable mathematical model into nonlinear dynamical systems. Oscillations in the "real world of archaeology" arise because of nonlinear feedbacks between different interacting components of the system in which the enterprise is embedded (e.g., independent developments within the distribution markets).[17] In this context, datasets with huge quantities of archaeological material can be helpful, since the high numbers will temper oscillations considerably, bringing the original, mathematically describable patterns more to the foreground. In this case, the form of the curve is similar to a normal frequency curve, among others studied by Gauß, which can be mathematically described by the formula (cf. matematica)

$$y = \sqrt{2\pi} \exp\left(-\frac{x^2}{2}\right)$$

Below I discuss five different interacting components underlying product cycles – knowledge and specialized staff, management, unqualified employees, capital and external suppliers, and services – which are clearly determinable within the archaeological evidence of the major terra sigillata production sites in the western Roman Empire.

Technical Knowledge and the Availability of Specialized Staff

As with most of the ancient pottery industries, the emergence of the Gaulish Samian potteries cannot be explained without bearing in mind their forerunners, in this case the Italian Arretine potteries. The advanced and complex techniques for firing and the manufacturing of the engobe could not have been developed on the spot at a new production site without massive knowledge transfer from forerunner production centers. Such demanding engobes were made of different clays than the body of the pots.[18] Advanced knowledge of the exact properties of these illite clays and, above all, the locations where they could be found were required (Fig. 2). It remains to be determined, however, where the knowledge of the latter may have originated.

The Italian forerunners from Arezzo and Pisa were deeply involved in the establishment of quality standards for the engobes within the South Gaulish industry around 20 A.D. The knowledge of how to make high-end engobes was not available in southern Gaul up to then, and the process can only have been overseen by experts from Italy.[19] Whereas Service I pot forms were made throughout the early stage of making "proto-sigillata," the somewhat later forms of Service II were dominating the production from the moment when "real Samian" was first being made in Gaul.

An overview of the Italian and Gaulish potters who were involved in South Gaulish production centers demonstrates the engagement of local manpower combined with external Italian know-how. It is also interesting to note that typical Roman economical phenomena that are well known in the Italian production sites, such as lessor/lessee dependencies, are already discernible in this early Gaulish production phase (Tab. 1).[20]

17 Turchin (2020).
18 Sciau/Relaix *et al.* (2006).
19 Leon/Sciau *et al.* (2015).
20 Mees (2002) 267.

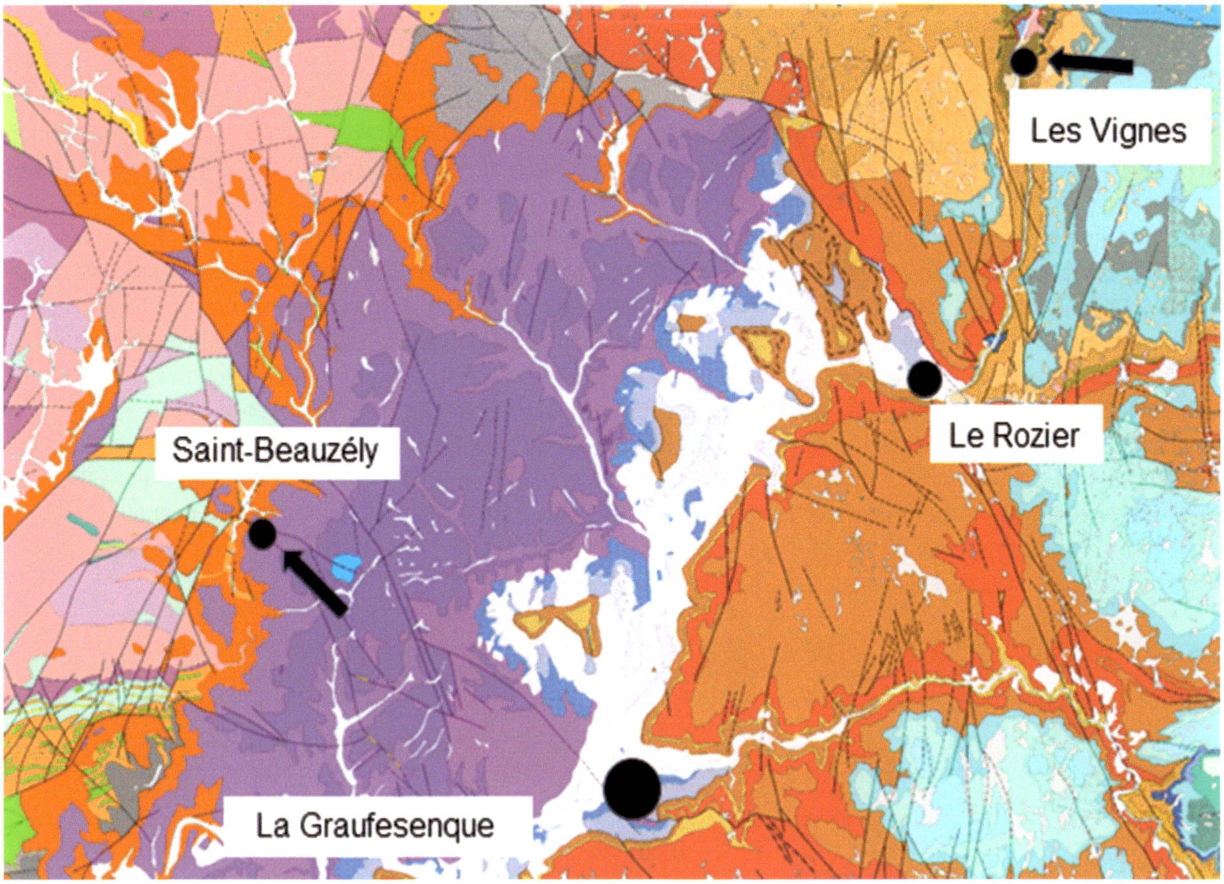

Fig. 2: Illite clay deposits of the engobes of La Graufesenque (red slip) and Le Rozier (marbled slip) in the modern departments of Aveyron and Lozère. After http://infoterre.brgm.fr/viewer (last access 08.07.2023).

Tab. 1: Italian influences on the emergence of the Gaulish terra sigillata industry. After Mees (2011) 231.

	Service I	Service II	Gaulish Potters	Italian Potters	Lessee	Calcareous Clay	Radial Stamp
Proto-Sigillata							
Montans	xx		Acutus			x	x
Lyon-Loyasse	xx			A Vibius Scrofula		x	x
Bram	xx					x	x
Narbonne	xx	x				x	x
Saint-Roman-en-Gal	xx	xx		Sex. Val. Niger	Amandus / Pomponi	x	x
La Graufesenque	x	xx	Ardacus; Bilicatos	Ateius	Seniseri / Ardaci	x	
Sigillata							
Montans			Acutus	L. Paconius		x	
Lyon-Muette	x	xx		Cn. Ateius; Rasinius; C. Sentius; Thyrsus		x	
Vienne				Mena / Avili		x	
La Graufesenque	x	xx	Ardacus / Bilicatos	Ateius / Vibius / Perennius		x	

Risk Management

In the early Gaulish production period, we can see a strong involvement of the Italian Samian industry, revealed by the appearance in Gaul of major Italian potter names, such as Ateius, Avilius, Perennius, Rasinius and Sentius (Tab. 1). However, during the Neronian period, at the height of the exportations from the La Graufesenque kiln site, a much more complex management arrangement started emerging.

In Italy, a single kiln docket suggests that the workers were organized in an *ordo* of dish makers – ο]ρδω κατιλαριω[ν.[21] This implies that there also must have been an *ordo* of cup makers. In fact, this organization is quite visible within the structure of South Gaulish potteries. For example, in the workshop of Aquitanus, certain dies were mostly used for dishes, others usually in combination with dishes and a third category mainly for decorated vessels and large, rouletted platters.[22] The latter were the most difficult to make, and it is therefore no surprise that these were made by more experienced potters.[23]

Although the membership list of an association of potters suggests some kind of self-organization,[24] a staggering number of kiln dockets have been found at La Graufesenque from its period of peak exports. These were presumably related to the management issue of mitigating the economic risks of an output which must have run into millions of vessels. Precise potter lists with numbers and sizes of vessels delivered by potters to individual kilns may have had the character of an insurance system. A few of these dockets mention "potter priests," *flamines*, who apparently also held religious authority when up to 30000 vessels were being stacked in a single kiln load (Fig. 3).[25]

(a) (b)

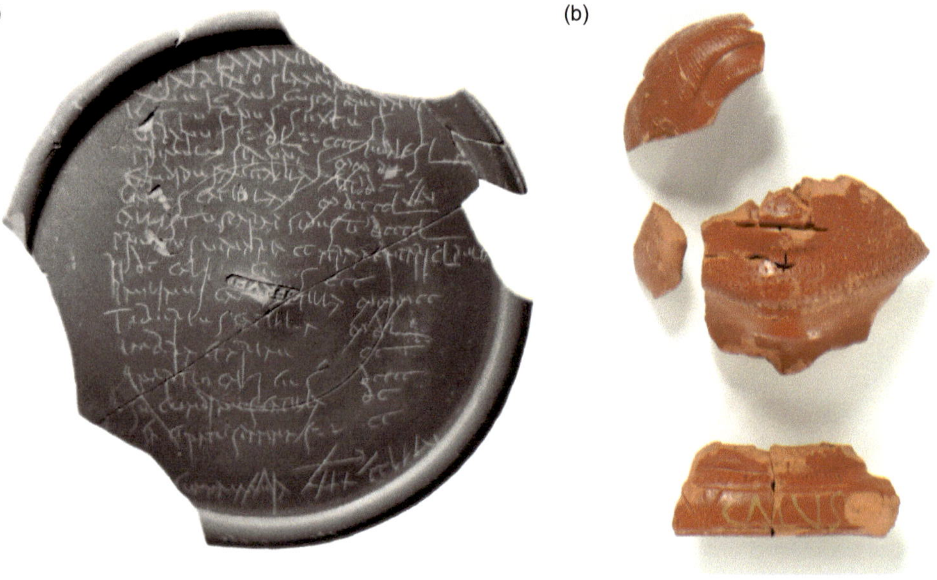

Fig. 3: (a) Docket list of the *flamines* P and Sabinus, found in La Graufesenque. After Vernhet/Bémont (1990–1991) 12;
(b) Head of the potter priest Calvus. Lettering in marbled engobe. After Mees (2012) 42, fig. 6.1. Photo: Iserhard/RGZM/LEIZA.

Availability of Unskilled Workers

From the early and middle production periods at La Graufesenque, virtually no misfired or misformed vessels appeared in the distribution chain. The huge kiln pit *Fosse de Gallicanus* in La Graufesenque, dating from the Neronian period, has

21 Mees (2002) 232; Zimmermann (2002) 157.
22 Mees (2013a) 81.
23 Mees (2013a) 80.
24 Mees (2002) 232; Zimmermann (2002) 157.
25 Mees (2012).

Fig. 4: Carelessly made vessel Drag. 37 from La Graufesenque made in the Hadrianic period, found in Mursa. Photo: Leleković.

yielded many vessels with minor defects, such as cracks in foot rings. Such defects are not known from contemporaneously exported vessels. However, especially during the latest production phase, "B Quality" vessels have been found on a very regular basis, suggesting that quality standards had dropped significantly, perhaps as the result of employing unqualified workers to produce such large quantities (Fig. 4).

Capital

It is clear that the conversion from "proto-sigillata" to "real" terra sigillata at La Graufesenque, which involved the application of extremely advanced technology, such as firing temperatures above 1000°C and the introduction of muffle-kilns, required considerable investment sums.[26] Moreover, whereas the "proto-sigillata" of La Graufesenque and Lezoux was mainly sold within the previous Iron Age distribution area of southern Narbonensis and the Allier basin, the situation changed when both production centers were integrated into the capital-intensive, long-distance trade.

Contrary to what has often been argued since the beginning of Samian research, La Graufesenque was not primarily targeting the Rhine provinces and their abundant military markets ca. 20 A.D.[27] In its early stage, this production center came to be connected to the already existing long-distance trade zone of the products from ancient Pisa in Aquitania (Fig. 5a), eventually profiting from the Roman crackdown on the indigenous revolt of 21 A.D.[28] This can be demonstrated by the distribution of pottery by the potter "Cocus i", who was working in the early La Graufesenque period (Fig. 5b).

A very similar phenomenon can be seen in the conversion from "proto-sigillata" to "real" terra sigillata at Lezoux. Whereas the "proto-sigillata" made during the 1st century A.D. in Lezoux was primarily sold in the Allier basin, after the technological change at Lezoux to "real" terra sigillata and becoming connected to the long-distance trade, Britannia became the first market for Lezoux products in the 2nd century A.D. (Fig. 6).

26 Polak (2000) 17, Fig. 2.2.
27 Dragendorff (1895) 107.
28 Tassaux (1984) 150.

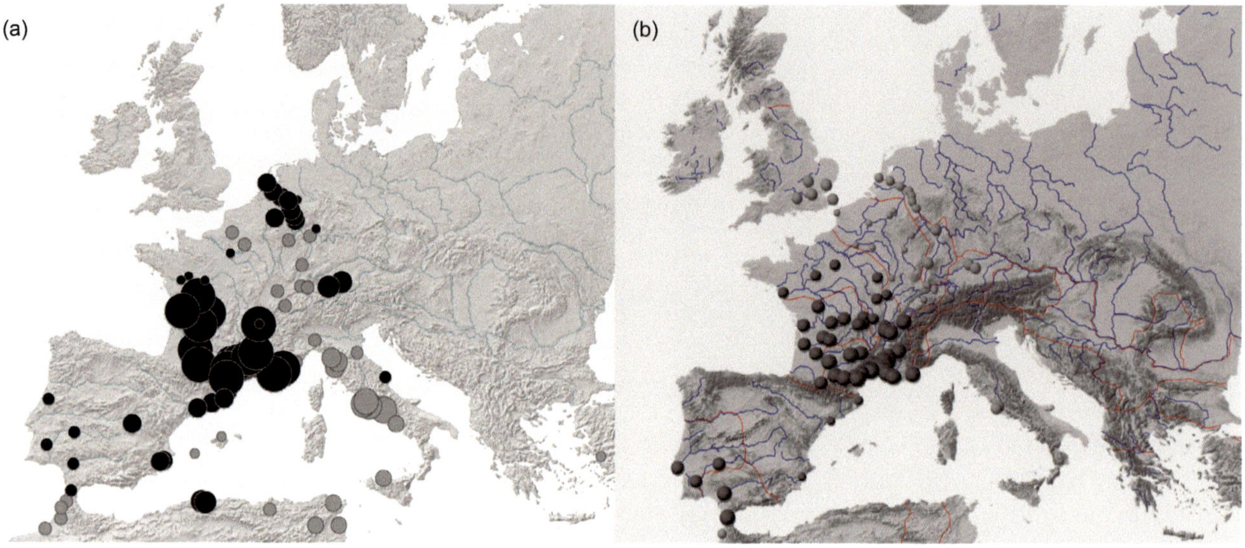

Fig. 5: (a) Primary markets of Pisa 20-10 B.C. Black dots: frequencies above the expected normal distribution. After Mees (2011) 85, fig. 88; **(b)** Primary market the La Graufesenque potter "Cocus i", ca. 20–30 A.D. Black dots: frequencies above the expected normal distribution. Map generated at Samian Research (http://www.rgzm.de/samian [last access 12.03.2024]).

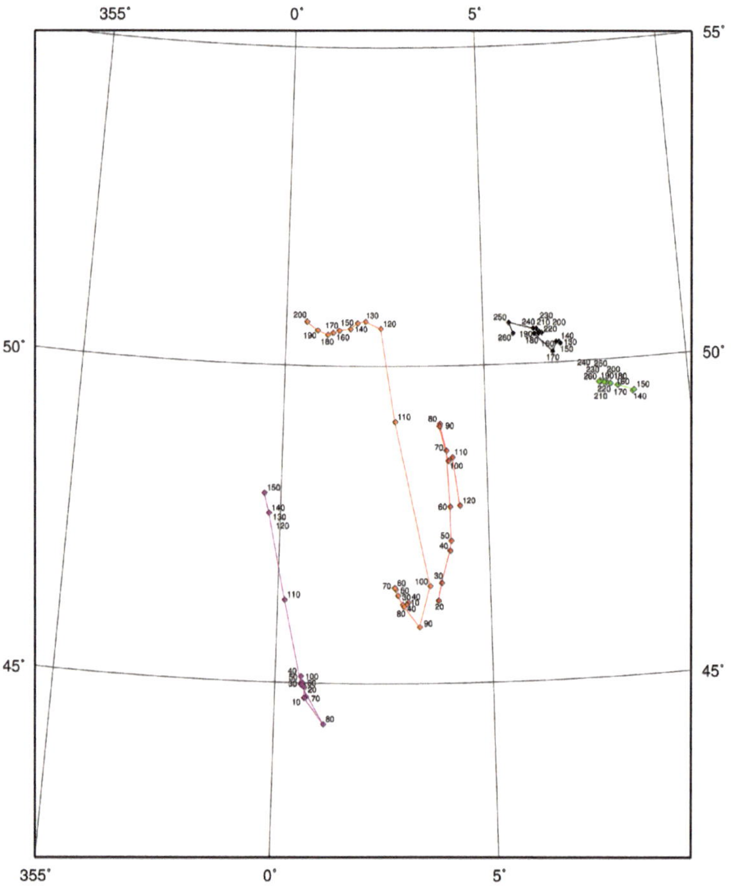

Fig. 6: Average dating and the average geographical distribution of terra sigillata from Lezoux (yellow line). A huge geographical jump is discernable after 100 A.D., when Britannia became the major market for Lezoux. After Mees/Polak (2013) 45, fig. 4.13.

In the case of Lezoux, there are no potter or investor names attached to the technical changes and sudden increase in the diffusion of its pottery. But for La Graufesenque, the senatorial Italian family of the Ateii seems to have been significantly involved in setting up this spinoff production center within the inaccessible mountainous region of the Gorges du Tarn (Tab. 1). It is impossible, however, to say exactly which member(s) of this widely branched family were investing in this enterprise.[29] The ban on senators being directly involved in commerce may explain why the potters' stamps did not explicitly state which member of this large family was investing in southern Gaul.

External Suppliers and Services

The influence of the Italian Samian production centers in the early phase of La Graufesenque is particularly noticeable. Indeed, from the discernable Italian potter names active there during this period, it can be deduced that Italian standards of quality were imposed on the South Gaulish manufacturing processes (Tab. 1).

As already has been noted with regard to the existing knowledge and availability of specialized staff, this Italian involvement was by no means based on charity. It was, after all, in the pure interest of the producers at Arezzo that they participated in establishing this South Gaulish Samian pottery-making center in 20 A.D., because they had recently lost the Gaulish market to their competitors at Pisa. Whereas Pisa had taken over the Gaulish market, Arezzo was able to maintain its market shares in the Mediterranean basin. Hence this move into southern Gaul can be interpreted as intended to weaken Pisa's Samian industry, which was geographically much more advantageously situated in the Arno estuary, gaining access to any western market, whereas Arezzo was located high in the mountains, which obviously caused it competitive difficulties in terms of the transport costs necessary to reach the attractive Gaulish and Germanic markets, where the competition with Pisa was fierce.[30]

Critical Mass

In business, "critical mass" is the point at which a growing producing entity becomes self-sustaining and no longer needs additional investment to remain economically viable.[31] In the context of business itself, it normally relates to the amount of revenues from customers, which is determined largely by the number of customers. These revenues are used to finance all business needs, both in terms of capital and human resources, to the point that they reach a size at which they can sustain themselves Secondly, when resources (such as available manpower) are not fully utilized and processes underperform, they are consuming more funds and thereby increasing the magnitude of business-critical mass (Fig. 7).[32]

Generally, before the "technological take-off" period emerges, the diffusion proceeds slowly or is restricted to the actual region of the production in question. The success of production sites like La Graufesenque depended mainly on the adoption of the new technologies by multiple producers. This was the cornerstone of the whole development. None of the major kiln sites in antiquity was operated by one single potter. Apparently, it was the technology adoption among an increasing number of potters (for the actual multiplication of the new technology, cf. Table 1) that functioned as a foundation for the commercial success. But once the "technological take-off" was achieved, diffusion sped up and the number of new technology adopters started to expand rapidly, usually at an exponential rate.[33]

Thus, the underlying factor is what can be called the technology replication coefficient (Φ).[34] It is also possible to identify growth in technology adoption (Ω), critical year ($Y_{crit,i,y}$), and critical penetration rate ($critTechnology_{i,y}$), where i denotes country, y indicates year, and N stands for the level of technology adoption in y–year. By definition, $N_{i,y} > 0$, as

29 Sangrisa (1998).
30 Mees (2011) 255.
31 Lechman (2015); Kenton (2023).
32 Kenton (2023).
33 Lechman (2015).
34 Lechman (2015) 3.

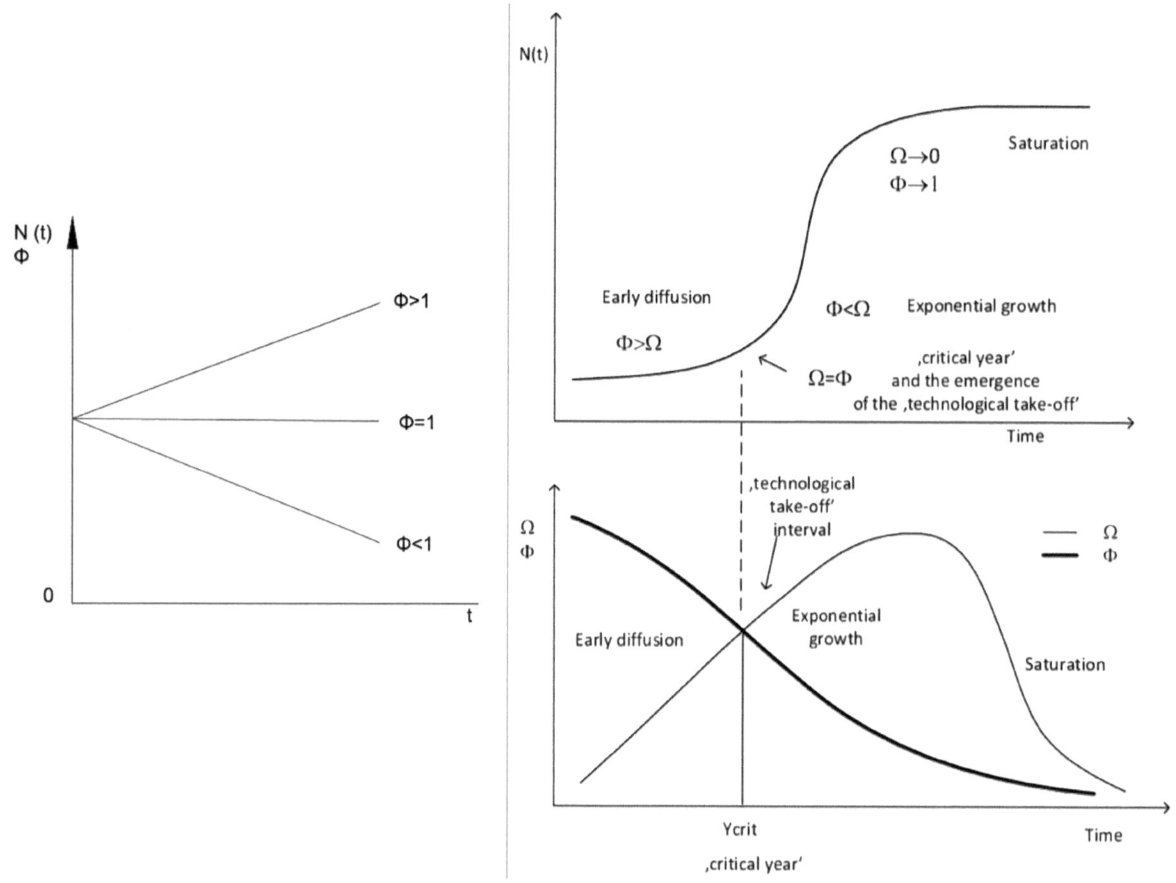

Fig. 7: Left: illustration of how respective values of $\Phi_{i,y}$ determine the $N_{i,y}$ over time. Right: Relationships between technology replication coefficient (Φ), technology adoption growth (Ω), critical year (*yCrit*) along the S-shaped technology diffusion trajectory. After Lechman (2015).

negative adoption is not possible, and if $N_y = 0$, the diffusion process would be non-existent or the kiln site would be operated by a single potter.[35] Following from this, the technology replication coefficient ($\Phi_{i,y}$) can be described as:

$$\Phi_{i,y} = \frac{Ni,y}{N(i,y-1)} \text{ resulting in } N_{i,y} = \Phi_{i,y}[N_{(i,y)-(i,y-1)}]$$

It remains a matter of debate, whether the "critical mass" point of no return (Fig. 7, right) already occurred when La Graufesenque's production was switched from "proto-sigillata" to "real sigillata" or instead whether it should be sought in the production phase during which the exports of La Graufesenque's products went beyond its primary early market in Aquitania (Fig. 5b) and mass production was achieved. Following from that, the business was connected to the long-distance trade that reached the military markets along the Rhine and in Britannia, which happened from 43 A.D. onwards. Also, in the case of Lezoux, the technological change from "proto-sigillata" towards "real sigillata" caused a tremendous increase in the diffusion of its products (Fig. 6, yellow line).

The end of the increase in the technological takeoff interval can be recognized in the development of the La Graufesenque production cycles of pot forms. Whereas until the Neronian period a general quantitative increase in production can be determined, along with the general maintenance of the high-quality standards, a major market shakeout took place from the Flavian period onwards. Not only was the general quality drastically lowered, as seen in the shiny engobe, weights of the vessels and crackles surfaces, but also the hitherto mainly Italian repertoire of forms was abandoned and standardized new shapes appeared. A correspondence analysis of the occurrences of pot forms (shapes) in sites dated to

35 Lechman (2015) 3.

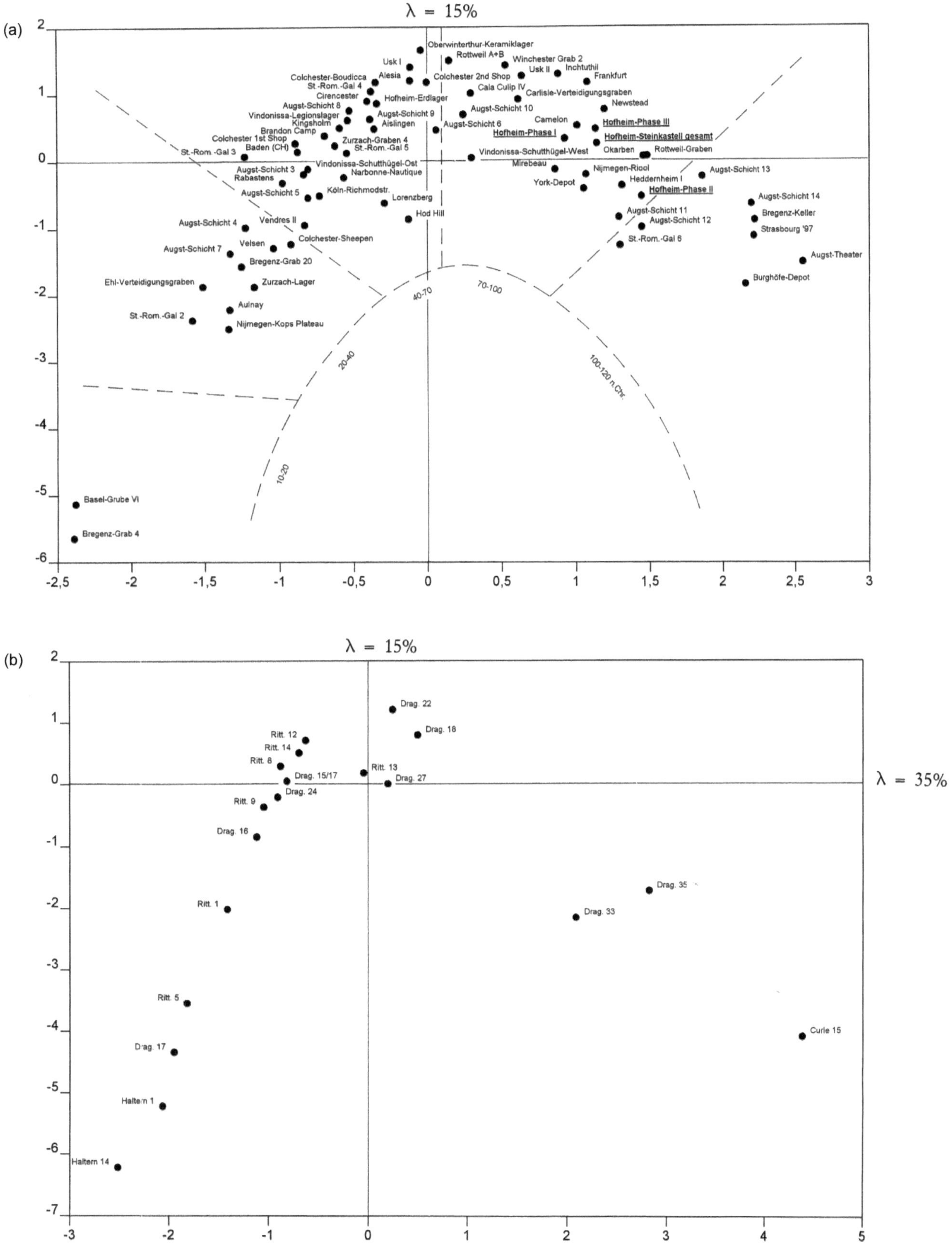

Fig. 8: (a) Correspondence analysis of findspots with plain Samian in 1st century A.D. findspots. After Mees (2013b) 26, fig. 8; **(b)** Correspondence analysis of plain Samian within the 1st century A.D. findspots. After Mees (2013b) 27, fig. 9.

the 1st century A.D. demonstrates the clear chronological development within the repertoire of forms (Fig. 8a–b). The repertoire shakeout from the Flavian period onwards resulted in a diminished spectrum of only four major types within the category of plain undecorated ware (e.g., Drag. 18, Drag. 27; Drag. 33 and Drag. 35).

Marketing Principles

Several marketing principles visible in the distribution of the industrially organized Samian industry can be observed within the archaeological evidence. I will discuss below four main parameters: directional trade; consumer driven markets; differential marketing; and supply chain continuity.

Directional Trade

A distribution map of 2nd-century A.D. Samian ware in Gaul reveals the striking occurrences of terra sigillata in the major regional towns. Outside these regional centers, the quantities of these wares found are surprisingly low (Fig. 9a).

The large quantities recorded at mainly civil financial hubs *en route* (e.g., Narbonne, Lyon, Augst, London), where safety of capital could be assured, strongly suggest that private entrepreneurship was behind this trading model: most of the added value generated during at least the first half of the export distances was realized at such sites.[36] This pattern apparently determined the character of the capital-intensive, long-distance trade involving Samian ware during both the 1st and 2nd centuries. The importance of a major hub like London for the distribution of 2nd-century A.D. Samian ware in Britannia can be demonstrated with the help of network analysis as well (Fig. 9b).[37] As far as trading models are concerned, we observe what could be considered a "post-industrial" model, in which differentiated regional hierarchical hubs were set up. At the end of a long economic development from self-sustaining local economies, finally even secondary hubs were appearing to cover the wider environments of the major hubs (Fig. 9c).[38] It is therefore rather surprising to discover that post-industrial economic structures may already have been employed in Roman times.

The strong divergence from an expected "down-the-line" pattern – the further away from the production center, the lower the quantities – strongly points to a "directional trade" model of commodities distribution, in which commercial hubs are redirecting the flow of trading goods (Fig. 9d).[39]

Consumer-driven Markets

The development of the Samian market was based on the repertoire of potforms on offer. Within each terra sigillata kiln site, a strict quality-control system must have been established, not only regarding the special red slip being used as engobe, but also the potform profiles. The concept behind this was the establishment of services. The standardization of ceramic form types already in antiquity made it fairly easy for modern archaeology to establish typologies, although the ancient concept of services was often overlooked by the seminal studies of Samian pottery.[40]

The proportions and dimensions of the vessels were chronologically developing during Roman times which involved standardization throughout all the Samian pottery kiln sites. For this phenomenon itself, the term "evolutive services" has been coined,[41] but it remains unclear how this process was enforced. This phenomenon itself is also recognizable in Roman amphora production, where it also remains unclear *how* this standardization was achieved, or *what* was driving

36 Reddé/Mees (2022) 66.
37 Reddé/Mees (2022) 78.
38 Haas/Neumair (2015) 67.
39 Zeitlin (1982).
40 Mees (forthcoming).
41 Polak (2000) 65.

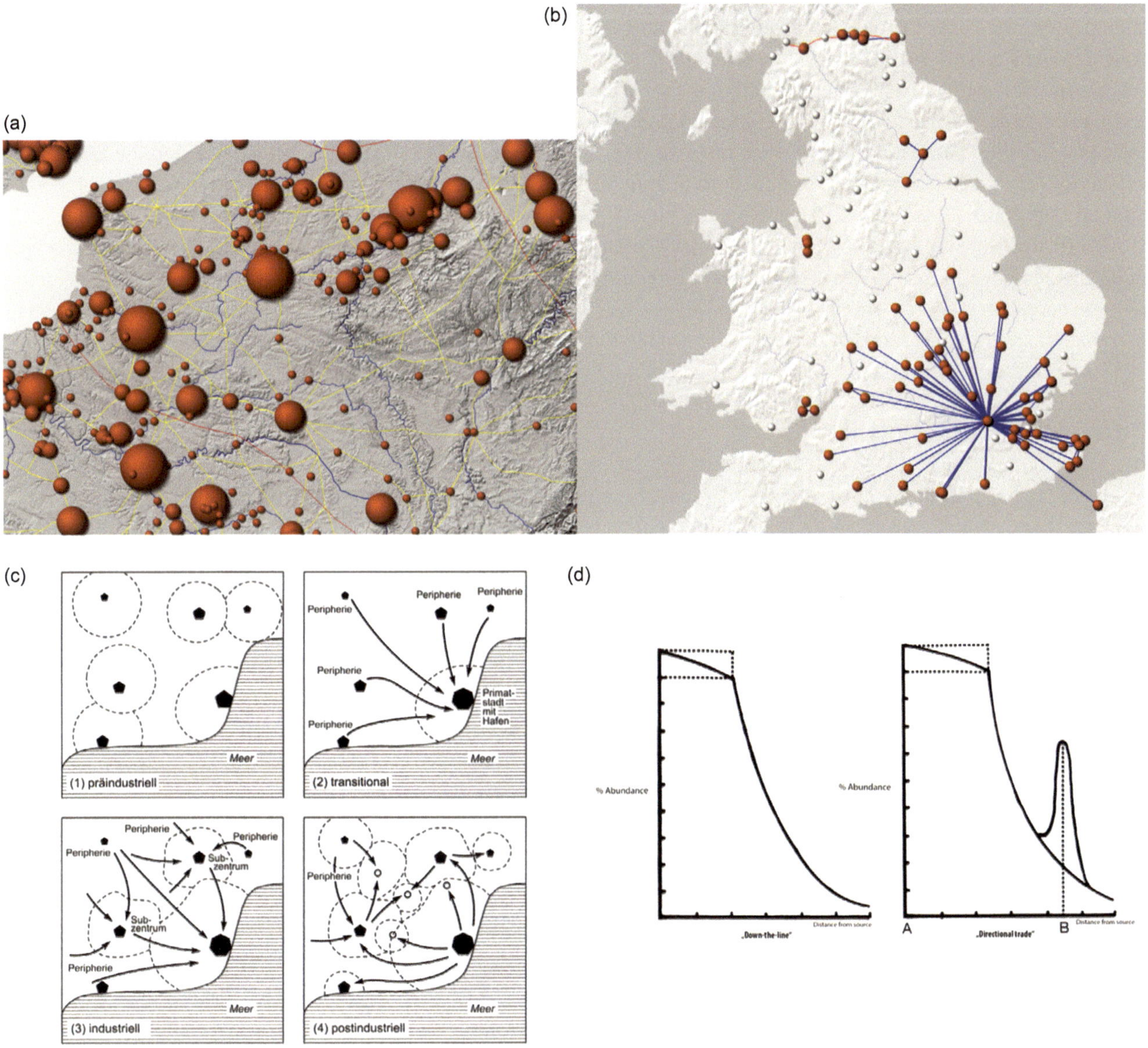

Fig. 9: (a) Distribution of 2nd century A.D. Lezoux Samian in Northern Gaul. Map generated at Samian Research (http://www.rgzm.de/samian [last access 07.03.2024]); **(b)** Network analysis of sites with an observed number of pots greater than the expected value and a 'Network Density Value' of 0.8 are indicated with a red dot. A blue line is drawn between two significant sites. Hubs of Samian trade with high correlations with adjacent sites comprising significant frequencies of products of the terra sigillata potter "Cinnamus ii" can be distinguished at Carlisle and Corbridge adjacent to Hadrian's Wall and Camelon by the Antonine Wall, and at Chester, Caerleon, London, Colchester, Richborough and Gloucester. Sites with fewer than five occurrences are statistically ignored and displayed as grey dots. After Reddé/Mees (2022) 78, fig. 20; **(c)** Spatial structures in center-periphery models. After Haas/Neumair (2015) 66, fig. 3.13; **(d)** Model of "directional trade", where commercial hubs are redirecting trading goods. After Renfrew (1972) figs. 20.9–20.12; Zeitlin (1982) figs. 1, 3.

this focus on standardization. But the suspicion arises that the merchants or investors, and not the potters, were actually the major agents enforcing this standardization.

Studies of the Chinese porcelain industry and its production for the European markets in the 17th–18th centuries clearly show a direct correlation between the prices of tea and sizes of vessels. It is important to note that these interactions took place over ca. 10000 km.[42] In the Roman Empire, distances between a Samian or amphorae kiln site and a consumption site were seldom more than 1500 km.[43]

We can observe clear developments within the repertoire of Samian forms, in which individual forms, such as Drag. 27 (cup), were disappearing during the 2nd century A.D., whereas other forms, such as Drag. 31 (large dish), became increasingly popular (Fig. 10a). This pattern is also recognizable, for example, when we analyze in a second step the aggregated forms (e.g., cups, dishes) from this period: cups in general were slowly disappearing, while dishes were becoming very popular (Fig. 10b).

It remains unclear, however, whether the decline and popularity trends described above were caused by changes in dining behavior. Moreover, we also have very little information on prices for Samian ware. Therefore, if we keep the Chinese porcelain market in mind as a likely parallel, we could suggest that this development was most likely a reaction to market demands and presumably not driven by production factors themselves.

Differential Marketing

Although virtually all of the northwestern markets of the Roman Empire were supplied with massive amounts of terra sigillata, there were – seen from an Empire-wide perspective – regional differences, which point towards differential marketing. For example, La Graufesenque products like marbled Samian were mainly sold into the Mediterranean basin, where they reached a market share of ca. 5–10%. The advanced technique required to produce the marbled slip must have also caused high percentages of failed products, elevating the successfully-produced vessels to upper market commodities. It is probably not accidental that these products were mainly found in the more thoroughly Romanized market areas, where more capital was available than in the provinces north of the Alps that were conquered from 122 B.C. onwards (Fig. 11a).

Supply Chain Continuity

Whereas in modern Europe the economy has switched to a just-in-time production with an adapted supply-chain (causing long fleets of trucks on the motorways), the finds related to Roman Samian trade suggests only a small amount of discontinuity within the supply chain. Although trade in southern Gaul was mainly connected to the organizations of the *nautae* and *navicularii*,[44] the more or less exclusive appearance of negotiatores cretariae north of the Alps suggests a more granular flow of goods, since these were individuals rather than groups.[45] This coincides with the appearance of shops / depots (Fig. 11b, yellow dots) in the northern Samian markets, which regularly featured homogenous stacks of vessels. Since these stacks often show large numbers of identical stamps on identical forms, it is thought that they were at least partly originating from a single kiln load.[46] This suggests that creating stocks of up to a couple of hundred vessels was pivotal to a continuously functioning market reacting to demand and supply, but the quantities involved are rarely higher than one would expect for normal shops (Fig. 11b). Although the La Graufesenque dockets only recorded kiln firings during the summer season, the quantities were apparently enough to secure a continuous influx of Samian goods into these markets. The supply chain buffering in the warehouses in Narbonne and London was evidently enough to guarantee a full season's supply of these fine ceramics (Fig. 11b, pink dots).

42 Pitts (2013).
43 Reddé/Mees (2022) 68, figs. 7–8.
44 Reddé/Mees (2022) 72.
45 Reddé/Mees (2022) 71, fig. 11.
46 Weber (2013).

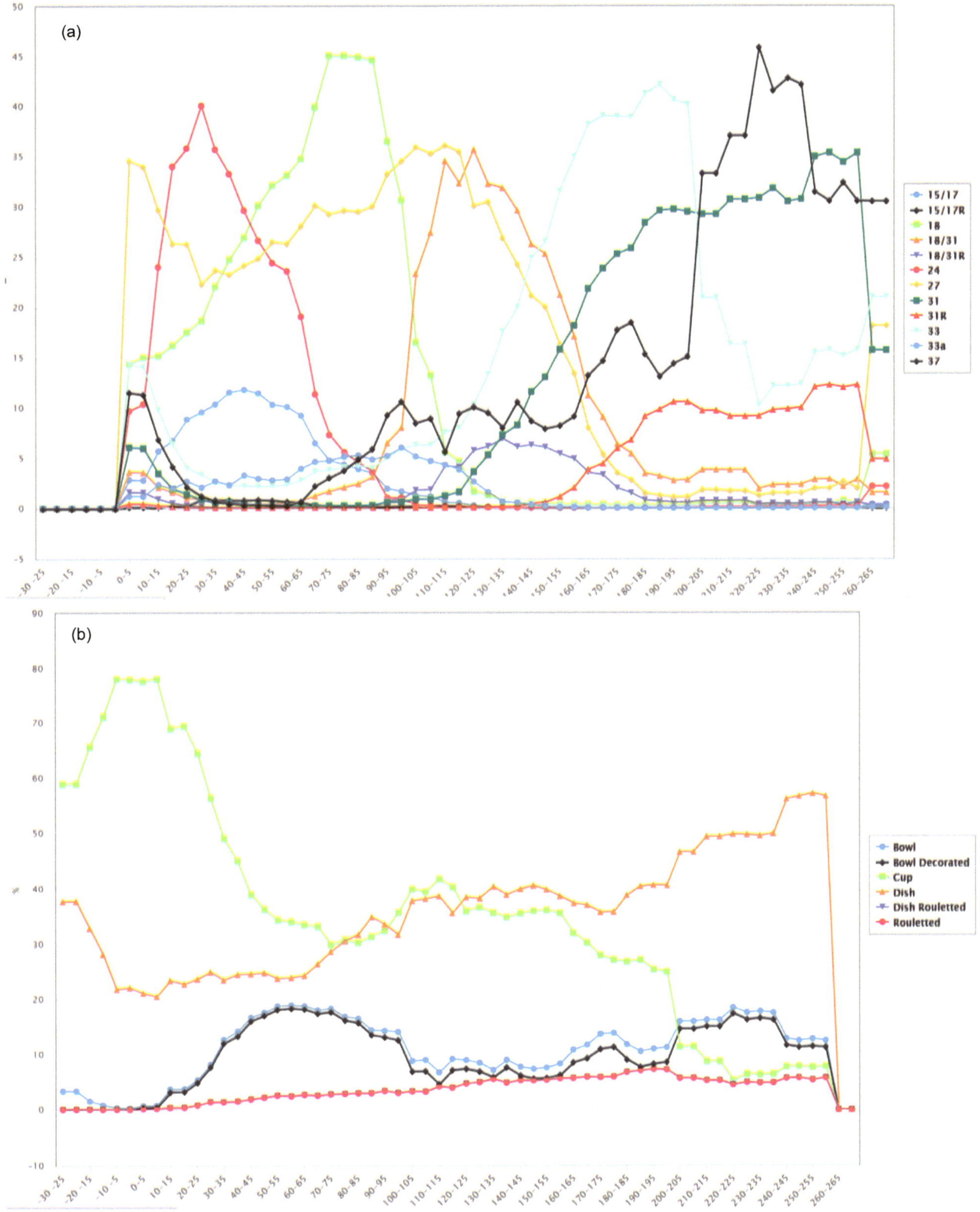

Fig. 10: (a) Quinquennial chronological development of market shares of individual Samian pot form types. Graph generated at Samian Research (https://www.rgzm.de/samian [last access 02.03.2024]); **(b)** Quinquennial chronological development of market shares of aggregated Samian pot forms. Graph generated at Samian Research (https://www.rgzm.de/samian [last access 02.03.2024]).

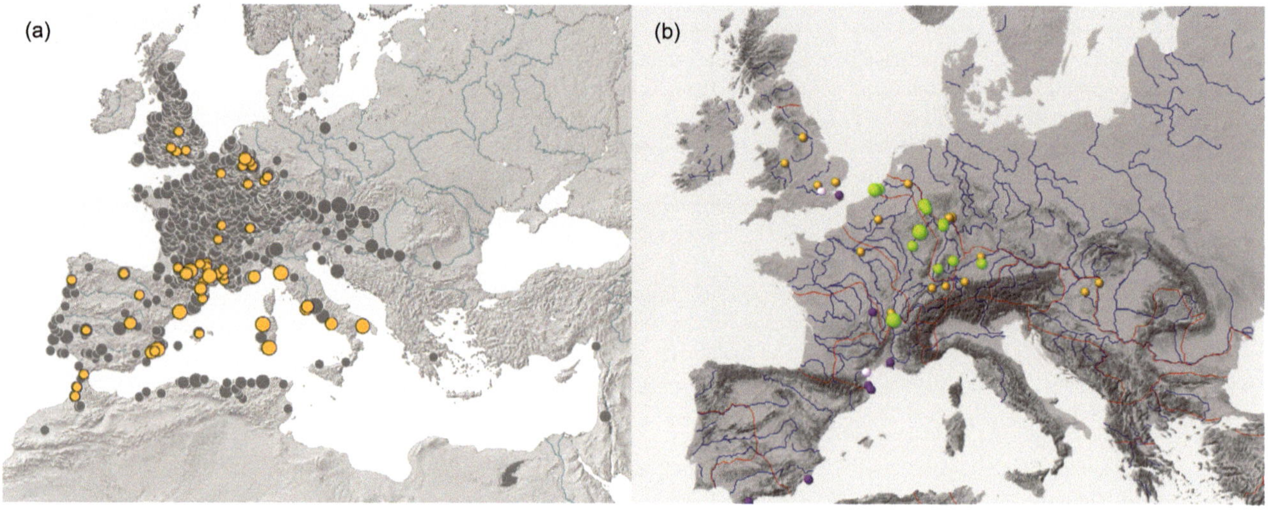

Fig. 11: (a) Distribution of marbled Samian from La Graufesenque. Yellow dots: marbled Samian. Grey dots: standard red slip Samian from La Graufesenque. Map generated with data from Samian Research (http://www.rgzm.de/samian [last access 02.03.2024]); **(b)** Distribution map of Samian depots (yellow), warehouses (pink), *negotiatores cretariae* (green) and shipwrecks (purple). Map generated with the data available at Samian Research (https://www.rgzm.de/samian [last access 02.03.2024]).

Conclusions

In this paper I have attempted to demonstrate that it may be possible to describe important economic processes related to the ancient pottery industry by economic principles which are valid up to modern times. In cases where our archaeological record is incomplete, one can eventually construct a more informative analysis of the economy of ancient potteries by adding elements of the industrial development from the Renaissance period or even from modern economy models. By adding these economic principles to our analytical toolbox for studying pottery industries, we are enhancing our "modernist" views on the ancient economy. This is different from the "primitivist" views that consider the ancient economy as wildly different from how the economies of the Western world work today. Far from developing standalone new theories, what I advocate in this paper is to recognize well-known modern economic principles, which are embedded within our observations from the archaeological records. With the increased granularity of modern datasets there is a unique opportunity to develop deep insights into how competitive marketing operated within workshops and how careers of highly individualized potters developed in the ancient pottery industry.

Sigle

Matematica: https://www.matematica.pt/en/useful/list-curves.php (last access 02.03.2024).

Bibliography

Arrington (2017): Nathan T. Arrington, "Connoisseurship, Vases, and Greek Art and Archaeology", in: J. Michael Padgett (ed.), *The Berlin Painter and his World. Athenian Vase-Painting in the Early Fifth Century B.C*, Princeton N.J., 21–40.

Bowman/Thomas (1994): Alan K. Bowman and J. David Thomas, *The Vindolanda Writing-Tablets, Tabulae Vindolandenses II*, London.

Bowman/Thomas/Tomlin (2010): Alan K. Bowman, J. David Thomas and Roberta Tomlin, "The Vindolanda Writing-Tablets (Tabulae Vindolandenses IV, Part 1)", in: *Britannia* 41, 187–224.

Dragendorff (1895): Hans Dragendorff, Terra sigillata. *Ein Beitrag zur Geschichte der griechischen und römischen Keramik*, Bonner Jahrbücher Beihefte 96/97, Berlin.

Fink (1971): Robert O. Fink, *Roman Military Records on Papyrus*, American Philological Association Monograph 26, Cleveland.

Finley (1987): Moses Finley, *Ancient History: Evidence and Models*, Berkeley/Los Angeles.

Franconi/Brughmans *et al.* (2023): Tyler Franconi, Tom Brughmans, Ekaterina Borisova and Laura Paulsen, "From Empire-wide Integration to Regional Localization: A Synthetic and Lechman Quantitative Study of Heterogeneous Amphora Data in Roman Germania Reveals Centuries-long Change in Regional Patterns of Production and Consumption", in: *PLoS ONE* 18, e0279382. https://doi.org/10.1371/journal.pone.0279382.

Fulford/Durham (2013): Michael Fulford and Emma Durham (eds.), *Seeing Red: New Economic and Social Perspectives on Terra Sigillata*, Institute of Classical Studies, Supplement 102.10, Reading.

Gialdroni (2022): Stefania Gialdroni, "Roman Law, Commercial Law and Levin Goldschmidts's Legacy", in: Verboven/Erdkamp (2022b) 248–264.

Green (2000): Kevin Greene, "Technological Innovation and Economic Progress in the Ancient World: M.I. Finley Re-Considered", in: *Economic History Review* 53, 29–59.

Haas/Neumair (2015): Hans-Dieter Haas and Simon-Martin Neumair, *Wirtschaftsgeographie*, Darmstadt.

Haggard/MacIntyre/Tiede (2008): Stephan Haggard, Andrew MacIntyre and Lydia Tiede, "The Rule of Law and Economic Development", in: *Annual Review of Political Science* 11, 205–234. https://doi.org/10.1146/annurev.polisci.10.081205.100244.

Hasaki/Bentz (2020): Eleni Hasaki and Martin Bentz (eds.), *Reconstructing Scales of Production in the Ancient Greek World: Producers, Processes, Products, People* (Panel 3.4, Proceedings of the 19th International Congress of Classical Archaeology, vol. 8, Cologne/Bonn, May 22–26, 2018), Heidelberg. https://doi.org/10.11588/propylaeum.639.c9977.

Kenton (2023): Will Kenton, "Critical Mass: Overview, Special Considerations, and Examples", in: *Investopedia* (version January 31, 2023). https://www.investopedia.com/terms/c/critical-mass.asp (last access 02.06.2023).

Lechman (2015): Ewa Lechman, "The 'Technological Take-off' and the 'Critical Mass'. A Trial Conceptualization", in: *GUT Faculty of Management and Economics. Working Paper Series A (Economics, Management, Statistics)*, 1/2015. https://hdl.handle.net/10419/173322.

Leon/Sciau *et al.* (2015): Yoanna Leon, Philippe Sciau, Michel Passelac, Corinne Sanchez, Robert Sablayrolles, Philippe Goudeau and Nobumichi Tamura, "Evolution of Terra Sigillata Technology from Italy to Gaul Through a Multi-technique Approach", in: *Journal of Analytical Atomic Spectrometry* 30, 658–665. https://doi.org/10.1039/C4JA00367E.

Lidow (2022), Derek Balfour Lidow, "The Prehistoric Entrepreneur: Rethinking the Definition", in: *Journal of Management History* 28, 458–475. https://doi.org/10.1108/JMH-11-2021-0058.

Mataix Ferrándiz (2022): Emilia Mataix Ferrándiz, "Goods, Law, and Trade: Material Evidence for Lease and Hire Contrats (locatio conductio) and a Grain Sample Recorded in CIL 4.9591", in: Verboven/Erdkamp (2022b) 131–157.

Mees (2002): Allard W. Mees, *Organisationsformen römischer Töpfer-Manufakturen am Beispiel von Arezzo und Rheinzabern unter Berücksichtigung von Papyri, Inschriften und Rechtsquellen.* Monographien Römisch-Germanisches Zentralmuseum 52, Mainz, https://doi.org/10.11588/propylaeum.519.

Mees (2011): Allard W. Mees, *Die Verbreitung von Terra Sigillata aus den Manufakturen von Arezzo, Pisa, Lyon und La Graufesenque. Die Transformation der italischen Sigillata-Herstellung in Gallien.* Monographien des Römisch-Germanischen Zentralmuseums 93, Mainz.

Mees (2012): Allard W. Mees, "The Portrait of the Potter Calus", in: David Bird (ed.), *Dating and Interpreting the Past in the Western Roman Empire. Essays in honour of Brenda Dickinson*, Oxford, 41–57.

Mees (2013a): Allard W. Mees, "The Internal Organisation of Terra Sigillata (Samian) Workshops", in: Fulford/Durham (2013) 66–96.

Mees (2013b): Allard W. Mees, *Hofheim II, Terra Sigillata. Mit einem Beitrag von Dirk Allgaier. Freiburger Beiträge zur Archäologie und Geschichte des ersten Jahrtausends* 18, Freiburg.

Mees/Polak (2013), Allard Mees and Marinus Polak, "Scattered Pots. Exploring Spatial and Chronological Aspects of Samian Ware", in: Fulford/Durham (2013), 36–48.

Mees (forthcoming): Allard W. Mees (ed.), *Von Typo-Chronologie zur Ontologie – Hans Dragendorffs und Joseph Déchelettes Methodik und ihr Einfluss auf die heutige Terra-Sigillata-Forschung* (Symposium RGK. Mehr als nur Scherben, 26.11.-27.11, 2020), forthcoming.

Pitts (2013): Martin Pitts, "Pots and Comparative History: The Case of Imported Roman Fine Wares and Chinese Porcelain in NW Europe", in: Fulford/Durham (2013) 381–390.

Polak (2000): Marinus Polak, *South Gaulish Terra Sigillata with Potters' Stamps from Vechten*, Rei Cretariae Romanae Fautorum Acta, Supplement 9, Nijmegen.

Reddé/Mees (2022): Michel Reddé and Allard W. Mees, "Hadrian's Wall and its Continental Hinterland", in: *Britannia* 53, 55–84.

Renfrew (1972): Colin Renfrew, *The Emergence of Civilization. The Cyclades and the Aegean in the Third Millennium B.C.*, London.

Sangrisa (1998): Paolo Sangrisa, "Terra sigillata e politica augustea: alcune note su Cn. Ateius," in: *Studi Classici e Orientali* 46, 919–932.

Sciau/Relaix *et al.* (2006): Philippe Sciau Sabrina Relaix, Christian Roucau, Yolande Kihn and Delhia Chabanne, "Microstructural and Microchemical Characterization of Roman Period Terra Sigillate Slips from Archeological Sites in Southern France", in: *Journal of the American Ceramic Society* 89.3, 1053–1058. https://doi.org/10.1111/j.1551-2916.2005.00827.x.

Tassaux *et al.* (1984): Danielle Tassaux, Francis Tassaux, Pierre Caillat, Louis Maurin, Marie-Hélène Santrot, Jacques Santrot, Paola Starakis and Pierre Tronche, "Un camp militaire augusto-tibérien en Aquitaine, 2e partie", in: *Aquitania* 2, 105–157.

Temin (2001): Peter Temin, "A Market Economy in the Early Roman Empire," in: *Journal of Roman Studies* 91, 169–181.

Turchin (2020): Peter Turchin, "Cliodynamics is Not 'Cyclical History'", in: *Blog of Peter Turchin* (entry September 18, 2020). https://peterturchin.com/cliodynamics-is-not-cyclical-history (last access 12.03.2024).

Verboven/Erdkamp (2022a): Koen Verboven and Paul Erdkamp, "Introduction: Legal Systems and Economic Development", in: Verboven/Erdkamp (2022b) 1–24.

Verboven/Erdkamp (2022b): Koen Verboven and Paul Erdkamp (eds.), *Law and Economic Performance in the Roman World, Impact of Empire* 44, Leiden/Boston.

Vernhet/Bémont (1990–1991): Alain Vernhet and Colette Bémont, "Un nouveau compte de potiers de La Graufesenque portant mention de flamines", *Annales de Pegasus* 1990–1991, 12–14.

Weber (2013): Meike Weber, “A Reassessment of Pre-consumption Deposits and Samian Export in the Antonine Period”, in: Fulford/Durham (2013) 188–209.
Zeitlin (1982): Robert N. Zeitlin, “Toward a More Comprehensive Model of Interregional Commodity Distribution: Political Variables and Prehistoric Obsidian Procurement in Mesoamerica”, in: *American Antiquity* 47, 260–275.
Zimmermann (2002): Carola Zimmermann, *Handwerkervereine im griechischen Osten des Imperium Romanum.* Römisch-Germanisches Zentralmuseum Monographien 57. Mainz.

Maiwenn Tonna

25 Looking for the Lost Decoration of an Apulian Volute Krater by the Painter of Sèvres 1

Abstract: The Apulian volute krater MNC 1, created by the Sèvres 1 Painter and housed in the Sèvres Museum, is a rich source of information—due to its size, iconography, and preservation—for enhancing our understanding of production techniques and decorative stages for Apulian vases produced around Bari during the last quarter of the 4th century B.C. It also provides insights into the context in which the vase was discovered and has been preserved until today. Indeed, only the preparatory motifs have survived; no overpaints are visible. The combination of archival research, photogrammetry, comparisons with other vases by the same painter, close observations using raking light, and multispectral analysis enables us to propose a virtual reconstruction of what the decoration of the MNC 1 might have looked like when it left the workshop.

Introduction

The Apulian volute krater MNC 1 of the Sèvres Museum shows a few evanescent traces of overpainted decorations, but its lavish and elaborate, ornately styled overpainted details are mostly lost and illegible (Fig. 1). This vase was decorated by the Sèvres 1 Painter between 325 and 320 B.C. and is indeed the painter's eponymous vase. Because of its monumental size and its iconography, it was also the masterpiece of Dominique Vivant Denon's collection. D. V. Denon (1747–1825) was a *chargé d'affaires* at the French embassy of Naples between December 1778 and 1785. Before that, he spent fourteen months traveling through southern Italy and Sicily: from October 1777 to December 1778, he was working as the manager of a small team of artists hired by Jean Benjamin De Laborde and Jean–Claude Richard Abbé de Saint-Non to make picturesque sketches of South Italian landscapes in order to illustrate Saint-Non's forthcoming *Itinéraire pittoresque*. It has also been proved that the accompanying text of the *Itinéraire pittoresque* was mostly by Denon's hand and based on his partly lost travel diaries. It was during this trip through southern Italy and during the long stay in Naples that followed that D. V. Denon acquired a huge collection of ancient vases, 525 of which were sold to the French king Louis XVI in 1785.

MNC 1 has a huge and obvious crack running under its left handle; this happened during its firing and was apparently due to the handle's weight. Because of this huge crack, we know for certain that MNC 1 was never a functional piece for banquets and that it was designed from the very beginning as an adornment for a tomb. The good state of preservation of the shape, which is almost entire, also speaks for a funerary context. However, we do not know precisely where MNC 1 was discovered, since our indirect sources on Denon's travels do not mention its *estrazione*; but it is possible to make assumptions based on the known areas of provenance of other vases by the same painter and on what we know of Denon's itinerary in Puglia. The reasons for Denon's silence on this discovery are probably linked to his understanding of contemporary laws on archaeology. King Ferdinand IV of Naples had indeed issued a decree forbidding excavations conducted by unauthorized private individuals and tried to prevent foreigners from taking archaeological artifacts out of the kingdom. Denon's silence was probably a measure of precaution to avoid accusations of theft.

Even though MNC 1 may look somewhat unimpressive or poorly decorated because of its bad state of preservation, this vase in fact provides us with a wealth of information on the artist's technique and raises many challenging questions. How did the artist create the decoration of this vase, and what steps did he take to do so? How are we to account for the almost complete disappearance of the vase's overpainted decoration? This paper will attempt to answer these questions by offering a fresh study of the context of production of this vase and a close observation of the barely legible remains of its decoration; it will also document the evolution of the state of preservation of this vase in modern times, between its discovery in the 18th century and today. These observations will lead to a tentative reconstruction of the lost designs and to a virtual or digital restoration of this vase.

Note: I am grateful to the organizers for inviting me to present a poster in their Conference. I owe warm thanks to Anaïs Boucher, Viviane Mesqui, Dr. Marlène Nazarian, Prof. Claude Pouzadoux, Prof. Évelyne Prioux, Dr. Rozenn Douaud, Jérôme Louvet and Louise Détrez for their availability and invaluable help with the research and the translation. Research conducted in the frame of the ACuTe *(Antiquity in a Cup of Tea)* project of the LabEx "Les Passés dans le Présent" (Investissement d'Avenir, réf. ANR-11-LABX-0026-01).

https://doi.org/10.1515/9783111189635-025

(a) (b)

Fig. 1: Apulian red-figured volute krater, attributed to Sèvres 1 Painter (325–320 B.C.). H. 68.5 cm, diam. 35 cm. Sèvres, Musée National de Céramique, inv. no. MNC 1. **(a)** Side A; **(b)** Side B © Photo: Louise Détrez, with the kind permission of Sèvres – Manufacture et Musée.

Context of Production: The Technique of the Sèvres 1 Painter and the Decoration of the Volute Krater MNC 1

Great technical virtuosity in the use of tetrachromy is a recurring feature of the painters and workshops of the last third of the 4th century B.C. This trend was very well identified by A. D. Trendall and was labeled "the ornate style" of Apulian vase painting. The ornate style consists in applying a wealth of overpainted details to red-figured vases. This technique, which relies on the use of various shades of overpaint (white, yellow ochre, brown, red), consists in reinforcing the ornamentation but also the modeling and tridimensionality of objects, figures, and architectures. The illusion of volume is created through effects of chiaroscuro or camaïeu of varied hues. The graphic black lines of the design over the orange-red clay surface are thus completed with a masterful technique based on the use of several shades of yellow, brown, and red. By the end of the 4th century B.C., the technique of red-figured painters and of the painters who specialized in polychrome *ceramica sovradippinta* shared many features. Red-figured painters tended to use a wealth of overpainted details to add depth and brightness to their designs; their preparatory designs were sometimes mere sketches, entirely covered and hidden from viewers' eyes by the richly colored overpainted decorations.

The loss of all overpainted details on the volute krater MNC 1 enables us to gain a better understanding of the painter's technique and of the steps he followed when decorating a vase. He started by applying *miltos* to the vase in order to redden the clay. He would then draw a quick sketch into diluted clay on the surface of the vase and apply a so-called black glaze around the figures and objects intended. This painter also made preparatory sketches with highly diluted clay in order to mark the location of future overpainted elements: this is the case, for instance, with the body of the central female character of side A, with the wings of the swan she holds, and with plant decoration. Finally, the painter would add overpainted elements and zones to the vase and fire it. The reason why he used highly diluted clay for his sketches is that they would become invisible after the addition of overpaint and the firing.

According to A. D. Trendall, the Painter of Sèvres 1 belonged to a coterie of artists who have been described as the Copenhagen Group. His activity is related to that of other painters of this group, such as the Bitonto Group and the Copenhagen Head Painter.[1] The Copenhagen Group was active during the last quarter of the 4th century B.C. in Peucetia and Daunia; it can be recognized as one of the most characteristic groups of Late Apulian painting because of its abundant production, its penchant for color, its habit of representing figures within a *naiskos*, and the use of female heads as a main motif on monumental vases. All these tendencies are notable characteristics of MNC 1.

Even though its overpainted decoration has almost completely disappeared, one may note the quality of the iconography that adorns the entire vase. On side A (Fig. 1) the body of the vase shows a tall *naiskos* with its roof beams in perspective. Within the *naiskos*, a woman is seated on a huge cubic case. She is seen in three-quarter profile, but it seems difficult at first to decide what direction she is facing in because her head has been sketched extremely schematically. She is apparently looking at a large bird lying or standing on her lap. The sketchy lines on her body seem to suggest that she is wearing a light chiton and that she is unveiling herself with a gesture of *anakalypsis*. The bird is apparently a large swan with open wings. One also notes a small, unclear, roughly triangular motif in the lower left corner of the *naiskos*. On each side of the building can be seen two candelabra like plants. The shoulder of the vase has a frieze of *languettes* running on both sides. On side A, the neck of the vase is decorated with a female head emerging in profile from a blooming anthemion and facing left. Higher on the neck, there is a frieze of alternating paterae and roses. The mouth is adorned with two friezes: one of posts and another of *ovulae*. On both sides the krater's volutes enclose *gorgoneia* mascarons. Side B displays a less detailed decoration (Fig. 1). The body of the vase has one main motif: a female head in profile facing left. This figure wears an embroidered *kekryphalos* and has thick, curly hair done up at the back of her head in a stiff, voluminous ponytail. The iconographic field is completed by plant decoration (volutes), a writing tablet. and a *haltēr*, which last may seem surprising since it represents an attribute only rarely connected with women. The neck is decorated with a huge palmette flanked by volutes on each side of the vase. Farther up on the neck, one can see a frieze of alternating volutes and dots. The mouth has two friezes, as noted above: one of posts and another of *languettes*. On the shoulder are preserved three of an original four plastic swans' heads. On side B, one of these was wrongly restored as a mere ring.

Toward a Virtual Restoration of the Lost Decoration

One may question the identification of the seated figure on side A. Are we to think of Leda herself? Or of an anonymous woman represented more or less in the guise of Leda? First, the swan is huge, which may indicate that it is no mere pet bird but Zeus metamorphosed into a swan. Second, the nude leg and the gesture of *anakalypsis* encourage the viewer to recognize a seduction scene: *anakalypsis* is indeed frequently connected with such scenes in Italiote iconography. These elements suggest that the figure could be Leda herself or at least an anonymous woman willfully presented as a new Leda, with a conscious allusion to mythological images. On side B, the writing tablet and the *haltēr* are striking details, both of which point to the level of education of the deceased, possibly a woman, for whose tomb this vase was initially designed.

The hand of the Sèvres 1 Painter is associated with very distinctive characteristics: this painter represented female heads with a round eye, a large pupil, an eyelid consisting of two converging lines,[2] a round bit of hair and a bump on the forehead. Several vases have been attributed to this same artist because of such elements.[3] One of them is an amphora

1 *RVAp* II, 945–954; Riccardi (2009) 28; Detrez/Pouzadoux (2022) 47.

2 *RVAp* II, 945–946, pl. 371; Riccardi (2009) 29; Detrez/Pouzadoux (2022) 47.

3 Riccardi (2009) 28–29; Maiellaro (2019) 20.

kept in the Bibliothèque Nationale (Fig. 2). Further characteristics of this artist's work are the wide locks of hair brought together behind the ear and the tall, candelabra like plants around the *naiskos*: the same elements are also noticeable, for instance, on a volute krater kept in a private collection in Geneva.[4] It is particularly interesting that four vases attributed to this painter were discovered together in one and the same tomb in Bitonto (tomb 7-2003). This attested provenance tends to suggest that the Sèvres 1 Painter's workshop was possibly located in the vicinity of Bari.

(a) (b)

Fig. 2: Apulian red-figured amphora attributed to the Sèvres 1 Painter, female profiles, last quarter of the 4th century B.C. H. 32.1 cm, diam. 14.4 cm. Paris, Bibliothèque Nationale de France, inv. no. De Ridder.896. **(a)** Side A; **(b)** Side B. © All rights reserved.

4 Détrez/Pouzadoux (2022) 50.

The decoration of the vase that we perceive today has little to do with what the original viewers of this vase were able to see as it was deposited in the tomb. A recurring feature of the Sèvres 1 Painter's recovered work is indeed the very poor state of preservation of overpainted zones and details. It is possible, for instance, to compare MNC 1 with another krater, in Strasbourg, by the same hand (Fig. 3). Once again, the vase seems to have lost all its overpainted details. The loss could be due to the technique of the painter and to specific technical procedures that made his overpaint more fragile, but it could also be due to the chemical characteristics of the soil in which several of his vases were buried: a high level of humidity or acidity, or both, may have caused the overpainted decorations to disappear.

Fig. 3: Apulian red-figured volute krater, attributed to the Sèvres 1 Painter, ca. 340-320 B.C. H. 49.8 cm, diam. 29.5 cm. Side A: youth under a *naiskos*. Strasbourg, Musee Archeologique, inv. no. 11.002.1.16. © Musée Archéologique de Strasbourg / Photo: Musées de Strasbourg.

Our first attempts to virtually reconstruct the lost decorations of this vase (Fig. 4) were mainly based on close observations of its surface under a raking light or with a small UV lamp. This method has already provided us several hints and answers: with the raking light we were able to perceive slight traces, "phantoms," of motifs under certain angles, and the UV light also allowed certain traces of overpaint to glow on the neck of the vase, especially on the anthemion of side A, thus confirming that this vase had once been adorned with a wealth of overpainted motifs. Photogrammetric views, combined with chromatic saturation obtained thanks to Affinity software, allowed us to make further lost details reappear: for instance, we found that the female figure of side B was wearing not one but two necklaces, and we were able to perceive a series of patterns in overpaint on the frieze of paterae and roses of side A. Even more details appeared as we tentatively used a multispectral camera to analyze the surface of the vase: this allowed us to see further traces of overpaint on the *naiskos*, the seated female figure, and the swan of side A—that is to say, on parts of the vase where the naked eye no longer sees any trace of overpainted decoration. Sadly, the results remained somewhat inconclusive for side A. For side B, by contrast, observation in infrared revealed that the female head originally had a round earring and a hair clip. The multispectral camera also allowed us to gain information on the original appearance of the mascarons: both mascarons of side A certainly had overpainted details. It seems that the skin and eyes, lips, and hair of the side A *gorgoneia* were white. On side B, the skin was probably orange-red (cf. Fig. 1). Finally, we also noed a few traces of white on the plastic swan heads. Apart from these phantom traces, it was possible to draw many comparisons between MNC 1 and other vases decorated by the Sèvres 1 Painter or other artists of the ornate style: such comparisons were crucial for understanding the slight traces that we perceived on MNC 1 and besides for tentatively reconstructing the colors of the lost overpainted details.

As this analysis has shown, MNC 1 was once a very elaborate example of the ornate style. Thankfully, modern sources allow us to understand a little better why so little of this initial decoration still persists today. First, we are lucky to have a testimony on the appearance of this vase and its state of preservation as it arrived in Sèvres. Jean-Jacques Lagrenée (1739–1821), the adjunct head designer of the Sèvres Manufacture, indeed made an album of gouaches representing a hundred vases of the Denon collection in1786–1787: that is to say, immediately after the vases arrived in Sèvres. His painting of the volute krater MNC 1 was completed with a quick comment stating that this vase was initially made "to be seen

(a)

(b)

Fig. 4: Proposed reconstruciton of the decoration on MNC 1. **(a)** Side A; **(b)** Side B (© Photo: Rozenn Douaud; Preparatory drawings: Maiwenn Tonna).

from a considerable distance" ("pour être vu de très loin"). Obviously, Lagrenée, who had traveled in Italy and was a good connoisseur of red-figured vases, was rather disappointed by the appearance of MNC 1 and implicitly criticized what he interpreted as a sketchy, rough, careless decoration.[5] It is intriguing that MNC 1 also appears on a calotype or talbotype—one of the very first photographs made by the École de Sèvres and currently kept in the Sèvres archives. On what appears to be a talbotype made by L. Robert (Fig. 5), the Sèvres chemist, sometime between 1851 and 1858, one can see MNC 1 with astonishing traces of white overpaint that build a quadrangular frame inside the walls of the *naiskos* or enhance the candelabra like plants of side A and the roses of the neck frieze.[6] This photograph thus proves that fugitive traces of white overpaint, which still existed in the mid-19th century, were willfully erased at a later time, probably because they made no sense to the modern viewers. This willful erasure of the few remaining white traces of overpaint must have happened before the 1930s, when MNC 1 was once again photographed for the purpose of M. Massoul's publication of a volume of the *Corpus Vasorum Antiquorum* on the ancient ceramics of the Sèvres Museum. The *CVA* photographs show none of the white traces seen on the talbotype of the 1850s, and Madeleine Massoul appears to have been completely unaware of their existence: accordingly she wrote that the painter of this vase had simply forgotten to add in the details before the firing. The same idea was expressed, in an identical way, by A. D. Trendall, who never saw this vase but was probably amazed

5 Lagrenée (le Jeune) (1786–1787), notation in the margin: "Ce vase dans l'original a 2 pieds 5 pouces de haut avec les anses sur 18 pouces de large C'est le plus grand de la collection. Tous ses ornements et figures sont très grossièrement ébauchés il a sûrement été fait pour être vu de très loin".
6 Sèvres, Musée National de Céramique, inv. no. AMNS 1996 D95.4.

by the complete absence of traces of overpaint in the photographs published by Madeleine Massoul in her volume of the *CVA*.[7] Evidently, most of the overpainted details had already disappeared by the time MNC 1 arrived in Sèvres, but the few remaining traces that would have helped to understand the nature of its decoration at an earlier point were erased during restoration sometime between the 1850s and the 1930s.

Fig. 5: Tablotype by Louis Robert, ca. 1851–1858. Sèvres, Musée National de Céramique, inv. no. AMNS 1996 D95.4 © Archives de la Manufacture Nationale de Sèvres.

Conclusions

The volute krater MNC 1 is a very valuable artifact for understanding the technique of painters operating in the Apulian ornate style. Because of the almost complete disappearance of its traces of overpaint, it provides us a rare example of what a sketch could look like and allows us to glimpse the preparatory stages of the decorative process. A tentative reconstruction of the lost overpainted elements has been made possible by the combined use of several methodologies and sources: close observation under raking light, chromatic saturation of photogrammetric pictures in image-processing software, multispectral photography, comparison with other vases attributed to the Sèvres 1 Painter and to related painters sharing common features with him, analysis of archival documents, photographs, and modern comments on this krater's decoration. The virtual restoration that we have proposed, we hope, will allow modern viewers to appreciate the quality and complexity of the iconography of MNC 1. It also shows that this vase fully belongs among the elaborate examples of the Apulian ornate style. Far from being careless or producing quick, sketchy works, the Sèvres 1 Painter was keen on adding tridimensionality, brightness, and a wealth of details to his compositions.

7 *CVA* France 13 – Musée de Sèvres, Paris, Honoré Champion (1936) 75, pl. 37: "ce vase semble avoir été inachevé ; les rehauts blancs auraient été omis".

Abbreviations

CVA *Corpus Vasorum Antiquorum*.
RVAp II Arthur Dale Trendall and Alexander Cambitoglou, *The Red-figured Vases of Apulia. Late Apulian*, VOL. II, Oxford, 1982.

Bibliography

Détrez/Pouzadoux (2022): Louise Détrez and Claude Pouzadoux, "Retour à la Terre: un partenariat entre le musée National de la Céramique de Sèvres et le Centre Jean Bérard de Naples pour la recontextualisation du cratère à volutes apulien MNC 1", in: Alexandra Attia, Daniela Costanzo, Christian Mazet and Valeria Petta (eds.), *«Infinito sarà il tempo dell'Ade». L'archéologie funéraire en Italie du Sud (fin VIe – début IIIe siècle av. J.-C.)* (Actes de la rencontre franco-italienne, Paris, 24–25 mars 2017 – Atti dell'incontro italo-francese, Parigi, 24–25 marzo 2017), Lavello, 43–54.
Lagrenée (le Jeune) (1786–1787): Jean-Jacques Lagrenée, *Cent vases étrusques dessinés par Lagrenée d'après les vases originaux appartenant au Roy*, album of gouaches kept in the Nationalmuseet of Copenhagen.
Maiellaro (2019): Giuseppe Maiellaro (ed.), *Il Grand Mausolée di Polignano. Riscoperta di un contesto peuceta del IV secolo a.C.*, Foggia.
Riccardi (2009): Ada Riccardi (ed.), *Donne e Guerrieri da Ruvo a Bitonto. Le scoperte del III millennio*, Bari.

Section V: **Craftspeople's Mobility and Knowledge Transfer: Distribution and Statistics**

Xenia Charalambidou

26 Technology Transfer and Mobility During the Period of Greek Migrations to Italy (8th to 6th Centuries B.C.): Insights from Coarse ware Pottery

Abstract: This chapter highlights intra- and interregional artisanal interactions and mobility, with an emphasis on the Geometric and Archaic periods – the time of the Greek migrations to Italy and creation of the Greek diaspora. It explores artisanal interplay and movements of products and people through examples of coarse pottery in order to demonstrate the need for a new balance between the study of fine wares and coarse wares, one which has long been anticipated in the field of Classical archaeology and is now gaining ground. These examples come from Cycladic Naxos in the Aegean, as well as Sicilian Naxos, a significant colonial settlement on the eastern shore of Sicily that, based on later colonization mythography, was founded by Euboea (especially Chalcis) and Cycladic Naxos in 734 B.C.

Introduction

In terms of traditional chronology,[1] the 8th century B.C. can be viewed as a point of acceleration in the Mediterranean, evident in patterns of intensified connectivity and mobility, including the founding of settlements and/or traveling mainly for professional reasons by poets, physicians, diplomats and emissaries, sailors, merchants, cooks, and artisans, among others.[2] Intensification and expansion of mobility and trade led to intra- and interregional interactions that were realized through the exchange of people, goods, technologies, and consumption practices.

The phenomenon of intra- and interregional artisanal mobility at the time of the Greek migrations to Italy and establishment of a diaspora there, which occurred during the Geometric and Archaic periods, is central to my "Pots and Pans" research.[3] I have stressed in an earlier paper the importance of investigating the phenomenon of the mobility of artisans (here potters), not only through macroscopic approaches, but also through interdisciplinary methodologies, placing equal emphasis on *both* fine and coarse wares.[4] As scholarship concerning the Early Iron Age and Archaic periods to date has mainly focused on the work of potters on the move with respect to fine wares – and this, thus far, has been realized mainly through macroscopic approaches – in this chapter I would like to highlight the importance of coarser – in fabric – vessels as similarly strong evidence for mobility of people and products.[5] Such a shift in perspective is required to counterbal-

1 I consider absolute dates as 'provisional', also in anticipation of new research on Iron Age Mediterranean chronology, see recently Doumet-Serhal *et al.* (2023).

2 Traveling professionals are not solely an Early Iron Age and Archaic phenomenon in the Mediterranean. Professionals on the move preceded this era already in the Bronze Age – and lasted long after it too, right up until preindustrial times in various regions. See, for example, Papadopoulos (1997) and more recently Charalambidou (forthcoming a), with bibliography.

3 The full name of this research project is *Pots and Pans: pottery production and consumption in ancient Greek mother-cities and in indigenous and colonial communities in Italy [ca. 800-550 B.C.]* for which the present author is a coordinator and senior postdoctoral researcher. It is part of the wider *Melting Pot* Project (the NWO-funded project, no. VC.GW17.136, directed by Prof. J. P. Crielaard), which is entitled *What went into the melting pot? Land-use, agriculture, and craft production as indicators for the contributions of Greek migrants and local inhabitants to the so-called Greek colonization in Italy (ca. 800-550 B.C.)*. The whole campaign investigates ancient Mediterranean migration by promoting the trans-disciplinary integration of macroscopic and scientific analyses of ancient craft-products with research on bio- and geo-archaeology and landscape modelling.

4 Charalambidou (forthcoming a). For the purposes of this chapter, I group semi-coarse and coarse vessels together under the general term "coarse". The same applies for the fine and semi-fine vessels, which are here grouped under the general term "fine".

5 Some very important advancements towards the inclusion of the interdisciplinary (macroscopic and science-based) study of coarser ceramic categories in relation to artisanal mobility have, however, already been made in the prehistoric Aegean and the Mediterranean: see, for example, Müller/Kilikoglou/Day (2015); Gauß *et al.* (2015); Gauß *et al.* (2017) (the last two papers, with discussion also about the Iron Age); Lis *et al.* (2020); Day *et al.* (2022).

Note: I would like to thank especially Maria Costanza Lentini, Peter Day, Beatrice McLoughlin, and. Gloria London for their valuable feedback. This chapter was written during my research at the Vrije Universiteit Amsterdam. My heartfelt thanks to the Dutch Research Council (NWO-Nederlandse Organisatie voor Wetenschappelijk Onderzoek)- funded project in which I participate as a senior postdoctoral researcher, *What Went into the Melting Pot?* (see full title in note 3), and the project director, Prof. Jan Paul Crielaard of Vrije Universiteit Amsterdam.

https://doi.org/10.1515/9783111189635-026

ance the focus of scholarship on ceramic fine (painted) table wares, which still assigns less value as to how other ceramic categories, such as storage and cooking wares, can illuminate for us the wide spectrum of activities they reflect (*i.e.*, from production, provision and procurement to cooking and consumption).[6]

Approach and Methodology

To understand the interplay and mobility of artisans, here potters active at a local/regional and transregional level, my efforts concentrate on characterizing and documenting the different pottery *chaînes opératoires* (or production sequences)[7] in order to postulate the degree and range of social dynamics and interactions of the communities in contact. My particular focus is on pottery production and consumption in ancient Greek mother-cities and the colonial and indigenous communities of Italy between ca. 800 and 550 B.C. The main goal is to determine what share Greek and native traditions had in these cultures in Italy, with an emphasis on their ceramic technologies; the second objective is to examine how material culture evolved in the light of the newly-formed identities and newly-developed socio-economic structures.

The "Pots and Pans" research through the lens of pottery therefore seeks to clarify how knowledge, technology, and consumption were transferred between mother-cities, colonial establishments, and the local/regional native milieux in Italy. Within this research, I, with my collaborators, study the selection and preparation of raw clay materials, forming, finishing operations, surface treatments, and firing techniques of regional ceramic traditions in Greece, southern Italy, and Sicily. We have adopted a wide range of approaches to elucidate ceramic *chaînes opératoires*: using macroscopic study (morphological/typological, stylistic/iconographic), science-based analyses of pottery (petrographic, elemental, and micro-structural), geological prospection, and drawing on ethnoarchaeological research.

For the purposes of this chapter, case studies representing the project are taken from Cycladic Naxos and Sicilian Naxos, the latter one of the most important colonies on the eastern shore of Sicily that, according to tradition, was founded by Euboea, especially Chalcis, and Cycladic Naxos in 734 B.C.[8]

Short-Distance Interaction and Mobility

As certain categories of artifacts now reveal, potters were already on the move during the Bronze Age, an example of which is to be found in the spread of Aeginetan-type cooking pots in the Euboean and Pagasetic gulfs towards the end of the Late Bronze Age.[9] In the Early Iron Age and Early Archaic/Protoarchaic period, there are many vessel categories that deserve attention when exploring artisanal interplay and mobility. Among them, a group of vessels that can potentially point to both local and traveling potters is the coarse pithoi that often carried relief decoration, a popular type from the last third of the 8th through the 7th centuries B.C. in the Aegean (including Cycladic Naxos).[10] This could be so because such products, given their substantial size and weight, would often have been made locally, even by itinerant potters, as ethnoarchaeological research suggests.[11]

The importance of ceramic coarse ware traditions in the investigation of artisanal interactions and movement of goods and people can be highlighted on Iron Age Cycladic Naxos for instance, by means of various other categories. A good illustration of this is the presence of the coarse storage jars with tripod feet or a plain base, which appear (probably

6 Important steps have been taken to readdress this bias: e.g, McLoughlin (2000), (2011); Strack (2007); Gros (2007); Spataro/Villing (2015); Charalambidou (forthcoming b).
7 On the concept of *chaîne opératoire*: Leroi-Gourhan (1993). More recently, see Dobres (2000) especially 164–211; Gosselain (2000); Roux (2019), *passim*.
8 On colonization mythographies about Sicilian Naxos as well as on excavations and research at the site, see Lentini (1990), (1998), (2004a), (2004b), (2009), (2011), (2012), (2015a), (2015b), (2016), (2017), (2020a), (2020b), (2022).
9 More recently, Lis/Rückl/Choleva (2015), Lis *et al.* (2020). See also note 5.
10 See, for example, the neck-handled relief decorated pithoi: Simantoni-Bournia (1990), (2001), (2004); McLoughlin (2000), (2011).
11 Cf. Hampe/Winter (1962); Voyatzoglou (1984); Blitzer (1990); Leontidis (1996); Psaropoulou (1996); Giannopoulou (2010); London (2020).

in secondary use) in funerary contexts both inland, at the Iron Age cemetery of Tsikalario[12] as well as in a possible burial context at Agios Ioannis Apeiranthou, and in the cemeteries of the harbor town of Naxos (Fig. 1).[13] Preliminary results of macroscopic, petrographic, and elemental analyses (by the author of this chapter, Evangelia Kiriatzi and Noémi S. Müller) indicate that the majority of analyzed jars from the harbor town in the western part of the island (from the Plithos cemetery) and some of the analyzed storage jars from the Tsikalario cemetery were made in a coarse fabric associated with the dominant granite-granodiorite intrusion in western Naxos (Fig. 1b, 1c, 1d). Equally significant is the presence of at least one distinctly different fabric group among the analyzed jars from Tsikalario (Naxos, Archaeological Museum, inv. no. MN 3839; Fig. 1f), which is, in contrast, compatible with the local geology of central and eastern Naxos. Such variation in compositional data may indicate the existence of pottery production sites in different parts of the island.[14] The analyzed jars could have been produced in several towns or villages throughout the island, whereas some from western Naxos may have reached Tsikalario via intraregional interactions and mobility: the jars made in the first fabric presumably reached the island's interior through trade or exchange, or were being brought by people who moved there (perhaps as a result of intermarriages?).[15]

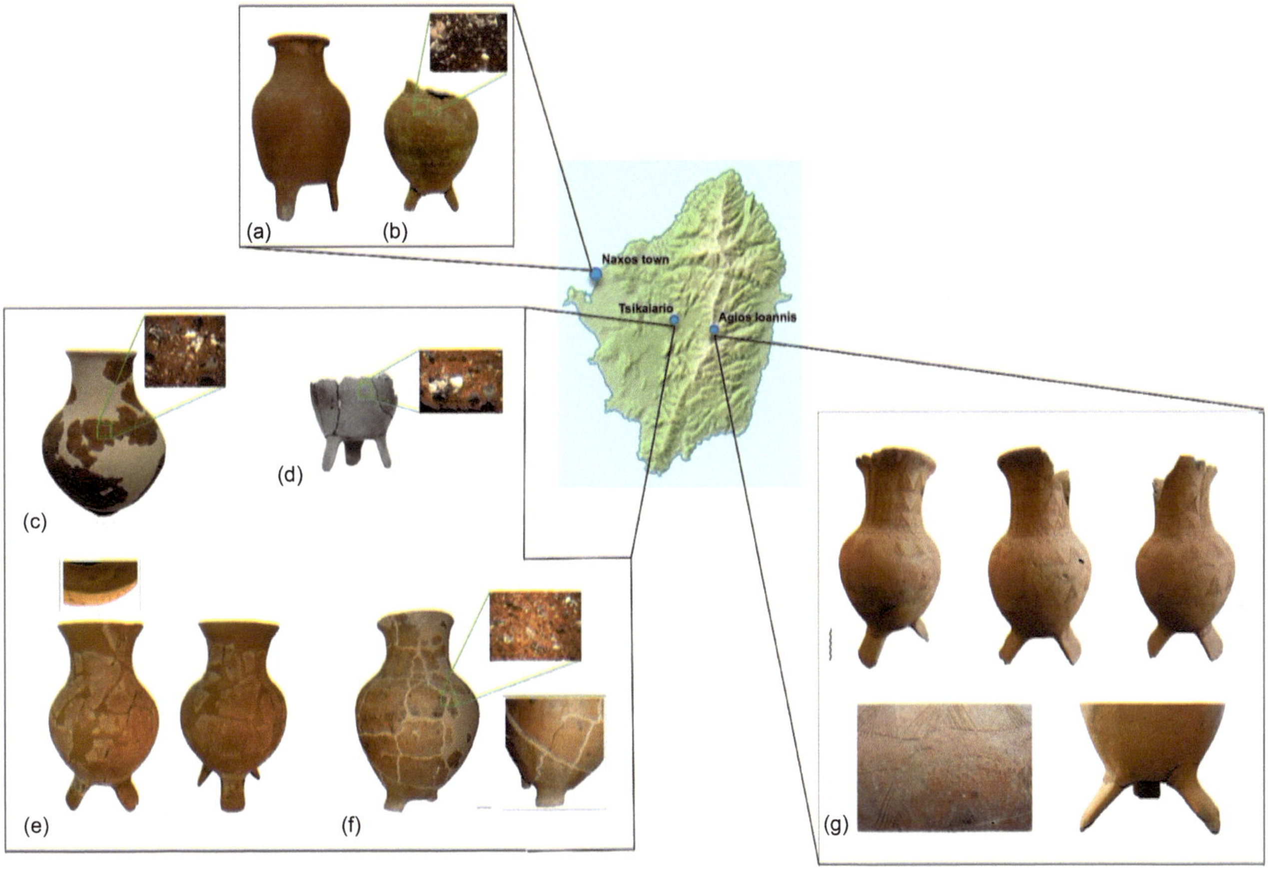

Fig. 1: Naxian coarse storage jars unearthed both inland (Tsikalario cemetery, Agios Ioannis Apeiranthou), and in the cemeteries of the harbour town of Naxos. Photos: Xenia Charalambidou.

The storage vessels as well as the cooking pots (with either a horizontal/flat base or round-bottomed; see below) from the Tsikalario cemetery reflect multiple chaînes opératoires. It is worth noting that after the disappearance of wheel-shaped

12 Zaphiropoulou (1983), (2001) 285–292, (2008–2009); more recently, Charalambidou (2018), (2021a), (2021b), (2022) 61–65, with further bibliography on Tsikalario.

13 See references and bibliography in Charalambidou (2021a), (2021b), (2022) 63.

14 Charalambidou/Kiriatzi/Müller (2017) 123–128.

15 Charalambidou (2021a) 325.

coarse pots, probably following the Late Helladic IIIC Grotta settlement phase, the reintroduction of rotational devices for the forming and/or finishing of coarse pots is first detected at this site in the Geometric period. To date, the earliest of this evidence from the site, on two coarse amphorae, dates to around the middle or the third quarter of the 8th century B.C. (e.g., Fig. 2a–c).[16] The cemetery's finds furthermore prove that a simplistic evolutionary conception of a shift from hand-built to wheel-shaped pots (meaning here: shaped with a rotational device) does not do justice to the complexities and coexistence of such traditions: even after the reoccurrence of wheel-shaped pots at the cemetery of Tsikalario, the hand-built traditions continued.

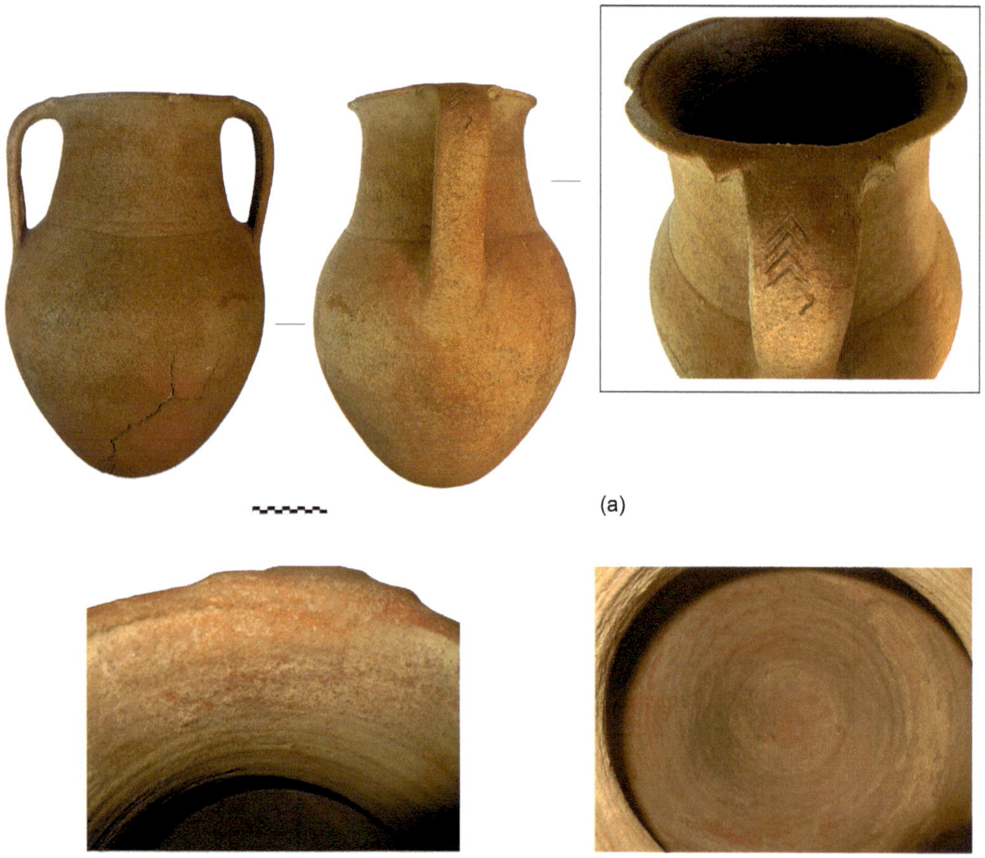

Fig. 2: (a–c) Coarse amphora (Naxos, Archaeological Museum, inv. no. 3825) from the centre of low tumulus 6, Tsikalario, Cyclades. Photos: Denitsa Nenova and Xenia Charalambidou (© Charalambidou and the *Tsikalario Archaeological Project*).

In the Tsikalario cemetery we encounter an interesting shape, a two-handled cooking pot with a horizontal rim, vertical neck marked off from the shoulder, and rounded or ovoid body with horizontal base (e.g., Figs. 3a–d; 4a–f). It is difficult to ascribe a definitive name to these Naxian vessels: neither amphorae (due to their two handles) nor chytrai (which is a term best reserved for round-bottomed cooking pots) seems appropriate in this case. At least some of these pots in the cemetery, either employed as burial containers or intended for funerary rituals, perhaps were first used in the household.[17] Since these vessels do not have a rounded base, as noted above, in their primary function they would not necessarily have required being propped up in some fashion while being used for cooking. Due to their flat base, for slow cooking they could have simply been set in the ashes and coals of a hearth. This particular form, based on the date of the associated finds in the cemetery, probably dates from the Late Geometric to the Early Archaic/Protoarchaic period (*i.e.*, second half of the 8th to the 7th century B.C.). This proposed date is further supported from other Aegean parallels, for example in Athens

16 Charalambidou (2021a).
17 Charalambidou (2021b).

and Eretria on Euboea.[18] What probably differentiates the Naxian two-handled pots from these parallels is the use of a slow rotational device for the neck of these Naxian vessels explaining why this part was quite cylindrical and straight-sided, in relation to the other Aegean examples.[19] The Naxian pots were most likely made using the coil technique, as is demonstrated by some marks or seams on their surfaces. In some cases (Fig. 4b–c), one can observe a clear seam around the maximum perimeter of the body, where clay was added when work on the upper part (re)commenced after the lower body had been allowed to dry out (and so become stronger).[20] It is the upper part of these vessels in particular that was further elaborated, being finished on a slow rotational device, whereas evidence for the use of such device cannot be observed either in the interior of the vessel or in the lower parts of its exterior.

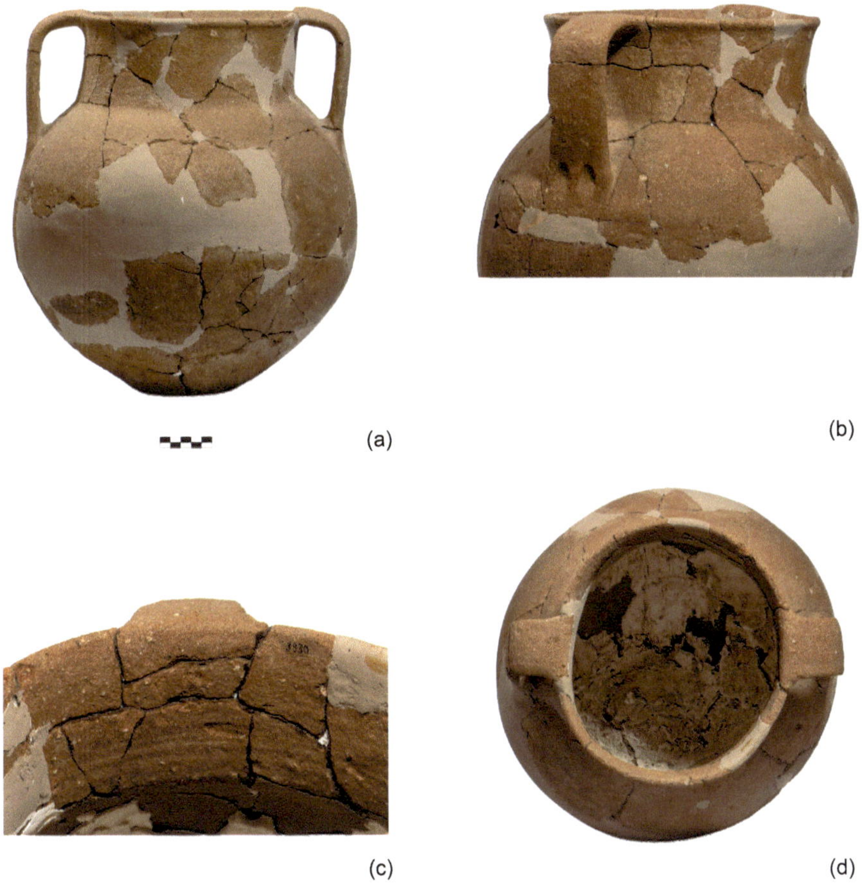

Fig. 3: (a–d) Two-handled cooking pot (Naxos, Archaeological Museum inv. no. 3930, burial context 13), Tsikalario, Cyclades. Photos: Robert Miller, Xenia Charalambidou and Nikos Sepetzoglou (© Charalambidou and the *Tsikalario Archaeological Project*).

18 Gros (2007) 112–113 (type 9). Our Naxian pots probably belong to Gros type 9c (pp. 113, cat. nos. 181–188, annexes pp. 54–56).

19 I would like to thank Beatrice McLoughlin for her valuable feedback on this issue.

20 It is a common practice in interrupted techniques around the world: the lower body is created, allowed to dry out for a while, and then the potter adds wet clay to the top of the semi-dried lower body – as a coil – in this case capping the top seam on the exterior and then builds the shoulder, neck and rim from that point on: pers. comm. McLoughlin.

Fig. 4: (a–f) Two-handled cooking pot (MN 5086, burial context 4/low tumulus A), Tsikalario, Cyclades, with a microphotograph (f) of its Naxian fabric (XPL, 500 µm). Photos: Robert Miller, Xenia Charalambidou and Nikos Sepetzoglou (© Charalambidou and *the Tsikalario Archaeological Project*).

Such a phenomenon—the incorporation of rotational devices used in existing potting traditions—offers us a great opportunity to highlight interactions among potters transcending Cycladic communities.[21] Preliminary comparisons of the potting traditions on Naxos and another Cycladic island, Andros, which had developed independently, indicate that wheel shaping became an integral part of the Iron Age Naxian tradition earlier, from at least the middle or the third quarter of the 8th century onward (see above), whereas on Andros at the Geometric settlement of Zagora wheel rotation for coarse wares is documented later, during the last quarter of the 8th century B.C.[22] The Andriot tradition was often grounded in the drawing-up technique, while the incorporation of wheel shaping into existing practices came only at the end of the Early Iron Age. Since the Zagora potters had achieved the same skill set, and for a restricted range of shapes, in the last decades of the 8th century, this raises the question of artisanal interplays (and even artisanal mobility) between Naxos and Andros. The importance of this practice also lies in the fact that during this period neighboring communities, such as on Euboea, seem not to have adopted wheel shaping for coarse wares.

Interregional Interactions and Mobility Across the Mediterranean

As argued elsewhere,[23] cooking pots as well as storage vessels comprise significant categories by which one can explore mobility and cultural transmission across the Mediterranean. Cooking vessels, for example, can significantly reflect economic, political, and social developments as much as painted fine wares do. Furthermore, they offer information on culinary behavior and cultural diffusion.[24] The circulation of Greek-type cooking pots during the time of Greek mobility and migration into southern Italy between the 8th and the 6th centuries B.C. in colonial and indigenous contexts, existing alongside coarse indigenous cooking products, makes the point well. In addition to the circulation of fine painted tablewares and spread of the associated drinking practices, the presence of Greek-type cooking shapes may even indicate new cooking habits. For instance, Greek-type chytrai[25] circulated in southern Italy especially from the 7th century onward, with examples found at Incoronata, Metapontum-Andrisani, Siris-Polieion, and Taranto.[26] They may also reflect the arrival of newcomers who brought cooking vessels with them or manufactured familiar forms using locally available clays.

From among the available case studies of the "Pots and Pans" research, I will choose a pertinent example from Sicilian Naxos (modern Giardini Naxos).[27] Sicilian Naxos shows a diversified ceramic repertoire, with a wide range of imported vessels, Greek-type pottery made locally, and indigenous ceramics found in varying proportions over time. In contrast to the lack of substantial amounts of Euboean and Euboeanizing pottery at Pithekoussai,[28] Sicilian Naxos, especially during the first phase of the settlement, exhibits a wide range of Euboean (metropolitan) and locally made Euboeanizing shapes,[29] alongside various other Greek vessel types as well as indigenous and mixed elements. Based on close similarities in form, style, forming and finishing techniques with Aegean, and in particular Euboean, products, but with the use of local/regional raw materials for the manufacture of these Greek-type forms, I argued elsewhere that many of the colony's Greek-type artifacts may have been made by migrant Aegean potters working there.[30] Here, I would also like to stress that any past

21 See also Charalambidou (2021a).
22 McLoughlin (2011); McLoughlin/Paspalas (forthcoming).
23 Charalambidou (forthcoming a).
24 Indicatively, Joyner (2007); Spagnoli (2010); Spataro/Villing (2015); Hruby/Trusty (2017a).
25 In the Aegean, some of the earliest chytrai come from late Middle Geometric II contexts at Corinth: Pfaff (1988) 78–79, nos 115–116, pl. 32. The form circulated widely, with examples from Athens – Rotroff (2015) – and Aegina – Gauß *et al.* (2015); Klebinder-Gauß/Strack (2015); Gauß/Klebinder-Gauß (2017) 29, fig. 7; Cycladic Naxos – Tsikalario: Charalambidou (2018) 161, fig. 21. See also Gros (2007) pl. 53. Such deep containers could facilitate stirring, contain vigorously boiling liquids or semi-liquids, and minimize evaporation. They would probably be practical for the preparation of soups, broths, and porridges containing vegetables, legumes, cereals, or meat, as well as to boil water: see Manakidou (2022) 433. Residue analyses are pending to explore with science-based methods their contents.
26 Quercia (2015a) 314–316, fig. 2, (2015b) 205–207, figs. 18.2–3, fig. 18.4 (top left).
27 I am grateful to Dr Maria Costanza Lentini and the Direction of Parco Archeologico di Naxos for our excellent collaboration.
28 Boardman (1996) 156; Kotsonas (2015) 249.
29 See references above to Lentini's publications as well as Coldstream (2004) and Charalambidou (forthcoming a).
30 Charalambidou (forthcoming a), with a summary of our macroscopic and science-based results.

discussions in connection with the pottery discovered in Sicilian Naxos have thus far revolved around fine wares.[31] The potting traditions of coarse wares in this colonial establishment and any possible ties with those of the Aegean mother-cities have not been considered before the "Pots and Pans" research.

Among coarse wares found in Sicilian Naxos, there are numerous cooking pots with multiple *chaînes opératoires*, reflecting indigenous, Greek-style, and hybrid traditions. For the purpose of this discussion, I would like to place emphasis on a well-preserved, two-handled cooking pot representing a form described above (when describing cooking pots from Cycladic Naxos) – one with a horizontal rim, vertical neck marked off from the shoulder, and rounded body with horizontal base – which is dated to the 7th century B.C. and was used as a burial vessel in the North Cemetery of Sicilian Naxos (Fig. 5a–e). This pot was probably made locally/regionally (it has a coarse metamorphic and igneous fabric, proving a Sicilian origin), but appears to recall Aegean cooking forms and technology, as a detailed examination shows. It is not only its form that connects it with the Aegean (compare with Figs. 3–4 from Cycladic Naxos; cf. [Gros] Type 9c above). A link with the Naxian two-handled cooking pots discussed above is also seen in the area around the maximum diameter of the body, where a "projection" can be observed on the exterior surface: as with the Tsikalario cooking pots, this is an articulation of the clay coil seam, which caps the lower part of the body when starting the construction of the upper body to form the shoulder. This forming element may represent a gestural link between potters – on Cycladic Naxos and in Sicilian Naxos – in particular because this element is not intentionally visible from all sides. In other words, the potters in both sites, while not trying to create a visibly articulated shoulder, accidentally achieved an articulated ridge simply due to the way they created the shoulders of the pots. Other links with the Cycladic Naxian vessels may include the more elaborate finishing of the pot's neck and the manner in which the lower half is left in a slightly rougher state than the upper, where the efforts of the potter to smooth the surface were concentrated. Such preliminary observations on these technological phenomena can show affinities with the forming and finishing procedures followed at Cycladic Naxos, giving us a case to interrogate and compare their *chaînes opératoires*. None of these technological features is thus far known in the indigenous Sikel coarse ware traditions. It is therefore likely that the potter who fashioned this cooking vessel in Sicilian Naxos used local materials, while following the technology of their Aegean homeland.

This presents a particularly compelling situation for a number of reasons. First, it can be recognized as an example of technological transfer from a Greek mother-city to its colonial settlement in Italy, especially when we consider that Cycladic Naxos was traditionally one of the two founders of Sicilian Naxos. As the role of Chalcis, the other founder, has been established mainly through imported Euboean and locally made Euboeanizing vessels (many of which are fine table ware),[32] this newly-detected Aegean impact (possibly from Cycladic Naxos) could have been a welcome addition to our understanding of the interplays among far-flung communities in contact.

Cooking pots as well as textile production and weaving,[33] both belonging to the women's sphere, can be used as indicators of cultural relations and identity signifiers.[34] Furthermore, cooking pots can also reflect the arrival of various people, men as well as women, who brought cooking vessels with them or manufactured their homeland cooking forms by using clays from the areas they had moved into. Such pots could even potentially suggest the presence of immigrant women potters or users from the Aegean at the Sicilian colony, since cooking pots were especially connected with women's preferences in the kitchen.

This proposal, in turn, brings to the fore the question of whether we possess archaeological evidence of immigrant women at colonial settlements – still a highly debated topic.[35] Irad Malkin's recent research argues for the presence of wives accompanying their husbands in these far-flung travels: it is a view worth considering carefully.[36] His opinion is based on the simple, but solid, argument that women from metropolitan Greece were, for example, needed in the performance of cult in colonial contexts. Indigenous women would have had no knowledge either of the rites and rituals

31 Coldstream (2004); Kourou (2004).
32 See above, including note 30.
33 Quercia/Foxhall (2014), with bibliography.
34 Cf. Spagnoli (2010).
35 See, for example, Shepherd (2012), with earlier bibliography. Other variables of female recognition include the analysis of grave goods and patterns in their distribution; these can also contribute to the identification of women in the archaeological record of the "Greek West", especially when used in conjunction with advances in skeletal analysis (when the latter evidence is preserved): Liston (2012); Shepherd (2012) 223.
36 Malkin (2021) with bibliography. See also Malkin (2020).

Fig. (5a–e): Two-handled cooking pot (North Cemetery of Giardini Naxos, tomb 265), Giardini Naxos, Sicily, with a microphotograph of its local/regional fabric (XPL, 1mm). Photos: Robert Miller, Xenia Charalambidou and Nikos Sepetzoglou. Courtesy: Maria Costanza Lentini.

of the *oikos* or of the religion of the new settlement, so essential to the very existence of the new habitation sites. Gillian Shepherd, furthermore, reminds us that even if the initial threat of danger was a decisive factor in deciding who joined the newly established settlements, women could have arrived from Greece at a later, safer, stage in the settlement process.[37]

37 Shepherd (2012) 219.

Such cooking pots as these may not have been manufactured in large quantities, in the area of Sicilian Naxos, and may even have been produced in the context of the household. Their production may well have been in the hands of male as well as of female potters, as documented for example in the ethnoarchaeological work by Gloria London on traditional women potters producing hand-built coarse ware pots on Cyprus since the 19th century A.D. (Fig. 6).[38] In different places and circumstances, women as well as men could have been active agents in pottery manufacture: certain coarse ware categories, especially those that were often connected to household activities, could engage both genders in pottery production. Likewise, scholarly discussion of ancient female artisans in fine painted pottery has begun to gain ground recently.[39] Furthermore, the anthropological documentation of one female ceramicist is now a strong possibility for the Early Iron Age: in Eleutherna, on Crete, a woman between 45 and 50 years old was inhumed at the Orthi Petra cemetery among funerary pyres dating to the Late Geometric and Archaic periods. According to Anagnostis Agelarakis, the skeletal-anatomical and palaeopathological evidence indicates that she could have worked at a potter's wheel.[40]

Fig. 6: A woman potter, Christinou, from Agios Demetrios (Marathasa), Cyprus (July 1986), while coil-building a coarse hand-built pot. Photos: Courtesy of Gloria London.

Conclusions

I have argued that both technology and consumers' behavior were transmitted between Greece and Italy as a phenomenon of the Iron Age Greek diaspora, and this can be revealed not only by the study of fine wares, but also the documen-

38 London (2000), (2020).
39 Murray/Chorghay/MacPherson (2020).
40 Agelarakis (2020).

tation of coarse wares there. Coarse potting traditions can advance our understanding of the range of social dynamics and interactions among neighboring and far-flung communities that were in contact. New research on coarse wares can significantly enhance our knowledge of the complexity of economic, political, and social conditions of the past, especially during periods of Mediterranean history marked by the mobility and circulation of people, goods, production, and consumption practices. Coarse vessels, cooking pots in particular, can offer valuable information on issues of technological transfer, gendered roles in production and use of ceramics, cultural diffusion, culinary behavior, and cultural identity. They definitely have the potential to help us make the focus of scholarship devoted to Mediterranean archaeology more inclusive.[41] It is crucial to highlight the multiplicity of agents involved in various sociocultural processes, especially individuals and social groups that often have been excluded from discussions about the Mediterranean past.

Bibliography

Agelarakis (2020): Anagnostis Agelarakis, "Bioarchaeological Evidence Reflective of the Use of the Kick-wheel by a Female Ceramicist in Ancient Eleutherna", in: Nikolaos Ch. Stampolidis and Mimika Giannopoulou (eds.), *Eleutherna and the Outside World*, University of Crete, Eleutherna, 206–218.

Blitzer (1990): Harriet Blitzer, "Κορωνέϊκα: Storage-jar Production and Trade in the Traditional Aegean", in: *Hesperia* 59, 675–711.

Boardman (1996): John Boardman, "Euboeans Overseas: A Question of Identity", in: Don Evely, Irene S. Lemos and Susan Sherratt (eds.), *Minotaur and Centaur: Studies in the Archaeology of Crete and Euboea Presented to M.R. Popham*, Oxford, 155–160.

Charalambidou (2018): Xenia Charalambidou, "Iron Age Mortuary Practices and Material Culture at the Inland Cemetery of Tsikalario on Naxos: Differentiation and Connectivity", in: *Annual of the British School at Athens* 113, 143–198.

Charalambidou (2021a): Xenia Charalambidou, "Contextualizing Artisanal Interplay and Technological Changes on Iron Age Naxos (Cyclades). Some Preliminary Observations", in: *Interdisciplinaria Archaeologica – Natural Sciences in Archaeology (IANSA)* 12 (2), 311–329. https://doi.org/10.24916/iansa.2021.2.15.

Charalambidou (2021b): Xenia Charalambidou, "Style and Function of Geometric Coarse Jars from Naxos: Examples from Inland and Coastal Regions of the Island", in: Vassilis Lambrinoudakis, Lina Mendoni, Maria Koutsoumpou, Tania Panagou, Alexandra Sfyroera and Xenia Charalambidou (eds.), *Έξοχος άλλων: Τιμητικός τόμος για την Καθηγήτρια Ε. Σημαντώνη-Μπουρνιά*, Athens, 147–162.

Charalambidou (2022): Xenia Charalambidou, "Migrating from Greece to Italy: Features of Euboean and Naxian Settlements before and during the Early Iron Age Greek Diaspora to Italy", in: Rodolfo Brancato, Luigi Maria Caliò, Marianna Figuera, Gian Michele Gerogiannis, Eleonora Pappalardo and Simona Todaro (eds.), *Schemata. La città oltre la forma. Per una nuova definizione dei paesaggi urbani e delle loro funzioni: urbanizzazione e società nel Mediterraneo pre-classico. Età arcaica* (Atti del Convegno Internazionale, organizzato dall'Università degli Studi di Catania, dall'Università degli Studi della Campania Luigi Vanvitelli e dal Consorzio Universitario Archimede, Siracusa, 26–28 febbraio 2020), *Cronache di Archeologia* – Monografie, Catania, 43–82.

Charalambidou (forthcoming a): Xenia Charalambidou, "Technological Transmission and Potters on the Move during the Iron Age Greek Diaspora to Italy (8th–6th centuries BC): Approaches and Methodological Considerations", in: Arianna Esposito (ed.), *Itinerances des artisans potiers et pratiques inter-artisanales en Italie centro-meridionale (VIIIe-VIe s. av. J.-C.). Tradition, innovation et contacts techno-culturels* (Colloque International, Naples, 17 September 2021), forthcoming.

Charalambidou (forthcoming b): Xenia Charalambidou, "The Social Roles of Feasting in the Central and Southern Aegean from the Postpalatial to the Protoarchaic Period", in: Catherine Morgan, Jan Paul Crielaard and Xenia Charalambidou (eds.), *Feasting with the Greeks: Towards a Social Archaeology of Ritual Consumption in the Greek World*, Oxford.

Charalambidou/Kiriatzi/Müller (2017): Xenia Charalambidou, Evangelia Kiriatzi and Noémi S. Müller, "Scales of Ceramic Analysis on Naxos (Cyclades)", in: Søren Handberg, Anastasia Gadolou (eds.), *Material Koinai in the Greek Early Iron Age and Archaic Period*, Monographs of the Danish Institute at Athens 22, Aarhus, 109–132.

Coldstream (2004): John Nicolas Coldstream, "The Various Aegean Affinities of the Early Pottery from Sicilian Naxos", in: Lentini (2004a) 40–49.

Day *et al.* (2022): Peter M. Day, Elina Kardamaki, Vasco Hachtmann and Adamantia Vasilogamvrou, "Pottery Production and Consumption at Ayios Vasileios during LH II-IIIA and the Cretan Connection", in: Anna Lucia D'Agata, Luca Girella, Eleni Papadopoulou and Davide G. Aquini (eds.), *One State, Many Worlds. Crete in the Late Minoan II-IIIA2 Early Period* (Proceedings of the International Conference, Khania, Μεγάλο Αρσενάλι, 21st-23rd November 2019), SMEA Supplement, Rome, 365–386.

Dobres (2000): Marcia-Anne Dobres, *Technology and Social Agency. Outlining a Practice Framework for Archaeology*, Oxford/Malden, Massachusetts.

Gauß/Klebinder Gauß/von Róden (2015): Walter Gauß, Gudrun Klebinder-Gauß and Constance von Róden (eds.), *The Transmission of Technical Knowledge in the Production of Ancient Mediterranean Pottery*, Vienna.

41 Cf. more recently, see Hamilakis/Greenberg (2022), as well as a recent colloquium and workshop organized by N. Papadimitriou, A. Kotsonas, L. Bhalerao and A. Anagnostopoulos, at the Museum of Paul and Alexandra Canellopoulos Museum in Athens with theme "Decoloniality, Inclusivity and the Greek Past", https://camu.gr/en/decoloniality-inclusivity-and-the-greek-past/?fbclid=IwAR3iAqWUCXAj87dDm8VKKzwHodgfnyXxxV8pBSAnXGbUFD28Tp69AUu4FN8 (last access 01.03.2024).

Gauß *et al.* (2015): Walter Gauß, Gudrun Klebinder-Gauß, Evangelia Kiriatzi, Areti Pentedeka and Myrto Georgakopoulou, "Aegina: An Important Centre of Production of Cooking Pottery from the Prehistoric to the Historic Era", in: Spataro/Villing (2015) 65–74.

Gauß *et al.* (2017): Walter Gauß, Evangelia Kiriatzi, Michael Lindblom, Bartek Lis and Jerolyn E. Morrison, "Aeginetan Late Bronze and Early Iron Age Cooking Pottery", in: Hruby/Trusty (2017b) 46–56.

Gauß/Klebinder-Gauß (2017): Walter Gauß and Gudrun Klebinder-Gauß, "Äginetische Keramik: Produktion und Verbreitung in prähistorischer und historischer Zeit", in: Felix Lang and Wolfgang Wohlmayr (eds.), *50 Jahre Archäologie an der Paris Lodron-Salzburg Universität* (Workshop Salzburg, 14 Dezember 2016. ArchaeoPlus – Schriften zur Archäologie und Archäometrie der Paris Lodron), Universität Salzburg Band 9, Salzburg, 23–34.

Giannopoulou (2010): Mimika Giannopoulou, *Pithoi. Technology and History of Storage Vessels through the Ages*, BAR International Series 2140, Oxford.

Gosselain (2000): Olivier P. Gosselain, "Materialising Identities. An African Perspective", in: *Journal of Archaeological Method and Theory* 7, 187–217.

Gros (2007): Jean-Sebastien Gros, *Η χρηστική κεραμεική στην Κεντρική Ελλάδα κατά την Πρώιμη Εποχή του Σιδήρου (περ. 1100-675 π.Χ.), Τυπολογία, παραγωγή, διακίνηση, κατανάλωση*, PhD Dissertation, University of Thessaly and Université de Montpellier III.

Hamilakis/Greenberg (2022): Yiannis Hamilakis and Rafael Greenberg, *Archaeology, Nation and Race. Confronting the Past, Decolonizing the Future in Greece and Israel*, Cambridge.

Hampe/Winter (1962): Roland Hampe and Adam Winter, *Bei Töpfern und Töpferninnen in Kreta, Messenia und Zypern*, Mainz.

Hruby/Trusty (2017a): July Hruby and Debra Trusty, "Approaches to Bronze Age Greek Cooking Vessels", in: Hruby/Trusty (2017b) 1–13.

Hruby/Trusty (2017b): July Hruby and Debra Trusty, *From Cooking Vessels to Cultural Practices in the Late Bronze Age Aegean*, Oxford.

James/Dillon (2012): Sharon L. James and Sheila Dillon (eds.), *A Companion to Women in the Ancient World*, Malden, MA.

Joyner (2007): Louise Joyner, "Cooking Pots as Indicators of Cultural Change: A Petrographic Study of Byzantine and Frankish Cooking Wares from Corinth", in: *Hesperia* 76, 183–227.

Klebinder-Gauß/Strack (2015): Gudrun Klebinder-Gauß and Sara Strack, "An Aeginetan Potter's Workshop in Athens?", in: Gauß/Klebinder Gauß/von Róden (2015) 77–96.

Kotsonas (2015): Antonis Kotsonas, "What Makes a Euboean Colony or Trading Station? Zagora in the Cyclades, Methone in the Thermaic Gulf, and Aegean Networks in the 8th Century BC", in: Jean-Paul Descoeudres and Stavros A. Paspalas (eds.), *Zagora in Context. Settlements and Intercommunal Links in the Geometric Period (900-700 BC)* (Proceedings of the Conference held by the Australian Archaeological Institute at Athens and the Athens Archaeological Society, Athens, 20-22 May 2012), Mediterranean Archaeology 25, Sydney, 243–257.

Kourou (2004): Nota Kourou, "Cycladic Naxian Late Geometric Pottery and History", in: Lentini (2004a) 77–85.

Lentini (1990): Maria Costanza Lentini, "Le oinochoai 'a collo tagliato'. Un contributo alla conoscenza della ceramica di Naxos di VIII e VII secolo a.C.", in: *Bollettino d'Arte* 75, 67–82.

Lentini (1998): Maria Costanza Lentini, "Nuovi rinvenimenti di ceramica euboica a Naxos di Sicilia", in: Michel Bats and Bruno d'Agostino (eds.), *Euboica. L'Eubea e la presenza euboica in Calcidica e in Occidente*, Naples, 377–386.

Lentini (2004a): Maria Costanza Lentini, *Le due città di Naxos* (Atti del Seminario di Studi Giardini, Naxos, 29-31 ottobre 2000), Milan.

Lentini (2004b): Maria Costanza Lentini, "L'abitato proto-arcaico di Naxos di Sicilia (scavi 1998–1999)", in: Lentini (2004a) 28–39.

Lentini (2009) Maria Costanza Lentini, "Ultime indagini nell'area urbana dell'antica Naxos. Scavi 2003–2006. Rapporto preliminare", in: M.C. Lentini (ed.), *Naxos di Sicilia. L'abitato coloniale e l'arsenale navale. Scavi 2003–2006*, Messina, 9–37.

Lentini (2011): Maria Costanza Lentini, "Sicilian Naxos: Evidence from the Dark Age", in: Alexander Mazarakis Ainian (ed.), *The "Dark Ages" Revisited* (Acts of an International Symposium in Memory of William D.E. Coulson, University of Thessaly, Volos, 14–17 June 2007), Volos, 529–540.

Lentini (2012): Maria Costanza Lentini, "Resti di capanne della tarda età del Ferro a Naxos di Sicilia", in: *Aristonothos. Scritti per il Mediterraneo antico* 7, 157–174.

Lentini (2015a): Maria Costanza Lentini, "Recent Investigation on the Early Settlement Levels at Sicilian Naxos", in: Jean-Paul Descoeudres and Stavros A. Paspalas (2015) 309–315.

Lentini (2015b): Maria Costanza Lentini, "Some Late Geometric and Early Orientalising Tableware from Sicilian Naxos", in: Vicky Vlachou (ed.), *Pots, Workshops and Early Iron Age Society. Function and Role of Ceramics in early Greece*, Brussels, 241–250.

Lentini (2016): Maria Costanza Lentini, *Le origini di Naxos. Nuovi dati sulla fondazione*, in: Lieve Donnellan, Valentino Nizzo and Gert-Jan Burgers (eds.), *Contexts of Early Colonization*, Rome, 311–321.

Lentini (2017): Maria Costanza Lentini, "An Early Orientalizing Spouted Krater from Naxos on Sicily", in: Xenia Charalambidou and Catherine Morgan (eds.), *Interpreting the Seventh Century BC: Tradition, and Innovation*, Oxford, 349–354.

Lentini (2020a): Maria Costanza Lentini, "Le cinte murarie di Naxos", in: Luigi M. Caliò, Gian Michelle Gerogiannis and Maria Kopsacheili (eds.), *Fortificazioni e società nel Mediterraneo occidentale. Sicilia e Italia*, Rome, 39–53.

Lentini (2020b): Maria Costanza Lentini, "Una brocchetta bronzea a collo tagliato da Naxos di Sicilia", in: Rosalba Amato, Gioacchino Barbera and Concetta Ciurcina (eds.), *Siracusa, la Sicilia, l'Europa. Scritti in onore di Giuseppe Voza*, Palermo, 183–186.

Lentini (2022): Maria Costanza Lentini, "Naxos di Sicilia: l'inizio della città", in: Rodolfo Brancato, Luigi M. Caliò, Marianna Figuera, Gian Michele Gerogiannis, Eleonora Pappalardo and Simona Todaro (eds.), Schemata. *La città oltre la forma*, Rome, 151–168.

Leontidis (1996): Anastasios Leontidis, "Αγγειοπλαστικά κέντρα και περιοδεύοντες τεχνίτες. Η περίπτωση των Μαργαριτών Μυλοποτάμου Ρεθύμνης", in: Eirini Gavrilaki (ed.), *Κεραμικά εργαστήρια στην Κρήτη από την Αρχαιότητα ως Σήμερα* (Πρακτικά Ημερίδας, Μαργαρίτες, 30 Σεπτεμβρίου 1995), Rethimno, 69–77.

Leroi-Gourhan (1993): Andre Leroi-Gourhan, *Gesture and Speech*, Cambridge.

Lis/Rückl/Choleva (2015): Bartek Lis, Štěpán Rückl and Maria Choleva, "Mobility in the Bronze Age Aegean. The Case of Aeginetan Potters", in: Gauß/Klebinder Gauß/von Róden (2015) 63–75.

Lis *et al.* (2020): Bartek Lis, Evangelia Kiriatzi, Anthi Batziou and Štěpán Rückl, "Dealing with the Crisis: Mobility of Aeginetan-Traditions Potters around 1200 BC", in: *Annual of the British School at Athens* 115, 269–327.

Liston (2012): Maria A. Liston, "Reading the Bones: Interpreting the Skeletal Evidence for Women's Lives in Ancient Greece", in: James/Dillon (2012) 125–140.
London (2000): Gloria London, *Women Potters of Cyprus* [DVD: 26 minutes], Nicosia.
London (2020): Gloria London, *Wine Jars and Jar Makers of Cyprus: The Ethnoarchaeology of Pitharia*, Nicosia.
Malkin (2020): I. Malkin, "Women and the Foundation of Greek Colonies", in: Michela Costanzi, Madalina Dana (eds.), *Another Way of Being Greek. Interactions and Cultural Innovations of the Greeks in a Colonial Milieu*, Louvain, 235–255.
Malkin (2021): Iran Malkin, "Greek Women and Greek Colonies", in: *Ancient West and East* 20, 51–69.
Manakidou (2022): Eleni Manakidou, "Clay Cooking Ware and Kitchen Equipment in the Ancient Greek Household", in: Soultana Maria Valamoti, Anastasia Dimoula and Maria Ntinou (eds.), *Cooking with Plants in Ancient Europe and Beyond, Interdisciplinary approaches to the Archaeology of Plant Foods*, Leiden, 431–442.
McLoughlin (2000): Beatrice McLoughlin, *Searching for the Potter's Intention: The Pithos Makers of Zagora. A Ceramic Analysis of the Pithoi from the Geometric Settlement Site Zagora, Andros*, MA thesis, University of Sydney.
McLoughlin (2011): Beatrice McLoughlin, "The Pithos Makers at Zagora: Ceramic Technology and Function in an Agricultural Settlement Context", in: Alexander Mazarakis Ainian (ed.), *The "Dark Ages" Revisited* (Acts of an International Symposium in Memory of William D.E. Coulson, University of Thessaly, Volos, 14–17 June 2007), Volos, 913–928.
McLoughlin/Paspalas (forthcoming): Beatrice McLoughlin and Stavros A. Paspalas, "Ninth-and Early Eighth-century Zagora, Andros: Indications of Central Aegean Networks and Engagements", in: Demetris Athanasoulis (ed.), *Περὶ τῶν Κυκλάδων νήσων, Το Αρχαιολογικό Έργο στις Κυκλάδες, Διεθνές Συνέδριο, Βυζαντινό και Χριστιανικό Μουσείο, Αθήνα, 22 – 26 Νοεμβρίου 2017.*
Müller/Kilikoglou/Day (2015): Noemi S. Müller, Vassilis Kilikoglou, Peter Day, "Home-made Recipes: Tradition and Innovation in Bronze Age Cooking Pots from Akrotiri, Thera", in: Spataro/Villing 2015, 37–48.
Murray/Chorghay/MacPherson (2020): Sarah C. Murray, Irum Chorghay and Jennifer MacPherson, "The Dipylon Mistress: Social and Economic Complexity, the Gendering of Craft Production, and Early Greek Ceramic Material Culture", in: *American Journal of Archaeology* 124, 215–244.
Papadopoulos (1997): John K. Papadopoulos, "Innovations, Imitations and Ceramic Style. Modes of Production and Modes of Dissemination", in: Robert Laffineur and Philip B. Betancourt (eds.), *Τέχνη. Craftsmen, Craftswomen and Craftsmanship in the Aegean Bronze Age* (Proceedings of the 6th International Aegean Conference, Philadelphia, Temple University, 18–21 April 1996), Aegaeum 16, Liège, 449–462.
Pfaff (1988): Christopher A. Pfaff, "A Geometric Well at Corinth: Well 1981–6", in: *Hesperia* 57, 21–80.
Psaropoulou (1996): Betty Psaropoulou, "Τα κεραμικά εργαστήρια των τελευταίων 100 χρόνων στην Κρήτη", in: Eirini Gavrilaki (ed.), *Κεραμικά εργαστήρια στην Κρήτη από την Αρχαιότητα ως σήμερα* (Πρακτικά ημερίδας, Μαργαρίτες, 30 Σεπτεμβρίου 1995), Rethimno, 101–138.
Quercia (2015a): Alessandro Quercia, "The Production and Distribution of Early Greek-style Cooking Wares in Areas of Cultural Contact: The Case of Southern Italy and Sicily", in: Gauß/Klebinder Gauß/von Róden (2015) 311–332.
Quercia (2015b): Alessandro Quercia, "Forms of Adoption, Adaptation and Resistance in the Cooking Ware Repertoire of Lucania, South Italy (8th–3rd centuries BC)", in: Spartaro/Villing (2015) 203–212.
Quercia/Foxhall (2014): Alessandro Quercia and Lin Foxhall, "Weaving Relationships in Areas of Cultural Contacts: Production, Use and Consumption of Loom Weights in Pre-Roman Sicily", in: Sanna Lipkin and Krista Vajanto (eds.), *Focus on Archaeological Textiles: Multidisciplinary Approaches*, MASF 3, 88–101.
Rotroff (2015): Susan Rotroff, "The Athenian Kitchen from the Early Iron Age to the Hellenistic Period", in: Spataro/Villing (2015) 180–189.
Roux (2019): Valentine Roux, in collaboration with M.-A. Courty, *Ceramics and Society. A Technological Approach to Archaeological Assemblages*, Switzerland.
Shepherd (2012): Gillian Shepherd, "Women in Magna Grecia", in: James/Dillon (2012) 215–228.
Simantoni-Bournia (1990): Eva Simantoni-Bournia, *Ανασκαφές Νάξου. Οι ανάγλυφοι πίθοι*, Βιβλιοθήκη της εν Αθήναις Αρχαιολογικής Εταιρείας 113, Athens.
Simantoni-Bournia (2001): Eva Simantoni-Bournia, " Ξώμπουργο Τήνου. Ό ανάγλυφος πιθαμφορέας της γεννήσεως. Έκδοση δεύτερη, βελτιωμένη", in: Alexandra Alexandri and Iphigeneia Leventi (eds.), *Καλλίστευμα. Μελέτες προς τιμήν της Όλγας Τζάχου-Αλεξανδρή*, Athens, 69–84.
Simantoni-Bournia (2004): Eva Simantoni-Bournia, *La céramique grecque à reliefs, ateliers insulaires du VIIIe au VI e siècle av. J.-C.*, Hautes études du monde gréco-romain 32, Genève.
Spagnoli (2010): Federica Spagnoli, *Cooking Pots as an Indicator of Cultural Relations between Levantine Peoples in Late Bronze and Iron Ages. Origins, Diffusion, and Typological Development of Cooking Ware in Levantine and Cypriot Repertoires (14th – 7th Centuries BC)*, Rome.
Spartaro/Villing (2015): Michela Spartaro and Alexandra Villing (eds.), *Ceramics, Cuisine and Culture*, Oxford.
Strack (2007): Sara Strack, *Regional Dynamics and Social Change in the Late Bronze and Early Iron Age: A Study of Handmade Pottery from Southern and Central Greece*, PhD Dissertation, University of Edinburgh.
Voyatzoglou (1984): Maria Voyatzoglou, "Thrapsano, Village of Jar Makers", in: Philip P. Betancourt (ed.), *East Cretan White-On-Dark Ware. Studies on a Handmade Pottery of the Early to Middle Minoan Periods*, Philadelphia, 130–142.
Zaphiropoulou (1983): Photini Zaphiropoulou, "La necropoli geometrica di Tsikalario a Naxos", in: *Magna Grecia* 18 (5–6), 1–4.
Zaphiropoulou (2001): Photini Zaphiropoulou, "Καύσεις στις γεωμετρικές Κυκλάδες. Οι περιπτώσεις της Νάξου και της Πάρου", in: Nikolaos Ch. Stampolidis (ed.), *Καύσεις στην Εποχή του Χαλκού και την Πρώιμη Εποχή του Σιδήρου*, Athens, 285–299.
Zaphiropoulou (2008–2009): Photini Zaphiropoulou, "The Tumulus Necropolis at Tsikalario on Naxos", in: *AION Archeologia e Storia antica* 15–16, 49–55.

Adrien Delahaye

27 Regional Markets and Distribution Patterns of Laconian Pottery in Crete and Beyond: The Aeginetan Hypothesis (6th–5th Centuries B.C.)

Abstract: While Crete has been strangely absent from distribution maps of Laconian pottery in the Mediterranean, the renewal of archaeological evidence on the island appears to show that it played a pivotal role in the organization of Laconian exchanges as part of a larger regional market that extended from the Peloponnese to Cyrenaica but was also connected to Egypt and the West via Kythera. The aim of this chapter will be to propose an overview of the underlying distribution networks, their actors, and their successive reorganizations. Beyond the use of the concepts of networks and connectivity, the chapter emphasizes the value of the regional market scale, the still-useful Braudelian long-term perspective, and the need for a careful consideration of physical parameters and human *savoir-faire* in mercantile interactions. In the end, the goals will be, first, to put Crete on the distribution map of Laconian pottery and, second, to propose a new vision of a Spartan Mediterranean, connecting regional markets at different geographical scales and drawn extensively from archaeological data.

Introduction

The so-called austerity turn in Sparta during the Archaic period and the Cretan Dark Ages gap of the 6th century B.C. may theoretically prevent any research about the diffusion of Laconian pottery in Crete.[1] Crete was indeed a blank spot on the diffusion maps of Laconian vases in the Mediterranean until recently. The recent progress in the available archaeological data, though, has shed new light on the island and its role in the circulations of wares that formed a much larger regional market extending from the Peloponnese to Cyrenaica, Egypt, and the West via Kythera, the southern Cyclades, and the Dodecanese.[2]

Beyond ceramology *stricto sensu*, imported pottery can be used as a primary source to reintegrate Laconian and Cretan territories into the Aegean exchange system. Even if pottery can potentially be considered a residual commodity, and even if Laconian productions are of secondary importance compared with the volumes of Corinthian, Attic, and even Ionian wares circulating in the Mediterranean at the time, these objects are nevertheless what remains of exchanges and can therefore be used as a proxy in approaching wider phenomena.[3]

In order to be used as proxies, these artifacts need to be interpreted beyond their place of discovery, within the framework of theoretical concepts such as trade, market, connectivity, or networks.[4] Although useful, we know that these tools must be used with caution in the study of the past, especially in archaeology. Apart from the problem of representativeness inherent to archaeological artifacts, another difficulty is to establish the exact nature and extent of widescale-structured institutions, or the information that the actors had at their disposal to make rational decisions. These limits shall therefore prevent us from elaborating overly generalizing models; otherwise, we risk attributing mercantilist behaviors to ancient actors.[5] The use of these concepts has fortunately become much more refined in recent years, acknowledging

1 I would like to express my gratitude to the organizers in Turin (Diego Elia, Eleni Hasaki, and Marco Serino), the École française d'Athènes, which funded this research, Isabella Bossolino and the Scuola archeologica italiana di Atene, the Connected Past network, the Museum and Ephorate of Heraklion who offered me the opportunity to improve my work, the Ephorate of Western Attica, Piraeus and the Islands which granted me the permission to study material in Aegina, and to Gudrun Klebinder-Gauß, who provided useful comments as well as the Fig. 5. Heartfelt thanks also to Athena Tsingarida, Didier Viviers, Conor Trainor, Antonis Kotsonas, Carl Knappett, Thomas Brogan, Tom Brughmans, Leonidas Souchleris, Katerina Tzanakaki, Pascal Arnaud, Franck Wojan, Donald Haggis, Margaret Mook, Gerald Schaus, Maria Pipili, Yannis Nakas, Catherine Bouras, Roberto Melfi and Despoina Nika for their help, their time and their helpful remarks.

2 Erickson (2005), (2010) for a synthesis about this renewal.

3 Cook (1959) 114–123 questioned the value of using pottery as an indicator trade. Vallet/Villard (1961) 295–318 have developed the link between pottery and trade. *Contra* Osborne (2007); Chankowski (2013) about the value of pottery and its importance in trade.

4 For a reflexive synthesis of the uses of these concepts in archaeology: Gras (2012); Moatti (2012); Zurbach (2012); Collar *et al.* (2014), (2015); Leidwanger/Knappett (2018); Foxhall (2021).

5 Gras (2012) 13.

https://doi.org/10.1515/9783111189635-027

the diversity of economic behaviors and the variety of the scales of analysis. In other words, complexity and variability can now be integrated against or within more flexible generalizing models. They can especially be of great use to study smaller pottery production, which also has its place within the web of the Archaic and Classical Mediterranean exchanges.

This chapter uses the regional market scale to understand the modalities of distribution of a secondary production, Laconian vases, in a specific and generally overlooked area: Crete. Since it allows for the possibility of multiple scales and actors, which can be potentially entangled and embedded the concept of a regional market allows us to think at the regional scale.[6] It is an easy concept to work with, one both limiting the risks of overinterpretation and allowing the results of other studies to be added later. By investigating the connection between Laconia and Crete, the aim here is to expand the results at a larger scale in order to integrate the circulations of these objects within the exchange system of the Aegean and identify the patterns and interconnections between regional markets and wider maritime exchange systems.

The tools of network analysis are used here only at a theoretical level and in an exploratory fashion in order to highlight some specific aspects of the distribution of Laconian pottery, but not in depth.[7] The homogeneity of the dataset is undermined by too many uncertainties about quantities, contexts, and dating. Such gaps in documentation are inherent to the field of archaeology,[8] but they are particularly important in the case of Laconian pottery, as its wide geographical distribution amplifies the risks of discrepancies, and these cannot be smoothed out sufficiently due to the relatively low quantities involved.[9] Any modeling using the tools of Network Science Techniques[10] faces the risk of producing suspect discrepancies. Another limitation lies in the difficulty of modeling the actors and their roles. In this respect, Latour's Actor-Network Theory (ANT) is useful in that it considers the interactions between objects, actors (individuals as well as collective organizations), and discourses, the *chaîne opératoire* from the producer to the end user, and also studies their effects on the stability of a network.[11] But its precise application in the field of archaeology has to be refined, and it remains difficult to "incorporate complexity, messiness and fluidity of relationships and social networks in ancient societies" into models.[12]

Any network analysis of artifacts' diffusion is, then, doomed to be partial. To overcome these limitations, it is necessary to choose an entry point—Laconian pottery in our case—to conduct a network analysis and study of its modalities of distribution in a specific regional market (that is, Crete) and the economic agents involved.[13] Laconian pottery is particularly suitable for such an analysis. Reticular logics and their history may help to explain its fragmented distribution, which is quite different from that of Corinthian and Attic products, but somehow close to that of Ionian vases. The approach adopted here is pragmatic, relying first and foremost on cross-referencing material and written data. This approach makes it possible to draw up a preliminary inventory and reduce uncertainties, a necessary step before any modeling and visualization, and then provides the necessary historical context in order to determine the role of the traders.[14] In the end, the challenge will be to propose an image of a Spartan Mediterranean other than I. Malkin's, one based not on colonial but on exchange networks.[15]

6 See Tsingarida/Viviers (2013).

7 A network is not a reality *per se*, but a tool for structuring and understanding flows. Zurbach (2012) 266.

8 The problems of incomplete quantitative datasets and their incorporation in SNA for archaeology have been highlighted by Knappett/Evans/Rivers (2008) 1012; Knappett (2011); Brughmans (2013). It appears possible to smooth out discrepancies by using big and homogenous datasets, which is not the case for the Laconian pottery, at least in Crete. Also Rivers (2018).

9 This statement is even truer in Crete, where Erickson (2005) 633, no. 99 notes that "the spotty excavation record for post-Minoan Crete, inadequate publication of excavated material, and the difficulty of comparing finds from different contexts limit the usefulness of scientific forms of analysis".

10 Social Network Analysis (SNA) or Agent Based Computational Models (ABCM). For a reflective summary on the use of these methods, see Foxhall (2021).

11 Latour (2005).

12 Foxhall (2021) 9, 13–14.

13 Zurbach (2012) 266. In an Archaic world – even considered as a "small world" – it is necessary to rely on case studies of regional markets before being able to propose complex models. On the "small Greek world", see Malkin (2011).

14 The Socio-Technical Configuration Analysis (STCA) could be an interesting path to explore in the future developments of this research in order to propose a network analysis both of the artifacts and the actors, through a two-mode network. At this stage, the formatting of the data was not sufficient to propose a network visualization of the data. See Miörner *et al.* (2022).

15 Malkin (1994) esp. 8–13, highlighted the existence of Spartan colonial networks on a Mediterranean scale, in contrast to the image of a landlocked and isolated Sparta. His demonstration was based above all on the dialectic between territorial challenges and the uses of myth, leaving aside archaeological *realia*. The next edition of the book shall integrate more of the latter (pers. comm. I. Malkin, whom I thank here).

From the Spartan Mediterranean to Laconia within the Mediterranean

The Spartan mirage of an austere, isolated Sparta may fit poorly within a paradigm of the Aegean Sea as a space of exchanges and circulations par excellence.[16] This incompatibility is, however, more historiographic than historical and lacks archaeological support. Irad Malkin has highlighted the existence of a Spartan Mediterranean, taking Spartan studies out of the Peloponnese: a paradox, though, in that it perpetuates the singularity of Sparta. The Mediterranean Sea has been made Spartan, but Sparta and its territory remained little integrated into the Mediterranean networks of exchange. The reason lies in the main use of mythical foundation sources, which limit the approach to the colonies, real or supposed, of Sparta. However, relying first on the archaeological data, a cartography of a completely different scale emerges. Sparta, far from having been isolated, appears on the contrary to have been integrated into several systems of exchange at different scales.

Our knowledge of the diffusion of Laconian pottery in the Mediterranean is based primarily on the work of C. M. Stibbe, P. Pelagatti, M. Nafissi, and F. Coudin.[17] Besides the new discoveries to be added to the diffusion maps, which will gradually become obsolete, these studies are largely based on funerary contexts from Etruria and Magna Graecia; the data from the Aegean were generally underestimated and therefore overlooked. In this respect, the conclusions about the distribution of Laconian vases can nowadays be enriched and nuanced, especially thanks to new discoveries in the Aegean.

The choice of Laconian pottery as the main documentary category (as a proxy) to approach the wider exchange networks is motivated by the volume of the archaeological evidence, its wide geographical distribution, its broad chronological spectrum (from the beginning of the 6th to the middle of the 5th century B.C.), and the variety of contexts, all of which make it the only material that can be used to quantify, compare, and identify developments in Laconian trade. Three categories of exported Laconian pottery can be analyzed: black-figured vases, black-glazed vases, and transport amphorae. The black-figured vases were the most exported and therefore are the most studied. As these products are absent from Crete, they will not be considered within the scope of this chapter. The Laconian black-glazed vases, exported from the beginning of the 6th to the end of the 5th century B.C., are the Laconian vases exported over the longest chronological period, and have a typology that is now well identified. Nevertheless, the majority of the Aegean material has not been the subject of any synthetic analysis. It should be noted that the study of figured and glazed vases and their distribution must be carried out together. Indeed, the method established by J. D. Beazley continues to determine our approach to figured vases, isolating artificially their study from the rest of their production. However, progress in the morphotypological approach has demonstrated the close relationship between glazed and figured vases; in fact, if pottery were compared to a wave, the figured vases would be no more than the foam at its crest.[18] Finally, the Laconian transport amphorae have a special position. Poorly published, they can be studied as containers, their content making it possible to trace the circulation of perishable goods such as oil or wine and broaden our understanding of the Laconian pottery market by integrating it into wider exchange flows.[19]

So far, the quantification of this study includes 2955 Laconian vases (MNI) spread over 80 sites throughout the Aegean and the Black Sea (Tab. 1). This number can be compared to the 750 artifacts (MNI) counted in the same area by F. Coudin at 22 sites.[20]

16 For a recent synthesis based on archaeological data: Delahaye (2019).
17 Nafissi (1989); Stibbe (1989); Pelagatti (1992); Stibbe (1994), (2000); Coudin (2009).
18 The expression refers to "L'écume de l'histoire", invented by Duby (1973) 8, according to whom "events are like the froth of history, bubbles, large or small, which burst on the surface, and whose bursting creates ripples which spread further and further. This one has left very lasting traces [. . .] These traces give it existence. Apart from them, the event is nothing". Figured vases are to material culture what the event is to history: what truly matters is the wave below it. In the case of Laconian pottery overseas: the black-glazed wares, the transport amphorae and the socio-economic and cultural phenomena which can explain their distribution. See also Lynch in this volume.
19 Pelagatti (1992) 10; Johnston (1992), (1993), (2005); Johnston *et al.* (2000).
20 MNI = Minimum Number of Individuals. Calculations based on Coudin (2009) Graph. 2.

Tab. 1: Comparative counts of Laconian pottery in the Aegean, the Black Sea and the Mediterranean. © Adrien Delahaye. Based on Coudin (2009) Graph. 2 and the author's research.

Site	K	KI	TrA	TabA	CS	CSI	OS	C	CI	S	SI	A	P	U	Laco.	Copies	TOTAL
Aptera	1														1	0	1
Azoria	2													x	2	0	2
Eleutherna	21	1									1				21	2	23
Itanos		1	2	1	1		2								6	1	7
Kastello Varypetrou	1	1													1	1	2
Knossos	3	2			1	1			1		1			x	4	5	9
Kommos	2		21	1											24	0	24
Kydonia	4	1						1	1			1			6	2	8
Lappa	1														1	0	1
Lyktos	3														3	0	3
Priniatikos Pyrgos	4	4													4	4	8
Tarra													1		1	0	1
TOTAL	**42**	**10**	**23**	**2**	**2**	**1**	**2**	**1**	**2**	**0**	**2**	**1**	**1**	**0**	**74**	**15**	**89**

K = Krater
KI = Krater imitation
TrA = Transport amphora
TabA = Table amphora
CS = Closed shape
CSI = Closed shape imitation
OS = Open shape
C = Cup
CI = Cup imitation
S = Skyphos
SI = Skyphos imitation
A = Aryball
P = Pyxid
U = Unquantified material

No Laconian Pottery in Crete?

The historiography and pottery of the 6th and 5th centuries B.C. in Crete bear many features similar to Laconia's. This period has indeed sometimes been referred to as the Cretan Dark Ages because of the scarcity of documentation, both written and material. The relative archaeological invisibility of the 6th and 5th centuries B.C. has been linked to the lack of documentation in Laconia from the second half of the 6th century B.C. onward and has been used as an argument to justify a shared Dorian austerity. However, the recent discoveries of Archaic and Classical Cretan pottery, as well as of imports in Crete, has challenged this picture.

Crete is absent from the 3383 Laconian vases collected in Coudin's corpus.[21] C. M. Stibbe, in his three-volume catalog of Laconian black-glazed pottery, listed 21 artifacts found in two locations, Eleutherna and Knossos, completed by the material from Kommos published by A. Johnston.[22] It is always tempting to discuss gaps in order to reconstruct trade routes. But these numerous gaps reflect, above all, the sites that have been excavated and published, and therefore the general state of knowledge at any given time.[23] This scarcity was quite suspicious from a maritime and logistical point of view. Indeed, if the diffusion

21 Coudin (2009).
22 Stibbe (1989) B28, K20; Johnston (1992), (1993) no. 53, 81–99; Johnston *et al.* (2000) 244; Johnston (2005) no. 198–199.
23 Gras (2012) 17–19.

of Laconian pottery in the past could be considered part of an east-west route between Ionia and the western Mediterranean, with no strong links to the Aegean area, this assumption is no longer valid. The picture has become much more complex, thanks to the discoveries of Laconian pottery in the northern Aegean, the Black Sea, and Cyrenaica and Egypt, which have established the existence of a north-south axis (Fig. 5). In the context of coastal sailing, it then becomes difficult to explain the presence of these artifacts in Cyrenaica without any connection to the Aegean, and therefore presumably via Crete.

After a preliminary investigation, at least 74 Laconian vases and 15 probable local imitations (MNI, so 89 in total; Fig. 1) were found in Crete, at Aptera, Azoria, Eleutherna, Itanos, Kastello Varypetrou, Knossos, Kommos, Kydonia, Lappa, Lyktos, Priniatikos Pyrgos and Tarra (Figs. 1–2). B. Ericksson has already listed the Laconian kraters found in Crete, the most well attested shape on the island, totaling 33 artifacts.[24] With 21 examples, Eleutherna "offers the most evidence on the island for 6th-century B.C. Lakonian imports."[25] For transport amphorae, A. Johnston published 21 items from Kommos. To these counts can be added other published and unpublished vases.[26]

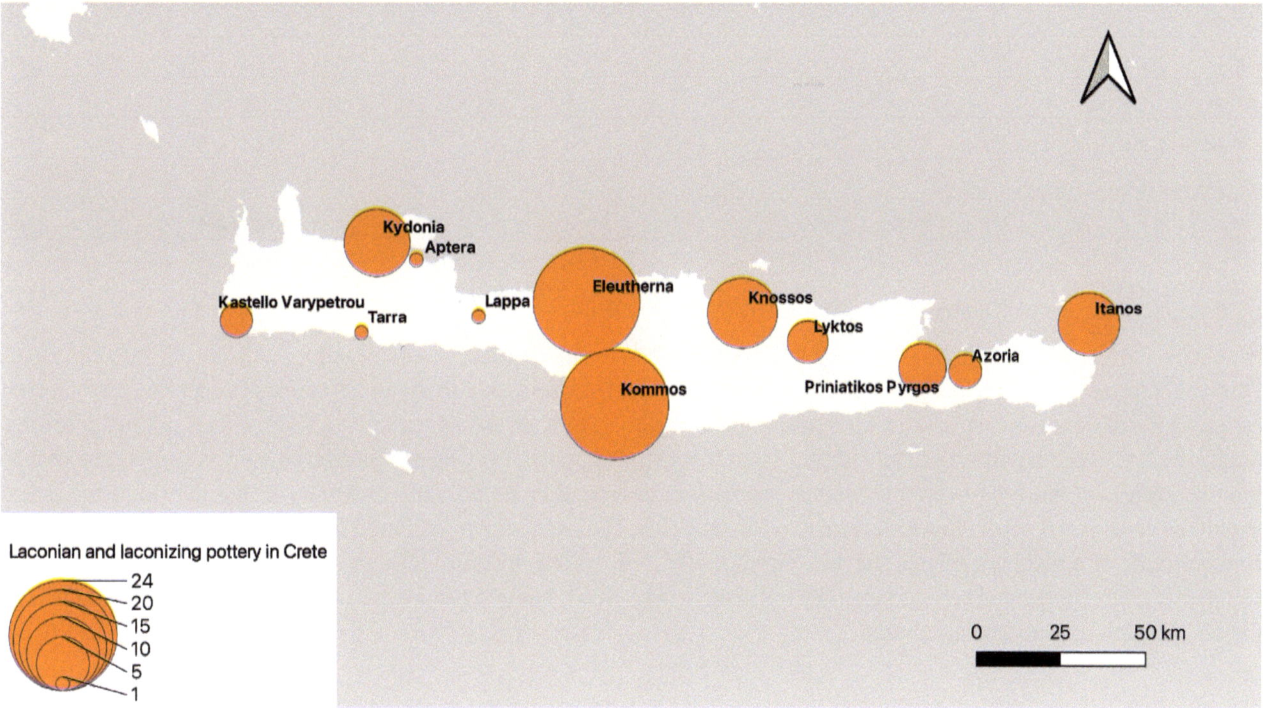

Fig. 1: Counts of Laconian pottery in Crete. © Adrien Delahaye.

24 Erickson (2010) 56–62, fig. 3.4–5, 119, 124, 251, tab. 10.1, 285–287, nos. 78–79, with the relevant references; the inventory is updated until 2005.
25 Erickson (2010) 56–57, figs. 3.4–5.
26 Kydonia: Hallager *et al.* (1997) 213, no. 77–P 1849, pls. 62, 96: d.5.; 209 no. 71–P 1375, no. 77–P 1860, pls. 63, 68: d.11; 211, no. 82–P 0091/0112, pls. 63, 83: d.1; Kanta/Tzedakis (1978) 38, no. 68, fig. 41, pl. 13, published as Geometric, this krater is Archaic, as observed by Erickson (2010) 286, no. 78, also no. 47, fig. 44, pl. 16 (cup), and no. 23, fig. 107, which could be a handle from a Lakonian krater. Kastello Varypetrou and Lappa: Erickson (2010) 251, tab. 10.1, 287, no. 78. Knossos: Stibbe (1989) no. K20; Coldstream (1973) 43, no. 37; 51, no. 23, fig. 7, pl. 19; 53, no. 26, fig. 7; 54, no. 46; Coldstream (1992) 75, no. 7; Erickson (2010) 122, no. 239, fig. 4.2. One Laconian krater was found in the well GG in the Unexplored Mansion (pers. comm. Conor Trainor; recently published by Paizi (2023) 16, Supplement Cat. 22, fig. 29). The KULP (*Knossos Urban Landscape Project*) has gathered through survey several fragments of Lakonian kraters and local imitations, along with imports from mainland Greece, which remain unquantified and unpublished for now (pers. comm. Conor Trainor); see Trainor (2019); also Paizi (2023) 4–5, no. 7. Kommos: Johnston (1993) 340, no. 6; 351, no. 53; 358–359, nos. 81–98; Johnston (2005) 364, nos. 198–199, 374, no. 241; Johnston *et al.* (2000) 244, nos. 331–332; 245, no. 338; Shaw/Shaw (2000) 374, no. 341. Priniatikos Pyrgos: Erickson (2010) 191, no. 11. Azoria: Erickson (2010) 286, no. 79 (unpublished) and pers. comm. Donald Haggis and Margaret Mook. Itanos: pers. comm. Athena Tsingarida and Didier Viviers. Lyktos: Chaniotis *et al.* (2021a) 220; Chaniotis/Kotsonas (2021b) 44 and pers. comm. Antonis Kotsonas. Tarra: Tzanakaki (2013) 207–208, fig. 15, no. 2426. Aptera: Tzanakaki (2013) 212, no. 51 and pers. comm. Katerina Tzanakaki.

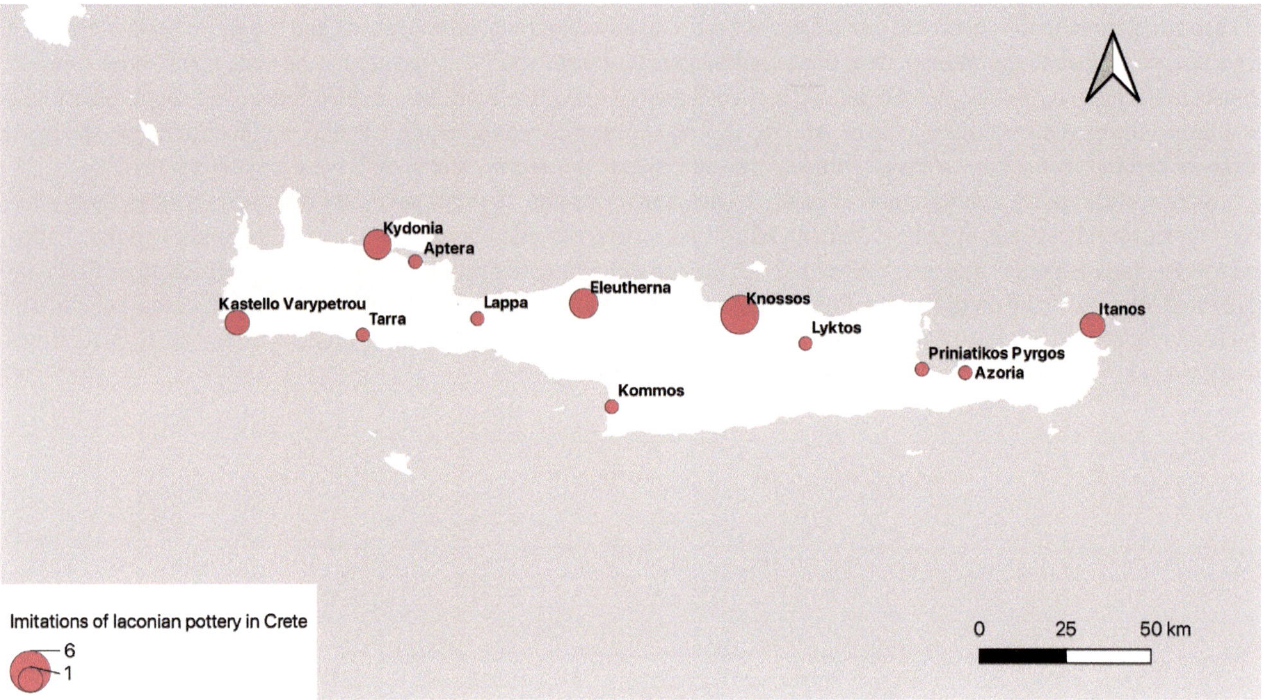

Fig. 2: Map of Cretan sites with Laconian or Laconizing pottery. © Adrien Delahaye.

This total may seem of little importance in quantitative terms, but in relation to the total corpus of Laconian vases, it represents 2.6% of Coudin's corpus, and 1.6% when updated with data from the Aegean (5588 MNI). This is not insignificant, especially for a region that was totally absent from the diffusion maps. Previously considered a blind spot, Crete should now be seen as a market—even a secondary one—and, beyond this, its possible centrality in the distribution system should be considered, especially with regard to North Africa. The Laconian vases found in Crete feature several elements characteristic of a distinct regional market. First, the figured vases are totally absent; second, black-glazed kraters are constantly overrepresented and occasionally imitated locally; third, Kommos appears as a privileged context for studying the distribution of transport amphorae.

The Laconian Kraters of Eleutherna

The largest assemblage of Laconian vases is found in Eleutherna (Figs. 1–3). Although the majority of the published finds are dated from the 8th and 7th centuries B.C., the necropolis of Orthi Petra also provided pottery from the 6th and early 5th centuries B.C., including a set of 21 Laconian kraters.[27] The funerary nature of the context of these kraters remains a matter of debate: admittedly discovered near funerary clusters, they actually come from superficial layers.[28] While almost all the Laconian material discovered in the Aegean comes from cultic contexts, the potential funerary deposition of those kraters raises the question of a local demand for this specific shape.

27 Published by Erickson (2010) 56–62, fig. 3.4–5.

28 Erickson (2005) 633, (2010) 45–50 considers that they could come from disturbed graves. Kotsonas (2002) 42–43, no. 71 suggests that they come from residual layers of the Pyrgi settlement or from traces of ritual activities linked to ancestor worship, possibly including ritual banquets.

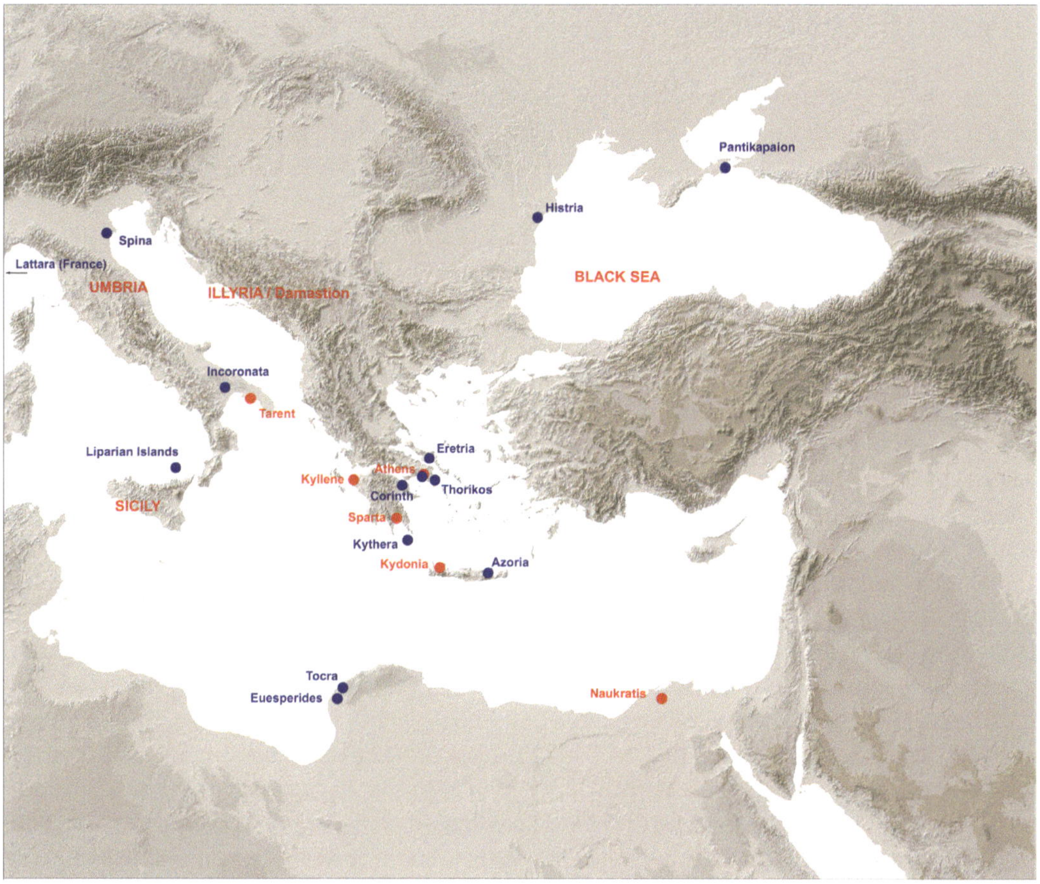

Fig. 3: Map of the diffusion of Aeginetan cooking pottery. The findspots of Aeginetan cooking ware are marked in blue, written sources mentioning Aeginetan presence beyond Aegina are marked by red dots. The classification is not always based on macroscopic or scientific analysis, but on references by other authors. © Anavasi, with additions by Gudrun Klebinder-Gauß.

The krater fragments can be categorized into groups E and F of the typology established by C. M. Stibbe.[29] Group F entirely consists of black-glazed vases. They were produced from the beginning of the 6th to the middle of the 5th century B.C. and are the most widespread Laconian kraters throughout the Mediterranean. While Stibbe's typology proves a concentration of their production during the 6th century, Eleutherna is of particular interest because many of the fragments come from contexts dated with a *terminus post quem* around 500-480 B.C.[30] Group E is smaller in number but is one of the oldest types and can be dated back to the end of the 7th and the beginning of the 6th centuries B.C.[31] It is attested in Sparta, Samos, Miletus, Sicily, Etruria, Naucratis, Aegina, and Cyrenaica.

Beyond Crete, a comparative approach shows marked similarities between Laconian pottery in Crete and the assemblages of Aegina, Gravisca in Etruria, and Sicily. These sites reveal a strong demand for the shape of the krater (50% in Eleutherna), the existence of local imitations, and a long duration of exports and their resilience until the middle of the 5th century B.C., despite the marked decline of Laconian pottery exports elsewhere, except in a minority of outlets that could be described as "safe" markets, such as Crete, Cyrenaica, Sicily, Gravisca, and Aegina.[32] These characteristics suggest the existence of a specific Cretan market for Laconian pottery.[33]

29 Stibbe (1989).

30 Stibbe (1989) 37–43, cat. F; Erickson (2010) 57, no. 36.

31 Stibbe (1989) 33–37, cat. E.

32 Williams (1993) 595 already foresaw this pattern for pottery in Crete, thinking that "much of which may well belong to the second half of the century, perhaps got there as a result of the trade with Cyrene and Tocra".

33 This paper does not include a detailed contextual analysis. Contextual data in Crete is often incomplete or remain unpublished in detail. Where such data is available, it reveals only secondary depositions, as at Knossos.

Laconian Pottery and Safe Markets

Among the 111 Laconian vases discovered in the Etruscan port of Gravisca, 57 are kraters; one is a local imitation, and eight are transport amphorae.[34] The site's chronological arc stretches from the early 6th to the early 5th century B.C.

In Sicily, P. Pelagatti counts 53 sites that yielded Laconian pottery, 49 of which provided a total of 546 black-glazed kraters, including 444 from groups E and F, 46 krateriskoi, and 41 local imitations, as well as 26 transport amphorae.[35] Nafissi noticed that after the middle of the 6th century B.C. the imports of Laconian pottery into Italy tended to concentrate in southern Italy and Sicily while decreasing in Etruria, and to shift away from black-figured kylikes in favor of black-glazed kraters.[36]

In Cyrene, 222 pieces (MNI) are published from Demeter and Kore's sanctuary; Tocra provided 108 artifacts.[37] O. Mei recently provided a useful updated count for Cyrene (338 MNI) and the whole of Cyrenaica (466).[38] Laconian pottery from Cyrenaica shows a far greater morphological variety than does that of Crete and Sicily, with a large number of forms and figured wares that outside of Laconia are found only in Samos, Miletus, Olympia, and Taranto.[39] After cups, kraters are the most represented shape, with 94 examples, or 20% of the total, including only 5% in Tocra. The presence of the rarely exported lakainai may be seen as an indication of cultural proximity to Laconia, perhaps partly due to the colonial ties between Sparta and Cyrene.[40] The chronology stretches from the period of 650-620 B.C., with the 6th century B.C. seeing a strong development that peaked in the third and fourth quarters, before ultimately fading in the 5th century B.C., with the last imports around 475 B.C.[41] But krater imports seem to have faded from the 3rd quarter of the 6th century onward in Cyrene.[42]

These similar pottery assemblages provide a distribution pattern between the Peloponnese, Crete, and Cyrenaica, even though there are some specific features associated with the latter, such as a higher proportion of imports and the presence of Laconian black figured wares. In order to explain these, a link with Egypt has long been theorized, on the basis of the numerous Laconian and Corinthian fine wares found at Cyrene, Tocra, and Naucratis.[43] The *emporion* of Naucratis provides a slightly different picture, which probably reflects the wider variety of its suppliers.[44] 70 sherds from 55 artifacts (MNI) have been counted so far, representing less than 1% of the total painted Greek pottery from the site.[45] Kylikes are the largest group, with approximately 50 sherds belonging to 34 vases (MNI); kraters represent 16 artifacts, or 29% of the total. Most are black-figured, which raises the question of selection criteria during the excavations. Laconian pottery is attested from the early phases of the *emporion*, starting around 600 B.C., with a takeoff and peak between 575 and 550 B.C., characterized by a strong increase in black-figured kylikes, before a sharp decline during the third quarter of the century. A shrinking number of vessels continued to arrive until the early 5th century B.C.

Among kraters found in Naucratis, most are from Group E. The diachronic evolution of imports of Laconian vases combined the typological spectrum dominated by black-figures tends to differentiate Naucratis from Crete, Aegina, and Sicily, Cyrenaica representing an intermediate case, with both black-figured and black-glazed wares. Naucratis does not seem to have imported Laconian kraters during the first half of the 5th century. One krater, though, is dated to the last quarter of the 5th century, and the two black-glazed kraters have been dated over a broad period and could be assigned a late date.[46]

---

34 Boitani (1992); Williams (1993) 591, no. 20 for the list.

35 Pelagatti (1992).

36 Nafissi (1989) 71–72.

37 Boardman/Hayes (1966) nos. 2109–2124, pl. 21; (1973) 81–95, nos. 932–1023, pls. 57–69; Schaus (1985) 24–28, nos. 84–124.

38 Mei (2013) 86–93. To be completed with the new material to be published from Euesperides. See Zimi *et al.* (2019) 27, where Laconian wares (including black-glazed kraters) are said to represent 3% of the total quantification of fine wares, a work still in progress.

39 Schaus (1985) 15.

40 Coudin (2009) 98–99.

41 Erickson (2010) 290. The 7th century B.C. date of several Laconian vases and other categories of artifacts is proposed by Schaus (1985) 100 due to the presence of Laconian settlers at the time of the colonial foundation. In other terms, they may belong to a different distribution phenomenon.

42 Coudin (2009) 88.

43 Roebuck (1950) 136–247; Johnston *et al.* (2000) 211; Erickson (2005) 628–629, no. 88.

44 Bergeron (2017) 16.

45 Bergeron (2017) 2.

46 But Bergeron (2017) 4, fig. 1, no. 17 considers that the absence of other Laconian pottery dated to the first half of the 5th century B.C. "might suggest that the kraters should best be dated earlier to the second half of the 6th c. BC".

Aegina is the last refuge market and may be the missing link between the rest. The Sanctuary of Aphaia yielded 77 vases, completed by 27 from Kolonna, for a grand total of 104 objects (MNI).[47] In this total, the krater predominates, with no fewer than 41 examples at Aphaia (44.5%) and 8 at Kolonna (29.6%). To this number many unquantified fragments may also be added.[48] The earliest pieces date back to the last third of the 7th century B.C., with the bulk belonging to the 6th century. One aspect of this group that makes it unusual is the concentration of kraters not in the first half of the 6th century B.C., but in the second, with some examples even belonging to the first quarter of the 5th century B.C. The taste for Laconian kraters is also evident in the total absence of Corinthian krater fragments, and the 17 black-figured Attic kraters that have been found at Aegina all date from the mid-6th century B.C. at the latest, indicating a switch to Laconian kraters.[49]

The Aeginetans: The Common Denominator?

This typological and chronological proximity between the aforementioned sites—even though Cyrenaica and Naucratis present more specific features—may indicate an involvement of the Aeginetans in distribution.[50] One must be careful not to identify traders involved in exchanges prematurely from archaeological remains, as a dataset can be compatible with different kinds of social structures.[51] But traders were an essential component of circulation and must be included in the discussion. If objects travel, it is thanks to the mobility of individuals (or groups of individuals) endowed with circulatory skills and a culture of mobility, which presuppose maritime, commercial, social, and legal knowledge.[52]

One must therefore ask how the Laconian vases got into Crete. In Kydonia, four Laconian kraters and one local imitation have been discovered so far. This case is particularly interesting, since Herodotus tells us that in 519 B.C. the city was the theater of a confrontation between Aeginetans, allied to Cretans, against Samians installed at the site since 524 B.C.[53] As discussed below, this event can be interpreted in a new light when one considers the supposed role of the Samians in the distribution of Laconian products.

The high concentration of Laconian black-figured vases on Samos had led E. A. Lane, and then P. Cartledge and M. Pipili, to postulate a central role of Samian merchants in the distribution of Laconian pottery until the middle of the 6th century B.C. and a massive drop in exports of Laconian vases that occurred in the Mediterranean.[54] This hypothesis, though, has been partly undermined by the publication of the Laconian pottery found at the Artemision of Samos, which contradicts the account of an absolute break in the middle of the 6th century B.C., and even more by G. Schaus' publication of the abundant Laconian pottery found at Miletus, which makes the merchants from the two Ionian cities at least equal candidates as intermediates of the distribution of Laconian pottery.[55]

Even though the theory of Samian involvement has been reconsidered, it cannot be totally dismissed, and allows us to connect some dots between Laconia, Crete, Samos, and Aegina. The Lacedemonians answered the call of the Samian aristocratic factions against Polycrates in 525 B.C.[56] The expedition was a failure, and the rebels, forced into exile, "afterwards founded Kydonia in Crete." After expeditions in Siphnos and Hydra, they settled in Kydonia around 524 B.C, where they "prospered for five years" before Aeginetans and Cretans overcame them in a sea battle.[57] Herodotus attributes the cause of the conflict between Aeginetans and Samians to a hypothetical quarrel dating back to the reign of King Amphicrates of Samos. Rather than revenge, it seems that the exiled Samians did not sail erratically, choosing clearly identifiable reloca-

47 Furtwängler *et al.* (1906) 455, no. 236a, pl. 128, 32; 458, no. 268; 466, nos. 367–373, pls. 121.59–63, 129.14; 458, no. 268a, pl. 122.40; Felten (1982); Williams (1993); Klebinder-Gauß (2012) nos. 29–33, 95–97, 326–327, 448–451.
48 Williams (1993) 572, no. 2: "in the garden of the old museum at Aigina port there were, at least until 1990, trays full of fragments of Laconian kraters. It is not clear whether these are Furtwängler's finds from Aphaia, or the discoveries of others from elsewhere on the island".
49 Williams (1993) 589.
50 Especially with the distribution of black-figured pottery.
51 Reed (2003) 63.
52 Clifford (1997); Tarrius (2001). For an introduction, Moatti (2012) 44.
53 Hdt. 3.44, 59.
54 Lane (1933) 179; Cartledge (1982); Pipili (2000).
55 Pipili (2001); Schaus (2020). Finally, although we cannot totally rule out the possibility of the involvement of Peloponnesian transporters – whatever their status – there are currently no sources confirming this hypothesis; see Delahaye (2019) 226–233.
56 Hdt. 3.44–57. Cartledge (1982) 247–248.
57 Hdt. 3.59.

tion sites: Hydra, at the tip of the Argolid and the crossing point between the Saronic and Argolic gulfs, and Kydonia, at the northwestern tip of Crete.

The discovery of three funerary stelae in Kydonia that each bears an epigram in the Aeginetan alphabet and date from the end of the 6th to the beginning of the 5th century, may confirm the Aeginetan installation.[58] The Kydonia episode reveals two things: first, the centrality of Kydonia as a strategic stopover to secure on trading routes; second, the existence of an economic rivalry in the area bounded by Aegina to the north, the east coast of the Peloponnese to the west, the Dodecanese to the east, and Crete to the south, in what can be called a Dorian Sea, with a branch extending towards northern Africa.[59]

Moreover, like the Samian exiles in 525 B.C., some of the Aeginetans driven out of their island by the Athenians in 431 B.C. chose to take refuge in Kydonia.[60] This second settlement may underline the importance of Kydonia as a trading post for Aegina. Others were settled by their Spartan allies in Thyreatis, located on the borders of Laconia but also on the western shoreline of Peloponnese, even fighting alongside the Lacedaemonians against the Athenians during the Peloponnesian War.[61]

Commercial links between Aegina and Laconia have been theorized for a long time. D. Williams suggests that some of the numerous abbreviated names attested in the graffiti on Laconian vases discovered at the Sanctuary of Aphaia may be those of traders.[62] Among them, he and A. Johnston identified a group of Aeginetan and Rhodian traders, accompanied by one potential Etruscan name, inscribed either in Laconia or in Aegina, which could connect the three points of a Laconia-Aegina-Etruria traffic.[63] Similarly, Johnston considers a graffito on a black-glazed Laconian amphora in the British Museum as the trademark of the well-attested Aeginetan trader Sostratos, and M. Torelli evokes potential family connections between the Sostratos active in Gravisca and another of the same name in Naucratis.[64] Lastly, besides Aphaia in Aegina, D. Williams noted that inscribed dedications on Laconian black-glazed kraters can be found at Amyklai, Kamiros, Aphytis, Tocra, Gela, and Gravisca.

In Naucratis, Aegina appears as one of the most active poleis in the management of the *emporion* and is the only one not located in eastern Greece.[65] While nine poleis from Asia Minor collectively founded the sanctuary of the Hellenion, Aegina, Miletus, and Samos built a sanctuary of their own.[66] Unsurprisingly, all the major protagonists of Aegean trade are to be found in Naucratis, including the three main candidates for a central role in the distribution of Laconian pottery: Samos, Miletus, and Aegina. The foundation of its own sanctuary by each of these three poleis on territory granted by a foreign power is not insignificant: commercial travel does not require the foundation of a sanctuary at every stopover.[67] The foundation reveals both the significant investment of these trade and civic communities in the *emporion* of Naucratis and the privileged state-to-state links between these poleis and Egypt.[68] It is of interest that the Aeginetan presence in Naucratis seems, once again, to be entangled in a thread of relations with Rhodes.[69] A Rhodian decree relating to Naucratis mentions a potential Aeginetan *proxenos* resident in Naucratis.[70] This interweaving of Aeginetan and Rhodian trading networks, together with the wide variety of pottery found on site, leads to a comparison between Naucratis and Gravisca, the Etruscan *emporion* founded at the same time, between the end of the 7th and the beginning of the 6th centuries B.C.

---

58 Guarducci (1939) 120–123; Jeffery (1990) 112, 314.

59 Stefanakis (1999) 254–255; Williams (1993) 595; Erickson (2005) 634–635, no. 117. While the open sea can to some extent be considered a lawless zone, this is not the case for ports. Ports are control points that can close off a navigation zone to certain players. This is the model that seems to apply to Kydonia. See Arnaud (2011) 64–65.

60 Thuc. 2.27.2.

61 Thuc. 4.56; 57.1–4.

62 Williams (1993) 593

63 Williams (1993) 576, fig. 17, B11, 591.

64 Johnston (2013) 110, no. 30; Torelli (1982) 318; *contra* Möller (2000) 56–57.

65 On *emporia*, see most recently Demetriou (2012), especially 64–152 on Gravisca and Naucratis.

66 Hdt. 2.178–182.3.

67 Bresson (2000) 48, no. 138.

68 Another reason, which does not exclude the first, could be their exclusion at the founding of Hellenion and the granting of compensatory privileges by Amasis. Bresson (2000) 18–24, 41, 49.

69 Whose central role in the management of the Hellenion has been demonstrated by Bresson (2000) 14–63.

70 *IG* XII.1. Austin (1970) 26, 29, 31, no. 3; Figueira (1988) 543–551. *Contra* Bresson (2000) 31–34, 67, no. 16 (with bibliography) who considers that the restitution Αἰγ[ινάταν] is incorrect and should read Αἰγ[ύπτιος].

B. Erickson corroborates this hypothesis of a "direct commercial link between the Peloponnese and Crete" partly controlled by Aeginetans, by means of the presence in Crete of some Argive vases, a product very rarely exported.[71] He also convincingly interprets as an Aeginetan trading feature the influx of Attic pottery from the mid-6th century B.C. onward in sites where Laconian wares were also discovered; this assertion seems to be confirmed by the disappearance of Attic wares from the island during the second half of the 5th century B.C., when Athens definitively ended the maritime power of Aegina.[72]

The diffusion of Laconian pottery, the graffiti found on it, comparison with the diffusion of other pottery, and the literary and epigraphic evidence all seem to draw pattern indicating the centrality of Aegina to the distribution of Laconian pottery between the mid-6th and the mid-5th centuries B.C. This makes sense, given Aegina's intense commercial activity, its role as a central clearing point, its capacity to operate both at a regional and at a long-distance scale, and its privileged links with the Peloponnese.[73]

Aeginetan Proxies

The wide diffusion of Aeginetan cooking pottery during the Archaic and Classical periods also seems to draw a pattern corresponding partly to this general picture (Fig. 4), with a presence archaeologically attested in Cyrenaica, Crete, Kythera, Attica, Corinth, Eretria, the Black Sea, the Liparian islands, and Incoronata in southern Italy.[74] This high-quality cooking ware was produced in Aegina but appears to have been exported, probably along with other goods, making it an important indicator of the Aeginetan exchange system.[75]

The circulation of Aeginetan pottery to seek Aeginetan networks can also be crossed with the circulation and use of the Aeginetan coins. The ancient sources refer to the famous coin depicting a turtle as the Peloponnesian coin.[76] In addition to hoards, both coins from excavations and local mints using the Aeginetan standard in the Peloponnese and Crete provide a fuller picture of the Aeginetan sphere of influence. The Aeginetan monetary and weight systems were widely used during the Archaic and Classical eras in the whole of the Peloponnese (except Corinth), Crete, Central Greece, a large part of the Cyclades, and a few cities in Asia Minor.[77]

Immediately south of Kommos, a treasure of 70 Aeginetan staters was discovered in Matala.[78] Buried around the end of the 6th century B.C., its earliest coin is dated around 530 B.C. More than evidence of the payment of Cretan mercenaries, M. Stefanakis and B. Ericksson suggest that it may indicate the Aeginetans' commercial activities in Crete.[79] Combined with the Laconian amphorae found in Kommos, it could emphasize that the area may have acted as a trading station linked to Cyrenaica. Local minting of pseudo-Aeginetan staters is attested in central and western Crete, notably in Kydonia after the victory against the Samians.[80] At the beginning of the Hellenistic period, while the Aeginetan standard remained in use, the monetary types were influenced by Peloponnesian coins, which were widely circulated throughout Crete. Finally, M. Stefanakis has shown that Cyrenaic coinage circulated in Crete in large quantities during the 4th and the early 3rd centuries B.C.[81] Like pottery, coins seem to tell the same story of enduring exchanges between Crete, the Peloponnese, and Cyrenaica from the Archaic to the Hellenistic period, at least, with the involvement of Aeginetan traders. The centrality of the latter is revealed by a basic network analysis (Tab. 2) whose purpose was to compare the available data presented above in order to visualize the continued presence of Aeginetans in the distri-

71 Erickson (2005) 631–632, no. 98 (Argive imports).

72 Erickson (2005) 635, no. 121–122, 137 (Attic imports), (2010) 63, 64, 274, 292, no. 115. Such a pattern (of an increase of Attic imports) is also proposed by Johnston (1979) 53.

73 Jennings (1988) 140–145, 151–155; Klebinder-Gauß (2015) 76.

74 Gauß/Kiriatzi (2011); Klebinder-Gauß (2012) 197–198; Gauß *et al.* (2015) 65–74. See also Klebinder-Gauß in this volume.

75 Klebinder-Gauß (2012) 74.

76 Poll. *Onom.* 9, 74; Hsch. s.v. "χελώνη".

77 Doyen/Apostolou (2018).

78 *IGCH* 1. 70 Aeginetic staters and 1 obol and 1 stater from Thera. Stefanakis (1999) 250–251.

79 Stefanakis (1999) 251; Erickson (2010) 278, no. 27.

80 Stefanakis (2019) 250–251.

81 Stefanakis (2013).

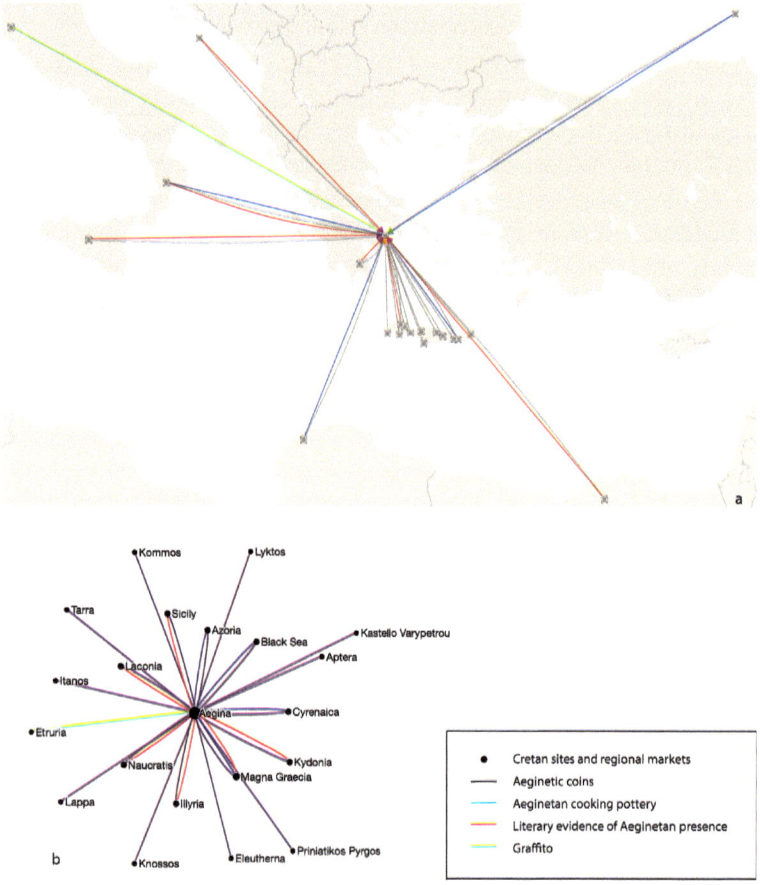

Fig. 4: Preliminary visualizations of Aeginetan trading activities within the network distribution of Laconian pottery. **(a)** Map visualization, **(b)** Node-link visualization. Realized with *The Vistorian* (https://vistorian.net [last access 02.07.2024]). © Adrien Delahaye.

bution system for Laconian pottery beginning in the second half of the 6th century B.C. It represents Aeginetan trading activities within the network distribution of Laconian pottery. The main regional markets for Laconian pottery where Aeginetan presence is attested constitute the nodes. Among regional markets, Cretan sites are represented individually by nodes, for the purpose of this article. The edges represent the indications of attested Aeginetan trading activities (coinage, cooking pottery, literary sources, graffiti) within this network of nodes. It contains 21 nodes and 30 links, all of them connected to Aegina which is the targeted source.[82]

Kythera as a Stepping Stone

The existence of a north-south maritime axis and its ramifications are attested on several occasions in ancient sources. When describing the Athenian maritime expeditions to the Laconian coast during the Peloponnesian War, Thucydides mentions the island of Kythera and its links with North Africa and Crete.[83] Kythera appears as a stepping stone (Fig. 5) between Laconia, Egypt, and Libya, also looking towards Crete and Sicily.[84] The term ὁλκάδες means "merchant ships shipping grain", indicating perhaps the circulation of perishable goods, aboard which pottery could have been included

82 This network analysis was realized with https://vistorian.net and analyzed with https://networknarratives.github.io (both last access 02.07.2024).

83 Thuc. 6.44. See also Erickson (2005) 630.

84 Broodbank (1999) 191.

Tab. 2: North-South and East-West maritime routes between Peloponnese, Crete, Cyrenaica and Asia. After Arnaud (2005) 183–188, 217.

N° Arnaud	Route	Distance	Source
North-South maritime routes between Peloponnese, Crete and Cyrenaica			
54	Peloponnesian capes – Berenice (Euesperides)	/	Str. 17.3.20
55	Cape Taenaros (Matapan) – Cape Phycus (Ras-Hamama)	2800 stadia	Str. 17.3.20; Plin. *HN* 5.32
56	Crete – Cape Phycus (Ras-Hamama)	1000 stadia (1 day and 1 night)	Plin. *HN* 4.60; 5.32
57	Apollonia – Kriou Metopon Cape	2000 stadia (2 days and 2 nights against the wind)	Eratosth.; Str. 10.4.5, 17.3.21
58	Cape Chersonesus (Ras-et-Tyn) in Cyrenaica – Kyklos island (Gavdos)	1500 stadia (with a south wind)	Str. 17.3.22
59	Petra Megale – Cape Chersonesos in Crete	3000 stadia	Str. 17.3.22
60	Cape Kimaros (Vouchas) – Cape Taenaros	700 stadia (1 day)	Str. 10.4.5; Ps. Skyl. 47
61	Cape Cadistos (Cape Zephyrium / Chersonisos) – Cape Malea	650 stadia	Plin. *HN* 4.60
East-West maritime routes between Peloponnese, Crete and Asia			
102a	Cape Malea – Rhodes via Crete	1000 stadia (1 day and 1 night)	Ps. Skyl. 113

as a secondary cargo.[85] The Egyptian War of 459-454 B.C. can be seen as an attempt to gain control of this regional market of the Aeginetans; the Athenian expedition assisted by the Milesians in 424 B.C. against Kythera may also have been partly motivated by commercial considerations.[86]

Even though Laconian pottery from Kythera remains largely unpublished, the excavations carried out by J. N. Coldstream and G. L. Huxley at Kastri have nevertheless yielded 30 Laconian fragments (MNI) dated from the late-5th century B.C., found in a rubbish dump.[87] While this assemblage is later than the Cretan contexts and consists mainly of domestic wares, it nevertheless includes imported fine wares from Laconia (including one krater), Attica, and Corinth (mainly drinking vessels), as well as transport amphorae from eastern Greece. The material was reviewed by Johnston, who mentions 46 new Laconian fine-ware artifacts (MNI, 1 krater) and extends the chronological range of the deposits from the early 5th century B.C. (including the krater) to the 4th century B.C.[88] Cretan imports are also attested, which can be compared to the few Cretan and Theran artifacts found in Cyrene and Tocra.[89] The scarcity of earlier sherds, going back to the 7th or 8th century B.C., as well as the unpublished data from the survey "reinforce the view that presence on the site was sporadic and thin, or both."[90]

The largely unpublished material from Paliokastro includes a fragment of a Geometric Laconian skyphos, several burial offerings, among them a subgeometric lakaina, and a Laconian aryballos along with some Corinthian wares.[91] Neither the typology nor the chronology of these objects offers any point of comparison with the Cretan case. This scarce evidence seems to indicate that Kythera was at best a stopover, but certainly not an outlet, for Laconian pottery during the 6th century B.C.

---

85 Roebuck (1950) 136–247 linked these circulations to the grain trade from Egypt. Also, Johnston *et al.* (2000) 211. Reed (2003) 69 questions the mechanical association often assumed in scholarship between pottery and grain circulations. *Contra* Erickson (2005) 628–629, no. 88 reminds us that "grain is the most likely candidate" as main cargo, by a process of elimination, and because of the written evidence (Hdt. 4.198–199; 2.154).

86 Egypt: Thuc. 1.104.2; 109.1–4; 110.1–4; Diod. 11.74.4–75.3; 77.1–2. Kythera: Thuc. 4.53.3–4; cf. Erickson (2010) 296.

87 Coldstream/Huxley (1972) 161–163, pl. 45, nos. 29–54 (Deposit σ); 202–203, pl. 58, nos. 318–321 (unstratified). The domestic common wares, not concerned by overseas exports, are not included here.

88 Johnston *et al.* (2014) 5–19 identifies one Cretan import, from another deposit (ω). Potentially Coldstream/Huxley (1972) 267–268, 270, nos. Q5, Q9, R2, pls. 86–88, according to Erickson (2010) 102, no. 76.

89 Boardman/Hayes (1966) 4–5, 10–14, 79–85, (1973) 39–41. The earliest objects date back to the 7th century B.C.

90 Johnston *et al.* (2014) 18–19. Coudin (2009) 240, map 1 mentions 17 Laconian vases dated of the end of the 7th century B.C., without further details.

91 Petrocheilos (2016) 267, fig. 26; Fragou/Kroustalis (2016) 120, fig. 13b-c.

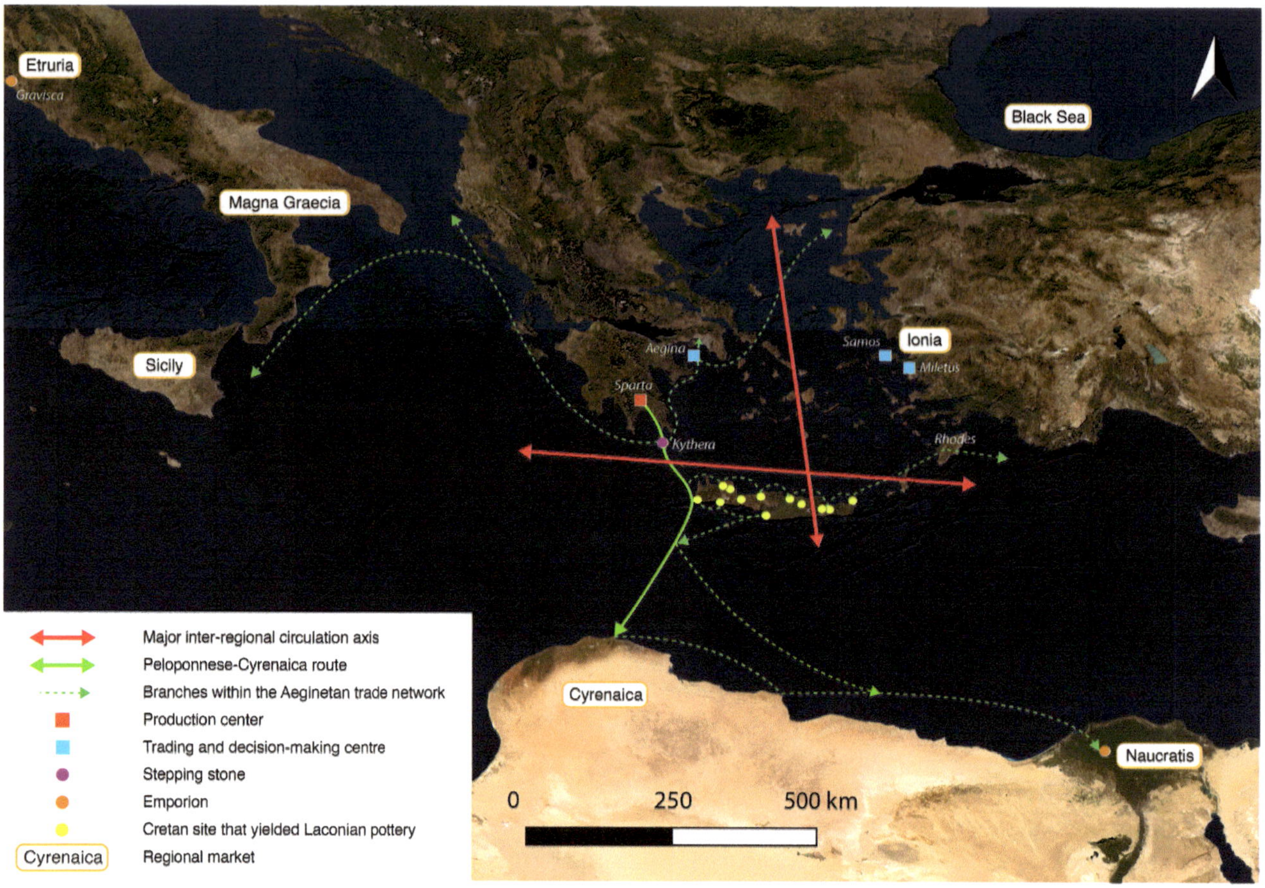

Fig. 5: Synthetic map of the distribution patterns of Laconian pottery in Crete, the Aegean and the Mediterranean. © Adrien Delahaye.

Memory of Networks and Long-Term Navigation Routes

During the Bronze Age, Crete already appears to have had a pivotal role on a similar navigation route between the south of the Peloponnese (via Kythera and Antikythera) and North Africa (Fig. 5; Tab. 2). Analysis of regional interrelation networks carried out by Cyprian Broodbank and Carl Knappett in the Aegean has demonstrated both the centrality of Crete and the role of intermediary played by Kythera between Crete and the Peloponnese.[92] The Minoan settlement of Kastri in Kythera during the Early Bronze Age II can be interpreted as the materialization of the opening of this route between Crete and the Peloponnese, evident from the modeling of maritime interactions in the Aegean Bronze Age.[93] In Crete, linkage between the Minoan port of Kommos, at the outlet of the Messara Plain, and North Africa has been demonstrated by L. Watrous, and the pottery evidence reveals other Cretan areas: the Peloponnese, Kythera, the Cyclades, Boiotia-Attica, Cyprus, Egypt, the Levant, western Anatolia, and Sardinia.[94]

This raises the question of the memory of networks and routes, and of geographical parameters that would lead to a route being used over the long term, according to a Braudelian approach to the study of the Mediterranean Sea, whose value can still be relevant today, even since the advent of the concept of connectivity.[95] A sea route exists first and foremost because it provides favorable navigation conditions. To understand the central role of Crete, it is indeed necessary to

92 Broodbank (2000) using Proximal Point Analysis (PPA) whose results are then compared with the archaeological data. Knappett (2011), Knappett/Evans/Rivers (2008) tried to optimize the modelling by balancing the proximity criterion with geographical distances and the cost.
93 Knappett/Evans/Rivers (2008) 1015–1017, figs. 3–5.
94 Watrous (1992) 176–177, 180; Tomlinson *et al.* (2010); Day *et al.* (2011); Rutter (2017); Gheorghiade/Spencer (2024) 37–38.
95 Braudel (1949); Horden/Purcell (2000).

understand maritime navigation in its practical aspects, by including parameters such as winds, currents, reefs, visibility, seasonality, and sailing technology, factors that have a major impact on mobility.[96] Using the testimonies of ancient geographers, Pascal Arnaud highlighted the existence of a route between Cape Malea, Crete, and Cyrenaica characterized by the regularity of the winds, which allow exceptionally fast north-south connections during the winter (Fig. 5).[97]

Stopping in one or several Cretan harbors seems therefore inevitable, at least on the northern shoreline, especially on return journeys against the wind (i.e., westward), when coastal sailing was surely necessary.[98] The south coast is indeed exposed to the south winds during winter, which create difficult sea conditions.[99] This would explain the lack of attested stopovers, except for Kommos at the gateway to the Messara Plain, and the southwestern site of Tarra, which can be connected to the route between the Peloponnese and Crete. On the opposite side, the north coast possesses many sites with anchorages easily accessible by coastal sailing and benefits from the decelerating effect of the Cretan mountains on the northern winds.[100] These anchorages may have been used as redistribution points for the Cretan domestic market, as well as shifting points from one sailing basin to another. Crete lies at a pivotal point on the east-west axis connecting Asia Minor to the western Mediterranean, through a connection between Rhodes and Cape Malea.[101] While the route between Messina and Alexandria seems to have been directly accessible to Egypt with a favorable northwest wind,[102] stopovers in either direction at Cape Malea, Kythera, Antikythera, Crete, Kasos, Karpathos, and Rhodes are well documented.[103] The Dorian arc, which stretches from the southern Peloponnese through Crete to the Dodecanese, could have acted as a hinge between north-south and east-west Mediterranean axes (Fig. 5).

But natural conditions are not enough to characterize a navigation route: it needs to be used by sailors. Maritime navigation was based on empirical knowledge of the space and conditions mentioned above, acquired through repeated journeys and oral transmission within groups over several generations.[104] Of course, the memory of a route can be lost. But since the intrinsic conditions of navigation in the Mediterranean have changed little over time, by learning the route, the possibility of maritime navigation rests primarily on the empirical knowledge of the aforementioned conditions shared by sailors. A hybrid navigation system favoring coastal sailing (cabotage) and tramping, potentially entangled with some direct sailing routes, allowed the possibility of picking up some complementary cargo during stopovers along the main route (Fig. 5).[105]

In this system, the choice of the route and the stopovers were under the supervision of the pilot and may have varied according to the season or trade patterns and geopolitics. The opportunity costs varied according to this shared knowledge that established the price of the pilot and the central role of the carriers.[106] Hence the influence that Cretan sailors may have exerted on navigation between the Peloponnese and Crete because of their empirical knowledge of winds, surface currents, and gusts from inland mountains. For instance, the famous story reported by Herodotus about the colonization of Cyrene by the Therans highlights how the latter "sent messengers to Crete to seek out any Cretan or sojourner there

---

96 See Alberti (2018) and, most recently, Gheorghiade/Spencer (2024) on modeling seasonal average wind regimes and their consequences for mobility from and to Kommos during LBA; specific coastal conditions are not modelled there. On the strategic location of Crete, its harbour networks (during Late Antiquity), and local environmental conditions and topographical features, see Roussos (2021) 331–334, 336, no. 8, figs. 5–7.

97 Arnaud (2005) 183–188.

98 Forsyth (2022); Gheorghiade/Spencer (2024) 43.

99 Roussos (2021) 332.

100 The wind flow is deflected around the island towards its edges. Roussos (2021) 332–333, nos. 8, 10.

101 This route is described by Ps.-Skylax, 113. Bouras (2016) 203, figs. 1, 216, 218.

102 Plin. *HN* 19.3.4.

103 Arnaud (2005) 217.

104 Arnaud (2012) 118, who notes that to annihilate a maritime power, all that's needed is to erase its memory of the roads by killing its pilots, as Aemilius Paullus did after his victory over the ligurian tribe of the Ingauni from Albingaunum in 184–181 A.D.

105 As theorized by Tchernia (2007) for the Roman Era. Trapero Fernández/Aragón (2022) defines define this mobility as "maritime transhumance".

106 Arnaud (2012) 61–62 with references, 118. The term pilot does not refer here to the modern sense of maritime pilot (person external to the crew who takes over the ship in difficult waters to carry out approach manoeuvres in the port of call), but in the ancient terms of κυβερνήτης (*Il.* 19.43; *Od.* 3.279; 9.78; Aesch. *Suppl.* 770; Eur. *Suppl.* 880; Pl. *Phdr.* 247c) or ἀρχικυβερνήτης (Str. 698; Plut. *Alex.* 66) which can be translated by 'chief pilot' or 'helmsman' and designate the member of the crew responsible for navigation and manoeuvres, who is not necessarily the captain (even though the ships of the Classical and Archaic era being quite small in general and operated with a limited crew, possibilities of qualified personnel were limited). I thank here Yannis Nakas for his valuable suggestions.

who had traveled to Libya."[107] They finally hired a *murex* fisher from Itanos named Corobios, who had already traveled to Cyrenaica.[108]

We also need to consider whether this navigation route, already known in Minoan times, continued to be used during the Archaic and Classical periods, and even the Hellenistic times, as the available evidence suggests. Cretan exports such as "Plakettenvasen", produced in western Crete, were indeed found in the Peloponnese (Geraki and Argos), Melos, Rhodes, Cyrenaica, Carthage, Naucratis, Alexandria, Illyria, and southern Italy.[109] The distribution of these later productions confirms that western Crete was part of a long-term trade route connecting Cyrenaica to the Peloponnese and beyond, towards the west and Egypt (Fig. 5).

Conclusions

This chapter offers a glimpse of broad patterns of the distribution networks of Laconian pottery in the Aegean and beyond. The Cretan case contributes to our understanding of these by providing a much more complex image of the distribution system than the hypothesis of an east-west route controlled by Samian merchants. The artifacts appear to be integrated into larger-scale circulations generated by a variety of actors, and partly structured into networks, which may be articulated and interwoven at different scales; this greatly complicates the overall picture we have of these circulations.

Within this emerging picture, Crete appears as one of the missing links between Laconia, Aegina, and, beyond them, northern Africa and Magna Graecia. It underlines how Aeginetans' trading activities played a major role in the distribution of Laconian pottery and the structuration of Mediterranean exchanges. The exact role of Cretan harbors and the interconnections between them has yet to be characterized more precisely. It is, for example, to be hoped that the excavations at Phalasarna will provide new insights into the role of this eastern port.

The Cretan case demonstrates that Laconian vases, even if they represented only a marginal element in two respects (Aegean and Mediterranean trade in general and pottery circulation in particular), can be used as an exploitable proxy to explore the wider exchange systems that underpin them.[110] But proxies of what? In our case, Laconian pottery can be used to highlight wider circulations of artifacts and of people (the Aeginetan traders). Pottery is what remains from them, like exchanges' dust.[111] These artifacts have no agency of their own, but they are what remains from ancient exchanges and can therefore be considered key proxies from which we can improve our understanding of networks that include human agents.[112] It is generally accepted that commodities in which the Aeginetans traded or acted as middlemen were probably, for the most part, perishable goods such as grain, ore, and eventually slaves,[113] along with which a variety of artifacts were carried as secondary cargo, hence the expression Αἰγιναία ἐμπολή.[114]

By integrating these agents and their motivations into the equation, a network can be seen as a "continual process of formation,"[115] which enables us to give them a diachronic dimension. Several reconfigurations of the distribution patterns of Laconian pottery are indeed noticeable during the period considered here and tend to make Crete in turn a commercial stopover, a secondary market, or a refuge market. It underlines, then, the heuristic interest of microregions and regional markets, which can be entangled with each other and embedded into systems of exchange polarized around more important actors and flows. In particular, this case study allows us to reintegrate Crete and Laconia into the concert of Archaic and Classical exchanges in a cartography of exchanges that partially overlaps with that of I. Malkin's Spartan Mediterranean.

---

107 Hdt. 4.151.
108 As an interesting fact, Corobios while left alone on the island of Platea and starving while the Thereans had temporarily sailed back home, was rescued by a Samian ship on his way to Egypt.
109 Massar (2022) 504–506.
110 Gras (2012) 13.
111 On this question: Bresson (2000) 71–72.
112 Foxhall (2021) 4–6.
113 Aegina is described as a "central clearing point" by Jennings (1988) 54, 221.
114 Which can be translated as Aeginetan commodity, but frequently meaning "junk". Klebinder-Gauß (2015) 76, with the relevant bibliography.
115 Foxhall (2021) 5.

If Phoenician east-west trade routes until the early 6th century B.C. and the network constituted by the Ionian poleis from the end of the 7th to the middle of the 6th century B.C. seem to have been predominant parts of the Mediterranean exchanges,[116] the Aeginetans seem to have played an increasing role afterwards, until the middle of the 5th century B.C. They successfully established themselves as trading middlemen between Attica, the Argolic Gulf, and the eastern Peloponnese at a regional scale, and between the Aegean, northern Africa, the Black Sea, and the western Mediterranean at an interregional scale.[117]

We cannot suggest a simple and complete replacement of the Ionians by the Aeginetans as main traders of Laconian pottery, since the Samians in Crete and the long-lasting presence of Chian wares at Naucratis seem to indicate a form of resilience in the Ionians' activities.[118] Based on the Cretan case, however, and its contextualization within the Laconian pottery assemblages of Sicily, Cyrenaica, and Aegina, it is possible to suggest that the sharp decline in exports of Laconian pottery, the abandonment of black-figured wares, and the loss of several overseas markets around the middle of the 6th century B.C. were followed by an adaptation in which Aeginetan traders may have played a central role.[119]

Bibliography

Agouridis (1997): Christos Agouridis, "Sea Routes and Navigation in the Third Millennium Aegean", in: *Oxford Journal of Archaeology* 16 (1), 1–24.

Alberti (2018): Gianmarco Alberti, "TRANSIT: a GIS Toolbox for Estimating the Duration of Ancient Sail-Powered Navigation", in: *Cartography and Geographic Information Science* 45, 510–528.

Arnaud (2005): Pascal Arnaud, *Les routes de la navigation antique: itinéraires en Méditerranée*, Paris.

Arnaud (2011): Pascal Arnaud, "Ancient Sailing-Routes and Trade Patterns: The Impact of Human Factors", in: Andrew I. Wilson and Damian Robinson (eds.), *Maritime Archaeology and Ancient Trade in the Mediterranean*, Oxford, 61–80.

Arnaud (2012): Pascal Arnaud, "La mer, vecteur des mobilités grecques", in: Laurent Capdetrey and Julien Zurbach (eds.), *Mobilités grecques: mouvements, réseaux, contacts en Méditerranée de l'époque archaïque à l'époque hellénistique*, Bordeaux, 89–135.

Austin (1970): Michel M. Austin, *Greece and Egypt in the Archaic Age*, Cambridge.

Bergeron (2017): Marianne Bergeron, "Laconian Pottery", in: *Naukratis: Greeks in Egypt*, London, 2–16.

Boardman/Hayes (1966): John Boardman and John Hayes, *Excavations at Tocra 1963–1965. The Archaic Deposits, I*, London.

Boardman/Hayes (1973): John Boardman and John Hayes, *Excavations at Tocra 1963–1965. The Archaic Deposits II and Later Deposits*, London.

Boitani (1992): Francesca Boitani, "Le ceramiche laconiche a Gravisca", in: Paola Pelagatti and Conrad M. Stibbe (eds.), *Lakonika: Ricerche e nuovi materiali di ceramica laconica*, Rome, 19–72.

Bouras (2016): Catherine Bouras, "The Geography of Connections: A Harbour Network in the Aegean Sea During the Roman Imperial Period?", in: Kerstin Höghammar, Brita Alroth and Adam Lindhagen (eds.), *Ancient Ports, the Geography of Connections*, Uppsala, 202–223.

Braudel (1949): Fernand Braudel, *La Méditerranée et le monde méditerranéen à l'époque de Philippe II*, Paris.

Bresson (2000): Alain Bresson, *La cité marchande*, Bordeaux.

Broodbank (1999): Cyprian Broodbank, "Kythera Survey: Preliminary Report on the 1998 Season", in: *Annual of the British School at Athens* 94, 191–214.

Broodbank (2000): Cyprian Broodbank, *An Island Archaeology of the Early Cyclades*, Cambridge.

Brughmans (2013): Tom Brughmans, "Thinking Through Networks: A Review of Formal Network Methods in Archaeology", in: *Journal of Archaeological Method and Theory* 20, 623–662.

Cartledge (1982): Paul Cartledge, "Sparta and Samos. A Special Relationship?", in: *The Classical Quarterly* 32, 243–265.

Chaniotis/Kotsonas/Sythiakak (2021): Angelos Chaniotis, Antonis Kotsonas and Vassiliki Sythiakak, "Ανασκαφή Λύκτου", in: Πρακτικά της εν Αθήναις Αρχαιολογικής Εταιρείας 176, 213–264.

Chaniotis/Kotsonas (2021): Angelos Chaniotis and Antonis Kotsonas, "Λύκτος Ηρακλείου Κρήτης", in: *ΕΡΓΟΝ* 68, 43–47.

116 Although Ionian traders continued to play an important, but probably declining, role in Aegean trade until the Ionian revolt of 499 B.C., or even 478 and the creation of the League of Delos, at which point traffic between the Greek and Achaemenid worlds seems to have fallen dramatically. Bresson (2000) 73–74.

117 For references and comments, see Figueira (1981) 276; Jennings (1988) 140–145, 151–155; Weilhartner (2010) nos. 722, 750, 885, 954, 1233, 1271 (Crete); 390 (Egypt); 661–662 (Sicily); 885 (Umbria, Illyria); 902 (Tarentum); 470 (Hellespont); numerous occurrences about Peloponnese; Klebinder-Gauß (2015) 76.

118 Nafissi (1989) 73–74 stood up against such a theory. *Contra* Williams (1993) 595 and Erickson (2005) 635–636, nos. 121–122, (2010) 63, 64, 274, 292, no. 115, who consider the possibility.

119 Williams (1993) 595 and Erickson (2010) 293 support the takeover of the distribution of Laconian pottery by the Aeginetans from 550 onwards. *Contra* Nafissi (1989) 73.

Chankowski (2013): Véronique Chankowski, "La céramique sur le marché: l'objet, sa valeur et son prix. Problèmes d'interprétation et de confrontation des sources", in: Tsingarida/Viviers (2013) 25–38.
Clifford (1997): James Clifford, *Routes: Travel and Translation in the Late Twentieth Century*, Cambridge.
Coldstream (1973): John Nicolas Coldstream, "Knossos 1951–61: Orientalizing and Archaic Pottery from the Town", in: *Annual of the British School at Athens* 68, 33–63.
Coldstream/Huxley (1972): John Nicolas Coldstream and George L. Huxley, *Kythera: Excavations and Studies*, London.
Coldstream (1992): John Nicolas Coldstream, "Early Hellenic Pottery", in: Leyland Hugh Sackett (ed.), *Knossos: From Greek City to Roman Colony. Excavations at the Unexplored Mansion II*, British School at Athens, Supplement 21, Athens, 67–88.
Collar *et al.* (2014): Anne Collar, Tom Brughmans, Fiona Coward and Claire Lemercier, "Analyser les réseaux du passé en archéologie et en histoire", in: *Nouvelles de l'archéologie* 135, 9–13.
Collar *et al.* (2015): Anna Collar, Fiona Coward, Tom Brughmans and Barbara J. Mills, "Networks in Archaeology: Phenomena, Abstraction, Representation", in: *Journal of Archaeological Method and Theory* 22, 1–32.
Cook (1959): Robert Manuel Cook, "Die bedeutung der bemalten keramik für den griechischen handel", in: *Jahrbuch des Deutschen Archäologischen Instituts* 74, 114–123.
Coudin (2009): Fabienne Coudin, *Les Laconiens et la Méditerranée à l'époque archaïque*, Paris.
Day *et al.* (2011): Peter M. Day, Patrick S. Quinn, Jeremy B. Rutter and Vassilis Kilikoglou, "A World of Goods: Transport Jars and Commodity Exchange at the Late Bronze Age Harbor of Kommos, Crete", in: *Hesperia* 80, 511–558.
Delahaye (2019): Adrien Delahaye, *Archéologie et images de l'austérité spartiate. L'apport de l'iconographie et de la culture matérielle laconiennes à l'histoire de Sparte (VIe -Ve siècles av. J.-C.)*, Ph.D. dissertation, Paris 1 Panthéon-Sorbonne.
Demetriou (2012): Denise Demetriou, *Negotiating Identity in the Ancient Mediterranean: The Archaic and Classical Greek Multiethnic Emporia*, Cambridge.
Doyen/Apostolou (2018): Charles Doyen and Eva Apostolou, ""Τὸ Πελοποννησίων νόμισμα": Παρατηρήσεις στη νομισματική παραγωγή της Πελοποννήσου κατά την αρχαϊκή και κλασική περίοδο", in: Charles Doyen and Eva Apostolou (eds.), *Obolos 10. La monnaie dans le Péloponnèse*, Athens, 33–42.
Duby (1973): Georges Duby, *Le dimanche de Bouvines: 27 juillet 1214*, Paris.
Erickson (2005): Brice Erickson, "Archaeology of Empire: Athens and Crete in the Fifth Century BC", in: *American Journal of Archaeology* 109 (4), 619–663.
Erickson (2010): Brice Erickson, *Crete in Transition: Pottery Styles and Island History in the Archaic and Classical Periods*, Princeton N.J.
Felten (1982): Vassiliki Felten, "Lakonische Keramik", in: Elena Walter-Karydi and Walter Hans (eds.), *Alt-Ägina 2.1*, Mainz, 19–22.
Figueira (1981): Thomas J. Figueira, *Aegina, Society and Politics*, New York.
Figueira (1988): Thomas J. Figueira, "The Chronology of the Conflict between Athens and Aegina in Herodotus Bk. 6", in: *Quaderni Urbinati di Cultura Classica* 28, 49–89.
Forsyth (2022): Douglas Forsyth, "Upwind Sailing Capabilities of Square-Rigged Ships in Late Antiquity and the Ramifications for Trade Networks", in: Anti Lampinen and Emilia Mataix Ferrándiz (eds.), *Seafaring and Mobility in the Late Antique Mediterranean*, London, 149–165.
Foxhall (2021): Lin Foxhall, "Materiality, Methodologies and the Agency of Things in Archaeological Networks", in: Lin Foxhall (ed.), *Interrogating Networks: Investigating Networks of Knowledge in Antiquity*, Oxford, 1–20.
Fragou/Kroustalis (2016): Gely Fragou, Evangelos Kroustalis, "Excavations on the Acropolis of Paliocastro on the Island of Kythera, Greece: Investigating the Relationship Between the Settlement and the Cemetery", in: Valeriu Sîrbu (ed.), *Istros XXII. Interdisciplinary Methods of Research for Prehistoric and Protohistoric Funerary Monuments*, Braila, 109–128.
Furtwängler *et al.* (1906): Adolf Furtwängler, Ernst Robert Fiechter and Hermann Thiersch, *Aegina: das heiligtum der Aphaia*, Munich.
Gauß *et al.* (2015): Walter Gauß, Gudrun Klebinder-Gauß, Evangelia Kiriatzi, Areti Pentedeka and Myrto Georgakopoulou, "Aegina: An Important Centre of Production of Cooking Pottery from the Prehistoric to the Historic Era", in: Michela Spataro and Alexandra Villing (eds.), *Ceramics, Cuisine and Culture*, Oxford, 65–74.
Gauß/Kiriatzi (2011): Walter Gauß and Evangelia Kiriatzi, *Pottery Production and Supply at Bronze Age Kolonna, Aegina*, Vienna.
Gheorghiade/Spencer (2024): Paula Gheorghiade and Christine Spencer, "Modelling the Cost of the Wind: A Preliminary Reassessment of Networks of Mobility in the Late Bronze Age Mediterranean", in: *Journal of Computer Applications in Archaeology* 7, 36–53.
Gras (2012): Michel Gras, "Avant les réseaux. Les stratigraphies conceptuelles de la Méditerranée", in: Laurent Capdetrey and Julien Zurbach (eds.), *Mobilités grecques: mouvements, réseaux, contacts en Méditerranée de l'époque archaïque à l'époque hellénistique*, Bordeaux, 13–24.
Guarducci (1939): Margherita Guarducci, *Inscriptiones creticae 1: Tituli Cretae Occidentalis*. Rome.
Hallager *et al.* (1997): Erik Hallager, Birgitta Pålsson Hallager and Margrete Hahn, "Archaic to Byzantine Periods: The Pottery and the Small Finds", in: Maria Andreadaki-Vlasaki and Erik Hallager (eds.), *The Greek-Swedish Excavations at the Agia Aikaterini Square, Kastelli, Khania, 1970–1987. Vol. I*, Stockholm, 209–212.
Horden/Purcell (2000): Peregrin Horden and Nicholas Purcell, *The Corrupting Sea. A Study of Mediterranean History*, Oxford.
Jeffery (1990): Lilian Hamilton Jeffery, *The Local Scripts of Archaic Greece*, Oxford.
Jennings (1988): James E. Jennings, *Aeginetan Trade, 650-457 BC: A Reexamination*, Ann Harbor.
Johnston (1979): Alan W. Johnston, *Trademarks on Greek Vases*, Warminster.
Johnston (1992): Alan W. Johnston, "Anfore laconiche a Kommos", in: Paola Pelagatti and Conrad M. Stibbe (eds.), *Lakonikà: ricerche e nuovi materiali di ceramica laconica*, Rome, 115–116.
Johnston (1993): Alan W. Johnston, "Pottery from Archaic Building Q at Kommos", in: *Hesperia* 62, 339–382.
Johnston (2005): Alan W. Johnston, "Kommos: Further Iron Age Pottery", in: *Hesperia* 74, 309–393.

Johnston (2013): Alan W. Johnston, "Naukratis, Aegina and Laconia: Some Individuals and Pottery Distribution", in: Tsingarida/Viviers (2013) 103–113.
Johnston *et al.* (2000): Alan W. Johnston, Peter J. Callaghan, Patricia Maynor Bikai and John W. Hayes, "The Iron Age Pottery from Kommos", in: Joseph W. Shaw and Maria C. Shaw (eds.), *Kommos IV. The Greek Sanctuary, Part I*, Princeton/Oxford, 210–301.
Johnston *et al.* (2014): Alan W. Johnston, Kathleen Slane and Joanita Vroom, "Kythera Forty Years On: The Pottery from Historical Kastri Revisited", in: *Annual of the British School at Athens* 109, 3–64.
Kanta/Tzedakis (1978): Athanassia Kanta and Giannis Tzedakis, *Καστέλλι Χανίων 1966: αναλυτική μελέτη της κεραμεικής από τη στρωματογραφημένη τάφρο Β και το πηγάδι*, Rome.
Klebinder-Gauß (2012): Gudrun Klebinder-Gauß, *Keramik aus klassischen kontexten im Apollon-heiligtum von Ägina-Kolonna. Lokale produktion und importe*, Vienna.
Klebinder-Gauß (2015): Gudrun Klebinder-Gauß, "Opportunity in Scarcity: Environment and Economy on Aegina", in: Constance Rüden and Achim Lichtenberger (eds.), *Multiple Mediterranean Realities. Current Approaches to Spaces, Resources and Connectivities*, Paderborn, 67–91.
Knappett (2011): Carl Knappett, *An Archaeology of Interaction: Network Perspectives on Material Culture and Society*, Oxford.
Knappett/Evans/Rivers (2008) Carl Knappett, Tim Evans and Ray Rivers, "Modelling Maritime Interaction in the Aegean Bronze Age", in: *Antiquity* 82 (318), 1009–1024. https://doi.org/10.1017/S0003598X0009774X.
Kotsonas (2002): Antonis Kotsonas, "The Rise of the Polis in Central Crete", in: *Eulimene* 3, 37–74.
Lane (1933): Edward Arthur Lane, "Lakonian Vase-painting", in: *Annual of the British School at Athens* 34, 99–189.
Latour (2005): Bruno Latour, *Reassembling the Social: An Introduction to Actor-Network-Theory*, Oxford.
Leidwanger/Knappett (2018): Justin Leidwanger and Carl Knappett, "Maritime Networks, Connectivity, and Mobility in the Ancient Mediterranean", in: Justin Leidwanger and Carl Knappett (eds.), *Maritime Networks in the Ancient Mediterranean World*, Cambridge.
Malkin (1994): Irad Malkin, *Myth and Territory in the Spartan Mediterranean*, Cambridge.
Malkin (2011): Irad Malkin, *A Small Greek World: Networks in the Ancient Mediterranean*, Oxford.
Massar (2022): Natacha Massar, "Cretan Fine Wares in and out of Crete. Trade Patterns and Distribution Networks", in: Laura Rembart and Alice Waldner (eds.), *Manufacturers and Markets: The Contribution of Hellenistic Pottery to Economies Large and Small*, Vienna, 494–515.
Mei (2013): Oscar Mei, *Cirene e la ceramica laconica*, Rome.
Miörner *et al.* (2022): Johan Miörner, Bernhard Truffer, Christian Binz, Jonas Heiberg and Xiao-Shan Yap, "Guidebook for Applying the Socio-Technical Configuration Analysis method", in: *GEIST Working Paper* 1, 1–19.
Moatti (2012): Claudia Moatti, "Mobilités et circulations: approches historiographique et conceptuelle", in: Laurent Capdetrey and Julien Zurbach (eds.), *Mobilités grecques: mouvements, réseaux, contacts en Méditerranée de l'époque archaïque à l'époque hellénistique*, Bordeaux, 39–52.
Möller (2000): Astrid Möller, *Naukratis: Trade in Archaic Greece*, Oxford.
Nafissi (1989): Massimo Nafissi, "Distribution and Trade", in: Stibbe (1989) 68–88.
Osborne (2007): Robin Osborne, "What Travelled with Greek Pottery?", in: *Mediterranean Historical Review* 22, 85–95.
Paizi (2023): Eirini Paizi, "Overseas Connections of Knossos and Crete in the Sixth and Fifth Centuries BC: Insights from the Unexplored Mansion", in: *Annual of the British School at Athens* 1–26.
Pelagatti (1992): Paola Pelagatti, "Ceramica laconica in Sicilia e a Lipari", in: Paola Pelagatti and Conrad M. Stibbe (eds.), *Lakonikà: ricerche e nuovi materiali di ceramica laconica*, Rome, 123–247.
Petrocheilos (2016): Ioannis Petrocheilos, "Η πρώτη ανασκαφή στο Παλιόκαστρο Κυθήρων (1999–2010)", in: *Δωδώνη: επιστημονική επετηρίδα Τμήματος Ιστορίας και Αρχαιολογίας* 45–46, 261–286.
Pipili (2000): Maria Pipili, "Vases from the Samian Heraion: Shapes and Iconography", in: Pascale Linant de Bellefonds (ed.), *Αγαθός Δαίμων. Mythes et cultes. Études d'iconographie en l'honneur de Lilly Kahil*, Bulletin de correspondance hellénique, Supplément 38 Athens, 409–421.
Pipili (2001): Maria Pipili, "Samos, the Artemis Sanctuary: The Laconian Pottery", in: *Jahrbuch des Deutschen Archäologischen Instituts* 116, 17–102.
Reed (2003): Charles M. Reed, *Maritime Traders in the Ancient Greek World*, Cambridge.
Roebuck (1950): Carl D. Roebuck, "The Grain Trade Between Greece and Egypt", in: *Classical Philology* 45, 236–47.
Roussos (2021): Konstantinos Roussos, "Port and Harbour Networks in Crete during Late Antiquity (4th–mid-7th c. AD): A Modern Approach", in: *Journal of Greek Archaeology* 6, 329–358.
Rivers (2018): Ray Rivers, "New Approaches to the Theran Eruption", in: Justin Leidwanger and Carl Knappett (eds.), *Maritime Networks in the Ancient Mediterranean World*, Cambridge, 39–60.
Rutter (2017): Jeremy B. Rutter, *House X at Kommos: A Minoan Mansion Near the Sea: Part 2. The Pottery*, Philadelphia.
Schaus (1985): Gerald P. Schaus, *The Extramural Sanctuary of Demeter and Persephone at Cyrene. Final Reports, II: The East Greek, Island, and Laconian Pottery*, Philadelphia.
Schaus (2020): Gerald P. Schaus, *Funde aus Milet. Teil 4: Laconian and Chian Fine Ware Pottery at Miletus*, Wiesbaden.
Shaw/Shaw (2000): Joseph W. Shaw and Maria C. Shaw (eds.), *Kommos IV. The Greek Sanctuary*, Princeton N. J.
Stefanakis (1999): Manolis Stefanakis, "The Introduction of Coinage in Crete and the Beginning of Local Minting", in: Angelos Chaniotis (ed.), *From Minoan Farmers to Roman Traders. Sidelights on the Economy of Ancient Crete*, Stuttgart, 247–268.
Stefanakis (2013): Manolis Stefanakis, "Looking Towards the North: The Circulation of Cyrenaic Coins on Crete", in: Michele Asolati (ed.), *Le monete di Cirene e della Cirenaica nel Mediterraneo*, Padua, 65–81.
Stefanakis (2019): Manolis Stefanakis, "Από την Πελοπόννησο στην Κρήτη: νομισματική κυκλοφορία και εικονογραφικές επιρροές", in: Charles Doyen and Eva Apostolou (eds.), *Obolos 10. La monnaie dans le Péloponnèse*, Athens, 217–245.
Stibbe (1989): Conrad M. Stibbe, *Laconian Black-glazed Pottery 1: Laconian Mixing Bowls*, Amsterdam.

Stibbe (1994): Conrad M. Stibbe, *Laconian Black-glazed Pottery 2: Laconian Drinking Vessels and Other Open Shapes*, Amsterdam.
Stibbe (2000): Conrad M. Stibbe, *Laconian Black-glazed Pottery 3: Laconian Oil Flasks and Other Closed Shapes*, Amsterdam.
Tarrius (2001): Alain Tarrius, *Les nouveaux cosmopolitismes*, La Tour d'Aigues.
Tchernia (2007): Pierre Tchernia, "Entrepôt et cargaisons comlpémentaires sur la route du blé d'Alexandrie", in: Jorge Pérez Ballester and Guillermo Pascual Berlanga (eds.), *Comercio, redistribución y fondeaderos: la navegación a vela en el Mediterráneo*, Valencia, 57–63.
Tomlinson *et al.* (2010): Jonathan E. Tomlinson, Jeremy B. Rutter and Sandra M.A. Hoffmann, "Mycenaean and Cypriot Late Bronze Age Ceramic Imports to Kommos: An Investigation by Neutron Activation Analysis", in: *Hesperia* 79, 191–231.
Torelli (1982): Mario Torelli, "Per la definizione del commercio greco-orientale: il caso di Gravisca", in: *I Focei dall'Anatolia all'Oceano*, in: *La Parola del Passato* CCIV-CCVII, 304–326.
Trainor (2019): Conor Trainor, "Polis, Colony, and Beyond: Urban Knossos from Archaic to Late Antique Times", in: Grigoris Siphakis and Alexis Kalokairinos (eds.), *Proceedings of the 12th International Congress of Cretan Studies,* Heraklion.
Trapero Fernández/Aragón (2022): Pedro Trapero Fernández and Enrique Aragón, "Modelling Cabotage. Coastal Navigation in the Western Mediterranean Sea During the Early Iron Age", in: *Journal of Archaeological Science: Report* 41. https://doi.org/10.1016/j.jasrep.2021.103270.
Tsingarida/Viviers (2013): Athena Tsingarida and Didier Viviers (eds.), *Pottery Markets in the Ancient Greek World: (8th – 1st Centuries B.C.)* (Proceedings of the International Symposium, Bruxelles, 19–21 June 2011), Brussels.
Tzanakaki (2013): Katerina Tzanakaki, "Αγία Ρουμέλη 1970 (Τάρρα). Οι αρχαϊκές ταφές", in: Wolf-Dietrich Niemeier, Oliver Pilz and Ivonne Kaiser (ed.), *Kreta in der geometrischen und archaischen Zeit*, Munich, 201–220.
Vallet/Villard (1961): Gérard Vallet and François Villard, "Céramique et histoire grecque", in: *Revue historique* 225, 295–318.
Watrous (1992): Vance L. Watrous, *Kommos III: The Late Bronze Age Pottery*, Princeton N.J.
Weilhartner (2010): Jörg Weilhartner, *Testimonia: die literarischen zeugnisse über das antike Aigina von Homer bis in byzantinische Zeit*, Vienna.
Williams (1993): Dyfri Williams, "Aegina, Aphaia-Tempel, XVII: The Laconian Pottery", in: *Archaölogischer Anzeiger* 4, 571–598.
Zimi *et al.* (2019): Eleni Zimi, Kristian Göransson and Keith Swift, "Pottery and Trade at Euesperides in Cyrenaica: An Overview", in: *Libyan Studies* 50, 21–33.
Zurbach (2012): Julien Zurbach, "Mobilités, réseaux, ethnicité. Bilan et perspectives", in: Julien Zurbach and Laurent Capdetrey (eds.), *Mobilités grecques: mouvements, réseaux, contacts en Méditerranée de l'époque archaïque à l'époque hellénistique*, Bordeaux, 261–273.

Mark Stansbury-O'Donnell

28 Consumers and Vases in Early South Italian Vase-Painting

Abstract: For the first generation of red-figured pottery workshops in Lucania and Apulia, we have a case of painters establishing a market for their wares both in the Greek cities of the Ionian Sea and in the Italic communities of Lucania and Apulia. A question that can be addressed is whether there is a strong association of certain shapes or subjects with specific regions, or whether the market for the red-figured wares were generalized across southern Italy. Using a database based on the lists of Trendall and supplemented with published and displayed works post-Trendall, the data on frequency of shapes, subjects, and regional findspots for the Pisticci-Amykos Group and the groups associated with the Sisyphus Painter and the Painter of the Birth of Dionysos were subjected to statistical tests, including the chi-square test and Cramer's V. The results suggest that there are cases of specific associations between variables; for example, the bell-krater being strongly associated with Lucania outside of the Greek cities. This paper seeks to demonstrate how statistical testing can point to situations where differing consumer preferences were met by the first generation of Lucanian and Apulian workshops.

Introduction

As A. D. Trendall noted, the Lucanian and Apulian painters who established the production of figure-decorated pottery in southern Italy were primarily serving local markets, in contrast to the Attic workshops that were their sources.[1] Many of the vases have been found in the Greek cities on the coast of the Ionian Sea, Taranto, Metapontum, and Herakleia/Policoro, but more were consumed inland in Apulia and Lucania where the population consisted of native Italic peoples. The question that I would like to address is whether there is any difference in the shapes of the vases or their subject matter between those chosen by the consumers in the cities and in the different interior regions of southern Italy that could be tested and be statistically significant?[2]

To put it another way, is it possible to associate certain shapes and subjects with one group/area or another? It is possible to approach this question statistically, although the difficulties posed by the majority of Lucanian and Apulian vases being without findspots and the irregular pattern of excavation and publication makes it challenging to develop a sample group. As part of another project on pursuit scenes, I began to develop a database of early Lucanian vases based initially on the lists of Trendall, to which were added early Apulian vases and vases published post-Trendall, which includes a higher proportion of vases with findspots compared to his original lists. As of spring 2023, with the addition of later Lucanian painting and the first chapter of Middle Apulian, there are about 4200 records in the database. Some initial statistical analysis of the Pisticci Painter and the Sisyphus Group, the first painters of the new schools, showed that the Sisyphus Group's choice of subject matter and vase shape was more aligned with their Attic model of the Polygnotan Group, whereas the Pisticci Painter showed far greater divergence in those variables from the workshop of the Niobid Painter that was its model.[3] The data also showed that the Sisyphus Group and the Pisticci Painter can be associated with different primary markets, Taranto and Peucetia in the case of the former and Metapontum and Lucania in the latter, with some overlap of the two in Messapia as a tertiary market.[4]

1 *LCS* 3-7; *RVAp* XLVIII-XLIX and 4–5.

2 This paper began as a response to a question from Ted Robinson from my paper at the *CVA* Brussels conference in October 2022, wondering about differences between Metapontum and the interior for Lucanian vase-painting. I would like to thank him both for his feedback at the conference and the question. The study of Schierup (2014) is an important foundation for this analysis. A number of additional vases from Lucania were listed in Bottini (2015).

3 Stansbury-O'Donnell (2024).

4 See Stansbury-O'Donnell (2024) Tables 9–10.

Note: I would like to thank Eleni Hasaki and Marco Serino for inviting me to serve on the Scientific Committee and to submit this paper. On the statistical analyses, I am grateful for the feedback of Efthymia Nikita of the Cyprus Institute and to my colleagues of the Pottery Research Network, especially to Armand Leroi, Imperial College London.

https://doi.org/10.1515/9783111189635-028

Constructing Sample Groups and Statistical Testing

To consider consumer choices in the markets in the cities and in the countryside in the selection of vase shapes and subject matter and to take account of the dearth of findspots, it is necessary to expand the sample beyond the work of the Pisticci Painter and Sisyphus Group. Of the 218 vases attributed to the Pisticci Painter in the database, only 84 have a recorded findspot, many only a town or site rather than a specific archaeological context. These can be grouped into the regional categories in Tab. 1.[5]

Tab. 1: Regional findspots by the Pisticci Painter and Protolucanian/Pisticci-Amykos Group (*LCS*, chapter 1).

REGION	Pisticci Ptr #	Pisticci Ptr %	Pisticci-Amykos Group #	Pisticci-Amykos Group %
Apulia C. (Peucetia)	10	11.9%	54	24.1%
Apulia S. (Messapia)	7	8.3%	17	7.6%
Apulia N. (Daunia)	0		1	0.4%
Apulia (general)	1	1.2%	3	1.3%
Greek Cities (Taranto, Metaponto, Herakleia)	23	27.4%	74	33.0%
Basento-Brandano Basin	31	36.9%	46	20.5%
Lucania	8	9.5%	13	5.8%
Campania	2	2.4%	10	4.5%
Calabria	0		2	0.9%
Sicily	1	1.2%	1	0.4%
Albania	1	1.2%	2	0.9%
Etruria	0		1	0.4%
Total	84		224	
No findspot	134		405	

With 20 vases from Metapontum (and 3 others from Taranto), 31 from the Basento–Brandano basin, and 8 from the rest of Lucania, there is not a sufficient quantity to address the question about shapes and subjects in the markets for the Pisticci Painter alone, as the number of categories for shape and subject are potentially high and the number of examples in many categories too low for statistical testing.[6] The close relationship of the Pisticci Painter to the Amykos Painter and workshop, as well as the Cyclops Painter and the PKP Group, forms a grouping of workshops similar to those of the Sisyphus Group, with consequently a larger number of vases with recorded findspots.[7] These workshops, named broadly the Pisticci-Amykos Group by Trendall, have a total of 629 vases currently in the database, of which 224 have a findspot, 74 from the Greek cities of Taranto, Metapontum, and Herakleia/Policoro and 59 from Lucania (hereafter this term includes vases from the Basento-Brandano basin as well as the highlands to the north and west).[8]

Before analyzing the data, we should consider two issues with this sample. First, of the 26 vases with findspots for the PKP Group, 8 come from the Policoro Painter tomb in Herakleia.[9] The tomb is highly unusual for its large number of figured vases, 6 hydriai, 4 pelikai, and 2 skyphoi. The types of vases are common for tombs in the Greek cities, but their

5 Stansbury-O'Donnell (2022) 470, Table 21.1 counts 223 vases by the Pisticci Painter based on the first version of the database that included some entries listed as either Pisticci or Amykos Painters or were reattributed in post-Trendall publications. The full data as of 25 April 2023 have been made available in an Excel spreadsheet at: https://docs.google.com/spreadsheets/d/13JpzQ5r-1lNVtTATOPd8nnwFPG02sSvB/edit?usp=share_link &ouid=113441084278290443126&rtpof=true&sd=true (last access 01.03.2024). The full data set includes vases without findspots, but the sheet "DATA-Findspots" includes only those with provenance by region.

6 Usually a minimum of 5 examples for each category is required for reliability.

7 These are the painters in Chapter 1 of *LCS* with the number of vases and those with findspot: Pisticci Painter and Workshop: 217 vases, 84 with findspot; Amykos Painter and Workshop: 297, 99 with findspot; Cyclops Painter: 41, 14 with findspot; PKP Group: 73, 26 with findspot. There is one additional vase attributed to both the Cyclops and Pisticci Painters with a findspot. Schierup (2014) does not include the PKP Group in her lists of vases with findspots, but given Denoyelle (2018) 57–58 linking the Cyclops and Palermo Painters as the same, it seems prudent to include the PKP Group in the sample.

8 Basento-Brandano: 46; Lucania: 13. Two of these are fragments of undetermined shape and so are not part of the analysis of vase shapes below, leaving 57 as the sample size for shapes.

9 Degrassi (1967). In addition to the six vases by the PKP Group, there was a hydria by the Amykos Painter. Two skyphoi by the Intermediate Group and a hydria by the Creusa Painter are not included in the sample.

focus on mythological and heroic subjects is highly unusual. Similarly, there are a dozen figured vases from Tomb 1/1968 at Marina Soprano in Pisticci, including works by the Pisticci and Cyclops Painters.[10] As we will see later, there are also examples of tombs in Rutigliano and Gravina in Peucetia with multiple vases, some with unusual subjects.[11] While these large assemblages are unusual, it seems best to include them in the sample.[12]

A second consideration is that a large number of vases were found in the excavation of the Kerameikos in Metapontum, showing that it was the production site for much of early Lucanian painting with most of the fragments attributed to the Creusa and Dolon Workshops that formed the second generation of Lucanian painting.[13] There are three examples attributed to the Amykos Painter in the excavation finds, and since this type of context does not reflect the potential destination and end-use of vases by the consumer, it is prudent to exclude these from the data for Metapontum and the Greek cities. Excluding these artifacts results in 71 vases for the Greek cities sample and 59 for Lucania.

In order to explore whether there is an association or independence of vase shape or of subject matter between the two areas, we can use a chi-square test, which takes the values of the categories and calculates the differences between the observed and expected values to see if the difference is statistically significant or not. For example, in a hypothetical, simple comparison of two variables, region (Greek or Lucania) and shape (krater and pelike), the data (fictional) could be set up in a 2x2 contingency table (Tab. 2):

Tab. 2: Hypothetical table of vase shapes and region.

	Greek Cities	**Lucania**
Krater	25	30
Pelike	37	25
Total	62	55
Chi-square	2.3668 w 1df	P=0.1239
Cramer's V	0.1422	

It would appear that the pelike might be associated more with urban areas and the krater slightly more with Lucania, but is this difference statistically significant, or possibly a result of the irregularity of vases with reported findspots? For categorical data such as this, we can use a chi-square test to determine whether the variable of vase shape here is independent of the variable of findspot (the null hypothesis); or whether there is some association between particular shapes and particular regions (alternative hypothesis). We can test the data with a chi-square calculator; the resulting value is 2.3668 with 1 degree of freedom.[14] The P value, which measures statistical significance, is 0.1239. Usually, one looks for a P value of 0.05 or lower, so in this case, the results are not considered statistically significant in themselves and we cannot reject the null hypothesis, meaning that the data does not indicate an association of specific shape with specific area, even though there is a majority of pelikai in the cities and kraters in Lucania. That does not mean there is not an association, but based only on this hypothetical sample, the data would not support the hypothesis that specific markets preferred specific shapes.

---

10 One can also cite the later Tomb 2/1986 at S. Maria del Casale as unusual with 7 figured vases, two by the Brooklyn-Budapest Painter and others by Apulian painters; see Bottini/Lecce (2013) 60, note 71 and fig. 10.

11 See, for example, Montanaro (2018) and Montanaro (2021).

12 The vases from the Tomb at Martina Soprano and the Amykos Painter vase in the Policoro Painter tomb are also included in Schierup (2014).

13 See D'Andria (1975); Silvestrelli (2005), (2019). On the organization of the Metapontine Kerameikos from the 5th into the 4th centuries, see Silvestrelli (2008) and (2016).

14 For a discussion of the chi-square test and Cramer's V see Drennan (2009) ch. 14. For the following analyses, I follow a standard compromise discussed in Drennan that no more than 20% of categories have less than 5 examples. The use of these tests for exploring the association of workshop with shape and subject matter is in Stansbury-O'Donnell (2024).

Lucanian Data and Analysis

There are significantly more categories in the variable vase shape to consider than the two in the demonstration, as the results in Tab. 3 show.[15]

Tab. 3: Distribution of vase shapes of the Pisticci-Amykos Group by region.

SHAPE	Greek Cities	Lucania
Bell-krater	7 (9.5%)	26 (45.6%)
Calyx Krater	1 (1.4%)	3 (5.3%)
Column Krater	0	1 (1.8%)
Undetermined Krater	3 (4.1%)	1 (1.8%)
Volute Krater	0	0
Pelike	20 (27.0%)	6 (10.5%)
Hydria	22 (29.7%)	6 (10.5%)
Panathenaic Amphora	7 (9.5%)	2 (3.5%)
Lebes	5 (6.8%)	1 (1.8%)
Skyphos	2 (2.7%)	4 (7.0%)
Lekythos	7 (9.5%)	1 (1.8%)
Oinochoe	0	6 (10.5%)
Total	74	57
Undetermined Fragments	0	2

Tab. 4: Distribution of vase shape by Region. Degrees of Freedom = 5.

SHAPE	Greek Cities	Lucania
Bell-krater	7 (9.9%)	26 (45.6%)
Other Kraters	4 (5.6%)	5 (8.9%)
Pelike	20 (28.2%)	6 (10.5%)
Hydria	20 (28.2%)	6 (10.5%)
Panathenaic Amphora	6 (8.5%)	2 (3.5%)
Other Shapes	14 (19.7%)	12 (21.1%)
Total	71	57
Chi-Square	**27.0739 w 5df**	**P=0.000056**
Cramer's V	**0.4599**	

As can be seen, there are a number of shape categories that have values of less than 5, and there are four cells with 0. Normally, a minimum value of 5 for each cell is regarded as necessary for statistical analysis, but a useful compromise is to limit this to no more than 20% of cells with less than 5 and no cells with 0. Since the bell-krater is so numerous, we can place the other, more complex types of kraters into a category of "other kraters." The pelike and hydria have sufficient quantity, and we can include the pseudo-panathenaic amphora as a fifth category with one cell having 2 examples. We can place the remaining shapes into a category of "other," resulting in a 6x2 contingency table (Tab. 4).

The data show bell-kraters are particularly numerous in Lucania, where they constitute almost half of the total, in comparison to about 10% in the Greek cities. The pelike and hydria combined are 56.4% of the total vases in the Greek cities, compared to 20.0% for those shapes in Lucania. There appears, then, to be an association of the bell-krater with Lucania, and the pelike, hydria, and also pseudo-panathenaic amphora with the Greek cities, but is this statistically significant given the size of the samples?

The resulting chi-square value for the data is 27.0739 with 5 degrees of freedom and a P value of 0.000056, well below the threshold of 0.05 indicating statistical significance.[16] It is, however, difficult to judge the strength of the result given the geometric curve for the P value below 0.01. A second statistical measure, Cramer's V, takes the chi-square value and adjusts for the sample size and degrees of freedom to yield a number from 0 to 1, with 0 signifying the independence of the two variables (distribution of shapes is the same in all regions, so no association of market and shape), and 1 signifying complete association (one region has only kraters and the other only hydriai in our hypothetical example).[17] For a 6x2 table, a Cramer's V of 0.1-0.3 would suggest a small degree of association between the two variables, a value of 0.3-0.5 a moderate degree of association, and above 0.5 a strong degree of association. At 0.4599, the association is bordering on the strong, and suggests that consumers in the two areas had strong, although not exclusive preferences for certain, select shapes, the bell-krater for Lucania and the pelike and hydria for the Greek cities.

We can also examine subject matter to see if there is a preference for certain themes in each of the two areas. In this case, the number of potential categories is huge if a depiction of a specific scene, such as Eos pursuing Kephalos, is counted

15 The two undetermined fragments have been excluded from the 57 examples from Lucania.

16 If we include Panathenaic amphorae in the other category, the resulting chi-square result is 25.9656 with 4 degrees of freedom, a P value of 0.000032, and a Cramer's V of 0.4504. This is very close to the result with six categories, but obscures the development of the Panathenaic amphora as a new shape in south Italy that was particularly important in Peucetia more than in Lucania as we will see later in Table 8.

17 Drennan (2009) ch. 14.

as a category, as the number of examples for most categories would be below 5 and many would be 0.[18] For the purposes of the database, specific subjects were coded into broader categories, so that a representation of Achilles, or Bellerophon slaying the chimaera, would each be a "heroic" scene.[19] Scenes primarily consisting of non-specific men and youths could be broken down into categories such as athletics, warrior departure, the symposion, but could also be grouped together if there were only a few examples of each. Group scenes include the so-called conversation scenes and have women, youth, and/or men present, but could also be coded more specifically if the action involved a ritual or funerary scene.[20] The categories used in the database are in Tab. 5.

Tab. 5: Subject matter categories.

1 Athletic	Scenes of athletes,sometimes includes a woman with wreath, strigil, aryballos, could include deity (esp. Eros, Nike)
2 Arming	Youth or man putting on or receiving arms, could include deity (esp. Eros, Nike)
3 Departing	Warrior departing, can include woman, phiale, oinochoe
4 Warrior	Warrior(s), standing or conversing, can include rider with horse or chariot, could include deity (esp. Eros, Nike)
5 Symposion	Banquet, can include women, kline
6 Komos	Dance-like processional scenes, can be mixed gender, sometimes instruments
7 Youth	Undifferentiated scenes with only youths, usually with some nudity, could include deity (esp. Eros, Nike)
8 Men/Male	Undifferentiated scenes with adult males, can include youths, could include deity (esp. Eros, Nike). Hunt.
9 Domestic	Scenes with woman with indication of domestic setting (such as chair, column), could include deity (esp. Eros, Nike) or sometimes male observers
10 Women	Undifferentiated scenes of woman, can include weaving, conversation, could include deity (esp. Eros, Nike)
11 Group	Mixed-gender undifferentiated gatherings, could include deity (esp. Eros, Nike)
12 Figure	Single figures of uncertain gender (mostly pottery fragments)
13 Music	Scenes with instruments or performance
14 Ritual	Ritual at altars, shrines, includes offerings and cult scenes.
15 Dionysiac	Scenes with satyrs/silens and maenads, including thiasos, can include Dionysos or youth, includes thrysos, tymbales, situla
16 Pursuit	One figure chasing another, can be deity or mortal
17 Heroic	Narrative scenes with heroes and mortals, can include gods as viewers, duels and battles
18 Myth	Narrative scenes with gods as protagonists
19 Gods	Non-narrative scenes of gods
20 Funerary	Scenes at tombs, usually includes naiskos, column, or stele
21 Theatrical	Includes phlyax scenes, actors
22 Other	Often animals

In looking at the subject matter data for the Greek cities, three scenes dominate (more than 20%): scenes of women, group scenes of women and men (almost all youths), and pursuit scenes, either with a divinity or youth as pursuer (Tab. 6). Scenes with women are also common in Lucania, but there is a strong preference for Dionysiac scenes that are more uncommon in the Greek cities.[21]

18 A similar approach was taken in a project by F. and I. Giudice; see Giudice (1999) and Giudice/Giudice Rizzo (2009). There are a number of categories in their project, such as hunting, courting, and making love have proportionally few examples and, whereas categories combine scenes that could be distinguished by gender. The scenes of draped youths that are found on the majority of vases are not counted as a subject unless there is some distinctive action like an offering. Some vases do have scenes on both sides of the vase, leading to the count for subjects to be higher than the count for vases.

19 In developing the categories, I used several Attic and Lucanian groups/painters and the broad categories found in J. D. Beazley and A. D. Trendall but favoring those categories for which at least two workshops had at least 5% of their work in the category.

20 The scenes of standing, draped youths that are standard on the reverse of most south Italian vases have not been classified as subjects unless there is some action or attribute such as nudity that places the scene in one of the categories.

21 The number of scenes is slightly larger than the number of vases as some have additional subjects on the reverse other than draped youths. If a reverse scene falls into the same category as the obverse, it is only counted once.

Tab. 6: Distribution of subject matter by region. Degrees of Freedom = 5.

SHAPE	Greek Cities	Lucania
Men and Youths (1–8)	8 (10.1%)	8 (13.3%)
Women (9–10)	19 (23.8%)	12 (20.0%)
Group (11)	19 (23.8%)	5 (8.3%)
Dionysiac (15)	6 (7.5%)	19 (31.7%)
Pursuit (16)	16 (20.0%)	8 (13.3%)
Other Scenes	12 (15.0%)	8 (13.3%)
Total	80	60
Chi-Square	**17.4734 w 5df**	**P=0.0037**
Cramer's V	**0.3533**	

Tab. 7: Appearance of Eros by region. Degrees of Freedom = 1.

SHAPE	Greek Cities	Lucania
Vases With Eros	20 (28.2%)	5 (8.8%)
Vases Without Eros	51 (71.8%)	52 (91.2%)
Total	71	57
Chi-Square	**7.5690 w 1df**	**P=0.005938**
Cramer's V	**0.2432**	

The chi-square value is less than that for shape, and Cramer's V at 0.3533 is moderate in terms of the strength of association for specific subject matter with each area. There would seem to be a statistically notable association of Dionysiac scenes with Lucania, and a greater interest in group scenes and scenes with women in the Greek cities.[22] There is somewhat more interest in the Greek cities in pursuit scenes, and there is further some distinction between the two areas in terms of the pursuer, with all 3 Eos-youth pursuits found in Lucania and 5 of the 6 Eros-woman pursuits in the Greek cities, as well as 4 of the 6 youth-woman pursuits. As Angelo Bottini has noted, the elites of Lucania appear to have been familiar with the movements concerning the afterlife, including Pythagoreanism, Orphism, and Dionysiac cults; he associates interest in winged deities with these cults, and the strong interest in Dionysiac scenes lends support to that hypothesis.[23] The association of subject matter with region, however, is not as strong the association of shape with region.

There are other possible iconographic questions to explore statistically, such as the representation of specific figures or objects in the scenes, particularly in the future as more vases with archaeological contexts and findspots are known. For example, Eros is a popular figure in early Lucanian painting generally, in non-narrative scenes with women, groups of youths and women, or pursuit narratives.[24] Among the 71 vases from the Greek cities, he appears on 20, whereas he appears on 5 of the 57 vases found in Lucania. Setting up a 2x2 table (Tab. 7), the chi-square value is 7.5690 with 1 degree of freedom and a P value of 0.005938, suggesting that this is a significant association of Eros with one region over the other. Cramer's V, however, shows that this is at best a small association, with a value of 0.2432. The result would be very different if Eros were to appear on half of the vases from the Greek cities, so one might say that Eros was a figure of interest there, but within the context of other figures or scenes.[25]

Apulian Data and Analysis

One can ask the same questions about shape and subject with regions for Apulia, but here the number of surviving, published vases is more challenging.[26] As Trendall noted, the four painters of the Sisyphus Group and the Gravina Painter and Painter of the Birth of Dionysos form the Pioneers of Apulian painting, with the latter two painters a bit later, like the PKP Group in Early Lucanian.[27] There are 230 vases in the database for these workshops, 99 of them with provenance: 40

22 For an overview of the iconography in Lucania see Söldner (2007), and more specifically with Eros, see Cassimatis (2014) 147–238.

23 Bottini (2012); Bottini/Lecce (2013) 50; Robinson (2011) esp. 59–63.

24 Söldner (2007) 117–122; Cassimatis (2014) 178–194.

25 See discussion of Söldner (2007) 85–91; Cassimatis (2014) 206–208, and 210–224.

26 Robinson (2014) 226–227 reasonably proposes that Apulian potters/painters made specific shapes for Italic clients as well as specific scenes and this discussion is intended to explore the data on that. Robinson also notes the problems of stylistic variation, attribution, and chronology and includes painters of the beginning of the Ornate style from chapter 2 of *RVAp*, as does Fontannaz (2005), who re-organizes some of Trendall's groupings in *RVAp*. For descriptive statistics of Apulian red-figured pottery broadly, see Herring (2018).

27 See *RVAp* chapter 1 (Painter of the Berlin Dancing Girl, Hearst Painter, Sisyphus Painter, Parasol Painter, and Ariadne Painter) and chapter 2: Gravina Painter, Painter of the Birth of Dionysos and connected vases.

(40.4%) are from Peucetia and 41 (41.4%) are from Taranto, making two dominant markets, as many experts have noted.[28] Most of the works from Taranto are fragmentary and appear to have served as tomb markers, whereas most of the vases from Peucetia were from inside tombs.[29] As mentioned earlier, there are several tombs in Peucetia with a large number of vases, similar to the Policoro Painter tomb in Herakleia/Policoro. Tomb 24/1976 in the Contrada Purgatorio excavations in Rutigliano included 7 vases of six different shapes, all by the Painter of the Berlin Dancing Girl; there are also 13 vases from two tombs (1/1974, 2/1994) in Gravina-Botromagno, and these assemblages point to different patterns of usage compared to Taranto.[30] The sample sizes are, however, smaller than in the Lucanian comparisons and subject to caution, as Fontannaz reported about 300 fragments in the Taranto museum, many of which have not been published, and many of the Rutigliano tombs have yet to be published.[31]

Turning first to shape and excluding fragments of undetermined types, we get the following distribution in Tab. 8.

As can be see, kraters are important in both regions, accounting for at least 60% of all vases. Bell and volute kraters are far more common in Peucetia, whereas calyx kraters are the most common type in Taranto, accounting for at least half of all kraters, where they served as *semata* above the tombs, leading to their very fragmentary condition compared to vases found in Peucetia.[32] Additionally, pseudo-panathenaic amphorae turn up frequently in Peucetia, but not in Taranto, which generally has less diversity in shapes than Peucetia.

The size of the sample means that taking the top five categories results in more than 20% of the cells with values of less than 5. In this case we can use the categories for bell-kraters and other kraters and place the remaining vases into the 'other category,' resulting in two degrees of freedom (Tab. 8). This results in a chi-square value of 11.0555 and a Cramer's V of 0.3918. This would recognize the association of bell-kraters and the array of other non-krater shapes with Peucetia as significant, but it blurs the strong association of calyx kraters with Taranto and volute kraters with Peucetia seen in Tab. 9. This differentiation would suggest an even stronger association between shape and markets in Taranto and Peucetia. Given the number of unpublished vases, we might be able to demonstrate a stronger, or perhaps weaker association of shape and market in the future.

Tab. 8: Distribution of Protoapulian vase shapes in Taranto and Peucetia.

SHAPE	Greek Cities (Taranto)	Peucetia
Bell-krater	3 (9.4%)	10 (25.0%)
Calyx Krater	17 (53.1%)	1 (2.5%)
Column Krater	0 (0.0%)	1 (2.5%)
Krater Undetermined Fragment	6 (18.8%)	0 (0.0%)
Volute Krater	0 (0.0%)	11 (27.5%)
Panathenaic Amphora	2 (6.3%)	7 (17.5%)
Hydria	2 (6.3%)	5 (12.5%)
Pelike	0 (0.0%)	2 (5.0%)
Oinochoe	0 (0.0%)	1 (2.5%)
Skyphos	2 (6.3%)	1 (2.5%)
Lekanis	0 (0.0%)	1 (2.5%)
Total	32 (100%)	40 (100%)
Undetermined Fragments/Test Piece	9	0

Tab. 9: Distribution of Protoapulian vase shapes in Taranto and Peucetia. Degrees of Freedom = 2.

SHAPE	Greek Cities (Taranto)	Peucetia
Bell-krater	3 (9.4%)	10 (25.0%)
Other Kraters	23 (71.9%)	13 (32.5%)
Other Shapes	6 (18.8%)	17 (42.5%)
Total	32	40
Chi-Square	**11.0555**	**P=0.003975**
Cramer's V	**0.3918**	

28 The full data as of 27 April 2023 have been made available in an Excel spreadsheet at: https://docs.google.com/spreadsheets/d/1yDp-Mx7o-qL0PTaU9YXtxO3do45-CICXN/edit?usp=share_link&ouid=113441084278290443126&rtpof=true&sd=true (last access 01.03.2024). It is worth noting that there are 6 vases (6.1%) from Messapia, the third largest market. On Messapia, see Giannotta (2014) esp. figs. 9.3–9.5.

29 See Robinson (2014) 218–220 and 224–228 on contexts for Taranto and Peucetia and Riccardi (2014) for both Lucanian and Apulian pottery in Peucetia.

30 See Montanaro (2018) 26–30 for a summary of the Gravina and Rutigliano tombs. There were also a Lucanian bell-krater and hydria by the Creusa Painter in tomb 24/1976 in Rutigliano, and it should be noted that Peucetia became a market for both Apulian and Lucanian wares in the early 4th century.

31 Fontannaz (2005) 126. Most of these were not listed by Trendall, making the sample of Apulian pottery very problematic. Some additional works have been published in Fontannaz (2014) and D'Amicis (2018).

32 Fontannaz (2005) 130 and n. 31 and 137; Robinson (2014) 219.

There are similar problems of small numbers with subject matter of Apulian pottery in Taranto and Peucetia. The five largest subject categories in Taranto are heroic scenes (14), Dionysiac (7), mythological narratives (6), and scenes with women, gods, and groups (5 each). In Peucetia, heroic scenes are also the largest category (24), followed by group scenes (8), athletic scenes (5), and Dionysiac and mythological scenes (4 each). Using the top five categories across the two regions (heroic, Dionysiac, group, mythological, and gods) along with other scenes results in one-third of the cells with less than 5 examples. We can consolidate the narrative and non-narrative scenes with gods (18–19) and also combine the scenes with men and youths (1–8) as we did in Table 6, resulting in six total categories and five degrees of freedom (Tab. 10).

Tab. 10: Distribution of Protoapulian subject matter in Taranto and Peucetia. Degrees of Freedom = 5.

SUBJECT	Greek Cities (Taranto)	Peucetia
Men and Youths (1–8)	4 (8.7%)	18 (26.5%)
Group (11)	5 (10.9%)	8 (11.8%)
Dionysiac (15)	7 (15.2%)	4 (5.9%)
Heroic and Divine (17)	14 (30.4%)	24 (35.3%)
Myth and Gods (18–19)	11 (23.9%)	6 (8.8%)
Other Scenes	5 (10.9%)	8 (11.8%)
Total	46	68
Chi-Square	**11.3927 W 5 df**	**P=0.0441**
Cramer's V	**0.3161**	

As can be seen, scenes of mortal males are about three times as many in Peucetia compared to Taranto, while scenes of myth and gods (18–19) and Dionysiac (15) scenes are three times higher proportionally in Taranto; other scenes show smaller differences, and the chi-square value of 11.3927 results in a Cramer's V of 0.3161, at the low end of the range of moderate strength of association. Shape, then, would seem to have a somewhat stronger association with region than subject matter in Protoapulian pottery.

Comparison of Lucanian and Apulian Pottery in the Greek Cities

Finally, we can also compare subject matter between the Apulian wares found in Taranto (46 examples) and the Lucanian wares found mostly in Metapontum (51), as well as Herakleia/Policoro (16) and Taranto (13) (Tab. 11).

Tab. 11: Distribution of subject matter in Greek cities: Apulian Wares (Taranto) and Lucanian (Metaponto, Herakleia/Policoro), Taranto. Degrees of Freedom = 5.

SHAPE	Apulian wares in Taranto	Lucanian wares in Metaponto, Herakleia, and Taranto
Men and Youths (1–8)	4 (8.7%)	8 (10.1%)
Women (9–10)	5 (10.9%)	19 (23.8%)
Group (11)	5 (10.9%)	19 (23.8%)
Dionysiac (15)	7 (15.2%)	6 (7.5%)
Pursuit (16)	0	16 (20.0%)
Heroic (17)	14 (30.4%)	7 (8.8%)
Myth and Divine (18–19)	11 (23.9%)	4 (5.1%)
Other Scenes	5 (10.9%)	1 (1.3%)
Total	46	80
Chi-Square (9-10/11/16/17/18-19/other)	**30.3090 w 5df**	**P=0.0000128**
Cramer's V	**0.4905**	

Heroic (17), myth (18), and divine (19) scenes account for 54.3% of all Apulian scenes in Taranto, but 15.2% for Lucanian ware in the Greek cities.[33] Lucanian scenes with women and groups are much more numerous than they are in Apulian ware found in Taranto, and pursuit scenes are one-fifth of the production for Lucanian but absent in Apulian ware in Taranto.[34] Using the five categories with the highest number of total examples from both fabrics (women (9–10), group (11), pursuits (16), heroic (17), and myth/gods (18–19), we get a chi-square value of 30.3090 and a Cramer's V of 0.4905, at the border of moderate/strong association. In other words, it appears that there are stronger associations of some scene categories in Lucanian pottery distributed in the Greek cities (women, groups, and pursuits) and other categories in Apulian pottery distributed in Taranto (heroic, mythological/gods, and Dionysiac). It would, then, appear that there was greater differentiation between Lucanian pottery and Apulian pottery distributed in the Greek cities than in either ware distributed to Lucania or Peucetia respectively.

My goal in this paper is to consider whether statistical analysis can help us to understand whether there might be regional variations in the distribution of shapes and subject matter by the early Apulian and Lucanian workshops that would indicate a stronger degree of association of certain categories with the different regions. A stronger association between certain shapes or subjects and specific regions cannot be proven through these tests, but they can point to cases where there is the potential for the painting workshops being selective in choosing shape and subject based on their consumers in the different markets of southern Italy. As new vases with provenance become known and published, statistical analyses will need updating and reexamination. It is possible that large numbers of unpublished vases could alter the picture found here, or permit a finer analysis of individual workshops rather than groups of workshops, but a statistical approach can allow us in the meantime to substantiate current theories and explore other areas of divergence and diversity.

Abbreviations

LCS Arthur Dale Trendall, *The Red-Figured Vases of Lucania, Campania, and Sicily*, Oxford, 1967.
RVAp Arthur Dale Trendall and Alexander Cambitoglou, *The Red-Figured Vases of Apulia. Early and Middle Apulian, Vol. I, Oxford.*
CVA *Corpus Vasorum Antiquorum.*

Bibliography

Bottini (2012): Angelo Bottini, "Tiresia e Pitagora fra Greci e Italici: la *nekyia* del Pittore di Dolone", in: *Mélanges de l'École française de Rome* 124, 461–474.
Bottini (2015): Angelo Bottini, "La ceramica a figure rosse di Metaponto nella *mesogaia* lucana fra V e IV secolo a. C. Un bilancio", in: *Mélanges de l'École française de Rome* 127, 1–22.
Bottini/Lecce (2013): Angelo Bottini and Lucia Lecce, "La Mesogaia lucana e il caso di Pisticci", in: Luigi Todisco (ed.), *La comunicazione verbale tra Greci e indigeni in Apulia nel V-IV secolo a.C.: quali elementi?* (Atti del Seminario di Studi linguistici, archeologici e storici, Bari, Università degli Studi "Aldo Moro", 30 ottobre 2012), Naples, 45–60.
Cassimatis (2014): Hélène Cassimatis, *Éros dans la céramique à figures rouges Italiote. Essai d'interprétation iconographique et iconologique*, Paris.
D'Amicis (2018): Amelia D'Amicis, "Materiali dall'Arsenale di Taranto. Un rinvenimento di cui parlare", in: Martine Denoyelle, Claude Pouzadoux and Francesca Silvestrelli (eds.), *Mobilità dei pittori e identità delle produzioni*, Cahier du Centre Jéan Berard 25, Naples, 189–204.
D'Andria (1975): Francesco D'Andria, "Scavi nella zona del *Kerameikos*", in: *Notizie degli scavi di antichità* 29, Supplement, 355–452.
Degrassi (1967): Nevio Degrassi, "Meisterwerke frühitaliotischer Vasenmalerei aus einem Grab in Policoro", in: Bernhard Neutsch (ed.), *Archäologische Forschungen in Lukanien. II. Herakleiastudien*, Heidelberg, 193–245.
Denoyelle (2018): Martine Denoyelle, "Trendall and After: Some News from the Palermo Painter", in: Ursula Kästner and Stephan Schmidt (eds.), *Inszenierung von Identitäten. Unteritalische Vasenmalerei zwischen Griechen und Indigenen*, Munich, 52–60.
Drennan (2009): Robert D. Drennan, *Statistics for Archaeologists. A Common Sense Approach*, 2nd edition, New York. https://doi.org/10.1007/978-1-4419-0413-3_14.

33 Of the seven Lucanian heroic scenes, 6 are from the Policoro Painter tomb in Herakleia/Policoro and 1 from Taranto. The most common subject in early Lucanian vases found in Taranto are group scenes (6 out of 13), and this is the most common scene in Herakleia Policoro after the heroic scenes in the Policoro Painter tomb (3 out of 16), making group scenes common in all three cities.
34 12 of the pursuit scenes are found in Metapontum, and 2 each in Herakleia/Policoro and Taranto.

Fontannaz (2005): D. Fontannaz, "La céramique proto-apulienne de Tarente: problems et perspectives d'une recontextualization", in: M. Denoyelle, E. Lippolis, M. Mazzei and C. Pouzadoux (eds.), *La céramique apulienne. Bilan et perspectives* (Actes de la Table Ronde organisée par l'École française de Rome en collaboration avec la Soprintendenza per i Beni Archeologici della Puglia et le Centre Jean Bérard de Naples – Napoli, Centre Jean Bérard, 30 novembre – 2 décembre 2000), Naples, 125–142.

Fontannaz (2014): Didier Fontannaz, "Production and Functions of Apulian Red-Figure Pottery in Taras: New Contexts and Problems of Interpretation", in: Thomas H. Carpenter, Kathleen M. Lynch and Edward G. D. Robinson (eds.), *The Italic People of Ancient Apulia. New Evidence from Pottery for Workshops, Markets, and Customs*, Cambridge, 71–95.

Giannotta (2014): Maria Teresa Giannotta, "Apulian Pottery in Messapian Contexts", in: Thomas H. Carpenter, Kathleen M. Lynch and Edward G. D. Robinson (eds.), *The Italic People of Ancient Apulia. New Evidence from Pottery for Workshops, Markets, and Customs*, Cambridge, 186–210.

Giudice (1999): Filippo Giudice, "Il viaggio delle immagini dall'Attica verso l'Occidente e il fenomeno del rapporto tra 'prodigi' e 'fortuna iconografica'", in: *Le myth grec dans l'Italie antique. Fonction et image* (Actes du Colloque International, Roma, 13–16 Novembre 1996), Rome, 267–327.

Giudice/Giudice Rizzo (2009): Filippo Giudice and Innocenza Giudice Rizzo, "Seeing the Image: Constructing a Data-Base of the Imagery on Attic Pottery from 635 to 300 BC", in: John H. Oakley and Olga Palagia (eds.), *Athenian Potters and Painters* II, Oxford, 48–62.

Herring (2018): Edward Herring, *Patterns in the Production of Apulian Red-Figure Pottery*, Newcastle upon Tyne.

Montanaro (2018): Andrea C. Montanaro, "Death is not for me: Funerary Contexts of Warriors Chiefs from Pre-Roman Apulia", in: Ursula Kästner and Stefan Schmidt (eds.), *Inszenierung von Identitäten. Unteritalische Vasenmalerei zwischen Griechen und Indigenen*, Munich, 25–38.

Montanaro (2021): Andrea C. Montanaro, "Rutigliano (Bari). La necropoli di contrada Purgatorio. Le tombe del settore settentrionale (scavi 1976–77). Riflessioni preliminari", in: *BABesch. Annual Papers on Mediterranean Archaeology* 96, 1–44.

Riccardi (2014): Ada Riccardi, "Apulian and Lucanian Pottery from Coastal Peucetian Contexts", in: Thomas H. Carpenter, Kathleen M. Lynch and Edward G. D. Robinson (eds.), *The Italic People of Ancient Apulia. New Evidence from Pottery for Workshops, Markets, and Customs*, Cambridge, 133–151.

Robinson (2011): Edward G. D. Robinson, "Identity in the Tomb of the Diver at Poseidonia", in: Margarita Gleba and Helle W. Horsnaes (eds.), *Communicating Identity in Italic Iron Age Communities*, Oxford, 50–72.

Robinson (2014): Edward G. D. Robinson, "The Early Phases of Apulian Red-figure", in: Stine Schierup and Victoria Sabetai (eds.), *The Regional Production of Red-figure Pottery: Greece, Magna Graecia and Etruria*, Aarhus, 217–234.

Schierup (2014): Stine Schierup, "Patterns of Use in Early Metapontine Red-figure Pottery: Distribution, Shapes and Iconography", in: Stine Schierup and Victoria Sabetai (eds.), *The Regional Production of Red-figure Pottery: Greece, Magna Graecia and Etruria*, Gösta Enbom Monograph 4, Aarhus, 191–216.

Silvestrelli (2005): Francesca Silvestrelli, "Le fasi iniziali della ceramica a figure rosse nel *kerameikos* di Metaponto", in: Martine Denoyelle, Enzo Lippolis, Marmina Mazzei and Claude Pouzadoux (eds.), *La céramique apulienne. Bilan et perspectives* (Actes de la Table Ronde organisée par l'École française de Rome en collaboration avec la Soprintendenza per i Beni Archeologici della Puglia et le Centre Jean Bérard de Naples, Napoli, Centre Jean Bérard, 30 novembre – 2 décembre 2000), Naples, 113–123.

Silvestrelli (2008): Francesca Silvestrelli, "La distribuzione della ceramica a figure rosse dei Pittori di Creusa, di Dolone e dell'*Anabates*", in: *Le perle e il filo. A Mario Torelli per i suoi settanta anni*, Verona, 271–292.

Silvestrelli (2016): Francesca Silvestrelli, "Potters and Painters in Metapontine Red-Figure Workshops: Some Preliminary Observations", in: Norbert Eschbach and Stefan Schmidt (eds.), *Töpfer-Maler-Werkstatt. Zuschreibungen in der griechischen Vasenmalerei und die Organisation antiker Keramikproduction*, *CVA* Deutschland Beihefte 7, Munich, 176–189.

Silvestrelli (2019): Francesca Silvestrelli, "La *polis* di Metaponto e il suo Territorio dalle prime esperienze alla metà del IV secolo a.C.", in: *Produzioni e committenze in Magna Grecia* (Atti del LV Convegno di Studi sulla Magna Grecia, Taranto, 24–27 settembre 2015), Taranto, 536–555.

Söldner (2007): Magdalene Söldner, *Bios Eudaimon. Zur Ikonographie des Menschen in der rotfigurigen Vaesnmalerei Unteritaliens. Die Bilder aus Lukanien*, Möhnsee.

Stansbury-O'Donnell (2022): Mark D. Stansbury-O'Donnell, "Winged Figures and Mortals at a Crossroad", in: Judith M. Barringer and François Lissarrague (eds.), *Images at the Crossroads. Media and Meaning in Greek Art*, Edinburgh, 459–475.

Stansbury-O'Donnell (2024): Mark D. Stansbury-O'Donnell, "From First to Second Generation: Adapting Subject Matter and Shapes to Local Markets in South Italy", in: Anne Coulié, Athena Tsingarida and Stefan Schmidt (eds.), *The Corpus Vasorum Antiquorum (1922–2022). A Century of Exploring Greek Vases: Typologies, Readings and Debates* (*CVA* Belgium, Supplement 1), 355–369.

Elisabetta Di Virgilio

29 Hybridity and Exchange Models in Campania: The Case of Ionian Cups from Ancient Cumae

Abstract: This chapter presents a typological study and macroscopic observations of the so-called Ionian cups coming from three sanctuaries at Cumae, on the Bay of Naples (Italy). The presence of hundreds of these cups in the sites of Magna Graecia and Sicily, generally manufactured in the East Greek area, has led scholars to hypothesize a Western production of them during the Archaic age, confirmed by the occasional discovery of production markers, especially in southern Italy. Workshops have not been identified so far at Cumae, but the coexistence of different fabrics and morphological models suggests a possible local production of these cups. Morphological study has shown a higher percentage of the Type B2 cup, while macroscopic analysis has revealed a remarkable difference in fabrics and banded decoration, attesting the existence of imitative products at local and regional levels, according to the colonial trends of the Archaic period.

Introduction

The existence of colonial workshops specialized in the production of so-called Ionian cups in Magna Graecia and Sicily is now a widely accepted phenomenon.[1] However, the lack of workshop indicators in the region of Campania[2] has made it difficult to demonstrate their presence, the more so since the clay sources of certain zones are very similar to each other, as is the case with Cumae and Pithekoussai.[3] In any event, the Bay of Naples can generally be assigned as the material source for most local pottery categories, and archaeometric research is increasingly improving in this connection.[4]

Through morphological and macroscopic investigation of a significant sample of Ionian cups, this study aims at a wider understanding of the exchange networks of objects and knowledge in Campania.[5]

Background

The ancient city of Cumae is located at the northern end of the Bay of Naples,[6] between two hills: to the west, the promontory of Cumae, overlooking the Tyrrhenian Coast, while to the east stands Mount Grillo (Fig. 1). Geomorphologically, this territory is part of the volcanic area of the Campi Flegrei, which lies to the west of Naples. This zone is characterized by pyroclastic flows such as the Campanian Ignimbrite and the Neapolitan Yellow Tuff accumulated over time, after several eruptions.[7] The promontory of Cumae was chosen by Greek settlers in the 8th century B.C. as the city's acropolis because of its strategic position, and it soon became the seat of the main local cults: those of Apollo and Jupiter. From the Archaic period on, the city had an excellent spatial organization, attested also by widespread distribution of the cult areas inside and along the city walls. The materials examined come from three of the main sanctuaries of the ancient city: the so-called

1 A general overview about this issue is given by Van Compernolle (1996) 298–302, (2000) 89–100.
2 Except for the case of ancient Trebula, in northern Campania, where a Late-Archaic kiln with waste of B2 Ionian cups, Fine Banded Ware and black-glazed pottery has been lately identified: Albore Livadie (2012) 187–204.
3 This issue has been exhaustively discussed in Gassner/Traplicher (2012).
4 Archaeometric analyses on pottery from Cumae have lately carried out: Bossière/Frère (2007) 176–220; Greco *et al.* (2012); Munzi *et al.* (2012).
5 This preliminary study is part of an ongoing doctoral research entitled *Ceramiche fini di età tardo-arcaica e classica da Cuma. Evidenze da contesti santuariali*, carried out at the University of Campania "Luigi Vanvitelli" under the supervision of C. Rescigno.
6 Bay of Naples is intended here as the volcanic area that extends from Cumae to Stabiae: Morra *et al.* (2014) 30.
7 For the geomorphological background of Campi Flegrei, see *supra*.

Note: I wish to thank the Parco Archaeologico of Campi Flegrei for permission to study the artifacts, the Laboratory of Archaeometry (University of the Peloponnese), the Institut für Klassische Archäologie (University of Vienna) and the FACEM project for providing the necessary facilities for this study.

https://doi.org/10.1515/9783111189635-029

Temple of Jupiter on the upper terrace of the acropolis, which has lately been reinterpreted as the Temple of Apollo;[8] an important sanctuary perhaps dedicated to Hera,[9] in a terraced area called Fondo Valentino southeast of the walls; nearby, on the site of the later Roman amphitheater, stood a temple whose cult has been associated with Zeus Meilichios.[10]

Fig. 1: Plan of the focus areas at Cumae. Editing: Elisabetta Di Virgilio, © Google Earth.

Methodology

In all 370 sherds of Ionian cups have been investigated. Most of them (N = 300) come from the Fondo Valentino; a smaller group comes from the Temple of Apollo, on the upper terrace (N = 60), with a few from the temple near the later Roman amphitheater (N = 10). The materials were first catalogued on site and divided into different types.[11] Then samples were observed first for the fresh breaks in hand examination and with a 10x magnifying lens. A digital microscope with camera (HD 1080P) was used for more accurate inspection and images. Some samples were selected for further investigation at the Laboratory of Archaeometry of the University of the Peloponnese, using a Leica S9i Digital Stereo Microscope.[12]

8 Excavations on the upper terrace are carried out by the University of Campania "Luigi Vanvitelli" since 2011. For a summary of the excavations and the new interpretations about the cult see: Rescigno (2021) 195–203; Rescigno/Parisi (2022) 35–59.

9 Based on epigraphic evidence. See La Rocca *et al.* (1995) 51–79; Munzi (2007) 109–130.

10 Caputo *et al.* (2016) 141–215.

11 Reference has been made to the traditional typology of Villard-Vallet (1955) 5–32, a still valid benchmark for the classification of Ionian cups in colonial contexts.

12 The attendance of the laboratory was funded by a Research Grant of the NSRF-funded project *Advancing Research Infrastructure & Scientific Techniques in Archaeological Science* of the University of Peloponnese (Kalamata, Greece).

Identification of the macroscopic groups was carried out with the following criteria: the color of the ceramic pastes and glaze (using the Munsell Color Chart), the hardness, the porosity, the surface treatment, and the inclusions. The results were compared with the fabric's description available in the online database FACEM,[13] which represented a necessary tool for the identification of the main macroscopic features and the trade patterns.

Results

Typological Study

Morphological features suggest a predominance of Type B2, as shown in the graph (Fig. 2a), whereas types A2, B1, and B3 are represented by a few sherds coming only from the Fondo Valentino. A general chronology can be given between the end of the 7th century B.C. (ca. 620 B.C.) and the end of the 6th century B.C. (ca. 520–500 B.C.), except for some more recent samples.

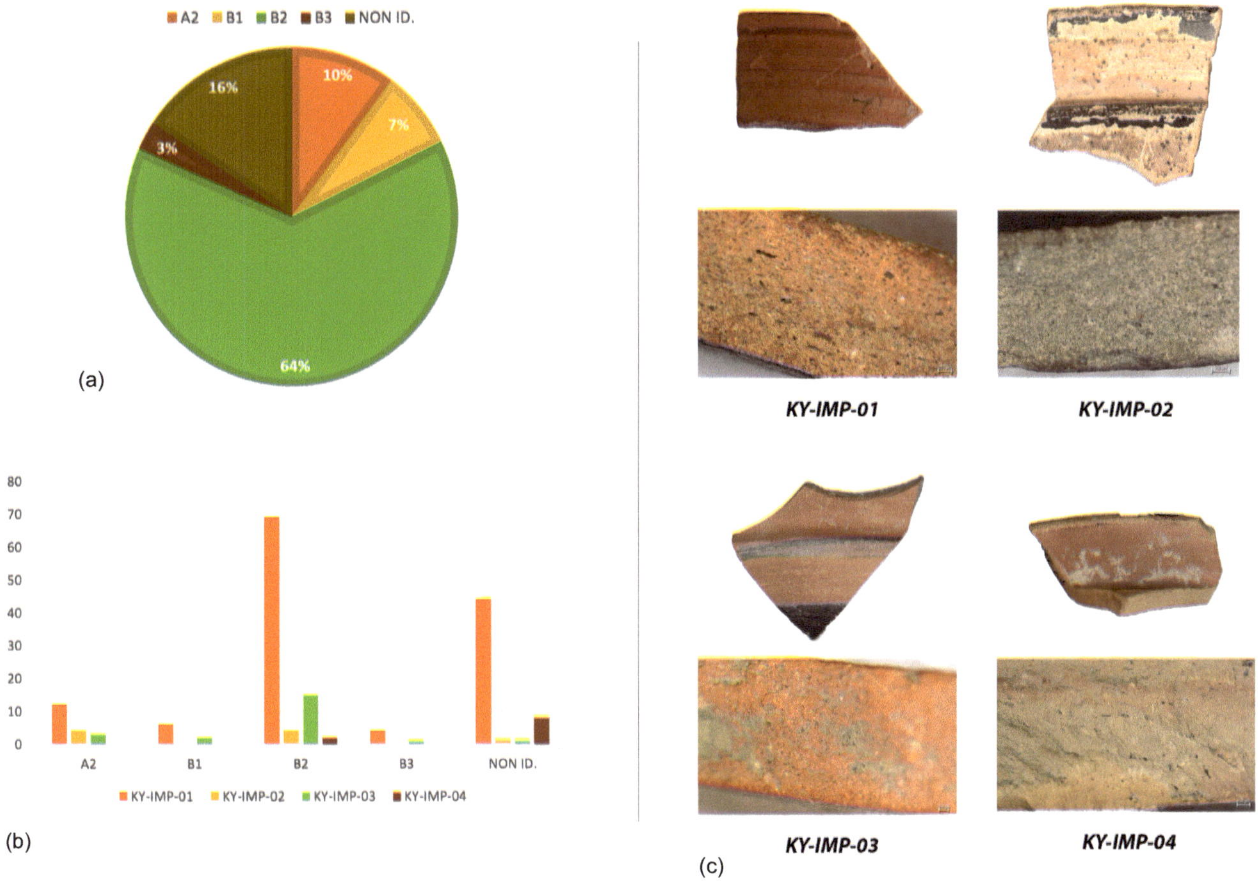

Fig. 2: (a) Pie chart showing Ionian cups' types; **(b)** Bar chart showing the distribution of the fabrics per type; **(c)** The four main fabric of Ionian cups. © Elisabetta di Virgilio.

The cups are significantly heterogeneous in both quality and profiles. Despite the very fragmentary state of the sherds, it is possible to outline morphological development among the identified types (Fig. 3).

13 FACEM is a database for pottery specialists that gives an overview of fabrics in the Central Mediterranean region: www.facem.at (last access 19.03.2024). See also Gassner in this volume.

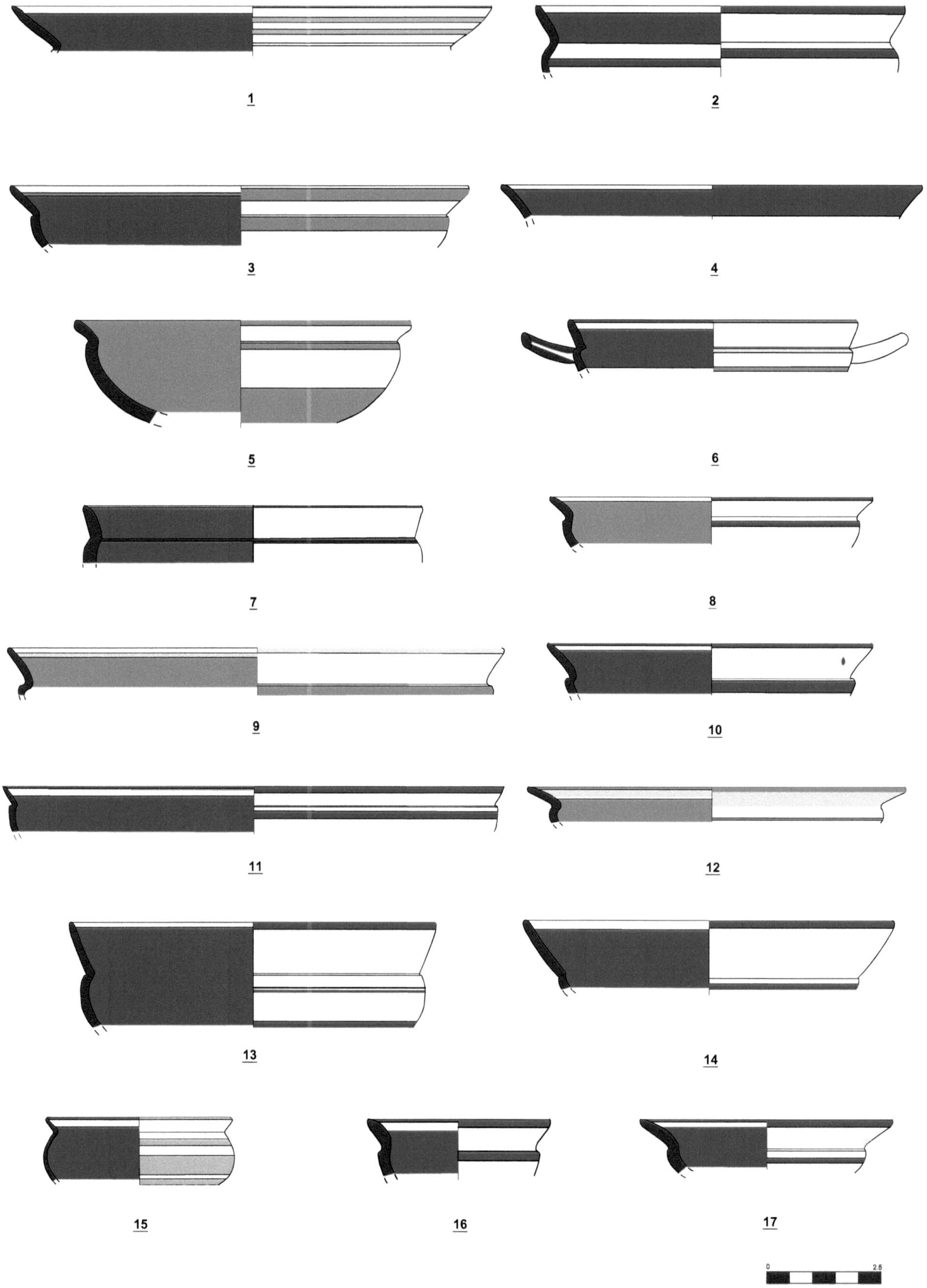

Fig. 3: Typology of Ionian cups found at Cumae. © Elisabetta di Virgilio.

Type A2 (620–550 B.C.) has a tall lip with convex profile, a pronounced shoulder attachment, and a short, conical foot. The decoration can be either the red-painted filetti on the external lip (Fig. 3.1) or painted bands on the rim and shoulder (Fig. 3.2, 3), with the inner lip slipped except for the reserved strip at the top.

Of Type B1 (600–550 B.C.) there are two variants. The first (Fig. 3.15) has a shorter, thinner lip, clearly distinct from the rounded body, and a smaller diameter (rim diam: 8–10 cm).[14] The other (Fig. 3.4), with a black-glazed lip on both surfaces, can be compared with an example already attested at Cumae, even though they differ in edge thickness.[15]

The formal repertoire of Type B2 cups (64% of the total) shows differences mainly in the lip's shape, whereas the banded decorative scheme remains unvaried. The diameter is generally between 11 and 22 cm.[16] The lip can be more or less flared (Fig. 3.5, 7, 12), quite thick, some with tapered edge. The shoulder offset, a specific trait of Campanian production, is common to only some examples (Fig. 3.8–11, 16, 17), whereas others have a distinct attachment without the offset (Fig. 3.5, 7). This type is dated between 570 and 520 B.C., whereas the samples with a grooved offset suggest a later chronology, up to the first years of the 5th century B.C.

Type B3 (560–550 B.C.), with high lip, slightly pronounced shoulder, deeper body, and stemmed foot (Fig. 3.13, 14), resembles the contemporary Attic Siana cups and can be compared with Hayes's Type X.[17]

Macroscopic Study of the Fabric

Four main groups have been identified through macroscopic observation (Fig. 2b–c):

- **KY-IMP-01**[18] is a wide group characterized by light orangish brown clay (5YR 6/5), granular fracture surface, and quite marked hardness. In hand examination, gold mica is visible both in the break and on the surface; some rounded whitish inclusions have been observed in the break. The glaze varies from orangish/reddish (2.5Y 4/5 – 10YR 2/1) to brownish and black (7.5YR 3/1 – 10YR 2/1). Some sherds show an orange slip on the external surface.
- **KY-IMP-02** is light pinkish brown (7.5YR 7/3), slightly compact and hard. Shiny mica, black rounded and white inclusions can be distinguished in the break. The quality is quite rough. The glaze is always grayish black (10YR 3/1), mainly opaque. These characteristics seem to match with those of the contemporary Fine Banded Ware from Cumae.[19]
- **KY-IMP-03**[20] is reddish brown, sometimes with a gray core (7.5YR 5/6 – 10YR 5/1), fine-grained, low in porosity, smooth in fracture. Inclusions are generally rare, small, and white.
- **KY-IMP-04** differs completely from the rest in clay and inclusions. It is yellowish light brown (7YR 8/4 – 10YR 8/2), homogeneous, with a smooth fracture surface and elongated, rounded voids. Some inclusions are visible in hand examination, such as quartz and white and reddish rounded minerals. It may be imported, compared also with other East Greek cups from the same contexts.

Discussion and Conclusion

The analysis of Ionian cups coming from three sacred areas at Cumae revealed a detailed picture of both the local production[21] and the exchange dynamics at the regional level along the Tyrrhenian routes. The general abundance of Type B2 is not surprising in colonial sanctuaries of southern Italy,[22] although we must emphasize the coexistence of hybrid models.

Typologically the local Ionian cups have well-outlined morphological features: the thick-walled and slightly flared lips with rounded edge, the not-concave profile, and the slight offset of the shoulder. Moreover, a signature of Cumaean

---

14 Comparison with a transition type between A2 and B1 coming from Paestum: Menard (1991) 737, fig. 47, no 53.

15 Tubelli (2006) 46, pl. 12.8. The shape resembles the *Panionion* type cups, of East Greek influence: Munzi (2007) 127, fig. 15.

16 A smaller cup from the temple of Apollo has a diameter of 8 cm (Fig. 3.16).

17 *Tocra I* 123–124.

18 Correspondence with the group 70 in Cuozzo *et al.* (2006) 13.

19 Correspondence with BNAP-IC-2 (FACEM).

20 Correspondence with PAE-G-3, VEL-G-2 (Glazed Ware from the region of Paestum) (FACEM).

21 I use the term 'local' to refer to an area within the Bay of Naples, as explained in Gassner/Traplicher (2012) 1–11.

22 The B2 type is the most widespread in various contexts from Southern Italy (Pithekoussai Velia, Paestum, Vico Equense).

production seems to be the presence on some samples of a little painted drop or circle on the external lip (Fig. 3.10, 17).[23] However, other formal traits, such as the more flaring, concave profile, recall the morphology of Paestan production,[24] detectable also in other contemporary materials.[25]

Analysis of technical features allowed us to attribute most samples to local craftmanship. Figure 2b shows a coexistence of the first three fabrics in all types, whereas the fourth, considered imported, appears with certainty only in Type B2. This may suggest increased importation beginning in the second quarter of the 6th century B.C., when local craft activity was already highly intensified, as is demonstrable from the high quality of the artifacts.[26] Technological signatures of the local cups may be the common application of a creamy or orangish slip and the sparkling surface deriving from the presence of shiny mica. However, macroscopic correspondence between the fabric KY-IMP-03 and the groups PAE-G/VEL-G (FACEM) may support a hypothesis of importing raw materials from the area of Poseidonia/Paestum, as already assumed for some cups coming from the Tempio con Portico at Cumae,[27] Such importation would help outline a framework of exchange along regional routes and between major production areas of Campania.

Abbreviation

FACEM Fabrics of the Central Mediterranean.
Tocra I John Boardman and John Walker Hayes, *Excavation at Tocra, 1963–1965: The Archaic Deposits I*, British School of Archaeology, Athens, 1966.

Bibliography

Albore Livadie (2012): Claude Albore Livadie, "Treglia (Comune di Pontelatone – Caserta): un impianto artigianale di epoca tardo-arcaica", in: Ugo Zannini (ed.), *Conference Proceedings Isti (Aurunci) graece Ausones nominantur* 10 (Sessa Aurunca, May 2009), Rome, 176–220.

Bossière/Frère (2007): Gérard Bossière and Dominique Frère, "Etude de la composition chimique des céramiques étrusques et campaniennes du Vie siècle. Distinction géographique et typologie chimique", in Dominique Frère (ed.), *Ceramiche fini a decoro sub-geometrico del VI sec. a. C. in Etruria e in Campania*, Collection de l'Ecole française de Rome 389, Rome, 176–220.

Caputo *et al.* (2016): Paolo Caputo, Cristina Regis and Carlo Rescigno, "Cuma. Il santuario nell'area dell'anfiteatro", in: *Notizie degli Scavi di Antichità*, 141–215.

Cuozzo *et al.* (2006): Mariassunta Cuozzo, Bruno d'Agostino and Laura Del Verme (eds.), *Cuma. Le fortificazioni 2. I materiali dai terrapieni arcaici*, Naples.

Franco (2010): Marianna Franco, "Coppe ioniche", in: Juliette De La Genière and Giovanna Greco (eds.), *Il santuario di Hera alla foce del Sele. Indagini e Studi 1987–2006*, Atti e Memorie della Società Magna Grecia, serie IV, vol. IV (2008–2010), Rome, 418–429.

Gassner (2003): Verena Gassner, *Velia Studien 2. Materielle Kultur und kulturelle Identität Eleas in spätarchaisch-frühklassischer Zeit. Untersuchungen zur Gefâss- und Baukeramik aus der Unterstadt (Grabungen 1987–1993)*, Vienna.

Gassner/Trapichler (2012): Verena Gassner and Maria Trapichler, "Pottery Production in the Bay of Naples. Problems, History of Research and Current Strategies", in: *FACEM* (version 06/12/2012), http://www.facem.at/project-papers.php (last access 19.03.2024).

Greco *et al.* (2012): Giovanna Greco, Bianca Ferrara and Antonella Tomeo, "L'area sud-occidentale del Foro di Cuma: le produzioni ceramiche", in: *FACEM* (version 06/12/2012), http://www.facem.at/project-papers.php (last access 19.03.2024).

La Rocca *et al.* (1995): Luigi La Rocca, Carlo Rescigno and Gianluca Sorricelli, "Cuma: l'edificio sacro di Fondo Valentino", in: *Studi sulla Campania preromana* III.2, 51–79.

Menard (1991): Theresa D. Menard, *The Archaic and Classical Pottery from the Sanctuary at località Santa Venera in Paestum, Italy*, Ann Arbor.

Morra *et al.* (2014): Vincenzo Morra, Alberto De Bonis and Celestino Grifa, "Geological Outlines of Studied Archaeological Sites", in: Giovanna Greco and Luigi Cicala (eds.), *Archaeometry. Comparing Experiences*, Naples, 29–36.

Munzi (2007): Priscilla Munzi, "Un contesto arcaico da Cuma. Le ceramiche decorate, non figurate, di produzione coloniale", in: Dominique Frère (ed.), *Ceramiche fini a decoro subgeometrico del VI sec. secolo a.C. in Etruria meridionale e in Campania*, Rome, 109–130.

23 A trait recognized also among the Archaic pottery found at Cumae: Munzi (2007) 126, fig. 15; Tubelli (2006) 43–51, pl. 12.23.
24 B2 Ionian cups from the Heraion at the Sele plain are an example: Franco (2010) 427, fig. 19.
25 A banded cup-skyphos from the area of Fondo Valentino, characterized by a very concave profile and a clear shoulder line, is not attested at Cumae, while it finds comparison with the *Schalenskyphos* 3 found at Velia, considered a Tyrrhenian variant: Gassner (2003) 48–50, Ib.14-Ic.4.
26 Contemporary imports from Attica and Eastern Mediterranean are widespread among the other materials.
27 Greco *et al.* (2012) 5.

Munzi *et al.* (2012): Priscilla Munzi, Vincenza Guarino, Alberto De Bonis, Celestino Grifa, Alessio Langella and Vincenzo Morra, "Le ceramiche ellenistiche del santuario periurbano settentrionale di Cuma", in: *FACEM* (version 06/12/2012), http://www.facem.at/project-papers.php (last access 19.03.2024).

Rescigno (2021): Carlo Rescigno, "Acropoli di Cuma. Nuove acquisizioni dalle ricerche condotte dall'Università degli Studi della Campania Luigi Vanvitelli. Estate 2019", in: *Puteoli, Cumae, Misenum. Rivista di Studi e Notiziario del Parco archeologico dei Campi Flegrei 1*, Pozzuoli, 195–203.

Rescigno/Parisi (2022): Carlo Rescigno and Valeria Parisi, *La colomba di Apollo. La fondazione di Cuma e il ruolo del culto apollineo nella colonizzazione euboica d'Occidente* (Atti del Convegno Internazionale, Santa Maria Capua Vetere, 16 novembre 2020), Pozzuoli.

Tubelli (2006): Amelia Tubelli, "Coppe ioniche", in: Mariassunta Cuozzo, Bruno d'Agostino and Laura Del Verme (eds.), *Cuma. Le fortificazioni 2. I materiali dai terrapieni arcaici*, Naples, 43–51.

Van Compernolle (1996): Thierry Van Compernolle, "Le produzioni ceramiche arcaiche. Coppe di tipo ionico", in: Enzo Lippolis (ed.), *I Greci in Occidente. Arte e artigianato in Magna Grecia*, Naples, 298–302.

Van Compernolle (2000): Thierry Van Compernolle, "Les céramiques ioniennes en Méditerranée centrale', in: Paloma Cabrera Bonet and Marta Santos Retolaza (eds.), *Ceràmiques jònies d'època arcaica: centres de producciò i coercialitzaciò al Mediterrani Occidental* (Actes de la taula rodona, Empùries, May 28-29, 1999), Monografies Emporitanes 11, Barcelona, 89–100.

Villard/Vallet (1955): François Villard and George Vallet, "Megara Hyblaea V: "Lampes de VIIe Siècle et Chronologie des Coupes Ioniennes", in: *Mélanges de l'École française de Rome* 67, 5–32.

Antoine Attout

30 East Greek Inspirations on Attic Black-figured Pottery: The Implications for Clients and Trade Networks

Abstract: In this chapter, we examine the distribution of selected Attic black-figured vases with elements inspired by eastern Greek ceramic traditions. Through four case studies, we show that those vases may have had very different distribution patterns, suggesting different distribution networks and clients. Naturally, this situation underlines the driving role that different networks of Attic ceramic customers and intermediaries may have played in the dynamics of inspiration that we are studying.

Introduction

Some fifty years ago, D. A. Jackson published the first, unique, in-depth study of Attic figured vases with eastern Greek inspiration.[1] Despite the great precision of Jackson's work, the subject now deserves a new, detailed analysis with updated data and innovative approaches. For instance, the distribution of the Attic vases concerned has hardly been examined. However, this approach could address the role of networks and customers in the complex phenomenon of the vases' distribution. Bearing this in mind, we hope to show by means of four specific cases that Attic black-figured vases with Ionian-inspired elements may have had very different distribution patterns, suggesting different distribution networks and clients.

Droop Cups with Special Ornaments

The first particular case concerns a couple of Droop cups recently discovered at Zeytintepe (near Miletus) with typical Ionian-inspired ornaments. The first cup (Fig. 1a) bears an uncommon succession of twin volutes with small lozenges or lotuses in the spandrels adorning the zone between the handles.[2] Interestingly enough, the cup also bears the signature of Amasis. Though N. Kunisch noticed that comparable features appear on a few other works by the Amasis Painter,[3] this pattern remains unparalleled in Attica. In contrast, numerous southern Ionian vases from the 6th century B.C. bear similar volutes. Such ornaments often appear on friezes located on the shoulders of Milesian amphorae,[4] or in a slightly different arrangement, on Milesian cups.[5] Here again, a series of miniature lozenges, buds, and lotus flowers often fills the spandrels between the volutes. Moreover, the existence of similar ornaments on other, older Milesian productions[6] clearly indicates that this kind of ornament belongs to Milesian tradition from the end of the 7th century B.C.

Two other special Droop cups,[7] also attributed to Amasis's workshop, do not bear so large a frieze of volutes, but the handle palmettes integrate a small complex of palmettes and volutes of a similar design (Fig. 1b). A fourth fragmentary cup bears a small palmette attached to the handle by a wavy stem. A little lozenge points to the base of the palmette in keeping with Ionian ornamental traditions.

All four special Droop cups were discovered in the sanctuary of Zeytintepe, near Miletus. Since so far we have not found another Attic Droop cup with this type of Ionian-inspired decoration, it seems that these peculiar cups were specially intended for southern Ionian clients or at least for people passing by Milesian sanctuaries. One of these cups bears a dedicatory inscription, most probably written by an Ionian dedicator named Protis.[8] We have noted elsewhere that the

1 Jackson (1976).
2 Kunisch (2016) cat. no. 980.
3 Kunisch (2016) 46.
4 E.g., Walter-Karydi (1973) cat. nos. 107, 613, 683 and 144.
5 E.g., Walter-Karydi (1973) cat. nos. 108 and 335.
6 E.g., Kerschner/Schlotzhauer (2005) cat. nos. 18 and 53.
7 Kunisch (2016) cat. nos. 981–982.
8 Kunisch (2016) 39.

https://doi.org/10.1515/9783111189635-030

(a) (b)

Fig. 1: Two Attic fragmentary Droop cups from Zeytintepe with Ionian-inspired decoration: **(a)** After Kunisch (2016) pl. 79, cat. no. 980; **(b)** After Kunisch (2016) pl. 88, cat. no. 981.

shape of these cups was probably intended to appeal to Ionian dedicators, since Droop cups were in fact highly prized in the southern Ionian sanctuaries—perhaps because this shape was derived from a type of Laconian cup, and Laconian pottery had long been imported into this context.[9]

Whatever the case, these Droop cups were very probably involved in a market-targeting strategy or a production commissioned by the dedicators of the southern Ionian sanctuaries. Such a dynamic echoes a small series of special cups attributed to the KX Painter and found at the Samian Heraion, whose shape and decoration also betray clear Ionian inspiration.[10]

The Shouldered Neck Amphorae

Attic shouldered neck amphorae also bear Ionian-inspired elements, but they show a completely different pattern of distribution. These amphorae appeared in Attic black-figure suddenly in the mid-6th century, being characterized mainly by a prominent shoulder and, in terms of decoration, by large complexes of volutes and palmettes around the handles.[11] D. A. Jackson showed that these features echo contemporary Ionian amphorae (see Fikellura-style amphorae) rooted in ancient Ionian traditions.[12] We now know that the Ionian amphorae in question came mainly from Milesian workshops[13] and that their earliest examples predate the first Attic shouldered neck amphorae.[14]

As M. Bentz[15] has shown (Fig. 2), these amphorae are rare around the Aegean, the overwhelming majority coming from southern Etruria (around 70% of the corpus, around 700 amphorae), particularly Vulci. Similar amphorae found in Athens are almost twenty times fewer than in Etruria (3% of the corpus, around 30 amphorae). Small concentrations of these amphorae are also found in Rhodes, Samos, and Miletus. The same diffusion pattern applies to the first generation of these amphorae (i.e., the Group E Amphorae, the Botkin Class, and the amphorae from the workshops of Exekias and Amasis). A few examples have been found in Samos and Miletus,[16] but the chief examples of these amphorae are mainly concentrated in Etruria.

9 Attout (forthcoming).
10 Brijder (1997) 1–5; Kreuzer (1998) cat. no. 206; Pipili (2000) 415.
11 Moore/Pease Philippides (1986) 11–12.
12 Jackson (1976) 13–36.
13 Cook/Dupont (2003) 77.
14 Kerschner/Schlotzhauer (2005).
15 Bentz (2009) 85–86; de La Genière (1987) 204–207.
16 Kreuzer (1998) cat. no. 27; Kunisch (2016) cat. nos. 9–10.

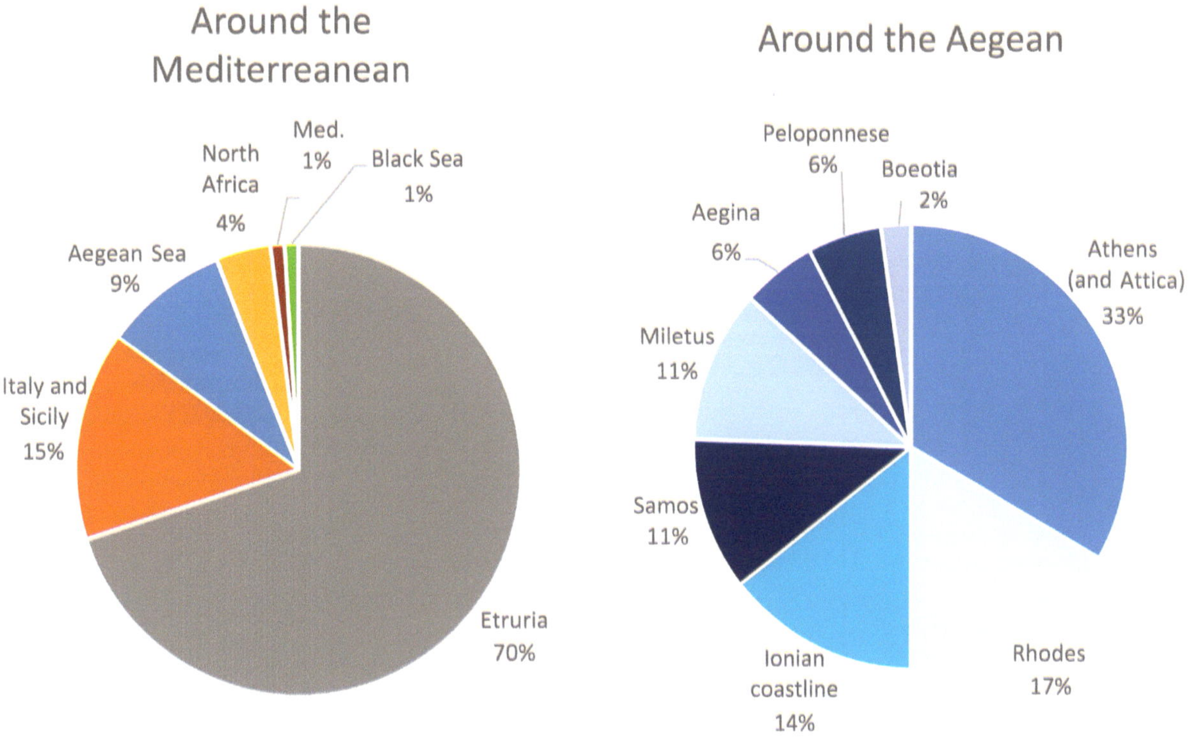

Fig. 2: Approximate distribution of the Attic standard amphora.

We may well wonder whether this situation is not simply due to the fact that the Etruscan necropoleis yielded very large quantities of Attic black-figured pottery and consequently an inordinate number of necked amphorae. In our opinion, this bias alone cannot explain such an imbalance. A number of authors have sought to identify the reasons for the success of shouldered neck amphorae in Etruria. S. Bundrick and J. de La Genière have pointed out that some shouldered neck amphorae had been used as cinerary urns and that their shape was perhaps reminiscent of older Etruscan urns.[17] For our part, we note that the necropoleis of southern Etruria have also yielded few Milesian amphorae produced between the second quarter and the end of the 6th century B.C.[18] These Attic amphorae may then reflect a desire to supply the Etruscan market with a product similar to Milesian amphorae but perhaps more easily available.

However, there is no reason to believe that Attic shouldered amphorae were produced specifically for Etruria. As already mentioned, several concentrations of these Attic amphorae also appear in southern Ionia and Rhodes. In these regions too, Attic shouldered neck amphorae may have served as an alternative to Milesian amphorae. The situation is particularly clear in the Rhodian necropoleis, where Attic amphorae were regularly used as cinerary urns much the same as southern Ionian amphorae.[19] In this respect, an Attic shouldered neck amphora whose body surprisingly was decorated with three simple black bands as in southern Ionian banded amphorae[20] (Fig. 3a) clearly illustrates the parallel between the two Attic and Ionian series (Fig. 3b).[21]

17 Bundrick (2019) 167–168; de La Genière (1987) 204–207.

18 Boldrini (1994) 105–114; Martelli Cristofani (1978) 184–188, 191–192; Pierro (1983) 103–107.

19 de La Genière (2010) 36–37.

20 Cook/Dupont (2003) 132–134.

21 *CVA* Greece 10 – Anna A. Lemos, *Rhodes, Archaeological Museum*, pl. 41.1.

Fig. 3: (a) Attic shouldered neck amphora with banded decoration Kamiros [Rhodes, Archaeological Museum, inv. no. 15643; drawn after *CVA* Greece 10, fig. 21]; **(b)** South-Ionian banded amphora [Munich, Antikensammlung, inv. no. 463; drawn after Cook/Dupont (2003) 133, fig. 19.1d].

Finally, unlike the droop cups mentioned above, Attic shouldered neck amphorae seem not to have been specifically intended for a southern Ionian clientele: they were much more widely distributed and seem to offer, at least initially, an alternative to contemporary Ionian amphorae in various regions such as Rhodes and southern Etruria.

Attic Chalices

Among the Attic black-figured vessels, a small group of chalices[22] directly derives from typical Chian chalices.[23] Attic versions of this shape are very rare and mainly found in Athens and Attica. Outside Attica, one chalix has been discovered in Gravisca,[24] and the Museum University College in Dublin holds another example reportedly from Naucratis.[25]

Even if the number of Attic chalices is small, the distribution pattern of these vases is significant: Naucratis and Gravisca were Ionian trading posts that according to the dedications on the Greek pottery from these sites seem to have functioned as nodes for a commercial network mostly frequented by Ionian merchants.[26]

22 Moore/Pease Philippides (1986) 57–58.
23 Lemos (1991) 4.
24 Iacobazzi (2004) cat. no. 527.
25 *CVA* Ireland 1 – Alan W. Johnston and Christina Souyoudzoglou-Haywood, *University College Dublin. University College Cork*, Dublin (2000) pl. 19.6.
26 Demetriou (2012) 77–83, with previous references. See also Krämer (2016) 82–84; Villing (2013) 83–86.

Bearing this in mind, it is worth pointing out that several cups from Amasis's workshop also have been found at Gravisca.[27] One of them[28] has an ornament under the handles that recalls the Milesian-inspired handle ornaments of the Attic shouldered neck amphorae. It is interesting that another of these cups,[29] attributed to the Amasis's workshop, also bears a dedicatory inscription on the lip, written in the Ionian dialect by a certain Hyblesios—this apparently being an Ionian name also found among Naucratis dedications.[30]

These Attic chalices suggest, then, that some Attic black-figured vases bearing Ionian-inspired elements were intended for a clientele of merchants or travelers crisscrossing the Mediterranean along a maritime network between important Ionian trade nodes.

Phialai in Polychrome Six Technique

Before concluding, we would like to briefly point out that the distribution of phialai decorated with the Polychome Six technique shows, once again, a very different picture. These phialai recall Chian phialai produced in the first half of the 6th century B.C.[31] whose inside is decorated with colorful vegetal motifs on a black background. As A. Tsingarida already showed,[32] most of these phialai are especially concentrated in Athens, mainly on the Acropolis. Other examples are distributed in several sanctuaries of Attica and around the Aegean to a lesser extent. This distribution clearly contrasts with the other series under consideration: these phialai seem to meet a very different kind of demand, one involving an important local clientele.

Conclusions

Finally, the diffusional analysis of the series under consideration suggests that the introduction of eastern Greek–inspired elements in Attic black-figured pottery is a complex phenomenon involving different series of vessels with different kinds of inspiration, responding to different distribution patterns and matching different demands around the Mediterranean. In particular, some of these vessels seem to have been involved in a target-marketing strategy aimed at a clientele associated with southern Ionian sanctuaries, whereas others seem to have been related to a wide network of trade routes crisscrossing the Mediterranean and frequented especially by Ionian merchants.

Abbreviation

CVA *Corpus Vasorum Antiquorum.*

Bibliography

Attout (forthcoming): Antoine Attout, "Triangle amoureux entre Athènes, Milet et Sparte. Références intriquées autour des coupes de Droop de l'atelier d'Amasis", in: Anastasia Paillard and Mitchka Shahryari (eds.), *Réemploi, Réutilisation et Référence dans les sociétés anciennes* (Workshop Proceedings, Lille, 5 February 2021), Lille, forthcoming.

Bentz (2009): Martin Bentz, "Maße, Form und Funktion: Die attisch-schwarzfigurigen Halsamphoren", in: Athena Tsingarida and Laurent Bavay (eds.), *Shapes and Uses of Greek Vases (7th–4th centuries B.C.)* (Symposium Proceedings, Brussels, April 27–29, 2006), Brussels, 79–87.

27 Iacobazzi (2004) 56–67.
28 Iacobazzi (2004) cat. nos. 69–70.
29 Iacobazzi (2004) cat. no. 58.
30 Demetriou (2012) 83, with previous references.
31 Lemos (1991) 85.
32 Tsingarida (2008) 191–192.

Boldrini (1994): Sabrina Boldrini, *Gravisca. Le ceramiche ioniche*, Bari.
Brijder (1997): Herman A. G. Brijder, “New Light on the Earliest Attic Black-Figure Drinking-Cups”, in: John H. Oakley, Williams D.E. Coulson and Olga Palagia (eds.), *Athenian Potters and Painters* I, Oxford, 1–15.
Bundrick (2019): Sheramy D. Bundrick, *Athens, Etruria, and the Many Lives of Greek Figured Pottery*, Madison.
Cook/Dupont (2003): Robert M. Cook and Pierre Dupont, *East Greek Pottery*, London.
de La Genière (1987): Juliette de La Genière, “Rituali funerari e produzione di vasi”, in: Maria Bonghi Jovino and Cristina Chiaramonte Treré (eds.), *Tarquinia: Ricerche, scavi e prospettive* (Conference Proceedings, Milano, 24–25 Giugno 1986), Rome, 203–208.
de La Genière (2010): Juliette de La Genière, “Barbares et grecs, des vases attiques pour les morts”, in: Filippo Giudice and Rosalba Panvini (eds.), *Il greco, il Barbaro e la ceramica Attica. Immaginario del diverso, processi di scambio e autorappresentazione degli* indigeni (Atti del convegno, Catania-Caltanissetta-Gela-Camarina-Vittoria-Siracusa, 14–19 Maggio 2001), Rome, 101–128.
Demetriou (2012): Denise Demetriou, *Negotiating Identity in the Ancient Mediterranean: The Archaic and Classical Greek Multiethnic Emporia*, Cambridge.
Iacobazzi (2004): Barbara Iacobazzi, *Gravisca. Le ceramiche attiche a figure nere*, Bari.
Jackson (1976): Donald A. Jackson, *East Greek Influence on Attic Vases*, London.
Kerschner/Schlotzhauer (2005): Michael Kerschner and Udo Schlotzhauer, “A New Classification System for East Greek Pottery”, in: *Ancient West and East*, 4, 1–56.
Krämer (2016): Robinson P. Krämer, “Trading Goods – Trading Gods. Greek Sanctuaries in the Mediterranean and their Role as Emporia and “Ports of Trade” (7th–6th Century BCE)”, in: *Distant Worlds Journal* 1, 75–98.
Kreuzer (1998): Bettina Kreuzer, *Samos. Die attisch schwarzfigurige Keramik aus dem Heraion von Samos*, Bonn.
Kunisch (2016): Norbert Kunisch, *Milet. Die Attische Importkeramik*, Berlin.
Lemos (1991): Anna A. Lemos, *Archaic Pottery of Chios: The Decorated Styles*, Oxford.
Martelli Cristofani (1978): Marina Martelli Cristofani, “La ceramica greco-orientale in Etruria”, in: *Les céramiques de la Grèce de l’Est et leur diffusion en Occident* (Colloquium Proceedings, Naples, July 6–9, 1976), Naples, 150–212.
Moore/Pease Philippides (1986): Mary B. Moore and Mary Z. Pease Philippides, *The Athenian Agora* XXIII. *Attic Black-Figured Pottery*, Princeton N.J.
Pierro (1983): Elena Pierro, *Ceramica “ionica” non figurata e coppe attiche a figure nere nel Museo di Tarquinia*, Rome.
Pipili (2000): Maria Pipili, “Vases from the Samian Heraion. Shapes and Iconography”, in: *Αγαθός Δαίμων. Mythes et cultes. Études d’iconographie en l’honneur de Lilly Kahil*, Bulletin de correspondance hellenique, Supplément 38 Athens, 409–421.
Tsingarida (2008): Athena Tsingarida, “Color for a Market? Special Techniques and Distribution Patterns in Late Archaic and Early Classical Greece”, in: Kenneth D. S. Lapatin (ed.), *Papers on Special Techniques in Athenian Vases* (International Symposium Proceedings in connection with the Exhibition *The Colors of Clay: Special Techniques in Athenian Vases*, Malibu, the Getty Villa, June 15–17, 2006), Los Angeles, 187–206.
Villing (2013): Alexandra Villing, “Trade and Traders: Value, Transport and Places of Exchange”, in: Athena Tsingarida and Didier Viviers (eds.), *Pottery Markets in the Ancient Greek World: 8th-1st centuries B.C.* (Proceedings of the International Symposium, Bruxelles, 19–21 June 2011), Brussels, 73–101.
Walter-Karydi (1973): Elena Walter-Karydi, *Samische Gefässe des 6. Jahrhunderts v. Chr. Landschaftsstile. Ostgriechischer Gefässe*, Bonn.

Epilogue

Eleni Hasaki, Marco Serino and Diego Elia

31 Greek and Roman Ceramic Producers: Operational Knowledge and Networked Mobilities

As the contributions in this volume clearly demonstrate, the study of Ancient Greek and Roman ceramics continues to flourish, improving our understanding of the workshop crew (potters and painters), and their work in all sectors (roof tiles, terracottas, plain and figured pottery) and periods (Archaic/Orientalizing, Classical, Hellenistic, and Roman) of the ceramic industry in various Mediterranean regions, including Attica, Laconia, Corinthia, the Cycladic islands, Magna Graecia and Sicily. Inspired by Social Network Analysis, we make an attempt to visualize below as a knowledge graph the various aspects of ancient life that ceramics shed light on. In this way we can better appreciate how the materiality of ceramics helps understand ancient communities and their technological, social, and economic perspectives, from workshop-level to intraregional and overseas levels (Fig. 1). The 29 papers in this volume represent two thematic horizons: a) how workshops developed a highly sophisticated operational knowledge, and b) how they, along with distributors and customers, maintained equally sophisticated networks for trading their wares, and for shaping and responding to customers' preferences at home, away from home, and while traveling.

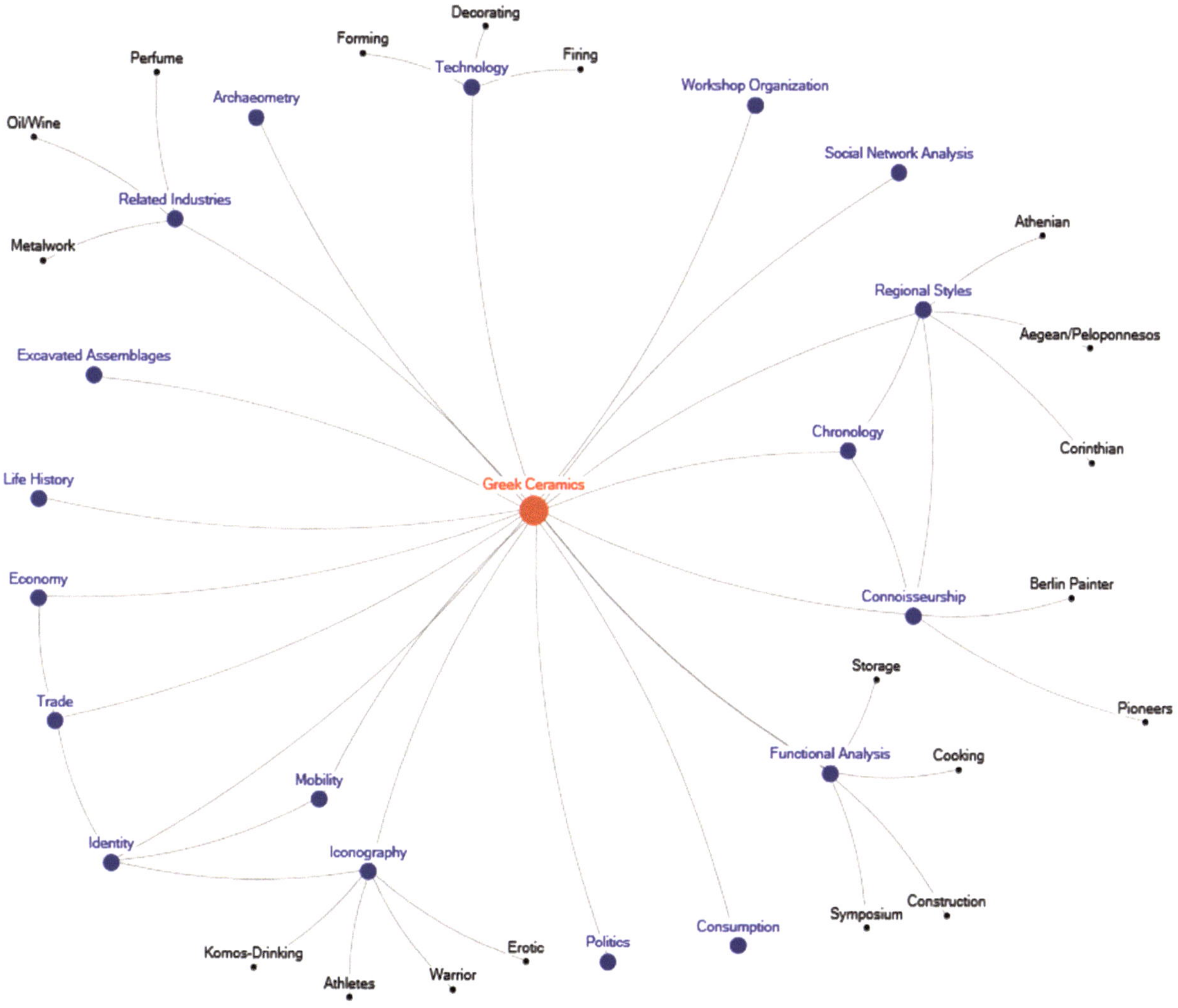

Fig. 1: Knowledge graph of select themes connected with the study of Greek ceramics. The nodes in black are listed only as examples. © Eleni Hasaki.

https://doi.org/10.1515/9783111189635-031

Operational Knowledge of Ceramic Production

The requisite knowledge underlying the *chaîne opératoire* in ancient pottery workshops was immense. All stages were of paramount importance: the tilemakers, coroplasts, vase potters and painters must have developed quality control tests for different stages including shaping, modeling, painting, drying, firing, and even repairing their artifacts. Papers in this volume highlight some of the most challenging aspects of the manufacturing process: slight changes in the colors of raw clays that become evident during the drying process (De Natale *et al.* through the case study of the Teanum Ware); changes in weight and size both during the drying and the firing stage; fingerprints on rooftiles (Giuliano); and the appropriate method for assembling in terracotta figurines and models to ensure that the ceramic object had undergone appropriate drying (Meirano). Any miscalculations, whether due to impatience or to inexperience, would have incurred the wrath of Omodamos (Conqueror of the Unbaked) in the kiln.[1]

The shelves and rows for drying vases in the pottery workshops at Pompeii underscore the importance of this phase (Cavassa *et al.*). It is interesting to consider whether workshops developed specific types of drying supports, perhaps similar to those for stacking in the kiln to ensure that all areas of the vases were adequately and evenly dried. The diversity of kiln supports – from Metapontum, Locri, Syracuse, and now Selinous – show how critical it was to stack the kiln with the appropriate spacers and supports. Tools for firing like tools that would have been used inside an ancient workshop are rarely recovered during excavation. The ones from the Kerameikos of the Greek colony of Selinous (Adorno) serve as bridges (or 'edges' in network terms) between artifacts and the hands of ancient craftspeople. From fingerprints to tools to drying shelves, to volumes of production we can better approximate the scale of production of vase-painting workshops both in Western Greece and in Athens (Sapirstein), one of the largest mainland Greek ceramic industries in the Archaic and Classical Mediterranean.

The production lines of pottery workshops may not correlate to the strict scholarly and strict divisions of slipped (glossed) and figured pottery. The workshops of slipped pottery may have diversified their production to attract customers from all economic strata and to satisfy their differentiated needs. The glossed equivalents of many figured shapes in the Athenian Agora reveal a blossoming industry of black-gloss ceramics (Lynch). Improving the composition of the slip and its application (through better tools and techniques) on those pots in turn led to advances in the decoration process of figured pottery, as the slip still constituted the primary background of the exterior (and interior) surfaces for the figural compositions. As some efforts are already under way to identify the potters who produced both slipped and figured pottery (Lynch, Silvestrelli), the increasing use of profile drawings in publications of figured pottery can help detect the potters.

Many papers in this volume deal with well-known artisanal personalities responsible for potting and painting in their respective corpora: Euphronios and Euxitheos (Hasaki); ΣΩΣΗΝΟΣ (Giuliano/Montana/Portale); Pisticci, Amykos, Dolon, Creusa and Anabates Painters in Metapontum (Silvestrelli); Locri Group and other local artisans from Locri Epizephyirii (Elia *et al.*); Himera, Santapaola, Chequer, and Dirce Painters from Sicily, Painter of the Geneva Orestes, and Assteas from Paestum (Serino). But the distinguishable 'hands' tell only part of the story and detailed reconstructions of the broader workforce in the Athenian Kerameikos have started shaping the sector(s) of the decorated pottery, not just those who dominated the ancient markets and modern scholarship. The case of Nikosthenes and his wider circle of collaborators (Sapirstein) represents a real and symbolic bridge between the Greek and Roman spheres of ceramic production and overseas marketing.

Layers of Technological Evidence: Between Materiality and Intangible Timetables

In the case of figured pottery, the advanced protocols of archaeometry (Mangone *et al.*) and diagnostic imaging with digital microscope have magnified (literally and metaphorically) the complexity of building up the figural composition, and how painters carefully planned the application of successive layers of treatment, whether this was the preliminary sketch, the background slip, the *miltos*, or the added colors (Elia *et al.*, Portale/Chirco, Serino). Thanks to experimental archaeology we can appreciate the experience of such layering, especially when the pre-firing coloring effects created much less contrast prior to firing (Hasaki, Serino). Such technical 'stratigraphies' lengthen and deepen the knowledge involved in the *chaîne*

1 Hasaki (2021), (2024).

opératoire, especially when one adds the important element of time, and the intervals between all stages and micro-stages of craft manufacturing. The dimension of time needs to be further emphasized, not only the broader time commitment to acquire and master the knowledge of each manufacturing stage, but also the more specific, workshop/based, timetable of organizing the workflow.[2] The sequence of layering could point to specific workshops; scholars have recently used this information to revisit older attributions (Portale/Chirco) or to better define the technological features of a workshop and their affinities, or even the different procedures adopted by different workshops that can help us better identify the various production traditions.

The operational knowledge becomes even more complex when a workshop sets out to create 'imitations' of a non-local shape. Western Greek ateliers imitated amphoras from mainland Greece, Cumaean workshops developed several lines of Ionian-inspired shapes, from chalices to neck amphoras, and Locrian workshops produced variants inspired both from Athens and from their neighboring potting cultures in Magna Graecia (Ercolin *et al.*). Several ceramic '*koinai*' had been developed and it is interesting to consider these local variations as constant reinforcers of their original cultural identity as pots and people moved around the Mediterranean. In terms of local production and local imitation of other productions, the case of Cuma in Magna Graecia and its Ionian cups, as well as that of Fregellae in Southern Lazio (Borgers/Diosono), can clearly demonstrate how the world outside the workshops is much more complicated and intricate, and that the dynamics of movements (not necessarily of people, but also of knowledge or even specific technological, morphological or aesthetic features) can be extremely fluid and eclectic (Di Virgilio). A similar vein can be seen in the case study of East Greek influences on Attic black-figured pottery, which also involves the analysis of clients and trade networks (Attout).

Archaeologists have developed strong interdisciplinary research agendas to recapture the complexity of that operational knowledge. First, archaeometry and computational imaging have been successful in extracting valuable information from the objects themselves and help us reconstruct the choices of ancient potters, coroplasts, and tile makers (D'Acri, Elia *et al.*, Ercolin *et al.*, Giuliano *et al.*, Mangone *et al.*, Serino). The archaeometric analyses through the elemental ratios in clay matrices have decoded what M. Rockman has termed "landscape learning" highlighting the agency of local communities of practice in understanding deeply their environmental resources "their fluctuations, their potentials, and their carrying capacity."[3] While this was previously discussed within the framework of ceramic ecology, Rockman's term allows for a wider applicability to several crafts.

Second, ethnoarchaeology and experimental archaeology have also been successfully integrated in research agendas. A rich but largely underutilized source of knowledge is the active community of modern artisans in several Mediterranean areas involved in the reproduction of ancient vases. Many of these artisans are involved in long-standing research and outreach collaborations with academic institutions and museums.[4] They often receive commissions by museums to produce replicas for educational programming. These replicas generate great momentum: they help the artisans expand their skills, enable museums to offer tactile experiences to their visitors, and create opportunities for curators and archaeologists to develop interdisciplinary research projects for museum collections.[5] Experimental archaeology can really help us recognize various levels of technological knowledge, and thereby assisting archaeologists' initial assessments of excavated materials. Archaeometry-based 'hard science' must be accompanied by handicraft-based wisdom and experience: the combination of the two approaches can truly advance our research (Cavassa *et al.*, Hasaki, Serino), as recent experiences with Cerveteri-based Roberto Paolini can well demonstrate.

Within the framework of the A.G.A.T.H.O.C.L.E.S project R. Paolini completed guest artist residencies at the Laboratory for Traditional Technology of the University of Arizona and at the Officina della ricerca e della Didattica of the University of

2 See the project *"Energetics of Potters and Painters in Ancient Greece"* discussed by E. Hasaki in this volume.

3 Rockman (2003) 5. Also, Rockman (2012). This term can be useful for studying a variety of Mediterranean crafts and it has been successfully applied to the esparto basketry craft for the imperial Roman mining industry by Gosner (2021). Our warm thanks to L. Gosner for introducing us to this concept.

4 Drawn from our experiences, we highlight Thanassis Katsaras from Eleusis who developed several educational programs for academic audiences and general public for the celebration of his hometown as Cultural Capital of Europe and created unique replicas of Protoattic amphoras at the Eleusis Museum. Dima Romanciuc and Maria Skliri-Starfa along with potter Dimitris Boutsalis held live demonstrations numerous times in the International Museum Day at the Archaeological Museum of Ancient Corinth bridging ancient and contemporary practices. As part of the dissemination strategy of the A.G.A.T.H.O.C.L.E.S. project, Roberto Paolini participated both in the European Research Night (2023 Edition) in Turin, and at a conference in Syracuse, at the Museo Archeologico Regionale "Paolo Orsi", with two different 'live painting' sessions.

5 As an example, A.G.A.T.H.O.C.L.E.S. collaborator Roberto Paolini has created commissioned vases like the Pyrgi Phiale (Rome, Villa Giulia Museum, inv. no. S13B-V6.; by the Brygos Painter; *BAPD* 23670), the Sarpedon krater (Cerveteri, Archaeological Museum, inv. no. 145139; by Euxitheos and Euphronios), for the Metropolitan Museum of Art and others. We thank Professor Guilia Rocco for further information on the Pyrgi phiale.

Turin in Spring and Fall 2023. He worked in tandem with archaeologists in both universities to decorate and fire a group of shapes in 1:1 scale from Corinth and Athens in black-figured, red-figured, and bilingual techniques and three South Italian red-figured vases (calyx krater, skyphoi, and lekane). The research questions included the time sequences of the manufacturing stages and the different gestures and types of coatings, including *miltos* in red-figured production (Fig. 2).[6]

Fig. 2: Roberto Paolini at work: **(a)** University of Arizona News on the workshop held at the University of Arizona with Roberto Paolini working on a South Italian krater (A.G.A.T.H.O.C.L.E.S. project). Photo: Kyle Mittan; **(b)** Selection of Athenian and Corinthian shapes made at the LTT (*Energetics of Potters and Painters in Ancient Greece project*). Photo: Eleni Hasaki; **(c)** Roberto Paolini working on a South Italian lekanis' lid at the Officina della didattica e della ricerca of the Department of Historical Studies, University of Turin; **(d)** European Research Night (Edition 2023) with live vase-painting by Roberto Paolini in the framework of the A.G.A.T.H.O.C.L.E.S. project (Giardini Reali, Turin). Photo: Marco Serino.

Networked Mobilities of Ceramic Workshops

The importance of specialized knowledge did not end in the technological realm. Pottery workshops and trading intermediaries had to capture the pulse of the market both at home and abroad. Trade networks, can shape and are influenced by a wide spectrum of factors relating to the choice of peculiar iconographies (Stansbury O'Donnell) or factors relating to technology, especially when a market leader industry spread in the Mediterranean (Klebinder-Gauß). Through trade networks we can trace migration of people/transfer of knowledge, following specific (sometimes hidden to the naked eye) technological procedures (Charalambidou, Di Virgilio, Serino) and we can better appreciate the power of such networks to connect markets with people (Attout, Delahaye, Mees).

The networked mobilities include the trading mechanisms that further underscore the concept of 'movement' we discussed in the introduction of this volume. Mobilities and overall 'movements' are keywords that cover trade markets (Klebinder-Gauß, Stansbury O'Donnell) and migration of people/transfer of knowledge of workshop's procedures and technological strategies, using the entire Mediterranean Sea as a network platform (e.g., Charalambidou, Gassner on amphorae). From this wide perspective, it is possible to follow the movements of iconographies in red-figured pottery and the specific market demands from indigenous people in Apulia (Stansbury O'Donnell), which imply cultural and mercantile links in a very restricted area between Lucania and Apulia (Magna Graecia). Magna Graecia and Sicily are the geographic focus of the artisanal production techniques related to the coroplastic world: this case study demonstrates

6 All vessels were made by M. Patarino, a Cerveteri-based potter. Among these vases we highlight a bilingual amphora, at a slightly smaller scale than the original (Boston, Museum of Fine Arts, inv. no. 99.538; attributed to the Andokides Painter/Lysippides Painter), a red-figured South Italian calyx krater (Syracuse, Museo Archeologico Regionale "Paoli Orsi", inv. no. 47102; attributed to the Prado-Fienga Group), two different versions of the red-figured skyphos by an early Sicilian workshop (Boston, Museum of Fine Arts, inv. no. 03.834; attributed to the Chequer Painter), and a lekane by the Locri Group (Private Collection: Elia (2010) FR 18).

the importance of hand modeling and the various strategies of clay manipulations that spread in Southern Italy, and how they can help us recognize regional trends and specific local markets demands (Meirano). These movements reflect the products' performance and consumer preferences in several communities whether in the long-established Western Greek cities or among the transient trading crews crisscrossing the Mediterranean and carrying a memory of home along on their travels (Attout).

Movements of pottery in trade markets are not only a matter of fine and figured vases; even cooking pottery can be the focus of exchange dynamics, if it has a peculiar fabric and it is supported by a strong and well-organized system, as the case of Aegina can demonstrate (Klebinder-Gauß). Movements entail not only a displacement of objects, but also of people, and of technological knowledge from one place to another. One can see mobilities of potters for coarseware in Iron Age Cycladic Naxos and Sicilian Naxos (Charalambidou); in Classical times, similar mobilities appear in the wide interconnections persisting both in terms of temporary relocation to enable workshop's collaborations and in permanent migration to another area between Magna Graecia and Sicily. The dynamics of these short- or long-term mobilities are presented in case-studies from Sicily, Lipari and the Tyrrhenian coast (Serino), and Locri Epizephyrii (Elia *et al.*), underscoring the Mediterranean networks of knowledge at work. Tile makers were also part of these mobilities as the 3rd century B.C. ateliers in Reggio Calabria demonstrate with the stamped tiles produced by the workshops of ΣΩΣΗΝΟΣ (Giuliano/Montana/Portale).

As they research the operational and networked knowledge of ancient producers, distributors, and customers, scholars now have powerful databases to aid in discerning trends and capturing nuances of ancient objects. The digital world also brings alive the ancient world of ceramics, whether through fascinating virtual reconstructions of workshops based on the archaeological records from excavations (Cavassa *et al.*, Xavier de Silva), or through restoration of the pottery itself, such as the case of the Apulian volute krater with iconography that can be 're-read' thanks to a virtual restoration of the entire scene (Tonna). The digital world can be also placed in a wider perspective, as the team publishing the two

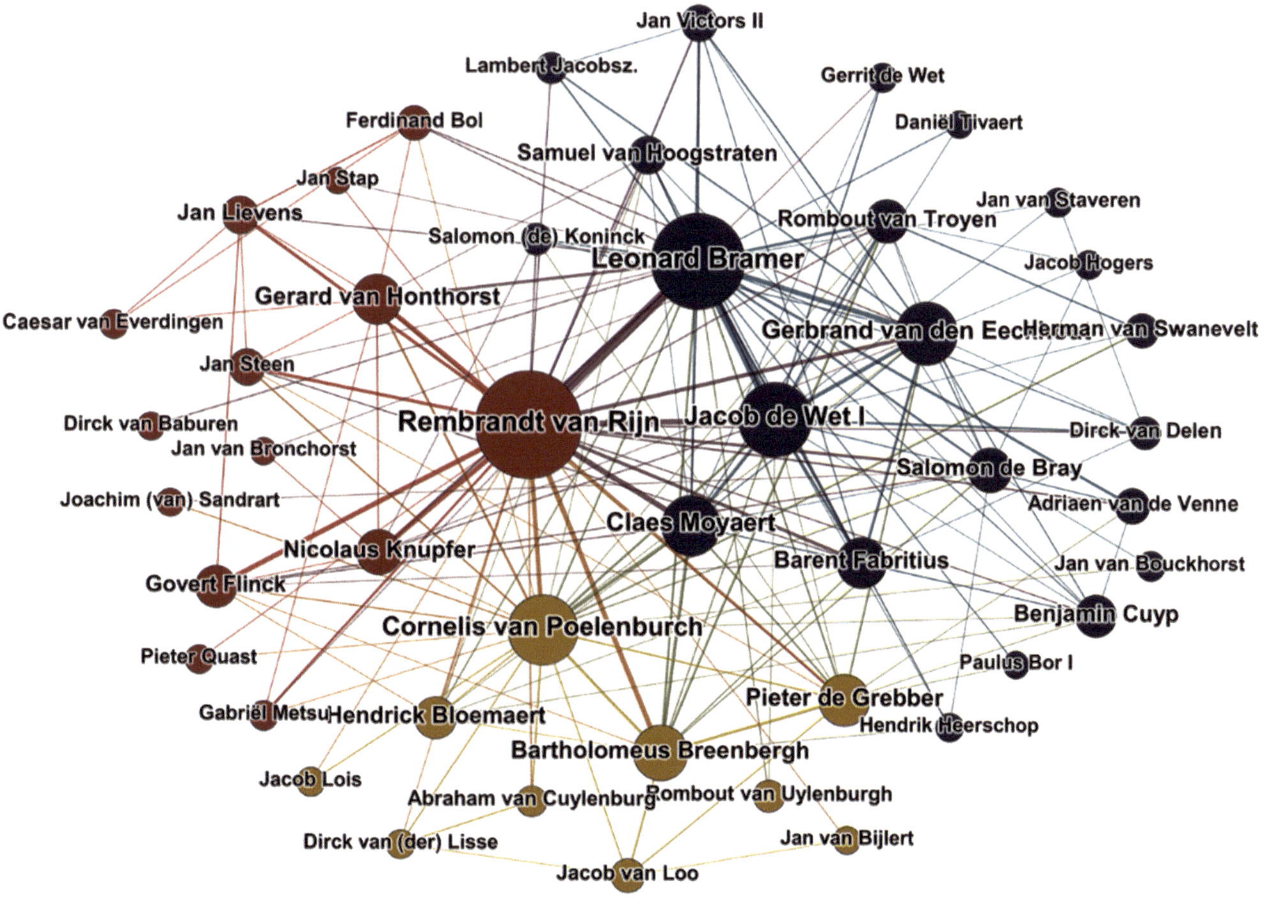

Fig. 3: Network of Dutch artists between 1620 and 1650, linked by at least three mutual subject matters. After Li (2021) 228, fig. 6.

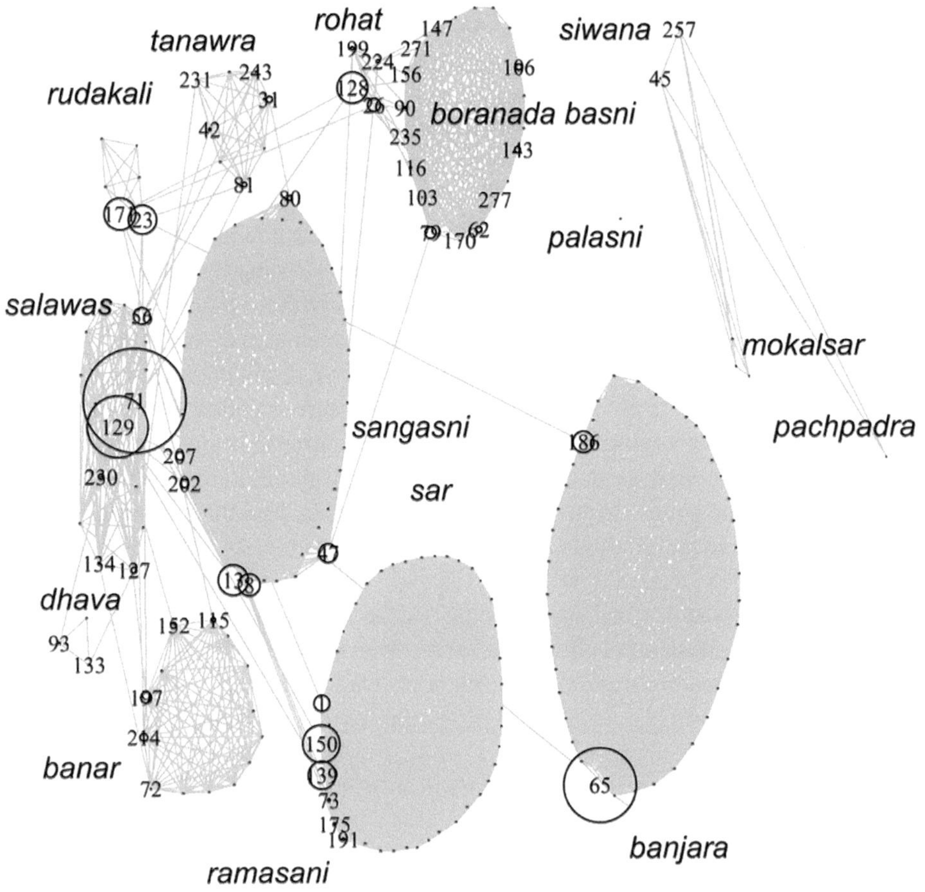

Fig. 4: Kinship ties among Muslim potters in Rajasthan. After Manzo *et al.* (2018) 1124, fig. 3.

workshops at Pompeii (Cavassa *et al.*) have done by adopting a multi-sensory immersion approach and connecting the tangible (and real) experience of the experimental archaeological sessions with the virtual reconstruction of a Pompeian workshop. Artificial intelligence (AI) will only advance existing processes and the potential of digital restorations.

Institutions and museum worldwide have made great strides in making their information and collections available online and facilitating inquiries on a larger scale. The open source format of those databases removes barriers of accessibility. The *Kerameikos.org* project represents 10 years' worth of metadata efforts to standardize and link separate databases. Consistency in terminology across collections is a major goal of this important project (Smith/Gruber/Harokopos).[7] Consistent terminology for descriptions of iconography is especially a desideratum for developing networks of iconographical themes. The full potential of this kind of standardization can be seen, for example, in networks of shared iconographies for Dutch paintings (Fig. 3).

The team led by G. Olcese (Olcese *et al.*) has undertaken a major project (*Immensa Aequora*) involving large datasets related to amphorae reaching Ostia, Rome's port. In connection with the research project on the amphorae and its link with a web-based information system, it is worth mentioning the FACEM project and its eighth release, with over 300 fabrics which highlights the importance of the fabric with the support of good images and archaeometric data that allows users to engage in analysis of distribution patterns and exchange modalities (Gassner).

7 For example, even in describing iconography is a desideratum for developing networks of iconographical themes. The full potential of standardized iconographical terms can be seen in networks of shared iconographies for Dutch paintings Li (2021). This could revolutionize preferred themes among contemporary artists, or regional preferences, and or chronological distributions. See also the digital poster by Eleni Hasaki and Diane Harris Cline (2022): Network Visualizations of Beazley's *ABV* and *ARV* Datasets: The Shape Sectors and Influential Artisans in the Athenian Kerameikoi (https://www.carc.ox.ac.uk/XDB/DMS/Hasaki%20Cline%20Oxford%202022%20final2.pdf; last access 03.06.2024).

Gradually our queries, relying on so many types of evidence (textual, epigraphical, stylistic, archaeometric, to name a few) will be best captured by knowledge graphs with incorporated ontologies (Smith/Gruber/Harokopos).[8] Quickly-generated results from statistical, digital, and other online applications are helpful but need always to be examined critically by archaeologists involved in such projects who will provide the necessary contextualization and the discussion of qualitative data along with the quantitative data. Another crucial aspect in data visualizations (whether graphs or network sociographs) is traceability and the ability to go back to the raw data.

From old data to contemporary data, archaeologists conducting ethnoarchaeological work in large potters' communities, continue their work of mapping social relationships among subgroups. Following Malkin's middle ground approach that a network could either simply be a metaphor or be the network of network theory, or both,[9] we have adopted different nuances of 'network approaches' in this volume. Networks have been discussed for the connections between Laconian pottery, Crete, and the whole Mediterranean area (Delahaye), for the hypothesized migration of red-figured painters from Sicily to Magna Graecia and vice versa (Serino), and for the production and marketing of terra sigillata (Samian ware) in the Roman empire, as explored by mathematical models (Mees). Drawing again on ethnoarchaeology, as for the operational knowledge, we can also turn to Social Network Analysis of contemporary potters' communities. A recent analysis of "emic" kinship ties and apprenticeship ties among potters of different religious groups in Rajasthan reminds us how differently (or similarly?) our "etic" stylistic ties in classical scholarship based on connoisseurship may map on ancient social relationships (Fig. 4).[10]

All of the contributions to this volume have allowed us to zoom in and out from local to Mediterranean horizons, and from individual makers to consumer communities. The authors have presented new material and have revisited known artifacts from new angles. From the digital microscope to networks of South Italian vases, and thanks to collaborations with artists, chemists, physicists and digital humanities specialists, these scholars have emphasized the importance of interdisciplinary collaborations, in a way creating modern networks of investigating teams to capture the dynamics of producers, promoters, and purchasers of ceramics in the ancient Greek and Roman world.

Bibliography

Dörpinghaus *et al.* (2022): Jens Dörpinghaus, Sonja Klante, Martin Chrisian, Christof Meigen and Carsten Düing, "From Social Networks to Knowledge Graphs: A Plea for Interdisciplinary Approaches," in: *Science Direct* 6.1, 100337. https://doi.org/10.1016/j.ssaho.2022.100337.

Elia (2010): Diego Elia, *Locri Epizefiri VI*. Nelle case di Ade. *La Necropoli in Contrada Lucifero, Nuovi Documenti*, Alessandria.

Gosner (2021): Linda Gosner, "Esparto Crafting under Empire: Local Technology and Imperial Industry in Roman Iberia", in: *Journal of Social Archaeology* 21, 329–352.

Hasaki (2021): Eleni Hasaki, *Potters at Work in Ancient Corinth: Industry, Religion, and the Penteskouphia Pinakes*, *Hesperia* Supplement 51, Princeton N.J.

Hasaki (2024): Eleni Hasaki, "Handling Greek Vases", in: Nikos Dimakis and Eirene Dimitriadou (eds.), *Καλλίνικος: Honorary Volume for Professor Panos Valavanis*, Athens, 25–36.

Horden/Purcell (2000): Peregrin Horden and Nicholas Purcell, *The Corrupting Sea. A Study of Mediterranean History*, Oxford.

Li (2021): Weixuan Li, "A Network of Iconography. Tracing the Evolution of Iconography in History Paintings in the Dutch Golden Age," in: *Early Modern Low Countries* 5, 216–249.

Malkin (2011): Irad Malkin, *A Small Greek World: Networks in the Ancient Mediterranean*, Oxford.

Manzo *et al.* (2018): Gianluca Manzo, Simone Gabriellini, Valentine Roux and Freda Nrikote M'm bogori, "Complex Contagions and the Diffusion of Innovations: Evidence from a Small-N Study," in: *Journal of Archaeological Method and Theory* 25, 1109–1154.

Rockman (2003): Merilyn Rockman, "Knowledge and Learning in the Archaeology of Colonization", in: Merilyn Rockman and James Steele (eds.), *Colonization of Unfamiliar Landscapes: The Archaeology of Adaptation*, London, 3–24.

Rockman (2012): Merilyn Rockman, "Apprentice to the Environment: Hunter-Gatherers and Landscape Learning", in: Willeke Wendrich (ed.), *Archaeology and Apprenticeship: Body Knowledge, Identity, and Communities of Practice*, Tucson, 99–118.

8 Dörpinghaus *et al.* (2022).

9 Malkin (2011) 16. The attitude of mobility in the ancient Mediterranean world was explored by Horden/Purcell (2000) and, especially in relation to the Greek world, by Malkin (2011): both were inspired by the Graph theories and network theories and they can be considered the main pillars for subsequent studies on network topics.

10 Li (2021); Manzo *et al.* (2018).

List of Contributors

Editors and Authors

Diego Elia is Professor of Classical Archaeology at the University of Turin, Department of Historical Studies. He is Field Director and Scientific Codirector of the archaeological missions of the University of Turin at Locri Epizephyrii, in southern Italy, and at the Roman villa of Costigliole Saluzzo, in Piedmont. Since 2017, he has been promoting interdisciplinary research on the House of the Ancient Hunt in Pompeii, and in 2018 was the Cocurator of the exhibition "Pompeiana Fragmenta: Conoscere e conservare (a) Pompei." His main research interests are craftsmanship and material culture in the Greek and Roman world, with a focus on Western red-figured pottery, funerary archaeology in ancient Greece, and urban planning and water management in Antiquity. In the last decade, as Director of S.U.S.C.O.R. (Struttura Universitaria in Scienze per la Conservazione, Restauro, Valorizzazione dei Beni Culturali) and President of the Master's Degree in Conservation and Restoration of Cultural Heritage, he has promoted numerous interdisciplinary projects involving archaeometry and conservation scientists. (diego.elia@unito.it)

Eleni Hasaki is Professor of Anthropology and Classics at the University of Arizona and the Codirector of the Laboratory for Traditional Technology. Her publications focus on Mediterranean technologies, apprenticeship systems, energetics, and workspaces of craft specialists in Classical Antiquity. In her most recent book, *Potters at Work in Ancient Corinth: Industry and Religion in the Penteskouphia Pinakes*, she contextualized the unique illustrations of potters at work within technological, religious, and social contexts. She employs experimental archaeology, digital humanities, and social-network analysis to study networks of technical expertise and the distribution of workshops in potters' communities of Ancient Greece. Her work has been recognized through senior fellowships at Harvard University, at the Center for Hellenic Studies, and at the National Gallery of Art, in Washington, D.C. She has also received generous funding including from the Archaeological Institute of America, the American School of Classical Studies at Athens, Harvard's Loeb Classical Library Foundation, the Samuel H. Kress Foundation, and the National Endowment for the Humanities. (hasakie@ arizona edu)

Marco Serino is Associate Professor of Classical Archaeology at the University of Turin, Department of Historical Studies. His research focuses on Greek and Roman pottery, computational imaging and digital humanities applied to ancient ceramics, Roman construction techniques in Pompeii, and on ancient Western Greek religious phenomena. His main area of interest is the early Western red-figured pottery, with a particular focus on the beginnings of the early Sicilian workshops and their productive and cultural backgrounds. He recently won a Marie Sklodowska–Curie Global Fellowship with a research project entitled A.G.A.T.H.O.C.L.E.S. (*The Archaeology of Gesture': Apprenticeship, Tools, Hands, Organizations, Collaboration, Learning Experience, and Social Network Analysis*), which leads him to conduct his research between Turin (Italy) and Tucson (Arizona, USA). (marco. serino@unito.it)

Authors

Linda Adorno earned her Ph.D. at Bonn University (Rheinische Friedrich Wilhelms, Philosophische Fakultät) in 2020. Since 2012 she has been collaborating with the German Archaeological Institute at Rome on various projects in Selinous. She participated in the research project of the University of Bonn on the Selinous Kerameikos. Since 2021 she has worked at the German Institute for the project *Ein neues Modell für die Stadt Selinunt*. She is also collaborating with the University of Augsburg on a project in Agrigento (the extraurban Sanctuary of Saint Anne). Her specialization is in pottery production, specifically the local production ceramic manufacture of Selinous, on which a monograph is forthcoming. (adornolinda@gmail.com)

Antoine Attout Antoine Attout is a Ph.D. candidate at the Université Libre de Bruxelles, under the supervision of Athena Tsingarida. He specializes in Attic pottery from the Archaic period and especially in the introduction of elements inspired by East Greek pottery traditions on Attic black-figure productions. (antoine.attout@ulb.be)

Barbara Borgers is postdoctoral researcher at the University of Vienna, Department of Classical Archaeology, an Honorary Research Fellow at the University of Kent. As a ceramic scientist, she has worked in various interdisciplinary research projects of heritage communities, museums, and commercial archaeology companies across Europe. She is currently Principal Investigator of the project "Cooking Ware as Indicator for Regional Trade in 4th–1st Century B.C. Central Italy". Adopting a multianalytical approach, her research considers technological variability and seeks to answer key questions about potters' transmission of technological knowledge and related social practices. (barbara.borgers@univie.ac.at)

Laetitia Cavassa is Archaeologist specializing in Hellenistic and Roman pottery. She worked in Aix-en-Provence (France) at the Centre Camille Jullian (AMU-CNRS) until 2014. Currently she works for the CNRS (Naples, Italy: Centre Jean Bérard). She directed the excavations of two potter's workshops in Pompeii. She earned her Ph.D. from the École Pratique des Hautes Etudes in Paris, with a thesis on pottery production in Pompeii. She has conducted numerous studies on pottery in the Mediterranean area (Egypt, Croatia, Italy, Greece, and Corsica). (laetitia.cavassa@univ-amu.fr)

Guilhem Chapelin is Architect and CNRS Agent in unit UAR 3155, Institut de Recherche sur l'Architecture Antique (AMU-CNRS) in Aix-en-Provence (France), specializing in ancient construction techniques and experimental archaeology. (guilhem.chapelin@univ-amu.fr)

https://doi.org/10.1515/9783111189635-032

Xenia Charalambidou is Senior Postdoctoral Researcher at Vrije Universiteit, Amsterdam. She has completed postdoctoral fellowships at the University of Athens, the University of Thessaly, and the University of Warsaw. For her science-based research she collaborates with NCSR Demokritos and the Fitch Laboratory of the British School at Athens. Her current research concerns social and geographical mobility, trade, technological transmission, cultural interaction, and consumer behavior in the Early Iron Age and the Archaic Mediterranean. Her publications include four co-edited volumes and a substantial number of articles in peer-reviewed journals. She has taught at the universities of Thessaly and of Warsaw, the Open University of Cyprus, and the Vrije Universiteit, Amsterdam. (xenia.charalambidou@gmail.com)

Gabriella Chirco is a Conservator who graduated from the University of Palermo in 2014 in Conservation and Restoration of Cultural Heritage. She specialized in painted artifacts on wood and textile support. Between 2014 and 2016, she obtained a second-level Master's Degree as an Expert Researcher in Nanomaterials and Nanotechnologies for Cultural Heritage at the University of Palermo, during which years she focused on microemulsions for cleaning painted surfaces. In 2022, she completed a Ph.D. at the University of Palermo in Human Sciences: Dynamics of Systems, Cultural Heritage, Cultural Studies, with an archaeological-archaeometric focus. Her research activity focuses on the characterization of materials and execution techniques, with particular attention to painted ceramic artifacts. (gabriella.chirco@unipa.it)

Patrizia Davit received her Ph.D. in Chemical Sciences at the University of Turin with a doctoral thesis entitled "Analytical Techniques Application in the Investigation of Ancient Pottery and Glass". Since 2006 she has been Research Technician at the University of Turin, Department of Chemistry. Her research activities are devoted to the characterization, technology of production investigation, provenance studies, and degradation-phenomena analysis of archaeological, historical, and artistic materials. Moreover, she is involved in several projects related to the teaching of chemistry and in the field of Public Engagement. (patrizia.davit@unito.it)

Mattia D'Acri is Ph.D. Candidate in Classical Archaeology at the University of Missouri, Columbia. He holds a graduate degree from the Scuola di Specializzazione in Archeologia in Matera, Italy. Specializing in ceramics, his research centers on the Protohistoric and Archaic periods in central and southern Italy. He has conducted extensive fieldwork at Rome, Gabii, Pompeii, Francavilla Marittima, and other sites, publishing insights into pottery assemblages. Since 2023, he has served as Codirector of the *Torre Mordillo Archaeological Project*. In 2023, he also coedited *Gabii Through Its Artefacts*, showcasing his commitment to archaeological exploration and scholarly contributions. (mattia.dacri@mail.missouri.edu)

Alberto De Bonis is Associate Professor at the University of Naples "Federico II," Department of Earth Sciences, Environment, and Resources. He specializes in the mineralogical and petrographic study of ancient ceramics and clay raw materials, as well as stones, mortars, plasters, mosaics, and wall paintings. (alberto.debonis@unina.it)

Adrien Delahaye is Associate Professor in Ancient Greek archaeology at the University Montpellier 3 – Paul Valéry. Former scientific member of the École française d'Athènes, he holds a doctorate in archaeology from the University Paris 1 Panthéon-Sorbonne. His research focuses on the history and archaeology of Sparta and Laconia in the Archaic and Classical periods, with particular emphasis on pottery and material culture, craft production, trade networks and the sanctuary of Apollo Amyklaios. (adrien.delahaye@efa.gr)

Alessandra De Natale is a Conservator specializing in the archaeological sector. She has assisted in numerous excavation campaigns with responsibility for emergency conservation and has directed and carried out restoration interventions on movable finds and decorated surfaces. Currently she collaborates with the MiC (Italian Ministry of Culture), department of Direzione Regionale Musei Molise. (alessandra.denatale@gmail.com)

Elisabetta di Virgilio is a Ph.D. Candidate at the University of Campania "Luigi Vanvitelli." She specializes in Greek and Roman pottery, and her current research is focused on the production and distribution of Late Archaic and Classical pottery from ancient Cumae, near Naples (Italy). (elisabetta.divirgilio@unicampania.it)

Francesca Diosono is Researcher at the Institute of Classical Archaeology of the Ludwig Maximilian University, Munich. Her fields of interest include Roman Archaeology, landscape archaeology, and Roman material culture. She has worked since the 2000s on a range of projects in Italy, Spain, France, and Jordan. She is the Director or Codirector of the archaeological excavations of the Sanctuary of Diana in Nemi (Rome), of the Sanctuary of Monte Sant'Angelo in Terracina (Latina, Italy), of Villa San Silvestro di Cascia (Perugia, Italy), and of the publications of the Fregellae excavations (Frosinone, Italy). (francescadiosono@virgilio.it)

Ludovica Xavier de Silva holds a Ph.D. in Classical Archaeology from University of Macerata. She is a member of the archaeological missions directed by Roberto Perna (University of Macerata, Italy) at the archaeological park of Pollentia–Urbs Salvia, in Albania (Drino Valley: Codirection with the Archaeological Institute of Tirana) and in Greece (Byzantine Houses Quarter of Gortyn, Crete: Italian Archaeological School of Athens). Her main research interest is the study of pottery production and its circulation in the Mediterranean from the Hellenistic to the Byzantine period. (ludovica.xavierdesilva@gmail.com)

Laura Di Siena holds Senior Archaeologist affiliated with the Danish Institute in Rome and with the University of Rome "Tor Vergata," where she is concluding her Ph.D. studies. Her research focuses on Classical and Medieval pottery manufacture. (laura.disiena91@gmail.com)

Elisa Ercolin holds Ph.D. in Archaeological Sciences from the University of Turin, Department of Historical Studies. Her scientific work focuses on Greek black-gloss pottery from southern Calabria and has developed interdisciplinary approaches aimed at the reconstruction of local repertoires, manufacturing processes, and craft contexts. She currently participates in the research projects of the University of Turin in Locri Epizephyrii and Pompeii and collaborates with the Soprintendenza Archeologia, Belle Arti e Paesaggio of Reggio Calabria and Vibo Valentia (Calabria, Italy) for the study of the black-gloss artifacts discovered by Paolo Orsi in the Sanctuary and Necropolis of *Medma*-Rosarno. (elisaercolin92@gmail.com)

Chiara Germinario is a Geologist specializing in the archaeometric study of archaeological materials (e.g., ceramics, pigments, mortars). She works at the University of Sannio (Benevento, Italy), where she was awarded her Ph.D. in 2015, and she is involved in interdisciplinary research projects in collaboration with national and international research institutes. She is author of publications in international scientific journals. (chiara.germinario@unisannio.it)

Lorena Carla Giannossa is Associate Professor in Analytical Chemistry at the University of Bari, Department of Chemistry. Her primary interests are in Cultural Heritage, Environment, and Characterization of Materials. She is the author of more than sixty papers in peer-reviewed international journals and textbooks, as well as two hundred communications at national and international conferences. (lorenacarla.giannossa@uniba.it)

Dario Giuliano is Ph.D. Candidate in Archaeology at the University of Palermo, in cotutorship with the Université Bordeaux-Montaigne. He is conducting a research project on roof tiles in ancient Sicily from their introduction to the Roman Conquest. He has worked in several archaeological excavations with both Italian and foreign universities. (giuliano.dar@gmail.com)

Celestino Grifa is Associate Professor at the University of Sannio (Benevento, Italy), Department of Science and Technology. He is a petrographer, and his core research area concerns the study of archaeological materials with a focus on stone, ceramic, and mortar-based materials. He is involved in national and international research projects and is the author of publications in leading scientific journals. (celestino.grifa@unisannio.it)

Monica Gulmini is Associate Professor at the University of Turin, Department of Chemistry, and a member of the AMAMASA team (ArchaeoMetry and Analytical Methods for hiStorical and Archaeological Materials). She teaches chemical methods for the study of tangible cultural heritage in undergraduate and graduate courses and is a frequent supervisor of thesis and doctoral research projects. Her research focuses on the chemical analysis of materials of artistic and archaeological interest (in particular, pottery, glass, historical musical instruments, and textiles) with the aim of providing information on production technologies and provenances. (monica.gulmini@unito.it)

Verena Gassner was Professor in Classical Archaeology at the University of Vienna, Institute of Classical Archaeology, from 2002 until her retirement in 2021. Her research focused on the study of the material culture in southern Italy and western Anatolia, especially on pottery, including interdisciplinary archaeometric research. During 2002–2021 she conducted the Austrian excavations at Elea/Velia (Campania) in southern Italy and explored the Greek fortifications, the sanctuaries on the central ridge, and the development of the eastern quarter of the town. (verena.gassner@univie.ac.at)

Gudrun Klebinder-Gauß studied Classical Archaeology and Ancient History at the University of Salzburg, Austria, where she completed her Ph.D. in 2002. She took part in numerous excavations in the Aegean, including Ephesus, Aegina, Aigeira, Kalapodi, and Didyma. As a specialist in Geometric and Archaic bronze art, she focuses on votive traditions and ritual practices. Another area of her research is pottery from the Archaic to the Hellenistic periods, with a particular interest in the production of cooking ware. (gudrun.klebinder-gauss@hotmail.com)

Ethan Gruber is Director of Data Science at the American Numismatic Society. With more than fifteen years' experience in digital humanities and cultural heritage web-development projects, he has developed digital collections, library, archives, and research-information systems, including *Online Coins of the Roman Empire* (OCRE) and PELLA Project. Specializing in Semantic Web Technologies, he is Chief Software Developer and Scientific Committee Member of both *Nomisma.org* and *Kerameikos.org*, both Linked Open Data projects. (ewg4xuva@gmail.com)

Nicholas A. Harokopos currently serves as Adjunct Faculty at the Department of History and Archaeology of the University of Athens. He has previously taught at the University of Patras and at the Athens School of Fine Arts, and since 2018 has been tutor of Classical Archaeology at the Athens Tourist Guide School of the Greek Ministry of Tourism. Specializing in Greek figured pottery and iconography, his research interests include representations of old men and women, the garments of musicians, and ancient religion. (nickchar@arch.uoa.gr)

Bastien Lemaire is in charge of archaeological operations at Mosaïques Archéologie. He is Associate Researcher at UMR 5140 ASM (Montpellier, France) and at the Centre Jean Bérard (Naples, Italy). His research aims to better understand the role and functions of periurban spaces in the socioeconomic history of antique cities. Far from being limited to the funerary world, it tests the nature of periurban spaces and their multifunctionality, thus defining the city more precisely and in its entirety. (b.lemaire@mosaiquesarcheologie.com)

Kathleen M. Lynch is Professor of Classics at the University of Cincinnati. Her research examines the cultural use of Archaic, Classical, and Hellenistic Greek pottery, especially from archaeological excavations. She has been a researcher at the Excavations of the Athenian Agora for over twenty years. She has also published Greek pottery from Troy and Gordion, in Turkey; Apollonia, in Albania; and Pylos, in Greece. Her book *The Symposium in Context: Pottery from a Late Archaic House Near the Athenian Agora* won the Archaeological Institute of America's Wiseman Award. (lynchkn@ucmail.uc.edu)

Annarosa Mangone is Associate Professor in Chemistry at the University of Bari, Department of Environment and Cultural Heritage. Her main research activity focuses on the development of innovative analytical methods for studying the environment and cultural heritage. She is the director of the Interdepartmental Research Centre "Laboratory for the Diagnostics of Cultural Heritage." Winner of numerous research awards, she is the author of more than 180 papers published in international peer-reviewed journals and textbooks, 1 patent, and 200 communications at national and international conferences. (annarosa.mangone@uniba.it)

Émilie Mannocci is a Ph.D. Candidate at Aix Marseille Université (Centre Camille Jullian, AMU-CNRS), in conjunction with the University of Padua, and in collaboration with the École française de Rome. She specializes in Roman ceramics, in particular the *terra sigillata* produced in northern Italy during the Augustan Period. (emiliemannocci@gmail.com)

Fabrizio Mastrorocco is Researcher at the Institute of Biomembranes, Bioenergetics, and Molecular Biotechnology (IBIOM), CNR Bari. His research activities include international centers such as the Institute of Astrophysics and Planetology CNRS-IRAP in Toulouse (France) and ETH Zurich (Switzerland). His main research activities focus on the analyses of trace and ultratrace in different matrices, analyses of organic molecules, and metabolomics analyses by liquid chromatography coupled with high-resolution mass spectrometry. The author of papers in peer-reviewed international journals and textbooks, he has also presented communications at numerous national and international conferences. (f.mastrorocco@ibiom.cnr.it)

Valeria Meirano teaches Classical Archaeology at the University of Turin, Department of Historical Studies. She is Scientific Codirector of the archaeological missions at Locri Epizephyrii, in southern Italy, and at the Roman site of Costigliole Saluzzo, in Piedmont. She has been a Visiting Professor and held fellowships in Greece, France, Denmark, and in the United States. Her research interests include funerary archaeology, sanctuaries, and rituals in ancient Greece, food iconographies and material culture in the Greek and Roman world (bronzes, terracotta figurines, ceramic productions). On behalf of the École française d'Athènes, she is responsible for the study of bronze vessels and other metal artifacts from Delphi. She recently developed a project on bronzes in Pompeii and cocurated an exhibition dedicated to the site. (valeria.meirano@unito.it)

Allard Mees is Archaeologist at the Leibniz-Zentrum für Archäologie, in Mainz, Germany, where he is Chair of the Scientific IT Department. He studied in Utrecht, worked at archaeological institutes in Budapest and Zürich, and received his Ph.D. in Freiburg im Breisgau. He has been heading major European-funded archaeological database projects on Roman ship archaeology and maintains a long-term European digital research framework on Roman ceramics. As a Fellow of the Society of Antiquaries, in London, he focuses his research on the Roman economy, statistics, and cultural astronomy. (allard.mees@rgzm.de)

Giuseppe Montana is Associate Professor at the University of Palermo, Department of Earth and Marine Sciences. He is the author of numerous papers published in international journals, as well as monographs and chapters in international volumes. His research topics range from Classical archaeometry (stones, ceramics, mortars) to mineropetrographic and geochemical applications for the Environment and Cultural Heritage. (giuseppe.montana@unipa.it)

Italo M. Muntoni has been Archaeologist at the Soprintendenza Archeologia, Belle Arti and Paesaggio per le province di Barletta–Andria–Trani e Foggia since 2010. He was awarded his Ph.D. at La Sapienza University, Rome, in 2001. Since 2020 he has been Adjunct Professor of Archaeology of Production at the Specialization School in Archaeology of Salento University (Lecce, Italy). (italomaria.muntoni@beniculturali.it)

Giovanni Murro is Senior Archaeologist affiliated with the Danish Institute in Rome and with the University of Groningen, where he is concluding his Ph.D. studies. He has directed and carried out numerous excavations as well as rescue archaeology interventions. His research focuses on landscape archaeology. (giomurro@hotmail.com)

Gloria Olcese is Professor of Methodology of Archaeological Research at the University of Milan. Her interests focus on the archaeology of production and trade, and multidisciplinary approaches integrating archaeology and archaeometry. She has been awarded fellowships in Germany (DAAD, Deutscher Akademischer Austauschdienst; DFG, Deutsche Forschungs Gemeinschaft) and Italy (University of Siena). She has coordinated several large-scale research projects: Ischia, financed by the Thyssen-Stiftung, Germany (1999–2003), and the FIRB project *Immensa Aequora* (2005–2011); "Archeologia e Ambiente delle 'Isole di Storia' del Tirreno," funded by the École française de Rome (2022–2026). She is Director of the Excavations of a large Roman Villa at Dragoncello/Acilia (near Ostia Antica, Rome). The author of several monographs and articles, she has also organized international conferences. (gloria.olcese@unimi.it)

Fabrice Paul is the founder of the Aix-en-Provence-based agency EDIKOM, which specializes in scientific, cultural, and heritage promotion through a range of multimedia and video media for museums, archaeological sites, and web platforms. Specializing in archaeology and digital technology, he often accompanies the narration of numerous reconstructions and 3D animations. (fabrice.paul@edikom.fr)

Marie Pawlowicz is a Ph.D. Candidate and Teaching Assistant in Art History and Archaeology at Aix Marseille University (Centre Camille Jullian, AMU-CNRS). She specializes in Roman harbor archaeology and ceramics from the western Mediterranean. (marie.pawlowicz@laposte.net)

Elisa Chiara Portale is Professor of Classical Archaeology at the University of Palermo. She has dealt with different areas and themes: Greek and Roman sculpture, painting, and mosaics; religion, sanctuaries, and votaries; Greek and Roman houses and decorations; cultural phenomena in provincial areas; post-Minoan Crete; Hellenistic ideology and art; history of archaeological research; terracotta figurines; Hellenistic ceramics; transport containers of the Hellenistic and Early Byzantine periods), with particular attention to the contexts of Sicily, Crete, and Tripolitania, and with openness to interdisciplinary approaches. She directs the archaeological missions of the University of Palermo in Greece (Gortyn) and in Libya (Sabratha, Leptis); in Sicily she coordinates archaeological research in Solunto, Agrigento, and Himera. (chiara.portale@unipa.it)

Andrea Razza has a Ph.D. (co-tutorship between the "Agriculture, Environment and Bioenergy" program of the University of Milan and the "Societat i Cultura" program of the Universitat de Barcelona). He has actively collaborated for many years in the research activities of the *Immensa Aequora* project (directed by Gloria Olcese) in several sites of central and southern Italy. His main lines of research are amphoras from the Late Republican and Early Imperial eras, with a specific focus on the central Tyrrhenian area; the material culture of the Rome and Ostia area between the Republican and Imperial periods; and the ancient structures for the production of wine (the so-called palmenti). (andrea.razza@hotmail.it)

Alessandro Re is Associate Professor at the University of Turin, Department of Physics. He is part of CHNet (Cultural Heritage Network), the network of the Italian National Institute for Nuclear Physics (INFN) devoted to Cultural Heritage, and he is Member of the Board of Directors of AIAr, the Italian Archaeometric Society. His main interest is the development and upgrading of new instruments and methods to be applied in the archaeometric field. Among the instruments he has developed there are devices to perform X-ray radiography and tomography of artworks and archaeological findings (from millimetric size up to some meters) and setups to detect the luminescence induced in various materials by X-ray or ion beams. He coordinates and participates in different projects and in the organization of national and international conferences. (alessandro.re@unito.it)

Philip Sapirstein received his doctorate in Art History and Archaeology from Cornell University (2008). Between 2013 and 2019 he was Assistant Professor in the School of Art, Art History, and Design at the University of Nebraska-Lincoln, and thereafter joined the faculty of Art History at the University of Toronto as Assistant Professor. His primary interests are Greek architecture, architectural terracottas, and ceramic production. Digital techniques for the 3D recording and analysis of ancient art and architecture, including photogrammetry, constitute another important area of his research. (phil.sapirstein@utoronto.ca)

Sine Grove Saxkjær is a Classical Archaeologist. She holds a Ph.D. from Aarhus University. Her research centers on ethnic and cultural identities in central and southern Italy across the 1st millennium B.C. (klasgs@cas.au.dk)

Francesca Silvestrelli is Assistant Professor of Classical Archaeology at Salento University (Lecce, Italy) and has worked on sites in central and southern Italy, Turkey, and Egypt. She is a specialist in the Archeology of production, Archeology of Magna Graecia, Pottery, and History of Archaeology, with a special focus on southern Italian red-figured and black-gloss ware production, distribution, and reception in settlements, sanctuaries, and necropoleis. A Trendall Fellow at the Institute of Classical Studies, University of London (UK, 2014), she is the author of a monograph (*Le duc de Luynes et la découverte de la Grande Grèce* [Paris, 2017]) and coauthor of a collective work (*The Chora of Metaponto 6. Sant'Angelo Vecchio* [Austin, 2016]). She is currently preparing the publication of a volume in the *Corpus Vasorum Antiquorum* of southern Italian red-figured vases from Metaponto. (francesca.silvestrelli@unisalento.it)

Tyler Jo Smith is Professor of Classical Art and Archaeology at the University of Virginia, where she also serves as Director of the Interdisciplinary Archaeology Program. As a specialist in ancient Greek vases and iconographies of performance, she is the author of *Komast Dancers in Archaic Greek Art* (2010) and of *Religion in the Art of Archaic and Classical Greece* (2021). Her current research is focused on ancient Greek festivals. She is the Cofounder of *Kerameikos.org*, a Linked Open Data project for figure-decorated Greek pottery based at the University of Virginia. (tjs6e@virginia.edu)

Mark Stansbury-O'Donnell is Professor of Archaeology and Art History at the University of St. Thomas, in Minnesota. He is the author of *Pictorial Narrative in Ancient Greek Art* (1999); *Vase Painting, Gender, and Social Identity in Archaic Athens* (2006); *Looking at Greek Art* (2011); and *A History of Greek Art* (2016), as well as articles and chapters exploring narrative and the contexts of Greek art. He is also Editor for the University of Wisconsin Press and Coeditor of *The Consumer's Choice: Uses of Greek Figure-Decorated Pottery* (2015). (m9stansburyo@stthomas.edu)

Domenico Michele Surace is Professor of Style, History of Art, and Costume at the Academy of Fine Arts in Reggio Calabria (Italy). His Ph.D. is in Archaeology (University of Barcelona) and in Agriculture, Environment and Bioenergy (University of Milan). He also teaches at University of Milan. He deals with study and dissemination in the historical-artistic field, and research through archaeological investigations and excavation campaigns in various sites with projection to the Mediterranean Basin. He has presented at various conferences and seminars and is also the author of several publications connected to the *Immensa Aequora* project. (dms1016@hotmail.com)

Maiwenn Tonna has received her Master's Degree at the University of Paris-Nanterre. She specializes in the iconography of music on southern Italian and Sicilian ceramics from the 5th to the 3rd century B.C. She is a member of the ACuTe (*Antiquity in a Cup of Tea*) project. (maiwenn.tonna84@orange.fr)

Bei Fragen zur Produktsicherheit wenden Sie sich bitte an:
If you have any questions regarding product safety, please contact:

Walter de Gruyter
Genthiner Straße 13
10785 Berlin
productsafety@degruyterbrill.com